T0271038

Analysis of Categorical Data with R

Analysis of Categorical Data with R, Second Edition presents a modern account of categorical data analysis using the R software environment. It covers recent techniques of model building and assessment for binary, multicategory, and count response variables and discusses fundamentals, such as odds ratio and probability estimation. The authors give detailed advice and guidelines on which procedures to use and why to use them.

The second edition is a substantial update of the first based on the authors' experience of teaching from the book for nearly a decade. The book is organized as before, but with new content throughout, and there are two new substantive topics in the advanced topics chapter—group testing and splines. The computing has been completely updated, with the `emmeans` package now integrated into the book. The examples have also been updated, notably to include new examples based on COVID-19, and there are more than 90 new exercises in the book. The solutions manual and teaching videos have also been updated.

Features:

- Requires no prior experience with R, and offers an introduction to the essential features and functions of R
- Includes numerous examples from medicine, psychology, sports, ecology, and many other areas
- Integrates extensive R code and output
- Graphically demonstrates many of the features and properties of various analysis methods
- Offers a substantial number of exercises in all chapters, enabling use as a course text or for self-study
- Is supplemented by a website with data sets, code, and teaching videos

Analysis of Categorical Data with R, Second Edition is primarily designed for a course on categorical data analysis taught at the advanced undergraduate or graduate level. Such a course could be taught in a statistics or biostatistics department, or within mathematics, psychology, social science, ecology, or another quantitative discipline. It could also be used by a self-learner and would make an ideal reference for a researcher from any discipline where categorical data arise.

CHAPMAN & HALL/CRC
Texts in Statistical Science Series

Recently Published Titles

Statistical Theory
A Concise Introduction, Second Edition
Felix Abramovich and Ya'acov Ritov

Applied Linear Regression for Longitudinal Data
With an Emphasis on Missing Observations
Frans E.S. Tan and Shahab Jolani

Fundamentals of Mathematical Statistics
Steffen Lauritzen

Modelling Survival Data in Medical Research, Fourth Edition
David Collett

Applied Categorical and Count Data Analysis, Second Edition
Wan Tang, Hua He and Xin M. Tu

Geographic Data Science with Python
Sergio Rey, Dani Arribas-Bel and Levi John Wolf

Models for Multi-State Survival Data
Rates, Risks, and Pseudo-Values
Per Kragh Andersen and Henrik Ravn

Spatio-Temporal Methods in Environmental Epidemiology with R, Second Edition
Gavin Shaddick, James V. Zidek, and Alex Schmidt

A Course in the Large Sample Theory of Statistical Inference
W. Jackson Hall and David Oakes

Statistical Inference, Second Edition
George Casella and Roger Berger

Nonparametric Statistical Methods Using R, Second Edition
John Kloke and Joesph McKean

Generalized Linear Mixed Models
Modern Concepts, Methods and Applications, Second Edition
Walter W. Stroup, Marina Ptukhina and Julie Garai

Analysis of Categorical Data with R, Second Edition
Christopher R. Bilder and Thomas M. Loughin

For more information about this series, please visit: https://www.routledge.com/Chapman--HallCRC-Texts-in-Statistical-Science/book-series/CHTEXSTASCI

Analysis of Categorical Data with R

Second Edition

Christopher R. Bilder and Thomas M. Loughin

CRC Press is an imprint of the
Taylor & Francis Group, an **informa** business
A CHAPMAN & HALL BOOK

Cover images: The golf and football photos are courtesy of Callum D. Bilder.

Second edition published 2025
by CRC Press
2385 Executive Center Drive, Suite 320, Boca Raton, FL 33431, U.S.A.

and by CRC Press
4 Park Square, Milton Park, Abingdon, Oxon, OX14 4RN

CRC Press is an imprint of Taylor & Francis Group, LLC

First edition published by Chapman and Hall, 2014

ISBN: 978-0-367-55323-4 (hbk)
ISBN: 978-0-367-55342-5 (pbk)
ISBN: 978-1-003-09309-1 (ebk)

DOI: 10.1201/9781003093091

Typeset in SFRM font
by KnowledgeWorks Global Ltd.

Publisher's note: This book has been prepared from camera-ready copy provided by the authors.

Contents

Preface to the second edition

We began the preface to the first edition with the claim, "We live in a categorical world!" This fact was glaringly apparent when we began writing the second edition during the COVID-19 pandemic. Determining who was positive or negative for SARS-CoV-2 was an immediate objective for everyone. Public health officials struggled to accurately estimate population proportions of cases, hospitalizations, and fatalities. Discussions about sensitivity and specificity of diagnostic tests that had once been limited to academic and medical circles now moved into public discourse. Incomplete reporting of vaccine efficacy led some people to misunderstand how *numbers* of deaths could be higher among vaccinated people than unvaccinated, because reports did not consider how many people were in the two groups or their respective age distributions.

Understanding the strengths and limitations of an analysis of categorical data has never been more important. This second edition reflects our experiences during the pandemic, our continued learning as academic and applied statisticians, and our development as teachers with diverse and varying audiences. It also highlights changes that have occurred with R and its packages since the first edition. We have reread and reconsidered every word in the text, every exercise in the chapters, and every line in the code. Hindsight has allowed us to improve discussions and examples in many places throughout the book to ensure that they are accessible and informative for all members of our broad target audience. In this process, we have endeavored to maintain what we felt made the first edition a valuable resource for students, self-learners, researchers, and instructors.

The book is organized in the same manner as the first edition. Chapters 1–4 examine categorical data analysis by response type: binary (Chapters 1–2), multicategory (Chapter 3), and count (Chapter 4). Chapter 5 unifies the previous chapters by examining model building techniques and diagnostic methods, while also providing two case studies. Chapter 6 explores more advanced topics important to categorical data analysis. Some of these topics are for specialized forms of data, while others provide alternative tools for analysis. We anticipate that an instructor for a one-semester course will be able to cover most of Chapters 1–5 and choose additional topics from Chapter 6 as desired.

There have been numerous additional changes to the book and its associated learning materials:

- The `emmeans` package has been completely integrated into the book. Since our first edition, this package has become a leader for estimation and inference in R. It is especially useful for problems with categorical explanatory variables, where existing solutions were potentially error prone and awkward to implement. Although not without its quirks, we anticipate `emmeans` will allow readers to more easily analyze important measures, such as odds ratios. We also continue to use the `mcprofile` package for this same purpose because it is flexible and can implement likelihood ratio methods, which `emmeans` cannot.

- Two new sections have been added to Chapter 6. The first, on group (pooled) testing, highlights what became a very hot research topic during the COVID-19 pandemic with over 1GB of research articles published during the pandemic's first two years. The reason for this great interest was that group testing greatly increased laboratory testing capacity worldwide. Our second new section, on splines, provides flexible tools to account for nonlinear associations in modeling. The content in this new section can be simultaneously taught with the regression modeling parts of Chapters 2–4.

- Section 6.5 has been extensively re-written to incorporate Hamiltonian sampling methods and their implementation through the `rstan` package. Hamiltonian sampling is a faster and less problematic way to implement Markov Chain Monte Carlo methods that have become very popular in recent years.

- Many COVID-19 examples and exercises have been added, including those that focus on testing accuracy, testing efficiency, vaccine efficacy, and antiviral treatment therapy. Emergency use authorizations, published papers, and our own experience were used as sources for these examples and exercises.

- More than 90 new exercises have been added to the book. Beyond those involving COVID-19, these exercises include applications from a wide range of areas, including air traffic controller errors, Ebola vaccine efficacy, food choices after exercise, golf putting, and squirrel habitats. Our exercises even examine how statistics can be used for science fiction applications involving *Star Trek* and *Battlestar Galactica*. Each application involves replicating categorical data analyses from published papers and/or those analyses that have occurred in our own work. We also point out parts of published analyses that were performed inefficiently or even incorrectly. This allows readers to learn from those mistakes while understanding how to apply better methods.

- Our additional exercises go beyond applications to data. We include more Monte Carlo simulation-based exercises to help readers learn abstract concepts. Exercises also include extensions of the content within the main text, like writing R functions to make coding more efficient, exploring other R packages, and implementing alternative statistical methods. We provide step-by-step guidance with these types of exercises. Not all new exercises involve R. We include exercises that allow readers to derive expressions to help reinforce topics discussed in the main text.

- A new set of instructional videos are available. These videos can supplement the teaching of a course or be used by self-learners interested in teaching themselves.

Our book's website continues to be available at `http://www.chrisbilder.com/categorical`. The code used to complete every calculation performed for the book is included in the R programs on this website. Many of these programs provide additional code showing different approaches for the same calculations that may not be mentioned within the book. Direct links to our videos and course notes from teaching a corresponding course are available from the website as well.

We continue to focus on those packages within the default installation of R. This is done because these packages are less likely to change drastically and will have the fewest bugs and/or errors. We take advantage of user-contributed packages when they provide a significantly better approach and have an established history of use. It is important to note that we generally avoid Posit's dialect of R known as the tidyverse. We find that the standard R approaches to most data management and presentation problems are sufficient and easier for readers to learn.

We owe a lot of people a lot of thanks for their continued support for both the first edition of this book and the work that went into the revision. Students in our courses have helped us immensely to understand the strengths and weaknesses of the first edition, and, through our interactions with them, we have grown as teachers and authors. Several students and self-learners who read the first edition identified errors or inconsistencies that we have aimed to resolve in this edition. Rob Calver and the staff at CRC Press were enormously patient and supportive of our efforts, despite our frequent requests for deadline extensions. The developers for all of the open-source software that went into this book—especially R, LyX, and LaTeX, and their many related packages—deserve much more recognition and gratitude than they receive. Most importantly, we must thank our families—especially our wives, Kimberly and Marie—who continued to take on more than their share of responsibilities at home while we holed up in our offices learning software, locating new data sets for exercises, and obsessing over proper use of adverbs.

Christopher R. Bilder and Thomas M. Loughin
Lincoln, NE and Burnaby, BC

Preface to the first edition

We live in a categorical world! From a positive or negative disease diagnosis to choosing all items that apply in a survey, outcomes are frequently organized into categories so that people can more easily make sense of them. However, analyzing data from categorical responses requires specialized techniques beyond those learned in a first or second course in Statistics. We offer this book to help students and researchers learn how to properly analyze categorical data. Unlike other texts on similar topics, our book is a modern account using the vastly popular R software. We use R not only as a data analysis method but also as a learning tool. For example, we use data simulation to help readers understand the underlying assumptions of a procedure and then to evaluate that procedure's performance. We also provide numerous graphical demonstrations of the features and properties of various analysis methods.

The focus of this book is on the analysis of data, rather than on the mathematical development of methods. We offer numerous examples from a wide range of disciplines—medicine, psychology, sports, ecology, and others—and provide extensive R code and output as we work through the examples. We give detailed advice and guidelines regarding which procedures to use and why to use them. While we treat likelihood methods as a tool, they are not used blindly. For example, we write out likelihood functions and explain how they are maximized. We describe where Wald, likelihood ratio, and score procedures come from. However, except in Appendix B, where we give a general introduction to likelihood methods, we do not frequently emphasize calculus or carry out mathematical analysis in the text. The use of calculus is mostly from a conceptual focus, rather than a mathematical one.

We therefore expect that this book will appeal to all readers with a basic background in regression analysis. At times, a rudimentary understanding of derivatives, integrals, and function maximization would be helpful, as would a very basic understanding of matrices, matrix multiplication, and finding inverses of matrices. However, the important points and application advice can be easily understood without these tools. We expect that advanced undergraduates in statistics and related fields will satisfy these prerequisites. Graduate students in statistics, biostatistics, and related fields will certainly have sufficient background for the book. In addition, many students and researchers outside these disciplines who possess the basic regression background should find this book useful both for its descriptions and motivations of the analysis methods and for its worked examples with R code.

The book does not require any prior experience with R. We provide an introduction to the essential features and functions of R in Appendix A. We also provide introductory details on the use of R in the earlier chapters to help inexperienced R users. Throughout the book, as new R functions are needed, their basic features are discussed in the text and their implementation shown with corresponding output. We focus on using R packages that are provided by default with the initial R installation. However, we make frequent use of other R packages when they are significantly better or contain functionality unavailable in the standard R packages. The book contains the code and output as it would appear in the R Console; we make minor modifications at times to the output only to save space within the book. Code provided in the book for plotting is often meant for color display rather than the actual black-and-white display shown in the print and some electronic editions.

The data set files and R programs that are referenced in each example are available from the book's website, `http://www.chrisbilder.com/categorical`. The programs include code used to create every plot and piece of output that we show. Many of these programs contain code to demonstrate additional features or to perform more detailed and complete analyses than what is presented in the book. We strongly recommend that the book and the website be used in tandem, both for teaching and for individual learning. The website also contains many "extras" that can help readers learn the material. Most importantly, we post videos from one of us teaching a course on the subject. These videos include live, in-class recordings that are synchronized with recordings of a tablet computer screen. Instructors may find these videos useful (as we have) for a blended or flipped classroom setting. Readers outside of a classroom setting may also find these videos especially useful as a substitute for a short-course on the subject.

The first four chapters of the book are organized by type of categorical response variable. Within each of these chapters, we first introduce the measurement type, followed by the basic distributional model that is most commonly used for that type of measurement. We slowly generalize to simple regression structures, followed by multiple regressions including transformations, interactions, and categorical explanatory variables. We conclude each of these chapters with some important special cases. Chapter 5 follows with model building and assessment methods for the response variables in the first four chapters. A final chapter discusses additional topics presented as extensions to the previous chapters. These topics include solutions to problems that are frequently mishandled in practice, such as how to incorporate diagnostic testing error into an analysis, the analysis of data from "choose-all-that-apply" questions, and methods for analyzing data arising under a complex survey sampling design. Many of these topics are broad enough that entire books have been written about them, so our treatment in Chapter 6 is meant to be introductory.

For instructors teaching a one-semester course with the book, we recommend covering most of Chapters 1–5. The topics in Chapter 6 provide supplemental material for readers to learn on their own or to provide an instructor a means to go beyond the basics. In particular, topics from Chapter 6 can make good class projects. This helps students gain experience in teaching themselves extensions to familiar topics, which they will face later in industry or in research.

An extensive set of exercises is provided at the end of each chapter (over 65 pages in all!). The exercises are deliberately variable in scope and subject matter, so that instructors can choose those that meet the particular needs of their own students. For example, some carry out an analysis step by step, while others present a problem and leave the reader to choose and implement a solution. An answer key to the exercises is available for instructors using the book for a course. Details on how to obtain the answer key are available through the book's website.

We could not have written this book without the help and support of many people. First and foremost, we thank our families, and especially our wives, Kimberly and Marie, who put in extra effort on our behalf so that we could reserve time to work on the book. We owe them a huge debt for their support and tolerance, but we are hoping that they will settle for a nice dinner. We thank Rob Calver and his staff at CRC Press for their continued support and encouragement during the writing process. We also thank the hundreds of students who have taken categorical courses from us over the last seventeen years. Their feedback helped us to hone the course material and its presentation to what they are today. We especially thank one of our students, Natalie Koziol, who wrote the MRCV package used in Section 6.4 and made the implementation of those methods available to R users. This book was written in LaTeX through LyX, and we are grateful to the many contributors to these open-source projects. Finally, we need to thank our past and present colleagues and mentors at Iowa State, Kansas State, Oklahoma State, Nebraska, and Simon Fraser Universities who have both supported our development and brought us interesting and challenging problems to work on.

Christopher R. Bilder and Thomas M. Loughin
Lincoln, NE and Surrey, BC

The Authors

Christopher R. Bilder is a Professor in the Department of Statistics at the University of Nebraska-Lincoln. Bilder has been the Principal Investigator for grants from the National Science Foundation and the National Institutes of Health involving research into categorical data analysis problems. His research has been published in a diverse set of outlets ranging from the *Journal of the American Statistical Association* to *Chance*. For his categorical data research, Bilder was awarded the American Statistical Association's Outstanding Statistical Application Award (twice) and the Best Paper in *Biometrics* by an International Biometric Society Member Award. Bilder is a Fellow of the American Statistical Association.

For over 20 years, Bilder has taught a course on categorical data analysis to students majoring in statistics and to students majoring in a wide variety of other fields of study. He also has been a pioneer in using technology in and outside the classroom through the use of class video capturing, course websites, distance learning, blended learning, and tablets during his career. Bilder's YouTube Channel at https://www.youtube.com/ChrisBilder hosts his videos from courses, workshops, and presentations.

Thomas M. Loughin is a Professor in the Department of Statistics and Actuarial Science at Simon Fraser University in Burnaby, BC, Canada. He was Chair of the department from 2014–2019 before coming to his senses. He previously held a faculty position at Kansas State University for 13 years. At K-State, he was partly funded by the College of Agriculture to provide statistical collaboration and consulting for faculty and students there. As a result, he has been active in statistical consulting, particularly the design and analysis of experiments, for most of his career. As a consultant and a teacher, he specializes in communication with subject-matter experts and students, re-expressing complex statistical concepts into language that is easy to understand.

Tom's research interests include categorical data analysis, statistical learning techniques, particularly tree-based ensembles, and sports analytics. He is a Fellow of the American Statistical Association and an accredited professional statistician, maintaining both P.Stat. (SSC) and PStat® (ASA). He has served on numerous committee positions within SSC and ASA and has held positions on the editorial boards of *Biometrics*, *Technometrics*, *The American Statistician*, and *Developmental Medicine & Child Neurology*.

Tom is an avid curler, playing in two leagues at the Cloverdale Curling Club. He is a craft beer aficionado and a member of the world's longest-running fantasy baseball league,[*] the Snedecor Baseball Mail League.

[*]Claim unconfirmed but unrefuted.

Chapter 1

Analyzing a binary response, part 1: introduction

Yes or no. Success or failure. Death or survival. For or against. Positive or negative. Binary responses may be the most prevalent type of categorical data. The purpose of this chapter is to show how to estimate and make inferences about a binary response probability and related quantities. We begin in Section 1.1 by examining a homogeneous population where there is one overall probability to be estimated. We generalize this situation in Section 1.2 to the setting where sampled items come from one of two groups.

Throughout this chapter, we emphasize the use of R with detailed code explanations. This is done on purpose because readers may have little R experience beyond the introduction in Appendix A. Future chapters will still emphasize the use of R but spend less time explaining code.

1.1 One binary variable

1.1.1 Bernoulli and binomial probability distributions

Almost every statistical analysis begins with some kind of statistical model. A statistical model generally takes the form of a probability distribution that attempts to quantify the uncertainty that comes with observing a new response. The model is intended to represent the unknown phenomenon that governs the observation process. At the same time, the model needs to be convenient to work with mathematically, so that inference procedures such as confidence intervals and hypothesis tests can be developed. Selecting a model is typically a compromise between two competing goals: providing a more detailed approximation to the process that generates the data and providing inference procedures that are easy to use.

In the case of binary responses, the natural model is the Bernoulli probability distribution (or just "Bernoulli distribution" for short). Let Y denote a random variable that has two possible outcomes, numerically represented as 0 and 1. Typically, we will say $Y = 1$ is a success and $Y = 0$ is a failure. For example, a success would be a basketball free throw attempt that is good or an individual who is cured of a disease by a new drug; a failure would be a free throw attempt that is missed or an individual who is not cured. We denote the probability of success as $P(Y = 1) = \pi$ and the probability of failure as $P(Y = 0) = 1 - \pi$. Then, Y has a Bernoulli distribution. The Bernoulli probability mass function (PMF) for Y combines the two probability statements into one expression:

$$P(Y = y) = \pi^y(1 - \pi)^{1-y}$$

for $y = 0$ or 1, where we use the standard convention that a capital letter Y denotes the random variable and the lowercase letter y denotes a possible value of Y. The expected value of Y is $E(Y) = \pi$, and the variance of Y is $Var(Y) = \pi(1 - \pi)$.

DOI: 10.1201/9781003093091-1

Often, one observes multiple Bernoulli random variable responses through repeated sampling or trials in identical settings, resulting in what are known as *Bernoulli trials*. This leads to defining separate random variables for each trial, $Y_1, \ldots, Y_n$, where n is the number of trials. If all trials are identical and independent, we can treat $W = \sum_{i=1}^{n} Y_i$ as a binomial random variable with PMF of

$$P(W = w) = \binom{n}{w} \pi^w (1 - \pi)^{n-w} \tag{1.1}$$

for $w = 0, \ldots, n$. The combination function $\binom{n}{w} = n!/[w!(n-w)!]$ counts the number of ways w successes and $n-w$ failures can be ordered. The expected value of W is $E(W) = n\pi$, and the variance of W is $Var(W) = n\pi(1-\pi)$. Notice that the Bernoulli distribution is a special case of the binomial distribution when $n = 1$.

We next show how R can be used to examine properties of the binomial distribution.

Example: Binomial distribution in R (Binomial.R)

The purpose of this example is to calculate simple probabilities using a binomial distribution and to show how these calculations are performed in R. We will be very basic with our use of R in this example. If you find its use difficult, we recommend reading Appendix A before proceeding further.

Consider a binomial random variable counting the number of successes from an experiment that is repeated $n = 5$ times, and suppose that the probability of success is $\pi = 0.6$. For example, suppose an individual has this success rate in a particular card game or shooting a basketball into a goal from a specific location. We can calculate the probability of each number of successes, $w = 0, 1, 2, 3, 4, 5$, using Equation 1.1. For example, the probability of 1 success out of 5 trials is

$$P(W = 1) = \binom{5}{1} 0.6^1 (1 - 0.6)^{5-1} = \frac{5!}{1!4!} 0.6^1 0.4^4 = 0.0768.$$

This calculation is performed in R using the `dbinom()` function.

```
> dbinom(x = 1, size = 5, prob = 0.6)
[1] 0.0768
```

Within the function, the `x` argument denotes the observed value of the binomial random variable (what we are calling w), the `size` argument is the number of trials (n), and the `prob` argument is π. We could have used `dbinom(1, 5, 0.6)` to obtain the same probability as long as the numerical values were in the same order as the arguments within the function (a full list of arguments and their order is available in the help for `dbinom()`). Generally, we will always specify the argument names in our code except with the most basic functions.

We find all of the probabilities for $w = 0, \ldots, 5$ by changing the `x` argument to `0:5`, which represents 0 to 5 by 1.

```
> dbinom(x = 0:5, size = 5, prob = 0.6)
[1] 0.01024 0.07680 0.23040 0.34560 0.25920 0.07776
```

To display these probabilities in a more descriptive format, we save them into an object and print from a data frame using the `data.frame()` function.

```
> pmf <- dbinom(x = 0:5, size = 5, prob = 0.6)
> save <- data.frame(w = 0:5, prob = round(x = pmf, digits = 4))
> save
  w   prob
1 0 0.0102
2 1 0.0768
3 2 0.2304
4 3 0.3456
5 4 0.2592
6 5 0.0778
```

Note that we could have used different names than `pmf` and `save` for our objects if desired. The `round()` function rounds the values in the `pmf` object to 4 decimal places.

We plot the PMF using the `plot()` and `abline()` functions.

```
> plot(x = save$w, y = save$prob, type = "h", xlab = "w", ylab =
    "P(W=w)", main = "Plot of a binomial PMF for n=5, pi=0.6",
    panel.first = grid(col = "gray", lty = "dotted"), lwd = 3)
> abline(h = 0)
```

Figure 1.1 shows the resulting plot. Appendix A.7.2 describes most of the arguments within `plot()`, so we provide only brief descriptions here. The `x` and `y` arguments specify the x- and y-axis values where we use the `$` symbol to access parts of the `save` data frame. The `type = "h"` argument value specifies that vertical bars are to be plotted from 0 to the values given in the `y` argument (the letter `h` is used to represent histogram-like bars). The `main` argument contains the plot title.[1] The `abline()` function plots a horizontal line at 0 to anchor the bottom of each vertical line.

The simpler specification `plot(x = save$w, y = save$prob, type = "h")` produces a plot similar to Figure 1.1, but our extra arguments make the plot easier to interpret.

Assumptions

The binomial distribution is a reasonable model for the distribution of successes in a given number of trials as long as the process of observing repeated trials satisfies certain assumptions. Those assumptions are

1. THERE ARE n IDENTICAL TRIALS. This refers to the process by which the trials are conducted. The action that results in the trial and the measurement taken must be the same in each trial. The trials cannot be a mixture of different types of actions or measurements.

2. THERE ARE TWO POSSIBLE OUTCOMES FOR EACH TRIAL. This is generally just a matter of knowing what is measured. However, there are cases where a response measurement has more than two levels, but interest lies only in whether or not one

[1] If it is desired to use a plot title with better notation, we could have used `main = expression(paste("Plot of a binomial PMF for ", italic(n) == 5, " and ", italic(pi) == 0.6))` to obtain "Plot of the binomial PMF for $n = 6$ and $\pi = 0.6$." In this code, `expression()` allows us to include mathematical symbols and `paste()` combines these symbols with regular text. Please see Appendix A.7.4 for more information.

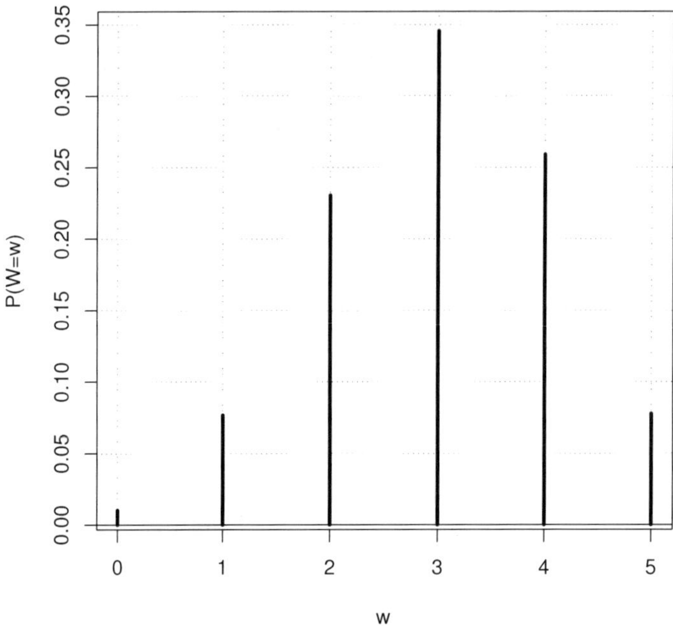

Figure 1.1: PMF for W.

particular level occurs. In this case, the special level can be considered "success" and all remaining levels "failure."

3. THE TRIALS ARE INDEPENDENT OF EACH OTHER. In particular, there is nothing in the conduct of the trials that would cause any subset of trials to behave more similarly to one another. Counterexamples include (a) measuring trials in "clusters," where the units on which success or failure is measured are grouped together somehow before observation, and (b) trials measured in a time series, where trials measured close together in time might react more similarly than those measured far apart.

4. THE PROBABILITY OF SUCCESS REMAINS CONSTANT FOR EACH TRIAL. Any factor that can affect the probability of success needs to be held constant from trial to trial. Because these factors are not always known in advance, this can be a very difficult condition to confirm. We often can only confirm that the "obvious" factors are not changing, and then merely assume that others are not as well.

5. THE RANDOM VARIABLE OF INTEREST W IS THE NUMBER OF SUCCESSES. Specifically, this implies that we are *not* interested in the order in which successes and failures occur, but rather only in their total counts.

The next two examples detail these assumptions with respect to applications, and they demonstrate how it can be difficult to ensure that these assumptions are satisfied.

Example: Field goal kicking

In American and Canadian football, points can be scored by a field goal, which involves kicking a ball through a target area (goal) at each end of the field. Suppose an experiment is conducted where a placekicker successively attempts five field goal kicks during practice. A success occurs on one trial when the football is kicked over the crossbar and between the two uprights of the goal posts. A failure occurs when the football does not achieve this result (the football is kicked to the left or right of both uprights or falls short of the crossbar). We want to use these results to estimate the placekicker's true probability of success, so we record how many kicks are successful.

To use the binomial distribution here, the experiment needs to satisfy the following conditions:

1. THERE ARE n IDENTICAL TRIALS. In this case, $n = 5$ field goals are attempted. The action is always kicking a football in the same way, and the measurement of success is made the same way each time.

2. THERE ARE TWO POSSIBLE OUTCOMES FOR EACH TRIAL. Each field goal attempt results in a success or a failure. Notice that we could further divide failures into other categories, such as too short, long enough but to the left of the uprights, or long enough but to the right of the uprights. If our interest is only in whether the kick is successful, then differentiating among the modes of failure is unnecessary.

3. THE TRIALS ARE INDEPENDENT OF EACH OTHER. Given that these field goals are attempted successively, this may be difficult to satisfy. For example, if one field goal attempt is missed to the left, the kicker may compensate by trying to kick farther to the right on the next attempt. On the other hand, the independence assumption may be approximately satisfied by a placekicker who tries to apply the exact same technique on each trial.

4. THE PROBABILITY OF SUCCESS REMAINS CONSTANT FOR EACH TRIAL. To make sure this is true, the field goals need to be attempted from the same distance under the same surrounding conditions. Weather conditions need to be constant. The same placekicker, ball, and goalposts are used each time. We assume that fatigue does not affect the placekicker for this small number of attempts. If the attempts occur close together in time, then it may be reasonable to assume that extraneous factors are reasonably constant as well, at least enough so that they do not have a substantial effect on the success of the field goals.

5. THE RANDOM VARIABLE OF INTEREST W IS THE NUMBER OF SUCCESSES. As we will see in Section 1.1.2, we need only to record W $(= 0, 1, 2, 3, 4, \text{or } 5)$, not the entire sequence of trial results, when estimating the probability of success.

Example: Disease prevalence

The *prevalence* of a disease is the proportion of a population that is afflicted with that disease. This is equivalent to the probability that a randomly selected member of the population has the disease. Many public health studies are performed to understand disease prevalence, because knowing the prevalence is the first step toward solving societal problems caused by the disease. For example, suppose there is concern that a new infectious disease may be transmitted to individuals through blood donations. One way to examine the disease prevalence would be to take a sample of 1,000 blood donations and test each for the disease.

To use the binomial distribution here, this setting needs to satisfy the following conditions:

1. THERE ARE n IDENTICAL TRIALS. In this case, $n = 1000$ blood donations are examined. Each blood donation needs to be collected and tested the same way. In particular, trials would *not* be identical if different diagnostic measures were used on different donations to determine presence of disease.

2. THERE ARE TWO POSSIBLE OUTCOMES FOR EACH TRIAL. Each blood donation is either positive or negative for the disease. Making this determination is not necessarily as straightforward as it may seem. Often, there is a numerical score reported from the results of an assay, such as the number of amplification cycles needed to cross a particular threshold. Also, for some assays, multiple thresholds may be used leading to responses such as positive, indeterminate, or negative.

3. THE TRIALS ARE INDEPENDENT OF EACH OTHER. This may be difficult to satisfy completely. For example, if married spouses are included in the sample, then presence of the disease in one spouse's donation may suggest a greater chance that the other spouse's donation will also test positive. Independence can be ensured by random sampling from a large population of donations but may always be in question when any non-random sampling method is used.

4. THE PROBABILITY OF SUCCESS REMAINS CONSTANT FOR EACH TRIAL. Each sampled donation needs to have the same probability of having the disease. This could be unreasonable if there are factors, such as risky behavior, that make donations for certain subpopulations more likely to have the disease than others. Similarly, if donations are collected over an extended period of time, prevalence may not be constant during that period.

5. THE RANDOM VARIABLE OF INTEREST W IS THE NUMBER OF SUCCESSES. There are $W = 0, \ldots, 1000$ possible positive blood donations. To estimate prevalence, we need to know how many positive donations there are but not specifically which ones are positive.

The previous examples show that it may be difficult to satisfy all of the assumptions for a binomial model. However, the binomial model may still be used as an approximation to the true model in a given problem, as long as violations are expected to be mild. In that case, the potentially violated assumptions then would need to be identified in any stated results. Alternatively, if assumptions are not satisfied, there are other models and procedures that can be used to analyze binary responses. In particular, if the probability of success does not remain constant for each trial—for example, if disease probability is related to certain risky behaviors—we may be able to identify and measure the factors causing the variations and then use a regression model for the probability of success (to be discussed in Chapter 2).

Simulating a binomial sample

What does a sample look like when a binomial distribution is the statistical model? Of course, the observed values only can be $0, 1, \ldots, n$. The proportion of observed values that are $0, 1, \ldots, n$ are governed by the PMF and the parameter π within it. The mean and variance of these observed values are also controlled by the PMF and π. These properties are easily derived mathematically using basic definitions of mean and variance. In more complex problems, however, properties of statistics are much harder to derive mathematically.

We show in the next example how to *simulate* a sample using R so that we can check whether theory matches what actually happens. This example also introduces Monte Carlo computer simulation as a valuable tool for evaluating a statistical procedure. All statistical procedures have assumptions underlying their mathematical framework. Monte Carlo simulation is especially useful in assessing how well procedures perform when these assumptions are violated. For example, we may want to know if a confidence interval that is designed to work in large samples maintains its stated confidence level when the sample size is small.

A Monte Carlo simulation works by creating a computer version of the population we are studying, sampling from this virtual population in a prescribed way, performing the statistical analysis that we are studying, and measuring how it performs. The details of each step vary from one problem to the next. In all cases, we draw a "large" number of samples from the virtual population. In so doing, the *law of large numbers* assures us that the average performance measured across the samples is likely to be close to the true performance of the procedure in this context (a more mathematical definition of the law of large numbers is contained on p. 232 of Casella and Berger, 2002). We follow this prescription in the example below.

Example: Simulation with the binomial distribution in R (Binomial.R)

Below is the code that simulates 1,000 random observations of W from a binomial distribution with $\pi = 0.6$ and $n = 5$.

```
> set.seed(4848)
> bin5 <- rbinom(n = 1000, size = 5, prob = 0.6)
> bin5[1:10]
[1] 3 2 4 1 3 1 3 3 3 4
```

The `set.seed()` function sets a seed number for the simulated observations. Without going into the details behind random number generation, the seed number specification allows us to obtain identical simulated observations each time the same code is executed.[2] The `rbinom()` function simulates the observations, where the `n` argument gives the number of observations for W that we wish to simulate (not n as in the number of trials). The `bin5` object contains 1,000 values, where the first 10 are printed.

The population mean and variance for W are

$$E(W) = n\pi = 5 \times 0.6 = 3$$

and

$$Var(W) = n\pi(1 - \pi) = 5 \times 0.6 \times 0.4 = 1.2.$$

We calculate the sample mean and variance of the simulated observations using the `mean()` and `var()` functions.

```
> mean(bin5)
[1] 2.991
> var(bin5)
[1] 1.236155
```

[2]It is best not to use the same seed number for other simulated data examples. There are many different ways that a seed number can be chosen. One way is to choose a number by running `runif(n = 1)` within R (this simulates one observation from a Uniform(0,1) distribution) and taking the first few significant digits. We will often simply use the first four digits for convenience. More digits should be chosen whenever a seed number is needed for research purposes.

The sample mean and variance are very similar to $E(W)$ and $Var(W)$, as expected. If a larger number of observations were simulated, say 10,000, we generally would expect these sample measures to be even closer to their population quantities because of the law of large numbers.

To examine how well the observed frequency distribution follows the PMF, we use `table()` to find the frequencies of each possible response and then use `hist()` to plot a histogram of the relative frequencies.

```
> table(x = bin5)
x
0   1   2   3   4   5
12  84 215 362 244  83
> hist(x = bin5, main = "Binomial with n=5, pi=0.6, 1000 bin.
    observations", probability = TRUE, breaks = -0.5:5.5, ylab =
    "Relative frequency", col = NA)
```

For example, we see that $w = 3$ is observed with a relative frequency of $362/1000 = 0.362$. We found earlier that $P(W = 3) = 0.3456$, which is very similar. The histogram is in Figure 1.2, and its shape is also very similar to the PMF plot in Figure 1.1. Within the `hist()` function, we used `probability = TRUE` to provide the relative frequencies (`probability = FALSE` gives the frequencies, which is the default), `breaks` to specify the classes for the bars to be -0.5 to 5.5 by 1 (the bars will not be drawn correctly here without specifying `breaks` in this situation), and `col = NA` so that the bars are not filled in with a color.

1.1.2 Inference for the probability of success

The purpose of this section is to estimate and make inferences about the probability of success parameter π from the Bernoulli distribution. We start by estimating the parameter using its maximum likelihood estimate because it is relatively easy to compute and has properties that make it appealing in large samples. Next, confidence intervals for the true probability of success are presented and compared. Many different confidence intervals have been proposed in the statistics literature. We will present the simplest—but worst—interval first and then offer several better alternatives. We conclude this section with hypothesis tests for π.

For those readers unfamiliar with estimation and inference procedures associated with the likelihood function, we encourage you to read Appendix B first for an introduction. We will reference specific parts of the appendix within this section.

Maximum likelihood estimation and inference

As described in Appendix B, a likelihood function is a function of one or more parameters conditional on the observed data. The likelihood function for π when $y_1, \ldots, y_n$ are observations from a Bernoulli distribution is

$$
\begin{aligned}
L(\pi|y_1, \ldots, y_n) &= P(Y_1 = y_1) \times \cdots \times P(Y_n = y_n) \\
&= \pi^w (1-\pi)^{n-w}. \qquad (1.2)
\end{aligned}
$$

Alternatively, when we only record the number of successes out of a number of trials, the likelihood function for π is simply $L(\pi|w) = P(W = w) = \binom{n}{w} \pi^w (1-\pi)^{n-w}$. The value of

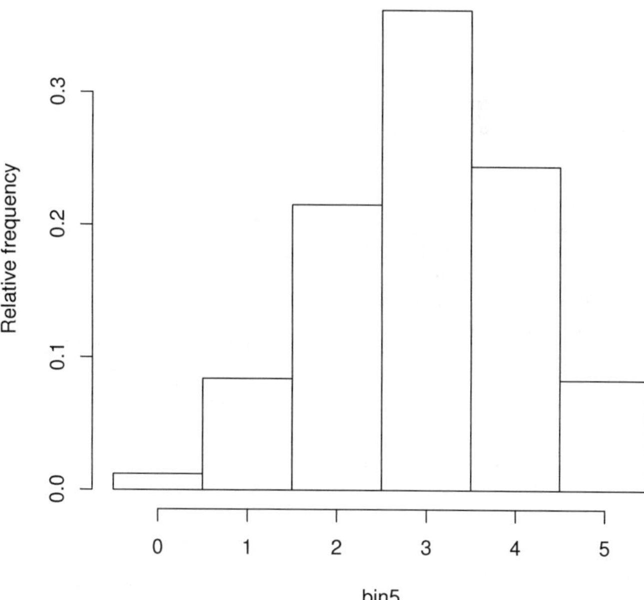

Figure 1.2: Histogram for the observed values of w.

π that maximizes the likelihood function is considered to be the most plausible value for the parameter, and it is called the maximum likelihood estimate (MLE). Appendix B.3 shows that the MLE of π is $\hat{\pi} = w/n$, which is simply the observed proportion of successes. This is true for both $L(\pi|y_1, \ldots, y_n)$ and $L(\pi|w)$ because $\binom{n}{w}$ contains no information about π.

Because $\hat{\pi}$ will vary from sample to sample, it is a statistic and has a corresponding probability distribution.[3] This probability distribution is called the *sampling distribution* of the statistic. As with all MLEs, $\hat{\pi}$ has an approximate normal distribution in large samples (see Appendix B.3.3). The mean of the normal distribution is π, and the variance is found from

$$\begin{aligned}
\widehat{Var}(\hat{\pi}) &= -E\left\{\frac{\partial^2 \log\left[L(\pi|W)\right]}{\partial \pi^2}\right\}^{-1}\Bigg|_{\pi=\hat{\pi}} \\
&= -E\left\{-\frac{W}{\pi^2} - \frac{n-W}{(1-\pi)^2}\right\}^{-1}\Bigg|_{\pi=\hat{\pi}} \\
&= \left[\frac{n}{\pi} + \frac{n}{1-\pi}\right]^{-1}\Bigg|_{\pi=\hat{\pi}} \\
&= \frac{\hat{\pi}(1-\hat{\pi})}{n},
\end{aligned} \tag{1.3}$$

[3] In order for $\hat{\pi}$ to have a probability distribution, it needs to be a random variable. Thus, we are actually using $\hat{\pi} = W/n$ in this case. We could have defined W/n as $\hat{\Pi}$ instead, but this level of formality is unnecessary here. It will be apparent from a statistic's use whether it is a random or observed quantity.

where log(·) is the natural log function (see Appendix B.3.4). We can write the distribution as $\hat{\pi} \dot{\sim} N(\pi, \widehat{Var}(\hat{\pi}))$ where $\dot{\sim}$ denotes "approximately distributed as." The approximation tends to be better as the sample size grows larger.

Wald confidence interval

Using this normal distribution, we can treat $(\hat{\pi} - \pi)/\widehat{Var}(\hat{\pi})^{1/2}$ as an approximate standard normal quantity (see Appendix B.5). Thus, for any $0 < \alpha < 1$, we have

$$P\left(Z_{\alpha/2} < \frac{\hat{\pi} - \pi}{\sqrt{\widehat{Var}(\hat{\pi})}} < Z_{1-\alpha/2}\right) \approx 1 - \alpha,$$

where Z_a is the ath quantile from a standard normal distribution (e.g., $Z_{0.975} = 1.96$). After rearranging terms and recognizing that $-Z_{\alpha/2} = Z_{1-\alpha/2}$, we obtain

$$P\left(\hat{\pi} - Z_{1-\alpha/2}\sqrt{\widehat{Var}(\hat{\pi})} < \pi < \hat{\pi} + Z_{1-\alpha/2}\sqrt{\widehat{Var}(\hat{\pi})}\right) \approx 1 - \alpha.$$

Now, we have an approximate probability that has the parameter π centered between two statistics. When we replace $\hat{\pi}$ and $\widehat{Var}(\hat{\pi})$ with observed values from the sample, we obtain the $(1 - \alpha)100\%$ confidence interval for π as

$$\hat{\pi} - Z_{1-\alpha/2}\sqrt{\hat{\pi}(1-\hat{\pi})/n} < \pi < \hat{\pi} + Z_{1-\alpha/2}\sqrt{\hat{\pi}(1-\hat{\pi})/n}.$$

This is the usual interval for a probability of success that is given in most introductory statistics textbooks. Confidence intervals based on the approximate normality of MLEs are called "Wald confidence intervals" because Wald (1943) was the first to show this property of MLEs in large samples.

When w is close to 0 or n, two problems occur with this interval:

1. Calculated limits may be less than 0 or greater than 1, which is outside the boundaries for a probability.
2. When $w = 0$ or n, $\sqrt{\hat{\pi}(1-\hat{\pi})/n} = 0$ for $n > 0$. This leads to the lower and upper limits to be exactly the same (0 for $w = 0$ or 1 for $w = n$).

We will discuss additional problems with the Wald interval shortly.

Example: Intervals for π (CIpi.R)

Suppose $w = 4$ successes are observed out of $n = 10$ trials. The 95% Wald confidence interval for π is $0.4 \pm 1.96\sqrt{0.4(1-0.4)/10} = (0.0964,\ 0.7036)$, where we use the shorthand notation within parentheses to mean $0.0964 < \pi < 0.7036$. The R code below shows how these calculations are carried out.

```
> w <- 4
> n <- 10
> alpha <- 0.05
> pi.hat <- w/n
> var.wald <- pi.hat*(1-pi.hat)/n
> lower <- pi.hat - qnorm(p = 1-alpha/2) * sqrt(var.wald)
> upper <- pi.hat + qnorm(p = 1-alpha/2) * sqrt(var.wald)
> round(data.frame(lower, upper), 4)
  lower  upper
```

```
1 0.0964 0.7036
```

In the code, we use the qnorm() function to find the $1-\alpha/2$ quantile from a standard normal distribution. We can calculate the interval quicker by taking advantage of how R performs vector calculations.

```
> round(pi.hat + qnorm(p = c(alpha/2, 1-alpha/2)) *
    sqrt(var.wald), 4)
[1] 0.0964 0.7036
```

See Appendix A.4 for a similar example.

The confidence interval is quite wide and may not be meaningful for some applications. However, it does give information on a range for π, which may be useful in hypothesis testing situations. For example, a test of $H_0 : \pi = 0.5$ vs. $H_a : \pi \neq 0.5$ would not reject H_0 because 0.5 is within this range. If the test was instead $H_0 : \pi = 0.8$ vs. $H_a : \pi \neq 0.8$, there is evidence to reject the null hypothesis. Other ways to perform these tests with a test statistic and a p-value will be discussed shortly.

The inferences for π from the Wald confidence interval rely on the underlying normal distribution approximation for the maximum likelihood estimator. For this approximation to work well, we need a large sample, and, unfortunately, the sample size in the last example was quite small. Furthermore, notice that $\hat{\pi}$ can take on only 11 different possible values in the last example: 0/10, 1/10, ..., 10/10, but a normal distribution is a continuous function. These problems lead the Wald conference interval to be *approximate*, in the sense that the probability that the interval covers (i.e., contains or includes) the parameter is not necessarily equal to the stated level $1-\alpha$. This probability is known as a confidence interval's *coverage* or *true confidence level*. The quality of the approximation varies with n and π, and as we will see later, the Wald interval generally has coverage less than $1-\alpha$. Such an interval is called a *liberal* interval. On the other hand, an interval with coverage in excess of the stated level is called *conservative*. While this latter property may seem to be a good quality, it can lead to intervals that are quite wide in comparison to others. We want confidence intervals that place the parameter within as narrow a range as possible, while maintaining at least the stated confidence level. If we wanted intervals that had greater coverage, we would have stated a higher confidence level!

Brown et al. (2001) present a thorough review of most competing intervals for π. We present their recommendations next along with our own thoughts on the best intervals.

Wilson confidence interval

When $n < 40$, Brown et al. (2001) recommend the Wilson interval or the Jeffreys interval because they maintain true confidence levels closer to the stated level than other intervals. The Wilson interval formula is found by examining the test statistic

$$Z_S = \frac{\hat{\pi} - \pi_0}{\sqrt{\pi_0(1-\pi_0)/n}},$$

which is a *score test* statistic often used for a test of $H_0 : \pi = \pi_0$ vs. $H_a : \pi \neq \pi_0$, where $0 < \pi_0 < 1$ (see Appendix B.5). The variance in the denominator of Z_S is computed assuming that the null hypothesis is true, rather than using the unrestricted estimate based on the data. This leads to the advantage that the denominator is not 0 whenever $w = 0$ or n.

We can approximate the distribution of Z_S with a standard normal to obtain $P(-Z_{1-\alpha/2} < Z_S < Z_{1-\alpha/2}) \approx 1 - \alpha$. Treating the approximation as an equality, the Wilson interval contains the set of all possible values of π_0 that satisfy the equation. Conversely, the set of all possible values for π_0 that lead to a rejection of the null hypothesis are outside of the confidence interval. The process of forming an interval from a hypothesis test procedure like this is often referred to as "inverting the test." See also Appendix B.5.2. Because the Wilson interval is based on a score test, it is often referred to as a *score interval*.

The interval endpoints are found by setting Z_S equal to $\pm Z_{1-\alpha/2}$, and applying the quadratic formula to solve for π_0 (see Exercise 12). Thus, the $(1-\alpha)100\%$ Wilson interval is

$$\tilde{\pi} \pm \frac{Z_{1-\alpha/2}\sqrt{n}}{n + Z_{1-\alpha/2}^2}\sqrt{\hat{\pi}(1-\hat{\pi}) + \frac{Z_{1-\alpha/2}^2}{4n}}, \quad (1.4)$$

where

$$\tilde{\pi} = \frac{w + Z_{1-\alpha/2}^2/2}{n + Z_{1-\alpha/2}^2}$$

can be thought of as an adjusted estimate of π. This interval is named after Wilson (1927) who first proposed finding an interval for π in this manner. Note that the Wilson interval always has limits between 0 and 1.

The Wald and Wilson confidence intervals discussed so far are *frequentist* inference procedures. The "confidence" associated with these types of inference procedures comes about through repeating the process of taking a sample and calculating a confidence interval each time. Thus, the confidence interval for π is interpreted as follows:

> If the sampling process were repeated many times and a confidence interval were computed with each sample, we would expect $(1-\alpha)100\%$ of all intervals to contain the parameter π.

This wording is somewhat awkward, so we often use a less formal interpretation,

> We have $(1-\alpha)100\%$ confidence that the interval from <lower limit> to <upper limit> contains the true value of the parameter π.

The appropriate lower and upper limits are inserted within the interpretation. Here, we can think of "confidence" as describing our faith in the procedure, as measured by the probability that the procedure succeeds in its task of including the parameter, not the probability that this particular interval contains the parameter. A single interval calculated from a sample either does or does not contain the parameter, so it is inappropriate to say it has a *probability* (other than 0 or 1) of containing the parameter. This is a confusing aspect to confidence intervals, causing them to be misinterpreted frequently in practice.

On the other hand, a Bayesian credible interval does have a $(1-\alpha)100\%$ probability of containing the parameter, because parameters are random variables in the Bayesian paradigm. A Jeffreys interval for π, also recommended by Brown et al. (2001), is one such Bayesian interval. We postpone its discussion until Section 6.6, when we describe Bayesian inference procedures in detail.

Agresti-Coull confidence interval

The Wilson interval is our preferred choice for a confidence interval for π with all sample sizes. However, Brown et al. (2001) recommend the Agresti-Coull interval (Agresti and Coull, 1998) for $n \geq 40$, primarily because it has a little simpler expression that closely resembles the popular Wald interval. Furthermore, the limits for this interval are often very similar to the Wilson interval. The $(1-\alpha)100\%$ Agresti-Coull interval is

$$\tilde{\pi} - Z_{1-\alpha/2}\sqrt{\frac{\tilde{\pi}(1-\tilde{\pi})}{n+Z_{1-\alpha/2}^2}} < \pi < \tilde{\pi} + Z_{1-\alpha/2}\sqrt{\frac{\tilde{\pi}(1-\tilde{\pi})}{n+Z_{1-\alpha/2}^2}}.$$

The interval is essentially the Wald interval where $Z_{1-\alpha/2}^2/2$ successes and $Z_{1-\alpha/2}^2/2$ failures are added to the observed data. Specifically, for $\alpha = 0.05$, this means that about two successes and two failures are added because $Z_{1-0.05/2} = 1.96 \approx 2$. Similar to the Wald interval, this interval has the undesirable property that it may have limits less than 0 or greater than 1.

Example: Intervals for π (CIpi.R)

Suppose again $w = 4$ successes are observed out of $n = 10$ trials. For a 95% confidence interval, the adjusted estimate of π is

$$\tilde{\pi} = \frac{w + Z_{1-\alpha/2}^2/2}{n + Z_{1-\alpha/2}^2} = \frac{4 + 1.96^2/2}{10 + 1.96^2} = 0.4278.$$

The 95% Wilson interval limits are

$$\begin{aligned}&\tilde{\pi} \pm \frac{Z_{1-\alpha/2}\sqrt{n}}{n+Z_{1-\alpha/2}^2}\sqrt{\hat{\pi}(1-\hat{\pi}) + \frac{Z_{1-\alpha/2}^2}{4n}}\\ &= 0.4278 \pm \frac{1.96\sqrt{10}}{10+1.96^2}\sqrt{0.4(1-0.4) + \frac{1.96^2}{4\times 10}}\end{aligned}$$

leading to an interval of $0.1682 < \pi < 0.6873$. The 95% Agresti-Coull interval limits are

$$\tilde{\pi} \pm Z_{1-\alpha/2}\sqrt{\frac{\tilde{\pi}(1-\tilde{\pi})}{n+Z_{1-\alpha/2}^2}} = 0.4278 \pm 1.96\sqrt{\frac{0.4278(1-0.4278)}{10+1.96^2}}$$

leading to an interval of $0.1671 < \pi < 0.6884$. Both confidence intervals have limits that are quite similar in this case, but are rather different from the Wald interval limits of (0.0964, 0.7036) that we calculated earlier.

Continuing from the last example, below is how the calculations are performed in R.

```
> pi.tilde <- (w + qnorm(p = 1-alpha/2)^2 / 2) / (n + qnorm(p =
   1-alpha/2)^2)
> pi.tilde
[1] 0.4277533

> # Wilson CI
> round(pi.tilde + qnorm(p = c(alpha/2, 1-alpha/2)) * sqrt(n) /
   (n + qnorm(p = 1-alpha/2)^2) * sqrt(pi.hat*(1-pi.hat) +
   qnorm(p = 1-alpha/2)^2/(4*n)), 4)
[1] 0.1682 0.6873

> # Agresti-Coull CI
> var.ac <- pi.tilde*(1-p.tilde) / (n + qnorm(p = 1-alpha/2)^2)
```

```
> round(pi.tilde + qnorm(p = c(alpha/2, 1-alpha/2)) *
    sqrt(var.ac), 4)
[1] 0.1671 0.6884
```

After calculating $\tilde{\pi}$, we calculate the Wilson and Agresti-Coull intervals. Note that executing part of a line of code can help highlight how it works. For example, one can execute `qnorm(p = c(alpha/2, 1-alpha/2))` to see that it calculates -1.96 and 1.96.

The `binom.confint()` function from the `binom` package can be used to simplify the calculations. Note that this package is not in the default installation of R, so it needs to be installed before its use (see Appendix A.5 for more information regarding package installation). Below is our use of the function.

```
> library(package = binom)
> binom.confint(x = w, n = n, conf.level = 1-alpha, methods =
    "all")
          method x  n      mean      lower     upper
1  agresti-coull 4 10 0.4000000 0.16711063 0.6883959
2     asymptotic 4 10 0.4000000 0.09636369 0.7036363
3          bayes 4 10 0.4090909 0.15306710 0.6963205
4        cloglog 4 10 0.4000000 0.12269317 0.6702046
5          exact 4 10 0.4000000 0.12155226 0.7376219
6          logit 4 10 0.4000000 0.15834201 0.7025951
7         probit 4 10 0.4000000 0.14933907 0.7028372
8        profile 4 10 0.4000000 0.14570633 0.6999845
9            lrt 4 10 0.4000000 0.14564246 0.7000216
10     prop.test 4 10 0.4000000 0.13693056 0.7263303
11        wilson 4 10 0.4000000 0.16818033 0.6873262
```

The function calculates 11 different intervals for π. Specific intervals can be requested by using their appropriate name with the `methods` argument. Alternatively, `methods = "all"` calculates each interval, which is the default. The first, second, and eleventh intervals are the Agresti-Coull, Wald ("`asymptotic`"), and Wilson intervals, respectively. Please see this function's help for more information on the other intervals. The end-of-chapter exercises discuss some of these in more detail.

Clopper-Pearson confidence interval

The Clopper-Pearson interval (Clopper and Pearson, 1934) is the last confidence interval for π that we discuss in this section. While Brown et al. (2001) remark that the interval is "wastefully conservative and it is not a good choice for practical use," this interval does have a unique property that the other intervals do not: the true confidence level is always equal to or greater than the stated level. To achieve this true confidence level, the interval is usually wider than most other intervals for π.

The interval uses the relationship between the binomial distribution and the beta distribution to guarantee that its true confidence level is maintained at or above the desired level (see #2.40 on p. 82 of Casella and Berger, 2002 for this distributional relationship). In fact, because the actual or *exact* distribution for W is used, the interval is called an *exact inference* procedure. There are many other exact inference procedures available for statistical problems, and some of these are discussed in Section 6.2.

To review the beta distribution, let V be a beta random variable. The probability density function (PDF) for V is

$$f(v; a, b) = \frac{\Gamma(a+b)}{\Gamma(a)\Gamma(b)} v^{a-1}(1-v)^{b-1},\ 0 < v < 1, \tag{1.5}$$

where $a > 0$ and $b > 0$ are parameters and $\Gamma(\cdot)$ is the gamma function, $\Gamma(c) = \int_0^\infty x^{c-1}e^{-x}dx$ for $c > 0$. Note that $\Gamma(c) = (c-1)!$ for an integer c. The parameters a and b control the shape of the distribution. The distribution is right-skewed for $a > b$ and left-skewed for $a < b$. When $a = b$, the distribution is symmetric about $v = 0.5$. Our program Beta.R gives a few example plots of the distribution. The α quantile of a beta distribution, denoted by v_α or beta$(\alpha; a, b)$, is found by solving

$$\alpha = \int_0^{v_\alpha} \frac{\Gamma(a+b)}{\Gamma(a)\Gamma(b)} v^{a-1}(1-v)^{b-1} dv$$

for v_α.

The limits of the $(1-\alpha)100\%$ Clopper-Pearson interval are simply quantiles from two beta distributions:

$$\text{beta}\left(\alpha/2;\ w,\ n-w+1\right) < \pi < \text{beta}\left(1-\alpha/2;\ w+1,\ n-w\right).$$

Because of the restriction $a > 0$, the lower endpoint cannot be computed if $w = 0$. In that case, the lower limit is taken to be 0. Similarly, the upper limit is taken to be 1 whenever $w = n$ because of the restriction $b > 0$. Because the remaining beta quantiles lie strictly between 0 and 1, the Clopper-Pearson interval respects the natural boundaries for probabilities.

Often, the Clopper-Pearson interval is given in terms of quantiles from an F-distribution rather than a beta distribution. This comes about through a relationship between the two distributions; see #9.21 on p. 454 of Casella and Berger (2002) for this relationship. Both formulas produce the same limits and beta quantiles are easy to compute, so we omit the F-based formula here.

There are a few variations on the Clopper-Pearson interval. The Blaker interval proposed in Blaker (2000, 2001) also guarantees that the true confidence level is equal to or greater than the stated level. An added benefit is that the interval is no wider than the Clopper-Pearson interval and is often narrower. A disadvantage is that the interval is more difficult to calculate and requires an iterative numerical procedure to find its limits. The CIpi.R program shows how to calculate the interval using the `binom.blaker.limits()` function of the `BlakerCI` package. Another variation on the Clopper-Pearson interval is the mid-p interval. This interval no longer guarantees the true confidence level to be at least the stated level, but it is always shorter than the Clopper-Pearson interval while performing relatively well with respect to the stated confidence level (Brown et al., 2001). The CIpi.R program shows how to calculate this interval using the `midPci()` function of the `PropCIs` package.

Example: Intervals for π (CIpi.R)

Suppose $w = 4$ successes are observed out of $n = 10$ trials again. The 95% Clopper-Pearson interval is $\text{beta}\left(0.025;\ 4,\ 7\right) < \pi < \text{beta}\left(0.975;\ 5,\ 6\right)$. The `qbeta()` function in R calculates these quantiles resulting in an interval $0.1216 < \pi < 0.7376$. Notice that this is the widest of the intervals calculated so far.

Table 1.1: Confidence intervals for the hepatitis C prevalence.

Method	Interval	Length
Wald	(0.0157, 0.0291)	0.0134
Agresti-Coull	(0.0165, 0.0302)	0.0137
Wilson	(0.0166, 0.0301)	0.0135
Clopper-Pearson	(0.0162, 0.0302)	0.0140

Below is the R code used to calculate the interval.

```
> round(qbeta(p = c(alpha/2, 1-alpha/2), shape1 = c(w, w+1),
   shape2 = c(n-w+1, n-w)),4)
[1] 0.1216 0.7376

> binom.confint(x = w, n = n, conf.level = 1-alpha, methods =
   "exact")
  method x  n mean     lower     upper
1  exact 4 10  0.4 0.1215523 0.7376219
```

Within the `qbeta()` function call, the `shape1` argument is a and the `shape2` argument is b. We use vectors within `qbeta()` to find the quantiles. R matches each vector value to produce the equivalent of separate `qbeta()` function runs for the lower and upper limits. The `binom.confint()` function is used as an alternative way to find the interval where `methods = "exact"` gives the Clopper-Pearson interval. Note that this interval was also given earlier when we used `methods = "all"`.

Example: Hepatitis C prevalence among blood donors (HepCPrev.R)

Blood donations are screened for diseases to prevent transmission from donor to recipient. To examine how prevalent hepatitis C is among blood donors, Liu et al. (1997) focus on 1,875 blood donations in Xuzhou City, China.[4] They observed that 42 donations tested positive for the antigen produced by the body when infected with the virus. The 95% Wilson interval is $0.0166 < \pi < 0.0301$, where we used the same type of R code as in the previous examples. Thus, with 95% confidence, the interval 0.0166 to 0.0301 contains the true hepatitis C antigen prevalence for the Xuzhou City blood donor population.

In practice, we would only calculate one of the intervals discussed in this section. For demonstration purposes, Table 1.1 displays additional 95% confidence intervals. Because of the large sample size, we see that the intervals are similar, with the Wald interval being the most different from the others. The lengths of the intervals are similar as well, with the Clopper-Pearson interval being a little longer than the others.

[4]The study's main purpose was to examine how well "group testing" would work to estimate overall disease prevalence. See Section 6.7 for an introduction to group testing.

Tests

When only one parameter is of interest, such as π here, we generally prefer confidence intervals over hypothesis tests, because the interval gives a range of possible parameter values. We can typically infer that a hypothesized value for a parameter can be rejected if it does not lie within the confidence interval for the parameter. However, there are situations where a fixed known value of π, say π_0, is of special interest, leading to a formal hypothesis test of $H_0 : \pi = \pi_0$ vs. $H_a : \pi \neq \pi_0$.

With regard to the Wilson interval, it was noted that the score test statistic

$$Z_S = \frac{\hat{\pi} - \pi_0}{\sqrt{\pi_0(1-\pi_0)/n}}$$

is often used in these situations. When the null hypothesis is true, Z_S should have approximately a standard normal distribution, where the approximation is generally better for larger samples. The null hypothesis is rejected when an unusual value of Z_S is observed relative to this distribution, namely something less than $-Z_{1-\alpha/2}$ or greater than $Z_{1-\alpha/2}$. The p-value is a measure of how extreme the test statistic value is relative to what is expected when H_0 is true. This p-value is calculated as $2P(Z > |Z_S|)$, where Z has a standard normal distribution. Note that this test is equivalent to rejecting the null hypothesis when π_0 is outside of the Wilson interval. If desired, the `prop.test()` function can be used to calculate Z_S (Z_S^2 is actually given) and a corresponding p-value; this is demonstrated in the CIpi.R program. The function also calculates the Wilson interval.

We recommend using the score test to perform a test for π. However, there are alternative testing procedures. In particular, a likelihood ratio test is a general way to perform hypothesis tests that are needed elsewhere in the textbook (see Appendix B.5 for an introduction). Informally, the likelihood ratio (LR) statistic is

$$\Lambda = \frac{\text{Maximum of likelihood function under } H_0}{\text{Maximum of likelihood function under } H_0 \text{ or } H_a}.$$

For the specific test of $H_0 : \pi = \pi_0$ vs. $H_a : \pi \neq \pi_0$, the denominator is $\hat{\pi}^w(1-\hat{\pi})^{n-w}$ (using Equation 1.2), because the maximum possible value of the likelihood function occurs when it is evaluated at the MLE. The numerator is $\pi_0^w(1-\pi_0)^{n-w}$ because there is only one possible value of the likelihood function if the null hypothesis is true. A simple transformation of the LR statistic, $-2\log(\Lambda)$, forms the likelihood ratio test (LRT) statistic. The extra word "test" is included because the transformed statistic has an approximate χ_1^2 distribution in large samples if the null hypothesis is true. The LRT statistic can be re-expressed for the hypotheses here as

$$-2\log(\Lambda) = -2\left\{w\text{log}\left(\frac{\pi_0}{\hat{\pi}}\right) + (n-w)\text{log}\left(\frac{1-\pi_0}{1-\hat{\pi}}\right)\right\}.$$

We reject the null hypothesis if $-2\log(\Lambda) > \chi^2_{1,1-\alpha}$, where $\chi^2_{1,1-\alpha}$ is the $1-\alpha$ quantile from a chi-square distribution with 1 degree of freedom (for example, $\chi^2_{1,0.95} = 3.84$ when $\alpha = 0.05$). The p-value is $P(A > -2\log(\Lambda))$ where A has a χ_1^2 distribution.

An alternative way to calculate a confidence interval for π is to invert the LRT in a similar manner as was done for the Wilson interval (see Exercise 15). This LR interval is automatically calculated by the `binom.confint()` function of the `binom` package, where the `methods = "lrt"` argument value is used. We provide additional code in CIpi.R that shows how to find the interval without this function. The interval is generally harder to compute than the intervals that we recommend in this section and has no advantages over them. LR confidence intervals often are used in some more complicated contexts where

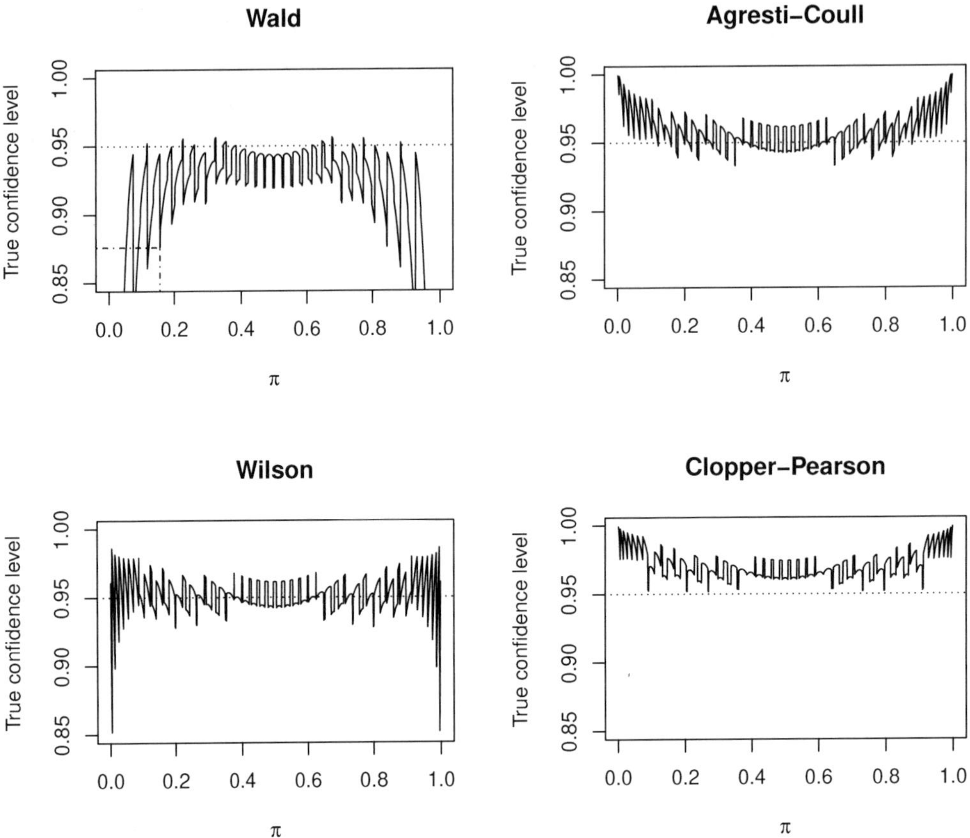

Figure 1.3: True confidence levels with $n = 40$ and $\alpha = 0.05$. The Wald plot includes a dot-dashed line illustrating that the true confidence level is 0.8760 at $\pi = 0.157$.

better intervals are not available (this will be the case in Chapters 2 to 4). The interval is better than the Wald interval in most problems.

1.1.3 True confidence levels for confidence intervals

As discussed on p. 11, a confidence interval method may not actually achieve its stated confidence level. The reasons for this are explained shortly. Figure 1.3 provides a comparison of the true confidence levels for the Wald, Wilson, Agresti-Coull, and Clopper-Pearson intervals. For each plot, n is 40 and the stated confidence level is 0.95 ($\alpha = 0.05$). The true confidence level (coverage) for each interval method is plotted as a function of π. For example, the true confidence level at $\pi = 0.157$ is 0.8759 for the Wald, 0.9507 for the Wilson, 0.9507 for the Agresti-Coull, and 0.9740 for the Clopper-Pearson intervals, respectively. Obviously, none of these intervals achieves exactly the stated confidence level on a consistent basis. Below are some general conclusions from examining this plot:

- The Wald interval tends to be the farthest from 0.95 the most often. In fact, the true confidence level is often below our lower plotting bound of 0.85 at extreme values of π.

- The Agresti-Coull interval does a much better job than the Wald with its true confidence level usually between 0.93 and 0.98. For values of π close to 0 or 1, the interval can be very conservative.

- The Wilson interval performs a little better than the Agresti-Coull interval with its true confidence level generally between 0.93 and 0.97; however, for very extreme π, it can be very liberal. Note that this performance for extreme π can be improved by changing the lower interval limit to $-\log(1-\alpha)/n$ when $w = 1$ and the upper interval limit to $1 + \log(1-\alpha)/n$ when $w = n - 1$; see p. 112 of Brown et al. (2001) for justification. This small modification is not implemented by `binom.confint()` or `prop.test()`.

- The Clopper-Pearson interval has a true confidence level at or above the stated level, generally oscillating between 0.95 and 0.98. For values of π close to 0 or 1, the interval can be very conservative.

Similar findings can be shown for other values of n and α. The R code used to construct Figure 1.3 is available in the ConfLevel4Intervals.R program, and it will be discussed shortly.

Why do these plots in Figure 1.3 have such strange patterns? It is because of the discreteness of a binomial random variable. For a given n, there are only $n+1$ possible intervals that can be formed, one for each value of $w = 0, 1, \ldots, n$. For a specific value of π, some of these intervals contain π and some do not. Thus, the true confidence level at π, say $C(\pi)$, is the sum of the binomial probabilities for all intervals that do contain π:

$$C(\pi) = \sum_{w=0}^{n} I(w) \binom{n}{w} \pi^{w}(1-\pi)^{n-w}, \tag{1.6}$$

where $I(w)$ is an indicator function with a value of 1 if the interval formed with w contains π and a value of 0 otherwise. Each of these binomial probabilities in Equation 1.6 changes slowly as π changes. As long as we do not move π past any interval limits, the true confidence level changes slowly too. However, as soon as π crosses over an interval limit, a mass of probability is suddenly either added to or subtracted from the true confidence level, resulting in the spikes that appear in all parts of Figure 1.3. We illustrate how to find the true confidence level and when these spikes occur in the next example.

<u>Example:</u> True confidence level for the Wald interval (ConfLevel.R)

We show in this example how to calculate a true confidence level for the Wald interval with $n = 40$, $\pi = 0.157$, and $\alpha = 0.05$. Below is a description of the process:

1. Find the probability of obtaining each possible value of w using the `dbinom()` function with $n = 40$ and $\pi = 0.157$,

2. Calculate the 95% Wald confidence interval that results from observing each possible value of w, and

3. Sum the probabilities corresponding to those intervals that contain $\pi = 0.157$; this is the true confidence level.

Below is the R code for the first two steps.

```
> pi <- 0.157
> alpha <- 0.05
> n <- 40
> w <- 0:n
> pi.hat <- w/n
> pmf <- dbinom(x = w, size = n, prob = pi)
> var.wald <- pi.hat*(1-pi.hat)/n
> lower <- pi.hat - qnorm(p = 1-alpha/2) * sqrt(var.wald)
> upper <- pi.hat + qnorm(p = 1-alpha/2) * sqrt(var.wald)
> round(lower[1:3], digits = 4)
[1] 0.0000 -0.0234 -0.0175
> round(upper[1:3], digits = 4)
[1] 0.0000 0.0734 0.1175
```

The code contains many of the same components that we have seen before. The main difference now is that we are calculating an interval for each possible value of w rather than for only one. This is accomplished by taking advantage of how R performs calculations on vectors and scalars (see Appendix A.4). For example, the vector `pi.hat` contains all 41 possible values of $\hat{\pi}$. The `1-pi.hat` portion of `var.wald` subtracts each of the these values from 1 to produce 41 new values in a vector. R performs multiplication in an *elementwise* manner for `pi.hat*(1-pi.hat)`, so that each of the 41 elements in `pi.hat` are multiplied by the corresponding 41 values in `1-pi.hat`. After dividing each resulting element by `n`, we obtain $\widehat{Var}(\hat{\pi})$ for $w = 0, \ldots, 40$.

The third step of the process involves summing the probabilities for those w that have an interval containing π. The code below performs these computations, showing the output for $\pi = 0.157$.

```
> save <- pi>lower & pi<upper
> data.frame(w, pi.hat, round(data.frame(pmf, lower, upper),4),
   save)[1:13,]
   w pi.hat    pmf   lower  upper  save
1  0  0.000 0.0011  0.0000 0.0000 FALSE
2  1  0.025 0.0080 -0.0234 0.0734 FALSE
3  2  0.050 0.0292 -0.0175 0.1175 FALSE
4  3  0.075 0.0689 -0.0066 0.1566 FALSE
5  4  0.100 0.1187  0.0070 0.1930  TRUE
6  5  0.125 0.1591  0.0225 0.2275  TRUE
7  6  0.150 0.1729  0.0393 0.2607  TRUE
8  7  0.175 0.1564  0.0572 0.2928  TRUE
9  8  0.200 0.1201  0.0760 0.3240  TRUE
10 9  0.225 0.0795  0.0956 0.3544  TRUE
11 10 0.250 0.0459  0.1158 0.3842  TRUE
12 11 0.275 0.0233  0.1366 0.4134  TRUE
13 12 0.300 0.0105  0.1580 0.4420 FALSE

> sum(save*pmf)
[1] 0.875905
> sum(dbinom(x = 4:11, size = n, prob = pi))
[1] 0.875905
```

The `pi>lower` portion of `save` performs logical checks. If π is above a lower confidence interval limit, a value of `TRUE` results; otherwise, a value of `FALSE` results. These logical checks are performed on all 41 lower limits. The `pi<upper` portion performs a similar check but now to determine if π is below each upper limit. The ampersand denotes "and" in a logical manner so that each element of `save` is `TRUE` only if `pi>lower` and `pi<upper` are both `TRUE`, which means the interval contains π. For example, the second interval is $(-0.0234, 0.0734)$, so it does not contain $\pi = 0.157$. The result is a `FALSE` value for the second element of `save`.

The data frame created at the end puts all of the calculated components together for printing. For example, we see that if $w = 3$, the corresponding interval does not contain π, but at $w = 4$ the corresponding interval does. By examining other parts of the data frame, we see that the intervals for $w = 4$ to 11 all contain $\pi = 0.157$. Probabilities for these values of w are summed by `sum(save*pmf)`. R converts the `TRUE` values in `save` to 1's and the `FALSE` values in `save` to 0's when performing elementwise multiplication between `save` and `pmf`. Thus, `save` essentially becomes $I(w)$ in Equation 1.6 so that each of its 41 elements is multiplied by the corresponding element in `pmf`. The results are summed to produce a true confidence level of 0.8759. More simply, the probability that a binomial random variable is between 4 and 11 with $n = 40$ and $\pi = 0.157$ can be calculated by the `sum()` and `dbinom()` functions. Obviously, the Wald confidence interval does not achieve its stated level of 95%.

Notice that the upper interval limit at $w = 3$ barely does not contain $\pi = 0.157$ and $P(W = 3) = 0.0689$. By using the same code with the change of `pi <- 0.156`, the upper limit at $w = 3$ now does contain $\pi = 0.156$, so that $P(W = 3) = 0.0706$ is included when summing probabilities for the true confidence level. Overall, $w = 3$ to 11 now have confidence intervals that contain $\pi = 0.156$, leading to a true confidence level of 0.9442! This demonstrates what we alluded to earlier as the cause for the spikes in Figure 1.3.

In simple problems like the present one, we can determine the probabilities for each interval that contains a given π, so that the plots like in Figure 1.3 can be made exactly. In other cases, we may have to rely on Monte Carlo simulation to estimate these probabilities. We explore the simulation approach next to compare an actual true confidence level to one estimated by simulation. This will be helpful later in the text when simulation is the only available method of assessment.

Example: True confidence level for the Wald interval (ConfLevel.R)

Suppose again that $n = 40$, $\pi = 0.157$, and $\alpha = 0.05$. Below is a description of the process to estimate the true confidence level through simulation:

1. Simulate 1,000 samples using the `rbinom()` function with $n = 40$ and $\pi = 0.157$,
2. Calculate the 95% Wald confidence interval for each sample, and
3. Calculate the proportion of intervals that contain $\pi = 0.157$; this is the estimated true confidence level.

Below is the R code.

```
> numb.bin.samples <- 1000  # Binomial samples of size n

> set.seed(4516)
> w <- rbinom(n = numb.bin.samples, size = n, prob = pi)
> pi.hat <- w/n
```

```
> var.wald <- pi.hat*(1-pi.hat)/n
> lower <- pi.hat - qnorm(p = 1-alpha/2) * sqrt(var.wald)
> upper <- pi.hat + qnorm(p = 1-alpha/2) * sqrt(var.wald)
> data.frame(w, pi.hat, lower, upper)[1:10,]
   w pi.hat        lower     upper
1  6  0.150  0.039344453 0.2606555
2  6  0.150  0.039344453 0.2606555
3  7  0.175  0.057249138 0.2927509
4  8  0.200  0.076040994 0.3239590
5  8  0.200  0.076040994 0.3239590
6  6  0.150  0.039344453 0.2606555
7  8  0.200  0.076040994 0.3239590
8  3  0.075 -0.006624323 0.1566243
9  5  0.125  0.022511030 0.2274890
10 4  0.100  0.007030745 0.1929693

> save <- pi>lower & pi<upper
> save[1:10]
 [1]  TRUE  TRUE  TRUE  TRUE  TRUE  TRUE  TRUE FALSE  TRUE
[10]  TRUE
> mean(save)
[1] 0.878
```

Again, we are using much of the same code as in the past. We see that sample #8 results in $\hat{\pi} = 0.075$ and an interval of $(-0.0066, 0.1566)$, which does not contain 0.157, so the corresponding value of `save` is `FALSE`. When using the `mean()` function, R converts the `TRUE` and `FALSE` values in `save` to 1's and 0's, respectively, to find the mean to be 0.878 (i.e., the proportion of `TRUE`'s in `save` is 0.878). This is our *estimated* true confidence level for the Wald interval at $n = 40$ and $\pi = 0.157$.

In this relatively simple simulation problem, we already know that the intervals for $w = 4, \ldots, 11$ contain $\pi = 0.157$ while the others do not. To see that the simulation is, indeed, estimating $P(4 \leq W \leq 11)$, the `table()` function is used to calculate the number of times each w occurs.

```
> counts <- table(w)
> counts
w
  1   2   3   4   5   6   7   8   9  10  11  12  13
  8  35  64 123 147 165 172 123  76  46  26  11   4

> sum(counts[4:11])/numb.bin.samples
[1] 0.878
```

For example, there were 64 out of the 1,000 observations that resulted in a $w = 3$. This is very similar to the $P(W = 3) = 0.0689$ that we obtained for the past example. Summing the counts for $w = 4, \ldots, 11$ and dividing by 1000, we obtain the same estimate of 0.878 for the true confidence level.

The estimate of the true confidence level here is almost the same as the actual true confidence level found in the previous example. Because of the large number of samples, the law of large numbers ensures that this will happen. We could even go as far as finding a confidence interval for the true confidence level! For this case, we have 878 "successes" out of 1,000 "trials." A 95% Wilson interval for the true confidence level itself is (0.8563, 0.8969), which happens to contain 0.8759, the actual true confidence level.

Figure 1.3 provides the true confidence levels for $\pi = 0.001, ..., 0.999$ by 0.0005, interpolating between consecutive values of π. To calculate true confidence levels at each π for a particular interval method, we repeat the same code as before, but now using a "for loop" over π values within R. The next example illustrates this process.

Example: True confidence level plot (ConfLevelWaldOnly.R, ConfLevel4Intervals.R)

With $n = 40$ and $\alpha = 0.05$, we calculate the true confidence levels for the Wald interval using the following code.

```
> alpha <- 0.05
> n <- 40
> w <- 0:n
> pi.hat <- w/n
> pi.seq <- seq(from = 0.001, to = 0.999, by = 0.0005)

> # Wald
> var.wald <- pi.hat*(1-pi.hat)/n
> lower.wald <- pi.hat - qnorm(p = 1-alpha/2) * sqrt(var.wald)
> upper.wald <- pi.hat + qnorm(p = 1-alpha/2) * sqrt(var.wald)

> # Save true confidence levels in a matrix
> save.true.conf <- matrix(data = NA, nrow = length(pi.seq),
   ncol = 2)

> # Create counter for the loop
> counter <- 1

> # Loop over each pi
> for(pi in pi.seq) {
    pmf <- dbinom(x = w, size = n, prob = pi)
    save.wald <- pi>lower.wald & pi<upper.wald
    wald <- sum(save.wald*pmf)
    save.true.conf[counter,] <- c(pi, wald)
    # print(save.true.conf[counter,])
    counter <- counter + 1
}

> plot(x = save.true.conf[,1], y = save.true.conf[,2], main =
   "Wald", xlab = expression(pi), ylab = "True confidence
   level", type = "l", ylim = c(0.85,1))
> abline(h = 1-alpha, lty = "dotted")
```

We create a vector `pi.seq` which is a sequence of numbers from 0.001 to 0.999 by 0.0005. The `for(pi in pi.seq)` function code (often referred to as a "for loop") instructs R to take one π value out of `pi.seq` at a time. The code enclosed by braces then finds the true confidence level for this π. The `save.true.conf` object is a matrix that is created to have 1,997 rows and 2 columns. At first, all of its values are initialized to be "NA" within R. Its values are updated then one row at a time by inserting the value of π and the true confidence level. Finally, the `counter` object allows us to change the row number of `save.true.conf` within the loop.[5]

After the for loop, we use use the `plot()` function to plot the value of π on the x-axis and the true confidence level on the y-axis using the appropriate columns of `save.true.conf`. The `type = "l"` argument instructs R to construct a line plot where each π and true confidence level pair is connected by a line. The `abline()` function draws a horizontal dotted line at 0.95, which is the stated confidence level. Please see the upper-left plot in Figure 1.3 for the final result. To construct all four plots in Figure 1.3, we insert the code for the other three intervals into the braces of the loop. We also add three columns to the `save.true.conf` matrix and use additional calls to the `plot()` function. Please see ConfLevel4Intervals.R for the code.

The `binom` package in R also can be used to calculate the true confidence levels. The `binom.coverage()` function calculates the true confidence level for one π at a time, and the `binom.plot()` function plots the true confidence levels over a set of different values of π. Examples of using these functions are in the programs for this example. We purposely demonstrated the calculations without `binom.coverage()` first, because convenient functions like it are not available for other situations examined in the textbook.

1.2 Two binary variables

We consider now the situation when the same Bernoulli trial is measured on units that can be classified into groups. The simplest such case is when a population consists of two groups, such as vaccinated and not vaccinated people, fresh- and salt-water fish, or American and foreign companies. Below are two examples with a binary response on trials that form two groups.

Example: Larry Bird's free throw shooting

A free throw is a shot in basketball where the shooter can shoot freely (unopposed by another player) from a specific location on the court. The shot is either made (a success) or missed (a failure). Most often during a National Basketball Association (NBA) game, free throws are shot in pairs. This means a free throw shooter has one attempt and then subsequently has a second attempt no matter what happens on the first.

The former NBA player Larry Bird was one of the most successful at making free throws during his career, with a success rate of 88.6%. By comparison, the NBA average during this time was about 75% (`http://www.basketball-reference.com`). Bird's outstanding success rate is among his many achievements, for which he has been recognized as one of the greatest players in NBA history. During the 1980–81 and 1981–82 NBA seasons, the outcomes from Bird's free throw attempts shot in pairs

[5] If desired, the call to the `print()` function can be uncommented to see the progress during the loop. If this is done, it is best to turn off the "buffered output" in R: select Misc > Buffered output from the R main menu.

Table 1.2: Larry Bird's free throw outcomes. Data source is Wardrop (1995).

		Second		
		Made	Missed	Total
First	Made	251	34	285
	Missed	48	5	53
	Total	299	39	338

Table 1.3: Pfizer-BioNTech COVID-19 vaccine clinical trial results. Data source is Polack et al. (2020).

		Test response		
		Positive	Negative	Total
Treatment	Vaccine	8	17,403	17,411
	Placebo	162	17,349	17,511
	Total	170	34,752	34,922

were recorded, and a summary is shown in Table 1.2. For example, Bird made both his first and second attempts 251 times. Also, Bird made the first attempt, but then subsequently missed the second attempt 34 times. Overall, he made the first attempt 285 times without regard to what happened on the second attempt. In total, Bird shot 338 pairs of free throw pairs during the season.

Basketball fans and commentators often speculate that the results of a second free throw might depend on the results of the first. For example, if a shooter misses the first attempt, will disappointment or determination lead to altering their approach for the second attempt? If so, then we should see that the probability of success on the second attempt is different depending on whether the first attempt was made or missed. Thus, the two groups in this problem are formed by the results of the first attempt, and the Bernoulli trials that we observe are the results of the second attempt. Given the data in Table 1.2, we will investigate if the second attempt outcome is dependent on what happens for the first attempt.

Example: COVID-19 vaccine clinical trial

Clinical trials are performed to determine the safety and efficacy of new drugs. Frequently, the safety and efficacy responses are categorical in nature; for example, the efficacy response may be simply whether a drug cures or does not cure a patient of a disease. To ensure that a new drug is indeed better than doing nothing (patients sometimes get better without intervention), it is essential to have a control group in the trial. This is achieved in clinical trials by randomizing patients into two groups: new drug or control. The control group is often a placebo, which is administered just like the new drug but contains no medication.

During the COVID-19 pandemic, pharmaceutical companies rushed to develop a vaccine to prevent infection from severe acute respiratory syndrome coronavirus 2 (SARS-CoV-2), the virus that leads to coronavirus disease 2019 (COVID-19). The Pfizer-BioNTech COVID-19 vaccine was the first to receive an Emergency Use Authorization from the U.S. Food and Drug Administration (FDA). Table 1.3 provides results from the definitive clinical trial used to examine the effectiveness of the vaccine. Out of 17,411 individuals who received the vaccine, only 8 individuals became infected with the virus. Out of 17,511 individuals who received the placebo, 162 individuals

Table 1.4: Probability and observed count structures for two independent binomial random variables.

		Response						Response		
		Success	Failure					Success	Failure	
Group	1	π_1	$1-\pi_1$	1		Group	1	w_1	n_1-w_1	n_1
	2	π_2	$1-\pi_2$	1			2	w_2	n_2-w_2	n_2
								w_+	n_+-w_+	n_+

became infected. The question of interest for the clinical trial was "Does the vaccine help to prevent infection?" We will develop comparison measures in this section to answer this question.

1.2.1 Notation and model

The model and notation follow those used for a single binomial random variable in Section 1.1. We start by considering two separate Bernoulli random variables, Y_1 and Y_2, one for each group. The probabilities of success for the two groups are denoted by π_1 and π_2, respectively. We observe n_j trials of Y_j leading to w_j observed successes, $j = 1, 2$.[6] We replace a subscript with "+" to denote a sum across that subscript. Thus, $n_+ = n_1 + n_2$ is the total number of trials, and $w_+ = w_1 + w_2$ is the total number of observed successes. This notation is depicted in Table 1.4. The table on the right side of Table 1.4 is called a *two-way contingency table*, because it gives a listing of all possible cross-tabulations of two categorical variables. We cover more general forms of contingency tables in Chapters 3 and 4.

We denote the random variable representing the number of successes in group j by W_j and write its PMF as

$$P(W_j = w_j) = \binom{n_j}{w_j}\pi_j^{w_j}(1-\pi_j)^{n_j-w_j},\ w_j = 0, 1, \ldots, n_j,\ j = 1, 2.$$

We assume that the two random variables Y_1 and Y_2 are independent, so that the outcome of one cannot affect the outcome of the other. In the COVID-19 vaccine clinical trial, for example, this means that people assigned to receive the vaccine cannot pass on immunity or disease to people in the placebo group and vice versa. This assumption is critical in what follows, and so this model is referred to as the *independent binomial model.* When independence is not satisfied, then other models need to be used that account for dependence between the random variables (e.g., see Section 1.2.6 for handling paired data).

When we want to simulate data from this model, we use R code like what is shown in the next example. This sampling procedure will be important shortly when we use these simulated counts to evaluate statistical inference procedures, like confidence intervals, to determine if they work as expected.

Example: Simulate counts in a contingency table (SimContingencyTable.R)

Consider the case of $\pi_1 = 0.2$, $\pi_2 = 0.4$, $n_1 = 10$, and $n_2 = 10$. The R code below shows how to simulate one set of counts for a contingency table.

[6] More formally, we could define y_{ij} as the observed value for the ith independent trial in the jth group. This leads to $w_j = \sum_{i=1}^{n_j} y_{ij}$.

```
> pi1 <- 0.2
> pi2 <- 0.4
> n1 <- 10
> n2 <- 10

> set.seed(8191)
> w1 <- rbinom(n = 1, size = n1, prob = pi1)
> w2 <- rbinom(n = 1, size = n2, prob = pi2)

> c.table <- array(data = c(w1, w2, n1-w1, n2-w2), dim = c(2,2),
   dimnames = list(Group = c(1,2), Response = c("Success",
   "Failure")))
> c.table
     Response
Group Success Failure
    1       1       9
    2       3       7

> list(Group = c(1,2), Response = c("Success", "Failure"))
$Group
[1] 1 2

$Response
[1] "Success" "Failure"
```

Similar to Section 1.1, we use `rbinom()` to simulate values for w_1 and w_2. To form the contingency table (what we name `c.table`), we use the `array()` function. Its `data` argument contains the counts within the contingency table. These counts are concatenated using the `c()` function. Notice that the data are entered by columns $(w_1, w_2, n_1 - w_1, n_2 - w_2)$. The `dim` argument specifies the contingency table's dimensions as (number of rows, number of columns), where the `c()` function is used again. Finally, the `dimnames` argument gives names for the row and column measurements. The names are given in a *list* format, which allows for a number of objects to be linked together (please see Appendix A.7.3 for more on lists, if needed). In this case, the objects are `Group` and `Response` that contain the levels of the rows and columns, respectively.

For this particular sample, $w_1 = 1$, $n_1 - w_1 = 9$, $w_2 = 3$, and $n_2 - w_2 = 7$. We access these values from within the contingency table by specifying a row and column number or name with `c.table`.

```
> c.table1[1,1]  # w1
[1] 1
> c.table["1", "Success"]  # w1
[1] 1
> c.table1[1,2]  # n1-w1
[1] 9
> c.table1[1,]  # w1 and n1-w1
Success Failure
      1       9
> sum(c.table1[1,])  # n1
[1] 10
```

For example, c.table[1,2] is equal to $n_1 - w_1$. We omit a column or row number within [ ] to have a whole row or column, respectively, displayed. Summed counts are found by using the sum() function with the appropriate counts. Alternatively, the functions addmargins(), colsums(), marginSums(), and rowSums() can be used to find summed counts for the margins of the table (i.e., the last row and column shown in Tables 1.2 and 1.3). These alternatives are demonstrated within our program.

If we wanted to repeat this process, say, 1,000 times, the n argument of the rbinom() functions would be changed to 1,000. Each contingency table would need to be formed separately using the array() function. The program for this example provides the code.

Likelihood and estimates

Our main interests are estimating the probabilities of success, π_1 and π_2, for each group and comparing these probabilities. Maximum likelihood estimation again provides a convenient and powerful solution. Because Y_1 and Y_2 are independent, so too are W_1 and W_2. The likelihood function formed by independent random variables is just the product of their respective likelihood functions. Hence, the likelihood function is $L(\pi_1, \pi_2|w_1, w_2) = L(\pi_1|w_1) \times L(\pi_2|w_2)$. Maximizing this function over π_1 and π_2 results in the intuitive estimates $\hat{\pi}_1 = w_1/n_1$ and $\hat{\pi}_2 = w_2/n_2$, the sample proportions in the two groups. In other words, when the random variables are independent, each probability is estimated using only the data from its own group.

Example: Larry Bird's free throw shooting (Bird.R)

The purpose of this example is to estimate the probability of successes using a contingency table structure in R. If the Bernoulli trial results are already summarized into counts as in Table 1.2, then a contingency table is created in R using the array() function.

```
> c.table <- array(data = c(251, 48, 34, 5), dim = c(2,2),
   dimnames = list(First = c("made", "missed"), Second =
   c("made", "missed")))
> c.table
        Second
First    made missed
made      251     34
missed     48      5
```

The estimates of the probability of successes (or sample proportions) are found by taking advantage of how R performs calculations.

```
> rowSums(c.table)  # n1 and n2
  made missed
   285     53

> pi.hat.table <- c.table/rowSums(c.table)
> pi.hat.table
         Second
First          made     missed
  made    0.8807018 0.11929825
```

```
missed 0.9056604 0.09433962
```

The `rowSums()` function finds the sum of counts in each row to obtain n_1 and n_2. By taking the contingency table of counts divided by these row sums, we obtain $\hat{\pi}_1$ and $\hat{\pi}_2$ in the first column and $1 - \hat{\pi}_1$ and $1 - \hat{\pi}_2$ in the second column. Notice that R does this division correctly by taking the counts in the first (second) row of `c.table` divided by the first (second) element of `rowSums(c.table)`.

Data are often available as measurements on each trial, rather than as summarized counts. For example, the Larry Bird data would originally have consisted of 338 unaggregated pairs of first and second free throw results. We created a hypothetical ordering of these data in the program and stored the data in a data frame named `all.data2`. We use the `head()` function to print the first six pairs below.

```
> head(all.data2)
  First Second
1  made   made
2  made missed
3  made missed
4  made   made
5  made   made
6  made   made
```

In this data frame, `First` represents the group and `Second` is the trial response. We call this data format the *raw data* because it represents how the data looked before being processed into counts.

To form a contingency table from the raw data, we can use the `table()` or `xtabs()` functions.

```
> bird.table1 <- table(all.data2$First, all.data2$Second)
> bird.table1
         made missed
made      251     34
missed     48      5
> bird.table1[1,1]  # w1
[1] 251

> bird.table2 <- xtabs(formula = ~ First + Second, data =
   all.data2)
> bird.table2
        Second
First     made missed
  made     251     34
  missed    48      5
> bird.table2[1,1]  # w1
[1] 251
```

In both cases, the functions produce a contingency table that is saved into an object to allow parts of it to be accessed as before. Note that the "xtabs" name comes about through an abbreviation of *crosstabulations*, which is a frequently used term to describe this joint summarization of multiple variables.

The estimated probability that Larry Bird makes his second free throw attempt is $\hat{\pi}_1 = 0.8807$ given that he makes the first and $\hat{\pi}_2 = 0.9057$ given he misses the first. In this sample, the probability of success on the second attempt is larger when the first attempt is missed rather than made. This is somewhat counterintuitive to many basketball fans' perceptions that a missed first free throw should lower the probability of success on the second free throw. However, this is only for one sample. We would like to generalize to the population of all free throw attempts by Larry Bird. To make this generalization, we need to use statistical inference procedures. We discuss these next.

1.2.2 Confidence intervals for the difference of two probabilities

We typically want to compare the binary responses across the two groups. For example, such a comparison allows us to determine if Larry Bird's second free throw attempt is affected by his first and if the COVID-19 vaccine is effective. There are three main approaches to make this comparison. The first is to examine the difference of the two probability of success parameters: $\pi_1 - \pi_2$. Later subsections will look at the other approaches.

Thanks to the invariance property of MLEs described in Appendix B.4, the corresponding estimate of $\pi_1 - \pi_2$ is simply the same function of the individual MLEs, $\hat{\pi}_1 - \hat{\pi}_2$. Each MLE has a probability distribution that is approximated by a normal distribution in large samples: $\hat{\pi}_j \dot{\sim} N(\pi_j, \widehat{Var}(\hat{\pi}_j))$, where $\widehat{Var}(\hat{\pi}_j) = \hat{\pi}_j(1-\hat{\pi}_j)/n_j$, $j = 1, 2$. Because linear combinations of normal random variables are themselves normal random variables (Casella and Berger, 2002, p. 156), the probability distribution for $\hat{\pi}_1 - \hat{\pi}_2$ is approximated by $N(\pi_1 - \pi_2, \widehat{Var}(\hat{\pi}_1 - \hat{\pi}_2))$, where $\widehat{Var}(\hat{\pi}_1 - \hat{\pi}_2) = \hat{\pi}_1(1-\hat{\pi}_1)/n_1 + \hat{\pi}_2(1-\hat{\pi}_2)/n_2$.[7] This distribution forms the basis for a range of inference procedures.

The easiest confidence interval to form for $\pi_1 - \pi_2$ uses the normal approximation for $\hat{\pi}_1 - \hat{\pi}_2$ directly to create a Wald interval:

$$\hat{\pi}_1 - \hat{\pi}_2 \pm Z_{1-\alpha/2}\sqrt{\frac{\hat{\pi}_1(1-\hat{\pi}_1)}{n_1} + \frac{\hat{\pi}_2(1-\hat{\pi}_2)}{n_2}}.$$

Unfortunately, the Wald interval for $\pi_1 - \pi_2$ has similar problems with achieving the stated level of confidence as the Wald interval for π. We will investigate this shortly.

Because of these problems, a number of other confidence intervals for $\pi_1 - \pi_2$ have been proposed. Inspired by the good general performance of the Agresti-Coull interval for a single probability, Agresti and Caffo (2000) investigate various intervals constructed as Wald-type intervals on data with different numbers of added successes and failures. They found that adding one success and one failure for each group results in an interval that does a good job of achieving the stated confidence level. Specifically, let

$$\tilde{\pi}_1 = \frac{w_1 + 1}{n_1 + 2} \text{ and } \tilde{\pi}_2 = \frac{w_2 + 1}{n_2 + 2}$$

be the amended estimates of π_1 and π_2. Notice that unlike $\tilde{\pi}$ for the Agresti-Coull interval, the estimates $\tilde{\pi}_1$ and $\tilde{\pi}_2$ do not change when the confidence level changes. The $(1-\alpha)100\%$

[7] This is an application of the following result: $Var(aU + bV) = a^2Var(U) + b^2Var(V) + 2abCov(U, V)$, where U and V are random variables and a and b are constants. If U and V are independent random variables, then $Cov(U, V) = 0$. See p. 171 of Casella and Berger (2002).

Agresti-Caffo confidence interval for $\pi_1 - \pi_2$ is

$$\tilde{\pi}_1 - \tilde{\pi}_2 \pm Z_{1-\alpha/2}\sqrt{\frac{\tilde{\pi}_1(1-\tilde{\pi}_1)}{n_1+2} + \frac{\tilde{\pi}_2(1-\tilde{\pi}_2)}{n_2+2}}.$$

Overall, we recommend the Agresti-Caffo method for general use.

Other confidence interval methods have been developed analogous to the single-parameter case discussed in Section 1.1.2. There is a score interval based on inverting the test statistic for a score test of $H_0 : \pi_1 - \pi_2 = d$ (i.e., determining for what values d of $\pi_1 - \pi_2$ that the null hypothesis is not rejected). This interval performs quite similarly to the Agresti-Caffo interval in terms of its true confidence level. A small disadvantage is that no closed-form mathematical expression exists, like is available for the Wald and Agresti-Caffo intervals. This is because the null hypothesis does not actually specify the values of π_1 and π_2. Iterative numerical procedures need to be used then to search for the interval's limits. Exercise 35 discusses how this interval is calculated. An additional confidence interval method is a two-group Bayesian credible interval similar to Jeffreys interval for π, but this involves distributions that are not as simple to compute as the standard normal. Details of this interval are available in Agresti and Min (2005a).

Example: Larry Bird's free throw shooting (Bird.R)

The purpose of this example is to calculate a confidence interval for the difference in the second free throw success probabilities given the first attempt outcomes. Continuing the code from earlier, we obtain the following intervals.

```
> alpha <- 0.05
> pi.hat1 <- pi.hat.table[1,1]
> pi.hat2 <- pi.hat.table[2,1]

> # Wald
> var.wald <- pi.hat1*(1-pi.hat1) / sum(c.table[1,]) +
   pi.hat2*(1-pi.hat2) / sum(c.table[2,])
> pi.hat1 - pi.hat2 + qnorm(p = c(alpha/2, 1-alpha/2)) *
   sqrt(var.wald)
[1] -0.11218742  0.06227017

> # Agresti-Caffo
> pi.tilde1 <- (c.table[1,1] + 1) / (sum(c.table[1,]) + 2)
> pi.tilde2 <- (c.table[2,1] + 1) / (sum(c.table[2,]) + 2)
> var.AC <- pi.tilde1*(1-pi.tilde1) / (sum(c.table[1,]) + 2) +
   pi.tilde2*(1-pi.tilde2) / (sum(c.table[2,]) + 2)
> pi.tilde1 - pi.tilde2 + qnorm(p = c(alpha/2, 1-alpha/2)) *
   sqrt(var.AC)
[1] -0.10353254  0.07781192
```

The 95% Wald interval is $-0.1122 < \pi_1 - \pi_2 < 0.0623$, and the 95% Agresti-Caffo interval is $-0.1035 < \pi_1 - \pi_2 < 0.0778$. These intervals are somewhat similar with the Wald interval being shifted to the left of the Agresti-Caffo interval. Using the Agresti-Caffo interval, with 95% confidence, the difference in the second free throw success probabilities given the outcome of the first is between -0.1035 and 0.0778. Because this interval contains 0, we cannot detect a change in Bird's probability of a successful second free throw following made and missed first attempts. This means that either there is no difference, or there is a difference, but it was not detected in this sample. The latter situation could be caused by either bad luck (an unusual sample) or too small of a sample size relative to the true size of the difference.

These confidence intervals can be obtained in other ways. The `wald2ci()` function of the `PropCIs` package takes the values of w_1, n_1, w_2, and n_2 as its arguments, where w_1 and w_2 are called `x1` and `x2` in the function. These values can be entered either directly as numbers, as we do below, or using previously assigned object names (e.g., `bird.table[1,1]` for w_1).

```
> # Wald
> wald2ci(x1 = 251, n1 = 285, x2 = 48, n2 = 53, conf.level =
    0.95, adjust = "Wald")

data:

95 percent confidence interval:
-0.11218742  0.06227017
sample estimates:
[1] -0.02495862

> # Agresti-Caffo
> wald2ci(x1 = 251, n1 = 285, x2 = 48, n2 = 53, conf.level =
    0.95, adjust = "AC")

data:

95 percent confidence interval:
 -0.10353254  0.07781192
sample estimates:
[1] -0.01286031
```

The `adjust` argument of `wald2ci()` specifies the Wald (`"Wald"`) or Agresti-Caffo (`"AC"`) interval. A quirk of the printed output is that there are no values printed next to `data:`.

The `prop.test()` function (first used in Section 1.1.2 to estimate and make inferences for π) also provides a convenient way to perform these calculations. We show how to use this function in Section 1.2.3 when we discuss hypothesis tests for $\pi_1 - \pi_2$.

To find a true confidence level for one of these confidence intervals, the joint probability distribution for all possible combinations of (W_1, W_2) is needed. This distribution is just the product of two binomial probabilities because these random variables are independent. For a given n_1 and n_2, there are $(n_1+1)(n_2+1)$ possible observed combinations of (w_1, w_2), and a confidence interval can be computed for each of these combinations. For set values of π_1 and π_2, some of these intervals contain $\pi_1 - \pi_2$ and some do not. Thus, the true confidence level at π_1 and π_2, $C(\pi_1, \pi_2)$, is the sum of the joint probabilities for all intervals that do contain $\pi_1 - \pi_2$:

$$C(\pi_1, \pi_2) = \sum_{w_2=0}^{n_2} \sum_{w_1=0}^{n_1} I(w_1, w_2) \binom{n_1}{w_1} \pi_1^{n_1}(1-\pi_1)^{n_1-w_1} \binom{n_2}{w_2} \pi_2^{w_2}(1-\pi_2)^{n_2-w_2}$$

where the indicator function $I(w_1, w_2)$ is 1 if the corresponding interval contains $\pi_1 - \pi_2$ and $I(w_1, w_2)$ is 0 otherwise. Calculation details are given in the next example.

Example: True confidence levels for the Wald and Agresti-Caffo intervals (ConfLevelTwoProb.R)

The true confidence level for the Wald interval can be found in a similar manner as discussed in Section 1.1.3. Consider the case of $\alpha = 0.05$, $\pi_1 = 0.2$, $\pi_2 = 0.4$, $n_1 = 10$, and $n_2 = 10$. To find all possible combinations of (w_1, w_2), we use the `expand.grid()` function, which finds all possible combinations of the arguments (separated by commas) within its parentheses. We repeat this same process to find all possible values of $(\hat{\pi}_1, \hat{\pi}_2)$ and $P(W_1 = w_1,\ W_2 = w_2)$. Below is the R code.

```
> alpha <- 0.05
> pi1 <- 0.2
> pi2 <- 0.4
> n1 <- 10
> n2 <- 10

> # All possible combinations of w1 and w2
> w.all <- expand.grid(w1 = 0:n1, w2 = 0:n2)

> # All possible combinations of pi^_1 and pi^_2
> pi.hat1 <- (0:n1)/n1
> pi.hat2 <- (0:n2)/n2
> pi.hat.all <- expand.grid(pi.hat1 = pi.hat1, pi.hat2 = pi.hat2)

> # Find joint probability for w1 and w2
> prob.w1 <- dbinom(x = 0:n1, size = n1, prob = pi1)
> prob.w2 <- dbinom(x = 0:n2, size = n2, prob = pi2)
> prob.all <- expand.grid(prob.w1 = prob.w1, prob.w2 = prob.w2)
> pmf <- prob.all$prob.w1 * prob.all$prob.w2

> # P(W1 = w1, W2 = w2)
> head(data.frame(w.all, pmf = round(pmf,4)))
  w1 w2    pmf
1  0  0 0.0006
2  1  0 0.0016
3  2  0 0.0018
4  3  0 0.0012
5  4  0 0.0005
6  5  0 0.0002
```

For example, the probability of observing $w_1 = 1$ and $w_2 = 0$ is $P(W_1 = 1,\ W_2 = 0) = 0.0016$. Using these probabilities, we calculate the true confidence level for the interval.

```
> var.wald <- pi.hat.all[,1] * (1-pi.hat.all[,1]) / n1 +
    pi.hat.all[,2] * (1-pi.hat.all[,2]) / n2
> lower <- pi.hat.all[,1] - pi.hat.all[,2] - qnorm(p =
    1-alpha/2) * sqrt(var.wald)
> upper <- pi.hat.all[,1] - pi.hat.all[,2] + qnorm(p =
    1-alpha/2) * sqrt(var.wald)
> save <- pi1-pi2>lower & pi1-pi2<upper
> sum(save*pmf)
```

```
[1] 0.9281274
> data.frame(w.all, round(data.frame(pmf, lower, upper),4),
   save)[1:15,]
   w1 w2    pmf   lower  upper  save
1   0  0 0.0006  0.0000 0.0000 FALSE
2   1  0 0.0016 -0.0859 0.2859 FALSE
3   2  0 0.0018 -0.0479 0.4479 FALSE
4   3  0 0.0012  0.0160 0.5840 FALSE
5   4  0 0.0005  0.0964 0.7036 FALSE
6   5  0 0.0002  0.1901 0.8099 FALSE
7   6  0 0.0000  0.2964 0.9036 FALSE
8   7  0 0.0000  0.4160 0.9840 FALSE
9   8  0 0.0000  0.5521 1.0479 FALSE
10  9  0 0.0000  0.7141 1.0859 FALSE
11 10  0 0.0000  1.0000 1.0000 FALSE
12  0  1 0.0043 -0.2859 0.0859  TRUE
13  1  1 0.0108 -0.2630 0.2630  TRUE
14  2  1 0.0122 -0.2099 0.4099  TRUE
15  3  1 0.0081 -0.1395 0.5395 FALSE
```

All possible Wald intervals are checked to determine if $\pi_1 - \pi_2 = 0.2 - 0.4 = -0.2$ is within each interval. The last data frame shows the first 15 intervals with the results from this check. The probabilities corresponding to the intervals that contain -0.2 are summed to produce the true confidence level of 0.9281, which is less than the stated level of 0.95.

We can also calculate the true confidence level while holding one of the probabilities constant and allowing the other to vary. Figure 1.4 shows a plot where $\pi_2 = 0.4$ and π_1 varies from 0.001 to 0.999 by 0.0005. We exclude the R code here because it is quite similar to what was used in Section 1.1.3, where now we use the `for()` function to loop over the different values of π_1. Both the Agresti-Caffo and Wald lines are drawn on the plot by using the `plot()` function first for the Wald true confidence levels and then using the `lines()` function for the Agresti-Caffo true confidence levels. The `legend()` function places the legend on the plot. Please see our program for the code.

Figure 1.4 shows that the Wald interval never achieves the true confidence level! The Agresti-Caffo interval has a true confidence level between 0.93 and 0.97. We encourage readers to change the pre-set value for π_2 in the program to examine what happens in other situations. For example, the Wald interval is extremely liberal and the Agresti-Caffo interval is extremely conservative for small π_1 when $\pi_2 = 0.1$ with $n_1 = 10$, $n_2 = 10$, and $\alpha = 0.05$.

We can also allow π_2 to vary by the same increments as π_1 in order to produce a three-dimensional plot with the true confidence level on the third axis. The R code is in the program for this example. Two `for()` function calls—one loop for π_2 and one loop for π_1—are used within the code. Once all true confidence levels are found, the `persp3d()` function from the `rgl` package of R produces an interactive three-dimensional plot. Using the left and right mouse buttons inside the plot window, the plot can be rotated and zoomed in, respectively. Figure 1.5 shows separate plots for the Wald (left) and Agresti-Caffo (right) intervals. For both plots, a plane is drawn at the 0.95 stated confidence level. We can see that the Wald interval never achieves the stated confidence level, but the Agresti-Caffo interval does a much better job. We encourage the reader to construct these plots to see the surfaces better by rotating them.

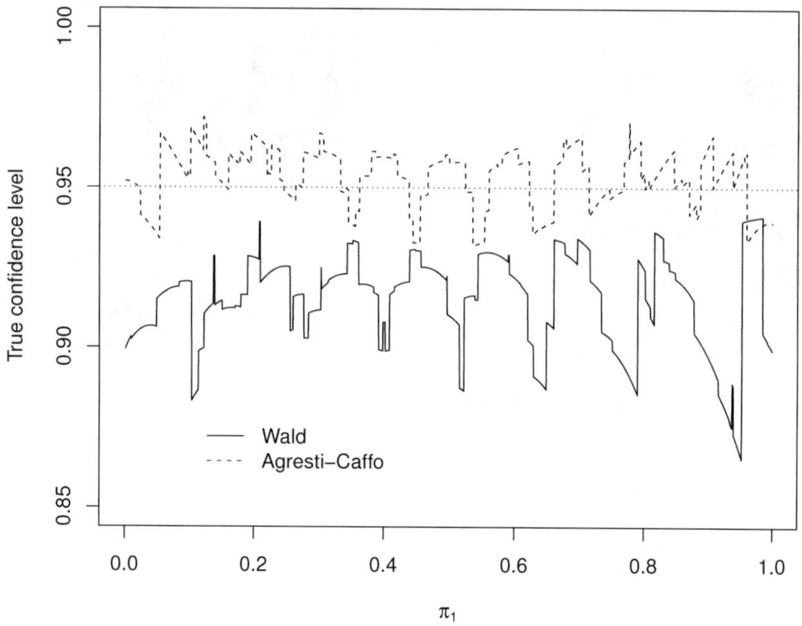

Figure 1.4: True confidence levels with $n_1 = 10$, $n_2 = 10$, $\pi_2 = 0.4$, and $\alpha = 0.05$.

These true confidence level calculations can also be made using Monte Carlo simulation. We provide an example showing how this is done in our program.

1.2.3 Test for the difference of two probabilities

The parameters π_1 and π_2 can also be compared using a hypothesis test of $H_0 : \pi_1 - \pi_2 = d$ vs. $H_a : \pi_1 - \pi_2 \neq d$, for some numerical value d. By far, the most common value of d is 0, leading to the test of $H_0 : \pi_1 - \pi_2 = 0$ vs. $H_a : \pi_1 - \pi_2 \neq 0$. This will be our focus in this subsection.

There are again several ways to perform this test. A Wald test uses a test statistic

$$Z_W = \frac{\hat{\pi}_1 - \hat{\pi}_2}{\sqrt{\frac{\hat{\pi}_1(1-\hat{\pi}_1)}{n_1} + \frac{\hat{\pi}_2(1-\hat{\pi}_2)}{n_2}}},$$

and compares this statistic against the standard normal distribution. Note that the denominator contains the estimated variance of $\hat{\pi}_1 - \hat{\pi}_2$ without regard to the null hypothesis.

Probability distributions for test statistics are computed assuming that null hypothesis is true. In the present context, that means that the two group probabilities are equal, and so a better estimated variance than what is in Z_W can be computed by assuming that $\pi_1 = \pi_2$. This condition implies that Y_1 and Y_2 have the same distribution. Thus, W_1 and W_2 are both counts of successes from the same Bernoulli random variable, and therefore w_1 and w_2 can be combined to represent w_+ successes in n_+ trials. Let $\bar{\pi} = w_+/n_+$ be the estimated probability of success when the null hypothesis is true. Then it can be shown

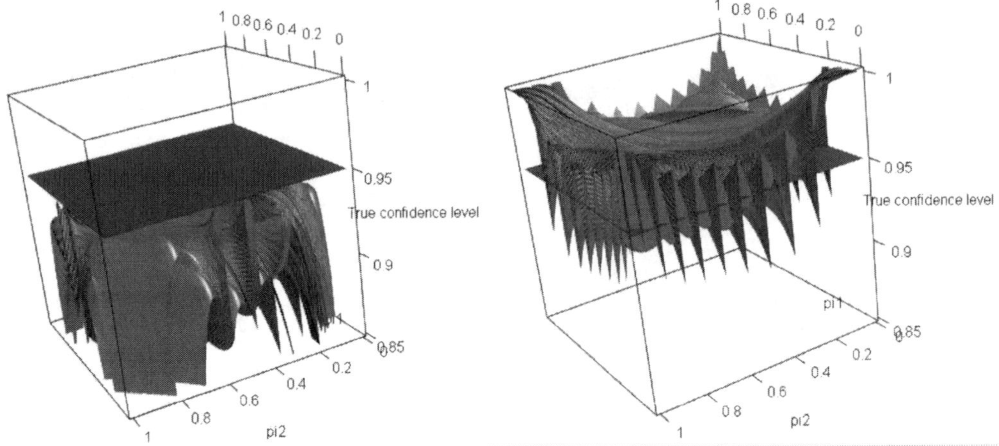

Figure 1.5: True confidence levels with $n_1 = 10$, $n_2 = 10$, and $\alpha = 0.05$. The left plot is for the Wald interval, and the right plot is for the Agresti-Caffo interval. Note that true confidence levels less than 0.85 are excluded from the Wald interval plot.

that $\widehat{Var}(\hat{\pi}_1 - \hat{\pi}_2) = \bar{\pi}(1-\bar{\pi})(1/n_1 + 1/n_2)$. This leads to the test statistic

$$Z_S = \frac{\hat{\pi}_1 - \hat{\pi}_2}{\sqrt{\bar{\pi}(1-\bar{\pi})(1/n_1 + 1/n_2)}}.$$

The statistic is compared with a standard normal distribution to judge whether there is sufficient evidence to reject the null hypothesis. This whole process is the score test for $\pi_1 - \pi_2$.

A more general procedure to compare observed counts to estimated expected counts from any hypothesized model is the Pearson chi-square test. This particular test is often used to perform hypothesis tests in a contingency table setting, and we will spend much more time discussing it in Section 3.2. The model that is implied by our null hypothesis is a single binomial distribution that has n_+ trials and probability of success $\bar{\pi}$. The test statistic is formed by computing (observed count − estimated expected count)^2/(estimated expected count) over *all* observed counts, meaning here both successes and failures in the two groups. The estimated expected number of successes in group j under the null hypothesis model is $n_j\bar{\pi}$, and similarly the expected number of failures is $n_j(1-\bar{\pi})$. Thus, the Pearson chi-square test statistic is

$$X^2 = \sum_{j=1}^{2} \left(\frac{(w_j - n_j\bar{\pi})^2}{n_j\bar{\pi}} + \frac{(n_j - w_j - n_j(1-\bar{\pi}))^2}{n_j(1-\bar{\pi})} \right). \tag{1.7}$$

This can be simplified to

$$X^2 = \sum_{j=1}^{2} \frac{(w_j - n_j\bar{\pi})^2}{n_j\bar{\pi}(1-\bar{\pi})}.$$

The X^2 statistic has a distribution that is approximately χ^2_1 when n_1 and n_2 are large and when the null hypothesis is true. If the null hypothesis is false, the observed counts tend not to be close to what is expected when the null hypothesis is true; thus, large values of X^2 relative to the χ^2_1 distribution lead to a rejection of the null hypothesis. It can be shown that the Pearson chi-square and score test results are identical for this setting, because $X^2 = Z_S^2$ (see Exercise 36) and the χ^2_1 distribution is equivalent to the distribution of a squared standard normal random variable (e.g., $Z^2_{0.975} = \chi^2_{1,0.95} = 3.84$). For those readers unfamiliar with the latter result, please see p. 53 of Casella and Berger (2002) for a derivation if desired.

An LRT can also be conducted. The test statistic can be shown to be

$$-2\log(\Lambda) = -2\left[w_1\log\left(\frac{\bar{\pi}}{\hat{\pi}_1}\right) + (n_1 - w_1)\log\left(\frac{1-\bar{\pi}}{1-\hat{\pi}_1}\right) + w_2\log\left(\frac{\bar{\pi}}{\hat{\pi}_2}\right) + (n_2 - w_2)\log\left(\frac{1-\bar{\pi}}{1-\hat{\pi}_2}\right)\right] \quad (1.8)$$

where we take $0 \times \log(\infty) = 0$ by convention. The null hypothesis is rejected if $-2\log(\Lambda) > \chi^2_{1,1-\alpha}$.

For all of these tests, the use of the standard normal or chi-square distribution is based on large-sample approximations. The tests are asymptotically equivalent, meaning that they will give essentially the same results in very large samples. In small samples, however, the three test statistics can have distributions under the null hypothesis that are quite different from their approximations. Larntz (1978) compares the score, the LRT, and three other tests in various small-sample settings and finds that the score test clearly maintains its size better than the others.[8] Thus, the score test is recommended here, as it was for testing the probability from a single group.

Example: Larry Bird's free throw shooting (Bird.R)

The purpose of this example is to show how to perform the score test, Pearson chi-square test, and LRT in R. We can use the `prop.test()` function to perform the equivalent score and Pearson chi-square tests.

```
> prop.test(x = c.table, conf.level = 0.95, correct = FALSE)

  2-sample test for equality of proportions without continuity
      correction

data:  c.table
X-squared = 0.2727, df = 1, p-value = 0.6015
alternative hypothesis: two.sided
95 percent confidence interval:
-0.11218742  0.06227017
sample estimates:
   prop 1    prop 2
0.8807018 0.9056604
```

The argument value for `x` is the contingency table. Alternatively, we could have assigned `x` a vector with w_1 and w_2 within it and used a new argument `n` with a vector value of n_1 and n_2 (see our program for an example). The `correct = FALSE` argument value guarantees that the test statistic is calculated as shown by Z_S; otherwise, a *continuity correction* is applied to help ensure that the test maintains its size at or below α.[9]

[8] The size of a testing procedure is the probability that it rejects the null hypothesis when the null hypothesis is true. A test that holds the correct size is one that rejects at a rate equal to the type I error level of α.

[9] Note that the test statistic Z_S is a discrete random variable but a continuous probability distribution is used to approximate its distribution. In general, continuity corrections to a test statistic are sometimes

The output gives the value of the Pearson chi-square test statistic as $X^2 = 0.2727$ and a p-value of $P(A > 0.2727) = 0.6015$, where A has a χ^2_1 distribution. The score test statistic is $Z_S = -\sqrt{0.2727} = -0.52$. The negative sign is included because $\hat{\pi}_1 < \hat{\pi}_2$ ($0.8807 < 0.9057$) and $\hat{\pi}_1 - \hat{\pi}_2$ is in the numerator of Z_S. The decision is to not reject the null hypothesis because of the large p-value. The conclusion is that there is not sufficient evidence to indicate a change in Bird's second free throw success probabilities over the possible outcomes of the first attempt. Note that the `chisq.test()` function and the `summary.table()` method function also provide ways to perform the Pearson chi-square test for 2×2 and larger contingency tables. We will discuss these functions in Section 3.2.

The code for the LRT is shown below.

```
> pi.bar <- colSums(c.table)[1]/sum(c.table)
> log.Lambda <- c.table[1,1] * log(pi.bar / pi.hat.table[1,1]) +
   c.table[1,2] * log((1-pi.bar) / (1-pi.hat.table[1,1])) +
   c.table[2,1] * log(pi.bar / pi.hat.table[2,1]) + c.table[2,2]
   * log((1-pi.bar) /(1-pi.hat.table[2,1]))
> test.stat <- -2*log.Lambda
> crit.val <- qchisq(p = 0.95, df = 1)
> p.val <- 1-pchisq(q = test.stat, df = 1)
> round(data.frame(pi.bar, test.stat, crit.val, p.val, row.names
   = NULL), 4)
  pi.bar test.stat crit.val p.val
1 0.8846    0.2858   3.8415 0.593
```

Under the null hypothesis, the estimate of the probability of success parameter is found using the sum of the counts in the first column of `c.table` divided by the total sample size, and the result is put into `pi.bar`. The code for the `log.Lambda` object shows how to convert most of Equation 1.8 into the correct R syntax. The test statistic is $-2\log(\Lambda) = 0.2858$, and the p-value is $P(A > 0.2858) = 0.5930$. These are nicely printed using the `data.frame()` function, where the `row.names = NULL` argument value prevents the printing of an errant row name. The overall conclusion is the same as for the score test. Note that the test statistic and p-value could have been calculated a little more easily using the `assocstats()` function of the `vcd` package, and this function also gives the Pearson chi-square test statistic as well. We show how to use this function in the program for this example.

We conclude this example with a few additional notes:

- Notice that the success probability conditioning on the first free throw being missed was subtracted from the success probability conditioning on the first free throw being made. The direction of the difference is especially important to know when a one-sided hypothesis test is performed or when 0 is outside of a confidence interval. For example, many basketball fans think that a missed first free throw reduces the chances of a made second free throw. If the lower limit of the interval had been positive (i.e., the whole interval is above 0), it would have confirmed this line of thinking with respect to Larry Bird.

made in these types of situations to improve the distributional approximation. These corrections can lead to very conservative tests (i.e., reject the null hypothesis at a rate less than α when the null hypothesis is true), so they are not often used. Alternative procedures are discussed in Section 6.2.

- As with other applications of statistics, care needs to be taken when interpreting the results with respect to the population. For example, suppose we wanted to make some claims regarding all of Larry Bird's past, present, and future free throw pairs when the data were collected. Strictly speaking, a random sample would need to be taken from this entire population to formally make statistical inferences. Random samples are often not possible in a sports setting, as in our example where we have data only from the 1980–1 and 1981–2 NBA seasons. Inference on a broader population of free throws may or may not be appropriate. For example, Larry Bird's free throw shooting success rate may have changed from year to year because of practice or injuries.

- In addition to the sampling problem, Larry Bird's career concluded in 1992. Therefore, the population data may be obtainable, and we could actually calculate population parameters such as π_1 and π_2. Statistical inference would not be necessary then.

1.2.4 Relative risks

The problem with basing inference on $\pi_1 - \pi_2$ is that it measures a quantity whose meaning changes depending on the sizes of π_1 and π_2. For example, consider two hypothetical scenarios where the probability of disease is listed for two groups of people, say for smokers (π_1) and for nonsmokers (π_2):

1. $\pi_1 = 0.51$ and $\pi_2 = 0.50$
2. $\pi_1 = 0.011$ and $\pi_2 = 0.001$.

In both cases, $\pi_1 - \pi_2 = 0.01$. But in the first scenario, an increase of 0.01 for smokers is rather small relative to the already sizable risk of disease in the nonsmoking population. On the other hand, in scenario 2 smokers have 11 times the chance of disease that nonsmokers have. We need to be able to convey the relative magnitudes of these changes better than differences allow.

In this instance, a preferred way to compare two probabilities is with the *relative risk*, $RR = \pi_1/\pi_2$ (assuming $\pi_2 \neq 0$). For the previous example, $RR = 0.011/0.001 = 11.0$ in the second scenario, meaning that smokers are *11 times as likely* to have the disease as nonsmokers. Alternatively, we could say that smokers are 10 times *more* likely to have the disease than nonsmokers. On the other hand, $RR = 0.51/0.50 = 1.02$ in the first scenario, indicating that smokers are just 2% more likely (or 1.02 times *as* likely) to have the disease. Notice that when $\pi_1 = \pi_2$, $RR = 1$.

These numerical values are based on population probabilities. To obtain an MLE for RR, we can use the invariance property of MLEs described in Appendix B.4 that allows us to substitute the observed proportions for the probabilities, $\widehat{RR} = \hat{\pi}_1/\hat{\pi}_2$, assuming $\hat{\pi}_2 \neq 0$. It is this estimate that is often given in news media reports that state risks associated with certain factors such as smoking or obesity.

Because $\widehat{RR}$ is an MLE, inference can be carried out with the usual procedures. It turns out, however, that the normal approximation is rather poor for MLEs that are ratios, especially when the estimate in the denominator may have non-negligible variability as is the case here. Therefore, inference based on a normal approximation for $\widehat{RR}$ is not recommended. However, the normal approximation holds better for $\log(\widehat{RR}) = \log(\hat{\pi}_1) -$

$\log(\hat{\pi}_2)$—the MLE for $\log(RR)$—so inference is generally carried out on the log scale. The variance estimate for $\log(\widehat{RR})$ can be derived by the delta method (Appendix B.4.2) as

$$\widehat{Var}(\log(\widehat{RR})) = \frac{1-\hat{\pi}_1}{n_1\hat{\pi}_1} + \frac{1-\hat{\pi}_2}{n_2\hat{\pi}_2} = \frac{1}{w_1} - \frac{1}{n_1} + \frac{1}{w_2} - \frac{1}{n_2}. \quad (1.9)$$

A $(1-\alpha)100\%$ Wald confidence interval for the population relative risk is found by first computing the confidence interval for $\log(RR)$,

$$\log(\widehat{RR}) \pm Z_{1-\alpha/2}\sqrt{\frac{1}{w_1} - \frac{1}{n_1} + \frac{1}{w_2} - \frac{1}{n_2}}.$$

The exponential transformation is then used to find the Wald interval for the relative risk itself:

$$\exp\left[\log(\widehat{RR}) \pm Z_{1-\alpha/2}\sqrt{\frac{1}{w_1} - \frac{1}{n_1} + \frac{1}{w_2} - \frac{1}{n_2}}\right],$$

where $\exp(\cdot)$ is the inverse of the natural log function ($b = \exp(a)$ is equivalent to $a = \log(b)$). When w_1 and/or w_2 are equal to 0, the confidence interval cannot be calculated. One ad-hoc adjustment is to add a small constant, such as 0.5, to the 0 cell count and the corresponding row total. For example, if $w_1 = 0$, replace w_1 with 0.5 and n_1 with $n_1 + 0.5$. A different adjustment is to add a small constant to all cells of the contingency table.

Example: COVID-19 vaccine clinical trial (COVID19vaccine.R)

The purpose of this example is to determine the effectiveness of the Pfizer-BioNTech COVID-19 vaccine by examining the relative risk of infection from SARS-CoV-2. Below is the R code used to enter the data into an array and to perform the necessary calculations.

```
> c.table <- array(data = c(8, 162, 17403, 17349), dim = c(2,2),
   dimnames = list(Treatment = c("Vaccine", "Placebo"), Outcome
   = c("Positive", "Negative")))
> c.table
         Outcome
Treatment Positive Negative
  Vaccine        8    17403
  Placebo      162    17349

> pi.hat.table <- c.table/rowSums(c.table)
> pi.hat.table
         Outcome
Treatment     Positive  Negative
  Vaccine 0.0004594796 0.9995405
  Placebo 0.0092513277 0.9907487

> pi.hat1 <- pi.hat.table[1,1]
> pi.hat2 <- pi.hat.table[2,1]

> round(pi.hat1/pi.hat2, 4)
[1] 0.0497
> round(1/(pi.hat1/pi.hat2), 4)   # inverted
[1] 20.1344

> # Wald confidence interval
```

```
> alpha <- 0.05
> var.log.RR <- 1/c.table[1,1] - 1/sum(c.table[1,]) +
   1/c.table[2,1] - 1/sum(c.table[2,])
> RR.CI <- exp(log(pi.hat1/pi.hat2) + qnorm(p = c(alpha/2,
   1-alpha/2)) * sqrt(var.log.RR))
> round(RR.CI, 4)
[1] 0.0244 0.1010
> rev(round(1/RR.CI, 4))  # inverted
[1] 9.9034 40.9345
```

Defining the index 1 to represent the vaccine group and the index 2 to represent the placebo group, we find $\widehat{RR} = 0.0497$. The estimated probability of infection is only 0.05 times (or 5%) as large for the vaccine group as for the placebo group. We used the word *estimated* with this interpretation because we are using parameter estimates. The 95% Wald confidence interval is $0.0244 < RR < 0.1010$. With 95% confidence, the probability of infection is only 0.02 to 0.10 times as large for people receiving the vaccine as for those receiving the placebo. Notice the use of `exp()` in the code to calculate the exponential function (for example, `exp(1)` is 2.718).

Rather than stating the relative risk, results of vaccine clinical trials are often reported in the form of vaccine efficacy, $VE = 1 - RR$. This represents the amount that the risk (probability) of infection is reduced by the vaccine. For example, when the COVID-19 vaccine clinical trial results were announced, a *New York Times* headline stated "New Pfizer results: Coronavirus vaccine is safe and 95% effective" (Thomas, 2020). This statement of effectiveness came from $\widehat{VE} = 1 - \widehat{RR} = 0.9503$. Estimates of vaccine efficacy should also include confidence intervals. Thus, with 95% confidence, the vaccine reduces the *population* risk of infection by 90% to 98%, where we subtract the lower and upper bounds of the interval for RR from 1.

Code from the previous example provides a template for how to perform the calculations for a new contingency table. This process can be simplified by using one function to perform the calculations. For example, Section 1.1.2 illustrated the simplicity of using the `binom.confint()` function from the `binom` package to estimate and calculate a confidence interval for π. Next, we show how to write a simple function for this same purpose by building on coding techniques shown in Appendix A.2 and A.7.3. Writing functions in this manner is very useful when coding in R. It helps to not only automate common tasks, but also to extend what one can do within the software environment.

Example: COVID-19 vaccine clinical trial (COVID19vaccine.R)

Our `RR()` function below has three arguments: 1) `data` for the contingency table, 2) `alpha` to specify α, and 3) `digits` to specify the amount of rounding for numerical values. The latter two arguments have default values set for them.

```
> RR <- function(data, alpha = 0.05, digits = 4) {
    n1 <- sum(data[1,])
    n2 <- sum(data[2,])
    pi.hat.table <- data/rowSums(data)
    pi.hat1 <- pi.hat.table[1,1]
    pi.hat2 <- pi.hat.table[2,1]
    RR.hat <- pi.hat1/pi.hat2
```

```
    var.log.RR <- 1/data[1,1] - 1/sum(data[1,]) + 1/data[2,1] -
        1/sum(data[2,])
    RR.CI <- exp(log(pi.hat1/pi.hat2) + qnorm(p = c(alpha/2,
        1-alpha/2)) * sqrt(var.log.RR))
    round(data.frame(Estimate = RR.hat, Lower = RR.CI[1], Upper
        = RR.CI[2], row.names = ""), digits)
  }

> RR(data = c.table, alpha = 0.05, digits = 4)
 Estimate  Lower Upper
   0.0497 0.0244 0.101
> RR(data = c.table)  # Use defaults
 Estimate  Lower Upper
   0.0497 0.0244 0.101
> save.res <- RR(data = c.table, alpha = 0.05, digits = 4)
> save.res[,1]
[1] 0.0497
> save.res[,2:3]
  Lower Upper
 0.0244 0.101
```

The code to calculate $\widehat{RR}$ and the confidence interval is essentially the same as in the previous example but now within the `RR()` function. Results returned by `RR()` are set by the function's last line, where each numerical value is rounded to four decimal places by default. These numerical values are organized into a data frame for convenience. Also, these numerical values can be saved into an object when the function is run, such as `save.res` here, and extracted as needed.

Modifications to this function are possible. For example, it may be of interest to have one function for VE. We discuss modifications like this in Exercises 46 and 48. Also, Exercise 49 discusses functions available in user-written packages to calculate the relative risk.

While the confidence interval presented for the relative risk is the most widely used, there are other intervals that have been proposed. The best of these other intervals is a score interval. This interval is formed again by inverting a hypothesis test (find all values of r for which $H_0 : RR = r$ is not rejected). Iterative numerical procedures are used to find the interval's limits. The `riskscoreci()` function of the `PropCIs` package performs the calculations. Exercise 35 examines this function further.

The ratio for the relative risk is often arranged so that $\widehat{RR} \geq 1$. This allows for an appealing interpretation that the group represented in the numerator has a risk that is, for example, "11 times as large" as the denominator group. This can be easier for a target audience to appreciate than the alternative, "0.091 times as large." However, the application may dictate which group should be the numerator regardless of the sample proportions. This was the case for the COVID-19 vaccine example, where it is natural to think of the vaccine in terms of its risk reduction.

The relative risk is generally a more useful measure than the difference between probabilities when the probabilities are fairly small. It is of limited use otherwise. For example, if $\pi_2 = 0.8$, then the maximum possible relative risk is $1/0.8{=}1.25$. It is therefore useful to have an alternative statistic for comparing probabilities that is applicable regardless of the sizes of the probabilities. This is part of the motivation for the next measure.

1.2.5 Odds ratios

We have focused so far on using probabilities to measure the chance that an event will occur. *Odds* can be used as a similar measure. Odds are simply the probability of a success divided by the probability of a failure, $Odds = \pi/(1-\pi)$. In some areas of application, such as betting, odds are used almost exclusively. For example, if a probability is $\pi = 0.1$, then the odds of success are $0.1/(1-0.1) = 1/9$. This is referred to as "9-to-1 odds *against*," because the probability of failure is 9 times the probability of success. Notice that there is a 1-to-1 relationship between probability and odds: if you know one, you can find the other. Also notice that odds have no upper limit, unlike probabilities. Like relative risks, odds are estimated with MLEs by replacing the probabilities with their corresponding estimates.

When there are two groups, we can calculate the odds separately in each group: $Odds_1 = \pi_1/(1-\pi_1)$ and $Odds_2 = \pi_2/(1-\pi_2)$. Then a comparison of the odds is made using an *odds ratio*. Formally, it is defined as

$$OR = \frac{Odds_1}{Odds_2} = \frac{\pi_1/(1-\pi_1)}{\pi_2/(1-\pi_2)} = \frac{\pi_1(1-\pi_2)}{\pi_2(1-\pi_1)}.$$

The odds ratio can be estimated by substituting the parameters with their corresponding estimates to obtain the MLE:

$$\widehat{OR} = \frac{\widehat{Odds}_1}{\widehat{Odds}_2} = \frac{\hat{\pi}_1(1-\hat{\pi}_2)}{\hat{\pi}_2(1-\hat{\pi}_1)} = \frac{(w_1/n_1)[(n_2-w_2)/n_2]}{(w_2/n_2)[(n_1-w_1)/n_1]} = \frac{w_1(n_2-w_2)}{w_2(n_1-w_1)}.$$

When counts are written in the form of a contingency table, the estimated odds ratio is a product of the counts on the "diagonal" (top-left to bottom-right) of the table divided by a product of the counts off of the diagonal.

Interpretation

Determining whether an odds ratio is equal to 1, greater than 1, or less than 1 is often of interest. An odds ratio equal to 1 means that the odds in group 1 are the same as the odds in group 2. Thus, we can say that the odds are not dependent on the group; i.e., the odds of success are *independent* of the group designation. An odds ratio greater than 1 means that the odds of success are higher for group 1 than for group 2. The opposite is true for an odds ratio less than 1. Note that $OR = 1$ implies that $RR = 1$ and $\pi_1 = \pi_2$ and vice versa; see Exercise 39. Because of this equivalence, the hypothesis test procedures of Section 1.2.3 can equivalently be used to test $H_0: OR = 1$ vs. $H_a: OR \neq 1$.

Because odds ratios are combinations of four probabilities, they are easy to misinterpret. The standard interpretation of an estimated odds ratio is:

> The estimated odds of success are $\widehat{OR}$ times as large for group 1 as for group 2.

For this statement, "success," "group 1," and "group 2" would be replaced with meaningful terms in the context of the problem. Also, $\widehat{OR}$ would be replaced with the calculated numerical value. The interpretation could be rephrased in terms of how much *larger* the odds are for group 1 than for group 2. In this case, the interpretation becomes

> The estimated odds of success are $\widehat{OR}-1$ times larger for group 1 than for group 2.

The small change in wording has a potentially important effect on the numerical value. Also, this alternative phrasing can be confusing if $\widehat{OR} < 1$, so we recommend against using it in those situations.

Changing the order of the ratio, so that we have $\widehat{Odds}_2/\widehat{Odds}_1$, can sometimes make the interpretation easier. We refer to this as *inverting* the odds ratio. With this inverted odds ratio, we can now say that

> The estimated odds of success are $1/\widehat{OR}$ times as large for group 2 as for group 1.

The ratio of the *odds of failure* for the two groups may also be of interest. Thus, $(1-\hat{\pi}_1)/\hat{\pi}_1$ is the odds of failure for group 1, and $(1-\hat{\pi}_2)/\hat{\pi}_2$ is the odds of failure for group 2. The ratio of group 1 to group 2 is

$$\frac{(1-\hat{\pi}_1)/\hat{\pi}_1}{(1-\hat{\pi}_2)/\hat{\pi}_2} = \frac{\hat{\pi}_2(1-\hat{\pi}_1)}{\hat{\pi}_1(1-\hat{\pi}_2)},$$

which is simply $1/\widehat{OR}$ from earlier. The interpretation becomes

> The estimated odds of failure are $1/\widehat{OR}$ times as large for group 1 as for group 2.

Finally, we can also invert this ratio of two odds of failure to obtain the interpretation of

> The estimated odds of failure are $\widehat{OR}$ times as large for group 2 as for group 1,

where the odds ratio is numerically the same as we had in the beginning.

This "symmetry" in the interpretation of odds ratios can be confusing at times, but it has a very distinct advantage. In *case-control studies*, measurements are taken retrospectively on fixed numbers of "cases" and "controls," which are equivalent to "successes" and "failures," respectively. That is, the proportion of cases relative to controls in the sample (e.g., patients who did or did not have a stroke) is chosen by the researcher and do not typically match the proportion of cases in the population. Whether they fall into group 1 or 2 (e.g., they did or did not take a particular medication in the past year) is the measurement of interest. In this context, the usual definitions of "group" and "response" are reversed: we want to know how taking medication might relate to the probability that someone had a stroke, but we measure how having a stroke is related to the probability that they took the medication. If one computes an odds ratio in this instance, then it is easy to see that if the odds of medication use are x times as large for those who had a stroke as for those who did not have a stroke, then the odds of having a stroke are also x times as large for medication users as for non-medication users! Thus, we can get an answer to the question of interest without actually measuring the odds of having a stroke for either medication group. Exercise 41 explores this example further.

It is incorrect to interpret an odds ratio as relating directly to the *probability* of success. We cannot say that "The estimated *probability* of success is $\widehat{OR}$ times as large for group 1 than for group 2." This is one substantial drawback for using odds ratios instead of relative risks, because most people are more comfortable discussing probabilities than odds. However, it is easy to see that $OR \approx RR$ if the two groups' probabilities of success are both fairly low. In this case, both probabilities of failure are nearly 1 and essentially cancel out in the odds ratio. For this reason, odds ratios can be used as a surrogate to relative risks when the successes being counted are relatively uncommon (e.g., both $\pi_j \leq 0.1$).

Confidence interval

Just as with relative risks, the probability distribution of $\log(\widehat{OR})$ is better approximated by a normal distribution than is the probability distribution of $\widehat{OR}$ itself. The variance of

$\log(\widehat{OR})$ is again found using the delta method and has a particularly easy form:

$$\widehat{Var}(\log(\widehat{OR})) = \frac{1}{w_1} + \frac{1}{n_1 - w_1} + \frac{1}{w_2} + \frac{1}{n_2 - w_2}.$$

The $(1-\alpha)100\%$ Wald confidence interval for OR becomes

$$\exp\left[\log\left(\widehat{OR}\right) \pm Z_{1-\alpha/2}\sqrt{\frac{1}{w_1} + \frac{1}{n_1 - w_1} + \frac{1}{w_2} + \frac{1}{n_2 - w_2}}\right]. \tag{1.10}$$

Lui and Lin (2003) show that the true confidence level of this interval is a little above the stated level most of the time. However, the interval is very conservative for small and large π_1 and π_2 and small n_+. These are cases that are likely to produce very small cell counts in the corresponding observed contingency table, and this leads to some instability in the estimates of OR and its variance.

Zero counts anywhere in the contingency table cause the estimated odds ratio to be 0 or undefined and the corresponding variance to be undefined. Small adjustments to the counts are often made in these situations. For example, we can use

$$\widetilde{OR} = \frac{(w_1 + 0.5)(n_2 - w_2 + 0.5)}{(w_2 + 0.5)(n_1 - w_1 + 0.5)}$$

as an estimate of the odds ratio. The addition of a 0.5 to each cell count leads the estimated odds ratio to be closer to 1 than without the adjustment. The estimated variance for $\log(\widetilde{OR})$ is then also changed to have 0.5 added to each count within it. An adjusted Wald confidence interval can then be formed with $\widetilde{OR}$ and this new variance by making the appropriate substitutions in Equation 1.10. Note that some people advocate making this adjustment at all times even when there are no 0 counts. The main advantage of this approach is the confidence interval is a little shorter in length (due to the "additional" observations) than without this adjustment. Lui and Lin (2003) show this adjusted Wald interval achieves a true confidence level very similar to the interval without the adjustment.

Example: COVID-19 vaccine clinical trial (COVID19vaccine.R)

The purpose of this example is to estimate the odds ratio describing the relationship between vaccine and infection from SARS-CoV-2. We have coded the data so that w_1 is the number of people contracting the virus in the vaccine group and w_2 is the same count for the placebo group. Continuing the code from earlier, we obtain the following results.

```
> OR.hat <- c.table[1,1] * c.table[2,2] / (c.table[2,1] *
   c.table[1,2])
> round(OR.hat, 4)
[1] 0.0492
> round(1/OR.hat, 4)
[1] 20.313

> alpha <- 0.05
> var.log.OR <- 1/c.table[1,1] + 1/c.table[1,2] + 1/c.table[2,1]
   + 1/c.table[2,2]
> OR.CI <- exp(log(OR.hat) + qnorm(p = c(alpha/2, 1-alpha/2)) *
   sqrt(var.log.OR))
> round(OR.CI, 4)
[1] 0.0242 0.1001
```

```
> rev(round(1/OR.CI, 4))
[1] 9.9851 41.3234
```

We find $\widehat{OR} = 0.0492$, so the estimated odds of infection from SARS-COV-2 are 0.05 times (or 5%) as large when the vaccine is given as when a placebo is given. The 95% Wald confidence interval is $0.02 < OR < 0.10$. Because 1 is not within the interval, there is sufficient evidence to indicate the vaccine decreases the true odds (and hence the probability) of infection. Notice that we did not include the word *estimated* in the last sentence because we interpreted OR rather than $\widehat{OR}$.

Alternatively, we could interpret the inverted odds ratios in terms of *protection* due to vaccination: The estimated odds of being infection-free are 20.31 times as large when the vaccine is given as when the placebo is given. The 95% Wald confidence interval for the odds ratio is (9.99, 41.32). Notice that the `rev()` function reverses the order of the elements within the object in parentheses. This is done so that the lower value of the confidence interval appears first in the resulting vector.

These values are essentially the same as what we calculated for the relative risk. This is because (a) the probability of infection is very low for both groups and (b) the sample size is very large. In smaller samples, the estimates $\widehat{RR}$ and $\widehat{OR}$ would still have been quite similar, but the confidence intervals would have differed. The difference in the confidence intervals stems from the difference in the variances: in very large samples, both $1/(n_j - w_j)$ and $-1/n_j$ are negligible relative to $1/w_j$, $j = 1, 2$.

Example: Larry Bird's free throw shooting (Bird.R)

The purpose of this example is to calculate an odds ratio comparing the odds of a successful second free throw attempt when the first is made to when the first is missed. To simplify the calculation process, we include our own function `OR()` below to calculate $\widehat{OR}$ and the corresponding confidence interval.

```
> OR <- function(data, alpha = 0.05, digits = 4) {
    OR.hat <- c.table[1,1]*c.table[2,2] /
        (c.table[2,1]*c.table[1,2])
    var.log.OR <- 1/c.table[1,1] + 1/c.table[1,2] +
        1/c.table[2,1] + 1/c.table[2,2]
    OR.CI <- exp(log(OR.hat) + qnorm(p = c(alpha/2, 1-alpha/2))
        * sqrt(var.log.OR))
    round(data.frame(Estimate = OR.hat, Lower = OR.CI[1], Upper
        = OR.CI[2], row.names = ""), digits)
  }

> OR(data = c.table, alpha = 0.05, digits = 2)
 Estimate Lower Upper
     0.77  0.29  2.07
> save.res <- OR(data = c.table, alpha = 0.05)
> 1/save.res[,1]
[1] 1.30039
> as.numeric(rev(1/save.res[,2:3]))  # as.numeric() removes
    labels
[1] 0.4840505 3.4940601
```

The estimated odds ratio is $\widehat{OR} = 0.77$, and the 95% confidence interval for OR is (0.29, 2.07). Because the estimated odds ratio is less than 1, we decided to invert it to help with its interpretation. The estimated odds of a successful second free throw attempt are 1.30 times as large when the first free throw is missed as when the first free throw is made. Also, with 95% confidence, the odds of a successful free throw attempt are between 0.48 and 3.49 times as large when the first free throw is missed as when the first free throw is made. Because 1 is within the interval, there is insufficient evidence to indicate an actual difference in the odds of success.

Exercises 47 and 49 discuss generalizations to `OR()` and other functions available in user-written packages to calculate an odds ratio.

The Wald confidence interval presented is the most widely used for odds ratios. Similar to previous comparisons of two probabilities, other intervals exist for OR. In particular, the score interval is formed by inverting a hypothesis test by finding all values of r for which $H_0 : OR = r$ is not rejected. The `orscoreci()` function of the `PropCIs` package uses iterative numerical procedures to find the interval's limits. Exercise 35 examines this function further. An LR-based confidence interval for OR can be found as well. The LR-based confidence interval is best introduced in the context of a regression model, so we postpone its introduction until Section 2.2.3. Exercise 10 of Chapter 2 examines this alternative interval for the Larry Bird free throw shooting data.

1.2.6 Matched pairs data

Section 1.2.2 examines a confidence interval for the difference of two probabilities. These intervals were developed using the fact that W_1 and W_2 are independent random variables. There are other situations where the two probabilities being compared correspond to dependent random variables. This occurs with *matched pairs* data, where two binary response observations, say X and Y, are made on each sampled unit, and a desired comparison between success probabilities for X and Y involves two correlated statistics. Below is an example to illustrate this problem.

Example: Prostate cancer diagnosis procedures

Zhou and Qin (2005) discuss a study that was used to compare the diagnostic accuracy of magnetic resonance imaging (MRI) and ultrasound in patients who had been established as having localized prostate cancer by a gold standard test. The data from one study location are given in Table 1.5. We are interested in comparing the probability of a localized diagnosis by an MRI to the probability of a localized diagnosis by an ultrasound. Notice that because both procedures are performed on the same patients, these probability estimates—10/16 and 7/16, respectively—are based on the same counts, and hence their corresponding statistics are not independent. This violates the assumption of independent binomial counts used in the model from Section 1.2.2.

To develop the appropriate analysis methods, note that any combination of two measurements can occur for each subject. That is, we may observe $(X = 1, Y = 1), (X = 1, Y = 2), (X = 2, Y = 1)$, or $(X = 2, Y = 2)$, where the values of X and Y correspond to row and column numbers, respectively. Each of these events has probability $\pi_{ij} = P(X = i, Y = j)$, $i = 1, 2$, $j = 1, 2$. These are called *joint probabilities.* As the previous example highlights, we are interested in summaries of these joint probabilities,

Table 1.5: Diagnoses established using MRI and ultrasound technology among individuals who truly have localized prostate cancer. Data source is Zhou and Qin (2005).

		Ultrasound		
		Localized	Advanced	
MRI	Localized	4	6	10
	Advanced	3	3	6
		7	9	16

rather than in the joint probabilities themselves. That is, we are interested in the *marginal probabilities*, $\pi_{1+} = \pi_{11} + \pi_{12}$ and $\pi_{+1} = \pi_{11} + \pi_{21}$, so called because they relate to counts in the margins of the table. With respect to the prostate cancer diagnosis example, π_{1+} is the probability of a localized diagnosis by MRI and π_{+1} is the probability of a localized diagnosis by ultrasound.

Suppose n units are cross-classified according to X and Y. We observe counts n_{11}, n_{12}, n_{21}, and n_{22}, with n_{ij} being the number of joint occurrences of $X = i, Y = j$. This is just like the binomial setting, except with four possible outcomes rather than two. The model that describes the counts produced in this manner is called the *multinomial distribution*. We cover this model in greater detail in Chapter 3. For now, it suffices to know that MLEs and variances follow the same pattern as in the binomial model: $\hat{\pi}_{ij} = n_{ij}/n$ with $\widehat{Var}(\hat{\pi}_{ij}) = \hat{\pi}_{ij}(1-\hat{\pi}_{ij})/n$, $\hat{\pi}_{1+} = (n_{11}+n_{12})/n$ with $\widehat{Var}(\hat{\pi}_{1+}) = \hat{\pi}_{1+}(1-\hat{\pi}_{1+})/n$, and $\hat{\pi}_{+1} = (n_{11}+n_{21})/n$ with $\widehat{Var}(\hat{\pi}_{+1}) = \hat{\pi}_{+1}(1-\hat{\pi}_{+1})/n$, where $n = n_{11}+n_{12}+n_{21}+n_{22}$.

Comparing marginal probabilities of success in matched pairs is usually done through the difference $\pi_{1+} - \pi_{+1}$, because this comparison is mathematically much easier to work with than other possible comparisons.[10] If $\pi_{+1} - \pi_{1+} = 0$, the two marginal probabilities are equal. Alternatively, if $\pi_{+1} - \pi_{1+} > 0$, this means that success (first response category) is more likely to happen for Y than for X. The opposite is true if $\pi_{+1} - \pi_{1+} < 0$.

Confidence interval for the difference of marginal probabilities

To develop a confidence interval for $\pi_{+1} - \pi_{1+}$, we start with the MLE $\hat{\pi}_{+1} - \hat{\pi}_{1+}$ and find its probability distribution. Once again we make use of the facts that

1. Each estimate, $\hat{\pi}_{1+}$ and $\hat{\pi}_{+1}$, has a distribution that is approximated in large samples by a normal distribution with variance as given above, and

2. Differences of normal random variables are normal random variables.

However, this problem differs from the one laid out in Section 1.2.2 because the two random variables $\hat{\pi}_{1+}$ and $\hat{\pi}_{+1}$ are dependent. Therefore, we have that[11]

$$\widehat{Var}(\hat{\pi}_{+1} - \hat{\pi}_{1+}) = \widehat{Var}(\hat{\pi}_{+1}) + \widehat{Var}(\hat{\pi}_{1+}) - 2\widehat{Cov}(\hat{\pi}_{+1}, \hat{\pi}_{1+}).$$

Using the underlying multinomial distribution for the cell counts, the covariance can be shown to be $\widehat{Cov}(\hat{\pi}_{+1}, \hat{\pi}_{1+}) = n^{-1}(\hat{\pi}_{11}\hat{\pi}_{22} - \hat{\pi}_{12}\hat{\pi}_{21})$.[12] Using the variances for the marginal

[10] In some situations, especially with very small marginal probabilities, the ratio of the two probabilities may be of more interest. See Bonett and Price (2006) for how to form intervals for ratios of marginal probabilities.

[11] This is another application of the following result: $Var(aU+bV) = a^2Var(U)+b^2Var(V)+2abCov(U,V)$, where U and V are random variables and a and b are constants. See p. 171 of Casella and Berger (2002).

[12] To show this result, note that $Cov(\hat{\pi}_{+1}, \hat{\pi}_{1+}) = Var(\hat{\pi}_{11})+Cov(\hat{\pi}_{11}, \hat{\pi}_{12})+Cov(\hat{\pi}_{11}, \hat{\pi}_{21})+Cov(\hat{\pi}_{12}, \hat{\pi}_{21})$, and the covariance of terms like $Cov(\hat{\pi}_{11}, \hat{\pi}_{12})$ is $-n^{-1}\pi_{11}\pi_{12}$. Please see Appendix B.2.2 or Section 3.1 for more on the multinomial distribution, if needed.

probabilities given earlier, we obtain

$$\widehat{Var}(\hat{\pi}_{+1} - \hat{\pi}_{1+}) = \frac{\hat{\pi}_{+1}(1-\hat{\pi}_{+1}) + \hat{\pi}_{1+}(1-\hat{\pi}_{1+}) - 2(\hat{\pi}_{11}\hat{\pi}_{22} - \hat{\pi}_{12}\hat{\pi}_{21})}{n}$$
$$= \frac{\hat{\pi}_{21} + \hat{\pi}_{12} - (\hat{\pi}_{21} - \hat{\pi}_{12})^2}{n}.$$

This leads to the $(1-\alpha)100\%$ Wald confidence interval for $\pi_{+1} - \pi_{1+}$:

$$\hat{\pi}_{+1} - \hat{\pi}_{1+} \pm Z_{1-\alpha/2}\sqrt{\widehat{Var}(\hat{\pi}_{+1} - \hat{\pi}_{1+})}.$$

Not surprisingly, the Wald interval has problems achieving the stated confidence level. There have been a number of other intervals proposed. We focus on the interval proposed by Agresti and Min (2005b) because of its simplicity and performance. Motivated by the Agresti-Caffo interval, Agresti and Min (2005b) propose to add 0.5 to each count in the contingency table and compute a Wald confidence interval using these adjusted counts. Specifically, let $\tilde{\pi}_{ij} = (n_{ij} + 0.5)/(n+2)$, so that $\tilde{\pi}_{+1} = (n_{11} + n_{21} + 1)/(n+2)$ and $\tilde{\pi}_{1+} = (n_{11} + n_{12} + 1)/(n+2)$. The $(1-\alpha)100\%$ Agresti-Min interval is

$$\tilde{\pi}_{+1} - \tilde{\pi}_{1+} \pm Z_{1-\alpha/2}\sqrt{\frac{\tilde{\pi}_{+1}(1-\tilde{\pi}_{+1}) + \tilde{\pi}_{1+}(1-\tilde{\pi}_{1+}) - 2(\tilde{\pi}_{11}\tilde{\pi}_{22} - \tilde{\pi}_{12}\tilde{\pi}_{21})}{n+2}}.$$

Agresti and Min (2005b) examine adding other constants to each cell, and they conclude that 0.5 is best in terms of the true confidence level and expected length.[13]

Example: Prostate cancer diagnosis procedures (DiagnosisCancer.R)

The purpose of this example is to compute confidence intervals for $\pi_{+1} - \pi_{1+}$. We begin by entering the data into R using the `array()` function.

```
> c.table <- array(data = c(4, 3, 6, 3), dim = c(2,2), dimnames
   = list(MRI = c("Localized", "Advanced"), Ultrasound =
   c("Localized", "Advanced")))
> c.table
           Ultrasound
MRI          Localized Advanced
  Localized          4        6
  Advanced           3        3

> n <- sum(c.table)
> pi.hat.plus1 <- sum(c.table[,1])/n
> pi.hat.1plus <- sum(c.table[1,])/n
> data.frame(pi.hat.plus1, pi.hat.1plus, diff = pi.hat.plus1 -
   pi.hat.1plus)
  pi.hat.plus1 pi.hat.1plus    diff
1       0.4375        0.625 -0.1875
```

[13]See Section 3 of their paper for further justification with respect to a score interval and a Bayesian credible interval. The main disadvantage of a score interval here is that it does not have a closed-form expression and it sometimes is longer than the Agresti-Min interval. The `scoreci.mp()` function in the `PropCIs` package calculates the score interval. We provide an example of its use in the program for the prostate cancer diagnosis example.

The estimate of $\pi_{+1} - \pi_{1+}$ is -0.1875. To find the Wald and Agresti-Min confidence intervals for $\pi_{+1} - \pi_{1+}$, we use the `diffpropci.Wald.mp()` and `diffpropci.mp()` functions, respectively, from the `PropCIs` package.

```
> library(package = PropCIs)
> diffpropci.Wald.mp(b = c.table[1,2], c = c.table[2,1], n =
    sum(c.table), conf.level = 0.95)

data:

95 percent confidence interval:
 -0.5433238  0.1683238
sample estimates:
[1] -0.1875

> diffpropci.mp(b = c.table[1,2], c = c.table[2,1], n =
    sum(c.table), conf.level = 0.95)

data:

95 percent confidence interval:
 -0.5022786  0.1689453
sample estimates:
[1] -0.1666667
```

The `b` and `c` arguments in the functions correspond to n_{12} and n_{21}, respectively, which leads to $\hat{\pi}_{+1} - \hat{\pi}_{1+} = \hat{\pi}_{21} - \hat{\pi}_{12}$ being calculated. The ordering of the estimates in this expression is important for determining that the interval is for $\pi_{+1} - \pi_{1+}$ rather than $\pi_{1+} - \pi_{+1}$. The 95% Wald and Agresti-Min intervals are $-0.5433 < \pi_{+1} - \pi_{1+} < 0.1683$ and $-0.5023 < \pi_{+1} - \pi_{1+} < 0.1689$, respectively. We show in the program for this example how to calculate these intervals by coding the formulas directly. Overall, we see that the interval is quite wide, leaving much uncertainty about the actual difference. In particular, note that 0 is within the interval, suggesting that the data provide no evidence that the two diagnostic tests have a different probability of correct (localized) diagnosis.

Test for the difference between marginal probabilities

Especially for the prostate cancer diagnosis example, knowing whether $\pi_{+1} - \pi_{1+}$ is different from 0 is a very important special case of the matched pairs data problem. To formally test $H_0 : \pi_{+1} - \pi_{1+} = 0$ vs. $H_a : \pi_{+1} - \pi_{1+} \neq 0$, we could use the Wald test statistic

$$Z_W = \frac{\hat{\pi}_{+1} - \hat{\pi}_{1+} - 0}{\sqrt{\widehat{Var}(\hat{\pi}_{+1} - \hat{\pi}_{1+})}} = \frac{\hat{\pi}_{21} - \hat{\pi}_{12}}{\sqrt{n^{-1}\left[\hat{\pi}_{21} + \hat{\pi}_{12} - (\hat{\pi}_{21} - \hat{\pi}_{12})^2\right]}}.$$

Instead, a score test statistic is often used that incorporates simplifications that occur under the null hypothesis. Because $\pi_{+1} - \pi_{1+} = 0$ implies that $\pi_{21} = \pi_{12}$, the variance in the denominator can be simplified to $n^{-1}(\hat{\pi}_{21} + \hat{\pi}_{12})$. If we square the resulting statistic, we obtain what is known as McNemar's test statistic (McNemar, 1947):

$$M = \frac{(\hat{\pi}_{21} - \hat{\pi}_{12})^2}{n^{-1}(\hat{\pi}_{21} + \hat{\pi}_{12})} = \frac{(n_{21} - n_{12})^2}{n_{21} + n_{12}}.$$

Under the null hypothesis, the statistic has an approximate χ^2_1 distribution for large samples. When $\pi_{21} \neq \pi_{12}$, the null hypothesis is false, and we would expect this to be reflected by large values of $(\hat{\pi}_{21} - \hat{\pi}_{12})^2$ relative to the denominator. Thus, the null hypothesis is rejected when M has an unusually large observed value with respect to the χ^2_1 distribution; in other words, reject H_0 if $M > \chi^2_{1,1-\alpha}$.

Example: Prostate cancer diagnosis procedures (DiagnosisCancer.R)

We formally perform a test of $H_0 : \pi_{+1} - \pi_{1+} = 0$ vs. $H_a : \pi_{+1} - \pi_{1+} \neq 0$ using the `mcnemar.test()` function from the `stats` package in R.

```
> mcnemar.test(x = c.table, correct = FALSE)

        McNemar's Chi-squared test

data:  c.table
McNemar's chi-squared = 1, df = 1, p-value = 0.3173
```

The `x` argument is for the contingency table. The `correct = FALSE` argument value prevents the numerator from being modified to $(|n_{21} - n_{12}| - 1)^2$, which is sometimes used to help the χ^2_1 distribution approximation in small samples. We examine better alternatives to these modifications in Section 6.2 with exact inference procedures. Because of the large p-value, there is insufficient evidence to indicate that the marginal probabilities are different.

1.2.7 Larger contingency tables

While Section 1.2 focuses on two groups with binary responses, there are many instances where more than two groups exist (i.e., there are more than two rows within a contingency table). For example, in addition to comparing a newly developed drug to a placebo in a clinical trial, prior approved drugs could be included as well. The usual goal in this case is to show that the new drug performs better than the placebo and performs similarly or better than any other prior approved drugs. Once more rows are added to a contingency table, it is often easier to perform the analysis with a binary regression model as covered in Chapter 2. Typically, the same types of questions as in Section 1.2 can be answered with these regression models. Plus, more complex scenarios can be considered, such as including additional variables in the analysis.

There may also be more than two response categories. Combined with more than two rows, a contingency table with I rows and J columns can be formed. Independent binomial sampling can no longer be used as the underlying probability distribution framework. Rather, we most often use multinomial or Poisson distributions. We postpone discussion of these types of models until Chapters 3 and 4.

1.3 Exercises

1. There are many simple experiments that can be performed to estimate π for a particular problem, and we encourage you to perform your own. The parts below discuss some experiments that we or our students have performed in the past. For each part, discuss conditions that are needed to satisfy the five assumptions for using the binomial distribution outlined in Section 1.1.1.

   (a) One of the first solid foods that infants can feed to themselves is the cereal Cheerios. Early on, infants often lack the dexterity in their fingers to pick up an individual Cheerio and put it into their mouth. To estimate the probability of success for one infant, an experiment was designed where a Cheerio would be set on an infant's food tray for 20 successive trials. If the Cheerio made it into the infant's mouth, the response for the trial was considered a success; otherwise, the response was considered a failure. Out of these 20 trials, 9 Cheerios made it into the infant's mouth.

   (b) In most billiards games, the person who breaks a rack of balls gets another turn if at least one ball is sunk, excluding the cue ball. This is advantageous for this person because there are a fixed number of balls that need to be sunk to win a game. To estimate one person's probability of success for sinking a ball on the break, there were 25 consecutive breaks performed in 8-ball billiards. Out of these 25, 15 breaks had a ball sunk.

   (c) The germination rate is the proportion of seeds that will sprout after being planted into soil. To estimate the germination rate for a particular type of sweet corn, 64 seeds were planted in a $3' \times 4'$ plot of land with fertile soil. The seed packet stated that sprouts should emerge within 7–14 days, and all planting and care guidelines given on the seed packet were followed as closely as possible. After waiting for three weeks, 48 seeds had sprouted out of the soil.

   (d) To estimate the proportion of residents in a town who drive electric vehicles, a person records the number of vehicles passing through a certain intersection for a one-half hour time period. A total of 125 vehicles passed through the intersection at this time, where 14 were electric.

2. Continuing Exercise 1, find the Wald, Agresti-Coull, Wilson, and Clopper-Pearson intervals using the given data. Interpret the intervals in the context of the experiments.

3. Richardson and Haller (2002) describe a classroom experiment where Milk Chocolate Hershey's Kisses are put into a cup and then poured onto a table. The purpose is to estimate the proportion of Kisses that land on their base. An example in the paper gives 39 out of 100 Kisses landing on their base (10 cups filled with 10 Kisses were used).

   (a) Describe the assumptions that would be needed to use one binomial distribution with the 100 Kisses.

   (b) Presuming the assumptions in part (a) are satisfied, compute the Wald, Agresti-Coull, Wilson, and Clopper-Pearson intervals to estimate the probability that a Kiss will land on its base. Interpret the intervals.

   (c) Which confidence interval method do you think is best for this type of setting? Why?

Table 1.6: Chlamydia diagnoses from two testing procedures. Data source is Rours et al. (2005).

		MagNA		
		Positive	Negative	Total
COBAS	Positive	27	4	31
AMPLICOR	Negative	17	702	719
	Total	44	706	750

(d) The experiment was repeated with the Milk Chocolate Kisses and also separately for Milk Chocolate with Almonds Kisses. In this experiment, 43 out of 100 Milk Chocolate Kisses landed on their base and 33 out of 100 Milk Chocolate with Almonds Kisses landed on their base. Compute the Wald and Agresti-Caffo intervals for the difference of two probabilities with these data. Is there sufficient evidence that a difference exists?

(e) Again, which confidence interval method for the setting in (d) do you think is best? Why?

4. Rours et al. (2005) collected urine specimens from 750 asymptomatic pregnant women in Rotterdam, Netherlands, to estimate the prevalence of chlamydia among pregnant women in the corresponding population. Of the 750 specimens, 48 tested positive for the disease.

   (a) Describe the assumptions that would be needed to use one binomial distribution for the sample.

   (b) Presuming the assumptions in part (a) are satisfied, find a confidence interval to estimate the prevalence of chlamydia. Use the most appropriate confidence interval procedure for this problem, and interpret the results.

5. Continuing Exercise 4, in addition to estimating the overall prevalence of chlamydia, the authors evaluated the accuracy of two different diagnostic procedures: an automated COBAS AMPLICOR PCR system and the MagNA DNA Pure LC DNA Isolation kit. Table 1.6 summarizes their diagnoses in a contingency table. Use the Agresti-Min confidence interval procedure to estimate the difference in probabilities of positive test results. Also, use McNemar's test to perform a test of equality for these two probabilities. Interpret the results.[14]

6. The Aptima Combo 2 Assay is used to simultaneously detect chlamydia and gonorrhea in individuals. Pages 19 and 22 of their product insert[15] provide information on the accuracy of the diagnostic test. We reproduce part of this information in the Aptima-Combo.csv file available on the textbook's website. Each row of our data set contains information on the testing accuracy for specific combinations of disease (chlamydia, gonorrhea), gender (male, female; no other genders reported), specimen type (swab, urine), and symptom status (symptomatic, asymptomatic). A table constructed from the first row of the data set is given below.

[14] The authors combined the $27 + 17 + 4 = 48$ positive test results to estimate the prevalence. They make the assumption then that there were no false negative tests given jointly by the two diagnostic procedures.

[15] Available from `https://www.fda.gov/files/vaccines,%20blood%20&%20biologics/published/Package-Insert---Aptima-Combo2-Assay.pdf`

```
> aptima <- read.csv(file = "C:\\data\\AptimaCombo.csv")
> c.table <- array(data = c(aptima[1,6], aptima[1,7],
   aptima[1,9], aptima[1,8]), dim = c(2,2), dimnames =
   list(True = c("+", "-"), Assay = c("+", "-")))
> c.table
    Assay
True   +   -
   + 190   7
   -  15 464

> Se.hat <- c.table[1,1]/sum(c.table[1,]) # Sensitivity
> Sp.hat <- c.table[2,2]/sum(c.table[2,]) # Specificity
> data.frame(Se.hat, Sp.hat)
    Se.hat    Sp.hat
1 0.964467 0.9686848
```

Listed in the data set's first row, there were 676 symptomatic males that provided swab specimens for chlamydia testing. Of these 676 individuals, 190 tested positive and were judged truly positive. Note that the results from a combination of other testing procedures were used to make the "true" diagnosis. Similarly, 464 individuals were truly negative and tested negative. There were 22 errors with 15 positive tests for truly negative individuals and 7 negative tests for truly positive individuals.

The purpose of this problem is to calculate confidence intervals for *sensitivity* (S_e) and *specificity* (S_p) accuracy measures. The sensitivity is defined as the proportion of truly positive individuals that test positive, and the specificity is defined as the proportion of truly negative individuals that test negative. Both of these proportions are equivalent to π defined in this chapter, but just for true positive or true negative individuals. Ideally, it is desirable to have S_e and S_p as close to 1 as possible. Point estimates ($\hat{S}_e$ and $\hat{S}_p$) for these measures are given in the previous output.

(a) The product insert gives Clopper-Pearson intervals for all disease, gender, specimen, and symptom combinations. Suggest possible reasons why it gives only the Clopper-Pearson intervals rather than the other intervals examined in Section 1.1.2.

(b) Compute the Clopper-Pearson intervals for S_e and S_p among the symptomatic males that provided swab specimens for chlamydia testing. Interpret the intervals.

(c) Calculate the Clopper-Pearson intervals for the other disease/gender/specimen/symptom combinations and display them in an organized manner (e.g., they can be put into one data frame with the appropriate labeling). Note that the `for()` function can be used to simplify the calculation process.

For more on the sensitivity and specificity of diagnostic tests, please see Section 6.1.

7. In a study of insect physiology, eggs from a beneficial species of moth were put in boxes placed in chambers at different temperatures. Each box had 30 eggs, and the number of eggs hatching after 30 days were counted. The box at 10°C had 0 hatch, the box at 15°C had 1 hatch, and the box at 20°C had 25 hatch. The data are courtesy of Jim Nechols, Department of Entomology, Kansas State University.

   (a) Construct appropriate confidence intervals for the probability that an egg hatches at each temperature.

(b) Assess informally whether the probabilities could be the same at each temperature. Explain your reasoning. Note that we will develop more formal ways of making these comparisons in Chapter 2.

8. Continuing Exercise 7, the researchers actually used 10 different boxes of 30 eggs at each temperature. The counts of hatched eggs for the 10 boxes at 15°C were 1, 2, 4, 1, 0, 0, 0, 12, 0, and 2.

   (a) Why do you suppose that the researchers used more than one box at each temperature?

   (b) Construct appropriate confidence intervals for the probability that an egg hatches in each box.

   (c) Based on the intervals, does it appear that the probability that an egg hatches is the same in all boxes? Explain your reasoning.

   (d) All 10 boxes were held in the same chamber at the same time, and the chamber was set to 15°C. How do you suppose it could happen that they give such different counts?

   (e) Do you think that it would be appropriate to consider the data as $w = 22$ successes coming from a binomial distribution with $n = 300$ trials? Why or why not?

9. Suppose that 79 out of 80 people contacted in a survey of city residents oppose a new tax. Let π represent the probability that a randomly selected resident opposes the tax.

   (a) Compute a 95% Wald confidence interval for π. Interpret the results.

   (b) Compute a 95% Wilson confidence interval for π. Interpret the results.

   (c) Is it possible that $\pi = 1$? Explain.

   (d) Which interval do you prefer? Explain.

10. The BD Veritor System is an antigen test used for COVID-19 detection. The clinical performance of the test is summarized at `https://www.fda.gov/media/139755/download` (see Table 1). This document was used by the FDA to grant an Emergency Use Authorization during the COVID-19 pandemic. The stated *sensitivity* (see Exercise 6) of the test is based on the correct identification of 26 positive specimens out of 31 known positive specimens.

    (a) Define $\hat{\pi}$ as the estimated sensitivity. Compute this value.

    (b) Calculate a 95% confidence interval for the sensitivity π. While only one interval would be calculated in practice, calculate the interval using the expressions for the Agresti-Coull, Clopper-Pearson, and Wilson intervals. Compare these intervals.

    (c) Interpret the Wilson interval from (b) in the context of the test.

    (d) The confidence interval for the sensitivity is quite wide. Suppose a larger sample size is taken and $\hat{\pi}$ remains the same to two decimal places. What is approximately the smallest sample size that would result in a 95% Wilson interval no wider than 0.05? To answer this question, one can use a trial and error method or a formal search method with R functions such as `uniroot()`.

(e) Examine the interval given for the positive percent agreement (PPA) measure in Table 1 of the document from the FDA. This document uses PPA as the sensitivity of the test. No confidence level or interval type is stated in the document. What confidence level and interval does it appear that BD use in their calculations? If you cannot decide on only one interval, list all intervals that it could be.

(f) Why is it poor practice for the document not to state a confidence level for their confidence interval?

(g) Suppose you are tested by the BD Veritor system. Would you be comfortable with the accuracy of the test?

11. There are three main types of COVID-19 tests: (1) reverse transcription polymerase chain reaction (RT-PCR), (2) antigen, and (3) antibody. Each type needs to have high sensitivity (see Exercise 6) to ensure that a truly positive individual receives an accurate test result. The RT-PCR type is considered to have the highest sensitivity, but it also takes the longest amount of time to perform. The antigen test is used as an alternative because it takes much less time. Unfortunately, this test can be less accurate than the RT-PCR. Fitzpatrick et al. (2021) warn readers of inflated accuracy claims by manufacturers of these antigen tests, including the BD Veritor System described in Exercise 10. Read this paper and summarize the problems described in it.

12. The Wilson interval is found by equating

$$Z_S = \frac{\hat{\pi} - \pi_0}{\sqrt{\pi_0(1-\pi_0)/n}},$$

to $Z_{1-\alpha/2}$ and solving for π_0. Show that the solutions obtained are the Wilson interval limits given in Section 1.1.2. Note that the quadratic formula will be helpful to complete this problem.

13. Describe one advantage and one disadvantage of the Clopper-Pearson interval in comparison to the Wald, Agresti-Coull, and Wilson intervals.

14. Figure 1.3 compares true confidence levels of confidence intervals for π at $n = 40$ and $\alpha = 0.05$. Reconstruct the plot for other values of n and α. Below are ideas for choosing possible values.

(a) Use levels of n and α found in other exercises. Discuss how your confidence level findings affect the particular type of confidence interval method that you chose for that exercise.

(b) Examine what happens to the true confidence levels as the sample size increases from $n = 40$. Conversely, examine what happens as the sample size decreases from $n = 40$.

(c) Examine what happens when α is not 0.05. Do some intervals achieve the corresponding stated level better?

Discuss your results with respect to the recommendations given in Brown et al. (2001).

15. There are many other proposed confidence intervals for π. One of these intervals mentioned in Section 1.1.2 was the LR interval. Complete the following using this interval.

(a) Verify the 95% confidence interval is $0.1456 < \pi < 0.7000$ when $n = 10$ and $w = 4$. Note that `binom.confint()` calculates this interval using the `methods = "lrt"` argument value. We also provide additional code to calculate the interval in CIpi.R.

(b) Construct a plot of the true confidence levels similar to those in Figure 1.3. Use $n = 40$ and $\alpha = 0.05$, and vary π from 0.001 to 0.999 by 0.0005.

(c) Compare the interval's true confidence level to those of the four other intervals discussed in Section 1.1.2. Which of these intervals is best? Explain.

16. Another interval for π is the logit confidence interval. This method finds a Wald confidence interval for $\log(\pi/(1-\pi))$, which is known as the *logit* transformation, and then transforms the endpoints back into an interval for π. The resulting interval is

$$\frac{\exp\left[\log\left(\frac{\hat{\pi}}{1-\hat{\pi}}\right) \pm Z_{1-\alpha/2}\sqrt{\frac{1}{n\hat{\pi}(1-\hat{\pi})}}\right]}{1+\exp\left[\log\left(\frac{\hat{\pi}}{1-\hat{\pi}}\right) \pm Z_{1-\alpha/2}\sqrt{\frac{1}{n\hat{\pi}(1-\hat{\pi})}}\right]}.$$

(a) What is the numerical range for $\log(\pi/(1-\pi))$? After transforming the endpoints back using the $\exp(\cdot)/[1+\exp(\cdot)]$ transformation, what is the new numerical range? Why is this new numerical range ideal?

(b) Verify the estimated variance of $\log(\hat{\pi}/(1-\hat{\pi}))$ is $[n\hat{\pi}(1-\hat{\pi})]^{-1}$ by using the delta method (see Appendix B.4.2).

(c) Verify the 95% confidence interval is $0.1583 < \pi < 0.7026$ when $n = 10$ and $w = 4$. Note that `binom.confint()` calculates this interval using the `methods = "logit"` argument value.

(d) Discuss the problems that occur when $\hat{\pi} = 0$ or 1. Propose a solution to solve the problems. How does `binom.confint()` account for the problem?

(e) Using your solution from (d) or the solution given by `binom.confint()`, construct a plot of the true confidence levels similar to those in Figure 1.3. Use $n = 40$ and $\alpha = 0.05$, and vary π from 0.001 to 0.999 by 0.0005.

(f) Compare the interval's true confidence level to those of the four other intervals discussed in Section 1.1.2. Which of these intervals is best? Explain.

17. In addition to the true confidence level of a confidence interval, the expected length of an interval is another important measure of its quality. The expected length is defined as

$$\sum_{w=0}^{n} L(\hat{\pi})\binom{n}{w}\pi^{w}(1-\pi)^{n-w}$$

where $L(\hat{\pi})$ denotes the interval length using a particular $\hat{\pi}$. For example, $L(\hat{\pi})$ for a Wald confidence interval when $w = 1$ and $n = 40$ is $0.0734 - (-0.0234) = 0.0968$ (see example on p. 19). In the expected length expression, each interval's length is multiplied by the corresponding probability that the interval is observed, and the expected length is obtained by summing these terms.

(a) We would like the expected length of a confidence interval to be as small as possible while maintaining the correct confidence level. Why?

(b) For $n = 40$, $\pi = 0.16$, and $\alpha = 0.05$, find the expected length for the Wald interval and verify it is 0.2215.

Table 1.7: Icing the kicker data. Data source is Berry and Wood (2004).

		Field goal		
		Success	Failure	Total
Strategy	No time-out	22	4	26
	Time-out	10	6	16
	Total	32	10	42

(c) For $n = 40$ and $\alpha = 0.05$, construct a similar plot to Figure 1.3, but now using the expected length for the y-axis on the plots. Compare the expected lengths among the intervals. To help with the comparisons, you will want to fix the y-axis on each plot to be the same (use the `ylim = c(0, 0.35)` argument value in each `plot()` function call). Alternatively to having four plots in one R Graphics window, overlay the expected lengths for all four intervals on one plot. This can be done by using one `plot()` function for the first interval's expected length. Three calls to the `lines()` function can then be used to overlay the remaining intervals' expected lengths.

(d) Using Figure 1.3 and the plot from (c), which interval is best? Explain.

(e) Outline the steps that would be needed to find the estimated expected length using Monte Carlo simulation.

18. Figure 1.3 shows the actual true confidence levels for the four intervals discussed in Section 1.1.3. Examine this same setting with Monte Carlo simulation and 1,000 samples taken for each π.

(a) Construct the analogous plot to Figure 1.3 using estimated true confidence levels rather than the actual true confidence levels. For a particular π, make sure to use the same 1,000 samples for each confidence interval method. This will allow for better direct comparisons between the confidence intervals.

(b) The plotted estimated true confidence levels should appear more jittery than in Figure 1.3. Why does this occur?

19. Before a placekicker attempts a field goal in a pressure situation, an opposing team may call a time-out. The purpose of this time-out is to give the kicker more time to think about the attempt in the hopes that this extra time will cause the kicker to become more nervous and lower his probability of success. This strategy is often referred to as "icing the kicker." Berry and Wood (2004) collected data from two National Football League seasons to investigate whether the strategy actually lowers the probability of success when implemented. Table 1.7 summarizes their data from the 31–40-yard field goals under pressure situations (attempt is in the last 3 minutes of a game and a successful field goal causes a lead change).

(a) Calculate the Wald and Agresti-Caffo confidence intervals for the difference in probabilities of success conditioning on the strategy. Interpret the intervals.

(b) Perform a score test, Pearson chi-square test, and LRT to test for the equality of the success probabilities. Perform two-sided tests here. Although coaches hope that icing the kicker will reduce the probability of success, it is also possible that the kicker could *gain* focus during the time-out and have *higher* probability of success.

(c) For the opposing team, "success" is when the kicker misses the kick. Estimate the relative risk for using the time-out strategy versus not using it, focusing on the

Table 1.8: Ethiopia HIV and condom use data. Data source is Aseffa et al. (1998).

		Test response		
		Positive	Negative	Total
Condom	Never	135	434	569
	Ever	15	9	24
	Total	150	443	593

probability of a missed kick ("failure" in Table 1.7). Calculate the corresponding confidence interval for this relative risk. Interpret the results.

(d) Estimate the odds ratio for the time-out strategy and calculate the corresponding confidence interval for it. Interpret the results.

(e) Is there sufficient evidence to conclude that the icing the kicker strategy works for the opposing team? Explain.

20. Aseffa et al. (1998) examine the prevalence of HIV among women visiting health care clinics in northwest Ethiopia. Along with testing individuals for HIV, additional information was collected on each woman such as condom use. Table 1.8 gives a contingency table summarizing the data.

    (a) Calculate the Wald and Agresti-Caffo confidence intervals for the difference in probabilities of being HIV positive based on condom use. Interpret the intervals.

    (b) Perform a score test, Pearson chi-square test, and LRT to test for the equality of the probabilities of HIV infection between condom users and non-users.

    (c) Estimate the odds ratio for HIV test response and calculate the corresponding confidence interval for it. Interpret the estimate and the interval.

    (d) Generally, it is thought that condom use helps to prevent HIV transmission. Do the results here agree with this? If not, what factors may have led to these results? Note that Aseffa et al. (1998) give the estimated odds ratio and a corresponding Wald interval, but fail to interpret them.

    (e) Table 3 of Aseffa et al. (1998) gives a large number of 2×2 contingency tables, each examining the relationship between a current disease status (HIV, syphilis, and chlamydia) and explanatory variables (literacy, married, age of first sexual intercourse, past sexually transmitted disease history, condom use, and genital ulcer presence). For each table, the estimated odds ratio and corresponding Wald confidence interval is calculated. If you have access to the paper, replicate some of their calculations and develop conclusions about the associations between the variables.

21. On September 24, 2009, news reports hailed the findings from an HIV vaccine clinical trial (RV 144) as being the first time that a vaccine worked to prevent infection. These news reports were often given as lead stories in the news media (e.g., Clarke, 2009; its corresponding video report is available on the textbook's website). The vaccine actually was a combination of two previous tested vaccines, ALVAC-HIV and AIDSVAX B/E, that were not shown to prevent HIV infection in previous clinical trials by themselves. The purpose of this problem is for you to analyze the research data so that you can draw your own conclusions. Table 1.9 summarizes the "modified intent-to-treat" data of Rerks-Ngarm et al. (2009) that was released to the media on September 24. Using the modified intent-to-treat data, complete the following parts.

Table 1.9: HIV vaccine results from the modified intent-to-treat data. Data source is Rerks-Ngarm et al. (2009).

		Test response		
		Positive	Negative	Total
Treatment	Vaccine	51	8,146	8,197
	Placebo	74	8,124	8,198
	Total	125	16,270	16,395

Table 1.10: HIV vaccine results from the intent-to-treat data. Data source is Rerks-Ngarm et al. (2009).

		Test response		
		Positive	Negative	Total
Treatment	Vaccine	56	8,146	8,202
	Placebo	76	8,124	8,200
	Total	132	16,270	16,402

(a) If you have access to Rerks-Ngarm et al. (2009), describe the population and sample used in the study. What is the population ultimately that the researchers would like to extend their conclusions to?

(b) Examine the benefits of using the vaccine over the placebo with the most appropriate measure(s) from Section 1.2. Use $\alpha = 0.05$.

(c) Suppose you were a reporter assigned to write a news report about the statistical results on September 24, 2009. Write a short news report that would be appropriate for a news media outlet. Remember that most readers will not be familiar with the previously calculated statistical measures, so be sure to explain what they are.

22. Continuing Exercise 21, the Rerks-Ngarm et al. (2009) paper was published on October 20, 2009, on the *New England Journal of Medicine*'s website. This paper contained two other versions of the data named the "intent-to-treat" data and the "per-protocol" data, and these data sets are shown in Tables 1.10 and 1.11. The intent-to-treat data contain seven additional individuals that were not in the modified intent-to-treat data. These additional individuals were enrolled in the clinical trial and treated, but later were found to have been HIV positive before the trial had started. The "per-protocol" data contain only those individuals that received all four treatments of ALVAC-HIV and both treatments of AIDSVAX B/E as specified in the treatment protocol for the clinical trial. Using each of these additional data sets, repeat (a) and (b) from Exercise 21.

Table 1.11: HIV vaccine results from the per-protocol data. Data source is Rerks-Ngarm et al. (2009).

		Test response		
		Positive	Negative	Total
Treatment	Vaccine	36	6,140	6,176
	Placebo	50	6,316	6,366
	Total	86	12,456	12,542

Table 1.12: HIV vaccine results for HVTN 505. Data source is Hammer et al. (2013).

		Test response		
		Positive	Negative	Total
Treatment	Vaccine	27	1,224	1,251
	Placebo	21	1,224	1,245
	Total	48	2,448	2,496

23. Continuing Exercises 21 and 22, there were a number of news media reports again regarding the HIV vaccine clinical trial when the Rerks-Ngarm et al. (2009) paper was published. Maugh (2009) said the following in a *Los Angeles Times* article:

    > A secondary analysis of data from the Thai AIDS vaccine trial—announced last month to much acclaim—suggests that the vaccine might provide some protection against the virus, but that the results are not statistically significant. In short, they could have come about merely by chance.

    Interestingly, news reports like this one were rarely publicized as much as the initial release of the modified intent-to-treat data results. Using this background, complete the following.

    (a) What might have been the reaction by the news media if reports on all three data sets had been released at the same time?

    (b) What if $\alpha = 0.10$ or $\alpha = 0.01$ were used instead of $\alpha = 0.05$ for the computations in Exercises 21 and 22? Would the conclusions from the clinical trial change? Discuss.

    (c) Why do you think the intent-to-treat and per-protocol data results were less publicized? Should it have been more or less publicized given the reaction to the modified intent-to-treat analysis? Discuss.

    (d) Suppose again that you were a reporter assigned to write a news report about the statistical results. Write a short news report that would be appropriate for a news media outlet that takes into account all of the available information. Remember that most readers will not be familiar with the previously calculated statistical measures, so be sure to explain what they are when writing the article.

24. The results of Rerks-Ngarm et al. (2009) (Exercise 21) gave new hope for finding an HIV vaccine and led to expanding efforts into other clinical trials. Unfortunately, efficacy has not been shown yet. Below are discussions regarding two of these trials.

    (a) HVTN 505 was a clinical trial that was terminated before its planned end date (McNeil, 2013). Table 1.12 displays the results for those who were in the study for at least 28 weeks. Using the available data, suggest reasons why this study was ended.

    (b) HVTN 702 was a clinical trial that was also terminated early (NIAID, 2020). The vaccine tested was based on the vaccine used for the RV 144 clinical trial discussed in Exercise 21. Table 1.13 displays the results for those who were in the study for at least 18 months. Using the available data, suggest reasons why this study was ended.

25. Textbook authors often use humor to illustrate their points, but does this humor make more sense to students for whom English is their first language (EFL) than to

Table 1.13: HIV vaccine results for HVTN 702. Data source is NIAID (2020).

		Test response		
		Positive	Negative	Total
Treatment	Vaccine	129	2,565	2,694
	Placebo	123	2,566	2,689
	Total	252	5,131	5,407

Table 1.14: Olestra side effects data. Data source is Cheskin et al. (1998).

		Response		
		Side effects	No side effects	Total
Potato chips	Olestra	89	474	563
	Regular	93	436	529
	Total	182	910	1,092

students for whom English is an additional language (EAL)? To address this question, students at one university were shown a passage from a textbook that was meant to be humorous. Of the 211 EAL students in the study, 118 thought the passage was humorous. Of the 206 EFL students in the study, 155 thought the passage was humorous. Analyze these data to answer the research question. Note that this data set is from only one passage of many others used to examine humor and language for the study. The data set is courtesy of Melody Geddert, English Language Studies, Kwantlen Polytechnic University.

26. Olestra is a fat substitute that was first used in the late 1990s. It is not used very often now because of gastrointestinal side effects that some people reported after consuming food with it. Cheskin et al. (1998) performed a controlled experiment to determine if side effects truly occurred when Olestra is used as a fat substitute in potato chips. Table 1.14 provides the resulting data.

    (a) If you have access to the paper, describe the sample used for the study and the intended population. What assumptions are needed to make this sample representative of the intended population?

    (b) Does Olestra cause side effects in people? Use the data and appropriate inference methods to support your answer.

    (c) Answer one of the two questions below according to your response in part (b).

        i. Olestra causes side effects: Would you recommend that people eat food that contains Olestra? Explain.

        ii. There is not sufficient evidence that Olestra causes side effects: The introduction of the problem said that Olestra is not often used in foods now because of side effects that some people reported. What could be a statistical reason for why your conclusion does not match those reports?

    (d) The news media will often report on results from medical research when it is first published. Suppose this paper on Olestra was just published. Write a one-paragraph description about the study that is at a level appropriate for a news media outlet (do not assume that a reader has taken any statistics courses). Focus only on the data analyzed for this problem.

27. People can experience acute mountain sickness (AMS) when they go to high altitude locations because of low levels of oxygen in the air. Symptoms include headache,

Table 1.15: Acute mountain sickness data. Data source is Gertsch et al. (2004b).

		Headache		
		Yes	No	Total
Treatment	Acetazolamide	23	95	118
	Placebo	63	56	119
	Total	86	151	237

nausea, and dizziness. There have been a number of studies to determine if particular drugs may prevent the onset of AMS. In particular, Gertsch et al. (2004b) examine the effects of the drug acetazolamide (taken prior to high attitude exposure) on the occurrence of AMS. Table 1.15 provides part of the data from the study.

(a) If you have access to the paper, describe the sample used and the specific population for the study. What may be a larger population that is of interest for the researchers?

(b) Does acetazolamide help to prevent headaches for people in the population? Use the following to help answer this question:

i. Difference of two probabilities

ii. Relative risk

iii. Odds ratio[16]

Calculate and interpret the confidence interval for each measure. Note that only one of these measures would typically be used in practice.

(c) Would you recommend that people take acetazolamide prior to ascending a tall mountain? Explain using results from part (b).

28. A recent rule change in golf allows golfers to either leave a flagstick in the hole while putting or take the flagstick out. Golfers had been required to take the flagstick out prior to this change. Now, with the choice, golfers would like to know which strategy—flagstick in or flagstick out—results in a larger probability of success for making a putt. The Edoardo Molinari Golf Academy (EMGA) performed an experiment[17] to determine the best putting strategy. A portion of their observed data for slightly off-center putts that approach the hole at a medium speed is given in Table 1.16.

(a) Find the 95% Agresti-Caffo confidence interval for the difference in the success probabilities. Interpret the interval in the context of the data problem.

(b) Bilder (2020) uses a score confidence interval rather than an Agresti-Caffo confidence interval to analyze EMGA's data. A score confidence interval for the difference between success probabilities in general is explained in Exercise 35. Using the information from that exercise, compute the 95% score confidence interval for the difference in the success probabilities. Compare this interval to what was obtained for part (a).

[16]The estimated odds ratio provided by the paper was 4.77 (see Table 2 within the paper). While this value is similar to what should be found for part (b), it is not exactly the same (inverting the estimate may be needed to see the similarity). Unfortunately, the authors do not provide the particular method that was used for their calculation.

[17]A video showing the experiment and data is available at `www.instagram.com/p/Bsdkf1blJPd` and `https://youtu.be/YgOWqFaQ1tU`.

Table 1.16: EMGA putting results for slightly off-center putts that approach the hole at a medium speed.

		Putt outcome		
		Success	Failure	Total
Flagstick	Out	73	27	100
	In	45	55	100
	Total	118	82	200

(c) Perform the score test for the difference in the success probabilities. Discuss how your conclusions agree with those from the score confidence interval.

(d) Estimate the relative risk and calculate the corresponding 95% confidence interval for it. Interpret the results.

(e) Estimate the odds ratio and calculate the corresponding 95% confidence interval for it. Interpret the results.

(f) What is the best putting strategy—flagstick in or flagstick out—when putting in this situation? Use the statistical inference results from this problem to explain your answer.

29. Continuing Exercise 28, Bilder (2020) includes an analysis of the full experiment performed by EMGA. This experiment used a Perfect Putter (a ramp that the ball rolls down) to control the speed of the ball as it reaches the hole (1 = low, 2 = medium, and 3 = high) and the entry line of the putt relative to the position of the flagstick if it was in the hole (1 = center, 2 = slightly off-center, 3 = grazing). For each combination of ball speed and entry line, 100 trials were run with the flagstick in and with the flagstick out the hole. The number of successful putts was recorded and is available in the file Flagstick.csv available on the textbook's website.

(a) Form 2×2 contingency tables like shown in Table 1.16 for each combination of ball speed and entry line.

(b) Calculate 95% Agresti-Caffo confidence intervals for the difference in the success probabilities with each table from part (a).

(c) Because part (b) requires the calculation of 9 separate intervals each with a confidence level of 95%, the overall confidence level that all 9 intervals contain the difference in the success probabilities is less than 95%. One approach to solve this problem is to use what is known as a *Bonferroni adjustment*. Section 2.2.3 discusses this adjustment in more detail, but it can applied in general using a confidence level of $1 - \alpha/g$ for each interval, where g is the number of intervals and $1-\alpha$ is the overall confidence level. For the situation here with a 95% overall confidence level, each interval would have a 99.44% confidence level. Repeat part (b) using the Bonferroni adjustment.

(d) Develop overall conclusions for the best putting strategy using the results from part (c). Your conclusions may be different for particular ball speeds and/or entry lines.

30. In addition to the Pfizer-BioNTech COVID-19 vaccine, many other vaccines were developed to prevent infection from SARS-CoV-2. Table 1.17 provides clinical trial results used to examine the effectiveness for some of these vaccines. Are these vaccines effective? Use vaccine efficacy VE with appropriate inference procedures to answer this question.

Table 1.17: COVID-19 vaccine clinical trial results. Data sources are Moderna (2021), Johnson & Johnson (2021), Oxford-AstraZeneca (2021), and Heath et al. (2021). These data were used to obtain approval for use in the United States and the United Kingdom. The clinical trials often continued after publication, so final counts likely changed.

Moderna vaccine

	Positive	Negative	Total
Vaccine	11	14,123	14,134
Placebo	185	13,888	14,073
Total	196	28,011	28,207

Johnson & Johnson vaccine

	Positive	Negative	Total
Vaccine	66	19,564	19,630
Placebo	193	19,498	19,691
Total	259	39,062	39,321

Oxford-AstraZeneca vaccine

	Positive	Negative	Total
Vaccine	84	8,513	8,597
Placebo	248	8,333	8,581
Total	332	16,846	17,178

Novavax vaccine

	Positive	Negative	Total
Vaccine	10	7,010	7,020
Placebo	96	6,923	7,019
Total	106	13,933	14,039

Table 1.18: Molnupiravir clinical trial results through day 29 of being infected with COVID-19; data source is Merck (2021).

		Outcome		
		Hospitalization and/or death	Neither hospitalization or death	Total
Treatment	Molnupiravir	28	357	385
	Placebo	53	324	377
	Total	81	681	762

31. Merck and Ridgeback Biotherapeutics developed the antiviral drug Molnupiravir. This drug was used in a clinical trial to examine its effectiveness in treating COVID-19 patients. An interim analysis led the companies to discontinue recruiting patients for the trial. The analysis was based on the data given in Table 1.18. Analyze the data to determine the reason for closing the trial to new patients. Compare your results to what is given in the press release by Merck (2021).

32. Uldbjerg et al. (2021) examine the relationship between antibiotic use by an expectant mother and subsequent development of asthma in their child. The purpose of this problem is to reproduce parts of the analysis results given in Tables 1 and 2 of their paper. Our Table 1.19 provides the portion of the data from the study to focus on here.

    (a) Estimate the odds ratio comparing the odds of asthma when the mother was or was not exposed to antibiotics during pregnancy. Interpret the estimate.

    (b) Calculate the corresponding 95% confidence interval for the odds ratio of interest in part (a). Interpret the interval.

    (c) Perform a Pearson chi-square test for independence with the data. Interpret the p-value.

33. One of the most famous and largest clinical trials ever performed was in 1954 to test the polio vaccine developed by Jonas Salk (Francis et al., 1955). Over 1.8 million children participated in the clinical trial to determine the efficacy of the vaccine. While the actual design of the trial sparked debate (Brownlee, 1955; Dawson, 2004), we forgo this discussion and focus on the data obtained from the randomized, placebo-controlled

Table 1.19: Childhood asthma data. Data source is Uldbjerg et al. (2021).

		Diagnosis		
		Asthma	No asthma	Total
Antibiotics	Exposed	804	4,718	5,522
	Not exposed	3,434	23,695	27,129
	Total	4,238	28,413	32,651

Table 1.20: Salk vaccine clinical trial results. Data source is Francis et al. (1955, p. 25).

		Diagnosis		
		Polio	Polio free	Total
Treatment	Vaccine	57	200,688	200,745
	Placebo	142	201,087	201,229
	Total	199	401,775	401,974

portion of the trial. Table 1.20 provides the data from the clinical trial. Does the vaccine help to prevent polio? Answer this question by completing the following parts.

(a) What is the estimated relative risk? Calculate the corresponding 95% Wald confidence interval.

(b) What is the estimated odds ratio? Calculate the corresponding 95% Wald confidence interval.

(c) The estimates from (a) and (b) are very similar. Why?

(d) Phrase interpretations of the relative risk and odds ratio that focus on the vaccine's effectiveness.

34. Matched pairs data often occurs when the same binary response is observed at two different time points, like what happens in the Larry Bird example. This example compares the success probabilities for the second free throw given the outcome of the first free throw attempt. The purpose of this exercise is instead to compare the probability of success on the second attempt to the probability of success on the first attempt. This is of interest to determine on which free throw attempt Larry Bird is better. For example, it is reasonable to hypothesize a "warming up" effect that may lead to a larger probability of success on the second attempt.

(a) Explain what $\pi_{+1} - \pi_{1+}$ means in the context of this problem.

(b) Calculate the Wald, score, and Agresti-Min intervals for $\pi_{+1} - \pi_{1+}$.

(c) Perform a hypothesis test of $H_0 : \pi_{+1} - \pi_{1+} = 0$ vs. $H_a : \pi_{+1} - \pi_{1+} \neq 0$ using McNemar's test.

(d) Interpret the intervals and the hypothesis test results in the context of the data.

(e) Why would a hypothesis test of $H_0 : \pi_{+1} - \pi_{1+} \leq 0$ vs. $H_a : \pi_{+1} - \pi_{1+} > 0$ more closely correspond to the warming up effect hypothesis? Perform this test. Note that you will want to use $(n_{21} - n_{12})/\sqrt{n_{21} + n_{12}}$ as the test statistic and $Z_{1-\alpha}$ as the critical value.

35. Similar to the Wilson interval for π, an interval for $\pi_1 - \pi_2$ can be found using a test statistic from a hypothesis test for the parameters. Consider the test $H_0 : \pi_1 - \pi_2 = d$

vs. $H_a: \pi_1 - \pi_2 \neq d$ where d denotes the hypothesized value of the $\pi_1 - \pi_2$ difference. The interval is the set of all possible values of d such that

$$\frac{|\hat{\pi}_1 - \hat{\pi}_2 - d|}{\sqrt{\hat{\pi}_1^{(0)}(1 - \hat{\pi}_1^{(0)})/n_1 + \hat{\pi}_2^{(0)}(1 - \hat{\pi}_2^{(0)})/n_2}} < Z_{1-\alpha/2}$$

is satisfied. Note that $\hat{\pi}_1^{(0)}$ and $\hat{\pi}_2^{(0)}$ denote the MLEs of π_1 and π_2 under the constraint that $\pi_1 - \pi_2 = d$. This interval is called a *score* interval for the difference of two probabilities. Unfortunately, there is no closed form solution and iterative numerical procedures must be used to find the confidence interval limits. The `diffscoreci()` function from the `PropCIs` package performs these calculations. Complete the following parts using this package.

(a) Calculate this score interval for $\pi_1 - \pi_2$ with data from the Larry Bird example. Compare the interval to the Wald and Agresti-Caffo intervals for $\pi_1 - \pi_2$.

(b) Score intervals can be calculated for other measures examined in this chapter. The following measures have functions in the `PropCIs` package that calculate these intervals: relative risk, `riskscoreci()`; odds ratio, `orscoreci()`; and difference of two probabilities in a matched pairs setting, `scoreci.mp()`. Calculate these intervals for the Larry Bird example (see Exercise 34 for more on matched pairs data and the Larry Bird example).

(c) Examine how well these score intervals achieve the stated confidence level using similar methods as in Section 1.2.2.

36. With regard to the Pearson chi-square test statistic in Section 1.2.3, complete the following parts.

(a) Show that

$$\sum_{j=1}^{2} \frac{(w_j - n_j\bar{\pi})^2}{n_j\bar{\pi}} + \frac{(n_j - w_j - n_j(1-\bar{\pi}))^2}{n_j(1-\bar{\pi})} = \sum_{j=1}^{2} \frac{(w_j - n_j\bar{\pi})^2}{n_j\bar{\pi}(1-\bar{\pi})},$$

(b) Show that $X^2 = Z_S^2$.

37. What is the numerical range for the relative risk? What does it mean for the relative risk to be equal to 1?

38. What is the numerical range for an odds of success? What does it mean for an odds to be equal to 1?

39. Show that $OR = 1$ when $\pi_1 = \pi_2$.

40. Show that $OR = \left(\frac{1-\pi_2}{1-\pi_1}\right) RR$. Under what situations would one expect $\widehat{OR}$ and $\widehat{RR}$ to have similar values?

41. Section 1.2.5 discusses the symmetry properties of odds ratios and how these properties are useful for case-control studies. For the example given on p. 44 with respect to stroke and medication status, answer the following questions. Use probabilities expressed as words, such as "P(took medication and had no stroke)."

(a) Write out the odds that a person took medication given that a person had a stroke. Similarly, write out the odds that a person took medication given that a person did not have a stroke.

(b) With these odds expressions, write out the odds ratio comparing the odds of taking medication for the stroke group vs. the no stroke group. Show that it can be reorganized into the odds of stroke for the medication group vs. the no medication group.

42. Below are three results used to find McNemar's test statistic. Show that these results are true.

(a) $\pi_{+1} - \pi_{1+} = \pi_{21} - \pi_{12}$

(b) $\pi_{+1}(1 - \pi_{+1}) + \pi_{1+}(1 - \pi_{1+}) - 2(\pi_{11}\pi_{22} - \pi_{12}\pi_{21}) = \pi_{21} + \pi_{12} - (\pi_{21} - \pi_{12})^2$

(c) $\pi_{21} + \pi_{12} + (\pi_{21} - \pi_{12})^2 = \pi_{12} + \pi_{21}$ under the null hypothesis of $\pi_{+1} - \pi_{1+} = 0$

43. Figures 1.4 and 1.5 in Section 1.2.2 show the true confidence levels for the Wald and Agresti-Caffo intervals for the difference of two probabilities. Reconstruct the plots for other values of n_1, n_2, π_1, π_2, and α. Below are ideas for choosing possible values.

(a) Use levels of n_1 and n_2 found in other exercises in this section. Discuss how your confidence level findings affect the particular type of confidence interval method that you chose for that exercise.

(b) Examine what happens to the true confidence levels as the sample size increases from $n_1 = n_2 = 10$. Conversely, examine what happens as the sample size decreases from those same values.

(c) Examine what happens when α is not 0.05. Does either interval achieve the corresponding stated level better?

Discuss your results with respect to the recommendations given in Section 1.2.2.

44. The purpose of this problem is to examine the true confidence level for the Wald confidence interval for the relative risk. We recommend using ConfLevelTwoProb.R as a basis for your R program. When w_1 and/or w_2 are equal to 0, add a small constant to their values when computing the interval as suggested in Section 1.2.4.

(a) Find the true confidence level when $\pi_1 = 0.2$, $\pi_2 = 0.4$, $n_1 = 10$, $n_2 = 10$, and $\alpha = 0.05$.

(b) Construct a plot of the true confidence levels similar to those in Figure 1.4. Use the same settings as in part (a), but allow π_1 to vary from 0.001 to 0.999 by 0.0005.

(c) Construct a three-dimensional plot of the true confidence levels similar to those in Figure 1.5. Use the same settings as in part (a), but allow both π_1 and π_2 to vary from 0.001 to 0.999 by 0.005.

(d) Discuss how well the confidence interval for the relative risk maintains the stated confidence level for appropriate values of π_1 and π_2 (remember that the usefulness of the relative risk decreases as π_1 and π_2 become large).

(e) Repeat this problem using values of n_1 and n_2 that correspond to other exercises where the relative risk was used. Discuss the certainty that you have in your conclusions for those other exercises after seeing the results here.

45. Repeat Exercise 44, but now for the odds ratio and its corresponding Wald confidence interval. Note that any counts within the contingency table structure which are equal to 0 require special attention when calculating the interval.

46. The `RR()` function was given in Section 1.2.4 to simplify the calculation of a relative risk. This function can be modified to perform calculations in the following manner:

    (a) Allow for 0 cell counts by adding a small constant to cells as needed,

    (b) Invert the statistic and corresponding confidence intervals so that $1/RR$ is estimated when requested, and

    (c) Focus the calculation on the probability of a failure ($(1 - \pi_1)/(1 - \pi_2)$) when requested.

    Construct a new function that allows for one or more of these modifications without a user needing to change the contents of the contingency table. For example, a call to the function could include an argument named `invert` that is `TRUE` or `FALSE` for if the relative risk should be inverted. With many of these modifications, an `if()` function can be useful. In particular, below shows how `if()` can be used to account for $w_1 = 0$ with an appropriate value for `constant`.

```
if (data[1,1] == 0) {
  data[1,1] <- constant
}
```

    Once the function is written, apply it to the COVID-19 vaccine (Pfizer-BioNTech) clinical trial data and verify the calculations are correct.

47. The `OR()` function was given in Section 1.2.5 to simplify the calculation of an odds ratio. This function can be modified to perform calculations in the following manner:

    (a) Allow for 0 cell counts by adding a small constant to cells as needed and

    (b) Invert the statistic and corresponding confidence intervals.

    Construct a new function that allows for one or both of these modifications without a user needing to change the contents of the contingency table. Apply the function to the COVID-19 vaccine (Pfizer-BioNTech) clinical trial example. See the suggestions in Exercise 46 for how to make these modifications.

48. Write a function that calculates VE and the associated Wald confidence interval.

49. Odds ratio and relative risk calculations are performed by functions within many R packages written for epidemiological applications. These packages can be found by searching for "epidemiol" on the web page `https://cran.r-project.org/web/packages/available_packages_by_name.html`, which provides an alphabetical listing of all R packages available from the Comprehensive R Archive Network (CRAN). Complete the following relating to these packages for epidemiological applications.

    (a) Provide a list of packages that may calculate odds ratios and/or relative risks. Examine the reference manuals for these packages to make these determinations.

    (b) One package that computes odds ratios and/or relative risks is `epitools`. Use its `oddsratio()` and `riskratio()` functions to reproduce the calculations for the COVID-19 vaccine (Pfizer-BioNTech) clinical trial example. In `riskratio()`, include the argument `rev = "both"` because the function assumes a reverse ordering of the rows (placebo, vaccine) and columns (negative, positive) than given in the example.

(c) For another epidemiology package other than `epitools`, use a function that computes odds ratios and/or relative risks to reproduce the calculations for the COVID-19 vaccine (Pfizer-BioNTech) clinical trial example.

Chapter 2

Analyzing a binary response, part 2: regression models

Section 1.1 discussed how to estimate and make inferences about a single probability of success π. Section 1.2 generalized this discussion to the situation of two probabilities of success that are now dependent on a level of a group. Chapter 2 completes the generalization to a situation where there are many different possible probabilities of success to estimate and perform inferences upon. Furthermore, we quantify how an explanatory variable with many possible levels (perhaps continuous rather than categorical) affects the probability of success. These generalizations are made through the use of binary regression models (also called binomial regression models).

2.1 Linear regression models

Before we introduce binary regression models, we need to review normal linear regression models for numerical responses. Let Y_i be the response variable for observations $i = 1, \ldots, n$. Also, consider the situation with p explanatory variables $x_{i1}, \ldots, x_{ip}$ that are each measured on the ith observation. We relate the explanatory variables to the numerical response variable through a linear model, which is customarily written as

$$Y_i = \beta_0 + \beta_1 x_{i1} + \cdots + \beta_p x_{ip} + \epsilon_i,$$

where ϵ_i for $i = 1, \ldots, n$ are independent and each has a normal probability distribution with mean 0 and variance σ^2. The $\beta_0, \ldots, \beta_p$ are *regression parameters* that quantify the relationships between the explanatory variables and response variable, which we will discuss shortly. An alternative way to write this model is

$$Y_i \sim N(\beta_0 + \beta_1 x_{i1} + \cdots + \beta_p x_{ip}, \sigma^2),$$

showing that each response has a normal distribution centered on its mean value

$$E(Y_i) = \beta_0 + \beta_1 x_{i1} + \cdots + \beta_p x_{ip},$$

for $i = 1, \ldots, n$.

The regression parameters quantify the relationship between the explanatory variables and the mean response $E(Y_i)$. For example, if $\beta_1 = 0$, then a linear relationship does not exist between the first explanatory variable and the mean response, given the other variables in the model. Alternatively, if $\beta_1 > 0$, a positive relationship exists (an increase in the explanatory variable leads to an increase in the mean response), and if $\beta_1 < 0$, a negative relationship exists (an increase in the explanatory variable leads to a decrease in the mean response). We use data to learn how the explanatory variables relate to the mean response

DOI: 10.1201/9781003093091-2

by estimating these regression parameters and using them for inference. In addition, we use these parameter estimates to estimate and make inferences about the mean response itself.

In the context of binary responses, we want to understand how explanatory variables relate to the probability of success, π. Let Y_i now be independent binary response variables for observations $i = 1, \ldots, n$, where a value of 1 denotes a success and a value of 0 denotes a failure. A normal distribution is obviously no longer an appropriate model for Y_i. Similar to Section 1.1, a Bernoulli distribution describes Y_i very well, but we now allow the probability of success parameter π_i to be different for each observation. Thus, the PMF for Y_i is $P(Y_i = y_i) = \pi_i^{y_i}(1-\pi_i)^{1-y_i}$ for $y_i = 0$ or 1.

To find the MLEs for π_i, the likelihood function is

$$\begin{aligned} L(\pi_1, \ldots, \pi_n | y_1, \ldots, y_n) &= P(Y_1 = y_1) \times \cdots \times P(Y_n = y_n) \\ &= \prod_{i=1}^{n} P(Y_i = y_i) \\ &= \prod_{i=1}^{n} \pi_i^{y_i}(1-\pi_i)^{1-y_i} \end{aligned} \tag{2.1}$$

Notice there are n different parameters for the n different observations. This means that the model is *saturated*: there are as many parameters as there are observations. The parameter estimates in this case turn out to be the same as the data ($\hat{\pi}_i = 0$ or 1), so there is nothing gained by working with this model.

Instead, we wish to relate the π_i's to the explanatory variables. We might propose a mathematical function to describe the structure or shape of the relationship, say $\pi_i = \pi(x_{i1}, \ldots, x_{ip})$, where this function contains unknown regression parameters that allow it to be used in many different problems. For example, in a simple setting with a single binary variable x_{i1} (x_{i1} could denote the group number for observation i as we had in Section 1.2), this function may designate two possible values of $\pi(x_{i1})$ depending on x_{i1}. For any case, we can substitute $\pi(x_{i1}, \ldots, x_{ip})$ into Equation 2.1 to estimate its regression parameters. This provides us with a model that can be used to estimate a probability of success as a function of any possible values for a set of explanatory variables. The next section shows that an appropriate choice of the function $\pi(x_{i1}, \ldots, x_{ip})$ leads to a full range of useful and convenient estimation and inference procedures.

2.2 Logistic regression models

The simplest type of function for $\pi(x_{i1}, \ldots, x_{ip})$ is the linear model $\pi_i = \beta_0 + \beta_1 x_{i1} + \cdots + \beta_p x_{ip}$. However, this model could lead to values of π_i less than 0 or greater than 1, depending on the values of the explanatory variables and regression parameters. Obviously, this is quite undesirable for estimating a probability. Fortunately, many nonlinear functions are available that force π_i to be between 0 and 1. The most commonly used function is the *logistic regression model*:

$$\pi_i = \frac{\exp(\beta_0 + \beta_1 x_{i1} + \cdots + \beta_p x_{ip})}{1 + \exp(\beta_0 + \beta_1 x_{i1} + \cdots + \beta_p x_{ip})}. \tag{2.2}$$

Notice that $\exp(\beta_0 + \beta_1 x_{i1} + \cdots + \beta_p x_{ip})$ is always positive and the numerator of Equation 2.2 is less than the denominator, which leads to $0 < \pi_i < 1$.

A logistic regression model can also be written as

$$\log\left(\frac{\pi_i}{1-\pi_i}\right) = \beta_0 + \beta_1 x_{i1} + \cdots + \beta_p x_{ip} \tag{2.3}$$

through some algebraic manipulations of Equation 2.2. The left side of Equation 2.3 is the natural logarithm for the odds of success, which we will use when interpreting the model in Section 2.2.3. This transformation of π_i is often referred to as the *logit transformation* (so named in analog to the "probit" model, discussed in Section 2.3), and the left side of the model is commonly written as $\text{logit}(\pi_i)$. The right side of Equation 2.3 is a linear combination of the regression parameters with the explanatory variables, and this is often referred to as the *linear predictor*. Equations 2.2 and 2.3 are equivalent statements of the logistic regression model.

We will often write the models without the i subscripts as

$$\pi = \frac{\exp(\beta_0 + \beta_1 x_1 + \cdots + \beta_p x_p)}{1+\exp(\beta_0 + \beta_1 x_1 + \cdots + \beta_p x_p)} \text{ and } \text{logit}(\pi) = \beta_0 + \beta_1 x_1 + \cdots + \beta_p x_p$$

when we are not referring to particular observations. It will be clear based on the context when the probability of success is a function of explanatory variables (rather than how π was used in Section 1.1.1), even though the symbol π does not mention them explicitly.

Example: Plot of the logistic regression model (PiPlot.R)

The purpose of this example is to examine the shape of the logistic regression model when there is a single explanatory variable x_1. Consider the model $\pi = \exp(\beta_0 + \beta_1 x_1)/[1+\exp(\beta_0+\beta_1 x_1)]$, which is equivalently expressed as $\text{logit}(\pi) = \beta_0 + \beta_1 x_1$. Suppose that $\beta_0 = 1$ and $\beta_1 = 0.5$. Figure 2.1 shows this model plotted on the left. The plot on the right is the same, but with $\beta_1 = -0.5$. We can make the following generalizations from examining the model and these plots:

- $0 < \pi < 1$
- When $\beta_1 > 0$, there is an increasing (positive) relationship between x_1 and π. When $\beta_1 < 0$, there is a decreasing (negative) relationship between x_1 and π.
- The shape of the curve is somewhat similar to the letter *s* (this shape is called "sigmoidal").
- The slope of the curve is dependent on the value of x_1. We can show this mathematically by taking the derivative of π with respect to x_1: $\partial\pi/\partial x_1 = \beta_1\pi(1-\pi)$.
- The shape of the curve is exactly the same if it is rotated 180 degrees about $\pi = 0.5$ (it is "rotationally symmetric").

We encourage readers to explore what happens when β_1 is increased or decreased by using the R program for this example. Specifically, with respect to the left plot, the curve becomes more steep in the middle as β_1 increases. Conversely, the curve becomes more flat in the middle as β_1 goes to toward 0.

Below is the R code used to create the plot on the left of Figure 2.1.

```
> par(mfrow = c(1,2))
> beta0 <- 1
> beta1 <- 0.5
> curve(expr = exp(beta0+beta1*x) / (1+exp(beta0+beta1*x)), xlim
```

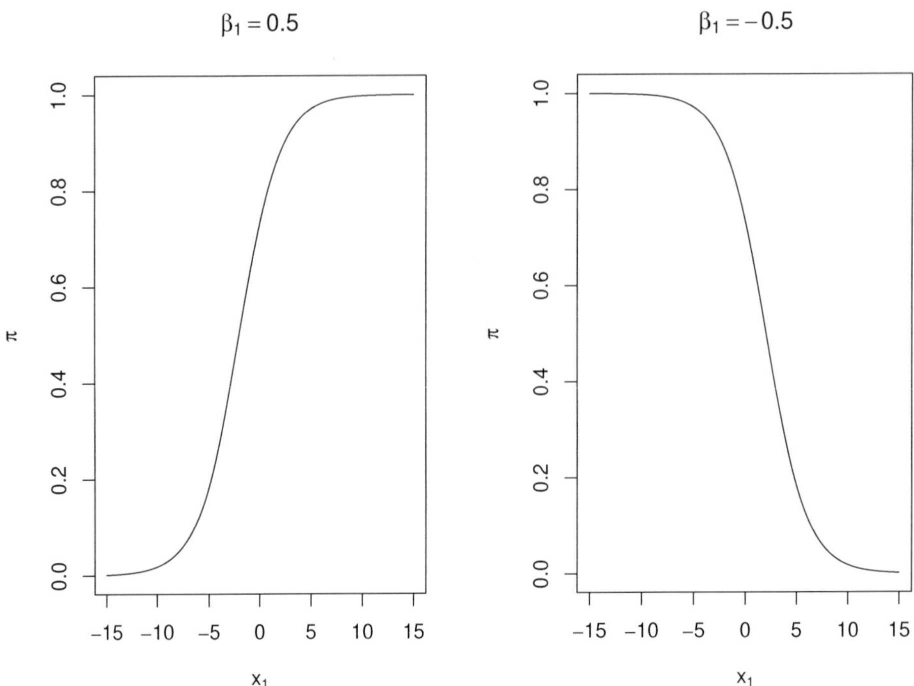

Figure 2.1: Logistic regression model for $\beta_0 = 1$ and $\beta_1 = 0.5$ and -0.5.

```
= c(-15, 15), col = "black", main = expression(beta[1] ==
0.5), xlab = expression(x[1]), ylab = expression(pi))
```

We use two new functions here. The `par()` function sets graphics parameters that control various plotting options for the `graphics` package, which is loaded when R starts up. In our code, we use it to partition the graphics window into 1 row and 2 columns using the `mfrow` argument, which stands for "make frame by row" (see Appendix A.7.3 for another example of its use). There are many other uses of the `par()` function, and we encourage readers to investigate these in the help for the function.

The second new function is the `curve()` function, which is used to plot the model. This is a very useful function for plotting mathematical functions that vary over one variable. In our example, the `expr` argument contains the logistic regression model where the letter `x` must be used as the variable name for the variable plotted on the x-axis. By default, the mathematical function is evaluated at 101 equally spaced x-axis values within the range specified by `xlim`. These resulting 101 points are then joined by straight lines. Also within `curve()`, we use the `expression()` function with the title and axis labels in order to include Greek letters and subscripts (see p. 637 in Appendix A for another example of `expression()`). We include a simpler use of the `curve()` function in the program for this example.

2.2.1 Parameter estimation

Maximum likelihood estimation is used to estimate the regression parameters, $\beta_0, \ldots, \beta_p$, of the logistic regression model. The initial estimation process is very similar to what is given in Appendix B.3.1 when there was a common value of π for all observations. Beginning from Equation 2.1, we substitute our model for π_i and take the natural logarithm to obtain the log-likelihood function:

$$\log\left[L(\beta_0, \ldots, \beta_p|y_1, \ldots, y_n)\right]$$

$$= \log\left(\prod_{i=1}^{n} \pi_i^{y_i}(1-\pi_i)^{1-y_i}\right) \tag{2.4}$$

$$= \sum_{i=1}^{n} y_i \log(\pi_i) + (1-y_i)\log(1-\pi_i)$$

$$= \sum_{i=1}^{n} y_i \log\left(\frac{\exp(\beta_0+\beta_1 x_{i1}+\cdots+\beta_p x_{ip})}{1+\exp(\beta_0+\beta_1 x_{i1}+\cdots+\beta_p x_{ip})}\right)$$

$$+(1-y_i)\log\left(\frac{1}{1+\exp(\beta_0+\beta_1 x_{i1}+\cdots+\beta_p x_{ip})}\right)$$

$$= \sum_{i=1}^{n} y_i\left(\beta_0+\beta_1 x_{i1}+\cdots+\beta_p x_{ip}\right) - \log\left(1+\exp(\beta_0+\beta_1 x_{i1}+\cdots+\beta_p x_{ip})\right). \tag{2.5}$$

We take the derivatives of Equation 2.5 separately with respect to each parameter $\beta_0, \ldots, \beta_p$, set these equal to 0, and solve them simultaneously to obtain the parameter estimates $\hat\beta_0, \ldots, \hat\beta_p$. When the parameter estimates are substituted into the model, we obtain the estimated probability of success as $\hat\pi = \exp(\hat\beta_0 + \hat\beta_1 x_1 + \cdots + \hat\beta_p x_p)/[1+\exp(\hat\beta_0+\hat\beta_1 x_1 + \cdots + \hat\beta_p x_p)]$.

Unfortunately, there are only a few simple cases where these parameter estimates have closed-form solutions; i.e., we cannot generally write out the parameter estimates in terms of the observed data like we could for the single probability estimate $\hat\pi$ in Section 1.1.2. Instead, we use iterative numerical procedures, as described in Appendix B.3.2, to successively find estimates of the regression parameters that increase the log-likelihood function. When the estimates change negligibly for successive iterations, this suggests that we have reached the peak of the log-likelihood function, and we say that they have *converged*. If the estimates continue to change noticeably up to a selected maximum number of iterations, the iterative numerical procedure has not converged, and those final parameter estimates should not be used. We discuss convergence and non-convergence in more detail in Section 2.2.7.

Within R and most statistical software packages, *iteratively reweighted least squares* (IRLS) is the iterative numerical procedure used to find the parameter estimates. This procedure uses the weighted least squares criterion, which is commonly used for normal linear regression models when there is non-constant variance. The IRLS algorithm alternates between updating the weights and updating the parameter estimates in an iterative fashion until convergence is reached. The `glm()` function ("glm" stands for "generalized linear model") within R implements this parameter estimation procedure, and we show how to use this function in the next example. Later in this section, we show an alternative method to maximize the likelihood function using the `optim()` function.

Example: Placekicking (Placekick.R, Placekick.csv)

As discussed in Chapter 1, points can be scored in American football by a placekicker kicking a ball through a target area at an end of the field. A success occurs when the football is kicked over the crossbar and between the two uprights of the goal posts. The placekicker's team receives 1 or 3 points, where a point after touchdown (PAT) provides 1 point and a field goal provides 3 points. A placekick that is not successful receives 0 points.

Pasteur and David (2017) summarize the papers that have investigated what factors affect the probability of success. We focus here on the work of Bilder and Loughin (1998), which was the first of these papers to investigate factors in addition to the distance of the placekick. The data available with this paper include:

- `week`: Week of the season
- `distance`: Distance of the placekick in yards
- `change`: Binary variable denoting lead-change (1) vs. non-lead-change (0) placekicks; lead-changing placekicks are those that have the potential to change which team is winning the game (for example, if a field goal is attempted by a team that is losing by 3 points or less, they will no longer be losing if the kick is successful)
- `elap30`: Number of minutes remaining before the end of the half, with overtime placekicks receiving a value of 0
- `PAT`: Binary variable denoting the type of placekick, where a PAT attempt is a 1 and a field goal attempt is a 0
- `type`: Binary variable denoting outdoor (1) vs. dome (0) placekicks
- `field`: Binary variable denoting grass (1) vs. artificial turf (0) placekicks
- `wind`: Binary variable for placekicks attempted in windy conditions (1) vs. non-windy conditions (0); we define windy as a wind stronger than 15 miles per hour at kickoff in an outdoor stadium
- `good`: Binary variable denoting successful (1) vs. failed (0) placekicks; this is our response variable

There are 1,425 placekick observations available from one National Football League (NFL) season. Below is how the data are read into R (see Appendix A.7.1 for more information on reading data into R).

```
> placekick <- read.csv(file = "C:\\data\\Placekick.csv")
> head(placekick)
  week distance change   elap30 PAT type field wind good
1    1       21      1 24.7167   0    1     1    0    1
2    1       21      0 15.8500   0    1     1    0    1
3    1       20      0  0.4500   1    1     1    0    1
4    1       28      0 13.5500   0    1     1    0    1
5    1       20      0 21.8667   1    0     0    0    1
6    1       25      0 17.6833   0    0     0    0    1
```

The data are saved in a comma-delimited (.csv) format, so we use the `read.csv()` function to read it in. The `file` argument specifies where the file is located on our computer. Note that double back slashes, not single back slashes, are required to separate folders and file names. Alternative ways to read the data into R are provided in the program.

The data are saved into a data frame called `placekick`, and the first six observations are printed using the `head()` function. For example, the first placekick is from the first week of the season and was attempted at 21 yards with 24.7 minutes remaining in the half. Also, this kick was a successful field goal attempted under lead-change conditions, on grass, in a domed stadium, and under non-windy conditions. Note that we do not consider the placekicker as a variable of interest (justification for this decision is given in the corresponding paper). We examine the possibility of a placekicker effect in Exercise 1 from Section 6.5.

For now, we use only the distance to estimate the probability of a successful placekick. Formally, Y is the response variable with a value of 1 for a success and 0 for a failure, and the explanatory variable x_1 denotes the distance in yards for the placekick. We use the observed data to estimate the parameters in the model

$$\text{logit}(\pi) = \beta_0 + \beta_1 x_1,$$

which can also be written with "`distance`" replacing x_1 to help make the model statement more descriptive.

Below is how we use R to fit the model with the `glm()` function.

```
> mod.fit <- glm(formula = good ~ distance, family =
    binomial(link = logit), data = placekick)
> mod.fit

Call:  glm(formula = good ~ distance, family = binomial(link =
    logit), data = placekick)

Coefficients:
(Intercept)      distance
      5.812        -0.115
Degrees of Freedom:1424 Total (i.e. Null); 1423 Residual
Null Deviance:       1013
Residual Deviance:775.7          AIC: 779.7
```

The results from `glm()` are saved into an object that we call `mod.fit`, which is a shortened version of "model fit" (we could choose a different name). The arguments within `glm()` are

- `formula`: Specifies the model with ~ separating the response and explanatory variables
- `family`: Gives the type of model to be fit where `binomial` states the response type (remember that Bernoulli is a special case of the binomial) and `logit` is the function on the left side of the model (we will discuss "link" functions in Section 2.3)
- `data`: Names the data frame containing the variables

By printing the `mod.fit` object through executing `mod.fit` at the command prompt, we see that the estimated logistic regression model is $\text{logit}(\hat{\pi}) = 5.8121 -$

0.1150`distance`. Because there is a negative parameter estimate corresponding to `distance`, the estimated probability of success decreases as the distance increases.

Additional information is printed in the output, but there are actually many more items, called *components*, that are stored within the `mod.fit` object. Using the `names()` function, we obtain a listing of these components.

```
> names(mod.fit)
 [1] "coefficients"        "residuals"          "fitted.values"

<OUTPUT EDITED>

[28] "method"              "contrasts"          "xlevels"
```

```
> mod.fit$coefficients
(Intercept)    distance
  5.8120798  -0.1150267
```

The `mod.fit` object is of a list format. As we have seen earlier, lists allow multiple objects to be linked together in one common place. To access components of a list, we use the format `<object>$<component>` without the $< >$. The example in the output shows `mod.fit$coefficients` contains $\hat{\beta}_0$ and $\hat{\beta}_1$. The results from many of the other components are obvious by their names. For example, `mod.fit$fitted.values` are the estimated probabilities for each observation ($\hat{\pi}_i$ for $i = 1, \ldots, n$). Other components are not as obvious, and we will discuss many of them as we progress through this chapter. Descriptions of all components are available in the "Value" section of the help for `glm()`.

A summary of what is inside `mod.fit` is obtained from the `summary()` function.

```
> summary(mod.fit)

Call:
glm(formula = good ~ distance, family = binomial(link = logit),
    data = placekick)

Deviance Residuals:
    Min       1Q   Median       3Q      Max
-2.7441   0.2425   0.2425   0.3801   1.6092

Coefficients:
             Estimate Std. Error z value Pr(>|z|)
(Intercept)  5.812080   0.326277   17.81   <2e-16 ***
distance    -0.115027   0.008339  -13.79   <2e-16 ***
---
Signif. codes:  0 '***' 0.001 '**' 0.01 '*' 0.05 '.' 0.1 ' ' 1

(Dispersion parameter for binomial family taken to be 1)

    Null deviance: 1013.43  on 1424  degrees of freedom
Residual deviance:  775.75  on 1423  degrees of freedom
AIC: 779.75
```

```
Number of Fisher Scoring iterations: 6
```

The above output displays a lot of information about the model that we will describe throughout this chapter. For now, note that the values of $\hat{\beta}_0$ and $\hat{\beta}_1$ are displayed in the `Coefficients` table under the `Estimate` header. Also, it took 6 iterations to obtain these estimates as given by the last line in the output (`Fisher scoring` is equivalent to IRLS for logistic regression models).

Additional background into how R works is helpful for understanding the results from `summary()`. There are *generic* functions, including `summary()`, that can be used on objects with different structures. When a generic function is run, it first finds the *class* of the object and subsequently runs a *method* function that is specifically written for objects of that class. The naming format for these method functions is `<generic function>.<class>`. In the case of `mod.fit`, its class is `glm` and `summary(mod.fit)` uses the method function `summary.glm(mod.fit)`. To see a list of method functions available for objects of a particular class, the `methods()` function is used.

```
> class(mod.fit)
[1] "glm" "lm"
> methods(class = "glm")
[1] add1            anova           coerce

<OUTPUT EDITED>

[22] show           slotsFromS3     summary
[25] vcov           weights
see '?methods' for accessing help and source code
```

Only the generic function name is given because R adds `.glm` to the end for each of them when they are applied to glm-class objects. For more information on the class of an object along with generic and method functions, please see p. 634 in Appendix A.7.3.

If more than one explanatory variable is included in the model, the variable names are separated by "+" symbols in the `formula` argument. For example, suppose we include the `change` variable in addition to `distance` in the model.

```
> mod.fit2 <- glm(formula = good ~ change + distance, family =
   binomial(link = logit), data = placekick)
> mod.fit2$coefficients
(Intercept)      change    distance
  5.8931814  -0.4477832  -0.1128888
```

The estimated logistic regression model is $\text{logit}(\hat{\pi}) = 5.8932 - 0.4478$`change` $-$ 0.1129`distance`.

The estimated variance-covariance matrix (hereafter referred to as a "covariance matrix" for simplicity) for a vector of parameter estimates $\hat{\boldsymbol{\beta}} = (\hat{\beta}_0, \hat{\beta}_1, \ldots, \hat{\beta}_p)'$, $\widehat{Var}(\hat{\boldsymbol{\beta}})$, has the usual form

$$\begin{bmatrix} \widehat{Var}(\hat{\beta}_0) & \widehat{Cov}(\hat{\beta}_0,\hat{\beta}_1) & \cdots & \widehat{Cov}(\hat{\beta}_0,\hat{\beta}_p) \\ \widehat{Cov}(\hat{\beta}_0,\hat{\beta}_1) & \widehat{Var}(\hat{\beta}_1) & \cdots & \widehat{Cov}(\hat{\beta}_1,\hat{\beta}_p) \\ \vdots & \vdots & \ddots & \vdots \\ \widehat{Cov}(\hat{\beta}_0,\hat{\beta}_p) & \widehat{Cov}(\hat{\beta}_1,\hat{\beta}_p) & \cdots & \widehat{Var}(\hat{\beta}_p) \end{bmatrix}$$

and can be found in R using the `vcov()` function.[1] This matrix is found by inverting a matrix of the second partial derivatives (often referred to as a "Hessian matrix") of the log-likelihood function evaluated at the parameter estimates and multiplying this resulting matrix by -1; see Appendix B.3.4 for details and Section 1.1.2 on p. 9 for a more simple case. In the end, the resulting covariance matrix simplifies to $(\mathbf{X}'\mathbf{V}\mathbf{X})^{-1}$ where

$$\mathbf{X} = \begin{bmatrix} 1 & x_{11} & x_{12} & \cdots & x_{1p} \\ 1 & x_{21} & x_{22} & \cdots & x_{2p} \\ \vdots & \vdots & \vdots & \ddots & \vdots \\ 1 & x_{n1} & x_{n2} & \cdots & x_{np} \end{bmatrix}$$

and $\mathbf{V} = \text{Diag}\left(\hat{\pi}_i(1-\hat{\pi}_i)\right)$, an $n \times n$ matrix with $\hat{\pi}_i(1-\hat{\pi}_i)$ on the diagonal and 0's elsewhere. This covariance matrix has the same form as the one resulting from weighted least squares estimation in a linear model. The individual elements of the matrix do not have a simple form, so we do not present them here; see Exercise 37 for a discussion.

Example: Placekicking (Placekick.R, Placekick.csv)

The purpose of this example is to show how to obtain the estimated covariance matrix for the regression parameter estimators. We use the `mod.fit` object from the model with `distance` as the only explanatory variable.

Below is the coefficients table again from the `summary(mod.fit)` output.

```
> round(summary(mod.fit)$coefficients, 4)
            Estimate Std. Error  z value Pr(>|z|)
(Intercept)   5.8121     0.3263  17.8133        0
distance     -0.1150     0.0083 -13.7937        0
```

We limit the displayed output by using the fact that `summary()` creates a list with `coefficients` as one component (if needed, please see the program for help). The `Std. Error` column gives the standard errors for the regression parameter estimators: $\widehat{Var}(\hat{\beta}_0)^{1/2}$ in the "`(Intercept)`" row and $\widehat{Var}(\hat{\beta}_1)^{1/2}$ in the "`distance`" row.

The `vcov()` function produces the estimated covariance matrix.

```
> vcov(mod.fit)
            (Intercept)      distance
(Intercept)  0.10645675 -2.606250e-03
distance    -0.00260625  6.953996e-05

> vcov(mod.fit)[2,2]   # Var-hat(beta-hat_1)
[1] 6.953996e-05
```

[1] Notice that `vcov` is listed in the `methods(class = "glm")` output given in the last example. When the generic function `vcov()` is applied to a `glm` class object, the `vcov.glm()` method function is run.

```
> summary(mod.fit)$coefficients[2,2]^2
[1] 6.953996e-05
```

We can extract the estimated variance for $\hat{\beta}_1$ by specifying the (2,2) element of the matrix. Thus, $\widehat{Var}(\hat{\beta}_1) = 0.0000695$, which is the square of 0.0083 given within the coefficients table.

For readers interested in the actual matrix calculations, we provide an example of how the covariance matrix is calculated by using $(\mathbf{X}'\mathbf{V}\mathbf{X})^{-1}$.

```
> pi.hat <- mod.fit$fitted.values
> V <- diag(pi.hat*(1-pi.hat))
> X <- cbind(1, placekick$distance)
> solve(t(X) %*% V %*% X)
            [,1]          [,2]
[1,]  0.10645678 -2.606250e-03
[2,] -0.00260625  6.953997e-05
```

We create the diagonal matrix $\mathbf{V} = \text{Diag}\left(\hat{\pi}_i(1-\hat{\pi}_i)\right)$ by using the `diag()` function, where we first extract $\hat{\pi}_i$ from `mod.fit`. Next, we create the $\mathbf{X}$ matrix by using the `cbind()` function to put a column of 1's before a column of the explanatory variable values.[2] Continuing, the `t()` function finds the transpose of the $\mathbf{X}$ matrix, the `%*%` syntax instructs R to perform matrix multiplication, and the `solve()` function finds the inverse of the matrix. The resulting matrix is practically the same as that produced by `vcov()`, where differences are due to rounding error.

As a reminder, we encourage readers to investigate individual parts of any set of code to help understand the coding process. For example, one could print part of `V` by executing `V[1:3, 1:3]` and see that the first three rows and columns of `V` are a diagonal matrix with values of $\hat{\pi}_1(1-\hat{\pi}_1)$, $\hat{\pi}_2(1-\hat{\pi}_2)$, and $\hat{\pi}_3(1-\hat{\pi}_3)$ as the (1,1), (2,2), and (3,3) elements, respectively. In addition, one can perform parts of the calculations within the `solve()` function to see the intermediate results.

While the `glm()` function will be used whenever we fit a logistic regression model, it is instructive to view the log-likelihood function in R while also fitting the model in a more general manner. In the next example, we illustrate how the log-likelihood function can be programmed directly into R and the general optimization function `optim()` can be used to find the parameter estimates. These programming techniques can be utilized for more complicated problems, such as when a function like `glm()` may not be available.

<u>Example</u>: Placekicking (Placekick.R, Placekick.csv)

We focus again on fitting the logistic regression model with only the `distance` explanatory variable. Below we create a function `logL()` to calculate the log-likelihood function for any given parameter values, explanatory variable values for x_1, and binary responses for Y.

[2] R *recycles* "1" n times so that the first column has the same number of elements in it as the second column.

```
> logL <- function(beta, x, Y) {
    pi <- exp(beta[1] + beta[2]*x)/(1 + exp(beta[1] + beta[2]*x))
    sum(Y*log(pi) + (1-Y)*log(1-pi))
}

> logL(beta = mod.fit$coefficients, x = placekick$distance, Y =
   placekick$good)
[1] -387.8725
> logLik(mod.fit)
'log Lik.' -387.8725 (df=2)
```

Inside the function, we find π_i through our model expression and then find

$$\log\left[L(\beta_0, \beta_1 | y_1, \ldots, y_n)\right] = \sum_{i=1}^{n} y_i \log(\pi_i) + (1 - y_i)\log(1 - \pi_i)$$

using the `sum()` and `log()` functions, while taking advantage of how R performs addition and multiplication on vectors in an elementwise manner. When we evaluate the function using the parameter estimates and the observed data, we obtain $\log\left[L(\beta_0, \beta_1 | y_1, \ldots, y_n)\right] = -387.87$. As a check, we obtain the same value from the generic function `logLik()`, which extracts the maximum log-likelihood function value from objects of various classes (the `logLik.glm()` method function is used here).

To maximize the log-likelihood function and find the corresponding MLEs, we use the `optim()` function. This is a very general optimization function that *minimizes* an R function with respect to a vector of parameters. Because we want to *maximize* the log-likelihood function instead, we specify `control = list(fnscale = -1)` in `optim()` to perform the equivalent of maximization by minimizing the negative of `logL()`. Below is our code.

```
> # Find starting values for parameter estimates
> reg.mod <- lm(formula = good ~ distance, data = placekick)
> reg.mod$coefficients
(Intercept)    distance
 1.25202444 -0.01330212

> mod.fit.optim <- optim(par = reg.mod$coefficients, fn = logL,
   hessian = TRUE, x = placekick$distance, Y = placekick$good,
   control = list(fnscale = -1), method = "BFGS")
> names(mod.fit.optim)
[1] "par"         "value"       "counts"      "convergence"
[5] "message"     "hessian"
> mod.fit.optim$par
(Intercept)    distance
  5.8112544  -0.1150046
> mod.fit.optim$value
[1] -387.8725
> mod.fit.optim$convergence
[1] 0
> -solve(mod.fit.optim$hessian)
             (Intercept)      distance
(Intercept)  0.106482867  -2.607258e-03
distance    -0.002607258   6.957463e-05
```

The optimization procedures within `optim()` need initial starting values for the regression parameters. We simply fit a linear regression model to the data using the `lm()` function and take the corresponding parameter estimates as these starting values (see Appendix A.7 for another example of using `lm()`). Within the call to `optim()`, we specify the initial parameter estimates using the `par` argument, and we specify the function to be maximized using the `fn` argument. Note that the first argument in the function named in `fn` must correspond to the initial parameter estimates; this is why `beta` was given as the first argument in `logL()`. The `TRUE` value for the `hessian` argument instructs R to obtain a numerical estimate of the Hessian matrix for the parameters. In other words, it estimates a matrix of second partial derivatives for the log-likelihood function, which can then be inverted to obtain the estimated covariance matrix for the parameter estimates (see Appendix B.3.4).[3] Finally, we specify the corresponding values of `x` and `Y` for the `logL()` function.

The object produced by `optim()` is a list that we save as `mod.fit.optim`. Components within the list include the parameter estimates (`$par`), the log-likelihood function's maximum value (`$value`), and the estimated covariance matrix (`$hessian`; we subsequently use `solve()` to find the estimated covariance matrix as the inverse of the Hessian matrix). All of these values are practically the same as those produced by `glm()`, where small differences are due to different convergence criteria for the iterative numerical procedures. The 0 value for `mod.fit.optim$convergence` means that convergence was achieved. For this implementation of `optim()`, we used the `method = "BFGS"` optimization procedure (similar to a Newton-Raphson procedure), but other procedures are available within the function (see Givens and Hoeting, 2013, or the `optim()` help for details on various iterative numerical procedures).

The log-likelihood function can be plotted for this example because there are only two parameters. The R code to create this plot is within the program for this example, and the code is similar to the three-dimensional confidence level plots on p. 36 used to produce Figure 1.5. Two `for()` function calls—one for loop for the β_0 values and one for loop for the β_1 values—are used to evaluate the log-likelihood function.[4] We use the `persp3d()` function from the `rgl` package to produce the three-dimensional log-likelihood function surface in Figure 2.2. We also use the `contour()` function from the `graphics` package to plot contours of the log-likelihood function in Figure 2.3.[5] We can see from the plots (a little clearer from the contour plot) that the values which maximize the log-likelihood function are 5.81 for β_0 and -0.115 for β_1.

An alternative form of the response variable is available when there are multiple Bernoulli trials recorded at some or all combinations of explanatory variables. In particular, if there

[3]The estimated covariance matrix obtained here is calculated as shown in Equation B.6, rather than as shown in Equation B.5 when an expectation is taken. For logistic regression, the resulting estimated covariance matrices are exactly the same no matter which equation is used. There are other models where the matrices will be different (e.g., see Exercise 31 and Section 6.1.2), but fortunately the two estimated covariance matrices are asymptotically equivalent (the same as $n \to \infty$).

[4]The `apply()` function can be used here instead of for loops. This code is available in the program for this example.

[5]For all possible values of β_0 and β_1, equal values of the log-likelihood function are joined with lines called *contours*. For example, the ellipse labeled -500 shows that $(\beta_0, \beta_1) = (2, -0.02)$, $(5.6, -0.07)$, and $(11.7, -0.27)$ all have values of $\log[L(\beta_0, \beta_1 | y_1, \ldots, y_n)]$ close to -500. Additional code in the program demonstrates this point.

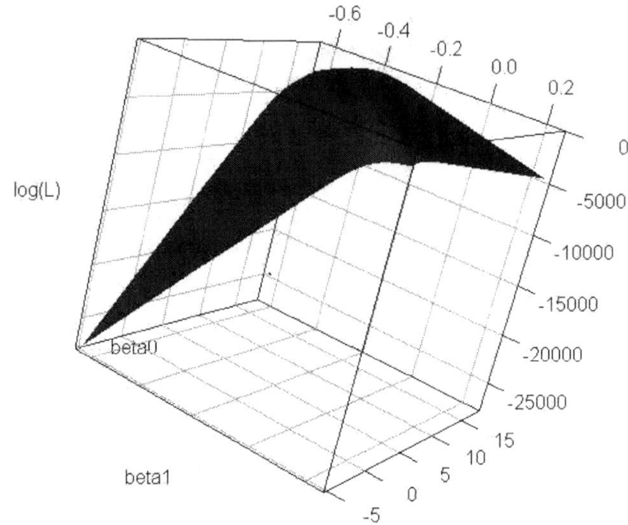

Figure 2.2: Three-dimensional plot of the log-likelihood function.

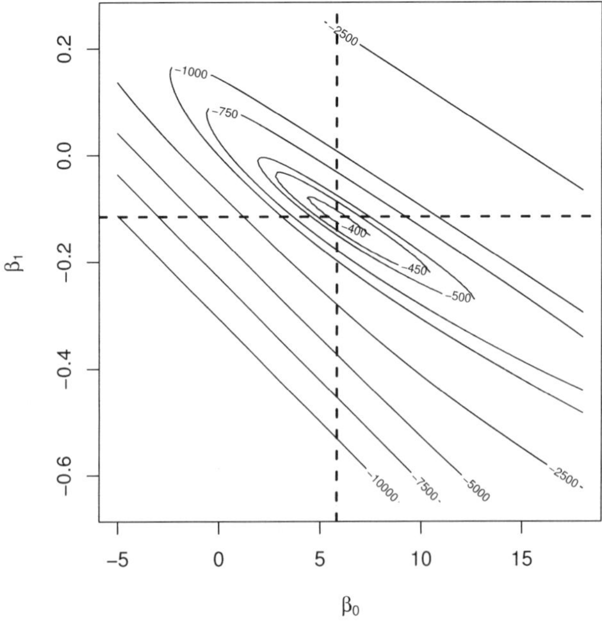

Figure 2.3: Contour plot of the log-likelihood function. Contours are drawn at −400, −450, −500, −750, −1000, −2500, −5000, −7500, and −10000.

are $J < n$ unique combinations of explanatory variables, we can aggregate the individual trial results as in Section 1.2, so that we observe w_j successes in n_j trials each with $x_{j1}, \ldots, x_{jp}$ explanatory variable values for $j = 1, \ldots, J$. In other words, we now have observations from J different binomial distributions with probabilities of success π_j, $j = 1, \ldots, J$. Alternatively, the data may naturally arise in the form of w_j successes out of n_j trials. For example, in an experiment to determine a lethal dose level, a particular amount of pesticide (x_{j1}) could be applied to an enclosure with a known number of ants (n_j), where the number of ants exterminated (w_j) would be recorded. By applying different amounts of pesticide to other enclosures, we have a binomial response setting for J treatment groups.

The same logistic regression model can be used to model the probability of success for the binomial response setting. The log-likelihood function is

$$\log\left[L(\beta_0, \ldots, \beta_p | w_1, \ldots, w_J)\right] = \sum_{j=1}^{J} \log\left[\binom{n_j}{w_j}\right] + w_j \log(\pi_j) + (n_j - w_j)\log(1 - \pi_j).$$

Because $\binom{n_j}{w_j}$ is a constant, this term does not change the estimated parameter values that result in the log-likelihood maximization. Therefore, the MLEs are the same as if the data instead consisted of $n = \sum_{j=1}^{J} n_j$ binary responses. We illustrate this in the next example.

Example: Placekicking (Placekick.R, Placekick.csv)

The purpose of this example is to show how to reform the placekicking data to a binomial response format and to show that the model parameter estimates are the same for this format. We focus only on the distance explanatory variable for this example.

The `aggregate()` function is used to find the number of successes and number of observations for each distance.

```
> w <- aggregate(x = good ~ distance, data = placekick, FUN =
   sum)
> n <- aggregate(x = good ~ distance, data = placekick, FUN =
   length)
> w.n <- data.frame(distance = w$distance, success = w$good,
   trials = n$good, proportion = round(w$good/n$good,4))
> head(w.n)
  distance success trials proportion
1       18       2      3     0.6667
2       19       7      7     1.0000
3       20     776    789     0.9835
4       21      19     20     0.9500
5       22      12     14     0.8571
6       23      26     27     0.9630
```

The `x` argument within `aggregate()` can contain different types of R objects depending on how the function is used. We provide here a formula expression giving the variable we want to aggregate on the left and the variable that we group by on the right. These variables are contained in the object listed in `data`. The `FUN` argument specifies a function that summarizes the response variable `good` within each group, where the `sum()` function adds the 0 and 1 responses for each distance and the `length()` function counts the number of observations for each distance. The final result is in the data frame `w.n`. For example, there are 2 successes out of 3 trials at a distance of 18 yards, which results in an observed proportion of successes of $2/3 \approx 0.6667$. The reason for a large number of observations at 20 yards is because this was the standard distance for PATs when the data were collected. NFL rules have changed since then and now most PATs are attempted at a longer distance.

The logistic regression model is estimated by the `glm()` function with two changes to our previous code.

```
> mod.fit.bin <- glm(formula = success/trials ~ distance,
    weights = trials, family = binomial(link = logit), data = w.n)
> summary(mod.fit.bin)
Call: glm(formula = success/trials ~ distance, family =
    binomial(link = logit), data = w.n, weights = trials)

Deviance Residuals:
    Min       1Q    Median       3Q      Max
-2.0373  -0.6449   -0.1424   0.5004   2.2758

Coefficients:
             Estimate Std. Error z value Pr(>|z|)
(Intercept)  5.812080   0.326277   17.81   <2e-16 ***
distance    -0.115027   0.008339  -13.79   <2e-16 ***
---
Signif. codes:  0 '***' 0.001 '**' 0.01 '*' 0.05 '.' 0.1 ' ' 1

(Dispersion parameter for binomial family taken to be 1)

    Null deviance: 282.181  on 42  degrees of freedom
Residual deviance:  44.499  on 41  degrees of freedom
AIC: 148.46

Number of Fisher Scoring iterations: 5
```

First, we specify the response in the `formula` argument as `success/trials`. Notice this finds the observed proportion of success for each distance. Second, we use the `weights` argument to specify the number of observations per distance. The estimated logistic regression model is $\text{logit}(\hat{\pi}) = 5.8121 - 0.1150$`distance`, which is the same as earlier using the binary response variable. It is possible that very small differences can occur in the parameter estimates because of the iterative numerical procedure.

2.2.2 Hypothesis tests for regression parameters

Hypothesis tests can be used to assess the importance of explanatory variables in a model. For example, a test of $H_0 : \beta_r = 0$ vs. $H_a : \beta_r \neq 0$ evaluates the rth explanatory variable term in the model $\text{logit}(\pi) = \beta_0 + \beta_1 x_1 + \cdots + \beta_r x_r + \cdots + \beta_p x_p$. If $\beta_r = 0$, this means the corresponding term is excluded from the model. If $\beta_r \neq 0$, this means the corresponding term is included in the model. Equivalently, we can state the hypotheses as

$$H_0 : \text{logit}(\pi) = \beta_0 + \beta_1 x_1 + \cdots + \beta_{r-1} x_{r-1} + \beta_{r+1} x_{r+1} + \cdots + \beta_p x_p$$
$$H_a : \text{logit}(\pi) = \beta_0 + \beta_1 x_1 + \cdots + \beta_r x_r + \cdots + \beta_p x_p,$$

which can be helpful to emphasize the comparison of two different models. The explanatory variables in the null-hypothesis model (also known as a *reduced model*) must all be in the alternative-hypothesis model (also known as a *full model*). The null-hypothesis model is said to be *nested* within the alternative-hypothesis model in this case.

Score inference is not readily available in statistical software packages for regression models, so we use a Wald test or an LRT to perform these tests. Appendix B.5.1 gives general details for how to perform them. We now focus on the specifics for the logistic regression setting.

Wald test

The Wald statistic

$$Z_W = \frac{\hat{\beta}_r}{\sqrt{\widehat{Var}(\hat{\beta}_r)}}$$

is used to test $H_0 : \beta_r = 0$ vs. $H_a : \beta_r \neq 0$. If the null hypothesis is true, Z_W has an approximate standard normal distribution for a large sample, and we reject the null hypothesis if Z_W has an unusual observed value for this distribution. We define unusual by $|Z_W| > Z_{1-\alpha/2}$. The p-value is $2P(Z > |Z_W|)$, where Z has a standard normal distribution.

More than one parameter may be tested for equality to 0 in the null hypothesis. However, we forgo discussion of it because Wald inference procedures here often encounter the same types of problems as those discussed in Chapter 1. In the context of hypothesis tests, this means that the stated type I error rate for a hypothesis test is not the same as the actual type I error rate (i.e., the size of the test is different from α). The LRT typically performs better than the Wald test, so we focus on this procedure next.

Likelihood ratio test

The LRT statistic can be written informally as

$$\Lambda = \frac{\text{Maximum of likelihood function under } H_0}{\text{Maximum of likelihood function under } H_0 \text{ or } H_a}. \tag{2.6}$$

In the context of testing whether some regression parameters are 0, the denominator in Equation 2.6 is the likelihood function evaluated at the MLEs for the model containing all $p+1$ regression parameters. The numerator in Equation 2.6 is the likelihood function evaluated at the MLEs for the model that excludes those variables whose regression parameters are set to 0 in the null hypothesis. For example, to test $H_0 : \beta_r = 0$ vs. $H_a : \beta_r \neq 0$, β_r would be held equal to 0, which means this corresponding explanatory variable would be excluded from the null-hypothesis model. Of course, the estimates for the remaining parameters need not be the same as with the alternative hypothesis model because of the differences between the two models.

If q regression parameters are set to 0 in the null hypothesis and if the null hypothesis is true, the $-2\log(\Lambda)$ statistic has an approximate χ^2_q distribution for a large sample. We reject the null hypothesis if $-2\log(\Lambda)$ has an unusually large observed value for this distribution. For example, if $\alpha = 0.05$ for the test of $H_0 : \beta_r = 0$ vs. $H_a : \beta_r \neq 0$, rejection occurs for $-2\log(\Lambda) > 3.84$, where $\chi^2_{1,0.95} = 3.84$ is the 0.95 quantile from a chi-square distribution. The p-value is $P(A > -2\log(\Lambda))$ where A has a χ^2_1 distribution when $q = 1$.

We most often use the generic functions `anova()` from the `stats` package and `Anova()` from the `car` package to perform these tests.[6] These functions test hypotheses of a similar

[6]Other common testing functions include the `drop1()` function from the `stats` package and the `lrtest()` function from the `lmtest` package.

structure to those typically seen in an analysis of variance (ANOVA), but can perform LRTs instead of the typical ANOVA F-tests. The two functions are based on model-comparison tests of different types. When used with one model-fit object resulting from `glm()`, the `anova()` function computes *type I* (sequential) tests, while `Anova()` computes *type II* (partial) tests. The difference between types lies in the null-hypothesis models used for each term tested. With type II tests, the null-hypothesis model for each tested term consists of all the other variables listed in the right side of the `formula` argument, ignoring any higher-order interactions that contain the term (see Section 2.2.5 for more on interactions). For type I tests, the null-hypothesis model contains only those variables listed in the `formula` argument *before* the tested term. Generally, type II tests are preferred unless there is a specific reason to consider a sequence of tests, such as in polynomial models. See Milliken and Johnson (2004) for further details.

The $-2\log(\Lambda)$ statistic has a simplified form. Suppose the estimated probability of successes under the null- and alternative-hypothesis models are denoted as $\hat{\pi}_i^{(0)}$ and $\hat{\pi}_i^{(a)}$, respectively, for $i = 1, \ldots, n$. Similarly, define the corresponding vectors of the regression parameter estimates as $\hat{\boldsymbol{\beta}}^{(0)}$ and $\hat{\boldsymbol{\beta}}^{(a)}$. We have then

$$\begin{aligned}
-2\log(\Lambda) &= -2\log\left(\frac{L(\hat{\boldsymbol{\beta}}^{(0)}|y_1,\ldots,y_n)}{L(\hat{\boldsymbol{\beta}}^{(a)}|y_1,\ldots,y_n)}\right) \\
&= -2\left[\log\left(L(\hat{\boldsymbol{\beta}}^{(0)}|y_1,\ldots,y_n)\right) - \log\left(L(\hat{\boldsymbol{\beta}}^{(a)}|y_1,\ldots,y_n)\right)\right] \\
&= -2\sum_{i=1}^{n} y_i\log(\hat{\pi}_i^{(0)}) + (1-y_i)\log(1-\hat{\pi}_i^{(0)}) - y_i\log(\hat{\pi}_i^{(a)}) - (1-y_i)\log(1-\hat{\pi}_i^{(a)}) \\
&= -2\sum_{i=1}^{n} y_i\log\left(\frac{\hat{\pi}_i^{(0)}}{\hat{\pi}_i^{(a)}}\right) + (1-y_i)\log\left(\frac{1-\hat{\pi}_i^{(0)}}{1-\hat{\pi}_i^{(a)}}\right). \qquad (2.7)
\end{aligned}$$

This formulation of the statistic can be easily coded into R. We show it and how to use the `anova()` and `Anova()` functions next.

Example: Placekicking (Placekick.R, Placekick.csv)

We begin this example by using the model that contains both `change` and `distance`,

$$\text{logit}(\pi) = \beta_0 + \beta_1\texttt{change} + \beta_2\texttt{distance},$$

which was originally estimated in Section 2.2.1 and its results saved in the `mod.fit2` object. Below is our code and output for Wald tests involving $H_0: \beta_r = 0$ vs. $H_a: \beta_r \neq 0$ with $r = 1, 2$.

```
> summary(mod.fit2)
Call: glm(formula = good ~ change + distance, family =
    binomial(link = logit), data = placekick)

Deviance Residuals:
    Min        1Q    Median        3Q       Max
 -2.7061    0.2282    0.2282    0.3750    1.5649

Coefficients:
             Estimate Std. Error z value Pr(>|z|)
(Intercept) 5.893181   0.333184   17.687   <2e-16 ***
```

```
change       -0.447783    0.193673   -2.312    0.0208 *
distance     -0.112889    0.008444  -13.370    <2e-16 ***
---
Signif. codes:  0 '***' 0.001 '**' 0.01 '*' 0.05 '.' 0.1 ' ' 1

(Dispersion parameter for binomial family taken to be 1)

    Null deviance: 1013.4 on 1424 degrees of freedom
Residual deviance:  770.5 on 1422 degrees of freedom
AIC: 776.5

Number of Fisher Scoring iterations: 6
```

Wald test statistics and p-values for each regression parameter are located in the `coefficients` table from `summary(mod.fit2)`. For example, to test the importance of the `change` explanatory variable, we use $H_0 : \beta_1 = 0$ vs. $H_a : \beta_1 \neq 0$ resulting in $Z_W = -0.4478/0.1937 = -2.312$ (`z value` column) and a p-value of $2P(Z > |-2.312|) = 0.0208$ (`Pr(>|Z|)` column). Using $\alpha = 0.05$, we would reject the null hypothesis. However, notice that we would not reject it using $\alpha = 0.01$. This indicates marginal evidence that `change` is important to include in the model given that `distance` is in the model. Note that it is important to include the "`distance` is in the model" part because the null and alternative hypotheses are equivalently stated as

$$H_0 : \text{logit}(\pi) = \beta_0 + \beta_1\texttt{distance}$$
$$H_a : \text{logit}(\pi) = \beta_0 + \beta_1\texttt{change} + \beta_2\texttt{distance}.$$

With respect to testing the parameter for `distance` in the model containing `change`, the p-value is given as `<2e-16` in the output, which means the p-value is less than 2×10^{-16}. Using any reasonable α level, there is strong evidence of the importance of `distance` given that `change` is in the model. In the program for this example, we show the R code needed to extract information from `mod.fit2` to perform the Wald tests without the help of `summary(mod.fit2)`.

Below is our code and output from `Anova()` for the LRTs.

```
> library(package = car)
> Anova(mod.fit2, test = "LR")
Analysis of Deviance Table (Type II tests)

Response: good
         LR Chisq Df Pr(>Chisq)
change      5.246  1    0.02200 *
distance  218.650  1    < 2e-16 ***
---
Signif. codes:  0 '***' 0.001 '**' 0.01 '*' 0.05 '.' 0.1 ' ' 1
```

The `Anova()` function produces LRTs using the `test = "LR"` argument value, which is the default. For the test of `change` with $H_0 : \beta_1 = 0$ vs. $H_a : \beta_1 \neq 0$, we obtain $-2\log(\Lambda) = 5.246$ with a p-value of $P(A > 5.246) = 0.0220$ (A is a random variable with a χ^2_1 distribution), and we reach the same conclusion as with the previous Wald test. The p-value for the `distance` test is given as `< 2e-16` in the output, which again indicates there is strong evidence that `distance` is important (given that the model includes `change`).

The anova() function with the test = "Chisq" argument value also performs LRTs, but in a sequential manner based on the ordering of the explanatory variables in the model fit.

```
> anova(mod.fit2, test = "Chisq")
Analysis of Deviance Table

Model: binomial, link: logit

Response: good

Terms added sequentially (first to last)

         Df Deviance Resid. Df Resid. Dev P(>|Chi|)
NULL                      1424    1013.43
change    1   24.277      1423     989.15 8.343e-07 ***
distance  1  218.650      1422     770.50 < 2.2e-16 ***
---
Signif. codes:  0 '***' 0.001 '**' 0.01 '*' 0.05 '.' 0.1 ' ' 1
```

The change row of the function's output gives a test of

$$H_0 : \text{logit}(\pi) = \beta_0$$
$$H_a : \text{logit}(\pi) = \beta_0 + \beta_1 \texttt{change},$$

which results in a p-value of 8.34×10^{-7}. The distance row of the function's output gives a test of

$$H_0 : \text{logit}(\pi) = \beta_0 + \beta_1 \texttt{change}$$
$$H_a : \text{logit}(\pi) = \beta_0 + \beta_1 \texttt{change} + \beta_2 \texttt{distance},$$

which results in a p-value of less than 2×10^{-16}. The hypotheses and calculations for the test of distance are exactly the same as those using the Anova() function.

Another way to use the anova() function is to fit both the null- and alternative-hypothesis models and then use the corresponding model-fit objects as argument values within the function. This will be especially useful later when we test null hypotheses that do not correspond to simple deletions of one term from the full model. To test the change variable given that distance is in the model, we can use the following code.

```
> # mod.fit: distance, mod.fit2: change and distance
> anova(mod.fit, mod.fit2, test = "Chisq")
Analysis of Deviance Table

Model 1: good ~ distance
Model 2: good ~ change + distance
  Resid. Df Resid. Dev Df Deviance P(>|Chi|)
1      1423     775.75
2      1422     770.50  1   5.2455   0.02200 *
---
Signif. codes:  0 '***' 0.001 '**' 0.01 '*' 0.05 '.' 0.1 ' ' 1
```

We obtain the same results as before with $-2\log(\Lambda) = 5.246$ and a p-value of 0.0220.

Equation 2.7 can be programmed directly into R to perform LRTs. While this generally is not needed for logistic regression applications, it is often helpful to go through the calculations in this manner to better understand what functions like `anova()` and `Anova()` are computing. We demonstrate the LRT for the test of $H_0 : \text{logit}(\pi) = \beta_0$ vs. $H_a : \text{logit}(\pi) = \beta_0 + \beta_1$`change`.

```
> mod.fit.Ho <- glm(formula = good ~ 1, family = binomial(link =
    logit), data = placekick)
> mod.fit.Ha <- glm(formula = good ~ change, family =
    binomial(link = logit), data = placekick)
> anova(mod.fit.Ho, mod.fit.Ha, test = "Chisq")
Analysis of Deviance Table

Model 1: good ~ 1
Model 2: good ~ change
  Resid. Df Resid. Dev Df Deviance P(>|Chi|)
1      1424    1013.43
2      1423     989.15  1   24.277 8.343e-07 ***
---
Signif. codes:  0 '***' 0.001 '**' 0.01 '*' 0.05 '.' 0.1 ' ' 1

> pi.hat.Ho <- mod.fit.Ho$fitted.values
> pi.hat.Ha <- mod.fit.Ha$fitted.values
> y <- placekick$good
> stat <- -2*sum(y*log(pi.hat.Ho/pi.hat.Ha) +
    (1-y)*log((1-pi.hat.Ho)/(1-pi.hat.Ha)))
> pvalue <- 1 - pchisq(q = stat, df = 1)
> data.frame(stat, pvalue)
      stat       pvalue
1 24.27703 8.342813e-07
```

We fit both the null- and alternative-hypothesis models using `glm()`. Note that we use `formula = good ~ 1` to fit a model with only an intercept.[7] The `anova()` function gives a $-2\log(\Lambda) = 24.277$ value along with a p-value equal to 8.343×10^{-7} as seen before. After creating objects for $\hat{\pi}_i^{(0)}$, $\hat{\pi}_i^{(a)}$, and y_i, we code Equation 2.7 using the fact that R performs mathematical operations on vectors elementwise.[8] To find the p-value, we use the `pchisq()` function and subtract it from 1 because we want the area to the right of the quantile but the function finds the area to its left. The resulting test statistic and p-value are the same as found earlier.

In the previous example's output, the word *deviance* and its abbreviation "dev" appeared a number of times. This word also appears in the `summary()` output as well. Deviance

[7]In contrast, if we wanted to exclude an intercept parameter from being estimated for a model, we would include a "`-1`" as part of a `formula` argument value along with any explanatory variable names.

[8]Remember that it is helpful to execute segments of code to determine what the code does. For example, one could first execute `y*log(pi.hat.Ho/pi.hat.Ha)` separately to see that a vector of zero and non-zero values is returned. When combined with `(1-y)*log((1-pi.hat.Ho)/(1-pi.hat.Ha))` and then using `-2*sum()`, we obtain Equation 2.7.

refers to the amount that a particular model deviates from another model as measured by the LRT statistic $-2\log(\Lambda)$. For example, the $-2\log(\Lambda) = 5.246$ value used for testing `change` (given that `distance` is in the model) is a measure of how much the estimated probability of successes for the model excluding `change` deviate from those for the model including `change`.

The *residual deviance* of a model is a particular form of the $-2\log(\Lambda)$ statistic that measures how much the probabilities estimated from the model deviate from the observed proportions of successes (which are simply 0/1 or 1/1 for binary response data).[9] These observed proportions are equivalent to estimating a logistic regression model where each observation is represented by one parameter, say $\text{logit}(\pi_i) = \gamma_i$ for $i = 1, \ldots, n$. This model is frequently referred to as the *saturated model,* because the number of parameters is equal to the number of observations, so that no additional parameters can be estimated. We will use the residual deviance as a measure of overall goodness of fit for a model in Chapter 5. The *null deviance* denotes how much the probabilities estimated from the model $\text{logit}(\pi_i) = \beta_0$ for $i = 1, \ldots, n$ deviate from the observed proportion of successes. Because $\text{logit}(\pi_i) = \beta_0$ contains only the intercept term (and thus just one parameter), every π_i is estimated to be the same value for this particular model (the MLE for π_i is the same as the MLE of a single probability success π that we estimated in Section 1.1.2).

Residual deviance statistics are often calculated as an intermediate step for performing an LRT to compare two models. For example, consider the hypotheses

$$\begin{aligned} H_0 &: \text{logit}(\pi^{(0)}) = \beta_0 + \beta_1 x_1 \\ H_a &: \text{logit}(\pi^{(a)}) = \beta_0 + \beta_1 x_1 + \beta_2 x_2 + \beta_3 x_3 \end{aligned}$$

where we use the superscript for π again to help differentiate between the two models, and suppose we are working with binary response data. The residual deviance for the model $\text{logit}(\pi^{(0)}) = \beta_0 + \beta_1 x_1$ tests $H_0 : \text{logit}(\pi^{(0)}) = \beta_0 + \beta_1 x_1$ vs. H_a : Saturated model, using the quantity:

$$-2\log(\Lambda) = -2\sum_{i=1}^{n} y_i \log\left(\frac{\hat{\pi}_i^{(0)}}{y_i}\right) + (1 - y_i)\log\left(\frac{1 - \hat{\pi}_i^{(0)}}{1 - y_i}\right). \tag{2.8}$$

Values of $y_i = 0$ or 1 in the fraction denominators do not cause problems because y_i is a coefficient on the log function (i.e., 0 multiplied by another quantity is 0). The residual deviance for the model $\text{logit}(\pi^{(a)}) = \beta_0 + \beta_1 x_1 + \beta_2 x_2 + \beta_3 x_3$ tests $H_0 : \text{logit}(\pi^{(a)}) = \beta_0 + \beta_1 x_1 + \beta_2 x_2 + \beta_3 x_3$ vs. H_a : Saturated model, using the quantity:

$$-2\log(\Lambda) = -2\sum_{i=1}^{n} y_i \log\left(\frac{\hat{\pi}_i^{(a)}}{y_i}\right) + (1 - y_i)\log\left(\frac{1 - \hat{\pi}_i^{(a)}}{1 - y_i}\right). \tag{2.9}$$

When Equation 2.9 is subtracted from Equation 2.8, we obtain Equation 2.7, which is the correct LRT statistic needed for the original set of hypotheses.

The residual deviance takes a slightly different form when we have binomial responses W_j for $j = 1, \ldots, J$. The statistic is

$$-2\log(\Lambda) = -2\sum_{j=1}^{J} \left[w_j \log\left(\frac{\hat{\pi}_j}{w_j/n_j}\right) + (n_j - w_j)\log\left(\frac{1 - \hat{\pi}_j}{1 - w_j/n_j}\right)\right], \tag{2.10}$$

[9]This is analogous to residual sum of squares in linear regression, which measures how much model-estimated means deviate from observed responses.

where $\hat{\pi}_j$ is the estimate of the probability of success at each $j = 1, \ldots, J$. Note that the saturated model in this setting estimates the probability of success as w_j/n_j. Residual deviances for null- and alternative-hypothesis models can still be used in hypothesis tests in the same manner as in the previous binary response case.

Example: Placekicking (Placekick.R, Placekick.csv)

To perform the test of

$$H_0 : \text{logit}(\pi) = \beta_0 + \beta_1\texttt{distance}$$
$$H_a : \text{logit}(\pi) = \beta_0 + \beta_1\texttt{change} + \beta_2\texttt{distance},$$

we used the `anova()` function earlier to obtain $-2\log(\Lambda) = 5.246$. Alternatively, we can use the residual deviances from the two models.

```
> # mod.fit: distance, mod.fit2: change and distance
> df <- mod.fit$df.residual - mod.fit2$df.residual
> stat <- mod.fit$deviance - mod.fit2$deviance
> pvalue <- 1 - pchisq(q = stat, df = df)

> # Test statistic, p-value
> data.frame(Ho.resid.dev = mod.fit$deviance, Ha.resid.dev =
   mod.fit2$deviance, df = df, stat = round(stat,4), pvalue =
   round(pvalue,4))
  Ho.resid.dev Ha.resid.dev df   stat pvalue
1      775.745     770.4995  1 5.2455  0.022
```

We use the `deviance` and `df.residual` components of the model-fit objects to obtain the residual deviances and corresponding degrees of freedom, respectively. The $-2\log(\Lambda)$ value and the p-value are both the same as earlier.

In this section, we illustrated a few different ways to perform an LRT. Generally, the `Anova()` and `anova()` functions are the easiest functions to use, and we recommend their use. We included other ways to perform the tests to help demonstrate more general methods and reinforce how the calculations are actually done. These more general methods will also be useful for other types of models where a `glm()`-like function does not exist for model fitting and no `Anova()` or `anova()` function exists for testing, which is the case for some parts of Chapters 3 and 6.

2.2.3 Odds ratios

In the linear model $E(Y) = \beta_0 + \beta_1 x_1 + \cdots + \beta_p x_p$, the regression parameter β_r is interpreted as the change in the mean response for each 1-unit increase in x_r, holding the other variables in the model constant. In a logistic regression model, the interpretation of regression parameters needs to account for the fact that they are related to the probability of success through $\text{logit}(\pi) = \beta_0 + \beta_1 x_1 + \cdots + \beta_p x_p$. Holding other variables constant, a 1-unit increase in x_r causes $\text{logit}(\pi)$ to change by β_r. This constant change in the log-odds of success leads to a convenient way to use odds in the interpretation.

To see how this works, consider a logistic regression model with only one explanatory variable x. The odds of success at a particular value of x are

$$Odds_x = \exp(\beta_0 + \beta_1 x).$$

If x is increased by $c > 0$ units, the odds of success become

$$Odds_{x+c} = \exp(\beta_0 + \beta_1(x + c)).$$

To determine how much the odds of success have changed by this c-unit increase, we find the odds ratio:

$$OR = \frac{Odds_{x+c}}{Odds_x} = \frac{\exp(\beta_0 + \beta_1(x + c))}{\exp(\beta_0 + \beta_1 x)} = \exp(c\beta_1).$$

Interestingly, the original value of the explanatory variable x is canceled out in the simplification; only the amount of increase c and the coefficient β_1 matter. When x is binary with a coding of 0 and 1, c is always 1. Otherwise, the value of c should be chosen in the context of the explanatory variable itself. We will discuss this further shortly and provide an example using the placekicking data.

To interpret the odds ratio, one could apply the same interpretation from Section 1.2.5:

> The odds of success are OR times as large for the explanatory variable at $x + c$ as for the explanatory variable at x,

where we treat x here as a particular numerical value of the explanatory variable. This interpretation works fine when the explanatory variable is binary with $x = 0$ and $c = 1$. However, the interpretation is somewhat cumbersome when x is a continuous variable because x was canceled out in our expression for the odds ratio (the actual value of x does not matter). For this reason, we recommend using the following interpretation when x is continuous:

> The odds of success change by $\exp(c\beta_1)$ times for every c-unit increase in the explanatory variable.

It is also fine to say that the odds "increase" instead of "change" when $\exp(c\beta_1) > 1$, and "decrease" when $\exp(c\beta_1) < 1$. Other interpretations are possible, and we provide examples shortly.

Estimates of the regression parameters can be substituted into OR to estimate the odds ratio. The estimated odds ratio becomes

$$\widehat{OR} = \exp(c\hat{\beta}_1),$$

and its interpretation follows from above with "estimated" included prior to "odds." Because the estimated odds ratio is a statistic, it will vary from sample to sample. Therefore, we need to find a confidence interval for OR to make inferences with a particular level of confidence. The likelihood-based procedures discussed in Appendix B.3.4 and B.5.2 provide the basis for the confidence intervals discussed next.

To find a Wald confidence interval for OR, we first find a confidence interval for $c\beta_1$ and then use the exponential function with the interval endpoints. To this end, we extract $\widehat{Var}(\hat{\beta}_1)$ from the estimated covariance matrix for the parameter estimators and form the $(1 - \alpha)100\%$ Wald confidence interval for $c\beta_1$ as

$$c\hat{\beta}_1 \pm cZ_{1-\alpha/2}\sqrt{\widehat{Var}(\hat{\beta}_1)},$$

where we used $\widehat{Var}(c\hat{\beta}_1) = c^2\widehat{Var}(\hat{\beta}_1)$.[10] The $(1 - \alpha)100\%$ Wald confidence interval for OR then becomes

$$\exp\left(c\hat{\beta}_1 \pm cZ_{1-\alpha/2}\sqrt{\widehat{Var}(\hat{\beta}_1)}\right). \tag{2.11}$$

[10]In general, if a is a constant and Y is a random variable, we have $Var(aY) = a^2Var(Y)$; see p. 60 of Casella and Berger (2002).

The standard interpretation of the confidence interval results from adapting the interpretation for OR:

> With $(1-\alpha)100\%$ confidence, the odds of a success change by an amount between <lower limit> to <upper limit> times for every c-unit increase in the explanatory variable,

where the appropriate numerical values replace $< >$ and the explanatory variable is continuous. A similar adaptation for the interpretation is made when the explanatory variable is binary.

Wald confidence intervals are relatively simple to calculate and generally perform well with large samples. Alternatively, profile LR confidence intervals (see Appendix B.5.2) tend to perform better, especially when the sample size is not large. The downside to their use is that calculation can be more difficult. To calculate a profile LR interval, we first find the set of β_1 values such that

$$-2\log\left(\frac{L(\tilde{\beta}_0, \beta_1|y_1, \ldots, y_n)}{L(\hat{\beta}_0, \hat{\beta}_1|y_1, \ldots, y_n)}\right) < \chi^2_{1,1-\alpha} \tag{2.12}$$

is satisfied, where $\tilde{\beta}_0$ is the MLE of β_0 following specification of a value of β_1. In most settings, there are no closed-form solutions for the lower and upper limits, so iterative numerical procedures are needed to find them. Once the confidence interval limits for β_1 are found, say, "lower" and "upper," we use the exponential function and take into account a value of c to find the $(1-\alpha)100\%$ profile LR confidence interval for OR:

$$\exp(c \times \text{lower}) < OR < \exp(c \times \text{upper}).$$

A few additional notes are needed about odds ratios before proceeding to an example:

1. Inverting odds ratios less than 1 can be helpful for interpretation purposes.

2. A value of $c = 1$ is sometimes used as a default. However, this may not make sense, depending on the context of the data. For example, if $0.1 < x < 0.2$, there is little sense in reporting a change in odds using $c = 1$. Similarly, if $0 < x < 1000$, a value of $c = 1$ may be too small to be meaningful. Instead, an appropriate value of c should be chosen in the context of the explanatory variable. Absent any other guidance, taking c to be the standard deviation of x can be a reasonable choice.

3. When there is more than one explanatory variable, the odds ratio relating to the effect of each variable on the odds of success can be shown to be $\exp(c\beta_r)$ for each $x_r, r = 1, \ldots, p$ in the model. The same interpretation of the odds ratio generally applies, with the addition of "holding the other explanatory variables constant." If the model contains interaction, categorical, or transformed explanatory variables, some care is required in applying this interpretation. We address how to approach interactions and transformations in Section 2.2.5 and categorical variables in Section 2.2.6.

4. Odds ratios resulting from a logistic regression model fit are widely used in research. These are sometimes referred to as *adjusted odds ratios* when the model contains more that one explanatory variable. This adjustment simply refers to calculating the odds ratio for a variable holding the other variables in the model constant, as described in the previous note. *Unadjusted odds ratios* are those calculated using each variable by itself in a model (or more simply using the tools in Section 1.2.5). The unadjusted odds ratio for a variable may be a poor estimate when other variables that have large effects on the odds of success are correlated with it. Exercise 32 provides an example where this terminology is used in research.

5. In many problems, more than one odds ratio is computed. Creating $(1-\alpha)100\%$ confidence intervals for each odds ratio results in a confidence level less than $(1-\alpha)100\%$ that *all* intervals contain (cover) their parameters. This is referred to as the problem of performing *simultaneous inference* or *multiple inference* on a *family* (i.e., a group of similar parameters). The problem becomes worse as the size of the family grows. The probability that all intervals contain their parameters is the *familywise confidence level.* To maintain a familywise confidence level of $(1-\alpha)100\%$, the individual intervals need to be adjusted so that each has a somewhat higher confidence level. The *Bonferroni adjustment* is one simple way to achieve this: if the size of the family is g, then each confidence interval is found using a $1-\alpha/g$ confidence level. This method is easy to use but tends to be somewhat conservative: it produces intervals with a familywise confidence level greater than $(1-\alpha)$, and hence are wider than necessary. Better methods of control exist and are available in certain functions within R (examples to be discussed shortly).

Example: Placekicking (Placekick.R, Placekick.csv)

For this example, we use the `mod.fit` object from the logistic regression model that includes only `distance` as an explanatory variable. Estimated odds ratios for a 1-yard increase and a 10-yard decrease in `distance` are calculated as follows.

```
> exp(mod.fit$coefficients[2])
  distance
 0.8913424
> exp(-10*mod.fit$coefficients[2])
 distance
 3.159035
```

We see that $\exp(\hat{\beta}_1) = 0.8913$ with $c = 1$. Because a 1-yard change is rather small (field-goal attempts range roughly from 20–60 yards), we instead focus on the change in the odds of success for a 10-yard change. Also, because the estimated odds of success are lower for an increase in distance ($\exp(c\hat{\beta}_1) < 1$ for $c > 0$), we focus on $c = -10$ for our primary interpretation. Thus, we find that the estimated odds of success change by 3.16 times for every 10-yard decrease in the distance of the placekick. If a football coach were given a choice between attempting a 50-yard placekick or trying to gain 10 more yards and attempting a 40-yard placekick (or a 10-yard decrease from anywhere else on the field), the fact that the estimated odds of success are $3.16 - 1 = 2.16$ times larger for the shorter distance could factor into the coach's decision.[11] Therefore, the preference would be for shorter placekicks. While football fans will not be surprised that shorter kicks have a higher chance of success, the amount by which the odds of success change is not widely known.

The `confint()` function provides a convenient way to calculate confidence intervals associated with regression parameters. This is a generic function that by default finds profile LR intervals for our setting. We use this function below to calculate the interval corresponding to the distance of the placekick.

[11]The subtraction of 1 from $\widehat{OR}$ was included in the calculation because we are examining how much *larger* one odds is compared to another odds. A similar subtraction of 1 was included in Section 1.2.4 for the relative risk ("as likely" vs. "more likely" discussion).

```
> beta.ci <- confint(object = mod.fit, parm = "distance", level
   = 0.95)
Waiting for profiling to be done...
> beta.ci
       2.5 %      97.5 %
 -0.13181435 -0.09907103
> rev(exp(-10*beta.ci))  # OR CI for c = -10
    97.5 %    2.5 %
  2.693147 3.736478
> # Remove labels with as.numeric()
> as.numeric(rev(exp(-10*beta.ci)))
[1] 2.693147 3.736478
```

The 95% profile LR confidence interval for the `distance` regression parameter is $-0.1318 < \beta_1 < -0.0991$. Using $c = -10$, the 95% profile LR interval for the odds ratio is $2.69 < OR < 3.74$ where $OR = \exp(-10\beta_1)$. With 95% confidence, the odds of success change by an amount between 2.69 to 3.74 times for every 10-yard decrease in the distance of the placekick. Because the interval is entirely above 1, there is sufficient evidence that a 10-yard decrease in distance increases the odds of a successful placekick. Note that we use `as.numeric()` in the previous code to prevent unnecessary labels from being printed.

To calculate a Wald interval, we need to use the specific method function `confint.default()`.

```
> beta.ci <- confint.default(object = mod.fit, parm =
   "distance", level = 0.95)
> beta.ci
              2.5 %     97.5 %
distance -0.1313709 -0.0986824
> as.numeric(rev(exp(-10*beta.ci)))  # OR CI for c = -10
[1] 2.682701 3.719946
```

The 95% Wald confidence interval for the `distance` parameter is $-0.1314 < \beta_1 < -0.0987$. Using $c = -10$, we obtain the 95% Wald interval $2.68 < OR < 3.72$. This interval is similar to the profile LR interval because of the large sample size.

To see how computations for the Wald interval are performed, the code below shows Equation 2.11 coded directly into R.

```
> beta.ci <- mod.fit$coefficients[2] + qnorm(p = c(0.025,
   0.975)) * sqrt(vcov(mod.fit)[2,2])
> beta.ci
[1] -0.1313709 -0.0986824
> rev(exp(-10*beta.ci))
[1] 2.682701 3.719946
```

The `vcov()` function calculates the estimated covariance matrix for the parameter estimates using the information within `mod.fit`. By specifying `vcov(mod.fit)[2,2]`, we extract $\widehat{Var}(\hat{\beta}_1)$ from the matrix. The `mod.fit$coefficients[2]` syntax extracts

$\hat{\beta}_1$ from the vector of parameter estimates. Putting these elements together, we calculate the confidence interval for β_1 and then the desired confidence interval for the odds ratio.

To see how the computations for the profile LR interval are performed, we include two sets of code in our program that compute the interval without the `confint()` function. Both rely on rewriting Equation 2.12 as

$$-2\left[\log\left(L(\tilde{\beta}_0, \beta_1|y_1,\dots,y_n)\right) - \log\left(L(\hat{\beta}_0, \hat{\beta}_1|y_1,\dots,y_n)\right)\right] - \chi^2_{1,0.95} < 0. \quad (2.13)$$

The first method sequentially calculates the left-side of Equation 2.13 for a range of possible β_1 values. The smallest and largest values of β_1 that satisfy the inequality correspond to the lower and upper interval limits for β_1. The second method replaces the less than sign in Equation 2.13 with an equal sign and iteratively solves for β_1 values using the `uniroot()` function. After the lower and upper limits are found by either method, the `exp()` function is used to find the interval limits for the odds ratio.

2.2.4 Probability of success

Once a logistic regression model is estimated, it is often of interest to estimate the probability of success for a set of explanatory variable values. This can be done simply by substituting the parameter estimates into the model:

$$\hat{\pi} = \frac{\exp(\hat{\beta}_0 + \hat{\beta}_1 x_1 + \cdots + \hat{\beta}_p x_p)}{1 + \exp(\hat{\beta}_0 + \hat{\beta}_1 x_1 + \cdots + \hat{\beta}_p x_p)}.$$

Because $\hat{\pi}$ is a statistic, it varies from sample to sample. Therefore, we need to find a confidence interval for π to make inferences with a particular level of confidence. Both Wald and profile LR intervals are discussed in this section.

To help explain how the Wald interval for π is calculated, consider again a logistic regression model with only one explanatory variable. The estimated probability of success is then $\hat{\pi} = \exp(\hat{\beta}_0 + \hat{\beta}_1 x)/[1 + \exp(\hat{\beta}_0 + \hat{\beta}_1 x)]$. The normal distribution is a better approximation to the distribution of $\hat{\beta}_0$ and $\hat{\beta}_1$ than it is for $\hat{\pi}$ (see Exercise 38), so we proceed in a manner similar to finding the Wald interval for the odds ratio. We first find a confidence interval for the linear predictor, $\text{logit}(\pi) = \beta_0 + \beta_1 x$, and then transform this interval's endpoints into an interval for π using the $\exp(\cdot)/[1 + \exp(\cdot)]$ transformation. The $(1 - \alpha)100\%$ Wald confidence interval for $\beta_0 + \beta_1 x$ is

$$\hat{\beta}_0 + \hat{\beta}_1 x \pm Z_{1-\alpha/2}\sqrt{\widehat{Var}(\hat{\beta}_0 + \hat{\beta}_1 x)}. \quad (2.14)$$

The estimated variance is found by

$$\widehat{Var}(\hat{\beta}_0 + \hat{\beta}_1 x) = \widehat{Var}(\hat{\beta}_0) + x^2\widehat{Var}(\hat{\beta}_1) + 2x\widehat{Cov}(\hat{\beta}_0, \hat{\beta}_1),$$

where each variance and covariance term is available from the estimated covariance matrix of the parameter estimates.[12] Using the interval limits for $\beta_0 + \beta_1 x$, the $(1 - \alpha)100\%$ Wald

[12]This is another application of the following result: $Var(aU+bV) = a^2Var(U)+b^2Var(V)+2abCov(U,V)$, where U and V are random variables and a and b are constants. See p. 171 of Casella and Berger (2002).

confidence interval for π is

$$\frac{\exp\left(\hat{\beta}_0 + \hat{\beta}_1 x \pm Z_{1-\alpha/2}\sqrt{\widehat{Var}(\hat{\beta}_0 + \hat{\beta}_1 x)}\right)}{1 + \exp\left(\hat{\beta}_0 + \hat{\beta}_1 x \pm Z_{1-\alpha/2}\sqrt{\widehat{Var}(\hat{\beta}_0 + \hat{\beta}_1 x)}\right)}. \tag{2.15}$$

Note that the lower (upper) limit for π uses the minus (plus) sign in the $\pm$ part of Equation 2.15.

When there are p explanatory variables in the model, the Wald interval for π is found in the same manner. The interval is

$$\frac{\exp\left(\hat{\beta}_0 + \hat{\beta}_1 x_1 + \cdots + \hat{\beta}_p x_p \pm Z_{1-\alpha/2}\sqrt{\widehat{Var}(\hat{\beta}_0 + \hat{\beta}_1 x_1 + \cdots + \hat{\beta}_p x_p)}\right)}{1 + \exp\left(\hat{\beta}_0 + \hat{\beta}_1 x_1 + \cdots + \hat{\beta}_p x_p \pm Z_{1-\alpha/2}\sqrt{\widehat{Var}(\hat{\beta}_0 + \hat{\beta}_1 x_1 + \cdots + \hat{\beta}_p x_p)}\right)},$$

where the variance expression is found in a similar way as in the single variable case:

$$\widehat{Var}(\hat{\beta}_0 + \hat{\beta}_1 x_1 + \cdots + \hat{\beta}_p x_p) = \sum_{i=0}^{p} x_i^2 \widehat{Var}(\hat{\beta}_i) + 2\sum_{i=0}^{p-1}\sum_{j=i+1}^{p} x_i x_j \widehat{Cov}(\hat{\beta}_i, \hat{\beta}_j), \tag{2.16}$$

with $x_0 = 1$.

While Equation 2.16 may be long, it is calculated automatically by functions in R. We focus on two main functions for this purpose. First, `predict()` of the `stats` package calculates the variance and also $\hat{\pi}$. These calculated values can be appropriately combined to find the Wald confidence interval for π. Second, `emmeans()` of the `emmeans` package (not in the default installation of R) performs these and many other calculations. This package is one of a number user-contributed packages that provide a convenient suite of functions for estimation and inference involving linear combinations of regression parameters. The name of the package is short for "estimated marginal means." We use it here to estimate and perform inference for the mean of Y, $E(Y) = \pi$.

Both functions lead to the same results. We introduce `predict()` of the `stats` package because it is in the default installation of R and the most predominantly used. This function will be our focus when it can produce the desired calculations. We introduce `emmeans()` here for comparison purposes. Later in the textbook, this function will perform other calculations more simply than they can be done otherwise. We next provide details on the use of both functions.

Example: Placekicking (Placekick.R, Placekick.csv)

We begin by using the `mod.fit` object from the logistic regression model that has only `distance` as an explanatory variable. Below are three ways that the estimated probability of success can be calculated for a distance of 20 yards.

```
> linear.pred <- mod.fit$coefficients[1] +
   mod.fit$coefficients[2] * 20
> linear.pred
(Intercept)
   3.511547
> as.numeric(exp(linear.pred)/(1 + exp(linear.pred)))
 0.9710145
```

```
> predict.data <- data.frame(distance = 20)
> predict(object = mod.fit, newdata = predict.data, type =
   "link")
       1
3.511547
> predict(object = mod.fit, newdata = predict.data, type =
   "response")
        1
0.9710145

> head(placekick$distance == 20)
[1] FALSE FALSE  TRUE FALSE  TRUE FALSE
> mod.fit$fitted.values[3]  # 3rd obs. has distance = 20
        3
0.9710145
```

The first way directly calculates the linear predictor as $\hat{\beta}_0 + \hat{\beta}_1 x = 5.8121 - 0.1150 \times 20 = 3.5115$, resulting in $\hat{\pi} = \exp(3.5115)/[1 + \exp(3.5115)] = 0.9710$. Note that the `(Intercept)` heading is a label left over from `mod.fit$coefficients[1]`, which can be removed by using `as.numeric()`.

The second way to calculate $\hat{\pi}$ is to use the `predict()` function, and we recommend using this function most of the time for these types of calculations. The function is a generic function that actually uses the `predict.glm()` method function to perform calculations. To use `predict()`, a data frame must contain the explanatory variable values at which the estimates of π are desired. This data frame is included then in the `newdata` argument of `predict()`. Additionally, the `object` argument specifies where the model-fit information from `glm()` is located, and the `type = "response"` argument value instructs R to estimate π. Alternatively, the `type = "link"` argument value instructs R to estimate $\beta_0 + \beta_1 x$.

If no data frame is given in the `newdata` argument of `predict()`, π is estimated for each observation in the data set. The result is exactly the same as given by `mod.fit$fitted.values`. Thus, a third way to estimate π is to look at values of `mod.fit$fitted.values` corresponding to the distance of 20 yards. Of course, estimates of π may be of interest for x values not in the data set, but for our current case, the third observation corresponds to $x = 20$.

To calculate a Wald confidence interval for π, we use the `predict()` function to calculate $\hat{\beta}_0 + \hat{\beta}_1 x$ and $\widehat{Var}(\hat{\beta}_0 + \hat{\beta}_1 x)$ first. We then calculate the confidence interval for π using the appropriate code for Equations 2.14 and 2.15.

```
> alpha <- 0.05
> linear.pred <- predict(object = mod.fit, newdata =
   predict.data, type = "link", se = TRUE)
> linear.pred
$fit
       1
3.511547
$se.fit
[1] 0.1732707
$residual.scale
[1] 1
```

```
> pi.hat <- exp(linear.pred$fit) / (1 + exp(linear.pred$fit))
> CI.lin.pred <- linear.pred$fit + qnorm(p = c(alpha/2,
   1-alpha/2)) * linear.pred$se
> CI.pi <- exp(CI.lin.pred)/(1+exp(CI.lin.pred))
> round(data.frame(predict.data, pi.hat, lower = CI.pi[1], upper
   = CI.pi[2]), 4)
  distance pi.hat  lower  upper
1       20  0.971 0.9598 0.9792
```

The `se = TRUE` argument value within `predict()` leads to the calculation of $\widehat{Var}(\hat{\beta}_0+\hat{\beta}_1 x)^{1/2}$, the "standard error" for $\hat{\beta}_0+\hat{\beta}_1 x$. Note that the use of this additional argument causes the resulting object to be a list of component vectors, whereas a single vector was produced when standard errors were not requested. The 95% Wald confidence interval for π is $0.9598 < \pi < 0.9792$; thus, the probability of success for the placekick is quite high at a distance of 20 yards.

When we want to estimate probabilities for more than one distance or there are additional explanatory variables, we must specify all values of all explanatory variables in `predict()` by including them in the data frame for `newdata`. For example, if we use the model with `distance` and `change` to estimate the probability of success for 20- and 30-yard lead-changing placekicks, we could create a data frame such as `predict.data <- data.frame(distance = c(20,30), change = c(1, 1))`. Our R program contains the full set of code for this additional case.

In contrast to `predict()`, the `emmeans()` function calculates Wald intervals without the need for additional code to form the interval.

```
> library(package = emmeans)
> predict.data2 <- list(distance = 20)
> emmeans(object = mod.fit, specs = ~ distance, at =
   predict.data2, type = "response", level = 0.95)
 distance  prob      SE  df asymp.LCL asymp.UCL
       20 0.971 0.00488 Inf      0.96     0.979
```

The `level` argument specifies the confidence level, and the `object` and `type` arguments represent the same arguments as for `predict()`. The `specs` argument specifies the variable(s) for which we wish to estimate probabilities. The format is similar to the right side of a `formula` argument, with a `+` separating different variables when needed, and does not have to include all variables in the model. The optional `at` argument lists the values of these variables at which we wish to do the computations, similar to `newdata` in `predict()`, but in a list format. For example, rather than using `newdata = data.frame(distance = c(20,30), change = c(1, 1))`, `emmeans()` uses `at = list(distance = c(20,30), change = 1)`, where all possible combinations of the explanatory variable values are created automatically. In the special case where just one probability estimate is needed, both list and data frame formats will work for `at` because a data frame is a special case of a list (see Appendix A.7.3).

Although the output from `emmeans()` appears to be a data frame, it is instead a calculation summary printed by an automatic invoking of a method function for `summary()`. This is a very important aspect to understand when the calculated values need to be saved to an object for future use, such as for performing additional calculations or for creating plots. We provide a simple illustration below of saving the results into an object and extracting important portions of it.

```
> pi.calc <- emmeans(object = mod.fit, specs = ~ distance, at =
    predict.data, type = "response")
> sum.pi.calc <- summary(pi.calc, level = 0.95)
> names(sum.pi.calc)
[1] "distance"  "prob"      "SE"        "df"
[5] "asymp.LCL" "asymp.UCL"
> sum.pi.calc$prob
[1] 0.9710145
> sum.pi.calc$asymp.LCL
[1] 0.9597647
> sum.pi.calc$asymp.UCL
[1] 0.9791871
```

Additionally, `as.data.frame()` can be used in the same way as `summary()`.

Objects created from `emmeans()`, like `pi.calc` in this example, generally will not be of direct interest to users. Rather, the calculations generated by method functions, such as `summary()`, will be of interest. Readers still interested in exploring the structure of objects returned by `emmeans()` may notice that `names()` does not work. This is because `emmeans()` creates objects using what is known as the S4 class system in R, which is different from the S3 class system used so far in the textbook.[13] The main differences that would be important to readers are 1) `names()` is replaced with `slotNames()` and 2) `$` is replaced with @. Section 3.4.3 discusses these class systems in more detail. Interestingly, despite the use of the S4 class system for `emmeans()`, the S3 class system is used by the method functions of the package.

When more than one confidence interval is calculated, it may be desirable to control the overall familywise confidence level. A number of adjustment methods are available with `summary()` by including an `adjust` argument and the name of the adjustment method. For example, a Bonferroni adjustment is requested by using `adjust = "bonferroni"`. We discuss these adjustments further in the examples of Section 2.2.6. The default for `summary()` is no control.

Although Wald intervals perform fine with large sample sizes, we recommend using profile LR intervals for π when possible because they generally better maintain their stated confidence levels. Like Wald intervals, profile LR intervals are calculated for $\text{logit}(\pi)$ first and then transformed into intervals for π using $\exp(\cdot)/[1 + \exp(\cdot)]$. The numerator of $-2\log(\Lambda)$ involves repeatedly maximizing the likelihood function while holding $\text{logit}(\pi) = \beta_0 + \beta_1 x_1 + \cdots + \beta_p x_p$ fixed at different values, which is more difficult than constraining just one parameter as was the case in Section 2.2.3. In some situations, the iterative numerical procedures may take excessively long to run or even fail to converge.

There are no functions within the default installation of R or in the `emmeans` package that compute profile LR intervals for π. For this reason, we use the `mcprofile` package for the computations. Like the `emmeans` package, this package provides a convenient suite of functions for estimation and inference involving linear combinations of regression parameters. Its one disadvantage compared to `emmeans` is that `mcprofile` requires more detailed code. In the end, this can be an advantage because it requires the user to understand their model better.

[13] The S3 class system is the predominant class system in R. There will be a few other occasions in the textbook where the S4 class system will be used because of user-contributed packages.

Because computations can be difficult, we recommend careful examination of the results produced by `mcprofile`. If warning messages occur or if Wald intervals are substantially different, stating both intervals or the Wald interval only may be preferred to stating the profile LR interval alone. In our experience with the package, we have encountered only a few situations (see Exercise 25) when there were any doubts about the profile LR interval calculations.

Example: Placekicking (Placekick.R, Placekick.csv)

To find a profile LR interval for π at a distance of 20 yards, we use the code below.

```
> library(package = mcprofile)

> # Coefficients in beta0 * 1 + beta1 * 20
> K <- matrix(data = c(1, 20), nrow = 1, ncol = 2)
> K
     [,1] [,2]
[1,]    1   20

> # Calculate -2log(Lambda)
> linear.combo <- mcprofile(object = mod.fit, CM = K)
> # CI for beta0 * 1 + beta1 * 20
> ci.logit.profile <- confint(object = linear.combo, level =
    0.95, adjust = "none")
> ci.logit.profile

   mcprofile - Confidence Intervals

level:              0.95
adjustment:         single-step

   Estimate lower upper
C1     3.51  3.19  3.87

> names(ci.logit.profile)
[1] "estimate"    "confint"     "CM"         "quant"
[5] "alternative" "level"       "adjust"
> # CI for pi
> exp(ci.logit.profile$confint)/(1 +
    exp(ci.logit.profile$confint))
      lower     upper
C1 0.9603165 0.979504
```

After the initial call to the `library()` function for the `mcprofile` package, we create a matrix `K` that contains coefficients for the linear combination of interest. In this case, we want to first find a confidence interval for $\beta_0 \times 1 + \beta_1 \times 20$, so our matrix contains a single row consisting of 1 and 20. This matrix of coefficients (also known as a *contrast matrix*) is used in the `mcprofile()` function as the argument value for `CM`. The `mcprofile()` function calculates $-2\log(\Lambda)$ using the maximized log likelihood function with constraints for a large number of possible values of logit(π). The `confint()` function finds the 95% profile LR interval to be $3.19 < \beta_0 + 20\beta_1 < 3.87$. We use the $\exp(\cdot)/[1+\exp(\cdot)]$ transformation then to find the interval for π as $0.9603 < \pi < 0.9795$, which is similar to the Wald interval because of the large sample size.

The matrix K should have as many columns as there are parameters in the regression model ($p+1$), and columns should follow the same order as their respective variables in the model. Also, there should be one row for each different combination of explanatory variable values at which a confidence interval is needed. For example, if we use the model with `distance` and `change` to estimate the probability of success for 20- and 30-yard lead-changing placekicks, we can create the K matrix and perform the calculations as shown below.

```
> mod.fit2$coefficients
(Intercept)      change    distance
  5.8931814  -0.4477832  -0.1128888

> K <- matrix(data = c(1, 1, 20,
                       1, 1, 30), nrow = 2, ncol = 3, byrow =
                         TRUE)
> K
     [,1] [,2] [,3]
[1,]    1    1   20
[2,]    1    1   30

> linear.combo <- mcprofile(object = mod.fit2, CM = K)
> ci.logit.profile <- confint(object = linear.combo, level =
   0.95, adjust = "none")
> exp(ci.logit.profile$confint)/(1 +
   exp(ci.logit.profile$confint))
      lower     upper
1 0.9411946 0.9741890
2 0.8505051 0.9167352
```

The `byrow = TRUE` argument in `matrix()` specifies that the numbers provided in `data` should be filled into the two rows of K by rows rather than by columns, which is the default. Thus, the first row is for a lead-change placekick at 20 yards ($\text{logit}(\pi) = \beta_0 \times 1 + \beta_1 \times 1 + \beta_2 \times 20$) and the second row is for a lead-change placekick at 30 yards ($\text{logit}(\pi) = \beta_0 \times 1 + \beta_1 \times 1 + \beta_2 \times 30$). Our spacing of the numerical values in `data` is done only to help view what the matrix should look like. R simply reads its contents as the vector `c(1, 1, 20, 1, 1, 30)`.

The `confint()` function includes the `adjust` argument for control of the familywise confidence level, where `adjust = "none"` is for no control. The 95% confidence intervals are $0.9412 < \pi < 0.9742$ for a lead-change placekick at 20 yards and $0.8505 < \pi < 0.9167$ for a lead-change placekick at 30 yards. To control the overall familywise confidence level, a Bonferroni adjustment can be requested by using `adjust = "bonferroni"`. The default for `confint()` is to use a *single-step adjustment*. We discuss this particular adjustment in Section 2.2.5.

We also demonstrate in our program how to find the same Wald intervals as earlier but now with the `mcprofile` package using the `wald()` function.

For a linear model with only one explanatory variable x, it is common to examine a scatter plot of the data with the estimated linear regression model plotted upon it. The purpose of this type of plot is to obtain a visual assessment of trends in the data and to assess the fit of the model. Because the response variable in logistic regression is binary, constructing a plot like this is not very informative because all plotted points would be at $y = 0$ or 1 on the y-axis. However, we can instead plot the observed proportion of successes at each unique x to obtain a general understanding of how well the model fits the data (more formal assessments are discussed in Chapter 5). The estimated model should be expected to pass "near" these proportions as long as each is based on a large enough number of observations. A key to assessment with these types of plots then is knowing the number of observations that exist at each x, so that one can truly distinguish between a poor and good fit. In the extreme case of a continuous x, the number of observations would be 1 at each unique x and this plot generally would not be very useful.

Example: Placekicking (Placekick.R, Placekick.csv)

In an earlier example from Section 2.2.1, we found the number of successes out of the total number of trials for each distance. Using the resulting data frame `w.n`, we use the `plot()` and `curve()` functions here to plot the observed proportion of successes at each distance and then overlay the estimated logistic regression model.

```
> head(w.n)
  distance  success  trials  proportion
1       18        2       3      0.6667
2       19        7       7      1.0000
3       20      776     789      0.9835
4       21       19      20      0.9500
5       22       12      14      0.8571
6       23       26      27      0.9630
> plot(x = w$distance, y = w$good/n$good, xlab = "Distance
   (yards)", ylab = "Estimated probability", panel.first =
   grid(col = "gray", lty = "dotted"))
> curve(expr = predict(object = mod.fit, newdata =
   data.frame(distance = x), type = "response"), col = "red",
   add = TRUE, xlim = c(18, 66))
```

The `curve()` function evaluates the `predict()` function at 101 distances within the x-axis limits by specifying `newdata = data.frame(distance = x)` for the `expr` argument. The resulting plot is given in Figure 2.4 (excluding the dot-dash lines).

We would also like to plot 95% Wald confidence intervals for π at each possible distance. These *bands* are added to the plot by using `curve()` again and a new function that we wrote named `ci.pi()`.

```
> ci.pi <- function(newdata, mod.fit.obj, alpha){
  linear.pred <- predict(object = mod.fit.obj, newdata =
      newdata, type = "link", se = TRUE)
  CI.lin.pred.lower <- linear.pred$fit - qnorm(p =
      1-alpha/2)*linear.pred$se
  CI.lin.pred.upper <- linear.pred$fit + qnorm(p =
      1-alpha/2)*linear.pred$se
  CI.pi.lower <- exp(CI.lin.pred.lower) / (1 +
      exp(CI.lin.pred.lower))
  CI.pi.upper <- exp(CI.lin.pred.upper) / (1 +
      exp(CI.lin.pred.upper))
  list(lower = CI.pi.lower, upper = CI.pi.upper)
}
```

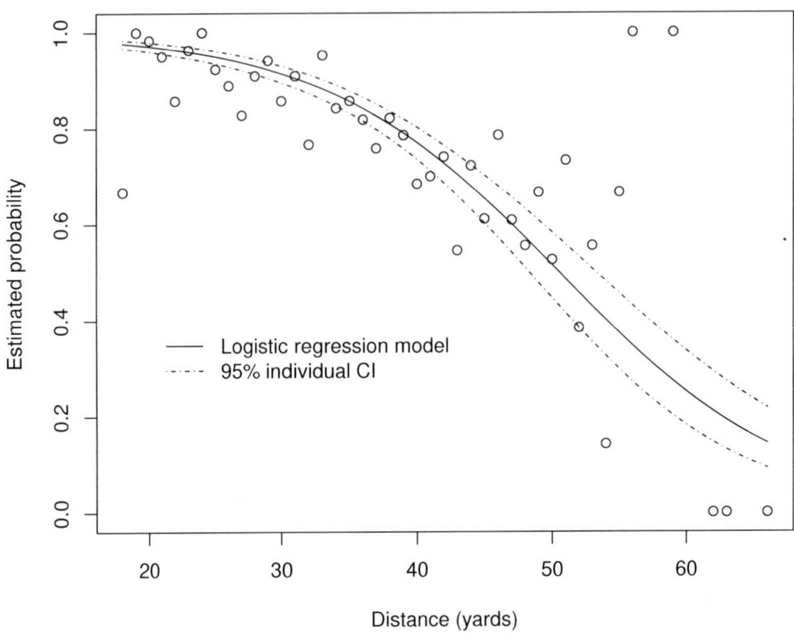

Figure 2.4: Estimated probability of success for a placekick.

```
> # Test case
> ci.pi(newdata = data.frame(distance = 20), mod.fit.obj =
   mod.fit, alpha = 0.05)
$lower
        1
0.9597647

$upper
        1
0.9791871

> # Plot C.I. bands
> curve(expr = ci.pi(newdata = data.frame(distance = x),
   mod.fit.obj = mod.fit, alpha = 0.05)$lower, col = "blue", lty
   = "dotdash", add = TRUE, xlim = c(18, 66))
> curve(expr = ci.pi(newdata = data.frame(distance = x),
   mod.fit.obj = mod.fit, alpha = 0.05)$upper, col = "blue", lty
   = "dotdash", add = TRUE, xlim = c(18, 66))

> # Legend
> legend(x = 20, y = 0.4, legend = c("Logistic regression
   model", "95% individual CI"), lty = c("solid", "dotdash"),
   col = c("red", "blue"), bty = "n")
```

Because the `expr` argument of `curve()` requires a function name as its value, the `ci.pi()` function performs our confidence limit calculations. This new function essentially contains the same code as used earlier to calculate a Wald confidence interval for π, and it stores the lower and upper limits as separate components in a list, so that each limit may be plotted with a separate call to `curve()`. The `ci.pi()` function can be used more generally to calculate Wald confidence intervals for π whenever they are needed. At the end of the code, we use the `legend()` function to clarify what each plotted line represents. The `x` and `y` arguments in `legend()` specify the x- and y-coordinate values for the upper-left corner of the legend box.

Notice that the estimated model in Figure 2.4 generally goes through the middle of the plotted proportions. However, there are a number of proportions that are far from the estimated model, for example at 18, 56, and 59 yards. These are distances at which very few attempts were recorded—three at 18 and one each at 56 and 59—so the sample proportions can vary considerably. The estimated model curve should not be expected to come as close to those points as it should to points based on large numbers of trials.

A better way to approach a plot like Figure 2.4 is to incorporate information about the number of trials n_j into every plotting point. This can be done with a bubble plot, where each plotting point is proportional in size to the number of trials. The `symbols()` function produces these plots in R.

```
> symbols(x = w$distance, y = w$good/n$good, circles =
   sqrt(n$good), inches = 0.5, xlab = "Distance (yards)", ylab =
   "Estimated probability", panel.first = grid(col = "gray", lty
   = "dotted"))
```

The `x` and `y` arguments within `symbols()` work the same way as in `plot()`. The `circles` argument controls the size of the plotting point where the maximum radius is specified in inches by the `inches` argument. For this plot, we use the number of observations at each unique distance to correspond to the plotting point size. The `sqrt()` function is used only to lessen the absolute differences between the values in `n$good`, and this does not need to be used in general.[14]

Figure 2.5 shows the final bubble plot with estimates and Wald confidence interval bands for π included. We can see that indeed the 18-yard placekicks should not cause concern because of the small size of the plotting point. There are similar points at the largest distances, where many of these include only one observation. As expected, the 20-yard placekicks have the largest plotting point, and it appears to be centered close to the estimated model. There are a few large plotting points, such as at 32 yards ($\hat{\pi} = 0.89$, observed proportion of $23/30 = 0.77$) and 51 yards ($\hat{\pi} = 0.49$, observed proportion of $11/15 = 0.73$), for which the model may not fit well. How to more formally assess these observations and others will be an important subject of Chapter 5 when we examine model diagnostic measures.

We provide additional code in the program that gives the same plot as shown in Figure 2.5 but with profile LR confidence interval bands rather than the Wald confidence interval bands. These two types of bands are indistinguishable on these plots

[14]There are 789 observations at a distance of 20 yards. The next largest number of observations is 30, which is at a distance of 32 yards. Because of the large absolute difference in the number of observations, this causes the plotting point at 20 yards to be very large and the other plotting points to be very small. To lessen this absolute difference, we use a transformation on `n$good`.

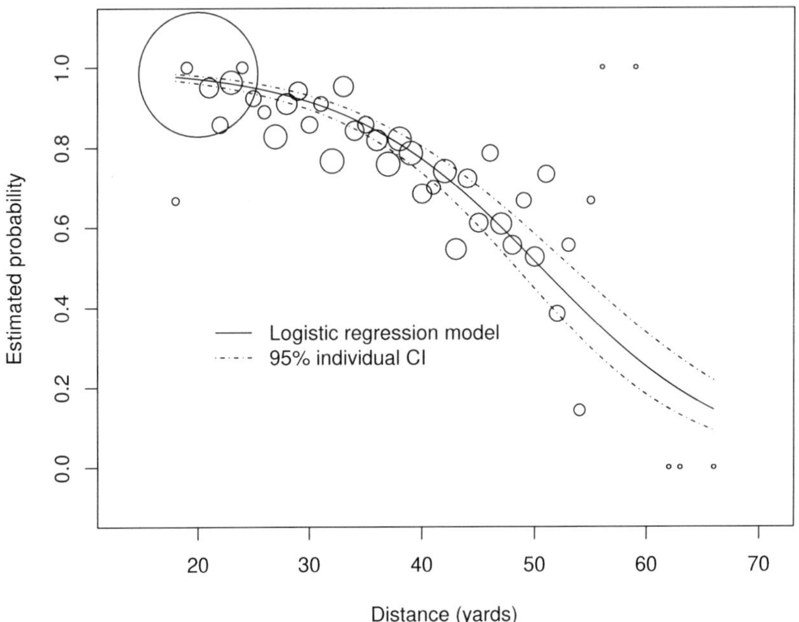

Figure 2.5: Bubble plot of the estimated probability of success for a placekick; the plotting point has a size proportional to the number of observations at a distance.

because of the large sample size. One small advantage for using Wald bands here is that they are computed much more quickly than those from profile LR intervals.

We have focused on understanding the effects of explanatory variables by examining odds ratios. This is largely because the use of the logit transformation in the logistic regression model makes computation and inference for odds ratios relatively straightforward. Other effect measures can be computed using similar processes to those used for estimating and making inferences for π. In particular, we can compare the probabilities of success for two groups using differences and ratios (relative risks), like in Sections 1.2.2 and 1.2.4, respectively, but now using the logistic regression model. However, these comparisons yield more complicated structures that do not simplify into basic linear combinations of regression parameters, and hence generally change depending on the values of the explanatory variable that are being compared. For example, consider again the logistic regression model with only one explanatory variable, $\pi = \exp(\beta_0 + \beta_1 x)/[1 + \exp(\beta_0 + \beta_1 x)]$. When x is increased by c-units, the relative risk is

$$\frac{\exp[\beta_0 + \beta_1(x + c)]/\{1 + \exp[\beta_0 + \beta_1(x + c)]\}}{\exp(\beta_0 + \beta_1 x)/[1 + \exp(\beta_0 + \beta_1 x)]} = \frac{\exp(c\beta_1)[1 + \exp(\beta_0 + \beta_1 x)]}{\{1 + \exp[\beta_0 + \beta_1(x + c)]\}}.$$

Therefore, the ratio has different values depending on the value of x. The same can be shown for the difference of two probabilities of success from the logistic regression model. Section 2.2.3 showed that odds ratios for many logistic regression models do not have this dependence. For this reason, odds ratios are used with logistic regression much more frequently than differences and ratios of probabilities.

To estimate differences and ratios of probabilities, estimates of π are first found at the chosen levels of the explanatory variable and used in the calculation of the corresponding differences and ratios. Wald and profile LR intervals follow in the same manner as with odds ratios. Exercises 14 and 15 provide examples of the process for models with a binary explanatory variable. Exercise 16 goes through the process using the placekicking data.

2.2.5 Interactions and transformations for explanatory variables

Interactions and transformations extend the variety of shapes that can be proposed for relating π to the explanatory variables. Terms are created and added to the linear predictor of a logistic regression model in the same way as for linear regression.[15] We focus in this section on pairwise (two-way) interactions and quadratic terms, because they are the most common and interpretable. Other types of terms, such as three-way interactions, are examined in the exercises. Section 6.8 explores the use of cubic splines as an alternative way to include nonlinear effects for an explanatory variable in a model.

Interactions between explanatory variables are needed when the effect of one explanatory variable on the probability of success depends on the value for a second explanatory variable. There are a few ways to include these terms in a `formula` argument for `glm()`. To describe them, suppose there are two explanatory variables named `x1` and `x2` in a data frame containing x_1 and x_2, and the goal is to fit the model $\text{logit}(\pi) = \beta_0 + \beta_1 x_1 + \beta_2 x_2 + \beta_3 x_1 x_2$. The $x_1 x_2$ term (x_1 is multiplied by x_2) represents the interaction. Thus, the effect of x_1 on $\text{logit}(\pi)$ depends not only on β_1 but also $\beta_3 x_2$. Similarly, the effect of x_2 on $\text{logit}(\pi)$ depends not only on β_2 but also $\beta_3 x_1$. Note that we adopt the convention that the explanatory variables themselves—their *main effects*—are always included in the model whenever the interaction is included. Models constructed in this way are said to respect *marginality*. Section 5.4.1 explains why marginality is a desirable property.

Below are equivalent ways to code the model:

1. `formula = y ~ x1 + x2 + x1:x2`

2. `formula = y ~ x1*x2`

3. `formula = y ~ (x1 + x2)^2`

In the first `formula` argument, the colon symbol denotes an interaction between `x1` and `x2`. In the second `formula` argument, the asterisk automatically creates all main effects and interactions among the connected variables. For the third `formula` argument, the `^2` creates all combinations of up to two variables from among the variables listed within the parentheses. If there is a third variable `x3`, then `(x1 + x2 + x3)^2` is equivalent to `x1*x2 + x1*x3 + x2*x3`. Also, `(x1 + x2 + x3)^3` is equivalent to `x1*x2*x3`.

Quadratic and higher-order polynomials are used to model a nonlinear relationship between an explanatory variable and $\text{logit}(\pi)$. There are two main ways to include a quadratic term in a `formula` argument. To describe them, suppose there is one explanatory variable named `x1` in a data frame containing x_1, and the goal is to fit the model $\text{logit}(\pi) = \beta_0 + \beta_1 x_1 + \beta_2 x_1^2$. Below are equivalent ways to code the model:

1. `formula = y ~ x1 + I(x1^2)`

2. `formula = y ~ poly(x1, degree = 2, raw = TRUE)`

[15] Linear regression textbooks that provide details include Kutner et al. (2004, Chapter 8), Weisberg (2013, Chapter 5), and Montgomery et al. (2021, Chapters 7 and 8). A full set of course notes and videos for a linear regression course is available from `www.chrisbilder.com`.

In the first `formula` argument, we want the ^ symbol to create an exponent as R normally would for mathematical operations. However, this symbol has a special meaning in `formula` arguments as we have just seen, which conflicts with its intended use here. The identity function `I()` tells R to treat its argument in the usual way, so that the ^ symbol is interpreted as we intend. In the second `formula` argument, the `poly()` function is used with a `degree` argument to specify the degree of the polynomial—the highest-order term to be included. All lower-order terms are included in `formula` as well. The `raw = TRUE` argument value specifies that x_1 is used in the calculations rather than an orthogonal polynomial representation (`raw = FALSE`).

Example: Placekicking (Placekick.R, Placekick.csv)

Generally, one would conjecture that the more time that a football is in the air, the more it is susceptible to the effect of wind. Also, longer placekicks generally have more time in the air than shorter kicks. Thus, it would be interesting to determine whether there is an interaction between `distance` and `wind`. The `wind` explanatory variable in the data set is a binary variable for placekicks attempted in windy conditions (1) vs. non-windy conditions (0), where windy conditions are defined as a wind stronger than 15 miles per hour at kickoff in an outdoor stadium.[16]

We next fit a model including this interaction term along with the main effects.

```
> mod.fit.Ho <- glm(formula = good ~ distance + wind, family =
    binomial(link = logit), data = placekick)
> mod.fit.Ha <- glm(formula = good ~ distance + wind +
    distance:wind, family = binomial(link = logit), data =
    placekick)
> summary(mod.fit.Ha)

Call: glm(formula = good ~ distance + wind + distance:wind,
    family = binomial(link = logit), data = placekick)

Deviance Residuals:
    Min       1Q   Median       3Q      Max
-2.7291   0.2465   0.2465   0.3791   1.8647

Coefficients:
               Estimate Std. Error z value Pr(>|z|)
(Intercept)    5.684181   0.335962  16.919   <2e-16 ***
distance      -0.110253   0.008603 -12.816   <2e-16 ***
wind           2.469975   1.662144   1.486   0.1373
distance:wind -0.083735   0.043301  -1.934   0.0531 .
---
Signif. codes:  0 '***' 0.001 '**' 0.01 '*' 0.05 '.' 0.1 ' ' 1

(Dispersion parameter for binomial family taken to be 1)
```

[16]While wind direction information was available when we collected these data, the stadium orientation was not, making it impossible to account for wind direction relative to the placekicks. Even if orientation had been available, it might not have been useful because of the swirling wind effects that can occur within a stadium. Also, note that wind speed was dichotomized into windy vs. non-windy conditions because of the unknown wind speed inside domed stadiums (only one domed stadium recorded this information). Because of these stadiums' ventilation systems, it is incorrect to assume the wind speed is 0.

```
Null deviance: 1013.43  on 1424  degrees of freedom
Residual deviance:  767.42  on 1421  degrees of freedom
AIC: 775.42

Number of Fisher Scoring iterations: 6

> # LRT
> anova(mod.fit.Ho, mod.fit.Ha, test = "Chisq")
Analysis of Deviance Table

Model 1: good ~ distance + wind
Model 2: good ~ distance + wind + distance:wind
  Resid. Df Resid. Dev Df Deviance P(>|Chi|)
1      1422     772.53
2      1421     767.42  1   5.1097   0.02379 *
---
Signif. codes:  0 '***' 0.001 '**' 0.01 '*' 0.05 '.' 0.1 ' ' 1

> # Alternative code for the LRT
> library(package = car)
> Anova(mod.fit.Ha, test = "LR")
Analysis of Deviance Table (Type II tests)

Response: good
              LR Chisq Df Pr(>Chisq)
distance       238.053  1    < 2e-16 ***
wind             3.212  1    0.07312 .
distance:wind    5.110  1    0.02379 *
---
Signif. codes:  0 '***' 0.001 '**' 0.01 '*' 0.05 '.' 0.1 ' ' 1
```

The estimated logistic regression model including the main effects and the interaction is

$$\begin{aligned}\text{logit}(\hat{\pi}) = 5.6842 &- 0.1103\texttt{distance} + 2.4700\texttt{wind} \\ &- 0.0837\texttt{distance} \times \texttt{wind}. \qquad (2.17)\end{aligned}$$

For testing $H_0 : \beta_3 = 0$ vs. $H_a : \beta_3 \neq 0$, the Wald test gives a p-value of 0.0531 and the LRT gives a p-value of 0.0238. Both tests suggest there is marginal evidence to support a `wind` and `distance` interaction. We show the use of `anova()` and `Anova()` here to emphasize that both produce the same results and to provide further comment on the use of type II tests by `Anova()`. The calculations provided by `Anova()` are for the hypothesis tests shown in Table 2.1. As discussed in Section 2.2.2, type II tests involve testing a particular term in the model, given all of the other terms in the model except for higher-order interactions that contain the term. Thus, the `distance` row of the output provides information for the test of $H_0 : \text{logit}(\pi) = \beta_0 + \beta_1\texttt{wind}$ vs. $H_a : \text{logit}(\pi) = \beta_0 + \beta_1\texttt{distance} + \beta_2\texttt{wind}$ without the interaction term in the null and alternative hypothesis models. The `Anova()` function provides tests in this manner so that they satisfy the principle of marginality described earlier. In particular, testing for a main effect using a model that includes the interaction in the null hypothesis would violate marginality.

Figure 2.6 displays plots of the estimated logistic regression models for the `wind` = 0 and 1 cases. The left plot uses the estimated logistic regression model without

Table 2.1: Linear predictor portion of the logistic regression models for the tests performed by `Anova()`.

	Null hypothesis	Alternative hypothesis
`distance`	$\beta_0 + \beta_1$`wind`	$\beta_0 + \beta_1$`distance` $+ \beta_2$`wind`
`wind`	$\beta_0 + \beta_1$`distance`	$\beta_0 + \beta_1$`distance` $+ \beta_2$`wind`
`distance:wind`	$\beta_0 + \beta_1$`distance` $+ \beta_2$`wind`	$\beta_0 + \beta_1$`distance` $+ \beta_2$`wind` $+ \beta_3$`distance` $\times$ `wind`

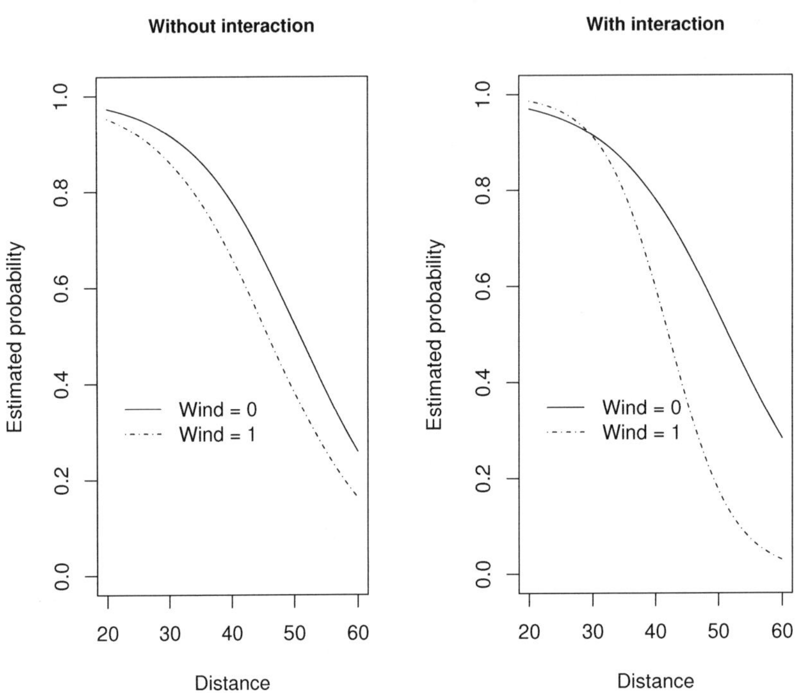

Figure 2.6: The estimated logistic regression model without (left) and with (right) a distance and wind interaction.

the interaction (`mod.fit.Ho` contains the model estimates), and the right plot uses the estimated logistic regression model with the interaction (`mod.fit.Ha` contains the model estimates). The effect of the interaction is evident in the right plot by the reduction in the estimated probability of success for the longer placekicks when it is windy. To create these plots, we use the `curve()` function in a similar manner as we did in Section 2.2.4 for Figure 2.4. The code is included in the program for this example.

The behavior illustrated in the right plot of Figure 2.6 is evident from closer examination of Equation 2.17. The negative coefficient on the distance main effect indicates that the probability of success decreases with increasing distance when `wind` $=$ `0`. The negative coefficient on the interaction indicates that this effect is exacerbated under windy conditions as the distance increases.

The inclusion of interactions and/or transformations causes odds ratios to depend on the numerical value of an explanatory variable. For example, consider the model $\text{logit}(\pi) = \beta_0 + \beta_1 x_1 + \beta_2 x_2 + \beta_3 x_1 x_2$. The odds ratio associated with changing x_2 by c units while holding x_1 constant is

$$OR = \frac{Odds_{x_2+c}}{Odds_{x_2}} = \frac{\exp(\beta_0 + \beta_1 x_1 + \beta_2(x_2 + c) + \beta_3 x_1(x_2 + c))}{\exp(\beta_0 + \beta_1 x_1 + \beta_2 x_2 + \beta_3 x_1 x_2)}$$
$$= \exp(c\beta_2 + c\beta_3 x_1) = \exp(c(\beta_2 + \beta_3 x_1)). \quad (2.18)$$

Thus, the increase or decrease in the odds of success for a c-unit change in x_2 depends on the level of x_1, which follows from the definition of an interaction. As another example, consider the model $\text{logit}(\pi) = \beta_0 + \beta_1 x_1 + \beta_2 x_1^2$. The odds ratio for a c-unit change in x_1 is

$$OR = \frac{Odds_{x_1+c}}{Odds_{x_1}} = \frac{\exp(\beta_0 + \beta_1(x_1 + c) + \beta_2(x_1 + c)^2)}{\exp(\beta_0 + \beta_1 x_1 + \beta_2 x_1^2)}$$
$$= \exp(c\beta_1 + 2cx_1\beta_2 + c^2\beta_2) = \exp(c\beta_1 + c\beta_2(2x_1 + c)),$$

which depends on the value of x_1.

Confidence intervals for these odds ratios can be more complicated to calculate, because they are generally based on linear combinations of regression parameters rather than on individual parameters. For the interaction example, where $OR = \exp(c(\beta_2 + \beta_3 x_1))$, the $(1 - \alpha)100\%$ Wald confidence interval is

$$\exp\left(c(\hat{\beta}_2 + \hat{\beta}_3 x_1) \pm cZ_{1-\alpha/2}\sqrt{\widehat{Var}(\hat{\beta}_2 + \hat{\beta}_3 x_1)}\right) \quad (2.19)$$

with

$$\widehat{Var}(\hat{\beta}_2 + \hat{\beta}_3 x_1) = \widehat{Var}(\hat{\beta}_2) + x_1^2\widehat{Var}(\hat{\beta}_3) + 2x_1\widehat{Cov}(\hat{\beta}_2, \hat{\beta}_3). \quad (2.20)$$

Other Wald intervals can be calculated in a similar manner. Unfortunately, the `confint.default()` function examined in Section 2.2.3 cannot be used for these computations. To program the interval expressions directly, the only challenging part is to find the variance for the linear combination of the parameter estimators, so we provide code to do this in the next example. The `emmeans` and `mcprofile` packages provide simpler ways to perform the calculations, where the latter package also computes profile LR intervals.

Example: Placekicking (Placekick.R, Placekick.csv)

Because of the interaction in Equation 2.17, we need to interpret the effect of `wind` at specific levels of `distance` and to interpret the effect of `distance` at specific levels of `wind`. For the former case, the odds ratio is given by Equation 2.18 with `distance` represented by x_1 and `wind` represented by x_2. Below is the corresponding R code and output to obtain odds ratio estimates and 95% Wald confidence intervals.

```
> beta.hat <- mod.fit.Ha$coefficients[2:4]
> c <- 1
> distance <- seq(from = 20, to = 60, by = 10)

> OR.wind <- exp(c*(beta.hat[2] + beta.hat[3]*distance))
> cov.mat <- vcov(mod.fit.Ha)[2:4,2:4]
> # Var(beta-hat_2 + distance*beta-hat_3)
> var.log.OR <- cov.mat[2,2] + distance^2*cov.mat[3,3] +
   2*distance*cov.mat[2,3]
```

```
> # Confidence intervals
> ci.log.OR.low <- c*(beta.hat[2] + beta.hat[3]*distance) -
   c*qnorm(p = 0.975)*sqrt(var.log.OR)
> ci.log.OR.up <- c*(beta.hat[2] + beta.hat[3]*distance) +
   c*qnorm(p = 0.975)*sqrt(var.log.OR)

> round(data.frame(distance = distance, OR.hat = 1/OR.wind,
   OR.low = 1/exp(ci.log.OR.up), OR.up = 1/exp(ci.log.OR.low)),2)
  distance OR.hat OR.low OR.up
1       20   0.45   0.09  2.34
2       30   1.04   0.40  2.71
3       40   2.41   1.14  5.08
4       50   5.57   1.54 20.06
5       60  12.86   1.67 99.13
```

We begin by extracting $\hat{\beta}_1, \hat{\beta}_2$, and $\hat{\beta}_3$ from `mod.fit.Ha$coefficients`. While this is not necessary, we do it here so that the resulting vector will have the same indices as the subscripts for the parameter estimates (i.e., `beta.hat[1]` is $\hat{\beta}_1$). The estimated odds ratios are within the `OR.wind` object, where we use $c = 1$ because `wind` is a binary variable and we examine distances of 20, 30, 40, 50, and 60 yards. We use Equations 2.19 and 2.20 to calculate $\widehat{Var}(\hat{\beta}_2 + \hat{\beta}_3 x_1)$ (or equivalently $\widehat{Var}(\log(\widehat{OR}))$, which is how we chose the name for this object) and the confidence interval for each odds ratio. The estimates and intervals for all odds ratios are inverted so that the numerator represents non-windy placekicks and the denominator represents windy placekicks.

The 95% confidence interval for the `wind` odds ratio at a distance of 30 yards is $0.40 < OR < 2.71$. Because 1 is inside the interval, there is insufficient evidence to indicate that windy conditions have an effect on the success or failure of a placekick at 30 yards. On the other hand, the 95% confidence interval for the `wind` odds ratio at a distance of 50 yards is $1.54 < OR < 20.06$. Because 1 is outside the interval, there is sufficient evidence that `wind` has an effect on the success or failure of a placekick at this distance. As originally conjectured, the longer the distance, the more a placekick is susceptible to the wind. In particular, the odds of success are smaller under windy conditions than non-windy conditions at these longer distances.

The `emmeans` package computes these Wald intervals a little more easily by taking advantage of its `contrast()` and `confint()` functions. The `contrast()` function enables comparisons between specific levels of an explanatory variable (windy vs. non-windy), and the `confint()` function calculates the needed intervals for these comparisons.

```
> calc.est <- emmeans(object = mod.fit.Ha, specs = ~ distance +
   wind, at = list(distance = c(20,30,40,50,60)), type =
   "response")
> summary(calc.est)
 distance wind   prob      SE  df asymp.LCL asymp.UCL
       20    0 0.9701 0.00518 Inf   0.95808     0.979
       30    0 0.9150 0.00907 Inf   0.89549     0.931
       40    0 0.7814 0.01717 Inf   0.74593     0.813
       50    0 0.5428 0.03632 Inf   0.47121     0.613
       60    0 0.2827 0.04421 Inf   0.20452     0.377
       20    1 0.9863 0.01110 Inf   0.93510     0.997
```

```
     30        1 0.9117 0.03803 Inf     0.80354        0.963
     40        1 0.5974 0.08827 Inf     0.41954        0.753
     50        1 0.1758 0.09236 Inf     0.05761        0.427
     60        1 0.0297 0.02940 Inf     0.00414        0.184

> test.info <- contrast(object = calc.est, method = "pairwise",
   simple = list("wind"), combine = TRUE)
> confint(object = test.info, adjust = "none", level = 0.95)
 distance contrast        odds.ratio     SE  df asymp.LCL asymp.UCL
 20          wind0 / wind1      0.451  0.379 Inf    0.0872      2.34
 30          wind0 / wind1      1.043  0.508 Inf    0.4019      2.71
 40          wind0 / wind1      2.410  0.917 Inf    1.1430      5.08
 50          wind0 / wind1      5.567  3.641 Inf    1.5447     20.06
 60          wind0 / wind1     12.860 13.401 Inf    1.6683     99.13

Confidence level used: 0.95
Intervals are back-transformed from the log odds ratio scale
```

The `wind0 / wind1` value under `contrast` in the output indicates that the odds for non-windy placekicks (`wind = 0`) are in the numerator and the odds for windy placekicks (`wind = 1`) are in the denominator of the odds ratio. This is a *contrast* because the log-odds for windy placekicks is subtracted from the log-odds for non-windy placekicks, prior to using the exponential function.[17] Overall, we see the calculated values here are the same as before.

While the **emmeans** package avoids the need to program mathematical expressions into R, it comes at a cost of understanding its many arguments. However, these arguments provide great flexibility for this and other data analyses. We next explain the arguments used by the code here.

- `emmeans()`: Model-fit information from `glm()` is given in the `object` argument. The `specs` argument is used to list only the explanatory variables we want to examine. By default, the function performs calculations for these explanatory variables at (1) the mean value of all numerical variables with more than two values (`distance`), (2) each level of numerical values with only two levels (`wind`), and (3) each level of factor variables (see Section 2.2.6). The linear predictor ($\hat{\beta}_0 + \hat{\beta}_1$`distance` $+ \hat{\beta}_2$`wind` $+$ $\hat{\beta}_3$`distance` $\times$ `wind`; i.e., the log-odds evaluated at the explanatory variable values) is calculated for at all combinations of these values. Transformations or interactions should not be given in `specs`, unlike the `formula` argument syntax for `glm()`; these are detected automatically by `emmeans()`. The `at` argument provides an alternative list of numerical values or specific levels at which we want to compute the linear predictor for any variable given in `specs`. In our example, we request distances of 20, 30, 40, 50, and 60 rather than using the default of the mean for `distance`. These five distances are combined with the two integer values for `wind`, producing 10 different evaluations of the linear predictor. The `type = "response"` argument value leads to the reporting of probabilities in the output rather the linear predictors in subsequent functions like `summary()` and `confint()`. This also subsequently leads `contrast()` to calculate odds ratios rather than differences of the linear predictors.

[17] Contrasts are used widely in statistics to compare quantities by linear combinations. For example, suppose there are two population means, μ_1 and μ_2. A common way to compare these two quantities is with the contrast $\mu_1 - \mu_2$. For an odds ratio, we could compare two odds, denoted as $Odds_1$ and $Odds_2$, by forming a contrast of the log-odds. This results in $\log(Odds_1) - \log(Odds_2) = \log(Odds_1/Odds_2) = \log(OR)$ using properties of logarithms.

- `contrast()`: This function performs contrast calculations using the results from `emmeans()` via the `object` argument. The `simple` argument restricts contrasts to be computed at levels for only those variables named in `list()`. In our example, we restrict attention to comparisons of `wind` separately at each `distance`. Different contrasts can be calculated as requested by the `method` argument. The `"pairwise"` argument value for `method` leads to all possible pairwise comparisons among the levels of `wind`. For this situation, there are only the 0 and 1 levels. Standard errors are computed and Wald tests are performed for each contrast. We do not print these values because our ultimate goal is to compute confidence intervals.

- `confint()`: Building on the results from `contrast()` via the `object` argument, Wald confidence intervals for the odds ratios are calculated. If control of the familywise confidence level is desired, this can be specified with the `adjust` argument. For example, `adjust = "bonferroni"` uses the Bonferroni adjustment for all distances because `combine = TRUE` is provided in `contrast()` to combine all odds ratios into one family. The `level` argument specifies the familywise confidence level or the level for each interval (`adjust = "none"`).

We encourage readers to examine results returned by the code if an argument is changed or omitted and to explore the help for these functions to discover other arguments. In particular, when `simple = list("distance")` is used in `contrast()`, comparisons are made among each pairwise combination of `distance` at each level of `wind`. Some of these odds ratio calculated (e.g., `distance` = 20 vs. `distance` = 30 is the same as a `distance` = 30 vs. `distance` = 40) will be identical because of the form of the model. Additionally, if the `method` argument value is changed to `"consec"`, odds ratios are computed only for consecutive values of `distance`. Thus, odds ratios are given for comparing 30-yard to 20-yard placekicks, 40-yard to 30-yard placekicks, and so forth at each level of `wind`. Adding `reverse = TRUE` to `contrast()` inverts these odds ratios. Finally, when `adjust = "none"` is omitted, a Bonferroni adjustment is applied by default when `simple` is included and a Tukey adjustment is made when `simple` is not included. This demonstrates that it is important to understand the default argument values in functions we use, and we recommend always specifying values for important arguments as part of good coding practice.

The `mcprofile` package computes both profile LR and Wald intervals for the odds ratios. To begin, we specify the coefficient matrix, which we call `K`.

```
> K.names <- list(c("Wind, distance = 20", "Wind, distance =
   30", "Wind, distance = 40", "Wind, distance = 50", "Wind,
   distance = 60"), c("beta0", "beta1", "beta2", "beta3"))
> K <- matrix(data = c(0, 0, 1, 20,
                       0, 0, 1, 30,
                       0, 0, 1, 40,
                       0, 0, 1, 50,
                       0, 0, 1, 60), nrow = 5, ncol = 4,
         byrow = TRUE, dimnames = K.names)
> K
                    beta0 beta1 beta2 beta3
Wind, distance = 20     0     0     1    20
Wind, distance = 30     0     0     1    30
Wind, distance = 40     0     0     1    40
Wind, distance = 50     0     0     1    50
Wind, distance = 60     0     0     1    60
```

We include row and column names to help understand the structure of `K`. These names are included in the same way that the `dimnames` argument of `array()` was used in Section 1.2 for contingency tables, except we do not specify overall names for the row and column groups. The profile LR intervals are computed next using `mcprofile()` and then `confint()`.

```
> linear.combo <- mcprofile(object = mod.fit.Ha, CM = K)
> ci.log.OR <- confint(object = linear.combo, level = 0.95,
   adjust = "none")
> exp(ci.log.OR)

  mcprofile - Confidence Intervals

level:             0.95
adjustment:        none

                     Estimate   lower   upper
Wind, distance = 20    2.2150 0.54627 16.061
Wind, distance = 30    0.9588 0.41399  2.930
Wind, distance = 40    0.4150 0.19430  0.876
Wind, distance = 50    0.1796 0.04387  0.594
Wind, distance = 60    0.0778 0.00764  0.497

> # Invert odds ratios
> round(data.frame(OR.hat = 1/exp(ci.log.OR$estimate), OR.low =
   1/exp(ci.log.OR$confint$upper), OR.up =
   1/exp(ci.log.OR$confint$lower)), 2)
                     Estimate OR.low  OR.up
Wind, distance = 20      0.45   0.06   1.83
Wind, distance = 30      1.04   0.34   2.42
Wind, distance = 40      2.41   1.14   5.15
Wind, distance = 50      5.57   1.68  22.79
Wind, distance = 60     12.86   2.01 130.85
```

We use the `mcprofile` package's method function for `exp()` to quickly apply the exponential function to the results saved as a list in `ci.log.OR`. Unfortunately, it is not as easy to invert the odds ratios from the object, so we extract components from `ci.log.OR` and form a data frame with the results.[18] These inverted odds ratios can be compared directly to the previous Wald intervals. The same general conclusions are reached for the two different types of intervals.

Wald confidence intervals can also be calculated by `mcprofile` with its `wald()` function.

```
> save.wald <- wald(object = linear.combo)
```

[18] Alternatively, we could have used `CM = -K` in `mcprofile()`. However, prior to calculating the odds ratios, one may not know that inverting them is preferred.

```
> ci.log.OR.wald <- confint(object = save.wald, level = 0.95,
    adjust = "none")
```

Applying the exponential function to the intervals in `ci.log.OR.wald` (and then inverting the intervals), yields the same Wald intervals as calculated earlier.

The familywise confidence level is controlled with the appropriate use of the `adjust` argument. For example, the argument value `adjust = "bonferroni"` performs a Bonferroni adjustment to the set of comparisons represented by the rows of `K`. A better form of control is available through the argument value `adjust = "single-step"`, which is based on logic comparable to Tukey's studentized-range statistic for pairwise comparisons in ANOVA as described in Westfall and Young (1993). Results of both methods are given below for profile LR intervals.

```
> ci.log.OR.bon <- confint(object = linear.combo, level = 0.95,
    adjust = "bonferroni")
> ci.log.OR.ss <- confint(object = linear.combo, level = 0.95,
    adjust = "single-step")
> round(data.frame(OR.hat = 1/exp(ci.log.OR.ss$estimate),
    bon.low = 1/exp(ci.log.OR.bon$confint$upper), bon.up =
    1/exp(ci.log.OR.bon$confint$lower), ss.low =
    1/exp(ci.log.OR.ss$confint$upper), ss.up =
    1/exp(ci.log.OR.ss$confint$lower)), 2)
                    Estimate bon.low bon.up ss.low  ss.up
Wind, distance = 20     0.45    0.03   2.63   0.04   2.36
Wind, distance = 30     1.04    0.22   3.03   0.25   2.84
Wind, distance = 40     2.41    0.90   6.59   0.97   6.11
Wind, distance = 50     5.57    1.19  37.92   1.32  32.35
Wind, distance = 60    12.86    1.19 312.81   1.39 237.94
```

It is good practice to present both adjusted and unadjusted results for an analysis involving many intervals. While adjusted intervals are wider, statistically significant findings are more likely to be reproducible in future investigations than those found without adjustments. The overall conclusions largely stay the same for the example here. The changes to interval limits above 1 are larger than those below 1 because of the asymmetric nature of odds ratios. Also, the interval for 40 yards now contains 1, although it is relatively close, as it was with the unadjusted interval.

The estimated odds ratios and corresponding confidence intervals are also found for `distance`, where `wind` is either 0 or 1. The R code is available in the program for this example. For 95% Wald intervals, the odds of a successful placekick increase by between 2.54 to 3.56 times for each 10-yard decrease in `distance` when it is not windy (`wind` = 0), and between 3.03 to 15.98 times when it is windy (`wind` = 1). For 95% profile LR intervals, the intervals are 2.55 to 3.58 (not windy) and 3.40 to 18.79 (windy) for a 10-yard decrease in `distance`.

While we do not provide a data example for quadratic terms in this section, the same principles apply for estimating and interpreting the model. For example, the model $\text{logit}(\pi) = \beta_0 + \beta_1 x_1 + \beta_2 x_1^2$ was shown earlier to have an odds ratio of $OR = \exp(d\beta_1 + d\beta_2(2x_1 + d))$ for a d-unit increase in x_1 (we use a d-unit, rather than a c-unit, increase to differentiate its code from the `c()` function in R). The coefficient matrix for `mcprofile()` can take the form of `matrix(data = c(0, d, d*(2*x1+d)), nrow = 1, ncol = 3)` with a value of `x1`

Table 2.2: Indicator variables and model representation for a 4-level categorical explanatory variable.

Level	x_1	x_2	x_3	Model
A	0	0	0	$\text{logit}(\pi) = \beta_0$
B	1	0	0	$\text{logit}(\pi) = \beta_0 + \beta_1$
C	0	1	0	$\text{logit}(\pi) = \beta_0 + \beta_2$
D	0	0	1	$\text{logit}(\pi) = \beta_0 + \beta_3$

for x_1 and a value of `d` for d. Alternatively, the form `matrix(data = c(0, 1, 2*x1+d), nrow = 1, ncol = 3)` can be used as long as the confidence interval limits for $\log(OR)$ are multiplied by d. Profile LR computations using this alternative version of the matrix tend to work better. Exercise 17 examines a squared distance term for the placekicking example.

2.2.6 Categorical explanatory variables

Categorical explanatory variables are represented in a logistic regression model the same way as they are in a linear regression model. If a variable has q categorical levels, then $q - 1$ indicator variables can be used to represent it. An indicator variable is just a special kind of binary variable that takes the value 1 when a particular outcome is observed and 0 otherwise. For example, we saw earlier that `change` in the placekicking data set was a categorical variable with two levels. It was represented in a logistic regression model with one indicator variable (1 for lead-change placekicks and 0 for all other placekicks).

R stores categorical variables as a *factor* object type. By default, R orders the levels of a factor using a numerical and then alphabetical ordering, where lowercase letters are ordered before uppercase letters.[19] To see the ordering of any factor, the `levels()` function can be used. This ordering is important because most functions in R use it to construct indicator variables with the "set first level to 0" method of construction.[20] For example, Table 2.2 shows the default R coding for indicator variables of a categorical explanatory variable with the $q = 4$ levels coded as "A", "B", "C", and "D." The first level "A" is the base level where all indicator variables are set to 0. The remaining levels each are assigned one indicator variable. In this case, the "B" level is represented by x_1 where $x_1 = 1$ for an observed level of "B" and $x_1 = 0$ otherwise. The "C" and "D" levels are defined in a similar manner for x_2 and x_3, respectively. The `relevel()` function in R can be used to define a new base level if desired, and this is discussed further in the next example.

A categorical explanatory variable is represented in a logistic regression model using all of its indicator variables. For example, a logistic regression model with the four-level categorical explanatory variable is written as

$$\text{logit}(\pi) = \beta_0 + \beta_1 x_1 + \beta_2 x_2 + \beta_3 x_3. \tag{2.21}$$

A somewhat less formal representation substitutes the level names for each of their corresponding indicator variables; i.e., $\text{logit}(\pi) = \beta_0 + \beta_1 B + \beta_2 C + \beta_3 D$. The model at each level of the explanatory variable is found by substituting the indicator variable values shown in

[19] There are exceptions to this ordering. First, a categorical explanatory variable could be coded numerically, like with levels 1, 2, and 3, and treated as if these are actual numerical values rather than codings. We discuss how to handle this type of situation later in this section. Second, a factor may be defined with a different ordering than the default. Examples are provided shortly for how to reorder levels of a factor. Exercise 11 provides an example when the reordering may occur unexpectedly.

[20] The "set last level to 0" construction is used sometimes by other statistical software, like SAS. Other forms of construction exist; see Section 6.3.5 for an example.

Table 2.2 into Equation 2.21. This results in the models shown in the last column of Table 2.2, illustrating that each level has a unique representation.

A question of interest with a categorical explanatory variable is "Does the probability of success change across its levels?" There is no change when the regression parameters for the indicator variables are all 0. Therefore, we can answer the question by testing the null hypothesis that all indicator variables have parameters equal to zero. For example, we would test all three regression parameters in Equation 2.21 simultaneously with H_0 : $\beta_1 = \beta_2 = \beta_3 = 0$ vs. H_a : Not all equal to 0, using an LRT or other appropriate testing procedure.

Example: Control of the Tomato Spotted Wilt Virus (TomatoVirus.R, TomatoVirus.csv)

Plant viruses are often spread by insects. This occurs when insects feed on plants already infected with a virus and subsequently become carriers of the virus themselves. When these insects then feed on other plants, they may transmit this virus to these new plants.

To better understand one particular virus, the Tomato Spotted Wilt Virus, and how to control thrips that spread it, researchers at Kansas State University performed an experiment in a number of greenhouses.[21] One hundred uninfected tomato plants were put into each greenhouse. Within each greenhouse, one of two methods was used to introduce the virus to the clean plants:

1. Additional infected plants were placed among the clean ones, and then "uninfected" thrips were released to spread the virus (coded as `Infest = 1`).

2. Thrips that already carried the virus were released onto the clean plants (`Infest = 2`).

To examine ways of controlling the spread of the virus to plants, the researchers used one of three methods:

1. Biological control: Use predatory spider mites to attack the thrips (`Control = "B"`).

2. Chemical control: Use a pesticide to kill the thrips (`Control = "C"`).

3. No control: Do not use any method to control the thrips (`Control = "N"`).

Among the plants that were originally clean, the number displaying symptoms of infection were recorded after 8 weeks for each greenhouse. Below is a portion of the data where each row of the `tomato` data frame represents a greenhouse.

```
> tomato <- read.csv(file = "C:\\data\\TomatoVirus.csv",
    stringsAsFactors = TRUE)
> head(tomato)
  Infest Control Plants Virus8
1      1       C    100     21
2      2       C    100     10
3      1       B    100     19
```

[21]Data courtesy of Drs. James Nechols and David Margolies, Department of Entomology, Kansas State University. Additional information regarding the virus is available at `https://youtu.be/9DiL-UQ6-Uw` and from the textbook's website.

```
4          1          N      100      40
5          2          C      100      30
6          2          B      100      30
```

From row 1 of the data set, we see that 21 out of 100 originally clean plants in a greenhouse showed symptoms after 8 weeks, and these plants had infestation method #1 applied while trying to control the thrips with a chemical application.

We include the `stringsAsFactors = TRUE` argument[22] in `read.csv()` so that R automatically recognizes any variable with letter values as a factor type.

```
> class(tomato$Control)
[1] "factor"
> levels(tomato$Control)
[1] "B" "C" "N"
> contrasts(tomato$Control)
  C N
B 0 0
C 1 0
N 0 1
```

The `contrasts()` function shows how R would represent `Control` with indicator variables in a model, similar to what was shown in Table 2.2. Thus, the B level of `Control` is the base level, and there are two indicator variables representing the levels C and N of `Control`.

The `Infest` variable is coded numerically rather than with letters. Because the variable has only two levels, there is no practical difference between using it as numerical or factor. Also, using numerical levels of 1 and 2, rather than 0 and 1, is fine as long as one properly tracks what each level represents. If `Infest` had more than two levels, we would need to redefine its class as `factor` using the `factor()` function.[23] Below is the code to explore the class of `Infest` and to replace its original version with a new factor version.

```
> class(tomato$Infest)
[1] "numeric"
> levels(tomato$Infest)
NULL

> tomato$Infest <- factor(tomato$Infest)
> class(tomato$Infest)
[1] "factor"
> levels(tomato$Infest)
[1] "1" "2"
> contrasts(tomato$Infest)
  2
1 0
```

[22]Starting with R version 4.0.0, the default value for `stringsAsFactors` was changed from `TRUE` to `FALSE`. If this new default was used instead, `Control` would have to be changed to a factor manually. Code showing the process is in the program for this example.

[23]An additional use of `factor()` is to change the level ordering, similar to `relevel()` but with more flexibility, and to change the level names. Examples are provided in the program for this example.

```
2 1
```

The `factor()` function can be used throughout the analysis in a similar manner as above. For example, we could leave `Infest` as numerical and use `factor(Infest)` in the `formula` argument of `glm()` if we want to model it as categorical. Again, using a factor version of a two-level explanatory variable is not needed, but we do it here for demonstration purposes and continue using it throughout our example.

Below is the output from estimating the logistic regression model with both the `Infest` and `Control` factors.

```
> mod.fit <- glm(formula = Virus8/Plants ~ Infest + Control,
   family = binomial(link = logit), data = tomato, weights =
   Plants)
> summary(mod.fit)

Call: glm(formula = Virus8/Plants ~ Infest + Control, family =
   binomial(link = logit), data = tomato, weights = Plants)

Deviance Residuals:
    Min       1Q   Median       3Q      Max
  -4.288   -2.425   -1.467    1.828    8.379

Coefficients:
            Estimate Std. Error z value Pr(>|z|)
(Intercept)  -0.6652     0.1018  -6.533 6.45e-11 ***
Infest2       0.2196     0.1091   2.013   0.0441 *
ControlC     -0.7933     0.1319  -6.014 1.81e-09 ***
ControlN      0.5152     0.1313   3.923 8.74e-05 ***
---
Signif. codes:  0 '***' 0.001 '**' 0.01 '*' 0.05 '.' 0.1 ' ' 1

(Dispersion parameter for binomial family taken to be 1)

    Null deviance: 278.69  on 15  degrees of freedom
Residual deviance: 183.27  on 12  degrees of freedom
AIC: 266.77

Number of Fisher Scoring iterations: 4
```

Because the response variable is given in a binomial form, we use the `weights` argument along with the success/trials formulation in the `formula` argument (see Section 2.2.1). The estimated logistic regression model is

$$\text{logit}(\hat{\pi}) = -0.6652 + 0.2196\texttt{Infest2} - 0.7933\texttt{ControlC} + 0.5152\texttt{ControlN},$$

where we use R's notation of `Infest2` for the indicator variable representing level 2 of `Infest` and `ControlC` and `ControlN` for the indicator variables representing levels `"C"` and `"N"` of `Control`, respectively.

Probability estimates for plants showing symptoms are computed using the `predict()` function. The process is the same as shown in Section 2.2.4, but now with the values for the categorical explanatory variables enclosed within quotes. Below is an example using the first infestation method and the three control methods.

```
> predict.data <- data.frame(Infest = c("1", "1", "1"), Control
   = c("B", "C", "N"))
> predict(object = mod.fit, newdata = predict.data, type =
   "response")
   1    2    3
0.34 0.19 0.46
```

The estimated probability of plants showing symptoms is largest when no control is used. This could be concluded as well based on the estimated regression parameters for `ControlC` and `ControlN`, because no control is the only one with a positive value. Also, based on the positive estimated regression parameter for `Infest2`, the probability of plants showing symptoms is estimated to be larger in greenhouses where infestation method #2 is used. This is somewhat expected because the thrips are already virus carriers with this method. This model and interpretation assume that there is no interaction between the infestation and control methods. We next examine how to include these interactions and evaluate their importance.

Interactions

Suppose that we have a numerical variable, say v, along with a four-level categorical variable. To include v with the model in Equation 2.21, we add the term $\beta_4 v$. However, the resulting model assumes that the effect of v on logit(π) (i.e., the log-odds of success) is the same for all levels of the categorical variable. If we want to allow the effects of v to change across these levels, we need to include the interaction. To represent an interaction between a categorical explanatory variable and a numerical explanatory variable in general, we need to add to a model all pairwise products between terms representing the two variables. For example, the interaction between the four-level categorical explanatory variable and v is created by adding three new terms to the model: $\beta_5 v x_1$, $\beta_6 v x_2$, and $\beta_7 v x_3$.

To create an interaction between two categorical explanatory variables in a model, we add the pairwise products between the two sets of indicator variables. For example, suppose that there is a three-level categorical variable X and a three-level categorical variable Z, represented by appropriate indicator variables x_1 and x_2 for X and z_1 and z_2 for Z. The logistic regression model with the interaction between X and Z is

$$\begin{aligned}\text{logit}(\pi) = \beta_0 + \beta_1 x_1 + \beta_2 x_2 + \beta_3 z_1 + \beta_4 z_2 + \beta_5 x_1 z_1 + \beta_6 x_1 z_2 \\ + \beta_7 x_2 z_1 + \beta_8 x_2 z_2.\end{aligned} \quad (2.22)$$

To test the interaction, the null hypothesis is that all of the regression parameters for interactions terms are 0 ($H_0: \beta_5 = \beta_6 = \beta_7 = \beta_8 = 0$ in our example) and the alternative is that at least one of these parameters is not 0. Examples are given next.

Example: Control of the Tomato Spotted Wilt Virus (TomatoVirus.R, TomatoVirus.csv)

It is important to investigate whether the three methods of thrip control have similar effects for both types of infestation. Thus, we want to know whether there is an interaction between `Control` and `Infest`. The code below shows how to include this interaction and evaluate its importance.

```
> mod.fit.inter <- glm(formula = Virus8/Plants ~ Infest +
   Control + Infest:Control, family = binomial(link = logit),
   data = tomato, weights = Plants)
> summary(mod.fit.inter)

Call: glm(formula = Virus8/Plants ~ Infest + Control +
   Infest:Control, family = binomial(link = logit), data =
   tomato, weights = Plants)

Deviance Residuals:
    Min       1Q   Median       3Q      Max
 -3.466   -2.705   -1.267    2.811    6.791

Coefficients:
                 Estimate Std. Error z value Pr(>|z|)
(Intercept)       -1.0460     0.1316  -7.947 1.92e-15 ***
Infest2            0.9258     0.1752   5.283 1.27e-07 ***
ControlC          -0.1623     0.1901  -0.854    0.393
ControlN           1.1260     0.1933   5.826 5.68e-09 ***
Infest2:ControlC  -1.2114     0.2679  -4.521 6.15e-06 ***
Infest2:ControlN  -1.1662     0.2662  -4.381 1.18e-05 ***
---
Signif. codes:  0 '***' 0.001 '**' 0.01 '*' 0.05 '.' 0.1 ' ' 1

(Dispersion parameter for binomial family taken to be 1)

    Null deviance: 278.69  on 15  degrees of freedom
Residual deviance: 155.05  on 10  degrees of freedom
AIC: 242.55

Number of Fisher Scoring iterations: 4

> library(package = car)
> Anova(mod.fit.inter)
Analysis of Deviance Table (Type II tests)

Response: Virus8/Plants
               LR Chisq Df Pr(>Chisq)
Infest            4.060  1    0.0439 *
Control          91.584  2  < 2.2e-16 ***
Infest:Control   28.224  2  7.434e-07 ***
---
Signif. codes:  0 '***' 0.001 '**' 0.01 '*' 0.05 '.' 0.1 ' ' 1
```

The estimated logistic regression model is

$$\begin{aligned}\text{logit}(\hat{\pi}) = & -1.0460 + 0.9258\texttt{Infest2} - 0.1623\texttt{ControlC} + 1.1260\texttt{ControlN} \\ & -1.2114\texttt{Infest2}\times\texttt{ControlC} - 1.1662\texttt{Infest2} \times \texttt{ControlN},\end{aligned}$$

where the coefficient names used by R again correspond to the indicators for the levels of the categorical variables. An LRT to evaluate the importance of the interaction term tests the regression parameters corresponding to Infest2 $\times$ ControlC and Infest2 $\times$ ControlN, say β_4 and β_5. The hypotheses are $H_0 : \beta_4 = \beta_5 = 0$ vs. $H_a : \beta_4 \neq$

0 and/or $\beta_5 \neq 0$. Equivalently, we could write the hypotheses in terms of model comparisons:

$$\begin{aligned} H_0 &: \text{logit}(\pi) = \beta_0 + \beta_1\texttt{Infest2} + \beta_2\texttt{ControlC} + \beta_3\texttt{ControlN} \\ H_a &: \text{logit}(\pi) = \beta_0 + \beta_1\texttt{Infest2} + \beta_2\texttt{ControlC} + \beta_3\texttt{ControlN} \\ &\quad +\beta_4\texttt{Infest2} \times \texttt{ControlC} + \beta_5\texttt{Infest2} \times \texttt{ControlN}. \end{aligned}$$

The test statistic is $-2\log(\Lambda) = 28.224$, and the p-value is 7.4×10^{-7} using a χ^2_2 approximation. Thus, there is strong evidence of an interaction between the infestation and control methods.

Odds ratios

Consider again the model given in Equation 2.21. The odds of success at level B are $\exp(\beta_0 + \beta_1)$ because $x_1 = 1, x_2 = 0$, and $x_3 = 0$, and the odds of success at level A are $\exp(\beta_0)$ because $x_1 = 0, x_2 = 0$, and $x_3 = 0$. The resulting odds ratio comparing level B to A is

$$\frac{Odds_{x_1=1,x_2=0,x_3=0}}{Odds_{x_1=0,x_2=0,x_3=0}} = \frac{\exp(\beta_0 + \beta_1)}{\exp(\beta_0)} = \exp(\beta_1).$$

In a similar manner, one can show that the odds ratio comparing C to A is $\exp(\beta_2)$ and the odds ratio comparing D to A is $\exp(\beta_3)$. We can use the same technique to compare non-base levels. For example, the odds ratio comparing level B to level C is

$$\frac{Odds_{x_1=1,x_2=0,x_3=0}}{Odds_{x_1=0,x_2=1,x_3=0}} = \frac{\exp(\beta_0 + \beta_1)}{\exp(\beta_0 + \beta_2)} = \exp(\beta_1 - \beta_2).$$

Estimates and confidence intervals are formed in the same way as discussed in Sections 2.2.3 and 2.2.5. For example, the estimated odds ratio comparing level B to C is $\exp(\hat{\beta}_1 - \hat{\beta}_2)$, and the $(1-\alpha)100\%$ Wald confidence interval for $\exp(\beta_1 - \beta_2)$ is

$$\exp\left(\hat{\beta}_1 - \hat{\beta}_2 \pm Z_{1-\alpha/2}\sqrt{\widehat{Var}(\hat{\beta}_1 - \hat{\beta}_2)}\right)$$

with

$$\widehat{Var}(\hat{\beta}_1 - \hat{\beta}_2) = \widehat{Var}(\hat{\beta}_1) + \widehat{Var}(\hat{\beta}_2) - 2\widehat{Cov}(\hat{\beta}_1, \hat{\beta}_2).$$

When there are two or more categorical explanatory variables in a model, odds ratios comparing two levels of a one variable are again formed by taking the ratio of the two odds. When there is an interaction between two variables, odds ratios depends on the level of the other categorical variable. Consider again Equation 2.22, which models the log-odds of success against two three-level categorical explanatory variables, each with levels A, B, and C, and their interaction. A typical analysis focusing on X is to compute odds ratios to compare B to A, C to A, and B to C. Because of the interaction, each of these odds ratios needs to be computed separately for each level of Z. Thus, there are nine different odds ratios among levels of X. These odds ratios are displayed in Table 2.3. For example, the comparison of levels B and C of X at level B of Z is found by

$$\frac{Odds_{x_1=1,x_2=0,z_1=1,z_2=0}}{Odds_{x_1=0,x_2=1,z_1=1,z_2=0}} = \frac{\exp(\beta_0 + \beta_1 + \beta_3 + \beta_5)}{\exp(\beta_0 + \beta_2 + \beta_3 + \beta_7)} = \exp(\beta_1 - \beta_2 + \beta_5 - \beta_7).$$

Odds ratios for comparing levels of Z at each level of X can be found in the same way if needed.

Table 2.3: Odds ratios comparing levels of X separately at each level of Z.

Z	X	OR
A	B to A	$\exp(\beta_1)$
	C to A	$\exp(\beta_2)$
	B to C	$\exp(\beta_1 - \beta_2)$
B	B to A	$\exp(\beta_1 + \beta_5)$
	C to A	$\exp(\beta_2 + \beta_7)$
	B to C	$\exp(\beta_1 - \beta_2 + \beta_5 - \beta_7)$
C	B to A	$\exp(\beta_1 + \beta_6)$
	C to A	$\exp(\beta_2 + \beta_8)$
	B to C	$\exp(\beta_1 - \beta_2 + \beta_6 - \beta_8)$

If there is no interaction, then the parameters that define the interaction between X and Z—$\beta_5, \beta_6, \beta_7$, and β_8—are all 0. It is then apparent from Table 2.3 for this situation that the odds ratios comparing levels of X are the same at all levels of Z, and therefore need to be computed only once.

Before proceeding to an example, we want to emphasize that developing these odds ratio expressions is a two-step process. First, identify the odds of interest and write them in terms of the model. Second, take the ratio of two odds and simplify the expression. When our own students have difficulty with odds ratios, it is usually because they attempt to immediately guess which parameters comprise an odds ratio without first finding the odds.

Example: Control of the Tomato Spotted Wilt Virus (TomatoVirus.R, TomatoVirus.csv)

Because the interaction between the infestation and control methods is significant, we would normally focus only on the model that includes the interaction. However, it is instructive first to see how calculations are performed using the model without the interaction. We later perform the more complicated investigations with the model including the interaction.

To understand the effect of `Control` on plants displaying symptoms of infection, we need the odds for

- Biological control: $\exp(\beta_0 + \beta_1 \texttt{Infest2})$
- Chemical control: $\exp(\beta_0 + \beta_1 \texttt{Infest2} + \beta_2)$
- No control: $\exp(\beta_0 + \beta_1 \texttt{Infest2} + \beta_3)$

Odds ratios are found by taking the ratio of each pairwise combination. In all cases, the $\beta_0 + \beta_1 \texttt{Infest2}$ portion is canceled out in the ratios, resulting in the odds ratios for

- Chemical control vs. biological control: $\exp(\beta_2)$
- No control vs. biological control: $\exp(\beta_3)$
- No control vs. chemical control: $\exp(\beta_3 - \beta_2)$

The estimated odds ratios are given below.

```
> # Compare control levels C and N each to B
> exp(mod.fit$coefficients[3:4])
ControlC ControlN
```

```
0.452342 1.674025

> # Control N vs. Control C
> exp(mod.fit$coefficients[4] - mod.fit$coefficients[3])
ControlN
3.700795
```

For example, the estimated odds ratio comparing no control to a biological control (the base level) is exp(0.5152) = 1.67. Thus, the estimated odds of plants showing symptoms are 1.67 times as large when using no control methods as when using a biological control, where the infestation method is held constant. We invert this odds ratio because we prefer to make a statement about the effects of biological control relative to no control, rather than vice versa.[24] Thus, the estimated odds of plants showing symptoms are 1/1.67 = 0.5973 times as large when using a biological control method as when using no control methods, where the infestation method is held constant. The use of spider mites (biological control) is estimated to reduce the odds that a plant shows symptoms by approximately 40%.

Below is the code to find the corresponding confidence intervals for the odds ratios.

```
> # Wald interval
> exp(confint.default(object = mod.fit, parm = c("ControlC",
   "ControlN"), level = 0.95))
              2.5 %    97.5 %
ControlC 0.3492898 0.5857982
ControlN 1.2941188 2.1654579

> # Profile LR interval
> exp(confint(object = mod.fit, parm = c("ControlC",
   "ControlN"), level = 0.95))
Waiting for profiling to be done...
              2.5 %    97.5 %
ControlC 0.3486313 0.5848759
ControlN 1.2945744 2.1666574

> # Wald interval for Control N vs. Control C
> beta.hat <- mod.fit$coefficients[-1]  # Match indices with [i]
> exp(beta.hat[3] - beta.hat[2])
ControlN
3.700795
> cov.mat <- vcov(mod.fit)[2:4,2:4]
> var.N.C <- cov.mat[3,3] + cov.mat[2,2] - 2*cov.mat[3,2]
> CI.betas <- beta.hat[3] - beta.hat[2] + qnorm(p = c(0.025,
   0.975)) * sqrt(var.N.C)
> exp(CI.betas)
[1] 2.800422 4.890649
```

For example, the 95% Wald confidence interval comparing level "N" to level "B" is 1.29 to 2.17. Thus, with 95% confidence, the odds that a plant shows symptoms are between 1.29 and 2.17 times as large when using no control method as when using a

[24]Alternatively, we could have set the base level to `Control = "N"` before estimating the model, which would lead to these types of comparisons more quickly.

biological control (holding the infestation method constant). Alternatively, we could also say with 95% confidence, the odds that a plant shows symptoms are between 0.46 and 0.77 times as large when using a biological control method as when using no control method (holding the infestation method constant). Thus, the use of spider mites (biological control) reduces the odds that a plant shows symptoms by approximately 23% to 54%.

To compare no control to chemical control, we needed to code $\exp(\hat{\beta}_3 - \hat{\beta}_2 \pm Z_{1-\alpha/2}\widehat{Var}(\hat{\beta}_2 - \hat{\beta}_3)^{1/2})$, where

$$\widehat{Var}(\hat{\beta}_2 - \hat{\beta}_3) = \widehat{Var}(\hat{\beta}_2) + \widehat{Var}(\hat{\beta}_3) - 2\widehat{Cov}(\hat{\beta}_2, \hat{\beta}_3),$$

into R. A simpler way to find the confidence interval is to use the `relevel()` function and change the base level to `"C"`. After re-estimating the model, we can use the `confint.default()` and `confint()` functions to obtain the corresponding confidence interval. This process is demonstrated by code in the program for this example. We also show examples of using the `emmeans` and `mcprofile` packages to find Wald and LR confidence intervals, respectively, without `relevel()`. An important difference from past code for `emmeans()` is that the `specs` argument is only `~ Control` rather than `~ Infest + Control`. This is because odds ratios for the control method do not change as a function of infestation (i.e., there is no interaction).

Next, we consider the estimated logistic regression model that includes the interaction. To understand the effect of `Control` on the response, we need to calculate odds ratios between pairs of `Control` levels, keeping the level of `Infest2` fixed at either 0 (`Infest = 1`) or 1 (`Infest = 2`). The odds are

- Biological control: $\exp(\beta_0 + \beta_1 \texttt{Infest2})$
- Chemical control: $\exp(\beta_0 + \beta_1 \texttt{Infest2} + \beta_2 + \beta_4 \texttt{Infest2})$
- No control: $\exp(\beta_0 + \beta_1 \texttt{Infest2} + \beta_3 + \beta_5 \texttt{Infest2})$

Odds ratios are found again by taking the ratio of each pairwise combination. These odds ratios are

- Chemical control vs. biological control: $\exp(\beta_2 + \beta_4 \texttt{Infest2})$
- No control vs. biological control: $\exp(\beta_3 + \beta_5 \texttt{Infest2})$
- No control vs. chemical control: $\exp(\beta_3 + \beta_5 \texttt{Infest2} - \beta_2 - \beta_4 \texttt{Infest2})$

because the $\beta_0 + \beta_1 \texttt{Infest2}$ portion is canceled out.

To complete finding the odds ratios, we need to include whether infestation method 1 or 2 is used. For example, the odds ratio comparing no control (`"N"`) to biological control (`"B"`) with infestation method 1 is

$$\frac{Odds_{\texttt{ControlC=0,ControlN=1,Infest2=0}}}{Odds_{\texttt{ControlC=0,ControlN=0,Infest2=0}}} = \frac{\exp(\beta_0 + \beta_3)}{\exp(\beta_0)} = \exp(\beta_3),$$

and the odds ratio comparing no control (`"N"`) to biological control (`"B"`) with infestation method 2 is

$$\frac{Odds_{\texttt{ControlC=0,ControlN=1,Infest2=1}}}{Odds_{\texttt{ControlC=0,ControlN=0,Infest2=1}}} = \frac{\exp(\beta_0 + \beta_1 + \beta_3 + \beta_5)}{\exp(\beta_0 + \beta_1)} = \exp(\beta_3 + \beta_5).$$

Other odds ratios are calculated in a similar manner. Below are all the estimated odds ratios for `Control` holding `Infest2` constant.

```
> beta.hat <- mod.fit.inter$coefficients[-1]
> C.B.Infest2.0 <- exp(beta.hat[2])
> C.B.Infest2.1 <- exp(beta.hat[2] + beta.hat[4])
> N.B.Infest2.0 <- exp(beta.hat[3])
> N.B.Infest2.1 <- exp(beta.hat[3] + beta.hat[5])
> N.C.Infest2.0 <- exp(beta.hat[3] - beta.hat[2])
> N.C.Infest2.1 <- exp(beta.hat[3] + beta.hat[5] - beta.hat[2]
   - beta.hat[4])

> comp <- c("C vs. B, Infest2 = 0", "C vs. B, Infest2 = 1",
            "N vs. B, Infest2 = 0", "N vs. B, Infest2 = 1",
            "N vs. C, Infest2 = 0", "N vs. C, Infest2 = 1")
> data.frame(Comparison = comp, OR.hat = round(c(C.B.Infest2.0,
   C.B.Infest2.1, N.B.Infest2.0, N.B.Infest2.1, N.C.Infest2.0,
   N.C.Infest2.1), 2))
            Comparison OR.hat
1 C vs. B, Infest2 = 0   0.85
2 C vs. B, Infest2 = 1   0.25
3 N vs. B, Infest2 = 0   3.08
4 N vs. B, Infest2 = 1   0.96
5 N vs. C, Infest2 = 0   3.63
6 N vs. C, Infest2 = 1   3.79
```

For example, the estimated odds ratio comparing level "N" to level "B" with Infest2 = 1 is $\exp(1.1260 - 1.1662) = 0.96$. Thus, the estimated odds of plants showing symptoms are 0.96 times as large with no control as with a biological control when infected thrips are released into the greenhouse. We can also see why the interaction between Infest and Control was significant. The "N" vs. "B" and "C" vs. "B" odds ratios each differ by a large amount between the two levels of Infest.

We can use similar methods as earlier to find Wald intervals. Because there are many intervals to calculate involving many variances and covariances, we focus on the emmeans package.

```
> library(package = emmeans)
> calc.est.inter <- emmeans(object = mod.fit.inter, specs = ~
   Infest + Control, type = "response")
> confint(object = contrast(object = calc.est.inter, method =
   "revpairwise", simple = list("Control")), adjust = "none",
   level = 0.95)
Infest = 1:
 contrast odds.ratio     SE  df asymp.LCL asymp.UCL
 C / B         0.850 0.1616 Inf     0.586     1.234
 N / B         3.083 0.5959 Inf     2.111     4.503
 N / C         3.627 0.7149 Inf     2.465     5.337

Infest = 2:
 contrast odds.ratio     SE  df asymp.LCL asymp.UCL
 C / B         0.253 0.0478 Inf     0.175     0.367
 N / B         0.961 0.1758 Inf     0.671     1.375
 N / C         3.795 0.7813 Inf     2.535     5.681
```

Our interest is comparing all levels of Control for all levels of Infest, so we omit the at argument for emmeans() and specify Control in the argument value for simple.

Also, our interest is in the confidence intervals, not the hypothesis tests, for the odds ratios, so we include the `constrast()` code as the `object` argument for `confint()`. The `method = "revpairwise"` argument value within `contrast()` specifies pairwise comparisons for the odds but the `Control` levels are in the reverse order of what `method = "pairwise"` provides. Alternatively, we could have used the `at` argument in `emmeans()` to specify an ordering of the levels there. This can be helpful for data analyses in general because particular orderings may be preferred for the comparisons of interest.

To control the familywise confidence level, the `adjust` argument for `confint()` can be set to an appropriate name. This would provide control for each level of `Infest`; however, one would generally like to control across all levels. The additional argument value of `combine = TRUE` in `contrast()` would provide the desired outcome.

The `mcprofile` package allows for the same comparisons by coding the matrix of contrast coefficients directly.

```
> library(package = mcprofile)
> K.names <- list(comp, c("beta0", "beta1", "beta2", "beta3",
    "beta4", "beta5"))
> K <- matrix(data = c(0, 0,  1,  0,  0,  0,
                       0, 0,  1,  0,  1,  0,
                       0, 0,  0,  1,  0,  0,
                       0, 0,  0,  1,  0,  1,
                       0, 0, -1,  1,  0,  0,
                       0, 0, -1,  1, -1,  1),
          nrow = 6, ncol = 6, byrow = TRUE, dimnames = K.names)
> K
                      beta0 beta1 beta2 beta3 beta4 beta5
C vs. B, Infest2 = 0      0     0     1     0     0     0
C vs. B, Infest2 = 1      0     0     1     0     1     0
N vs. B, Infest2 = 0      0     0     0     1     0     0
N vs. B, Infest2 = 1      0     0     0     1     0     1
N vs. C, Infest2 = 0      0     0    -1     1     0     0
N vs. C, Infest2 = 1      0     0    -1     1    -1     1

> linear.combo <- mcprofile(object = mod.fit.inter, CM = K)
> ci.log.OR <- confint(object = linear.combo, level = 0.95,
    adjust = "none")
> exp(ci.log.OR)

   mcprofile - Confidence Intervals

level:           0.95
adjustment:      none

                     Estimate lower upper
C vs. B, Infest2 = 0    0.850 0.585 1.234
C vs. B, Infest2 = 1    0.253 0.174 0.365
N vs. B, Infest2 = 0    3.083 2.116 4.517
N vs. B, Infest2 = 1    0.961 0.671 1.375
N vs. C, Infest2 = 0    3.627 2.472 5.357
N vs. C, Infest2 = 1    3.795 2.544 5.708
```

The columns of K are ordered corresponding to the six regression parameters estimated by the model. Each row identifies the linear combination of parameters needed for a particular odds ratio. For example, row 4 corresponds to estimating the odds ratio comparing level "N" to level "B" with Infest2 $= 1$ (equivalently, Infest = 2). This odds ratio is $\exp(\beta_3 + \beta_5)$, where β_3 and β_5 are the fourth and sixth parameters in the model, respectively.

The 95% profile LR interval comparing level "N" to level "B" with Infest2 $= 1$ is (0.67, 1.38). Thus, with 95% confidence, the odds of plants showing symptoms are between 0.67 and 1.38 times as large with no control as with a biological control when infected thrips are released in the greenhouse. Because 1 is within the interval, there is insufficient evidence to conclude a biological control is effective in this setting. Similar interpretations can be constructed for the other odds ratios. Notice the profile LR interval for comparing level "N" to level "B" with Infest2 $= 0$ is (2.12, 4.52). Because the interval is above 1, there is sufficient evidence to conclude the biological control reduces the odds of plants showing symptoms when interspersing infected plants with uninfected thrips.

Because there are many different odds ratios, we provide a set of intervals that maintain the familywise confidence level at 95% using the single-step adjustment.

```
> ci.log.OR.ss <- confint(object = linear.combo, level = 0.95,
   adjust = "single-step")
> exp(ci.log.OR.ss)

  mcprofile - Confidence Intervals

level:               0.95
adjustment:          single-step

                         Estimate lower  upper
C vs. B, Infest2 = 0      0.850 0.517  1.393
C vs. B, Infest2 = 1      0.253 0.153  0.410
N vs. B, Infest2 = 0      3.083 1.874  5.124
N vs. B, Infest2 = 1      0.961 0.596  1.546
N vs. C, Infest2 = 0      3.627 2.184  6.095
N vs. C, Infest2 = 1      3.795 2.236  6.536
```

The intervals are wider than their unadjusted counterparts given in the previous output, reflecting their larger individual confidence levels. The advantage of these adjusted intervals is that we can claim that we are 95% confident that the process has covered *all* of the odds ratios, rather than merely having 95% confidence separately for each one.

The code to compute the Bonferroni adjusted intervals is within our program. These intervals are similar, although the single-step intervals are narrower. Code to compute the Wald intervals and to control the familywise confidence level is also contained within the program.

2.2.7 Convergence of parameter estimation

The glm() function uses computational procedures that iterate until convergence is attained or until the specified maximum number of iterations is reached. The criterion used to determine convergence is the change in successive values of the residual deviance, rather than the change in successive estimates of the regression parameters. This creates a single, compact criterion that equitably balances the convergence of all parameter estimates simultaneously. If we let $G^{(k)}$ denote the residual deviance at iteration k, then convergence occurs when

$$\frac{\left|G^{(k)} - G^{(k-1)}\right|}{0.1 + \left|G^{(k)}\right|} < \epsilon,$$

where ϵ is some specified small number greater than 0. The numerator $\left|G^{(k)} - G^{(k-1)}\right|$ gives an overall measure of how much $\hat{\pi}_1, \ldots, \hat{\pi}_n$ change, which in the vast majority of cases corresponds to how much $\hat{\beta}_0, \ldots, \hat{\beta}_p$ change from the previous iteration to the current iteration (an upcoming example shows when this is not true). The denominator $0.1 + G^{(k)}$ converts this into an approximately proportional change.

The `glm()` function provides a few ways to control how convergence is determined. First, its `epsilon` argument allows the user to state the value of ϵ. The default value is `epsilon = 10^(-8)`. Second, the `maxit` argument states the maximum number of iterations allowed for the numerical procedure, where the default is `maxit = 25`. The `trace = TRUE` argument value can be used to see the actual $G^{(k)}$ values for each iteration. The default is `trace = FALSE`.

Example: Placekicking (Placekick.R, Placekick.csv)

Although the default values work fine for the logistic regression model using distance as the only explanatory variable, we explore the changes that occur when we alter the argument values for `epsilon`, `maxit`, and `trace`.

```
> mod.fit <- glm(formula = good ~ distance, family =
    binomial(link = logit), data = placekick, trace = TRUE,
    epsilon = 0.0001, maxit = 50)
Deviance = 836.7715 Iterations - 1
Deviance = 781.1072 Iterations - 2
Deviance = 775.8357 Iterations - 3
Deviance = 775.7451 Iterations - 4
Deviance = 775.745 Iterations - 5

> mod.fit$control
$epsilon
[1] 1e-04

$maxit
[1] 50

$trace
[1] TRUE

> mod.fit$converged
[1] TRUE
```

The `trace = TRUE` argument value causes the residual deviances to be printed along with the iteration number. The convergence criterion value for iteration $k = 5$ is

$$\frac{\left|G^{(k)} - G^{(k-1)}\right|}{0.1 + \left|G^{(k)}\right|} = \frac{|775.745 - 775.7451|}{0.1 + |775.745|} = 1.3 \times 10^{-7},$$

which is less than the stated $\epsilon = 0.0001$, so the iterative numerical procedure stopped. For iteration $k = 4$, the convergence criterion value is 0.00012, which is greater than 0.0001, so this is why the procedure continued. The `control` and `converged` components of the `mod.fit` object are also printed in the output showing the convergence control values and that convergence was obtained.

We include another example in our program with `maxit = 3`. Because convergence is not achieved in so few iterations, R prints `"Warning message: glm.fit: algorithm did not converge"` after iteration $k = 3$. While convergence is not achieved, R still allows access to the parameter estimates obtained from the last iteration. As expected, these parameter estimates are different than those when convergence was achieved, and they should not be used.

When convergence does not occur, the first possible solution is to try a larger number of iterations. For example, if the default 25 iterations are not enough, try 50 iterations. If this does not work, there may be some fundamental problem with the data or the model that is interfering with the iterative numerical procedures. The most common problem occurs when one or more explanatory variables perfectly separate the data between $y_i = 0$ and 1 values; this is often referred to as *complete separation.* In addition to a convergence warning message, `glm()` may also report `glm.fit: fitted probabilities numerically 0 or 1 occurred`; however, this statement alone is not always indicative of model-fitting problems. We illustrate the complete separation problem in the next example.

Example: Complete separation (Non-convergence.R)

We begin by creating a simple data set with one explanatory variable x_1 such that $y = 0$ when $x_1 \leq 5$, and $y = 1$ when $x_1 \geq 6$. Thus, x_1 perfectly separates the two possible values of y. Below is the R code to create these data and to attempt fitting a model.

```
> set1 <- data.frame(x1 = c(1,2,3,4,5,6,7,8,9,10), y =
    c(0,0,0,0,0,1,1,1,1,1))
> set1
  x1 y
1  1 0
2  2 0

<OUTPUT EDITED>

10 10 1

> mod.fit1 <- glm(formula = y ~ x1, data = set1, family =
    binomial(link = logit), trace = TRUE)
Deviance = 4.270292 Iterations - 1
Deviance = 2.574098 Iterations - 2

<OUTPUT EDITED>

Deviance = 7.864775e-10 Iterations - 25

> summary(mod.fit1)
```

```
Call: glm(formula = y ~ x1, family = binomial(link = logit),
    data = set1, trace = TRUE)

Coefficients:
            Estimate Std. Error z value Pr(>|z|)
(Intercept)   -245.8   337834.2  -0.001    0.999
x1              44.7    61172.1   0.001    0.999

(Dispersion parameter for binomial family taken to be 1)

    Null deviance: 1.3863e+01  on 9  degrees of freedom
Residual deviance: 7.8648e-10  on 8  degrees of freedom
AIC: 4

Number of Fisher Scoring iterations: 25

> mod.fit1$coefficients
(Intercept)          x1
 -245.84732    44.69951

Warning messages:
1: glm.fit: algorithm did not converge
2: glm.fit: fitted probabilities numerically 0 or 1 occurred
```

R indicates both that convergence did not occur and that the model is trying to create some estimates of π that are practically equal to 0 or 1. The left plot of Figure 2.7 shows the estimated model at the last iteration (code contained within our program). Because there is a separation between the $y = 0$ and 1 values, the estimated probability of success is approximately equal to 0 up to $x_1 = 5$ and 1 for $x_1 = 6$ and beyond. The slope of the line between $x_1 = 5$ and 6 will continue to get larger as the iterations are allowed to continue.

If the `maxit` argument is set to a higher value than the default of 25, `glm()` will indicate convergence is attained at iteration 26 with $\hat{\beta}_0 = -256.85$ and $\hat{\beta}_1 = 46.70$. Notice that these values are rather different from those at iteration 25, whereas we would expect parameter estimates to be very similar when the algorithm is close to convergence. If we additionally set `epsilon` to something much smaller, like `1e-15`, the algorithm converges at iteration 34 with $\hat{\beta}_0 = -333.02$ and $\hat{\beta}_1 = 60.55$. Estimates would continue to change for even more iterations, where R's precision leads to an eventual maximum number of iterations that can be implemented.

The `glm()` convergence criterion does not work completely here, because the best possible model fit would be a perfectly vertical line somewhere between $x_1 = 5$ and 6. Thus, the estimates for β_1 would keep increasing without limit if iterations continued indefinitely. The residual deviance for this perfect fit would be 0, which is a problem because it implies that decreases in the residual deviance must eventually become smaller than any criterion, suggesting algorithm convergence.

There are two additional warning signs in this example that one can use to identify when there are possible convergence problems. First, the residual deviance from the supposed convergence at iteration 26 is 2.9×10^{-10}, which is quite close to zero. This indicates that the model seems to be trying to fit the data perfectly, suggesting possible complete separation. Second, the standard errors for the regression parameter estimates are very large relative to these estimates, Wald test statistics are practically 0, and corresponding p-values are practically 1. This anomalous behavior is an

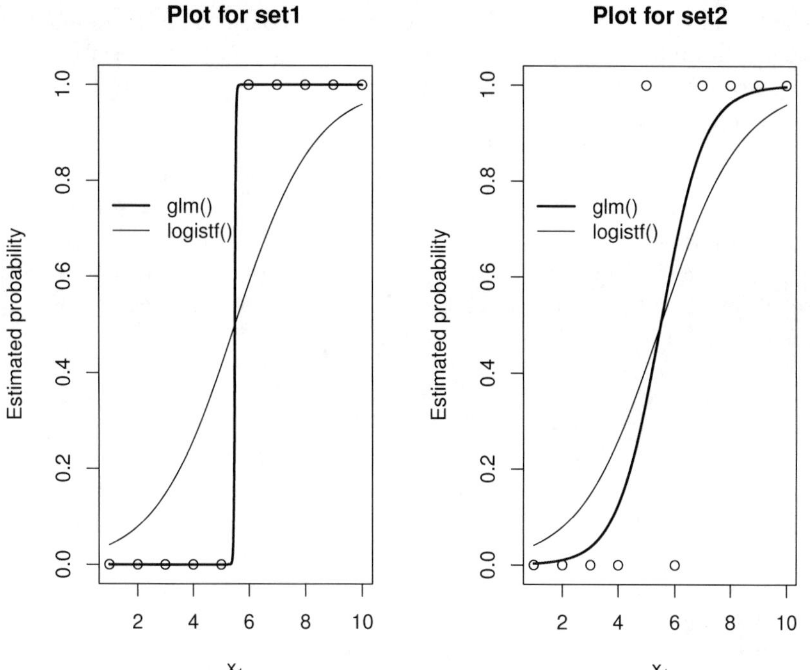

Figure 2.7: Plots showing when complete separation occurs (left) and when complete separation does not occur (right).

indication that there are problems with the reported estimates and convergence has not occurred.

To show that these phenomena are the results of complete separation, consider what happens if we reverse the y values at $x_1 = 5$ and 6 (see our program for code; data contained in `set2`). Convergence occurs in 6 iterations, and Figure 2.7 (right side) shows a plot of the estimated model. The estimated slope is much smaller than before, and the probability estimates are no longer essentially 0 or 1. Furthermore, the residual deviance is 5.0 rather than approximately 0, and standard errors, Wald statistics, and p-values have more typical values.

Heinze and Schemper (2002) outline a number of possible options for what to do when complete separation occurs. The most desirable options are to use either (1) exact logistic regression or (2) a modified likelihood function.[25] The first approach uses the exact distribution of each regression parameter estimator conditional on certain features of the data. We provide details of this procedure and other exact methods in Section 6.2. In the second, because the likelihood function increases without bound during estimation, the function can be modified (sometimes referred to as being "penalized") to potentially prevent this problem from happening. We detail this approach next.

[25]Another potential solution involves Bayesian methods. We explore this approach in Exercise 7 under "Bayesian methods for categorical data" in Chapter 6.

Firth (1993) proposes a modification to the likelihood function that lessens the bias of parameter estimators in fixed sample sizes.[26] Complete separation leads to an extreme case of bias because of infinite parameter estimates, so Firth's proposal is useful here. To help explain this procedure, we write the first derivative with respect to β_r of the log-likelihood function (Equation 2.5) as

$$\sum_{i=1}^{n}(y_i - \pi_i)x_{ir}$$

for $r = 0, \ldots, p$ and $x_{i0} = 1$. This equation is often referred to as the score function (see Appendix B). As discussed in Section 2.2.1, we normally would set the score function equal to 0 and solve for the regression parameters to produce their estimates. Firth suggests adding a penalty to the score function to produce

$$\sum_{i=1}^{n}(y_i - \pi_i + h_i(0.5 - \pi_i))x_{ir},$$

where h_i is the ith diagonal element from the *hat matrix* $\mathbf{H} = \mathbf{V}^{1/2}\mathbf{X}(\mathbf{X}'\mathbf{V}\mathbf{X})^{-1}\mathbf{X}'\mathbf{V}^{1/2}$ (see Section 2.2.1 for matrix definitions).[27] We again set this modified score function to 0 and solve for the regression parameters to produce their estimates. This penalty has the effect of changing y_i, $i = 1, \ldots, n$, from 0 or 1 to values slightly closer to 0.5. Therefore, it prevents any $\hat{\beta}_r$ from straying too far from 0, so that the iterative numerical procedures always converge to finite values (Heinze and Schemper, 2002). Wald and LR inference procedures are used in the same way as with ordinary maximum likelihood estimation. Heinze and Schemper (2002) show that convergence to finite parameter estimates always occurs using this modified score function.

Example: Complete separation (Non-convergence.R)

Two packages that implement Firth's modified score function are `logistf` and `brglm2`. We focus on the former in this example and provide comments about the latter at the end.

Below is our code used to fit the same model as in the example with complete separation but now using the modified score function.

```
> library(package = logistf)
> mod.fit.firth <- logistf(formula = y ~ x1, data = set1, pl =
    TRUE, alpha = 0.05)
> options(width = 60, digits = 4)  # Control output display
> summary(mod.fit.firth)
logistf(formula = y ~ x1, data = set1, alpha = 0.05)

Model fitted by Penalized ML
Coefficients:
                  coef se(coef) lower 0.95 upper 0.95 Chisq
(Intercept) -5.3386   3.3227    -30.7818     -0.8313 6.374
```

[26] Bias measures the difference between the expected value of a statistic and the parameter it is estimating. Ideally, we would like the bias to be 0. Maximum likelihood estimators are approximately unbiased (bias is 0) for large samples, but there may be significant bias present with small samples; see Appendix B for a discussion on bias.

[27] Note that h_i helps to measure the influence of the ith observation on the parameter estimates. We discuss influence and the hat matrix in more detail in Section 5.2.3.

```
x1              0.9706    0.5765         0.1962       5.5642 7.759
                       p method
(Intercept) 0.011579         2
x1          0.005345         2
Method: 1-Wald, 2-Profile penalized log-likelihood, 3-None

Likelihood ratio test=7.759 on 1 df, p=0.005345, n=10
Wald test = 2.834 on 1 df, p = 0.09226

> options(width = 80, digits = 7)  # Default
```

The `logistf()` function estimates the model to be $\text{logit}(\hat{\pi}) = -5.3386 + 0.9706x_1$ without experiencing any convergence problems. The method function associated with `summary()` provides the parameter estimates and other information, including $\widehat{Var}(\hat{\beta}_1)^{1/2} = 0.5765$, a 95% profile likelihood interval (`pl = TRUE`) of $0.20 < \beta_1 < 5.56$, and an LRT for $H_0: \beta_1 = 0$ vs. $H_a: \beta_1 \neq 0$ with a p-value $= 0.0053$.[28] Figure 2.7 plots the model (see our program for code).

In addition to `summary()`, other generic functions, like `anova()` and `predict()`, have method functions that can be used with an object created by `logistf()`. Some arguments within these functions are a little different than we have seen before. Also, the `logistftest()` function performs LRTs in much the same manner as `anova()` using the modified likelihood function. Please see the examples in our corresponding program for how to use these functions. Heinze and Schemper (2002) and Siino et al. (2018) show that these LR-based inferences are preferred over Wald-based inferences.

We provide examples for the `brglm2` package within our program for this example. The code to fit the model takes advantage of the `method` argument within `glm()`, which specifies a function to calculate the estimates of the regression parameters. By default, `glm.fit()` is used by `glm()` to calculate MLEs. For Firth's method, the `brglm2` package provides a replacement function named `brglmFit()`. The resulting estimated model is the same as found by `logistf()`. The main advantage of `brglm2` is that the output looks the same as with `glm()` and generic functions available to `glm()` can be used again. However, the LR statistics use the likelihood function without any modification, so we do not recommend using LR-based inferences with the `brglm2` package. The focus of this package and related papers (e.g., Kosmidis et al. 2020) is on Wald-based inferences. For example, `confint()` provides Wald intervals for the regression parameters.

We conclude this section with some general notes about complete separation and convergence:

1. Complete separation is not necessarily bad if you want to distinguish between the response levels of y. The problem is that the model estimated by maximum likelihood does not provide a good way to interpret the relationship between y and the explanatory variables.

2. It can be difficult to see complete separation graphically if there is more than one explanatory variable, because the separation may arise with respect to some linear

[28]The default formatting given by `summary()` is quite wide and uses an excessive number of digits after a decimal point. The `options()` function allows us to limit the width and digits using the `width` and `digits` arguments, respectively.

combination of the explanatory variables rather than just one. There may even be times when the `glm()` function does not provide a warning. When parameter estimates have very large magnitudes with extremely large estimated standard deviations, this is a sign that complete separation may exist. Another sign is that most or all estimated probabilities of success are essentially 0 or 1. Exercise 24 demonstrates this for a data set.

3. A solution that is often used to address complete separation is to add a pseudo observation to the data in an attempt to force convergence of the parameter estimates. This is somewhat similar to what was done in Section 1.2.5, where we estimated an odds ratio for a contingency table with zero cell counts by adding a small constant to the cell counts. Exercise 34 demonstrates this for a data set.

4. R may print the message of `fitted probabilities numerically 0 or 1 occurred` even when complete separation does not happen. This simply means that at least some of the $\hat{\pi}_i$'s are very close to 0 or 1, but others may not be. One can examine how often 0 and 1 estimates occur to determine if complete separation or quasi-separation (sets of explanatory variable values always lead to a 0 and/or 1 response; see Exercise 26) is the problem. Also, one can decrease ϵ to increase the number of iterations used by `glm()`. The resulting regression parameter estimates can be compared to those obtained with the default value of ϵ. If there is not a significant change in them, one can be confident that the estimates have converged.

2.2.8 Monte Carlo simulation

Through the use of Monte Carlo simulation, properties of the logistic regression model and associated inference procedures can be investigated. For example, in this section we examine whether a normal distribution is a good approximation for the probability distribution of the regression parameter estimates. This will help us to determine whether Wald confidence intervals are appropriate.

Example: Simulating data (SamplingDist.R)

Before we can investigate the distribution of our estimators, we need to be able to simulate responses in a logistic regression setting. For simplicity, we consider the logistic regression model $\text{logit}(\pi) = \beta_0 + \beta_1 x_1$, where we simulate x_1 from a uniform probability distribution that has a minimum and maximum value of 0 and 1, respectively. We choose values of β_0 and β_1 for our "true model" so that $\pi = 0.01$ when $x_1 = 0$ and $\pi = 0.99$ when $x_1 = 1$. Below is our code for finding the values of β_0 and β_1 that satisfy these requirements, and Figure 2.8 displays a plot of the true model.

```
> # x_1=0: logit(pi) = beta_0 + beta_1*0 = beta_0
> pi.x0 <- 0.01
> beta0 <- log(pi.x0/(1-pi.x0))
> beta0
[1] -4.59512

> # x_1=1: logit(pi) = beta_0 + beta_1*1 = beta_0 + beta_1
> pi.x1 <- 0.99
> beta1 <- log(pi.x1/(1-pi.x1)) - beta0
> beta1
```

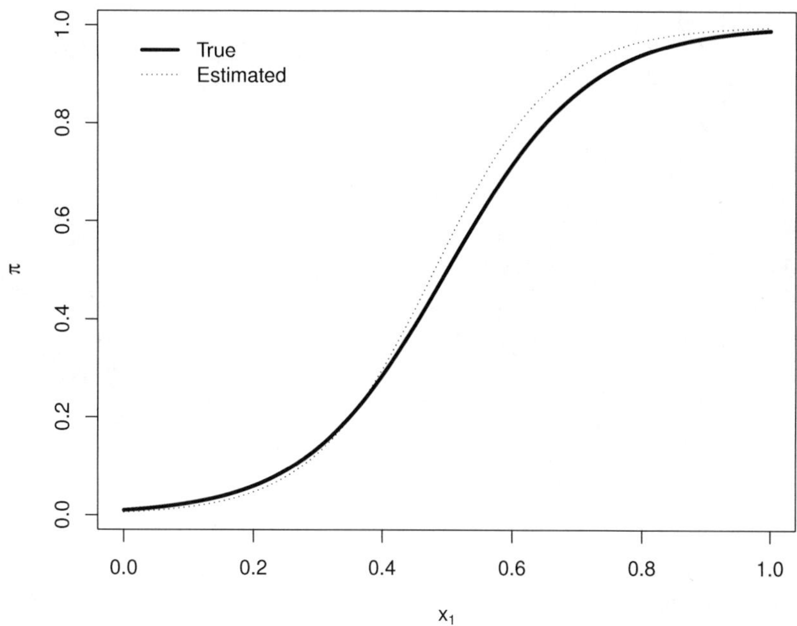

Figure 2.8: True and estimated model.

```
[1] 9.19024

> # Check
> exp(beta0 + beta1*c(0,1)) / (1 + exp(beta0 + beta1*c(0,1)))
[1] 0.01 0.99

> curve(expr = exp(beta0 + beta1*x) / (1 + exp(beta0 +
    beta1*x)), xlim = c(0,1), ylab = expression(pi), n = 1000,
    lwd = 3, xlab = expression(x[1]))
```

To simulate data from the model, we utilize much of the same methodology as in Chapter 1, where the `rbinom()` function was used to simulate observations from a Bernoulli distribution. The main difference now is that we allow the probability of success parameter to vary as a function of an explanatory variable. We first sample $n = 500$ values from the uniform distribution using the `runif()` function to find values for x_1. These x_1 values are inserted into our model to find π. We then use the `rbinom()` function to simulate one binary response y for each x_1. Below is our code.

```
> set.seed(8238)
> x1 <- runif(n = 500, min = 0, max = 1)
> pi <- exp(beta0 + beta1*x1) / (1 + exp(beta0 + beta1*x1))
> set.seed(1829)
> y <- rbinom(n = length(x1), size = 1, prob = pi)
> head(data.frame(y, x1, pi))
```

```
  y        x1         pi
1 1 0.4447197 0.37565323
2 1 0.9523950 0.98459621
3 1 0.5019858 0.50456242
4 0 0.1425656 0.03609259
5 0 0.2042915 0.06194091
6 0 0.2137520 0.06718933

> mod.fit <- glm(formula = y ~ x1, family = binomial(link =
    logit))
> mod.fit$coefficients
(Intercept)          x1
   -5.14340    10.71400
> beta.hat0 <- mod.fit$coefficients[1]
> beta.hat1 <- mod.fit$coefficients[2]
> curve(expr = exp(beta.hat0 + beta.hat1*x) / (1 + exp(beta.hat0
    + beta.hat1*x)), xlim = c(0,1), ylab = expression(pi), add =
    TRUE, col = "red", n = 1000)
> legend(x = 0, y = 1, legend = c("True", "Estimated"), lty =
    c(1,1), col = c("black", "black"), lwd = c(3, 1), bty = "n")
```

For example, the first simulated value for x_1 is 0.4447, which results in $\pi = 0.3757$. The simulated response for this x_1 is a success, $y = 1$. The logistic regression model is fit to the 500 (x_1, y) pairs, resulting in the estimate, $\text{logit}(\hat{\pi}) = -5.14 + 10.71x_1$. As expected, this is somewhat similar to the true model $\text{logit}(\pi) = -4.60 + 9.19x_1$. Note that we did not use a `data` argument in `glm()` because `x` and `y` are defined outside of a data frame. A plot of the estimated model is given in Figure 2.8.

The probability distributions of $\hat{\beta}_0$ and $\hat{\beta}_1$ can be found exactly by randomly generating infinitely many sets of data and then estimating the regression parameters separately on each set. Estimates can be summarized in a histogram or some other form that allows us to examine their distribution. Because this is impossible, we can instead approximate the sampling distributions using a large number of data sets, where "large" typically depends on the computational time required to complete the task and the desired precision with which the sampling distribution is to be estimated.

We next carry out this process for our logistic regression model. Using the data simulation process from the last example, the next example discusses how to repeat it. This allows us to investigate the distribution of our estimators.

Example: Probability distribution of a regression parameter estimator (SamplingDist.R)

We repeat the data set simulation process 10,000 times and fit the model for each simulated data set of size $n = 500$. Generally, a smaller number of replicates for the simulation process would be sufficient; however, a larger value is chosen for this setting to obtain good estimates of some extreme distributional quantiles. We do not include our code here because of space restrictions and similarities with the last example. Note that most of our code is embedded within a function that we created, named `sim.all()`, so that we can repeat the simulation process for different sample sizes.

We would like to point out the structure of the simulation, because the same structure is used frequently for simulations in R. Data are stored in a matrix, where each

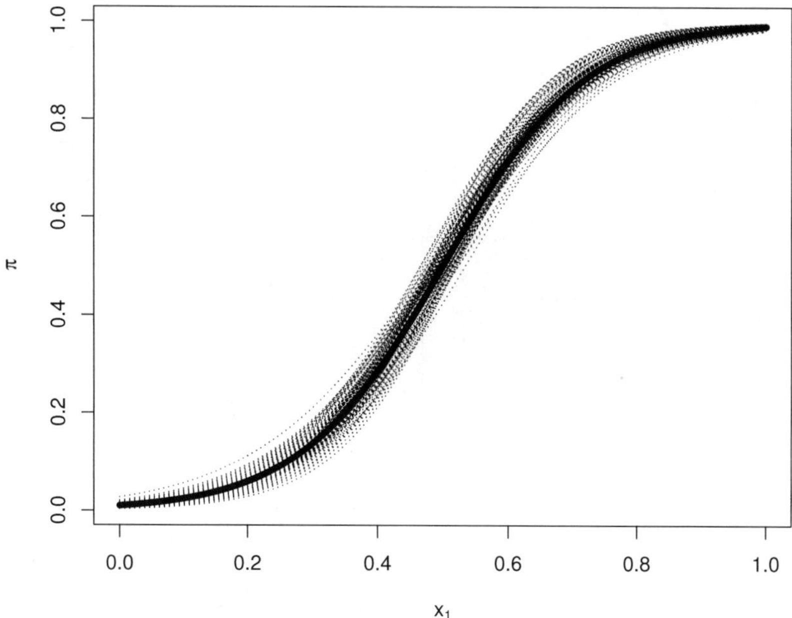

Figure 2.9: True model (black, thick line) and first 100 estimated models.

column of the matrix is one sample data set of n observations. Our data matrix, `y.mat`, stores the binary responses in 500 rows and 10,000 columns. An analysis function is applied to each column of the matrix in one step using the `apply()` function. The syntax for `apply()` is: `apply(X = , MARGIN = , FUN = , ...)`. The `X` argument is used for the data matrix, the `MARGIN` argument value specifies whether the analysis function is to be applied to rows (`=1`) or columns (`=2`) of the matrix, and `FUN` denotes the analysis function to be applied. Alternatively, we could have used a `for()` function to loop over each data set, but `apply()` is more convenient.

Figure 2.9 plots the true model (as in Figure 2.8) with the estimated models from the first 100 data sets. The variability among the estimated models is evident from the plot, but they clearly show the same general trend as the true model from which they were sampled. Note that this picture is highly dependent on the sample size. For example, readers can rerun the simulation with a smaller sample size and see how the variability increases.

The variability among estimated models is further displayed in Figure 2.10. The figure gives a histogram of the 10,000 different $z_0 = (\hat{\beta}_1 - \beta_1)/\widehat{Var}(\hat{\beta}_1)^{1/2}$ values, where we have included a standard normal distribution overlay. Table 2.4 provides a quantile comparison for the simulated z_0 values and for a standard normal distribution. Overall, the standard normal distribution approximation is not too bad. There is some left skewness in the simulated distribution, and this leads to small differences in the extreme quantiles.

Because the distribution of z_0 is reasonably approximated by a normal distribution, we would expect Wald-based inferences for β_1 to work well. Indeed, the Wald interval for β_1 has an estimated true confidence level of 0.9520 when the stated level is 0.95

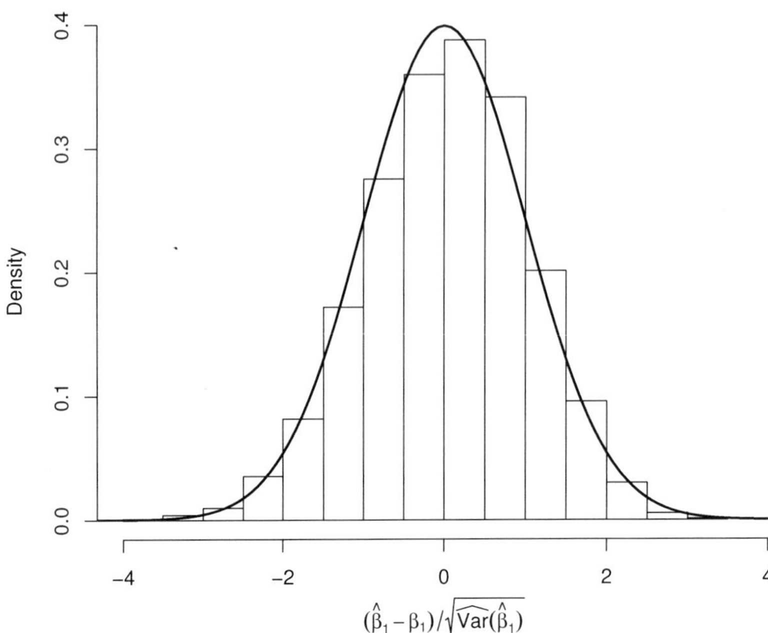

Figure 2.10: Histogram of z_0 with standard normal distribution overlay.

Table 2.4: Quantiles for the standard normal distribution and for the simulated distribution of z_0.

	0.005	0.025	0.05	0.95	0.975	0.995
Standard normal	−2.576	−1.960	−1.645	1.645	1.960	2.576
Simulated z_0	−2.691	−2.004	−1.666	1.615	1.897	2.382

(calculation details are given in our program). If instead there was a poor approximation, this would lead to an interval with a true confidence level quite different from the stated level.

Figure 2.11 displays a histogram of the 10,000 different values of $-2\log(\Lambda) = -2\log[L(\tilde{\beta}_0, \beta_1|y_1, \ldots, y_{500})/L(\hat{\beta}_0, \hat{\beta}_1|y_1, \ldots, y_{500})]$, where we have included a χ^2_1 distribution overlay. Table 2.5 provides a comparison of the quantiles for the simulated $-2\log(\Lambda)$ values and for a χ^2_1 distribution. Similar to the results from z_0, we see that the χ^2_1 distribution approximation is not too bad, but there is more right skewness in the simulated distribution than in a χ^2_1 distribution. This leads to small differences in the extreme quantiles. The estimated true confidence level for the profile LR interval for β_1 is 0.9482 when the stated level is 0.95.

We have written the program so that readers can explore other sample sizes. For example, the normal and χ^2 distributions do not work as well for smaller sizes and work better for larger sample sizes.

There are many other investigations that can be performed for logistic regression models and associated procedures. For example, we can investigate the *consistency* of the parameter

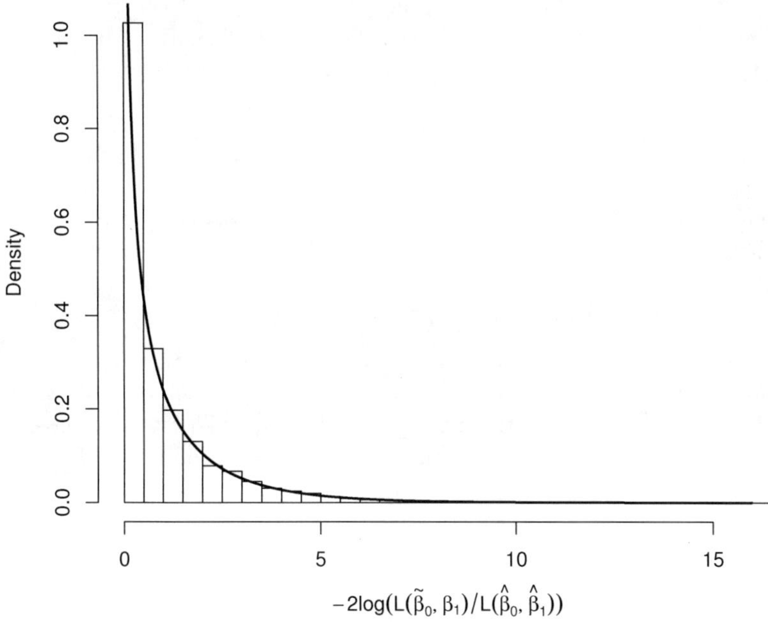

Figure 2.11: Histogram of $-2\log(\Lambda)$ with χ_1^2 distribution overlay.

Table 2.5: Quantiles for the χ_1^2 distribution and for the simulated distribution of $-2\log(\Lambda)$.

	0.90	0.95	0.995
χ_1^2	2.706	3.841	7.879
Simulated $-2\log(\Lambda)$	2.750	3.896	8.408

estimates. Estimators are consistent if they continue to get closer to their corresponding parameters as the sample size grows (Casella and Berger, 2002, Section 10.1). We can also investigate the performance of inference procedures for other parameters. For example, we can examine the Wald and profile LR intervals for π at a particular value of x_1. We encourage readers to perform these investigations on their own.

2.3 Generalized linear models

Logistic regression models fall within a family of models called *generalized linear models* (GLMs). Each generalized linear model has three different parts:

1. RANDOM COMPONENT. This specifies the distribution for Y. For the logistic regression model, Y has a Bernoulli distribution, which is often generalized as a binomial.

2. SYSTEMATIC COMPONENT. This specifies a linear combination of the regression parameters with the explanatory variables. It is the linear predictor $\beta_0+\beta_1x_1+\cdots+\beta_px_p$ that we have seen previously for a logistic regression model.

3. LINK FUNCTION. This specifies how the expected value of the random component $E(Y)$ is linked to the systematic component. For the logistic regression model, we have $\text{logit}(\pi) = \beta_0 + \beta_1 x_1 + \cdots + \beta_p x_p$ where $E(Y) = \pi$ and the logit transformation is the link function.

Note that "linear" in generalized linear models comes from using a linear combination of the regression parameters with the explanatory variables in the systematic component. Alternative systematic components can involve more complex functional forms such as x^β. These would then be called generalized *nonlinear* models.

Link functions used with binary regression models

Other generalized linear models are sometimes used to model binary responses. These binary regression models have the same random and systematic components as the logistic regression model, but their link functions are different than the logit. The most important aspect of the link function in these cases is that its inverse must guarantee $E(Y)$ is between 0 and 1. For example, we saw in Section 2.2 that

$$\pi = \frac{\exp(\beta_0 + \beta_1 x_1 + \cdots + \beta_p x_p)}{1 + \exp(\beta_0 + \beta_1 x_1 + \cdots + \beta_p x_p)}$$

is always between 0 and 1. More generally, this guarantee can be achieved by using the inverse of a *cumulative distribution function* (CDF) as the link function. In fact, the logit transformation is the inverse CDF of a logistic probability distribution, which results in the name for the model.

To review CDFs, suppose X is a continuous random variable with a *probability density function* (PDF) $f(x)$. The CDF $F(x)$ gives the area under the plotted PDF to the left of x. More formally, the CDF of X is $F(x) = P(X \leq x) = \int_{-\infty}^{x} f(u)du$.[29] Because all probabilities are between 0 and 1, $0 \leq F(x) \leq 1$ for $-\infty < x < \infty$. We have used CDFs many times in R already through functions such as `pnorm()` and `pchisq()`. For example, `pnorm(q = 1.96)` can be expressed as $F(1.96) = 0.975$ where $F(\cdot)$ is the CDF of a standard normal distribution. Equivalently, $Z_{0.975} = 1.96$.

Example: Logistic probability distribution (Logistic.R)

A random variable X with a logistic probability distribution has a PDF

$$f(x) = \frac{\sigma^{-1}\exp[-(x-\mu)/\sigma]}{\{1 + \exp[-(x-\mu)/\sigma]\}^2}$$

for $-\infty < x < \infty$ and parameters $-\infty < \mu < \infty$ and $\sigma > 0$. It can be shown that $E(X) = \mu$ and $Var(X) = \sigma^2\pi^2/3$. In comparison to a normal distribution with the same parameter values (μ as the mean and σ^2 as the variance), the distributions are symmetric about μ, but the logistic distribution is more spread out (fatter tails)

[29] $f(u)$ rather than $f(x)$ is used as the integrand to avoid confusion between what is being integrated over and the limits of integration.

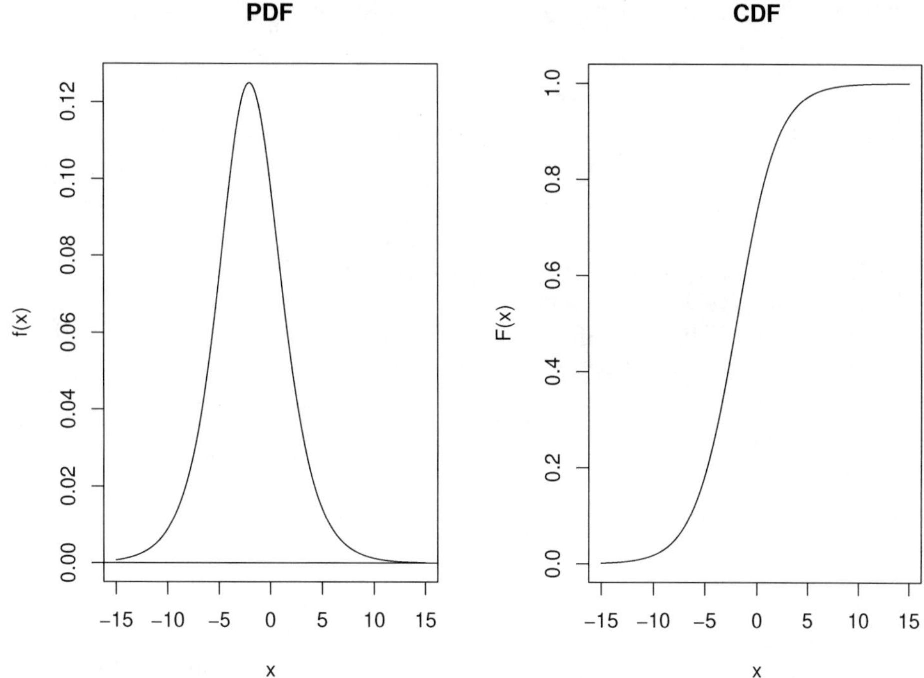

Figure 2.12: PDF and CDF for the logistic probability distribution with $\mu = -2$ and $\sigma = 2$.

because $Var(X) > \sigma^2$. The CDF is

$$\begin{aligned} F(x) &= \int_{-\infty}^{x} \frac{\sigma^{-1}\exp[-(u-\mu)/\sigma]}{\{1+\exp[-(u-\mu)/\sigma]\}^2} du \\ &= \frac{1}{1+\exp[-(x-\mu)/\sigma]} \\ &= \frac{\exp[(x-\mu)/\sigma]}{1+\exp[(x-\mu)/\sigma]}. \end{aligned}$$

Figure 2.12 displays plots of $f(x)$ and $F(x)$ where $\mu = -2$ and $\sigma = 2$, and the code used to produce the plots is below.

```
> mu <- -2
> sigma <- 2

> # Examples for f(-2) and F(-2)
> dlogis(x = -2, location = mu, scale = sigma)
[1] 0.125
> plogis(q = -2, location = mu, scale = sigma)
[1] 0.5

> par(mfrow = c(1,2))
```

```
> curve(expr = dlogis(x = x, location = mu, scale = sigma), ylab
   = "f(x)", xlab = "x", xlim = c(-15, 15), main = "PDF", col =
   "black", n = 1000)
> abline(h = 0)

> curve(expr = plogis(q = x, location = mu, scale = sigma), ylab
   = "F(x)", xlab = "x", xlim = c(-15, 15), main = "CDF", col =
   "black", n = 1000)
```

The `dlogis()` and `plogis()` functions evaluate the PDF and CDF for a logistic distribution, respectively.

The plot of the CDF in Figure 2.12 should look very familiar because it is the same as given in the left plot of Figure 2.1. Notice that

$$F(x) = \frac{\exp[(x-\mu)/\sigma]}{1+\exp[(x-\mu)/\sigma]} \text{ and } \pi = \frac{\exp(\beta_0+\beta_1 x)}{1+\exp(\beta_0+\beta_1 x)}$$

have the same mathematical form when $\beta_0 = -\mu\sigma^{-1}$ and $\beta_1 = \sigma^{-1}$. For example, with $\mu = -2$ and $\sigma = 2$, we obtain $\beta_0 = 1$ and $\beta_1 = 0.5$, which were the same values used for the left plot of Figure 2.1. This does *not* imply that the explanatory variable x has a logistic distribution. Rather, it is the mathematical function from a CDF for a logistic distribution that is used to guarantee $0 < \pi < 1$.

We can equivalently work with $Z = (X - \mu)/\sigma$ for a logistic distribution, similar to what is often done for a normal distribution.[30] This leads to the CDF $F(z) = \exp(z)/[1 + \exp(z)]$. Solving for z results in the inverse of this function: $z = \log\{F(z)/[1 - F(z)]\}$. This has the same mathematical form as the link function used with logistic regression. Thus, the link function is the *inverse CDF* for a logistic distribution.

Because the CDF always produces a value between 0 and 1, other inverse CDFs are used as link functions for binary regression models. While the logit link function is the one most prevalently used for binary regression, two others are common:

1. Probit regression: The inverse CDF of a standard normal distribution is used as the link function. If we denote the CDF of a standard normal distribution as $\Phi(\cdot)$, the model can be written as $\pi = \Phi(\beta_0 + \beta_1 x_1 + \cdots + \beta_p x_p)$ or $\Phi^{-1}(\pi) = \beta_0 + \beta_1 x_1 + \cdots + \beta_p x_p$. The function Φ^{-1} is known as the probit function[31], and the model is often written as $\text{probit}(\pi) = \beta_0 + \beta_1 x_1 + \cdots + \beta_p x_p$. Note that $\text{probit}(\pi)$ is equivalent to Z_π, the standard normal quantile at probability π.

2. Complementary log-log regression: The inverse CDF from a Gumbel distribution (also known as the extreme value distribution) is used to form the link function. The CDF is $F(x) = \exp\{-\exp[-(x - \mu)/\sigma]\}$ for $-\infty < x < \infty$ and parameters $-\infty < \mu < \infty$ and $\sigma > 0$. Rather than setting π equal to the CDF as done for the logistic and probit regression models, the complementary log-log model sets π to equal $1 - F(x)$, which is essentially the probability resulting from a complement of an

[30]The logistic and normal distributions are members of the location-scale family of distributions. These distributions allow for a transformation like $Z = (X - \mu)/\sigma$ to a standard PDF from which all corresponding probabilities can be found. See p. 116 of Casella and Berger (2002) for more information.

[31]The term probit is a shortened version of "probability unit" (Hubert, 1992). Please see Chapter 8 of Salsburg (2001) for an interesting account of how this model was developed by Chester Bliss.

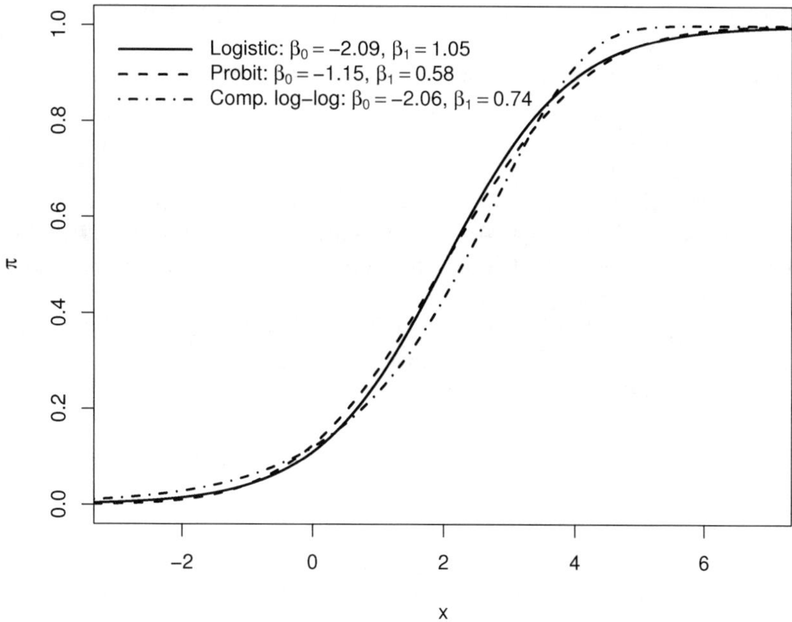

Figure 2.13: Logistic, probit, and complementary log-log models.

event. We can write our model then as $\pi = 1 - \exp[-\exp(\beta_0 + \beta_1 x_1 + \cdots + \beta_p x_p)]$ or $\log[-\log(1-\pi)] = \beta_0 + \beta_1 x_1 + \cdots + \beta_p x_p$.[32]

Example: Compare three binary regression models (PiPlot.R)

How different are the regression models with logit, probit, and complementary log-log link functions? Suppose the linear predictor has only one explanatory variable of the form $\beta_0 + \beta_1 x$. Figure 2.13 provides plots of these three models. We choose different values of β_0 and β_1 for each model so that the mean is 2 and the variance 3 for random variables with the corresponding CDFs. Please see the program for the code.

Both the logistic and probit regression models are rotationally symmetric about $\pi = 0.5$. The complementary log-log model does not have this same symmetry, because a Gumbel distribution is a skewed distribution. The main difference between the logistic and probit models is that the logistic model rises more quickly around $\pi = 0.5$, but tapers toward $\pi = 0$ and $\pi = 1$ a little more slowly than the probit model. This characteristic is because the logistic distribution has more probability in its tails than a normal distribution.

[32]By setting π equal to $1 - F(x)$, we are able to keep with standard conventions of what a positive and negative relationship mean with respect to the response variable. If π was set equal to $F(x)$ instead, the model would have a nonstandard interpretation. Namely, increases in an explanatory variable, say x_r, where its corresponding β_r is positive, would lead to *decreases* in the probability of success.

Estimation and inference for binary regression models

The probit and complementary log-log models are estimated using maximum likelihood estimation with `glm()` in a similar way as for the logistic model. The difference is that π_i is represented in the log-likelihood function of Equation 2.4 by the corresponding probit or complementary log-log model specification. Iterative numerical procedures are again needed to find the parameter estimates. Once the parameter estimates are found, the same inference procedures as used for the logistic model are available for the probit and complementary log-log models. These include Wald tests and LRTs from Section 2.2.1 and Wald and profile LR intervals from Sections 2.2.3 and 2.2.4. For example, we can use the `anova()` function for LRTs involving regression parameters, the `predict()` function for Wald confidence intervals for π, and functions from the `emmeans` and `mcprofile` packages for more general tests and confidence intervals.

One of the most important differences among the logistic, probit, and complementary log-log regression models arises when calculating odds ratios. In Section 2.2.3, we showed that the odds ratio for x in the model $\text{logit}(\pi) = \beta_0 + \beta_1 x$ is $OR = \exp(c\beta_1)$ for a c-unit increase in x. A very important aspect of this odds ratio was that it is the same for any value of x. Unfortunately, this result does not occur for probit and complementary log-log models. For example, consider the model $\text{probit}(\pi) = \beta_0 + \beta_1 x$. The odds of success are

$$Odds_x = \Phi(\beta_0 + \beta_1 x)/\left[1 - \Phi(\beta_0 + \beta_1 x)\right]$$

at a particular value of the explanatory variable x. If x is increased by $c > 0$ units, the odds of success become

$$Odds_{x+c} = \Phi(\beta_0 + \beta_1 x + \beta_1 c)/\left[1 - \Phi(\beta_0 + \beta_1 x + \beta_1 c)\right].$$

When the ratio of these two odds is taken, no algebraic simplification is possible, leading to an odds ratio that depends on the value of x. Odds ratios from the complementary log-log model also depend on x (see Exercise 30). This is one of the main reasons that logistic regression models are used the most among binary regression models.

Example: Placekicking (Placekick.R, Placekick.csv)

The purpose of this example is to compare the estimated logistic, probit, and complementary log-log regression models using the placekicking data set. We use `distance` as the only explanatory variable in the model. Our code to fit these models is shown below.

```
> mod.fit.logit <- glm(formula = good ~ distance, family =
    binomial(link = logit), data = placekick)
> round(summary(mod.fit.logit)$coefficients, 4)
            Estimate Std. Error  z value Pr(>|z|)
(Intercept)   5.8121     0.3263  17.8133        0
distance     -0.1150     0.0083 -13.7937        0

> mod.fit.probit <- glm(formula = good ~ distance, family =
    binomial(link = probit), data = placekick)
> round(summary(mod.fit.probit)$coefficients, 4)
            Estimate Std. Error  z value Pr(>|z|)
(Intercept)   3.2071     0.1570  20.4219        0
distance     -0.0628     0.0043 -14.5421        0
```

```
> mod.fit.cloglog <- glm(formula = good ~ distance, family =
    binomial(link = cloglog), data = placekick)
> round(summary(mod.fit.cloglog)$coefficients, 4)
            Estimate Std. Error  z value Pr(>|z|)
(Intercept)   2.3802     0.1184  20.1011        0
distance     -0.0522     0.0037 -14.0763        0
```

To fit the probit or complementary log-log models, we change the `link` argument value in `glm()` to its corresponding link function name. The estimated models are

- $\text{logit}(\hat{\pi}) = 5.8121 - 0.1150$`distance`
- $\text{probit}(\hat{\pi}) = 3.2071 - 0.0628$`distance`
- $\log[-\log(1-\hat{\pi})] = 2.3802 - 0.0522$`distance`

Each model results in a negative estimate of the effect that distance has on the probability of success, and the Wald test of $H_0 : \beta_1 = 0$ vs. $H_a : \beta_1 \neq 0$ in each case results in very small p-values indicating strong evidence for the importance of distance.

Estimates of the probability of success can be found using a model's parameter estimates and the corresponding CDF. For example, using the probit model and a distance of 20 yards produces the following estimate.

```
> lin.pred <- as.numeric(mod.fit.probit$coefficients[1] +
    mod.fit.probit$coefficients[2]*20)
> lin.pred
[1] 1.951195
> pnorm(q = lin.pred)
[1] 0.9744831
```

Thus, $\hat{\pi} = \Phi(3.2071 - 0.0628 \times 20) = \Phi(1.9512) = 0.9745$ when `distance` = 20.[33] These probability estimates can be found more easily using the `predict()` function.

```
> predict.data <- data.frame(distance = c(20, 35, 50))
> logistic.pi <- predict(object = mod.fit.logit, newdata =
    predict.data, type = "response")
> probit.pi <- predict(object = mod.fit.probit, newdata =
    predict.data, type = "response")
> cloglog.pi <- predict(object = mod.fit.cloglog, newdata =
    predict.data, type = "response")
> pi.hat <- data.frame(predict.data, logistic.pi, probit.pi,
    cloglog.pi)
> round(pi.hat, 4)
  distance logistic.pi probit.pi cloglog.pi
1       20      0.9710    0.9745     0.9777
2       35      0.8565    0.8436     0.8239
3       50      0.5152    0.5268     0.5477
```

[33]Using a different version of the notation, we have $Z_{0.9745} = 1.9512$.

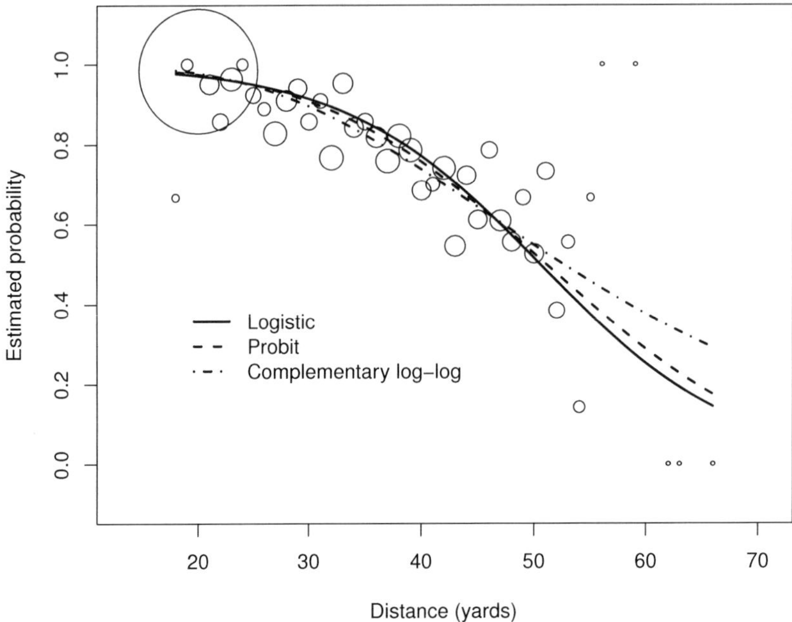

Figure 2.14: Bubble plot of the estimated probability of success for a placekick. The observed proportion of successes at each distance is plotted with the plotting point size proportional to the number of observations at a distance. The estimated logistic, probit, and complementary log-log regression models are all given in the plot.

Overall, we see that the differences in the estimated probabilities of success among the models are small, especially between the logistic and probit models. This may be surprising because the parameter estimates for the estimated models are quite different. For example, $\hat{\beta}_1$ for the complementary log-log model is less than half the $\hat{\beta}_1$ value for the logistic model. However, this is an inappropriate way to compare the models because of the differences in the way the regression parameters are used by their inverse link functions to form probabilities.

The similarities among the estimated models are seen further in Figure 2.14. In this figure, we provide a bubble plot with the plotting point proportional to the number of observations at each distance, which is the same as in Figure 2.5. Added to the plot are the three estimated models. Except for the very large distances where data are sparse, there is little difference among these models. Please see the program for the code.

We end this example by examining estimated odds ratios for the three models.

```
> odds.x20 <- pi.hat[1, 2:4]/(1 - pi.hat[1, 2:4])
> odds.x35 <- pi.hat[2, 2:4]/(1 - pi.hat[2, 2:4])
> odds.x50 <- pi.hat[3, 2:4]/(1 - pi.hat[3, 2:4])

> OR.20.35 <- odds.x20/odds.x35
```

```
> OR.35.50 <- odds.x35/odds.x50

> data.frame(OR = c("20 vs. 35", "35 vs. 50"),
   round(rbind(OR.20.35, OR.35.50), 2))
         OR logistic.pi probit.pi cloglog.pi
1 20 vs. 35        5.61      7.08       9.36
2 35 vs. 50        5.61      4.84       3.86
```

The code calculates the estimated odds of success for distances of 20, 35, and 50 yards. By forming ratios of consecutive odds, we obtain the estimated odds ratios for 15-yard increments of distance. As expected, the odds ratios are exactly the same for the logistic regression model, but are different for the probit and complementary log-log models.

Random components used with generalized linear models

The Bernoulli distribution is the standard random component for regression models of binary responses. There are many other probability distributions that can be used, depending on the response type. For example, the standard linear regression model

$$E(Y) = \beta_0 + \beta_1 x_1 + \cdots + \beta_p x_p$$

uses the normal distribution for Y along with an identity link function (i.e., $E(Y)$ equals the linear predictor). This type of model is appropriate with a continuous response variable that has an approximate normal distribution conditional on the explanatory variables. Also, in Chapter 4, we examine models for count responses that arise from mechanisms other than counting successes in repeated Bernoulli trials. In this setting, a Poisson random component is frequently used, where its mean is related to the explanatory variables through a log link,

$$\log[E(Y)] = \beta_0 + \beta_1 x_1 + \cdots + \beta_p x_p.$$

This link function ensures that $E(Y) > 0$ for any combination of regression parameters and explanatory variable values.

2.4 Exercises

1. Beginning with the logistic regression model written as

$$\pi = \frac{\exp(\beta_0 + \beta_1 x_1 + \cdots + \beta_p x_p)}{1 + \exp(\beta_0 + \beta_1 x_1 + \cdots + \beta_p x_p)},$$

   show that it leads to the model written as $\text{logit}(\pi) = \beta_0 + \beta_1 x_1 + \cdots + \beta_p x_p$ and vice versa.

2. Show $\partial\pi/\partial x_1 = \beta_1\pi(1-\pi)$ for the model $\pi = \exp(\beta_0 + \beta_1 x_1)/[1 + \exp(\beta_0 + \beta_1 x_1)]$. Discuss how the slope changes as a function of x_1.

3. By modifying the code within PiPlot.R, describe what happens to the logistic regression model when β_1 is increased or decreased in the case of $\beta_1 < 0$.

4. The failure of an O-ring on the space shuttle Challenger's booster rockets led to its destruction in 1986. Using data on previous space shuttle launches, Dalal et al. (1989) examine the probability of an O-ring failure as a function of temperature at launch and combustion pressure. Data from their paper is included in the Challenger.csv file. Below are the variables:

   - `Flight`: Flight number
   - `Temp`: Temperature (F) at launch
   - `Pressure`: Combustion pressure (psi)
   - `O.ring`: Number of primary field O-ring failures
   - `Number`: Total number of primary field O-rings (six total, three each for the two booster rockets)

   The response variable is `O.ring`, the number of trials per launch is `Number`, and the explanatory variables are `Temp` and `Pressure`. Note that more than one O-ring failed for some launches.

   (a) The authors use logistic regression to estimate the probability a randomly selected O-ring will fail. To use this model, the authors needed to assume for each launch that failure of any O-ring is independent of failure of any other O-ring. Discuss why this assumption is necessary and why it may not be true. Note that a subsequent analysis helped to alleviate the authors' concerns about independence.

   (b) Fit the logistic regression model using the explanatory variables in a linear form.

   (c) Perform LRTs to judge the importance of the explanatory variables in the model.

   (d) The authors chose to remove `Pressure` from the model based on the LRTs. Based on your results, discuss why you think this was done. Are there any potential problems with removing this variable?

5. Continuing Exercise 4, consider the simplified model $\text{logit}(\pi) = \beta_0 + \beta_1 \texttt{Temp}$, where π is the probability that a randomly selected O-ring fails.

   (a) Fit the model.

   (b) Construct two plots: (1) $\hat{\pi}$ vs. `Temp` and (2) Estimated expected number of O-ring failures per launch vs. `Temp`. Use a temperature range of 31° to 81° on the x-axis even though the minimum temperature in the data set was 53°.

   (c) Include the 95% Wald confidence interval bands for π on the plot. Why are the bands much wider for lower temperatures than for higher temperatures?

   (d) The temperature was 31° at launch for the Challenger in 1986. Estimate the probability that a randomly selected O-ring fails at this temperature. Using this estimate, compute the probability that *at least one* O-ring fails. What does this suggest about the potential safety of this launch?

   (e) Compute a 90% confidence interval for the first probability that was estimated in the previous part. Discuss what statistical assumptions need to be made for the estimation and the inference procedure.

   (f) Rather than using Wald or profile LR intervals for the probability of failure, Dalal et al. (1989) use a parametric bootstrap to compute intervals. Their process was to (1) simulate a large number of data sets from the estimated model of logit($\hat{\pi}$) = $\hat{\beta}_0 + \hat{\beta}_1$`Temp`, each with $n = 23$ for the same values of `Temp` as observed; (2) estimate new models for each data set, say logit($\hat{\pi}^\star$) = $\hat{\beta}_0^\star + \hat{\beta}_1^\star$`Temp`; and (3) compute $\hat{\pi}^\star$ at a specific temperature of interest. The authors used the 0.05 and 0.95 observed quantiles from the simulated distribution of $\hat{\pi}^\star$as their 90% confidence interval limits. Using the parametric bootstrap, compute 90% confidence intervals separately at temperatures of 31° and 72°.[34]

(g) Determine if a quadratic term is needed in the model for the temperature.

6. Continuing Exercise 5, investigate if the estimated probability of failure changes substantially if a probit or complementary log-log regression model is used instead of a logistic regression model.

7. Exercise 19 of Chapter 1 examined data from Berry and Wood (2004) to determine if an "icing the kicker" strategy reduces the probability of success for a field goal. Additional data collected for this investigation are included in the PlacekickBW.csv file. Below are descriptions of the variables available in this file:

   - `GameNum`: Identifies the year and game
   - `Kicker`: Last name of kicker
   - `Good`: Response variable (`"Y"` = success, `"N"` = failure)
   - `Distance`: Length in yards of the field goal
   - `Weather`: Levels of `"Clouds"`, `"Inside"`, `"SnowRain"`, and `"Sun"`
   - `Wind15`: 1 if wind speed is ≥15 miles per hour and the placekick is outdoors, 0 otherwise.
   - `Temperature`: Levels of `"Nice"` (40°F < temperature < 80° or inside a dome), `"Cold"` (temperature ≤ 40° and outdoors), and `"Hot"` (temperature ≥ 80° and outdoors)
   - `Grass`: 1 if kicking on a grass field, 0 otherwise
   - `Pressure`: `"Y"` if attempt is in the last 3 minutes of a game and a successful field goal causes a lead change, `"N"` otherwise
   - `Ice`: 1 if `Pressure` = `"Y"` and a time-out is called prior to the attempt, 0 otherwise

   These variables are similar but not all are exactly the same as given for the placekicking data described in Section 2.2.1. For example, `Good` is not coded with levels of 0 and 1, so this variable needs to be converted to a factor format or to 0/1 values directly (see Section 2.2.6). Other variables are not present, like a variable for PATs (only information on field goals is available). With this new data set, complete the following parts.

   (a) When using a `formula` argument value of `Good ~ Distance` in `glm()` with `Good` as a factor, how do you know whether R is modeling the probability of success or failure? Explain.

   (b) Fit the model from part (a). Plot this model using the `curve()` function.

[34]The bootstrap methods used here correspond to what is known as a percentile interval. There are better ways to form confidence intervals using the bootstrap, namely the BC_a and studentized intervals. Please see Davison and Hinkley (1997) for a discussion and the `boot()` and `boot.ci()` functions in the `boot` package for implementation.

(c) Add to the plot in part (b) the $\text{logit}(\hat{\pi}) = 5.8121 - 0.1150$`distance` model estimated in Section 2.2.1. Notice that the models are quite similar. Why is this desirable?

8. Continuing Exercise 7, use the `Distance`, `Weather`, `Wind15`, `Temperature`, `Grass`, `Pressure`, and `Ice` explanatory variables as linear terms in a new logistic regression model.

   (a) Fit the model and properly define the indicator variables used within it.

   (b) The authors use `"Sun"` as the base level category for `Weather`, which is not the default level that R uses. Describe how `"Sun"` can be specified as the base level in R.

   (c) Perform LRTs for all explanatory variables to evaluate their importance within the model. Discuss the results.

   (d) Estimate an appropriate odds ratio for distance, and compute the corresponding confidence interval. Interpret the odds ratio.

   (e) Estimate all six possible odds ratios for `Weather`, and compute the corresponding confidence intervals. Interpret the odds ratios. Is a value of 1 within any of the intervals? Relate this result to the LRT performed earlier for `Weather`.

   (f) Continuing part (e), discuss how the overall familywise error rate could be controlled for all of the confidence intervals. Carry out the necessary computations.

   (g) The purpose of this part is for you to estimate the probability of success for a field goal attempted by the Dallas Cowboys on December 4, 2011, in their game against the Arizona Cardinals. With the score tied and seven seconds remaining, the Dallas placekicker, Dan Bailey, attempted a 49-yard field goal that was successful. Unfortunately for Bailey, his coach, Jason Garrett, had called a time out right before the attempt, so Bailey had to re-kick. Bailey was not successful the second time, so Garrett essentially "iced" his own kicker, and the Cardinals went on to win the game in overtime.[35]

      i. Discuss the appropriateness of using this model to estimate the probability success.

      ii. The explanatory variable values for this placekick are: `Distance = 49` yards, `Wind15 = 0` (5 miles per hour), `Grass = 1`, `Pressure = "Y"`, and `Ice = 1`. The game was played in Arizona's retractable roof domed stadium, and information about whether the roof was open is unavailable. For this part, assume `Weather = "Sun"` and `Temperature = "Nice"`. Find the estimated probability of success for this field goal.

      iii. Compute the confidence interval for the probability of success using the explanatory variable values as given in ii.

   (h) Is there evidence to conclude that icing the kicker is a good strategy to follow?

9. Continuing Exercise 7, consider the model $\text{logit}(\pi) = \beta_0 + \beta_1$`Distance` $+ \beta_2$`Wind15` $+ \beta_3$`Distance` $\times$ `Wind15`, where π is the probability of success.

---

[35]A video of the field goal attempt is available at `https://www.youtube.com/watch?v=650ym30M7zI&t=93s`. The play-by-play description of the game can be examined at `https://www.espn.com/nfl/playbyplay?gameId=311204022`. Interestingly, if Dallas had won one additional game during the regular season, the New York Giants would not have made the playoffs and subsequently would not have won the Super Bowl.

(a) Fit the model with and without the `Distance` × `Wind15` interaction.

(b) Construct a plot similar to Figure 2.6 to see the effect of the interaction. Does the interaction appear to be important?

(c) Using the model with the interaction term, compute and interpret odds ratios to understand the effect of `Wind15` on the success or failure of a placekick.

(d) Perform an LRT to examine the importance of the interaction term. Compare these results to what was found earlier in Section 2.2.5. Suggest possible reasons for the similarities and/or differences.

10. The Larry Bird free throw shooting data discussed in Section 1.2 can be examined in a logistic regression context. The purpose here is to estimate the probability of success for the second free throw attempt given what happened on the first free throw attempt.

(a) The `c.table` object created in Bird.R contains the contingency table format of the data. For `glm()`, we need a data frame format instead. Because this data set is small, the easiest way to transform the data is to simply re-enter the data in a new data frame.

```
> bird <- data.frame(First = c("made", "missed"), success =
    c(251, 48), trials = c(285, 53))
> bird
   First success trials
1   made     251    285
2 missed      48     53
```

For larger contingency tables, like those examined in Chapters 3 and 4, it is better to transform the data directly from `c.table` using code.

```
> bird1 <- as.data.frame(as.table(c.table))
> trials <- aggregate(formula = Freq ~ First, data = bird1,
    FUN = sum)
> success <- bird1[bird1$Second == "made", ]
> bird2 <- data.frame(First = success$First, success =
    success$Freq, trials = trials$Freq)
> bird2
   First success trials
1   made     251    285
2 missed      48     53
```

Implement each line of code in the second method and describe what occurs.

(b) Fit a logistic regression model for the probability of success on the second attempt, where `First` is the explanatory variable.

(c) Estimate the odds ratio comparing the two levels of `First`. Calculate both Wald and profile LR intervals for the odds ratio. Compare these calculated values with those obtained in the Larry Bird example of Section 1.2.5. If you use `confint()` from the `stats` package to compute the intervals, make sure to use the correct value for the `parm` argument (same name as the indicator variable given by `summary()` when used with the model-fit object).

(d) Perform a hypothesis test of $H_0 : \beta_1 = 0$ vs. $H_a : \beta_1 \neq 0$ using Wald and LR test statistics. Compare these results to those found for the Larry Bird example in Section 1.2.3.

(e) Discuss why similarities and/or differences occur between the calculations here using logistic regression and the corresponding calculations in Chapter 1.

11. Exercise 33 of Chapter 1 analyzed data from the clinical trial for the Salk vaccine. This data also can be examined in a logistic regression context like was done for Exercise 10.

(a) Start with the contingency table format of the data shown below.

```
c.table <- array(data = c(57, 142, 200688, 201087), dim =
    c(2,2), dimnames = list(Treatment = c("vaccine",
    "placebo"), Result = c("polio", "polio free")))
```

Create a new data frame named `polio` that has a form appropriate for `glm()`.

```
> polio
  Treatment success trials
1   vaccine  200688 200745
2   placebo  201087 201229
```

Adapt the code used for `bird2` in Exercise 10 to create this data frame.

(b) Use the `levels()` function to examine ordering of the levels in `polio$Treatment`. Note that these are not in alphabetical order because of the ordering given originally in the `array()` function.

(c) Fit a logistic regression model for the probability of being polio free (second column of the contingency table) where the type of treatment is the explanatory variable. Specifically state the indicator variable coding for the explanatory variable.

(d) Estimate an odds ratio comparing the two treatments and calculate a corresponding confidence interval. Compare your results to those found in Exercise 33 of Chapter 1.

(e) Determine if there is a difference in the probability of being polio free based on the treatment.

(f) Suppose additional explanatory variables, such as geographical location and socioeconomic status, are available for each individual, and these are included in the model as well. Discuss why using a logistic regression model would be an easier way to assess the effectiveness of the treatment rather than a contingency table(s) format.

12. Similar to Exercise 10, the Olestra data discussed in Exercise 26 of Chapter 1 can be examined in a logistic regression context.

(a) Transform the contingency table to a data frame of the following format.

```
> olestra.data
  treatment side.effect trials
1   Olestra          89    563
2   Regular          93    529
```

This transformation can be performed with R code directly from the contingency table format of the data given in Chapter 1. Note that R may not order the levels of `treatment` in the expected manner depending on how the contingency table was originally constructed (see Exercise 11).

(b) Fit a logistic regression model using `treatment` as the explanatory variable and `side.effect` as the response variable. Remember to include `trials` as the number of trials for the response variable.

(c) Complete the following for an odds ratio involving `treatment`.

i. Compute the estimated odds ratio using the model.

ii. Find the 95% profile LR interval for the odds ratio. Show how the computations can be performed using both `confint()` and `mcprofile()`.

iii. Interpret the estimated odds ratio and confidence interval in the context of the potential for side effects with Olestra.

iv. Compare the estimated odds ratio and the profile LR interval to the estimate and interval found for the Chapter 1 exercise. Why are these numerical values similar or different?

(d) Perform an LRT using the model to determine if the treatment has an effect on whether there are side effects.

(e) Estimate the probability of having side effects given that a regular type of potato chip is consumed. Perform the same computation for Olestra potato chips. Compute 95% profile LR intervals for both cases and interpret the intervals.

13. Continuing Exercise 12, suppose the amount of potato chips that each individual consumed was available. This variable is denoted by `Chips` (measured in ounces) in the data file OlestraNotReal.csv. While its values are simulated for illustrative purposes, one may hypothesize that the amount of potato chips consumed could have an effect on whether side effects are observed. Also, one may further hypothesize that the response to the treatment may change as a function of the amount consumed.

The data file also contains the variables `Treatment` (with levels of Olestra or Regular) and `Response` (1 for side effects and 0 for no side effects). Using these data, complete the following parts.

(a) Use the `levels()` and `contrasts()` functions to show how R accounts for `Treatment` as a factor in any model. Describe in words how the indicator variable is constructed.

(b) Confirm that the estimated logistic regression model for `Response` using only `Treatment` as the explanatory variable is the same as obtained for Exercise 12.

(c) Fit the same model as in (b) but now with the amount of chips, the treatment, and their interaction. Use this model for the remainder of the exercise unless instructed to do otherwise.

(d) How many iterations does the `glm()` function need to reach convergence using the default convergence criterion? Show how the final convergence criterion value was calculated on the final iteration and confirm that convergence was obtained.

(e) Perform an LRT to determine if there is an interaction between the amount of chips and the treatment.

(f) Plot the estimated model with amount of chips on the x-axis and the estimated probability of side effects on the y-axis. Include two separate lines on the plot for the treatment levels. Describe the effect of the interaction term in the model as shown by the plot.

(g) Use odds ratios and their corresponding profile LR intervals to thoroughly interpret the effect the treatment has on whether side effects occur.

(h) Fit the same model as in part (c), but now using probit and complementary log-log regression models. Plot all three models on the same plot as in part (f). Use different combinations of line types, colors, and thicknesses to distinguish the models on the plot. Describe differences among the models.

14. The relative risk and the difference between two probabilities of success can be calculated using a logistic regression model. For this problem, consider the logistic regression model $\text{logit}(\pi) = \beta_0 + \beta_1 x_1$, where x_1 is 0 or 1.

    (a) Find the difference of the two probabilities of success in terms of the model.

    (b) Find the relative risk in terms of the model.

    (c) Through the delta method described in Appendix B, the estimated variances of the statistics in (a) and (b) can be found. Find these estimated variances in terms of $\widehat{Var}(\hat{\beta}_0)$, $\widehat{Var}(\hat{\beta}_1)$, and $\widehat{Cov}(\hat{\beta}_0, \hat{\beta}_1)$.

15. The deltaMethod() function of the car package provides a convenient way to apply the delta method to find variances of mathematical functions. The syntax of the function is given below.

```
deltaMethod(object = <model-fit object>, g = <mathematical
    function>, parameterNames = <Vector of parameter names>).
```

The parameterNames argument uses the same parameter names as given in the g argument. These names do not need to match those given by glm(), but they need to be in the same order as in the model-fit object. Apply deltaMethod() to the Larry Bird and Salk vaccine data of Exercises 10 and 11, respectively, as described below.

(a) Let $\pi_{\text{First=missed}}$ and $\pi_{\text{First=made}}$ denote the probability of success on the second attempt given what happened on the first attempt for Larry Bird. The following code estimates $\pi_{\text{First=missed}} - \pi_{\text{First=made}}$ and $Var(\hat{\pi}_{\text{First=missed}} - \hat{\pi}_{\text{First=made}})^{1/2}$, where mod.fit contains the results from fitting the model with glm().

```
library(package = car)
deltaMethod(object = mod.fit, g = "exp(beta0 + beta1*1)/(1 +
    exp(beta0 + beta1*1)) - exp(beta0)/(1+exp(beta0))",
    parameterNames = c("beta0", "beta1"))
```

Execute this code and use it to find the Wald confidence interval for the difference of the two probabilities. Compare your estimate and confidence interval to those for the Larry Bird example in Section 1.2.2.

(b) Let π_{vaccine} and π_{placebo} denote the probability of being polio free given the type of treatment received. Use the `deltaMethod()` function to estimate the relative risk given by $(1-\pi_{\text{vaccine}})/(1-\pi_{\text{placebo}})$ and its corresponding variance. Calculate a Wald confidence interval for the relative risk with these results. Compare your estimate and confidence interval to those for Exercise 33 of Chapter 1.

(c) Repeat part (b), but now with $\log[(1-\pi_{\text{vaccine}})/(1-\pi_{\text{placebo}})]$ to find an estimate and Wald confidence interval for the log of the relative risk. Use the exponential function to find the relative risk estimate and corresponding confidence interval.

16. Consider the model $\text{logit}(\pi) = \beta_0 + \beta_1$`distance` for the placekicking data set introduced in Section 2.2.1. Define π_x as the probability of success at a distance of x yards and c as a constant increment in the distance. Assuming the model estimation results from `glm()` are stored in `mod.fit`, complete the following parts.

    (a) Estimate $\pi_{x+c} - \pi_x$ and π_{x+c}/π_x for a few different values of x. Do the estimates change as a function of x for a constant c?

    (b) Compute 95% Wald confidence intervals for $\pi_{x+c} - \pi_x$ and π_{x+c}/π_x at the values of x and c chosen for part (a). Use the `emmeans` package for these computations. Below is an example of this process when $x = 20$ and $c = 10$.

```
library(package = emmeans)

calc.est1 <- emmeans(object = mod.fit, specs = ~ distance,
    at = list(distance = c(30,20)))

# Difference of probabilities
regrid1 <- regrid(object = calc.est1, transform = "response")
test.info2 <- contrast(object = regrid1, method = "pairwise")
confint(object = test.info2, adjust = "none", level = 0.95)

# Ratio of probabilities
regrid2 <- regrid(object = calc.est1, transform = "log")
test.info2 <- contrast(object = regrid2, method =
    "pairwise", type = "response")
confint(object = test.info2, adjust = "none", level = 0.95)
```

    The `regrid()` function transforms the linear predictor calculations of `emmeans()` to the probability and log-probability scales. The `contrast()` and `confint()` functions compute the appropriate tests and confidence intervals. The `type = "response"` argument value is needed for the second use of `contrast()` to display comparisons in terms of probabilities because of the log transformation by `regrid()`.

    (c) Show that $\widehat{Var}(\hat{\pi})^{1/2}$ is computed as the same value with `predict()` and `deltaMethod()` using the code below.

```
predict(object = mod.fit, newdata = data.frame(distance =
    x), type = "response", se.fit = TRUE)

library(package = car)
deltaMethod(object = mod.fit, g = "exp(beta0 + beta1*x)/(1 +
    exp(beta0 + beta1*x))", parameterNames = c("beta0",
```

```
    "beta1"))
```

where x is replaced by a particular value of distance.

(d) Complete part (b) using the deltaMethod() function for the computations instead.

17. Consider the model logit(π) = $\beta_0 + \beta_1$distance + $\beta_2$distance^2 for the placekicking data set introduced in Section 2.2.1. Estimate this model and interpret the effect that distance has on the success or failure of a placekick using odds ratios. Below is example code for the mcprofile and emmeans packages to make one set of comparisons between 20- and 30-yard placekicks. Assume model-fit information from glm() is stored in mod.fit.distsq.

```
library(package = mcprofile)
x <- 20
c <- 10
K <- matrix(data = c(0, 1, 2*x+c), nrow = 1, ncol = 3, byrow =
    TRUE)
linear.combo <- mcprofile(object = mod.fit.distsq, CM = K)
ci.log.OR <- confint(object = linear.combo, level = 0.95, adjust
    = "none")
1/exp(c*ci.log.OR$estimate)  # Estimate
rev(1/exp(c*ci.log.OR$confint))  # Interval

library(package = emmeans)
calc.est.sq <- emmeans(object = mod.fit.distsq, specs = ~
    distance, at = list(distance = c(20,30)), type = "response")
confint(object = contrast(object = calc.est.sq, method =
    "consec", combine = TRUE, reverse = TRUE), adjust = "none",
    level = 0.95)
```

18. Exercise 27 of Chapter 1 examined the effects of taking acetazolamide to prevent headaches that may occur because of AMS. The authors of the study also used ginkgo biloba, with or without acetazolamide, to prevent headaches. The data are provided in the code and output below.

```
> acet <- factor(x = c("no", "yes", "no", "yes"))
> ginkgo <- factor(x = c("no", "no", "yes", "yes"))
> headache <- c(63, 23, 72, 39)
> total <- c(119, 118, 124, 126)
> set1 <- data.frame(acet, ginkgo, headache, total)
> set1
  acet ginkgo headache total
1   no     no       63   119
2  yes     no       23   118
3   no    yes       72   124
4  yes    yes       39   126
```

Complete the following parts using occurrence of headaches as the response.

(a) Use the `levels()` and `contrasts()` functions to show how R accounts for the `acet` and `ginkgo` explanatory variables in a logistic regression model.

(b) Fit the logistic regression model with `acet`, `ginkgo`, and their interaction in the model. Use the binomial response format of the data directly with `glm()`. Explain why this is a saturated model.

(c) The data were actually observed in a Bernoulli response format rather than the binomial form given in the introduction to this exercise. This means that each row of the data set represented one individual and their corresponding `acet` and `ginkgo` treatment combination and `headache` response. Complete the following for data in this alternative format.

   i. Below is one way that `set1` can be converted to a binomial response format.

```
> index.headache <- rep(x = 1:nrow(set1), times =
    set1$headache)
> index.none <- rep(x = 1:nrow(set1), times = set1$total
    - set1$headache)
> set2 <- set1[c(index.headache , index.none),1:2]
> set2$headache <- c(rep(x = 1, times =
    length(index.headache)), rep(x = 0, times =
    length(index.none)))
> row.names(set2) <- NULL
```

   Execute this code and describe what each line of the code does.

   ii. Use the `aggregate()` function to verify that `set2` matches the data given in the binomial response format.

   iii. Fit the logistic regression model with `acet`, `ginkgo`, and their interaction in the model. What are the similarities and differences between the model here and the one estimated in part (b)?

Use the model and data format from part (b) for the remainder of this exercise.

(d) Perform an LRT to determine if there is evidence that an interaction between acetazolamide and ginkgo exists.

(e) Separately interpret the effects of taking acetazolamide or ginkgo with odds ratios. Because there is an interaction in the model, keep one of the explanatory variables fixed while examining the odds ratio for the other variable.

(f) Calculate the convergence criterion $\left(G^{(k)} - G^{(k-1)}\right) / \left(0.1 + \left|G^{(k)}\right|\right)$ for each iteration. Show that convergence occurs using the default settings in `glm()`.

(g) In studies like this one, it is important to control for other variables that may affect the response. For example, the paper provides a set of aggregated "baseline characteristics" in its Table 1 regarding gender, age, and oxygen saturation level. The authors indicated that these baseline characteristics were similar for each acetazolamide and ginkgo combination and then did not consider them further. Suggest an alternative way that these characteristics could be controlled through a logistic regression model.

(h) Elphick and Elphick (2004) and Adams (2004) wrote letters to the editor regarding Gertsch et al. (2004b). A reply by some of the original paper's authors was given in Gertsch et al. (2004a). If you have access to these references, summarize the problems described in the letters. Do you agree with the letters and/or the authors' reply? Explain. Note that the "Data supplement" link given in the reply contains information about the "subsequent clinical trial."

19. Thorburn et al. (2001) examine hepatitis C prevalence among healthcare workers in the Glasgow area of Scotland. These healthcare workers were categorized into the following occupational groups: (1) exposure prone (e.g., surgeons, dentists, surgical nurses), (2) fluid contact (e.g., non-surgical nurses), (3) lab staff, (4) patient contact (e.g., pharmacists, social workers), and (5) no patient contact (e.g., clerical). The collected data are available in the file HealthcareWorker.csv. Is there evidence of an occupational group effect on hepatitis status? If there is sufficient evidence of an effect, use the appropriate odds ratios to make comparisons among the groups. If there is not sufficient evidence of an effect, discuss why this may be a preferred result.

20. An example in Section 2.2.5 examined using $\text{logit}(\pi) = \beta_0 + \beta_1\texttt{distance} + \beta_2\texttt{wind} + \beta_3\texttt{distance} \times \texttt{wind}$ for the placekicking data set, and confidence intervals were calculated to interpret the odds ratio for wind. Because of the interaction, these intervals were found at a few specific distances. An alternative way to find these confidence intervals in R is through the `multcomp` package (Bretz et al., 2011). This is another user-contributed package that provides a convenient suite of functions for estimation and inference involving linear combinations of regression parameters. While the package only performs inferences using Wald-based methods, the `mcprofile` package is built upon it to perform profile LR inferences.

    (a) Implement the following continuation of code from the example. Verify the data frame contains the estimated `wind` odds ratio and the 95% Wald confidence interval when the `distance` is 20 yards.

    ```
    library(package = multcomp)
    K <- matrix(data = c(0, 0, 1, 1*20), nrow = 1, ncol = 4,
        byrow = TRUE)
    linear.combo <- glht(model = mod.fit.Ha, linfct = K)
    ci.log.OR <- confint(object = linear.combo, calpha =
        qnorm(0.975))
    data.frame(estimate = exp(-ci.log.OR$confint[1]), lower =
        exp(-ci.log.OR$confint[3]), upper =
        exp(-ci.log.OR$confint[2]))
    ```

    (b) Describe what each line of the code in (a) calculates given the argument values specified. You may want to examine the functions' help along with the on-line vignettes for the package at `http://cran.r-project.org/web/packages/multcomp/index.html`. Note that the elements of the `c()` function are put in an order corresponding to $\hat{\beta}_0$, $\hat{\beta}_1$, $\hat{\beta}_2$, and $\hat{\beta}_3$.

    (c) Modify `K` in (a) to also include distances of 30, 40, 50, and 60 yards. Complete the same odds ratio calculations for all of these distances.

    (d) The `multcomp` package contains numerous built-in adjustments for simultaneous inference. Like `mcprofile`, the default for both tests and confidence intervals is to use a single-step adjustment. Thus, the syntax to compute confidence intervals with a 95% familywise confidence level is `confint(object = linear.combo, level = 0.95)`. Compute these intervals for the distances (20 to 60 by 10), and explain what statement can be made about the familywise confidence level. Compare the intervals to the unadjusted intervals and comment on differences.

    (e) Implement the code below and compare it to what `summary(mod.fit.Ha)` gives.

```
K <- matrix(data = c(1, 0, 0, 0,  0, 1, 0, 0,  0, 0, 1, 0,
    0, 0, 0, 1), nrow = 4, ncol = 4, byrow = TRUE)
linear.combo <- glht(model = mod.fit.inter, linfct = K)
summary(object = linear.combo, test = adjusted(type =
    "none"))
```

21. Rather than finding the probability of success at an explanatory variable value, it is often of interest to find the value of an explanatory variable given a desired probability of success. This is referred to as *inverse prediction.* One application of inverse prediction involves finding the amount of a pesticide or herbicide needed to have a desired kill rate or probability, π, when applied to pests or plants. The *lethal dose level* x_π (commonly called "LDz", where $z = 100\pi$) is defined as

$$x_\pi = [\text{logit}(\pi) - \beta_0]/\beta_1$$

for the logistic regression model $\text{logit}(\pi) = \beta_0 + \beta_1 x$. The estimated value of x_π is $\hat{x}_\pi = [\text{logit}(\pi) - \hat{\beta}_0]/\hat{\beta}_1$, assuming that $\hat{\beta}_1 \neq 0$.

   (a) Show how x_π is derived by solving for x in the logistic regression model.

   (b) A $(1-\alpha)100\%$ confidence interval for x_π is the set of all possible values of x such that

$$\frac{\left|\hat{\beta}_0 + \hat{\beta}_1 x - \text{logit}(\pi)\right|}{\sqrt{\widehat{Var}(\hat{\beta}_0) + x^2\widehat{Var}(\hat{\beta}_1) + 2x\widehat{Cov}(\hat{\beta}_0, \hat{\beta}_1)}} < Z_{1-\alpha/2} \tag{2.23}$$

   is satisfied. Describe how this confidence interval for x_π is derived. Note that there is generally no closed-form solution for the confidence interval limits, which leads to the use of iterative numerical procedures.

   (c) The `dose.p()` function of the `MASS` package and the `deltaMethod()` function of the `car` package can calculate $\hat{x}_\pi$ and $\widehat{Var}(\hat{x}_\pi)^{1/2}$, where the estimated variance is found using the delta method. Using this statistic and its corresponding variance, show how an alternative $(1-\alpha)100\%$ Wald interval for x_π can be formed.

   (d) Derive the form of x_π using probit and complementary log-log models.

22. This exercise is an application of the concepts explored in Exercise 21. Turner et al. (1992) use logistic regression to estimate the rate at which picloram, a herbicide, kills tall larkspur, a weed. Their data were collected by applying four different levels of picloram (kg/ha) to separate land plots, and the number of weeds killed out of the number of weeds within each plot was recorded. The data are in the file Picloram.csv.[36]

   (a) Fit a logistic regression model using the picloram amount as the explanatory variable and the number of weeds killed as the response variable.

   (b) Plot the observed proportion of killed weeds vs. the picloram level. Include the estimated model on the plot. Describe how well the model fits the data.

[36] The experiment was performed in three separate runs, which are denoted by the `rep` variable in the data file. The authors fit a logistic regression model without accounting for the `rep` variable, and we do the same here. We include `rep` in a reanalysis of the data in Exercise 2 from Section 6.5.

(c) Estimate the 0.9 kill rate level (LD90) for picloram. Add a horizontal line at 0.9 and a vertical line at the LD90 to the plot in (b) to illustrate how it is found (the `segments()` function with `plot()` can be useful for this purpose).

(d) The code below uses Equation 2.23 to calculate a confidence interval for the dosage that yields a 0.9 kill rate.

```
root.func <- function(x, mod.fit.obj, pi0, alpha) {
  beta.hat <- mod.fit.obj$coefficients
  cov.mat <- vcov(mod.fit.obj)
  var.den <- cov.mat[1,1] + x^2*cov.mat[2,2] +
      2*x*cov.mat[1,2]
  abs(beta.hat[1] + beta.hat[2]*x - log(pi0/(1-pi0))) /
      sqrt(var.den) - qnorm(1-alpha/2)
}

lower <- uniroot(f = root.func, interval =
    c(min(set1$picloram), LD.x), mod.fit.obj = mod.fit, pi0
    = 0.9, alpha = 0.05)
lower  # lower$root contains the lower bound

upper <- uniroot(f = root.func, interval = c(LD.x,
    max(set1$picloram)), mod.fit.obj = mod.fit, pi0 = 0.9,
    alpha = 0.05)
upper  # upper$root contains the upper bound
```

Note that `set1` is a data frame containing the data and `LD.x` contains $\hat{x}_{0.9}$. Describe what each line of the R code does and run the code to find a 95% confidence interval for $x_{0.9}$.

(e) Use `dose.p()` or `deltaMethod()` to calculate $\hat{x}_{0.9}$ and $\widehat{Var}(\hat{x}_{0.9})^{1/2}$. The syntax for `dose.p()` is `dose.p(obj = <model-fit object>, p = <kill rate>)`. The syntax for `deltaMethod()` can be determined from Exercise 15. Form a 95% Wald interval for LD90 using the computed values from one of these functions and compare the interval to what was obtained in part (d).

(f) What amount of picloram should be used to have a 0.9 kill rate?

(g) The data for this problem consist of only four different dosage levels of picloram. What assumptions are needed for the model to provide a good estimate of x_π?

23. Repeat the analysis in Exercise 22 using the probit and complementary log-log models. Do the results change in comparison to using logistic regression?

24. Potter (2005) and Heinze (2006) examine a urinary incontinence study involving three explanatory variables and their relation to incontinence after a drug was administered. Unfortunately, the authors do not provide an exact description of what the response variable coding represents (most likely, the response $y = 1$ denotes continent and $y = 0$ denotes incontinent) and do not give detailed descriptions of the explanatory variables (they only say that x_1, x_2, and x_3 are measurements on the lower urinary tract). However, the main reason to examine these data here is to see an example where complete separation occurs when multiple explanatory variables are included in a logistic regression model. Using the data in the file Incontinence.csv, complete the following parts.

(a) Fit the logistic regression models $\text{logit}(\pi) = \beta_0 + \beta_1 x_r$ for $r = 1, 2, 3$ using $\pi = P(Y = 1)$ and maximum likelihood estimation. Plot the estimated models along with the observed values. Are there any problems with complete separation?

(b) Fit the logistic regression model $\text{logit}(\pi) = \beta_0 + \beta_1 x_1 + \beta_2 x_2 + \beta_3 x_3$ using maximum likelihood estimation. With the default number of iterations in `glm()`, obtain the parameter estimates from the last iteration. Use these to estimate the probability of success for each observation. Discuss the problems that occur.

(c) Fit the same model as in (b), but now using the modified likelihood function as discussed in Section 2.2.7. Use the `logistf()` function for these computations and include `alpha = 0.10` to compute 90% profile LR confidence intervals for the regression parameters. Do any problems occur with the estimation? Compare the results here to those found in (b).

(d) Using the model fit from (c), interpret the relationship between the explanatory variables and the response.

(e) Refit the model from (c) using `logistif()` function again, but now with `alpha = 0.05`. While the estimates of the regression parameters will converge, computational problems occur for the confidence interval limits. Include `plcontrol = logistpl.control(maxit = x)` in `logistif()` with a few different maximum number of iterations for `x` to show that a larger number of iterations does not help. The lower limit for `x2` and the upper limit for `x3` could be reported as infinite for this situation.

25. Pena et al. (2011) examine growth of the *Alicyclobacillus acidoterrestris* bacteria in apple juice as a function of four different variables. Their data are available in the AppleJuice.csv data file. The variables available are

- `pH`: Acidity level
- `nisin`: Nisin concentration (IU/ml)
- `temp`: Temperature (C)
- `Brix`: Concentration of soluble solids
- `growth16`: 1 for growth and 0 for no growth of the bacteria after 16 days of incubation

(a) The authors conclude the "best" model is $\text{logit}(\pi) = \beta_0 + \beta_1\texttt{pH} + \beta_2\texttt{nisin} + \beta_3\texttt{temp} + \beta_4\texttt{Brix} + \beta_5\texttt{Brix} \times \texttt{pH} + \beta_6\texttt{Brix} \times \texttt{nisin} + \beta_7\texttt{Brix} \times \texttt{temp}$. Estimate this model with `glm()` using the default values for `epsilon` and `maxit`. Use this model for the remainder of the exercise.

(b) There will be a warning message in the `glm()` output from part (a) stating `fitted probabilities numerically 0 or 1 occurred`. Show that convergence is actually obtained.

(c) Perform the appropriate LRTs and Wald tests to examine the importance of the interaction terms in the model. If you have access to the paper, compare your results to those of Table 2 in the paper. What type of tests does it appear that the authors use? Note that the authors omitted the `Brix`$\times$`pH` interaction results.

(d) Examine the odds ratios associated with `temp` as follows.

   i. Derive the odds ratio for the temperature holding `Brix` constant.

   ii. Why is $c = 5$ a reasonable choice for a temperature increase?

Table 2.6: Explanatory variable combinations of interest from Table 7 of Pena et al. (2011).

pH	3.7	4.0	5.0	3.7	4.0	5.0	3.7	4.0	5.0
nisin	5	5	5	40	40	40	5	5	5
temp	30	30	30	30	30	30	30	30	30
Brix	11.5	11.5	11.5	11.5	11.5	11.5	12.5	12.5	12.5

iii. For the levels of `Brix` in the data set, estimate the odds ratios for temperature with $c = 5$.

iv. Compute 95% profile LR and Wald confidence intervals for the odds ratios. Use the same values of `Brix` as in the previous part and $c = 5$. Compare the intervals. Note that the `mcprofile` package produced nonsensical results for us with some profile LR intervals not containing the corresponding odds ratio estimates.

v. Interpret the Wald confidence interval found in part iv. when `Brix` is 19 and $c = 5$.

vi. Discuss the effect that temperature has on the growth of bacteria in the apple juice.

(e) Table 7 in the paper presents a number of explanatory variable combinations and their estimated probabilities. We provide these combinations in Table 2.6. Find the corresponding estimated probabilities and 95% Wald confidence intervals for the true probabilities. Which of these combinations are best to prevent bacteria growth?

26. Continuing Exercise 25, the data set is an example of when *quasi-separation* occurs. This corresponds to when particular sets of explanatory variable values always lead to only a 0 or a 1 response.

(a) If complete separation occurred, what would a histogram of the $\hat{\pi}$ values look like? Explain your answer.

(b) Examine a histogram of the $\hat{\pi}$ values for the model estimated in Exercise 25. Also, determine the number of observations with $\hat{\pi}$ values that are less than 0.01 or greater than 0.99 and compare this to the overall number of observations in the data set. Discuss your findings here relative to your answer for part (a).

(c) Examine a binomial-response form of the data to determine what combination of values for some variables always lead to no bacteria growth. This data format shows the number of times bacteria growth was observed out of the total number of observations for each unique combination of `pH`, `nisin`, `temp`, and `Brix`. The `aggregate()` function can be used to transform the data to the specified format.

(d) Fit the same model as in Exercise 25, but now using the modified likelihood function as discussed in Section 2.2.7. We found that at least 50 iterations are needed to obtain convergence for the estimates of the regression parameters. This can be set by using `control = logistf.control(maxit = 50)` in `logistf()`. Warning messages will be printed for computational problems that occur for 95% profile confidence interval limits. However, we have found these limits do not change as the number of iterations increases.

(e) Find the 95% Wald confidence interval for the temperature odds ratio when `Brix` is 19 and $c = 5$. Compare your answer to what was found in Exercise 25.

27. Section 2.2.7 discusses the criteria for how R determines if and when the convergence of parameter estimates takes place. Discuss why the residual deviance is used for this purpose. Suggest other ways to determine convergence.

28. Continuing Exercise 13, construct a new hypothetical explanatory variable, say w, that will lead to non-convergence when included in a logistic regression model with amount of chips, the treatment, and their interaction. Illustrate this non-convergence through using the `glm()` function. Why does non-convergence also occur when estimating a probit or complementary log-log regression model?

29. What does the `na.action` argument of the `glm()` function control and what is its default? Discuss possible problems that may occur with this default value.

30. Consider the complementary log-log regression model $\log[-\log(1-\pi)] = \beta_0 + \beta_1 x_1$. Show that the odds ratio comparing different levels of x_1 remains a function of x_1.

31. Section 2.2.1 gives an example of how to fit a logistic regression model by creating an R function to calculate the log-likelihood function and then maximize it using `optim()`. Following this example, write an R function that calculates the log-likelihood function for the probit regression model. Use this function with `optim()` to estimate probit$(\pi) = \beta_0 + \beta_1$`distance` with the placekicking data set. Compare the estimated standard errors resulting from the use of `optim()` to those obtained from `glm()`. Repeat this process with the corresponding complementary log-log regression model.

32. Uldbjerg et al. (2021) examine the relationship between antibiotic use by an expectant mother and subsequent development of asthma in her child. This paper provides a standard example of how logistic regression is often used in scientific research. If you have access to the paper, read it and answer the questions below. Unfortunately, the entire data set used for the paper has not been made available by the authors.

    (a) What are the authors' stated reasons for not making the data available? Provide a discussion of the potential problems associated with making or not making the data available.

    (b) What is the population of interest?

    (c) "Unadjusted" and "adjusted" odds ratios are given in the paper. What does the adjustment represent? Note that Exercise 32 of Chapter 1 calculates one of the unadjusted odds ratios.

    (d) The authors present results for three different logistic regression models. Formally write out what the third model could be in terms of regression parameters and explanatory variables. Include definitions for the response variable and for the explanatory variables. Because the authors are not specific with their model, there is more than one way to answer to this question.

    (e) Is there discussion on the inclusion of interactions or transformations of explanatory variables in the logistic regression models? If the authors did not investigate these model features, how could this affect the conclusions from their data analysis?

    (f) Suppose explanatory variables are included in the logistic regression model in a linear form (no interactions or transformations). Using the results for Model 3 in Table 2, what is the estimated regression parameter for the antibiotic variable (any trimester)? Assume the variable value is 1 for exposure and 0 for no exposure.

(g) Using the results for Model 3, provide a formal interpretation of the estimated odds ratio for antibiotic exposure when this variable corresponds to no exposure or any exposure. Interpret the corresponding confidence interval as well.

(h) There are 95% confidence intervals given for odds ratios resulting from the models. Unfortunately, the type of confidence interval used is not given. Why would it be of interest to know the type? Explain.

(i) Is there sufficient evidence to indicate antibiotic exposure leads to a higher incidence rate of childhood asthma? Explain using the statistical evidence presented in the paper.

33. Show that the logistic regression model $\pi = \exp(\beta_0 + \beta_1 x)/[1 + \exp(\beta_0 + \beta_1 x)]$ can be rewritten as $\pi = 1 - 1/[1 + \exp(\beta_0 + \beta_1 x)]$. This alternative representation of the model can be useful for delta method calculations (see Exercise 34).

34. Exercises 28 and 29 of Chapter 1 examine data (Flagstick.csv) from a golf putting experiment. The ultimate goal was to determine whether taking the flagstick out or leaving the flagstick in results in a higher probability of success for making a putt. The purpose of this exercise is to re-examine the data using logistic regression.

(a) EMGA reached conclusions on the best approach by using the observed proportions only. No statistical inference methods were used. Why is this a poor way to develop conclusions?

(b) Bilder (2020) focuses on a logistic regression model that included terms representing flagstick (in or out), entry line, ball speed, and the interaction between flagstick and ball speed. The explanatory variables were included using the appropriate indicator variables to represent their levels. For example, the entry line variable was represented by two indicator variables to represent its three levels. Attempt to estimate this model using `glm()` with the default values for ϵ and the number of iterations. What aspects of the output suggest there is a convergence problem? Show that a larger number of iterations and stricter convergence criteria lead to different regression parameter estimates.

(c) What aspects of the data led to the convergence problem in (b)? Examine the data in a contingency table structure to help answer the question. For example, `xtabs(formula = Success ~ EntryLine + Flagstick + BallSpeed, data = set1)`, where the data is in a data frame named `set1`, provides an important view of the data.

(d) A small adjustment can be made to some data values so that the logistic regression model can be estimated properly. One approach is to add 0.5 to each 0 value for a number of successes and to subtract 0.5 from each 100 value for a number of successes.[37] This can be done using the `ifelse()` function as shown below.

```
const <- 0.5
Success2 <- ifelse(test = set1$Success == 0, yes = const, no
   = set1$Success)
```

[37] Other approaches include adding 0.5 successes and 0.5 failures to the counts for each combination of the explanatory variables. The data could also be left as is and a modified log-likelihood function (Section 2.2.7) or a Bayesian estimation approach (Section 6.6) could be used.

```
set1$Success2 <- ifelse(test = Success2 == 100, yes = 100 -
    const, no = Success2)
```

Fit the logistic regression model described in (b) with this adjusted data. Use the adjusted data and this model for the remainder of the exercise.

(e) The main explanatory variable of interest represents whether the flagstick is in or out of the hole. Use estimated odds ratios and corresponding confidence intervals to interpret this variable. Because of the interaction in the model, there is more than one odds ratio. Control the familywise error rate at 0.05 with a Bonferroni adjustment.

(f) Perform an LRT to assess the importance of the interaction between the flagstick and ball speed explanatory variables.

(g) For each ball speed and entry line combination, compute the difference between the estimated probabilities of success when the flagstick is out of the hole vs. when the flagstick is in the hole. Find the corresponding Wald confidence interval. Control the overall familywise confidence level for the nine intervals at 95% with a Bonferroni adjustment. Below is one way these calculations can be made using the `emmeans` package to estimate variances with the delta method.

```
# mod.fit contains the results from glm() when using formula
    = Success2/Trials ~ Flagstick * BallSpeed + EntryLine
library(package = emmeans)
calc.est1 <- emmeans(object = mod.fit, specs = ~ Flagstick +
    BallSpeed + EntryLine)
regrid1 <- regrid(object = calc.est1, transform = "response")
test.info1 <- contrast(object = regrid1, simple =
    list("Flagstick"), combine = TRUE, method =
    "revpairwise")
confint(object = test.info1, adjust = "Bonferroni", level =
    0.95)
```

Exercise 16 provides information about the use of `regrid()` for a similar situation. These same calculations can be made using the `deltaMethod()` function of the `car` package. Below is an example for one of the nine intervals.

```
parameterNames <- c("beta0", "beta1", "beta2", "beta3",
    "beta4", "beta5", "beta6", "beta7")
# pi_out - pi_in for BallSpeed = 2 and EntryLine = 2
g.part <- "1 - 1/(1 + exp(beta0 + beta1*1 + beta2*1 +
    beta3*0 + beta4*1 + beta5*0 + beta6*1*1 + beta7*1*0)) -
    1 + 1/(1 + exp(beta0 + beta1*0 + beta2*1 + beta3*0 +
    beta4*1 + beta5*0 + beta6*0*1 + beta7*0*0))"
calc.save <- deltaMethod(object = mod.fit, g = g.part, func
    = "BallSpeed = 2 EntryLine = 2", level = 0.9944,
    parameterNames = parameterNames)
round(calc.save, 4)
```

The above code for `g.part` uses the result from Exercise 33.

(h) Should the flagstick be left in or taken out of the hole? Using the results from this problem, develop overall recommendations.

(i) Exercise 29 of Chapter 1 analyzes the data without a logistic regression model. Discuss the advantages and disadvantages of this approach in comparison to using the logistic regression model.

35. Continuing Exercise 34, an alternative approach is to treat the ball speed and entry line variables as ordinal categorical variables that each contribute single terms to the model. For example, a ball speed of 3 is greater than a ball speed of 2, so this variable can be represented using one model term for these numerical values rather than using two separate indicator variables. Using the numerical values of these variables as the ordering and the original data set, complete the following parts.

    (a) Fit the logistic regression model that includes flagstick, ball speed, and entry line in the model as linear terms with their two-way interactions. There will be seven regression parameters estimated for this model.

    (b) To interpret the flagstick explanatory variable, multiple odds ratios are needed because of the interaction terms. Compute the estimated odds ratios with associated confidence intervals and provide a thorough interpretation of them. We recommend writing odds ratios as functions of the regression parameters as a first step for these calculations. Control the familywise confidence level at 0.95.

    (c) Plot the estimated probability of success from the model with ball speed on the x-axis. Include separate lines for the different combinations of flagstick and entry line. These lines should have different colors and/or line types, and include a corresponding legend. Interpret the plot relative to the results from the odds ratios in (b).

    (d) Section 2.2.4 discusses the use of a bubble plot to evaluate how well a model fits the data when there is a binomial response variable. Why would the bubble aspect of this plot not be useful to evaluate the model here?

    (e) Add the observed proportions for each explanatory variable combination to the plot in part (c). One way to make this addition is with the `points()` function when the original plot was constructed using `plot()`. Choose appropriate colors for these plotting characters that enable comparisons of the points to the model plotted. Assess how well the model fits the data relative to these observed proportions.

    (f) Compare the modeling approach for this problem to what was used in Exercise 34.

36. Show that when Equation 2.9 is subtracted from Equation 2.8, Equation 2.7 is the result.

37. For a simple model, $\text{logit}(\pi_i) = \beta_0 + \beta_1 x_{i1}$, one can show:

    - $\widehat{Var}(\hat{\beta}_0) = m^{-1}\sum_{i=1}^{n} x_{i1}^2 \hat{\pi}_i(1-\hat{\pi}_i)$
    - $\widehat{Var}(\hat{\beta}_1) = m^{-1}\sum_{i=1}^{n} \hat{\pi}_i(1-\hat{\pi}_i)$
    - $\widehat{Cov}(\hat{\beta}_0, \hat{\beta}_1) = -m^{-1}\sum_{i=1}^{n} x_{i1}\hat{\pi}_i(1-\hat{\pi}_i)$

    where $m = \left[\sum_{i=1}^{n} \hat{\pi}_i(1-\hat{\pi}_i)\right]\left[\sum_{i=1}^{n} x_{i1}^2\hat{\pi}_i(1-\hat{\pi}_i)\right] - \left[\sum_{i=1}^{n} x_{i1}\hat{\pi}_i(1-\hat{\pi}_i)\right]^2$. Complete the following:

    (a) Show how these expressions are found using $(\mathbf{X}'\mathbf{V}\mathbf{X})^{-1}$. Note that the inverse of a 2×2 matrix $\begin{bmatrix} a\ b \\ c\ d \end{bmatrix}$ is $m^{-1}\begin{bmatrix} d & -b \\ -c & a \end{bmatrix}$ where $m = ad - bc$.

(b) Using these formulas, calculate the estimated covariance matrix for the estimated model from the placekicking example, $\text{logit}(\hat{\pi}) = 5.8121 - 0.1150$`distance`. Verify that this matrix is the same as what `vcov()` produces.

38. Rather than using Equation 2.15 to form a confidence interval for π, a normal approximation can be used for the probability distribution of $\hat{\pi}$. This results in the Wald interval $\hat{\pi} \pm Z_{1-\alpha/2}\widehat{Var}(\hat{\pi})^{1/2}$, where $\widehat{Var}(\hat{\pi})$ is found using the delta method. For large samples and π not close to 0 or 1, this interval is very similar to the one produced by Equation 2.15. However, for smaller samples or when π is close to 0 or 1, this interval can have problems. Discuss what these problems are and why this interval is not desirable to use in general practice.

39. Suppose a logistic regression model with three binary explanatory variables is used to estimate the probability of success. This model includes all three linear terms, the three two-way interactions, and the one three-way interaction. Derive the odds ratio to examine the odds of success between two levels of one of the explanatory variables, holding the other two constant, and provide its interpretation. Give the variance of the odds ratio estimator that would be used in a Wald confidence interval.

Chapter 3

Analyzing a multicategory response

Chapters 1 and 2 investigated a response variable that had two possible category choices. Chapter 3 generalizes this to a setting where the response variable value is chosen from a fixed set of more than two category choices. For example, response options could have one of these forms:

1. Five-level Likert scale: Strongly disagree, disagree, neutral, agree, or strongly agree
2. Chemical compounds in drug discovery experiments: Positive, blocker, or neither
3. Cereal shelf-placement in a grocery store: Bottom, middle, or top
4. Canadian political party affiliation: Conservative, New Democratic, Liberal, Bloc Quebecois, or Green
5. Beef grades: Prime, choice, select, standard, utility, or commercial

For these examples, some responses are ordinal (e.g., Likert scale) and some are not (e.g., chemical compounds). We will investigate both ordinal and nominal (unordered) multicategory responses within this chapter.

In each of the above examples, an observed unit fits into exactly one category. For example, a chemical compound cannot be both a positive and a blocker. There are other situations where a unit may fit simultaneously into more than one category, such as in "choose-all-that-apply" survey questions. We will investigate these "multiple response" problems separately in Section 6.4.

3.1 Multinomial probability distribution

The multinomial probability distribution is the extension of the binomial distribution to situations where there are more than two categories for a response. Let Y denote the categorical response random variable with levels $j = 1, \ldots, J$, where each category has probability $\pi_j = P(Y = j)$ such that $\sum_{j=1}^{J} \pi_j = 1$. If there are n identical trials with responses $Y_1, \ldots, Y_n$, then we can define random variables N_j, $j = 1, \ldots, J$, such that N_j counts the number of trials responding with category j. That is, $N_j = \sum_{i=1}^{n} I(Y_i = j)$, where $I(\cdot) = 1$ when the condition in parentheses is true, and $I(\cdot) = 0$ otherwise. Let $n_1, \ldots, n_J$ denote the observed response count for category j with $\sum_{j=1}^{J} n_j = n$. The probability mass function (PMF) for observing a particular set of counts $n_1, \ldots, n_J$ is

$$P(N_1 = n_1, \ldots, N_J = n_J) = \frac{n!}{\prod_{j=1}^{J} n_j!} \prod_{j=1}^{J} \pi_j^{n_j}, \tag{3.1}$$

DOI: 10.1201/9781003093091-3

which is known as the *multinomial probability distribution*. Notice that when $J = 2$, the distribution simplifies to the binomial distribution as described in Section 1.1.1, where $n_1 = w$, $n_2 = n - w$, $\pi_1 = \pi$, and $\pi_2 = 1 - \pi$ in the notation of that section.

We use maximum likelihood estimation to obtain estimates of $\pi_1, \ldots, \pi_J$. The likelihood function is simply Equation 3.1, and the MLE for each π_j is $\hat{\pi}_j = n_j/n$ (i.e., the observed proportion for each category).

Example: Multinomial simulated sample (Multinomial.R)

To picture what data from a multinomial distribution might look like, we simulate a sample of $n = 1000$ trials from a multinomial distribution with $\pi_1 = 0.25$, $\pi_2 = 0.35$, $\pi_3 = 0.2$, $\pi_4 = 0.1$, and $\pi_5 = 0.1$. We use the `rmultinom()` function to simulate one set (`n = 1`) of the 1000 trials (`size = 1000`) with probabilities listed in the `prob` argument.

```
> pi.j <- c(0.25, 0.35, 0.2, 0.1, 0.1)
> set.seed(2195)
> n.j <- rmultinom(n = 1, size = 1000, prob = pi.j)
> data.frame(n.j, pihat.j = n.j/1000, pi.j)
  n.j pihat.j pi.j
1 256    0.256 0.25
2 365    0.365 0.35
3 195    0.195 0.20
4  82    0.082 0.10
5 102    0.102 0.10
```

The simulated counts are saved into an object named `n.j`. After dividing these counts by n, we see that the observed proportions are very similar to the actual parameter values.

If we wanted to simulate m sets of 1000 trials from the same multinomial distribution, we could change the `n` argument value to the numerical value for m. In this situation, there would be m columns of observations produced by `rmultinom()`. For example, the result for $m = 3$ is shown below.

```
> set.seed(9182)
> n.j <- rmultinom(n = 3, size = 1000, prob = pi.j)
> n.j
     [,1] [,2] [,3]
[1,]  259  259  237
[2,]  341  346  374
[3,]  200  188  198
[4,]   92  106   89
[5,]  108  101  102
```

As we can see, there is variability in the counts from one sample set to the next. This same type of variability should be expected in actual samples with multinomial responses.

The program for this example also shows how to use the `dmultinom()` function to evaluate the PMF for a given set of counts $n_1, \ldots, n_J$.

Table 3.1: $I \times J$ contingency table.

		Y				
		1	2	$\cdots$	J	Total
X	1	n_{11}	n_{12}	$\cdots$	n_{1J}	n_{1+}
	2	n_{21}	n_{22}	$\cdots$	n_{2J}	n_{2+}
	$\vdots$	$\vdots$	$\vdots$	$\ddots$	$\vdots$	$\vdots$
	I	n_{I1}	n_{I2}	$\cdots$	n_{IJ}	n_{I+}
	Total	n_{+1}	n_{+2}	$\cdots$	n_{+J}	n

3.2 $I \times J$ contingency tables and inference procedures

Section 1.2 discussed how counts from two binomial distributions can be analyzed in the form of a 2×2 contingency table. We expand this discussion now to allow for more than two rows and/or columns. We assume that two categorical variables, X and Y, are measured on each unit, where X has levels $i = 1, \ldots, I$, and Y has levels $j = 1, \ldots, J$. Our sample consists of n units that are cross-classified according to their levels of X and Y. The counts of units for which the combination $(X = i, Y = j)$ is observed are denoted by n_{ij}, and these represent observations of the random variables N_{ij}, $i = 1, \ldots, I$, $j = 1, \ldots, J$. Our goal is to use the observed counts to make inferences about the distribution of the N_{ij} and to identify any relationships between X and Y. To this end, we can create a contingency table summarizing these counts by letting X be the row variable and Y be the column variable. Table 3.1 provides the general contingency table. Note that we use a $+$ subscript to indicate a sum across one of the indices; for example, $n_{+j} = \sum_{i=1}^{I} n_{ij}$ is the total for column j. We let n_{++} be denoted simply by n.

In the discussion that follows, we examine two of the models that are typically used to represent counts in a contingency table. The first model assumes that the observed counts of the contingency table come from one multinomial distribution. The second model assumes that these counts arise independently from I multinomial distributions, where there is a separate distribution representing each row. In the end, these models lead to equivalent results.

3.2.1 One multinomial distribution

First, consider the situation where a fixed sample size of n units are sampled from a large population. We define $\pi_{ij} = P(X = i, Y = j)$. This is the same multinomial setting as in Section 3.1 with n trials and IJ possible responses, except now each possible response is a combination of two variables instead of just one. We assume that each sampled unit has one and only one X and Y category combination, so that $\sum_{i=1}^{I} \sum_{j=1}^{J} \pi_{ij} = 1$.

Maximum likelihood estimation for these probabilities proceeds analogously to what was done in Section 3.1, with minor modifications to the notation to account for the second variable. The PMF for $N_{11}, \ldots, N_{IJ}$ becomes

$$P(N_{11} = n_{11}, \ldots, N_{IJ} = n_{IJ}) = \frac{n!}{\prod_{i=1}^{I} \prod_{j=1}^{J} n_{ij}!} \prod_{i=1}^{I} \prod_{j=1}^{J} \pi_{ij}^{n_{ij}}, \quad (3.2)$$

which is also the likelihood function for a sample of size n. The MLE of π_{ij} is the estimated proportion $\hat{\pi}_{ij} = n_{ij}/n$.

Marginal distributions for X and Y may be found as well. The marginal probability for level i of X is $\pi_{i+} = P(X = i)$ for $i = 1, \dots, I$. The marginal distribution of X is therefore multinomial with n trials, probabilities $\pi_{1+}, \dots, \pi_{I+}$, and counts $n_{1+}, \dots, n_{I+}$. Similarly, the marginal probability for category j of Y is $\pi_{+j} = \sum_{i=1}^{I} \pi_{ij}$ for $j = 1, \dots, J$. The marginal distribution of Y is multinomial with n trials, probabilities $\pi_{+1}, \dots, \pi_{+J}$, and counts $n_{+1}, \dots, n_{+J}$. Note that $\sum_{i=1}^{I} \pi_{i+} = 1$ and $\sum_{j=1}^{J} \pi_{+j} = 1$. The MLEs of π_{i+} and π_{+j} are the corresponding row and column proportions, $\hat{\pi}_{i+} = n_{i+}/n$ and $\hat{\pi}_{+j} = n_{+j}/n$, respectively.

When the outcome of X does not have an effect on the probabilities for the outcomes of Y, we say that Y is *independent* of X. As a result of independence, the probability of any joint outcome $(X = i, Y = j)$ factors into the marginal probabilities for $X = i$ and $Y = j$: $\pi_{ij} = \pi_{i+}\pi_{+j}$. Independence simplifies the probability structure within a contingency table by reducing the number of unknown parameters to the $(I-1) + (J-1) = I + J - 2$ marginal probabilities. Note that the "-1" parts occur because of the $\sum_{i=1}^{I} \pi_{i+} = 1$ and $\sum_{j=1}^{J} \pi_{+j} = 1$ constraints on the probabilities. Without independence, there are $IJ - 1$ unknown probability parameters, where the "-1" part occurs because of the $\sum_{i=1}^{I} \sum_{j=1}^{J} \pi_{ij} = 1$ constraint on the probabilities. Thus, there is a reduction of $(IJ - 1) - (I + J - 2) = (I - 1)(J - 1)$ parameters compared to the same table without independence. We will examine how to perform a hypothesis test for independence shortly.

3.2.2 I multinomial distributions

Suppose each of the I levels of X represents a grouping, like year in high school (freshman, sophomore, junior, and senior) or blood type (A+, A–, B+, B–, AB+, AB–, O+, and O–). An alternative model is needed when samples of sizes n_{i+}, $i = 1, \dots, I$ are deliberately taken from each of I different groups for X. In this case, the marginal counts n_{i+} are fixed by design, so we have a separate J-category multinomial distribution in each of the I groups, where $n = \sum_{i=1}^{I} n_{i+}$. Each of these distributions has its own set of probability parameters. Define $P(Y = j|X = i) = \pi_{j|i}$ as the conditional probability of observing response category j of Y given that a unit is from group i of X. Note that $\sum_{j=1}^{J} \pi_{j|i} = 1$ for each $i = 1, \dots, I$. The conditional joint distribution of $N_{i1}, \dots, N_{iJ}$ has PMF

$$P(N_{i1} = n_{i1}, \dots, N_{iJ} = n_{iJ}|N_{i+} = n_{i+}) = \frac{n_{i+}!}{\prod_{j=1}^{J} n_{ij}!} \prod_{j=1}^{J} \pi_{j|i}^{n_{ij}}$$

for each $i = 1, \dots, I$. Assuming that I different samples are taken independently of one another, the likelihood for the parameters is just the product of I multinomial distributions,

$$\prod_{i=1}^{I} \frac{n_{i+}!}{\prod_{j=1}^{J} n_{ij}!} \prod_{j=1}^{J} \pi_{j|i}^{n_{ij}}. \tag{3.3}$$

As a result, this model is most often referred to as the *product multinomial* model. The MLE of $\pi_{j|i}$ is $\hat{\pi}_{j|i} = n_{ij}/n_{i+}$. Notice that the same estimates result from applying the definition of conditional probability, $P(Y = j|X = i) = P(X = i, Y = j)/P(X = i)$, to the estimates from the one multinomial model: $\hat{\pi}_{j|i} = \hat{\pi}_{ij}/\hat{\pi}_{i+} = (n_{ij}/n)/(n_{i+}/n) = n_{ij}/n_{i+}$. In fact, one can start with Equation 3.2, condition on $n_{1+}, \dots, n_{I+}$, and obtain the same PMF as given in Equation 3.3. See Exercise 3 for more details.

Independence of X and Y in the context of a product multinomial model means that the conditional probabilities for each Y are equal across the rows of the table. That is, for each j, $\pi_{j|1} = \cdots = \pi_{j|I}$. Equivalently, $\pi_{j|i} = \pi_{+j}$ for each i and j. To see this equivalence, note that $\pi_{ij} = \pi_{j|i}\pi_{i+}$ and $\pi_{+j} = \sum_{i=1}^{I} \pi_{ij} = \sum_{i=1}^{I} \pi_{j|i}\pi_{i+}$. Because $\pi_{j|1} = \cdots = \pi_{j|I}$, we refer to this constant value as $\pi_{j|i^*}$. This constant is factored out of the previous summation to leave us with $\pi_{+j} = \pi_{j|i^*} \sum_{i=1}^{I} \pi_{i+} = \pi_{j|i^*} \times 1 = \pi_{j|i^*}$.

Independence in the context of the product multinomial model also is mathematically equivalent to independence as defined for the one multinomial model. To show this equivalence, we again need to use the definition of conditional probability. Because X is not random in the product multinomial model, we define π_{i+} to be the fixed proportion of the total sample that is taken from group i. Then $\pi_{j|i} = \pi_{ij}/\pi_{i+}$ and $\pi_{j|i} = \pi_{+j}$ together imply that $\pi_{ij} = \pi_{i+}\pi_{+j}$.

We have now seen that parameter estimates from the one and product multinomial models are the same, the definitions of independence in the two models are equivalent, and the two models lead to exactly the same conditional distributions for Y given $X = i$. As a consequence, analyses conducted based on each model generally yield the same results. Therefore, when developing tests for independence and other analyses on contingency tables, we assume whichever model for the table is most convenient.

Example: Multinomial simulated sample (Multinomial.R)

We simulate samples here for both the one and product multinomial models to help readers understand the relationships between them. The data are simulated using `rmultinom()` as in Section 3.1. The main difference is in how we define probabilities to correspond to cells of an $I \times J$ contingency table.

Below is how we simulate a sample of size $n = 1000$ for a 2×3 contingency table corresponding to the one multinomial model with $\pi_{11} = 0.2$, $\pi_{21} = 0.3$, $\pi_{12} = 0.2$, $\pi_{22} = 0.1$, $\pi_{13} = 0.1$, and $\pi_{23} = 0.1$.

```
> # Probabilities entered by column for array()
> pi.ij <- c(0.2, 0.3, 0.2, 0.1, 0.1, 0.1)
> pi.table <- array(data = pi.ij, dim = c(2,3), dimnames =
   list(X = 1:2, Y = 1:3))
> pi.table  # pi_ij
   Y
X     1   2   3
  1 0.2 0.2 0.1
  2 0.3 0.1 0.1

> set.seed(9812)
> save <- rmultinom(n = 1, size = 1000, prob = pi.ij)
> c.table1 <- array(data = save, dim = c(2,3), dimnames = list(X
   = 1:2, Y = 1:3))
> c.table1
  Y
X   1   2   3
1 191 206  94
2 311  95 103
> c.table1/sum(c.table1)
   Y
X       1     2     3
  1 0.191 0.206 0.094
  2 0.311 0.095 0.103
```

For example, $n_{11} = 191$ and $n_{23} = 103$. We see that each $\hat{\pi}_{ij}$ is quite similar to its corresponding π_{ij} as would be expected for a sample this large. Also, notice that the row marginal totals vary—$n_{1+} = 491$ and $n_{2+} = 509$—and are very close to what would be expected considering that $\pi_{1+} = \pi_{2+} = 0.5$.

For the I multinomial setting, we again simulate a sample for a 2×3 contingency table (thus, $I = 2$). With this model, we need to draw samples of fixed size separately for each row. We also need to re-express the cell probabilities as conditional probabilities that sum to 1 in each row, which results in $\pi_{1|1} = 0.4, \pi_{2|1} = 0.4, \pi_{3|1} = 0.2$ and $\pi_{1|2} = 0.6, \pi_{2|2} = 0.2, \pi_{3|2} = 0.2$. Unlike the one multinomial model, the row totals are not specified by the probabilities on the table, so we can choose them to suit the needs of the problem. We arbitrarily select $n_{1+} = 400$ and $n_{2+} = 600$. Two separate calls to the `rmultinom()` function are used to generate samples independently from the two separate probability distributions.

```
> pi.cond <- pi.table/rowSums(pi.table)
> pi.cond  # pi_j|i
   Y
X     1   2   3
  1 0.4 0.4 0.2
  2 0.6 0.2 0.2

> set.seed(8111)
> save1 <- rmultinom(n = 1, size = 400, prob = pi.cond[1,])
> save2 <- rmultinom(n = 1, size = 600, prob = pi.cond[2,])

> c.table2 <- array(data = c(save1[1], save2[1], save1[2],
   save2[2], save1[3], save2[3]), dim = c(2,3), dimnames =
   list(X = 1:2, Y = 1:3))
> c.table2
   Y
X     1   2   3
  1 162 159  79
  2 351 126 123

> rowSums(c.table2)
  1   2
400 600
> c.table2/rowSums(c.table2)  # Estimate of pi_j|i
   Y
X       1      2      3
  1 0.405 0.3975 0.1975
  2 0.585 0.2100 0.2050

> round(c.table1/rowSums(c.table1), 4)  # From 1 multinomial
  Y
X     1      2      3
1 0.389 0.4196 0.1914
2 0.611 0.1866 0.2024
```

We again see that each $\hat{\pi}_{j|i}$ is quite similar to its respective $\pi_{j|i}$ as would be expected.

If we wanted to simulate the data under independence, we could find π_{ij}'s that satisfy $\pi_{ij} = \pi_{i+}\pi_{+j}$ for each i, j pair and use the `rmultinom()` function to simulate a sample from one six-category multinomial distribution. Equivalently, we could find π_{+j} for

$j = 1, 2,$ and 3, and simulate data from two three-category multinomial distributions using two calls to the `rmultinom()` function. We demonstrate this process in our program for this example.

3.2.3 Test for independence

A test for independence,

$$H_0 : \pi_{ij} = \pi_{i+}\pi_{+j} \text{ for each } i, j$$
$$H_a : \pi_{ij} \neq \pi_{i+}\pi_{+j} \text{ for some } i, j$$

can be performed using a Pearson chi-square test and an LRT. These tests were already shown in Section 1.2.3 as tests for the equality of two binomial success probabilities, which is equivalent to a test for independence in a product multinomial model with $I = J = 2$. The Pearson chi-square test statistic is again formed by summing (observed count − estimated expected count)^2/(estimated expected count) across all cells of the contingency table. The observed count is n_{ij}. The expected count under independence is $n\hat{\pi}_{i+}\hat{\pi}_{+j} = n_{i+}n_{+j}/n$ for the one multinomial model, or equivalently $n_{i+}\hat{\pi}_{+j} = n_{i+}n_{+j}/n$ under the product multinomial model. This leads to the test statistic

$$X^2 = \sum_{i=1}^{I}\sum_{j=1}^{J}\frac{(n_{ij} - n_{i+}n_{+j}/n)^2}{n_{i+}n_{+j}/n}.$$

Note that X^2 is equivalent to Equation 1.7 when $I = 2$ and $J = 2$ (see Exercise 1). The X^2 statistic has an approximate $\chi^2_{(I-1)(J-1)}$ distribution in large samples when the null hypothesis is true. When the null hypothesis is false, we expect large deviations between the observed and expected counts (relative to the size of $n_{i+}n_{+j}/n$), which lead to large values of the X^2 statistic. Therefore, we reject the null hypothesis of independence between X and Y when $X^2 > \chi^2_{(I-1)(J-1),1-\alpha}$.

The likelihood ratio is formed in the usual way as

$$\Lambda = \frac{\text{Maximum of likelihood function under } H_0}{\text{Maximum of likelihood function under } H_0 \text{ or } H_a}.$$

Computation of Λ is based on Equation 3.2, using $\hat{\pi}_{ij} = \hat{\pi}_{i+}\hat{\pi}_{+j} = n_{i+}n_{+j}/n$ in the numerator and $\hat{\pi}_{ij} = n_{ij}/n$ in the denominator to estimate each π_{ij}. Applying the usual transformation with Λ, we have

$$-2\log(\Lambda) = 2\sum_{i=1}^{I}\sum_{j=1}^{J} n_{ij}\log\left(\frac{n_{ij}}{n_{i+}n_{+j}/n}\right),$$

where a term with $n_{ij} = 0$ inside the summations is taken to 0. Like X^2, the LRT statistic has a large-sample $\chi^2_{(I-1)(J-1)}$ distribution when H_0 is true and uses the same rejection rule.

The degrees of freedom for the test of independence are found by calculating

$$(\text{Number of parameters under } H_a) - (\text{Number of parameters under } H_0).$$

This is a general way to find degrees of freedom for any model-comparison test. As shown in Section 3.2.1, we need to estimate $I + J - 2$ parameters when the null hypothesis of

Table 3.2: Bloating severity after eating a fiber-enriched cracker.

		Bloating severity			
		None	Low	Medium	High
Fiber source	None	6	4	2	0
	Bran	7	4	1	0
	Gum	2	2	3	5
	Both	2	5	3	2

independence holds and $IJ - 1$ when it does not. Thus, the degrees of freedom for the test of independence are $(IJ - 1) - (I + J - 2) = (I - 1)(J - 1)$.

Both the LRT and the Pearson chi-square test generally give similar results in large samples. However, at times, their values may differ considerably in smaller samples, leading to ambiguity if their values lie in opposite sides of the rejection region. There have been a number of recommendations given for what constitutes a "large enough" sample to obtain a good χ^2 approximation. The most common criteria are $n_{i+}n_{+j}/n > 1$ or > 5 for all cells of the contingency table. These criteria may not be satisfied when there are very small cell counts in many cells of the table. For example, a row for which the marginal count n_{i+} is not much larger than the number of columns cannot possibly have "large" expected cell counts in all columns. In these instances, a Monte Carlo simulation or the exact testing procedures described in Section 6.2 can provide a visual assessment to determine if the $\chi^2_{(I-1)(J-1)}$ distributional approximation is appropriate. This is our preferred approach whenever there is any doubt regarding the χ^2 approximation, and we provide an example of its implementation shortly.

Example: Fiber-enriched crackers (Fiber.R, Fiber.csv)

Dietary fiber is a healthful compound that is found in many vegetables and grains. Heavily processed foods are low in fiber, so it is sometimes added to such foods to make them more nutritious. Unfortunately, high-fiber foods can also have the side-effect of causing digestive bloating in some people. To investigate properties of a new fiber-enriched cracker, a study[1] was performed where individuals were asked to eat the cracker before having a meal. Shortly afterward, the participants were instructed to describe any bloating that they experienced. Table 3.2 gives the resulting data in a 4×4 contingency table. There were four different types of crackers based on their fiber source, and there were four levels of bloating severity reported. While not specified in the study description, we assume that each participant was assigned to only one fiber source group. Also, the study description does not mention whether the fact that there are 12 participants for each fiber source was by design or by chance, so it is not clear whether a one-multinomial or product-multinomial model might be more appropriate. Fortunately, as noted at the end of Section 3.2.2, this distinction has no bearing on the analysis results. Finally, while the data given with the study were in a 4×4 contingency table, we can see that fiber source could actually be split into two separate explanatory variables, one indicating whether bran is present and one indicating whether gum is present. We explore alternative analysis methods that account for this structure in Section 3.3.

[1]The study description is archived at `https://web.archive.org/web/20000620172010/http://lib.stat.cmu.edu/DASL/Stories/HighFiberDietPlan.html`. The original data source was The Data and Story Library.

We begin by reading the data into R and summarizing the data in a contingency table form using the `xtabs()` function.

```
> diet <- read.csv(file = "C:\\data\\Fiber.csv",
    stringsAsFactors = TRUE)
> head(diet)
  fiber  bloat count
1  bran   high     0
2   gum   high     5
3  both   high     2
4  none   high     0
5  bran medium     1
6   gum medium     3
```

```
> diet$fiber <- factor(x = diet$fiber, levels = c("none",
    "bran", "gum", "both"))
> diet$bloat <- factor(x = diet$bloat, levels = c("none", "low",
    "medium", "high"))
> diet.table <- xtabs(formula = count ~ fiber + bloat, data =
    diet)
> diet.table
      bloat
fiber  none low medium high
  none    6   4      2    0
  bran    7   4      1    0
  gum     2   2      3    5
  both    2   5      3    2
```

We use the `factor()` function to change the order of the levels for both variables to match that given in Table 3.2. We could instead have entered the contingency table directly into R using the `array()` function and arranged the array in the desired order. We provide an example of this in our program.

We would like to determine if bloating severity is related to the type of fiber. If there is a relation, this would indicate to the cracker manufacturer that some types of fiber may cause more severe bloating and should be avoided. We use a test for independence to make this determination. The hypotheses for the test are $H_0 : \pi_{ij} = \pi_{i+}\pi_{+j}$ for each i, j vs. $H_a : \pi_{ij} \neq \pi_{i+}\pi_{+j}$ for some i, j, where π_{ij} is the probability that a randomly selected person is assigned to fiber level i and experiences bloating level j. R provides a number of functions to test these hypotheses, including: (1) `chisq.test()`, (2) `assocstats()` from the `vcd` package, and (3) the generic function `summary()` to summarize a table object. Normally, we would only use one of these functions, but we show all three for illustrative purposes.

```
> ind.test <- chisq.test(x = diet.table, correct = FALSE)
Warning message:
In chisq.test(diet.table, correct = FALSE) :
  Chi-squared approximation may be incorrect
> ind.test

        Pearson's Chi-squared test
```

```
data:  diet.table
X-squared = 16.943, df = 9, p-value = 0.04962

> library(package = vcd)
> assocstats(x = diet.table)
                    X^2 df P(> X^2)
Likelihood Ratio 18.880  9 0.026230
Pearson          16.943  9 0.049621

Phi-Coefficient   : NA
Contingency Coeff.: 0.511
Cramer's V        : 0.343

> class(diet.table)
[1] "xtabs" "table"
> summary(diet.table)
Call: xtabs(formula = count ~ fiber + bloat, data = diet)
Number of cases in table: 48
Number of factors: 2
Test for independence of all factors:
        Chisq = 16.943, df = 9, p-value = 0.04962
        Chi-squared approximation may be incorrect

> qchisq(p = 0.95, df = 9)
[1] 16.91898
```

Note that we include the `correct = FALSE` argument value in `chisq.test()` to prevent a continuity correction from being applied. Please see Section 1.2.3 for why we avoid these corrections. The Pearson chi-square test statistic is $X^2 = 16.94$, the critical value at the $\alpha = 0.05$ level is $\chi^2_{9,0.95} = 16.92$, and the p-value is $P(A > 16.94) = 0.0496$ where $A \sim \chi^2_9$. Because the p-value is small, but not extremely so, we would say there is marginal evidence against independence (thus, there is some, but not strong, evidence of association) between fiber source and bloating. The LRT leads to a similar conclusion with $-2\log(\Lambda) = 18.88$ and a p-value of 0.0262.

There are a number of very small cell counts in the contingency table, so there may be some concern about the appropriateness of the χ^2_9 approximation. The expected cell counts under independence for this table satisfy the $n_{i+}n_{+j}/n > 1$ criterion given earlier, but not $n_{i+}n_{+j}/n > 5$.

```
> ind.test$expected
      bloat
fiber  none  low medium high
  none 4.25 3.75   2.25 1.75
  bran 4.25 3.75   2.25 1.75
  gum  4.25 3.75   2.25 1.75
  both 4.25 3.75   2.25 1.75
```

The `chisq.test()` and `summary()` functions even print a warning message about the appropriateness of the χ^2_9 approximation. This warning is given by the functions whenever $n_{i+}n_{+j}/n < 5$ for any cell.

One way to examine the distributional approximation is through a Monte Carlo simulation. This involves simulating a large number of contingency tables of size n

whose probabilities satisfy the null hypothesis of independence. Ideally, one should follow the same type of sampling method as the assumed model. For a one-multinomial model, we can use one call to `rmultinom()`, where we set each π_{ij} (`prob` argument) to $n_{i+}n_{+j}/n^2$ (the estimated expected count under the null hypothesis divided by the sample size). For a product-multinomial model, we can use I different calls to `rmultinom()` by setting $\pi_{j|i}$ equal to $\hat{\pi}_{+j}$. We chose to use the one multinomial model here because of its simplicity and the unknown nature of the actual sampling method.

The Pearson chi-square test statistic, say X^{2*}, is calculated for each contingency table of simulated counts. The estimated distribution for the statistic can be summarized through different plots, such as a histogram, an empirical CDF plot[2], and a quantile-quantile (QQ)-plot.[3] One can overlay the approximate χ^2_9 distribution (or examine the deviation from a straight line in the QQ-plot), to judge if the approximation is appropriate. The same procedure can also be applied to the LRT statistic to asses the quality of its approximate χ^2_9 distribution.

The Monte Carlo simulation using 10,000 simulated data sets is implemented with code in the program for this example. Figure 3.1 shows plots of the distribution for X^{2*} along with the χ^2_9 distribution. We see general agreement in both plots between the distribution of X^{2*} and a χ^2_9 distribution. The quantiles at $\alpha = 0.90, 0.95,$ and 0.99 are 14.57, 16.61, and 21.04 for X^{2*} and 14.68, 16.92, and 21.67 for χ^2_9, respectively. Thus, the χ^2_9 distribution appears to be a good approximation to the distribution of X^2 in this example.

It is interesting to note that the Monte Carlo simulation performed here is actually a form of the bootstrap. The bootstrap is applied here as a nonparametric procedure that provides estimates of a CDF for a statistic. For this example, we can obtain a bootstrap estimate of the p-value for the test by calculating the proportion of simulated X^{2*} values greater than or equal to $X^2 = 16.94$. The result is 0.0446, which is in general agreement with the p-value from the Pearson chi-square test. For a more in-depth discussion of the bootstrap, please see Davison and Hinkley (1997).

When independence is rejected, there are a number of ways to examine the association between X and Y. In Section 1.2 with 2×2 contingency tables, we could simply calculate an odds ratio to summarize the degree of association; i.e., the amount by which the odds ratio is greater than or less than 1. Now, with $I \times J$ contingency tables, one odds ratio does not summarize the association in the entire table. Instead, one approach is to calculate odds ratios for 2×2 parts of a contingency table that are of special interest. Unless there is firm guidance from the context of the problem, there is potential for creating many different odds ratios, and, hence, potential for inflated error rates for multiple tests or confidence intervals.

As an alternative to odds ratios, various forms of residuals may be calculated on each cell to measure how far the observed cell count deviates from what would be expected under independence. The basic residual in cell (i, j) has the form $n_{ij} - n_{i+}n_{+j}/n$. A different

[2]The empirical CDF is the sample analog of the CDF. For a sample of observations $w_1, w_2, \ldots, w_m$, the empirical CDF at w is the proportion of observations at or below w: $\hat{F}(w) = m^{-1}\sum_{j=1}^{m} I(w_j \leq w)$, where $I(\cdot)$ is an indicator function with a value of 1 when $w_j \leq w$ and a value of 0 otherwise.

[3]Ordered observed values from a statistic are plotted against the corresponding quantiles from a particular probability distribution of interest. For example, the observed median (provided the number of observed values is odd) would be plotted against the 0.5 quantile from the probability distribution. If the plotted points follow close to a straight line of $y = x$, where y (x) indicates what is plotted on the y-axis (x-axis), the statistic approximately follows the probability distribution.

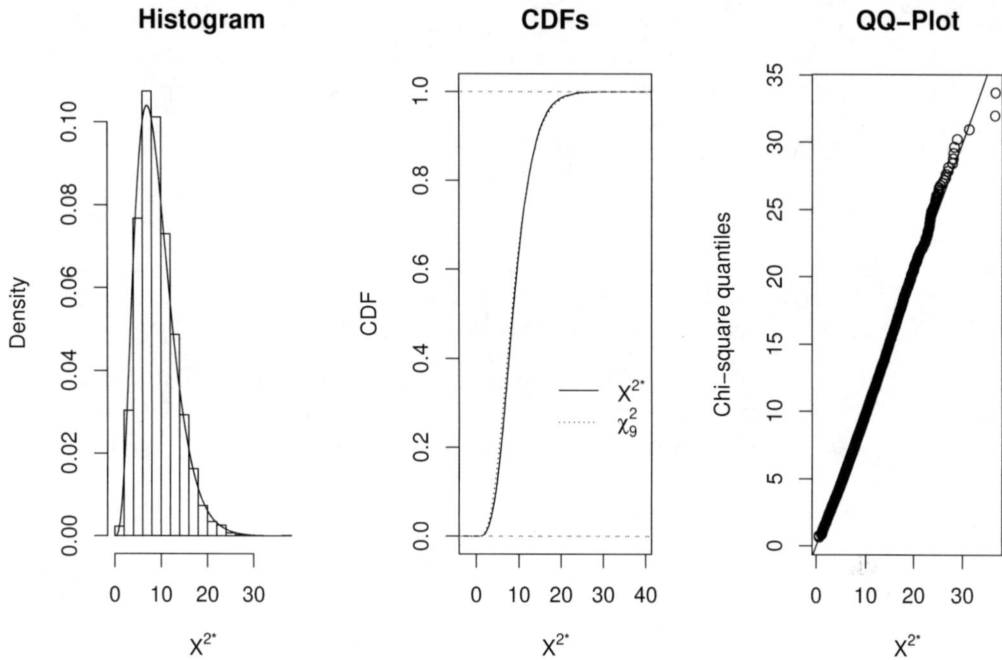

Figure 3.1: Plots of $X^{2\star}$ and a χ^2_9 distribution.

version of residual, the *standardized Pearson residual*, is

$$\frac{n_{ij} - n_{i+}n_{+j}/n}{\sqrt{(1 - n_{i+}/n)(1 - n_{+j}/n)n_{i+}n_{+j}/n}},$$

where the denominator is $\widehat{Var}(N_{ij} - N_{i+}N_{+j}/n)^{1/2}$. The `stdres` component of an object resulting from `chisq.test()` provides these standardized Pearson residuals. The advantage of a standardized Pearson residual is that its distribution can be approximated by the standard normal.[4] This means that values beyond, say, ± 2 can be considered "unusual" and should occur only in about 5% of cells. Cells where large standardized Pearson residuals occur indicate potential sources of the association between X and Y.

Residuals and odds ratios are somewhat ineffective in tables where I and/or J are much larger than 2, because they do not lead easily to a systematic understanding of the relationship between X and Y. However, displaying the residuals in the same rows and columns as the contingency table can sometimes reveal patterns of positive or negative values that hint at a larger cause for the association. For example, if residuals in a given row show an increasing (or decreasing) trend across increasing levels of an ordinal Y, then this indicates that this row's responses tend toward larger (or smaller) levels of Y than in other rows where this does not occur. Alternatively, if one cell has a residual that is large and of the opposite sign from others in the same row and column, then this one cell appears to violate independence. Unfortunately, patterns of residuals are not always apparent. Regression

[4]A *Pearson residual* has the same numerator, but the denominator is $\widehat{Var}(N_{ij})^{1/2} = \sqrt{(1 - n_{i+}/n)(1 - n_{+j}/n)}$, which results in a statistic that is not as well approximated by a standard normal distribution.

modeling is a more systematic approach to assess the association between variables. Sections 3.3 and 3.4 cover a variety of model structures that allow us to better understand the association between X and Y. Furthermore, ordinal properties of multinomial response categories can be taken into account more easily with these models and point to sources of association that may not be noticeable otherwise. Odds ratios and residuals can still be examined within these regression models.

3.3 Nominal response regression models

We now move to the problem of modeling the probabilities of a categorical response variable Y with response categories $j = 1, \ldots, J$ using explanatory variables $x_1, \ldots, x_p$. In Section 1.2.5, we defined odds for a binary response as $P(\text{success})/P(\text{failure})$. More generally, for a multinomial response, we can define odds to be a comparison of *any* pair of response categories. For example, $\pi_j/\pi_{j'}$ is the odds of category j relative to j'. A popular regression model for multinomial responses is developed then by selecting one response category as the base level and forming the odds of the remaining $J-1$ categories against this level. For example, let $j = 1$ represent the base level category, and form the odds π_j/π_1 for $j = 2, \ldots, J$. These odds are then modeled as a function of explanatory variables using a generalized form of logistic regression. Notice that if $J = 2$, we have $\log(\pi_2/\pi_1) = \log(\pi_2/(1-\pi_2))$, which is equivalent to $\log(\pi/(1-\pi))$ as in logistic regression (response category 2 becomes the "success" category).

Specifically, a *multinomial regression model* (also known as a *baseline-category logit model*) relates a set of explanatory variables to each log-odds according to

$$\log(\pi_j/\pi_1) = \beta_{j0} + \beta_{j1}x_1 + \cdots + \beta_{jp}x_p, \tag{3.4}$$

for $j = 2, \ldots, J$. This is like having $J-1$ logistic regression models using the same explanatory variables, $x_1, \ldots, x_p$. Each model has its own regression parameters, $\beta_{j0}, \beta_{j1}, \ldots, \beta_{jp}$ for $j = 2, \ldots, J$, which allow each response's log-odds to relate to the explanatory variables in a different way.

Also, notice that by subtracting the appropriate log-odds we can rewrite Equation 3.4 to compare any pair of response categories. For example, to find $\log(\pi_j/\pi_{j'})$ where $j' \neq 1$ and $j' \neq j$, we compute

$$\begin{aligned}
\log(\pi_j/\pi_{j'}) &= \log(\pi_j) - \log(\pi_{j'}) \\
&= \log(\pi_j) - \log(\pi_{j'}) + \log(\pi_1) - \log(\pi_1) \\
&= \log(\pi_j/\pi_1) - \log(\pi_{j'}/\pi_1) \\
&= (\beta_{j0} - \beta_{j'0}) + (\beta_{j1} - \beta_{j'1})x_1 + \cdots + (\beta_{jp} - \beta_{j'p})x_p. \qquad (3.5)
\end{aligned}$$

Thus, the choice of base level is not important and can be made based on convenience or interpretation.

While multinomial regression models are not GLMs as described in Section 2.3, they are part of larger model classes known as multinomial GLMs and vector GLMs. These classes allow for multiple parameters of a distribution, such as the multinomial distribution, to be functions of explanatory variables and regression parameters.

The probabilities for each individual category also can be found in terms of the model. We can rewrite Equation 3.4 as

$$\pi_j = \pi_1 \exp(\beta_{j0} + \beta_{j1}x_1 + \cdots + \beta_{jp}x_p)$$

using properties of logarithms. Noting that $\pi_1 + \pi_2 + \cdots + \pi_J = 1$, we have

$$\pi_1 + \pi_1 \exp(\beta_{20} + \beta_{21}x_1 + \cdots + \beta_{2p}x_p) + \cdots + \pi_1 \exp(\beta_{J0} + \beta_{J1}x_1 + \cdots + \beta_{Jp}x_p) = 1.$$

By factoring out the common π_1 in each term, we obtain an expression for π_1:

$$\pi_1 = \frac{1}{1 + \sum_{j=2}^{J} \exp(\beta_{j0} + \beta_{j1}x_1 + \cdots + \beta_{jp}x_p)}. \tag{3.6}$$

This leads to a general expression for π_j:

$$\pi_j = \frac{\exp(\beta_{j0} + \beta_{j1}x_1 + \cdots + \beta_{jp}x_p)}{1 + \sum_{j'=2}^{J} \exp(\beta_{j'0} + \beta_{j'1}x_1 + \cdots + \beta_{j'p}x_p)} \tag{3.7}$$

for $j = 2, \ldots, J$.

Parameters for the model are estimated using maximum likelihood. For a sample of m observations, we have y_i, the category response, and $x_{i1}, \ldots, x_{ip}$, the explanatory variables for $i = 1, \ldots, m$. The likelihood function is simply the product of m multinomial distributions with probability parameters as given by Equations 3.6 and 3.7. Iterative numerical procedures are used then to find the parameter estimates. The estimated covariance matrix for the parameter estimates is found using standard likelihood procedures as outlined in Appendix B.3.4. Wald and LR-based inference methods are performed in the same ways as for likelihood procedures in earlier chapters.

We will focus on two R functions for model estimation: `multinom()` from the `nnet` package and `vglm()` from the `VGAM` package. The `nnet` package is in the default installation of R, but one still needs to use the `library()` function to make its functions available. The origin of the package pre-dates the existence of R (package in S-Plus), so `multinom()` is widely used today. The algorithm used in `multinom()` can take many iterations to converge for some data sets, especially those where the explanatory variables are on different scales or are highly correlated. As mentioned in the help for `multinom()`, rescaling the explanatory variable values to be within 0 and 1 can help with convergence when needed, and this process is shown in Exercise 14. We have also found that increasing the number of iterations (`maxit` argument) and decreasing the convergence criterion (`reltol` argument) helps to achieve convergence and match estimates from other model-fitting functions. One main downside of using `multinom()` is that profile LR inference procedures are absent from the function and other functions that potentially could be used with its model fit. For example, neither a method function for `confint()` nor the `mcprofile` package has been extended to calculate profile LR intervals for objects resulting from `multinom()`. It is important to note that the `emmeans` package can be used for Wald-based inferences involving linear combinations of regression parameters.

The `VGAM` package provides an alternative way to fit the model and perform inferences. This package is user-contributed and not in the default installation of R, so it needs to be installed before use. Its `vglm()` function can fit a number of different vector GLMs, including multinomial regression models. We will use this function to estimate some of these other vector GLMs later in the textbook. Another benefit of using `vglm()` is profile LR intervals for regression parameters can be calculated with `confint()`. Unfortunately, the `mcprofile` and `emmeans` packages have not been extended to calculate confidence intervals involving linear combinations of regression parameters. Also, a downside of this package is it is much newer than the `nnet` package, so changes are still being made to it. In fact, the package's help warns "This package is undergoing continual development and improvement, therefore users should treat everything as subject to change." Since the first edition of our textbook, we have seen the positive aspects of these changes, such as the addition of profile

LR inference procedures, but also the negative aspects, such as arguments changing for `vglm()` that prevent our previous code from working.

The `VGAM` package is not alone among packages where changes prevent previous code from working or behaving as expected. Our statement here is not meant to alarm readers about user-contributed packages in general. Rather, there is a vast number of R packages contributed by users, and many of them can perform calculations that no other software can; however, these packages need to be used in a judicious manner. One needs to take into account many factors, such as age, stability, version number, and authors, when deciding whether to choose a user-contributed package. As mentioned in our Preface, this is why we focus in our textbook on functions in the default installation of R. When a package provides important capabilities beyond what is in the default installation, we present that package as well.

Example: Wheat kernels (Wheat.R, Wheat.csv)

The presence of sprouted or diseased kernels in wheat can reduce the value of a wheat producer's entire crop. It is important to identify these kernels after being harvested but prior to sale. To facilitate this identification process, automated systems have been developed to separate healthy kernels from the rest. Improving these systems requires better understanding of the measurable ways in which healthy kernels differ from kernels that have sprouted prematurely or are infected with a fungus ("Scab"). To this end, Martin et al. (1998) conducted a study examining numerous physical properties of kernels—density, hardness, size, weight, and moisture content—measured on a sample of wheat kernels from two different classes of wheat, hard red winter (`hrw`) and soft red winter (`srw`). Each kernel's condition was also classified as "Healthy," "Sprout," or "Scab" by human visual inspection. In the data provided by the authors of this paper, we have measurements from 275 wheat kernels. Below is a portion of the data.

```
> wheat <- read.csv(file = "C:\\data\\Wheat.csv",
    stringsAsFactors = TRUE)

> head(wheat, n = 3)
  class  density hardness    size  weight moisture    type
1   hrw 1.349253 60.32952 2.30274 24.6480 12.01538 Healthy
2   hrw 1.287440 56.08972 2.72573 33.2985 12.17396 Healthy
3   hrw 1.233985 43.98743 2.51246 31.7580 11.87949 Healthy

> tail(wheat, n = 3)
    class   density hardness    size  weight moisture type
273   srw 0.8491887 34.06615 1.40665 12.0870 11.92744 Scab
274   srw 1.1770230 60.97838 1.05690  9.4800 12.24046 Scab
275   srw 1.0305543 -9.57063 2.05691 23.8185 12.64962 Scab
```

We begin with useful plots to initially examine the data (the code is given in our program). Figure 3.2 shows a parallel coordinates plot constructed using the `parcoord()` function of the `MASS` package. The horizontal axis gives each variable name, and plotted above these names are the corresponding values for each kernel. These values are scaled so that the minimum and maximum values for each variable appear at the bottom and top, respectively, of the vertical axis. A line is drawn for each kernel indicating its position across the variables. For example, the kernel that has the smallest weight (lowest line for `weight`) also has the second smallest size (second lowest line

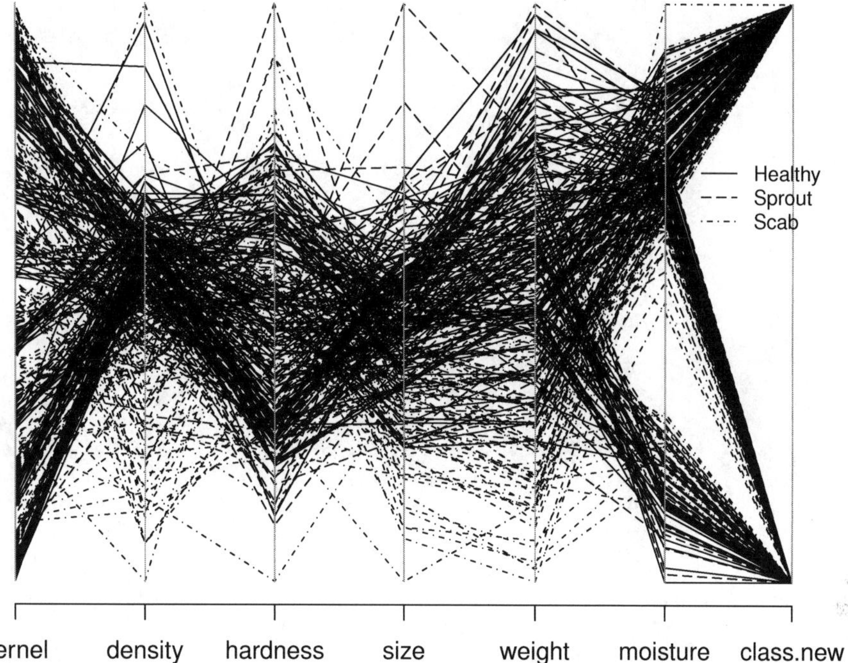

Figure 3.2: Parallel coordinate plot of the wheat data. The kernel variable corresponds to the observation number in the data set. The `class.new` variable is a 1 for soft red winter wheat and 0 for hard red winter wheat. We recommend viewing this plot in color (see program for code) if the version in your book is only in black and white.

for `size`).[5] We have used different line types to represent the three types of kernel, so that we can understand some of the features that differentiate the healthy, sprout, and scab conditions. We see that

1. Scab kernels generally have smaller density, size, and weight values.
2. Healthy kernels may have higher densities.
3. There is much overlap for healthy and sprout kernels.
4. The moisture content appears to be related to the hard or soft red winter wheat class.

To examine these data more formally, we fit a multinomial regression model using the wheat class and the five measurements on each kernel as explanatory variables. Below is our code using `multinom()`.

```
> levels(wheat$type)  # The 3 response categories
[1] "Healthy" "Scab"    "Sprout"

> library(package = nnet)
```

[5]Because it is difficult to identify particular kernel numbers, we provide code in our program that shows how to highlight a kernel. For example, one can see that kernel #269 is the kernel with the smallest weight.

```
> mod.fit <- multinom(formula = type ~ class + density +
    hardness + size + weight + moisture, data = wheat)
# weights:  24 (14 variable)
initial  value 302.118379
iter  10 value 234.991271
iter  20 value 192.127549
final  value 192.112352  converged

> summary(mod.fit)
Call: multinom(formula = type ~ class + density + hardness +
    size + weight + moisture, data = wheat)

Coefficients:
        (Intercept)   classsrw    density     hardness       size
Scab       30.54650 -0.6481277 -21.59715 -0.01590741 1.0691139
Sprout     19.16857 -0.2247384 -15.11667 -0.02102047 0.8756135
           weight    moisture
Scab   -0.2896482  0.10956505
Sprout -0.0473169 -0.04299695

Std. Errors:
        (Intercept)  classsrw  density    hardness      size
Scab       4.289865 0.6630948 3.116174 0.010274587 0.7722862
Sprout     3.767214 0.5009199 2.764306 0.008105748 0.5409317
           weight  moisture
Scab   0.06170252 0.1548407
Sprout 0.03697493 0.1127188

Residual Deviance: 384.2247
AIC: 412.2247
```

Output from `levels(wheat$type)` shows us that `"Healthy"` is stored as the first level of `type`, so `multinom()` uses it as the base level. For other applications, it may be desirable to use a different base level. This can be changed by using the `relevel()` or `factor()` functions in the same manner as shown in Section 2.2.6. Also, note that an indicator variable is created for `class`, where a 1 is for `"srw"` and a 0 for `"hrw"`. Based on this information and what is given in the `coefficients` table of the output, we can state the estimated model as

$$\begin{aligned}\log(\hat{\pi}_{\texttt{Scab}}/\hat{\pi}_{\texttt{Healthy}}) = 30.55 - 0.65\texttt{classsrw} - 21.60\texttt{density} - 0.016\texttt{hardness} \\ +1.07\texttt{size} - 0.29\texttt{weight} + 0.11\texttt{moisture}\end{aligned}$$

and

$$\begin{aligned}\log(\hat{\pi}_{\texttt{Sprout}}/\hat{\pi}_{\texttt{Healthy}}) = 19.17 - 0.22\texttt{classsrw} - 15.12\texttt{density} - 0.021\texttt{hardness} \\ +0.88\texttt{size} - 0.047\texttt{weight} - 0.043\texttt{moisture}.\end{aligned}$$

Hypothesis tests of the form $H_0 : \beta_{jr} = 0$ vs. $H_a : \beta_{jr} \neq 0$ for each parameter can be performed by Wald tests or LRTs. Such tests examine the effect of an explanatory variable on the log-odds comparing only one response level to the base response level, ignoring the variable's effects on other log-odds. For example, a Wald test of $H_0 : \beta_{21} = 0$ vs. $H_a : \beta_{21} \neq 0$, which corresponds to the `class` explanatory variable in the log-odds for scab vs. healthy, ignoring sprout, results in

$$Z_W = \frac{\hat{\beta}_{21}}{\sqrt{\widehat{Var}(\hat{\beta}_{21})}} = \frac{-0.6481}{0.6631} = -0.98$$

and a p-value of $2 \times P(Z > |-0.98|) = 0.33$. Thus, there is not sufficient evidence that hard and soft red winter wheat have different effects on the scab or healthy status of the kernels given the other explanatory variables in the model.

More interesting hypotheses explore the effects of a given explanatory variable on *all* response categories. For example, does class of wheat have *any* effect on the probabilities of healthy, scab, or sprout kernels? To investigate this, we need to test whether *all* parameters corresponding to the explanatory variable of interest, say x_r, are zero: $H_0 : \beta_{jr} = 0, j = 2, \ldots, J$. In the case of the `class` explanatory variable, we use $H_0 : \beta_{21} = \beta_{31} = 0$ vs. $H_a : \beta_{21} \neq 0$ and/or $\beta_{31} \neq 0$. This test is easily carried out with LR methods using the `Anova()` function.

```
> library(package = car)
> Anova(mod.fit)
Analysis of Deviance Table (Type II tests)

Response: type
         LR Chisq Df Pr(>Chisq)
class       0.964  2     0.6175
density    90.555  2  < 2.2e-16 ***
hardness    7.074  2     0.0291 *
size        3.211  2     0.2008
weight     28.230  2  7.411e-07 ***
moisture    1.193  2     0.5506
---
Signif. codes:  0 '***' 0.001 '**' 0.01 '*' 0.05 '.' 0.1 ' ' 1
```

The LRT statistic is $-2\log(\Lambda) = 0.964$, and the corresponding p-value is 0.6175 using a χ^2_2 distributional approximation. Because of the large p-value, there is not sufficient evidence to indicate that the class of wheat is important given that the other variables are in the model. Separate tests for `density`, `hardness`, and `weight` in the output all indicate at least marginal evidence of importance for these explanatory variables. Note that the `anova()` function may also be used to perform these tests one at a time. We demonstrate its use in the program for this example.

Below is our code using `vglm()`.

```
> mod.fit.v <- vglm(formula = type ~ class + density + hardness
   + size + weight + moisture, family = multinomial(refLevel =
   "Healthy"), data = wheat)
> summary(mod.fit.v)

Call: vglm(formula = type ~ class + density + hardness + size +
   weight + moisture, family = multinomial(refLevel =
   "Healthy"), data = wheat)

Coefficients:
                    Estimate Std. Error z value Pr(>|z|)
```

```
(Intercept):1   30.545829   4.289769    7.121 1.07e-12 ***
(Intercept):2   19.167800   3.767134    5.088 3.62e-07 ***
classsrw:1      -0.648242   0.663090   -0.978  0.32827
classsrw:2      -0.224810   0.500918   -0.449  0.65358
density:1      -21.596968   3.116126   -6.931 4.19e-12 ***
density:2      -15.116309   2.764258   -5.468 4.54e-08 ***
hardness:1      -0.015907   0.010275   -1.548  0.12157
hardness:2      -0.021021   0.008106   -2.593  0.00951 **
size:1           1.069246   0.772281    1.385  0.16620
size:2           0.875660   0.540932    1.619  0.10549
weight:1        -0.289652   0.061702   -4.694 2.67e-06 ***
weight:2        -0.047315   0.036975   -1.280  0.20066
moisture:1       0.109590   0.154840    0.708  0.47909
moisture:2      -0.042978   0.112718   -0.381  0.70299
---
Signif. codes:  0 '***' 0.001 '**' 0.01 '*' 0.05 '.' 0.1 ' ' 1

Names of linear predictors: log(mu[,2]/mu[,1]),
   log(mu[,3]/mu[,1])

Residual deviance: 384.2247 on 536 degrees of freedom

Log-likelihood: -192.1124 on 536 degrees of freedom

Number of Fisher scoring iterations: 6

Warning: Hauck-Donner effect detected in the following
   estimate(s): '(Intercept):1', 'weight:1'

Reference group is level  1 of the response
```

The parameter estimates are practically the same as those obtained from `multinom()` and displayed in a format much like what was seen in Chapter 2 with `glm()`. The labeling of "1" and "2" next to each coefficient name corresponds in order to the levels of the response factor that are not used as the base level. In this case, the first level of `type`, `"Healthy"`, has been set as the base level, so "1" and "2" refer to the second and third levels of `type`—`scab` and `sprout`, respectively. Thus, the coefficient `moisture:2`, $\hat{\beta}_{36} = -0.042978$, is from the $\log(\hat{\pi}_{\texttt{Sprout}}/\hat{\pi}_{\texttt{Healthy}})$ portion of the model.

Within `vglm()`, the `family` argument is similar to the corresponding argument for `glm()`. For the multinomial regression model, we specify `multinomial` for this argument's value and also indicate the base level (`refLevel`) for the response variable. While we used `"Healthy"` as the base level, the numerical position of a factor level can be used instead, like `1` here because the first level of `type` is `"Healthy"`. It is important to note that the default base level for `vglm()` is the *last* level of the response variable.

The output indicates concern about a "Hauck-Donner effect." This suggests a potential problem with complete or almost-complete separation (see Section 2.2.7), which can lead to low power for Wald tests (Yee, 2022). We are not concerned here about this potential issue for two reasons. First, we refit the model using a stricter convergence criterion (details provided in the program) and did not experience practical changes in the parameter estimates. Second, we will focus our examination on LRTs.

```
> anova(mod.fit.v)
Analysis of Deviance Table (Type II tests)

Model: 'multinomial', 'VGAMcategorical'

Link: 'multilogitlink'

Response: type

         Df Deviance Resid. Df Resid. Dev  Pr(>Chi)
class     2    0.964       538     385.19    0.6175
density   2   90.555       538     474.78 < 2.2e-16 ***
hardness  2    7.074       538     391.30    0.0291 *
size      2    3.211       538     387.44    0.2008
weight    2   28.230       538     412.45 7.411e-07 ***
moisture  2    1.193       538     385.42    0.5506
---
Signif. codes:  0 '***' 0.001 '**' 0.01 '*' 0.05 '.' 0.1 ' ' 1
```

The VGAM package has a method function for anova() that produces the Type II test p-values for LRTs by default. Notice this is different than how the method function for anova() worked using objects created by glm() (type I test p-values are given; see Section 2.2.2). The p-values match those given by Anova() on the fit produced by multinom() earlier in this example. It is very important to note that Anova(mod.fit.v)—with the capital first letter—produces output that is incorrect. There is not a method for this generic function in VGAM, so the default method function is invoked and incorrectly calculates test statistics and p-values. It is also important to note that this method function for anova() needs the response variable to be a factor type. While this is not an issue here because type is a factor, responses for other data problems can be coded numerically, like with levels of 1, 2, 3, and 4 as in Exercise 13.

Transformations and interaction terms are included within multinom() and vglm() in a similar manner as with glm(). We provide an example in our program that shows how to estimate parameters in $\log(\pi_j/\pi_{\texttt{Healthy}}) = \beta_{j0} + \beta_{j1}\texttt{classsrw} + \beta_{j2}\texttt{density} + \beta_{j3}\texttt{density}^2 + \beta_{j4}\texttt{classsrw} \times \texttt{density}$. For this example, multinom() does not achieve convergence with its default 100 iterations, so we needed to increase its maximum number of iterations. Furthermore, we needed to make its convergence criteria more strict to match the parameter estimates from vglm() using that function's convergence criteria defaults.

Estimates of $\pi_1, \ldots, \pi_J$ are calculated from Equations 3.6 and 3.7 using estimated regression parameters. Inferences for these probabilities are more complicated than for π with a logistic regression model. In logistic regression, the response variable has only two potential outcomes, and these outcomes are controlled by only one parameter, π. Inference for $1 - \pi$ follows from inference for π. In multinomial regression, the response variable has J potential outcomes, and these outcomes are controlled by J probability parameters that sum to 1, although we need to separately estimate only $J - 1$ of them because of the sum to 1 constraint. These parameters are clearly related to one another, but they are not as simply related as for logistic regression ($J = 2$). Thus, we may perform inference separately for each individual outcome probability, π_j, or we may perform inference about the entire probability distribution of a response, $\pi_1, \ldots, \pi_J$. The latter case is not easily addressed because a *joint confidence region* is needed in $J - 1$ dimensions. Computation methods are not readily available for these regions. For this reason, we focus on the former case: one-at-a-time confidence intervals for each π_j.

One-at-a-time Wald intervals are calculated as

$$\hat{\pi}_j \pm Z_{1-\alpha/2}\sqrt{\widehat{Var}(\hat{\pi}_j)}, \tag{3.8}$$

where $\widehat{Var}(\hat{\pi}_j)$ is found using the delta method and the estimated regression parameters. Profile LR intervals are not readily available in R, so we do not discuss them here.

The form of Equation 3.8 is somewhat different than what was given in Section 2.2.4. The reason is the multinomial regression model does not allow the parameters $\pi_1, \ldots, \pi_J$ to be expressed as a simple transformation of a linear combination of the estimated regression parameters. Unfortunately, Equation 3.8 can lead to an interval limit being below 0 or above 1. Also, because there are J different intervals, potential values of $\pi_1, \ldots, \pi_J$ from their intervals could be added together resulting in a sum not equal to 1. For this reason, an R function is not available within the `nnet` and `VGAM` packages to calculate Equation 3.8.[6] Despite these problems with confidence intervals for $\pi_1, \ldots, \pi_J$, we feel that it is better to offer some measure of uncertainty in the estimates using Equation 3.8 but interpret them with caution.

Example: Wheat kernels (Wheat.R, Wheat.csv)

For the first kernel of the data set, the estimated probability of being healthy is

$$\begin{aligned}\hat{\pi}_{\texttt{Healthy}} &= [1 + \exp(30.55 - 0.65 \times 0 + \cdots + 0.11 \times 12.02) \\ &\quad + \exp(19.17 - 0.22 \times 0 + \cdots - 0.043 \times 12.02)]^{-1} \\ &= 0.8552.\end{aligned}$$

The estimated probabilities of being scab (0.0464) and sprout (0.0984) are computed in a similar manner. We include code in our program to demonstrate how to program Equations 3.6 and 3.7 directly for these computations. Because $\hat{\pi}_{\texttt{Healthy}}$ is the largest probability, the model estimates that the first kernel is most likely healthy.

More convenient code is available for these calculations. Objects returned from `multinom()` and `vglm()` are of a list format and have a component named `fitted.values` that provide the probability estimates for each observation in the data set. Thus, `mod.fit$fitted.values` gives the estimated probabilities as calculated by `multinom()`. Because `vglm()` uses the S4 class system (see Section 2.2.4 for an introduction of S4), this same component is accessed with `mod.fit.v@fitted.values`. A more general approach to obtain estimated probabilities for any set of explanatory variable values is to use the generic function `predict()`.

```
> # Use object from multinom()
> predict(object = mod.fit, newdata = wheat[1,], type = "probs")
   Healthy        Scab     Sprout
0.85521096 0.04639683 0.09839221

> # Use object from vglm()
```

[6]See `https://stat.ethz.ch/pipermail/r-help/2004-April/048917.html` for an R listserv discussion of the issue. The author of the `nnet` package provides his own comments.

```
> predict(object = mod.fit.v, newdata = wheat[1,], type =
    "response")
    Healthy       Scab      Sprout
1 0.8552002 0.0464011 0.0983987
```

Different method functions are invoked by `predict()`, but the code is largely the same. For example, the `newdata` argument works like the same argument used for `predict()` in Section 2.2.4 when finding π with a logistic regression model. Thus, a new data frame of the explanatory variable values of interest could have been constructed rather than using a portion of the original data set with the argument. The one small difference between the implementations of `predict()` is the `type` argument values of `"probs"` and `"response"`, but both have the same meaning. When using objects created by `multinom()`, the `type` argument also allows for `type = "class"` to find the most probable response category for each kernel.

Unfortunately, neither method function for `predict()` provides the estimated variances needed to compute a Wald confidence interval for π_j. Instead, these variances and intervals can be calculated with the method function for `summary()` of the `emmeans` package.

```
> library(package = emmeans)
> calc.prob <- emmeans(object = mod.fit, specs = ~ type, at =
    wheat[1,1:6], mode = "prob")
> summary(object = calc.prob, df = Inf, level = 0.95)
 type     prob     SE   df  asymp.LCL asymp.UCL
 Healthy 0.8552 0.0600 Inf     0.7376    0.9728
 Scab    0.0464 0.0271 Inf    -0.0067    0.0995
 Sprout  0.0984 0.0429 Inf     0.0143    0.1825

Degrees-of-freedom method: user-specified
Confidence level used: 0.95
```

The 95% Wald interval for $\pi_{\texttt{Healthy}}$ is (0.7376, 0.9728) for kernels with the same explanatory variable values as the first kernel in the data set. The `emmeans()` function uses the object created by `multinom()` because the function will not accept objects created by `vglm()`. The other arguments for `emmeans()` are `specs` to specify the response variable, `mode` to indicate that probabilities should be calculated, and `at` to provide the explanatory variable values. A separate list was not needed for `at` because we have only one value for each explanatory variable and a data frame is a special form of a list. The `summary()` function includes `df = Inf` to represent an infinite degrees of freedom. This is needed because `summary()` incorrectly uses a t distribution with a finite degrees of freedom by default.[7] The calculations performed here can also be performed using the `deltaMethod()` function (introduced in Exercise 15 of Chapter 2) of the `car` package. We provide the code in our program. This code is more complex than using the `emmeans` package, so we do not recommend its use for this particular calculation here.

The estimated model can be plotted by including the estimated probability for each wheat type on the y-axis and one of the explanatory variables on the x-axis, while holding the other variables within the model constant. Because the plot is conditional

[7] A t distribution with an infinite degrees of freedom is a standard normal distribution.

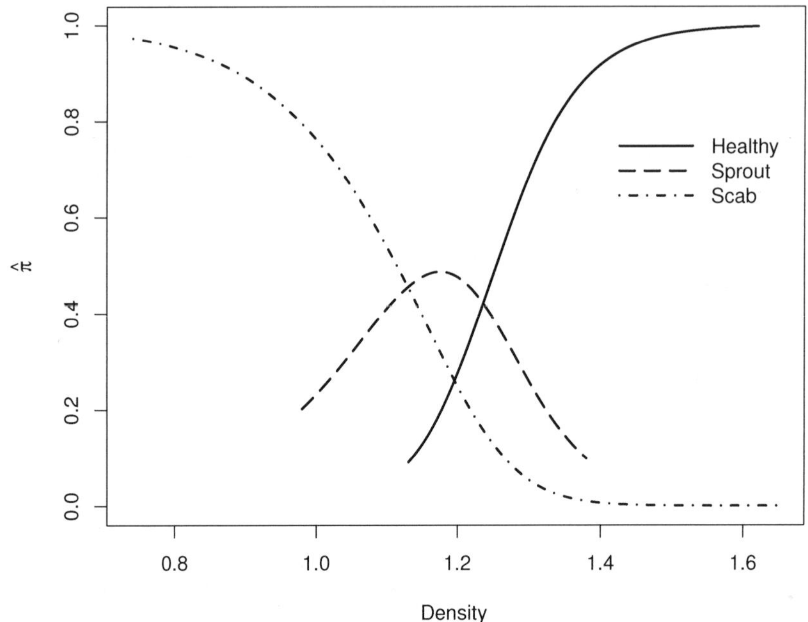

Figure 3.3: Estimated multinomial regression model for the wheat data with density as the only explanatory variable in the model.

on these variable values held constant, more than one plot should usually be made at different constant values. We discuss plots like this in Exercise 21. Instead for this example, we refit a model with `density` as the only explanatory variable:

$$\log(\hat{\pi}_{\texttt{Scab}}/\hat{\pi}_{\texttt{Healthy}}) = 29.38 - 24.56\texttt{density}$$

and

$$\log(\hat{\pi}_{\texttt{Sprout}}/\hat{\pi}_{\texttt{Healthy}}) = 19.12 - 15.48\texttt{density}.$$

The model is plotted in Figure 3.3 (see the program for the code), where the estimated probabilities for each kernel condition are drawn between the smallest and the largest observed density values for that condition. From the plot, we see that the estimated scab probability is the largest for the smaller density kernels. The estimated healthy probability is the largest for the high density kernels. For density levels in the middle, sprout has the largest estimated probability. The parallel coordinates plot in Figure 3.2 displays similar findings because density levels tend to follow the scab $<$ sprout $<$ healthy ordering.

3.3.1 Odds ratios

Because the log-odds are modeled directly in a multinomial regression model, odds ratios are useful for interpreting an explanatory variable's relationship with the response. As described in Section 2.2.3 for logistic regression models, odds ratios for numerical explanatory variables represent the increase/decrease in odds corresponding to a c-unit change in a particular explanatory variable. The only difference with multinomial models is that the odds are now formed as a comparison between two of the J response categories.

For example, suppose the terms in Equation 3.4 are distinct explanatory variables (not transformations or interactions). The odds of a category j response vs. a category 1 response are $\exp(\beta_{j0} + \beta_{j1}x_1 + \cdots + \beta_{jp}x_p)$. These odds change by $\exp(c\beta_{jr})$ times for every c-unit increase in x_r, holding the other variables constant. Similarly, from Equation 3.5, the odds of a category j vs. a category j' response ($j \neq j'$, $j > 1$, and $j' > 1$) change by $\exp(c(\beta_{jr} - \beta_{j'r}))$ times for every c-unit increase in x_r while holding the other variables in the model constant.

Maximum likelihood estimates of odds ratios are obtained by replacing the regression parameters with their corresponding estimates. Wald and LR-based inference methods for odds ratios are used in the same ways as discussed in earlier chapters.

Example: Wheat kernels (Wheat.R, Wheat.csv)

The odds for this example are constructed as $P(Y = j)/P(Y = 1)$, $j = 2, 3$, with categories 1 = healthy, 2 = scab, and 3 = sprout. The estimated odds ratios for scab vs. healthy or sprout vs. healthy are calculated for each explanatory variable as $\widehat{OR} = \exp(c\hat{\beta}_{jr})$ for $j = 2, 3$ and $r = 1, \ldots, 6$. To choose appropriate values of c, we set c to be equal to one standard deviation for each continuous explanatory variable. Each of these standard deviations is found using the `apply()` function, which applies the `sd()` function to specific columns (`MARGIN = 2`) of the `wheat` data frame. For `class`, we choose $c = 1$ because it is a binary variable for soft or hard red winter wheat. Below are the estimated standard deviations, followed by the odds ratios.

```
> sd.wheat <- apply(X = wheat[,-c(1,7)], MARGIN = 2, FUN = sd)
> c.value <- c(1, sd.wheat)  # class = 1 is first value
> round(c.value,2)
        density hardness      size    weight moisture
 1.00      0.13    27.36      0.49      7.92     2.03

> # beta.hat_jr for r = 1, ..., 6 and j = 2, 3, where
   beta.hat_jr = coefficients(mod.fit)[j-1,r+1]
> beta.hat2 <- coefficients(mod.fit)[1,2:7]
> beta.hat3 <- coefficients(mod.fit)[2,2:7]

> # Odds ratios for j = 2 vs. j = 1 (scab vs. healthy)
> round(exp(c.value*beta.hat2), 2)
        density hardness      size    weight moisture
 0.52      0.06     0.65      1.69      0.10     1.25
> round(1/exp(c.value*beta.hat2), 2)
        density hardness      size    weight moisture
 1.91     17.04     1.55      0.59      9.90     0.80

> # Odds ratios for j = 3 vs. j = 1 (sprout vs. healthy)
> round(exp(c.value*beta.hat3), 2)
        density hardness      size    weight moisture
 0.80      0.14     0.56      1.54      0.69     0.92
> round(1/exp(c.value*beta.hat3), 2)
        density hardness      size    weight moisture
```

```
  1.25      7.28      1.78      0.65      1.45      1.09
```

We use the `coefficients()` function to obtain the regression parameter estimates because there is no `coefficients` component in objects produced by `multinom()`. With objects produced by `vglm()`, the coefficients component does exist, so `mod.fit.v@coefficients` can be used to obtain the regression parameter estimates. Interpretations of the estimated odds ratios include:

- The estimated odds of a scab vs. a healthy kernel change by 0.06 times for every 0.13 increase in the density when holding the other variables constant. Equivalently, we can say that the estimated odds of a scab vs. a healthy kernel change by 17.04 times for every 0.13 *decrease* in the density when holding the other variables constant. The estimated odds of a sprout vs. a healthy kernel change by 7.28 times times for every 0.13 decrease in the density when holding the other variables constant.
- The estimated odds of a scab vs. healthy kernel change by 9.90 times for every 7.92 decrease in the weight when holding the other variables constant. The estimated odds of a sprout vs. healthy kernel change by 1.45 times for every 7.92 decrease in the weight when holding the other variables constant.

We see that the larger the density and weight, the more likely a kernel is healthy. This corresponds to what was shown in Figure 3.2. The plot shows that kernels with low density and weight are often of the scab condition rather than healthy condition. We see a similar separation between sprout and healthy for the density measurements, but not as much so for weight. Interpretations of the other variables are part of Exercise 12.

As with the `glm()` function, there are a number of method functions associated with objects resulting from `multinom()`.

```
> class(mod.fit)
[1] "multinom" "nnet"
> methods(class = multinom)
 [1] add1         anova        Anova        brief
 [5] coef         confint      Confint      deltaMethod
 [9] drop1        extractAIC   logLik       model.frame
[13] predict      print        S            summary
[17] vcov
see '?methods' for accessing help and source code

> sqrt(vcov(mod.fit)[2,2])  # sqrt(Var-hat(beta-hat_21))
[1] 0.6630948
```

Note that we previously used `coefficients()`, the equivalent to `coef.multinom()`, to extract the estimated regression parameters. There are also many method functions associated with objects resulting from `vglm()`. In a similar manner, these can be seen by using `class(mod.fit.v)` to show the class is `vglm` and `showMethods(classes = "vglm")` to display these functions.

For objects created by both `multinom()` and `vglm()`, there is a method function for `confint()` that calculates intervals for regression parameters. For `multinom()`, this function calculates Wald intervals. For `vglm()`, this function calculates both Wald and profile LR intervals, where `method = "profile"` needs to be specified for the latter case. Combined with the appropriate value of c for each variable and the `exp()` function, we calculate the 95% profile LR intervals for the odds ratios.

```
> conf.beta.v <- confint(object = mod.fit.v, level = 0.95,
    method = "profile")
There were 23 warnings (use warnings() to see them)
> round(conf.beta.v, 2)
                 2.5 % 97.5 %
(Intercept):1   22.63  39.46
(Intercept):2   12.21  27.01
classsrw:1      -1.97   0.65
classsrw:2      -1.21   0.76
density:1      -28.09 -15.86
density:2      -20.91 -10.05
hardness:1      -0.04   0.00
hardness:2      -0.04  -0.01
size:1          -0.45   2.59
size:2          -0.15   1.99
weight:1        -0.42  -0.17
weight:2        -0.12   0.02
moisture:1      -0.19   0.42
moisture:2      -0.26   0.18

> round(exp(rep(c.value, each = 2) * conf.beta.v[-c(1,2),]), 2)
            2.5 % 97.5 %
classsrw:1   0.14   1.91
classsrw:2   0.30   2.13
density:1    0.03   0.12
density:2    0.06   0.27
hardness:1   0.37   1.12
hardness:2   0.36   0.86
size:1       0.80   3.57
size:2       0.93   2.65
weight:1     0.04   0.25
weight:2     0.38   1.21
moisture:1   0.68   2.34
moisture:2   0.58   1.44

> # Density only
> data.frame(type = c("scab vs. healthy", "sprout vs. healthy"),
    low = round(1/exp(c.value[2] * conf.beta.v[5:6,2]), 2), up =
    round(1/exp(c.value[2] * conf.beta.v[5:6,1]), 2), row.names =
    NULL)
                type   low    up
1   scab vs. healthy  8.03 39.98
2 sprout vs. healthy  3.74 15.56
```

The warning messages are due to re-estimating the model under constraints to obtain the profile LR intervals. Some constraints make it difficult for model estimation, and similar warning messages often appear when using the `mcprofile` package in Chapter 2. Because these intervals are not unusually different than the Wald intervals (see the program for code), we are not concerned about the computations.

The density odds ratios can be interpreted as follows:

> With 95% confidence, the odds of a scab instead of a healthy kernel increase by between 8.03 and 39.98 times for every 0.13 decrease in density when holding the other variables constant. Also, with 95% confidence, the odds of

a sprout instead of a healthy kernel increase by between 3.74 and 15.56 times for every 0.13 decrease in density when holding the other variables constant.

Other odds ratios can be interpreted in a similar manner. For the scab vs. healthy comparison, only the density and weight odds ratio confidence intervals do not include 1. For the sprout vs. healthy comparison, only the density and hardness odds ratio confidence intervals do not include 1.

Odds ratio comparisons should be made for sprout vs. scab too. The simplest way to perform computations is to refit the model using sprout or scab as the base level, rather than healthy. Estimated odds ratios and corresponding confidence intervals are found using the same type of code used when healthy was the base level. Alternatively, the `emmeans` package can calculate Wald confidence intervals without needing to refit the model. These computations and interpretations are the subject of Exercise 12.

When the multinomial regression model includes an interaction or transformation, odds ratios for variables are found using the same process as described in Section 2.2.5. For example, consider the model

$$\log(\pi_j/\pi_1) = \beta_{j0} + \beta_{j1}x_1 + \beta_{j2}x_2 + \beta_{j3}x_1x_2$$

for $j = 2, \ldots, J$. The odds of a category j response vs. a category 1 response change by $\exp(c(\beta_{j1} + \beta_{j3}x_2))$ times for every c-unit increase in x_1 at a particular value for x_2. The derivation is left for Exercise 5. Wald and profile LR intervals for probabilities and odds ratios are created with the appropriate modifications to the processes used before, but only the Wald interval is readily available in R. For direct calculation with the previous odds ratio, the expression for the Wald interval is

$$\exp\left(c(\hat{\beta}_{j1} + \hat{\beta}_{j3}x_2) \pm cZ_{1-\alpha/2}\sqrt{\widehat{Var}(\hat{\beta}_{j1} + \hat{\beta}_{j3}x_2)}\right) \tag{3.9}$$

with

$$\widehat{Var}(\hat{\beta}_{j1} + \hat{\beta}_{j3}x_2) = \widehat{Var}(\hat{\beta}_{j1}) + x_2^2\widehat{Var}(\hat{\beta}_{j3}) + 2x_2\widehat{Cov}(\hat{\beta}_{j1}, \hat{\beta}_{j3}).$$

To calculate Equation 3.9 or other Wald intervals for these types of odds ratios, one can program their equations into R. The required variances and covariances are obtained from the estimated covariance matrix using `vcov()` with the object created by `multinom()` or `vglm()`. Alternatively, the `emmeans` package provides a somewhat simpler process that we demonstrate in the next example.

Example: Wheat kernels (Wheat.R, Wheat.csv)

Suppose `class`, `density`, and their interaction are included in a multinomial regression model. We estimate this model using `multinom()` rather than `vglm()` so that we can use the `emmeans` package.

```
> mod.fit.class.density <- multinom(formula = type ~
    class*density, data = wheat, reltol = 1e-12)
# weights:  15 (8 variable)
initial  value 302.118379
iter  10 value 230.236019
iter  20 value 228.419091
iter  30 value 228.317394
```

```
iter  40 value 228.222836
iter  50 value 228.218720
iter  60 value 228.218716
final  value 228.218715
converged

> coefficients(mod.fit.class.density)
       (Intercept) classsrw    density classsrw:density
Scab      29.41781 0.605879 -24.62440        -0.443696
Sprout    16.29562 6.322980 -13.15262        -5.196641

> # OR at density = 1
> density <- 1
> beta.hat2 <- coefficients(mod.fit.class.density)[1,-1]
> exp(beta.hat2[1] + beta.hat2[3]*density)
classsrw
1.176075
```

The estimated model is

$$\log(\hat{\pi}_{\texttt{Scab}}/\hat{\pi}_{\texttt{Healthy}}) = 29.42 + 0.61\texttt{classsrw} - 24.62\texttt{density} \\ -0.44\texttt{classsrw} \times \texttt{density}$$

and

$$\log(\hat{\pi}_{\texttt{Sprout}}/\hat{\pi}_{\texttt{Healthy}}) = 16.30 + 6.32\texttt{classsrw} - 13.15\texttt{density} \\ -5.20\texttt{classsrw} \times \texttt{density}.$$

We use a more strict convergence criterion than the default so that parameter estimates match more closely those from `vglm()` (the code is in our program). The estimated odds of a scab instead of a healthy kernel are $\exp(\hat{\beta}_{21} + \hat{\beta}_{23}\texttt{density})$ times as large for soft as for hard red winter wheat when holding density constant at a particular value. When the density is 1, this odds ratio is 1.18.

To use the **emmeans** package for this same odds ratio and the corresponding Wald confidence interval, we begin with `emmeans()` and include the response variable `type` and the explanatory variable `class` in its `specs` argument, because the odds are formed from different types of kernels and the ratios of odds are formed from the classes. It is important to note that `density` should not be included in `specs` because we will specify its value in the `at` argument. We next use `contrast()` and `confint()` in a similar manner as in Chapter 2 to compute estimates and intervals.

```
> calc.est.inter <- emmeans(object = mod.fit.class.density,
   specs = ~ type + class, at = list(density = 1), mode =
   "latent")
> test.info.inter <- contrast(object = calc.est.inter,
   interaction = list("revpairwise", "revpairwise"))
> confint(object = test.info.inter, df = Inf, type = "response",
   level = 0.95)
 type_revpairwise class_revpairwise ratio   SE  df
 Scab / Healthy   srw / hrw          1.18 1.64 Inf
 Sprout / Healthy srw / hrw          3.08 4.15 Inf
 Sprout / Scab    srw / hrw          2.62 1.66 Inf
 asymp.LCL asymp.UCL
```

```
    0.0763     18.14
    0.2212     43.00
    0.7576      9.08

Degrees-of-freedom method: user-specified
Confidence level used: 0.95
Intervals are back-transformed from the log scale
```

We needed to use mode with value of "latent" in emmeans() so that we can work with a rescaled version of the linear predictor. Included within contrast() is an interaction argument with two values in a list. This allows us to control the type of comparisons and their ordering relative to the levels of type and class. The code results in the calculation of the same odds ratio as found earlier. The corresponding 95% confidence interval is 0.0763 to 18.14 when the density is 1. Thus, there is not sufficient evidence that the class is important at this particular density value. Additional odds ratios are calculated comparing the odds of sprout vs. healthy and the odds of sprout vs. scab with the numerators of the ratio being for soft red winter wheat and the denominators being for hard red winter wheat.

The emmeans package performs many calculations in a straightforward manner once the correct arguments and their values are determined. However, there may be times when emmeans cannot perform a particular calculation of interest using the available pre-programmed argument values. For those situations, the package provides tools to manually specify contrasts of linear predictors that allow one to calculate the desired quantities. While not needed for the previous odds ratio calculation, we illustrate the process here because it will verify our results. To summarize the procedure, we need two contrasts for the log-odds of the form $\log(\hat{\pi}_{\texttt{Scab}}/\hat{\pi}_{\texttt{Healthy}}) = \log(\hat{\pi}_{\texttt{Scab}}) - \log(\hat{\pi}_{\texttt{Healthy}})$, where one log-odds is for soft red winter wheat and the second is for hard red winter wheat. We subtract the latter log-odds from the former to form the log of the odds ratio. The exponential function is applied with this result to obtain the odds ratio.

To begin, we determine how emmeans orders the linear predictors by using the coefficients() function in the package.

```
> calc.est.inter2 <- emmeans(object = mod.fit.class.density,
    specs = ~ type + class + density, at = list(density = 1),
    mode = "latent")
> find.order <- contrast(object = calc.est.inter2, method =
    "identity")
Note: Use 'contrast(regrid(object), ...)' to obtain contrasts of
    back-transformed estimates
> coefficients(find.order)
     type class density c.1 c.2 c.3 c.4 c.5 c.6
1 Healthy   hrw       1   1   0   0   0   0   0
2    Scab   hrw       1   0   1   0   0   0   0
3  Sprout   hrw       1   0   0   1   0   0   0
4 Healthy   srw       1   0   0   0   1   0   0
5    Scab   srw       1   0   0   0   0   1   0
6  Sprout   srw       1   0   0   0   0   0   1
```

Using this ordering information, we can form a contrast to calculate $-1\times\log(\hat{\pi}_{\texttt{Healthy}})+1\times\log(\hat{\pi}_{\texttt{Scab}})+0\times\log(\hat{\pi}_{\texttt{Sprout}}) = \log(\hat{\pi}_{\texttt{Scab}}/\hat{\pi}_{\texttt{Healthy}})$ for class = "hrw". Similarly, we can form a contrast to calculate the same log-odds for `class = "srw"`. The difference of these two provides the log of the odds ratio. Below is the code showing the process.

```
> hrw.dens1 <- c(-1,  1,  0,   0,  0,  0)
> srw.dens1 <- c( 0,  0,  0,  -1,  1,  0)
> test.info.inter2 <- contrast(object = calc.est.inter2, type =
    "response", method = list("Odds for hrw" = hrw.dens1, "Odds
    for srw" = srw.dens1, "OR for srw vs. hrw" = srw.dens1 -
    hrw.dens1))
> confint(object = test.info.inter2, df = Inf, level = 0.95)
 contrast             ratio     SE  df asymp.LCL asymp.UCL
 Odds for hrw        120.71 111.10 Inf   19.8772     733.1
 Odds for srw        141.97 148.98 Inf   18.1523    1110.3
 OR for srw vs. hrw    1.18   1.64 Inf    0.0763      18.1

Degrees-of-freedom method: user-specified
Confidence level used: 0.95
Intervals are back-transformed from the log scale
```

The `method` argument within `contrast()` includes a list for the contrasts. The `type = "response"` argument value uses the exponential function with the log of the odds ratio to obtain the odds ratio itself. It is important to note that the contrasts here are not the same as those used with the `mcprofile` package in Chapter 2. Those contrasts were made directly with the regression parameters. Instead, the `emmeans` package works with linear predictors (i.e., the log-odds here).

3.3.2 Contingency tables

The multinomial regression model provides a convenient way to perform the LRT for independence described in Section 3.2.3. We can treat the row variable X as a categorical explanatory variable (see Section 2.2.6) by constructing $I-1$ indicator variables $x_2,\ldots,x_I$ representing levels $2,\ldots,I$ of X. Note that we exclude the indicator x_1 for convenience so that the indicator variable indices match those used in the model by R. Using Y as the response variable with category probabilities $\pi_1,\ldots,\pi_J$, the multinomial regression model for the table is

$$\log(\pi_j/\pi_1) = \beta_{j0} + \beta_{j2}x_2 + \cdots + \beta_{jI}x_I, \tag{3.10}$$

for $j = 2,\ldots,J$. This model implies that $\log(\pi_j/\pi_1) = \beta_{j0}$ when $X = 1$ and $\log(\pi_j/\pi_1) = \beta_{j0}+\beta_{ji}$ when $X = i$ for $i = 2,\ldots,I$. Therefore, the log-odds between two columns depends on the row of the table. In addition, note that there are $I(J-1)$ regression parameters in total for the model. Because $\sum_{j=1}^{J}\pi_j = 1$ for each level of X (i.e., only $J-1$ response probabilities need to be estimated in each row), the model is saturated.

Independence between X and Y removes this row dependence, so that the log-odds between two columns is constant across rows. This implies the simplified model

$$\log(\pi_j/\pi_1) = \beta_{j0}.$$

Thus, testing for independence is equivalent to testing $H_0 : \beta_{j2} = \cdots = \beta_{jI} = 0$ for each $j = 2,\ldots,J$, vs. $H_a :$ At least one $\beta_{ji} \neq 0$ for some j and i. This test is easily performed using an LRT.

When the null hypothesis is rejected, the next step is to investigate the sources of dependence. As in Section 3.2.3, calculating odds ratios for selected 2×2 sections of the contingency table can sometimes help to explain the nature of the association. These odds ratios are easily obtained from the regression parameters as described in the previous section. For example, the estimated odds for comparing response categories $Y = j$ to $Y = 1$ are $\exp(\hat{\beta}_{ji})$ times as large for $X = i$ as for $X = 1$. Odds ratios involving rows and columns other than the first are found as exponentiated linear combinations of regression parameters. For example, the estimated odds for comparing response categories 2 and 3 are $\exp[(\hat{\beta}_{24} - \hat{\beta}_{34}) - (\hat{\beta}_{25} - \hat{\beta}_{35})]$ times as large in row 4 as they are in row 5. Confidence intervals can be formed for these odds ratios as described in Section 3.3.1. Changing the base level for X can help perform these computations.

Example: Fiber-enriched crackers (Fiber.R, Fiber.csv)

We use multinomial regression to repeat the calculations for the LRT for independence that were presented in the example of Section 3.2.3. We start by fitting the model using `multinom()` with bloating severity as the response variable and fiber source as the explanatory variable.

```
> library(package = nnet)
> mod.fit.nom <- multinom(formula = bloat ~ fiber, weights =
   count, data = diet)
# weights:  20 (12 variable)
initial  value 66.542129
iter  10 value 54.519963
iter  20 value 54.197000
final  value 54.195737
converged

> summary(mod.fit.nom)
Call: multinom(formula = bloat ~ fiber, data = diet, weights =
   count)

Coefficients:
        (Intercept)  fiberbran   fibergum fiberboth
low      -0.4057626 -0.1538545  0.4055575  1.322135
medium   -1.0980713 -0.8481379  1.5032639  1.503764
high    -12.4401085 -4.1103893 13.3561038 12.440403

Std. Errors:
        (Intercept)    fiberbran   fibergum  fiberboth
low       0.6455526    0.8997698   1.190217   1.056797
medium    0.8163281    1.3451836   1.224593   1.224649
high    205.2385583 1497.8087307 205.240263 205.240994

Residual Deviance: 108.3915
AIC: 132.3915
```

The function recognizes that `fiber` is a factor with levels ordered as `"none"`, `"bran"`, `"gum"`, and `"both"`. Therefore, it fits the model

$$\log(\pi_j/\pi_{\text{none}}) = \beta_{j0} + \beta_{j1}\texttt{bran} + \beta_{j2}\texttt{gum} + \beta_{j3}\texttt{both},$$

where bran, gum, and both represent indicator variables for the corresponding fiber source categories, and $j = 2, 3, 4$ represents response categories low, medium, and high, respectively.[8] The weights = count argument is used within multinom() because each row of diet represents contingency table counts (stored in the variable "count") rather than observations from individual trials. The vglm() function also uses a weights = count argument. However, vglm() will not work with 0 counts, so rows of the data frame that contain zero counts need to be removed. Our program provides code to implement vglm() with this removal.

The summary() generic function reports the parameter estimates and corresponding standard errors, $\widehat{Var}(\hat{\beta}_{ji})^{1/2}$. Note that the residual deviance is given as 108.3915 rather than 0 even though a saturated model is being fit. This is a side effect of how multinom() uses the weights argument to create a likelihood function. For this contingency table setting, the likelihood function should be a product of four multinomial PMFs, one for each fiber source, where the number of trials is 12 for each PMF. Then, the response category probabilities estimated by the model will be identical to the sample proportions in the table, which causes a residual deviance of zero. The multinom() function instead reformulates the counts into individual multinomial trials (i.e., the "raw data" format as examined in Section 1.2.1). This leads to 48 rows in the data frame, where each row represents one individual's fiber and bloating values. The likelihood function becomes the product of 48 multinomial PMFs with one trial for each PMF. Each individual's sample proportions for the bloating response categories are either 1, if the individual experienced that level of bloating, or 0 otherwise. The estimated probabilities from the model do not match these proportions, so the residual deviance differs from 0. However, despite this different formulation of the likelihood function by multinom(), we will not have any problems in the analysis that follows. For these data and in general, parameter estimates, standard errors, and LRT results for explanatory variables are the same no matter which multinomial likelihood function and data format is used. We will discuss these two data formats further in Chapter 5.

The test for independence of fiber source and bloating severity compares $H_0 : \beta_{j2} = \beta_{j3} = \beta_{j4} = 0$ for each $j = 2, 3, 4$ vs. H_a : Not all $\beta_{ji} = 0$ for some j and i. The LRT is performed by Anova().

```
> library(package = car)
> Anova(mod.fit.nom)
# weights:  8 (3 variable)
initial  value 66.542129
final  value 63.635876
converged

Analysis of Deviance Table (Type II tests)
Response: bloat
      LR Chisq Df Pr(>Chisq)
fiber    18.88  9    0.02623 *
---
Signif. codes:  0 '***' 0.001 '**' 0.01 '*' 0.05 '.' 0.1 ' ' 1
```

[8]We omit using "fiber" as part of the indicator variable names because fiber is the only explanatory variable in the model.

The LRT results in $-2\log(\Lambda) = 18.9$ with a p-value of 0.026. Thus, there is marginal evidence of dependence between type of fiber and severity of bloating. Notice that these values match what was found earlier using the `assocstats()` function.

To examine the sources of association, we can start by examining the estimated parameters of the model. For example, the estimated log-odds comparing the low bloating severity category to no bloating is

$$\log(\hat{\pi}_{\texttt{low}}/\hat{\pi}_{\texttt{none}}) = -0.41 - 0.15\texttt{bran} + 0.41\texttt{gum} + 1.32\texttt{both}.$$

The parameter estimates suggest that using gum, with or without bran, leads to a larger odds of low bloating relative to no bloating than does using no fiber. On the other hand, there appears to be a somewhat smaller effect in the opposite direction because of using bran instead of no fiber. We compute the estimated odds ratios as follows.

```
> round(exp(coefficients(mod.fit.nom)[,-1]), 2)
       fiberbran fibergum fiberboth
low         0.86      1.5      3.75
medium      0.43      4.5      4.50
high        0.02 631658.3 252812.49
```

For example, the estimated odds of having low bloating rather than none are $\exp(-0.15) = 0.86$ times as large for using bran as a fiber source than for using no fiber at all. Of course, final conclusions about the effects of fiber need to account for the variability of the parameter estimators, which we will do shortly.

Notice that the high bloating row in the table of odds ratios contains two very large estimates. Also, notice the table of standard errors from `summary(mod.fit.nom)` contains very large values in the high bloating row. In fact, if one were to decrease the convergence criterion value given by the `reltol` argument (default is `reltol = 10^(-8)`) for `multinom()`, one could see the parameter estimates are increasing toward infinity for the `fibergum` and `fiberboth` columns of the high bloating row despite the `converged` message printed in the output. These problems occur because of the 0 counts present in the corresponding cells of Table 3.2. We saw in Section 1.2.5 that 0 counts can cause difficulties with estimating odds ratios, and these problems occur here as well. An ad-hoc solution similar to what was used in Section 1.2.5 is to add 0.5 to the 0 cell counts. Alternatively, a small constant, such as 0.5, could be added to all cells of the contingency table. In either case, the regression parameter estimates are biased toward 0, and hence the odds ratios are biased toward 1. However, the reduction in the variance of the estimates is substantial, and there are not necessarily better solutions within the framework of using maximum likelihood estimation for the multinomial regression model.

The code below shows how 0.5 is added to the 0 cells using the `ifelse()` function. The `test` argument determines whether a count is equal to 0. *If* the count is 0, the `yes` argument is implemented to add 0.5 to the count. *Else*, the `no` argument is implemented to keep the count the same. This if-then-else approach is implemented separately for each element of `diet$count`. After adjusting the data, we refit the model.

```
> diet$count2 <- ifelse(test = diet$count == 0, yes = diet$count
   + 0.5, no = diet$count)
```

```
> mod.fit.nom2 <- multinom(formula = bloat ~ fiber, weights =
    count2, data = diet)
# weights:  20 (12 variable)
initial  value 67.928424
iter  10 value 58.549878
final  value 58.394315
converged

> sum.fit <- summary(mod.fit.nom2)
> round(sum.fit$coefficients, 4)
       (Intercept) fiberbran fibergum fiberboth
low        -0.4055   -0.1541   0.4055    1.3218
medium     -1.0986   -0.8473   1.5041    1.5040
high       -2.4849   -0.1542   3.4012    2.4849
> round(sum.fit$standard.errors, 4)
       (Intercept) fiberbran fibergum fiberboth
low         0.6455    0.8997   1.1902    1.0567
medium      0.8165    1.3452   1.2247    1.2247
high        1.4719    2.0759   1.6931    1.7795
```

We see that the estimated regression parameters and corresponding standard errors for low and medium bloating are very similar to before, but the highbloating values are much closer to 0 because of the added constant. Below are the estimated odds ratios and some of the corresponding 95% Wald intervals.

```
> round(exp(coefficients(mod.fit.nom2)[,-1]), 2)
       fiberbran fibergum fiberboth
low         0.86      1.5      3.75
medium      0.43      4.5      4.50
high        0.86     30.0     12.00

> conf.beta <- confint(object = mod.fit.nom2, level = 0.95)
> conf.beta , , low

                 2.5 %    97.5 %
(Intercept) -1.6706740 0.8596514
fiberbran   -1.9175361 1.6093729
fibergum    -1.9273405 2.7383320
fiberboth   -0.7493731 3.3929139

, , medium
                 2.5 %    97.5 %
(Intercept) -2.6988715 0.5016931
fiberbran   -3.4838153 1.7891901
fibergum    -0.8963637 3.9045092
fiberboth   -0.8963946 3.9044732

, , high

                  2.5 %    97.5 %
(Intercept) -5.36979896 0.4000565
fiberbran   -4.22290363 3.9145455
fibergum     0.08274864 6.7195650
```

```
fiberboth    -1.00285349 5.9725907

> round(exp(conf.beta[2:4,,3]), 2)  # compare high to no bloating
           2.5 % 97.5 %
fiberbran   0.01  50.13
fibergum    1.09 828.46
fiberboth   0.37 392.52
```

These estimated odds ratios are the same as what we found earlier, except for high bloating. The `confint()` function stores confidence limits for the regression parameters in a three-dimensional array using a [row, column, table] format. For example, to access the high vs. no bloating intervals, we specify the row and table indices within the array without giving a column index. To find the intervals for the odds ratios, we use the `exp()` function. The confidence intervals for high vs. no bloating are rather wide, calling their usefulness into question. This happens because the odds ratios and confidence intervals are based on such small counts that there is considerable uncertainty regarding the population values. Additional calculations for odds ratios comparing other levels of bloating are given in the program.

It is interesting to note that there is an apparent trend among the estimated odds ratios for gum and both when compared to none. We see that as the bloating severity increases, so do the estimated odds ratios. This may not be a coincidence. Notice that the bloating severity response is actually an ordinal variable. Section 3.4 will show how to adapt our modeling to ordinal response variables that allow a potentially more powerful, yet parsimonious, analysis to be conducted. An additional advantage to modeling the ordinal structure is that the impact of individual 0 counts is substantially lessened.

Interestingly, the estimated probabilities, odds, and odds ratios produced from Equation 3.10 are exactly the same as those obtained directly from the contingency table (as long as there are no 0 counts). For instance, in the last example, the estimated odds ratio of having low bloating rather than none is $n_{11}n_{22}/(n_{12}n_{21}) = 6 \times 4/(4 \times 7) = 0.86$ times as large for using bran as a fiber source than for using no fiber at all. This estimated odds ratio value is exactly the same as $\exp(-0.15) = 0.86$ computed using the estimated multinomial regression model. This happens because we are using a saturated model, so the parameter estimates produce odds ratios that match the data exactly.

In this sense, the multinomial model seems like a complicated way to perform the analysis. This is true for the relatively simple problem of relating a multinomial response to a single categorical explanatory variable. The strength of the modeling procedure is it can be used in more complex problems where the simple table-based analysis cannot. In particular, the multinomial regression model can be used to assess effects of arbitrarily many variables on the response (subject to adequate sample size). The variables can be numerical as shown earlier in this section, categorical as shown in this subsection, or any combination of these.

When additional categorical explanatory variables are available, we can examine the data in higher dimensional contingency tables. For example, consider a setting with a three-dimensional contingency table summarizing the counts for categorical variables X, Y, and Z, with levels $i = 1, \ldots, I$, $j = 1, \ldots, J$, and $k = 1, \ldots, K$, respectively. An analysis using the simpler tools of Section 3.2 requires that separate contingency tables summarizing any pair of variables (e.g., X and Y) be formed for each level of the third variable (e.g., Z). In some cases, there is an "obvious" way to do this, such as when X represents a treatment variable, Y a response variable, and Z a blocking or stratification variable. Then, modeling the relationship between X and Y controlling for the level of Z is the natural approach to the analysis. A Pearson or LR test for independence can be conducted in each table, and subsequently, odds ratios can be estimated and other follow-up analyses performed. However, this approach is limited in its scope. It allows assessments only of the association between X and Y separately on the level of Z. It makes no attempt to compare these associations across Z-levels, nor to combine them into a single association that applies to all Z-levels. Although table-based methods do exist to address the latter issue (see Section 6.4.5 of Agresti, 2013), they still do not allow consideration of the association between Z and Y, which is of interest when Z is another explanatory variable rather than a stratification variable.

The multinomial regression framework (and the Poisson regression model structure introduced in Section 4.2.4) allows these more general questions to be answered. For example, a general model allowing for association of both X and Z with Y is

$$\log(\pi_j/\pi_1) = \beta_{j0} + \beta^X_{j2}x_2 + \cdots + \beta^X_{jI}x_I + \beta^Z_{j2}z_2 + \cdots + \beta^Z_{jK}z_K,\ j = 2, \ldots, J$$

where we use $x_2, \ldots, x_I$ and $z_2, \ldots, z_K$ as indicator variables for the levels given in their indices. The superscript X or Z is added to the β parameters to make clear which variable's association with Y is described by the parameter. Excluding either the X and/or Z indicator variables from the model corresponds to independence between that explanatory variable and Y. An interaction between X and Z could be included in the model as well to allow each variable's association with Y to vary among the levels of the other explanatory variable. Adding additional categorical explanatory variables is a straightforward extension of this model, as is adding numerical explanatory variables.

Example: Fiber-enriched crackers (Fiber.R, Fiber.csv)

As mentioned in Section 3.2, the fiber explanatory variable is actually constructed from combinations of two separate variables: `bran` (`"yes"` or `"no"`) and `gum` (`"yes"` or `"no"`). This structure is often referred to as a 2×2 factorial.[9] It is typical in these structures to ask questions regarding separate effects of each constituent variable on the responses, as well as to look for interactions between the variables. To pursue this analysis, we first transform the explanatory variable `fiber` into `bran` and `gum` using the `ifelse()` function.

```
> diet$bran <- factor(ifelse(test = diet$fiber == "bran" |
   diet$fiber == "both", yes = "yes", no = "no"))
> diet$gum <- factor(ifelse(test = diet$fiber == "gum" |
   diet$fiber == "both", yes = "yes", no = "no"))
> head(diet, n = 4)
  fiber  bloat count bran gum
1  bran   high     0  yes  no
2   gum   high     5   no yes
3  both   high     2  yes yes
4  none   high     0   no  no
```

The single vertical line `|` joining conditions within the `test` argument means "or", so that the first line assigns `bran` a `"yes"` value if `fiber = "bran"` or `fiber = "both"`;

[9]In general an $a \times b$ factorial consists of all combinations of one variable with a levels and another with b levels. The variables making up the combinations are called "factors" (see, for example, Milliken and Johnson, 2004 for details).

otherwise, bran is set to "no". Also, we enclose ifelse() within factor() so that the result is a factor class type rather than character.

Next, we fit the multinomial model that includes effects for bran, gum, and their interaction.

```
> mod.fit.nom.inter <- multinom(formula = bloat ~ bran + gum +
    bran:gum, weights = count, data = diet)
# weights:  20 (12 variable)
initial  value 66.542129
iter  10 value 54.406806
iter  20 value 54.196639
final  value 54.195746
converged

> summary(mod.fit.nom.inter)
Call:
multinom(formula = bloat ~ bran + gum + bran:gum, data = diet,
    weights = count)

Coefficients:
       (Intercept)    branyes     gumyes branyes:gumyes
low     -0.4063036 -0.1532825  0.4076874      1.0674733
medium  -1.0973831 -0.8503013  1.5033271      0.8503872
high   -12.4349471 -2.0291752 13.3518350      1.1121073

Std. Errors:
       (Intercept)    branyes     gumyes branyes:gumyes
low      0.6456363   0.899801   1.190275        1.58416
medium   0.8160908   1.345541   1.224613        1.86471
high   204.7028906 561.368770 204.704601      561.37028

Residual Deviance: 108.3915
AIC: 132.3915

> Anova(mod.fit.nom.inter)
Analysis of Deviance Table (Type II tests)

Response: bloat
         LR Chisq Df Pr(>Chisq)
bran       2.5857  3   0.460010
gum       16.2897  3   0.000989 ***
bran:gum   0.4880  3   0.921514
---
Signif. codes:  0 '***' 0.001 '**' 0.01 '*' 0.05 '.' 0.1 ' ' 1

> logLik(mod.fit.nom.inter)
'log Lik.' -54.19575 (df=12)
```

The LRT for the interaction gives a p-value of 0.9215, indicating there is not sufficient evidence of an interaction. Furthermore, we see that bran has a large p-value, but gum has a small p-value. Thus, there is sufficient evidence to indicate that gum has an effect on the bloating severity.

Because the saturated model is being fit here, the problems with zero counts in the table that were noted in the previous example remain a problem in this model. In the end, these difficulties are not of concern for the LRT, because the $-2\log(\Lambda)$ statistic changes very little for a larger number of iterations.[10]

As an alternative to the multinomial structure of this chapter, analysis of contingency tables is frequently performed assuming the counts have a Poisson distribution (or variants of it). We will discuss these Poisson regression models in Chapter 4.

3.4 Ordinal response regression models

Many categorical response variables have a natural ordering to their levels. For example, a response variable may be measured using a Likert scale with categories "strongly disagree," "disagree," "neutral," "agree," or "strongly agree." If response levels can be arranged so that category 1 < category 2 < $\cdots$ < category J in some conceptual scale of measurement (e.g., amount of agreement), then regression models can incorporate this ordering through a variety of logit transformations of the response probabilities. In this section, we focus on modeling cumulative probabilities based on the category ordering.

The cumulative probability for category j of Y is $P(Y \leq j) = \pi_1 + \cdots + \pi_j$ for $j = 1, \ldots, J$. Note that $P(Y \leq J) = 1$. Regression models for ordinal multinomial responses can examine the effects of explanatory variables $x_1, \ldots, x_p$ on the log-odds of cumulative probabilities, also called *cumulative logits*,

$$\text{logit}(P(Y \leq j)) = \log\left(\frac{P(Y \leq j)}{1 - P(Y \leq j)}\right) = \log\left(\frac{\pi_1 + \cdots + \pi_j}{\pi_{j+1} + \cdots + \pi_J}\right).$$

In particular, the *proportional odds model* is a special type of vector GLM that assumes that the logit of these cumulative probabilities changes linearly as the explanatory variables change, and also that the slope of this relationship is the same regardless of the category j. Formally, the model is stated as

$$\text{logit}(P(Y \leq j)) = \beta_{j0} + \beta_1 x_1 + \cdots + \beta_p x_p,\ j = 1, \ldots, J - 1. \tag{3.11}$$

Notice that there are no j subscripts on the parameters $\beta_1, \ldots, \beta_p$. The model assumes that the effects of the explanatory variables are the same regardless of which cumulative probabilities are used to form the log-odds. Thus, the name "proportional odds" derives from each odds being a multiple of $\exp(\beta_{j0})$:

$$\begin{aligned}\frac{P(Y \leq j)}{1 - P(Y \leq j)} &= \exp(\beta_{j0} + \beta_1 x_1 + \cdots + \beta_p x_p)\\ &= \exp(\beta_{j0})\exp(\beta_1 x_1 + \cdots + \beta_p x_p).\end{aligned}$$

For any fixed j, increasing x_r by c units changes every log-odds in Equation 3.11 by $c\beta_r$ when holding other explanatory variables constant. On the other hand, the difference in the log-odds between response categories j and j' is constant, $\beta_{j0} - \beta_{j'0}$, and does not depend on the values of $x_1, \ldots, x_p$ when they are held fixed. These results relate directly to odds

[10]For the two 0 count occurrences, the multinomial regression model estimate of π_j is already very close to 0. The estimate can only get slightly closer to 0 if the convergence criterion was made more strict than the default used here.

ratios, as detailed in Section 3.4.1. Also, notice that the odds must increase as j increases, because we put progressively more probability in the numerator, $P(Y \leq j)$. This implies that $\beta_{10} < \cdots < \beta_{J-1,0}$.

Probabilities for observing a particular response category j are found by noting that

$$\begin{aligned} \pi_j &= P(Y = j) \\ &= P(Y \leq j) - P(Y \leq j-1), \end{aligned} \tag{3.12}$$

where $P(Y \leq 0) = 0$, $P(Y \leq J) = 1$, and

$$P(Y \leq j) = \frac{\exp(\beta_{j0} + \beta_1 x_1 + \cdots + \beta_p x_p)}{1 + \exp(\beta_{j0} + \beta_1 x_1 + \cdots + \beta_p x_p)}.$$

For example, the probability for category 1 is

$$\pi_1 = P(Y \leq 1) - P(Y \leq 0) = \frac{\exp(\beta_{10} + \beta_1 x_1 + \cdots + \beta_p x_p)}{1 + \exp(\beta_{10} + \beta_1 x_1 + \cdots + \beta_p x_p)},$$

and the probability for category J is

$$\pi_J = P(Y \leq J) - P(Y \leq J-1) = 1 - \frac{\exp(\beta_{J-1,0} + \beta_1 x_1 + \cdots + \beta_p x_p)}{1 + \exp(\beta_{J-1,0} + \beta_1 x_1 + \cdots + \beta_p x_p)}.$$

Example: Proportional odds model plots (CumulativeLogitModelPlot.R)

The purpose of this example is to examine features of the proportional odds model. Consider the model $\text{logit}(P(Y \leq j)) = \beta_{j0} + \beta_1 x_1$, where $J = 4, \beta_{10} = 0, \beta_{20} = 2, \beta_{30} = 4$, and $\beta_1 = 2$. Figure 3.4 plots the cumulative probabilities and individual category probabilities for this model. Notice that the cumulative probability curves in the left plot have exactly the same shape, because β_1 is the same for each response category j. The horizontal shift is due to the different values for β_{j0}. For example, $P(Y \leq j) = 0.5$ when $\beta_{j0} + \beta_1 x_1 = 0$, i.e., when $x_1 = -\beta_{j0}/\beta_1$.

Probabilities of response for individual categories, π_j, $j = 1, 2, 3, 4$, are found at each x_1 from the vertical distances between consecutive curves in the left plot, along with $P(Y \leq 0) = 0$ and $P(Y \leq 4) = 1$. These are shown in the plot on the right in Figure 3.4, which also illustrates the orderings among the categories for Y with respect to x_1. For example, the smallest values of x_1 lead to $Y = 4$ having the largest probability. As x_1 increases, the probability of observing $Y = 3$ becomes larger until eventually it has the largest probability for a particular range of x_1. Similar outcomes occur for $Y = 2$ and $Y = 1$ in this order.

The code for these plots is in the program for this example. We encourage readers to change elements of the code to see what happens as values of $\beta_{10}, \beta_{20}, \beta_{30}$, and β_1 change. For example, decreasing the absolute distance between β_{10} and β_{20} can lead to a case where π_2 is never the largest probability for any value of x_1 (see Exercise 6).

Parameters of the proportional odds model are estimated using maximum likelihood. Similar to Section 3.3, the likelihood function for a sample of size m is simply the product of m multinomial distributions with probability parameters expressed as functions of the explanatory variables as given by Equations 3.11 and 3.12. Iterative numerical procedures are used then to fit the model. The covariance matrix for the regression parameter estimates is found using standard likelihood procedures as outlined in Appendix B.3.4. Wald and LR-based inference procedures are performed in the usual ways as well.

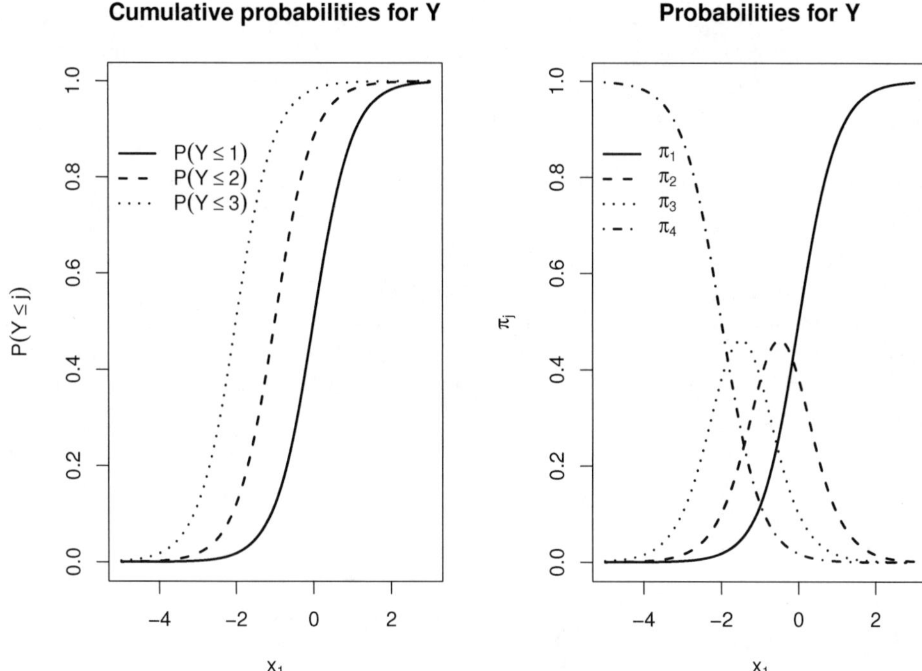

Figure 3.4: Proportional odds model for $\beta_{10} = 0, \beta_{20} = 2, \beta_{30} = 4$, and $\beta_1 = 2$.

We will focus on two R functions for model estimation: `polr()` from the `MASS` package and, once again, `vglm()` from the `VGAM` package. Similar to the `nnet` package, the `MASS` package is in the default installation of R, but `library(package = MASS)` needs to be executed to make its functions available. This package has a long history that pre-dates the existence of R, leading to the widespread use of `polr()` today. Profile LR intervals for model parameters can be calculated with `confint()`. However, for intervals involving linear combinations of model parameters, the `mcprofile` package has not been extended to models fit with `polr()`. Instead, the `emmeans` package can be used to calculate Wald intervals in this situation. With both `polr()` and `vglm()`, it is important to ensure that the levels of the categorical response are ordered in the desired way; otherwise, the ordering of the levels for Y will not be correctly taken into account. We will demonstrate how to check and, if necessary, change the ordering in the next example.

We will often consider hypotheses associated with tests of individual regression parameters, like $H_0 : \beta_r = 0$ vs. $H_a : \beta_r \neq 0$. If the null hypothesis is true, then the $J - 1$ log-odds comparing $P(Y \leq j)$ to $P(Y > j)$ do not depend on x_r, holding all other explanatory variables constant. If the alternative hypothesis is true, then the log-odds for each cumulative probability grow larger or smaller with x_r depending on the sign of β_r. In turn, this imposes an ordering on the individual category probabilities such as that seen in the right plot of Figure 3.4.

Contrast the previous test involving β_r to a corresponding test involving the multinomial regression model with a nominal response from Section 3.3, $H_0 : \beta_{2r} = \cdots = \beta_{Jr} = 0$ vs. H_a : At least one $\beta_{jr} \neq 0$. The alternative hypothesis places fewer constraints on how x_r relates to individual category probabilities by using more parameters. Generally, models with stricter assumptions can produce more biased estimates when the assumptions are wrong. However, when the assumptions are reasonably accurate, models that do not use them produce estimates that have more variability. This can lead to inferences that are unnecessarily conservative—for example, hypothesis tests have less power and confidence intervals are wider—compared to a model that makes accurate assumptions about the population. We will see this further in Section 3.4.2.

Example: Wheat kernels (Wheat.R, Wheat.csv)

Healthy kernels are the most desirable type of wheat kernel. One could also reason that presence of disease makes scab kernels less desirable than sprouted kernels. Thus, we use the ordering of scab ($Y = 1$) < sprout ($Y = 2$) < healthy ($Y = 3$) and fit a proportional odds model to these data.

We begin by creating the required ordering of the levels for the type variable. There are several ways to accomplish this, and we show one approach below that is fairly straightforward.

```
> levels(wheat$type)
[1] "Healthy" "Scab"    "Sprout"
> wheat$type.order <- factor(wheat$type, levels = c("Scab",
   "Sprout", "Healthy"))
> levels(wheat$type.order)
[1] "Scab"    "Sprout"  "Healthy"
```

The new variable of wheat named type.order contains the factor with the proper ordering of its levels. The proportional odds model is then fit using polr().

```
> library(package = MASS)
> mod.fit.ord <- polr(formula = type.order ~ class + density +
   hardness + size + weight + moisture, data = wheat, method =
   "logistic")
> summary(mod.fit.ord)

Re-fitting to get Hessian

Call: polr(formula = type.order ~ class + density + hardness +
   size + weight + moisture, data = wheat, method = "logistic")

Coefficients:
            Value Std. Error t value
classsrw  0.17370   0.391764  0.4434
density  13.50534   1.713009  7.8840
hardness  0.01039   0.005932  1.7522
size     -0.29253   0.413095 -0.7081
weight    0.12721   0.029996  4.2411
moisture -0.03902   0.088396 -0.4414

Intercepts:
               Value   Std. Error t value
Scab|Sprout    17.5724  2.2460      7.8237
Sprout|Healthy 20.0444  2.3395      8.5677

Residual Deviance: 422.4178
```

```
AIC: 438.4178

> library(package = car)
> Anova(mod.fit.ord)
Analysis of Deviance Table (Type II tests)

Response: type.order
         LR Chisq Df Pr(>Chisq)
class       0.197  1    0.65749
density    98.437  1  < 2.2e-16 ***
hardness    3.084  1    0.07908 .
size        0.499  1    0.47982
weight     18.965  1  1.332e-05 ***
moisture    0.195  1    0.65872
---
Signif. codes:  0 '***' 0.001 '**' 0.01 '*' 0.05 '.' 0.1 ' ' 1
```

The `method = "logistic"` argument value instructs R to use the logit transformation on the cumulative probabilities. The `polr()` function estimates the model

$$\text{logit}(P(Y \leq j)) = \beta_{j0} - \eta_1 x_1 - \cdots - \eta_p x_p, \tag{3.13}$$

where $-\eta_r$ is β_r in Equation 3.11. Because of these notational differences, we need to reverse the sign of all the estimated η_r values in the output to state the estimated model. This results in the model

$$\begin{aligned}\text{logit}(\widehat{P}(Y \leq j)) = \hat{\beta}_{j0} &- 0.17\texttt{classsrw} - 13.51\texttt{density} - 0.010\texttt{hardness} \\ &+0.29\texttt{size} - 0.13\texttt{weight} + 0.039\texttt{moisture},\end{aligned} \tag{3.14}$$

where $\hat{\beta}_{10} = 17.57$ and $\hat{\beta}_{20} = 20.04$. The $\hat{\beta}_{10}$ and $\hat{\beta}_{20}$ values are found in the `Intercepts` table with `Scab|Sprout` indicating $j = 1$ and `Sprout|Healthy` indicating $j = 2$.

The `t value` column in the `Coefficients` table provides the Wald statistics for testing $H_0 : \beta_r = 0$ vs. $H_a : \beta_r \neq 0$ for $r = 1, \ldots, 6$. The `Anova()` function provides the corresponding LRTs. Because of the large test statistic values for `density` and `weight`, there is sufficient evidence that these are important explanatory variables. There is marginal evidence that `hardness` is important too. As with other tests like these in previous chapters, each test is conditional on the other variables being in the model.

Below is our code using `vglm()`.

```
> library(package = VGAM)
> mod.fit.ord.v <- vglm(formula = type.order ~ class + density +
    hardness + size + weight + moisture, data = wheat, family =
    cumulative(parallel = TRUE))
> summary(mod.fit.ord.v)

Call:
vglm(formula = type.order ~ class + density + hardness + size +
    weight + moisture, family = cumulative(parallel = TRUE), data
    = wheat)

Coefficients:
```

```
                 Estimate Std. Error z value Pr(>|z|)
(Intercept):1   17.572326   2.245687   7.825 5.08e-15 ***
(Intercept):2   20.044351   2.339681   8.567  < 2e-16 ***
classsrw        -0.173694   0.384309  -0.452   0.6513
density        -13.505316   1.695674  -7.965 1.66e-15 ***
hardness        -0.010394   0.005964  -1.743   0.0814 .
size             0.292525   0.420863   0.695   0.4870
weight          -0.127213   0.030034  -4.236 2.28e-05 ***
moisture         0.039017   0.088564   0.441   0.6595
---
Signif. codes:  0 '***' 0.001 '**' 0.01 '*' 0.05 '.' 0.1 ' ' 1

Names of linear predictors: logitlink(P[Y<=1]),
   logitlink(P[Y<=2])

Residual deviance: 422.4178 on 542 degrees of freedom

Log-likelihood: -211.2089 on 542 degrees of freedom

Number of Fisher scoring iterations: 7

Warning: Hauck-Donner effect detected in the following
  estimate(s): '(Intercept):1'

Exponentiated coefficients:
    classsrw      density     hardness         size       weight
8.405541e-01 1.363691e-06 9.896601e-01 1.339807e+00 8.805459e-01
    moisture
1.039788e+00
Warning message:
In eval(slot(family, "initialize")) :
  response should be ordinal---see ordered()
```

The function directly estimates the parameters of Equation 3.11, leading to the model given in Equation 3.14. Within the call to `vglm()`, the `family = cumulative` argument value specifies a model for the cumulative logit, where `parallel = TRUE` denotes the proportional odds version. The warning message produced in the output is due to our ordering the values in `type.order` using the `factor()` function rather than the `ordered()` function. This will not affect the results though as long as the levels are in the correct order. The `anova()` function can be used with `mod.fit.ord.v` to calculate the same test statistics and p-values as was done with `mod.fit.ord`. As discussed in Section 3.3, the `Anova()` function does not produce the correct calculations with objects resulting from `vglm()`.

The `predict()` function finds the estimated probabilities for each response category.

```
> # Use object from polr()
> predict(object = mod.fit.ord, newdata = wheat[1,], type =
   "probs")
      Scab     Sprout    Healthy
 0.03661586 0.27384933 0.68953481

> # Use object from vglm()
```

```
> predict(object = mod.fit.ord.v, newdata = wheat[1,], type =
   "response")
        Scab      Sprout    Healthy
1 0.03661627 0.2738514 0.6895324
```

For example, the estimated probability of being healthy for the first observation is

$$\hat{\pi}_{\texttt{Healthy}} = 1 - \frac{\exp(20.04 - 0.17 \times 0 + \cdots + 0.04 \times 12.02)}{1 + \exp(20.04 - 0.17 \times 0 + \cdots + 0.04 \times 12.02)}$$
$$= 0.6895.$$

This probability is found using the `type = "probs"` argument for the object resulting from `polr()` and the `type = "response"` argument for the object resulting `vglm()`. We include code in our corresponding program showing how these probabilities are computed by coding Equation 3.12 directly using R's syntax. A `type = "class"` argument for the object from `polr()` (the default) identifies the response variable category with the largest estimated probability. For the first kernel, `Healthy` has the largest estimated probability, so the kernel is predicted to be `Healthy` (which is a correct prediction).

The `predict()` function does not provide the estimated variances needed to form Wald confidence intervals of the form $\hat{\pi}_j \pm Z_{1-\alpha/2}\widehat{Var}(\hat{\pi}_j)^{1/2}$ for π_j, similar to what was discussed in Section 3.3. These variances and intervals can be calculated with the method function for `summary()` of the `emmeans` package.

```
> library(package = emmeans)
> calc.prob <- emmeans(object = mod.fit.ord, specs = ~
   type.order, at = wheat[1,], mode = "prob")

Re-fitting to get Hessian

> summary(object = calc.prob, level = 0.95)
 type.order    prob     SE  df asymp.LCL asymp.UCL
 Scab        0.0366 0.0156 Inf   0.00612    0.0671
 Sprout      0.2738 0.0693 Inf   0.13797    0.4097
 Healthy     0.6895 0.0820 Inf   0.52882    0.8502

Confidence level used: 0.95
```

The 95% Wald interval for $\pi_{\texttt{Healthy}}$ is (0.5288, 0.8502) for kernels with the same explanatory variable values as the first kernel in the data set. The corresponding code is similar to what was used for multinomial regression models. In particular, we used the object resulting from `polr()` because `emmeans()` does not work with objects resulting from `vglm()`. The `specs` argument contains the response, the `at` argument specifies the explanatory variable values, and `mode = "prob"` specifies that calculations are to be performed directly on the probabilities and not on the linear predictors. However, we no longer need a `df` argument because a standard normal distribution is correctly used for the Wald intervals.

A plot of the estimated probabilities for each category vs. an individual explanatory variable (holding the other variables in the model constant) can be used to interpret the model. Exercise 21 examines plots of this form. We simplify the problem here

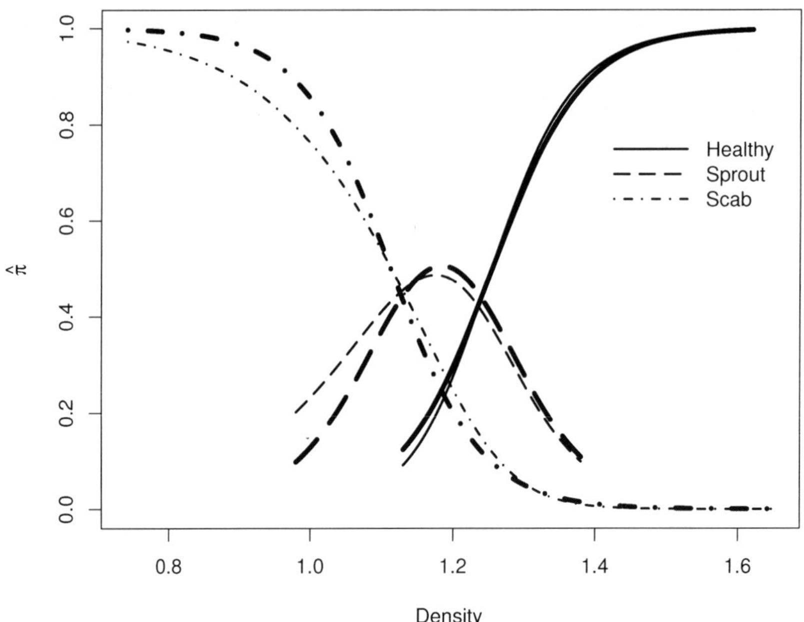

Figure 3.5: Estimated proportional odds (thicker line) and multinomial (thinner line) regression models for the wheat data, where density is the only explanatory variable included in the model.

by again using `density` as the only explanatory variable in fitting a new proportional odds model:

$$\text{logit}(\widehat{P}(Y \leq j)) = \hat{\beta}_{j0} - 15.64\texttt{density}$$

with $\hat{\beta}_{10} = 17.41$ and $\hat{\beta}_{20} = 19.63$ (see program for code). Figure 3.5 shows a plot of this model along with the multinomial regression model estimated earlier that treats the response as nominal. These models produce somewhat similar probability curves with the same scab < sprout < healthy ordering for density with both models. This provides some reassurance that the ordering used for the proportional odds model is appropriate and that the assumption of proportional odds among cumulative probabilities is reasonable.

There are a number of other functions that can fit a proportional odds model. We show how to use two of them—the `lrm()` function of the `rms` package and the `clm()` function of the `ordinal` package—in our program for this example.

3.4.1 Odds ratios

Odds ratios based on the cumulative probabilities are easily formed because the proportional odds model equates log-odds to the linear predictor. Thus, the odds ratio derivations from Chapter 2 apply here, but now with the cumulative probabilities. For example, the odds ratio involving one explanatory variable, say x_1, is

$$\frac{Odds_{x_1+c,x_2,\ldots,x_p}(Y \leq j)}{Odds_{x_1,\ldots,x_p}(Y \leq j)} = \frac{\exp(\beta_{j0} + \beta_1(x_1 + c) + \cdots + \beta_p x_p)}{\exp(\beta_{j0} + \beta_1 x_1 + \cdots + \beta_p x_p)} = \exp(c\beta_1),$$

where we let $Odds_{x_1,\ldots,x_p}(Y \leq j)$ denote the odds of observing category j or smaller for Y. Note that this odds ratio formulation assumes $x_2, \ldots, x_p$ are unchanged. The formal interpretation of the odds ratio is

> The odds of $Y \leq j$ vs. $Y > j$ change by $\exp(c\beta_1)$ times for every c-unit increase in x_1, holding the other explanatory variables in the model constant.

Interestingly, the odds ratio stays the same no matter what response category is used for j. This key feature of the proportional odds model occurs because the model assumes that the effects of explanatory variables are the same, as evidenced by the absence of a j subscript on the slope parameters $\beta_1, \ldots, \beta_p$.

Wald and profile LR intervals are formed in the same ways as in Chapter 2. Profile LR intervals are computed via a method function for `confint()` when using an object resulting from `polr()` or `vglm()`. Unfortunately, these intervals are limited to odds ratios based on only one regression parameter. For more complicated odds ratios, like those involving an interaction term, the `mcprofile` package cannot be used. Instead, Wald intervals can be computed using the `emmeans` package with the results from `polr()`.

Example: Wheat kernels (Wheat.R, Wheat.csv)

The estimated odds ratios for each explanatory variable are calculated as $\widehat{OR} = \exp(c\hat{\beta}_r)$ for $r = 1, \ldots, 6$. Similarly to Section 3.3, we set c to be equal to one standard deviation for each continuous explanatory variable and $c = 1$ for `class`. Recall that we also need to multiply the estimated regression parameters calculated with `polr()` by -1. Below are the calculations.

```
> round(c.value, 2)  # class = 1 is first value
       density hardness      size    weight moisture
  1.00    0.13    27.36      0.49      7.92     2.03

> round(exp(c.value * (-mod.fit.ord$coefficients)), 2)
       density hardness      size    weight moisture
0.84      0.17     0.75      1.15      0.37     1.08
> round(1/exp(c.value * (-mod.fit.ord$coefficients)), 2)
       density hardness      size    weight moisture
1.19      5.89     1.33      0.87      2.74     0.92
```

In the last line of code, we re-express the odds ratios with respect to decreases in their respective explanatory variables. Interpretations of the estimated odds ratios include:

- The estimated odds of a scab ($Y \leq 1$) vs. sprout or healthy ($Y > 1$) response are 0.84 times as large for soft red winter wheat as for hard red winter wheat, holding the other variables constant.
- The estimated odds of a scab vs. sprout or healthy change by 5.89 times for every 0.13 decrease in the density, holding the other variables constant.
- The estimated odds of a scab vs. sprout or healthy response change by 2.74 times for every 7.92 decrease in the weight, holding the other variables constant.

Because of the proportional odds assumption, each of these statements also applies to the odds of a scab or sprout vs. healthy response. For this reason, it is common to interpret odds ratios, such as for density, by saying:

> The estimated odds of kernel quality being below a particular level change by 5.89 times for every 0.13 decrease in the density, holding the other variables constant.

Overall, we see that the larger the density and weight, the more likely a kernel is from a better quality category. We can again relate these results back to Figure 3.2 on p. 187. The plot shows small density and weight kernels often have a scab condition. There is much more overlap among kernel classifications for some of the other explanatory variables in the plot, which corresponds to estimated odds ratios close to 1. Further interpretations of these other variables are part of Exercise 12.

The polr() function creates objects that have a polr class type. The methods() function with this class type shows what method functions are available.

```
> class(mod.fit.ord)
[1] "polr"
> methods(class = polr)
 [1] anova              Anova              brief
 [4] confint            Confint            deltaMethod
 [7] extractAIC         linearHypothesis   logLik
[10] model.frame        nobs               poTest
[13] predict            print              profile
[16] S                  simulate           summary
[19] vcov
see '?methods' for accessing help and source code
```

We have already used the summary.polr() and Anova.polr() functions. The corresponding method function for confint() constructs the profile LR confidence intervals for the parameters $\beta_1, \ldots, \beta_p$. Results from this function are used then to find profile LR intervals for the odds ratios.

```
> conf.beta <- confint(object = mod.fit.ord, level = 0.95)
Waiting for profiling to be done...

Re-fitting to get Hessian

> ci <- exp(c.value*(-conf.beta))
> round(data.frame(low = ci[,2], up = ci[,1]), 2)
          low   up
classsrw 0.39 1.81
density  0.11 0.26
hardness 0.55 1.03
size     0.77 1.72
weight   0.23 0.58
moisture 0.76 1.54

> round(data.frame(low = 1/ci[,1], up = 1/ci[,2]), 2)
          low   up
classsrw 0.55 2.57
density  3.87 9.36
hardness 0.97 1.83
size     0.58 1.29
weight   1.73 4.40
```

```
moisture 0.65 1.31
```

The density odds ratio interpretation is:

> With 95% confidence, the odds of kernel quality being below a particular level change by an amount between 3.87 to 9.36 times when density is decreased by 0.13, holding the other variables constant.

The confidence intervals for density and weight are the only intervals that do not contain 1. Thus, there is sufficient evidence that both of these explanatory variables are important.

The process using the resulting object from `vglm()` is essentially the same as shown for the multinomial regression model in Section 3.3. The main difference is that $\hat{\beta}_{10}$ and $\hat{\beta}_{20}$ are the first two elements of `mod.fit.ord.v@coefficients`, so we need to exclude them when computing the estimated odds ratios for each explanatory variable. Our code is in the program for this example.

Although we generally prefer LR intervals, Wald confidence intervals may be calculated as well. We demonstrate the process using the `emmeans` package and the kernel density variable.

```
> library(package = emmeans)
> calc.est <- emmeans(object = mod.fit.ord, specs = ~ density,
   at = list(density = c(1.25 + 0.13, 1.25)), mode =
   "linear.predictor")

Re-fitting to get Hessian

> confint(object = contrast(object = calc.est, method =
   "revpairwise", type = "response"), level = 0.95)
 contrast                    odds.ratio   SE  df asymp.LCL
 density1.25 / density1.38         5.79 1.29 Inf      3.74
 asymp.UCL
      8.95

Results are averaged over the levels of: class, cut
Confidence level used: 0.95
Intervals are back-transformed from the log odds ratio scale
```

The 95% Wald interval for the odds ratio is $(3.74, 8.95)$ when the density is decreased by 0.13. Because the odds ratio does not depend on a specific value of density, this is the interval for other potential values that differ by 0.13 in the `at` argument. It is also important to note that `emmeans()` correctly takes into account that `polr()` estimates a model (Equation 3.13) with positive/negative signs reversed for some of its regression parameters.

Within the `emmeans()` function, we set the `mode` argument to `"linear.predictor"`. This was done because Equation 3.11 has the log-odds on its left side when the right side involves only the linear predictor. There also is a special keyword `cut` that can be used within the `specs` argument of `emmeans()`. This word essentially represents the response, like what was included in `specs` for the multinomial regression model. For example, `specs = ~ density | cut` would estimate separate odds ratios for $j = 1$ and $j = 2$. As described earlier, a key feature of proportional odds models is that these odds ratios are the same, so this is why we do not use `cut` here. We mentioned the keyword because the output indicates that `cut` is averaged over. The output also indicates that `class` is averaged over, but this is not meaningful as well because `class` is not part of the odds ratio for the kernel density.

3.4.2 Contingency tables

Section 3.2 discussed how to test for independence in a contingency table using a multinomial regression model for a nominal response. A different test for independence can be performed using the proportional odds model, which specifies the type of dependence that may be present. This is because the proportional odds model assumes a specific structure for the association between a categorical explanatory variable X and a response variable Y, and hence uses fewer parameters than the nominal model to summarize this association.

In particular, for an $I \times J$ contingency table with ordinal categories of Y, the proportional odds model is

$$\text{logit}(P(Y \leq j)) = \beta_{j0} + \beta_2 x_2 + \cdots + \beta_I x_I,\ j = 1, \ldots, J-1 \qquad (3.15)$$

where $x_2, \ldots, x_I$ are indicator variables for rows $2, \ldots, I$, respectively. Any $\beta_i \neq 0$ for $i = 2, \ldots, I$ means that the odds involving cumulative probabilities for Y are not the same in rows 1 and i. Lower categories of Y are more likely to be observed in row i than in row 1 if $\beta_i > 0$, and less likely if $\beta_i < 0$. Thus, independence is tested as $H_0: \beta_2 = \beta_3 = \ldots = \beta_I = 0$ vs. $H_a:$ any $\beta_i \neq 0$, $i = 2, \ldots, I$. An LRT is performed using a χ^2_{I-1} distribution approximation for $-2\log(\Lambda)$. If the null hypothesis of independence is rejected, the association is summarized with the help of the signs and values of $\hat{\beta}_2, \ldots, \hat{\beta}_I$, along with confidence intervals for the corresponding parameters.

Notice that Equation 3.15 contains only $(I-1)+(J-1)$ parameters, compared to $I(J-1)$ in the corresponding multinomial regression model. The reduction in parameters comes from the proportional odds assumption, which specifies that the difference in the logits of cumulative probabilities for any two rows is controlled by just one parameter, β_i, regardless of j. Thus, the test for independence using the proportional odds model is more powerful than a test for independence using the multinomial regression model when the association does actually follow this structure. If it does not, the test using the proportional odds model may fail to detect any association, even when it is very strong.

Example: Fiber-enriched crackers (Fiber.R, Fiber.csv)

When examining a 4×4 contingency table for the fiber-enriched cracker data, we found previously that there was marginal evidence of an association between bloating severity and fiber source. Because bloating severity is measured in an ordinal manner (none < low < medium < high), a proportional odds model allows us to perform a potentially more powerful analysis. Using fiber as the explanatory variable with appropriately constructed indicator variables (see Section 3.3 example), our model is

$$\text{logit}(P(Y \leq j)) = \beta_{j0} + \beta_2 \texttt{bran} + \beta_3 \texttt{gum} + \beta_4 \texttt{both},$$

where j corresponds to levels 1 (none), 2 (low), and 3 (medium) of bloating severity. Notice that this model involves a smaller number of parameters than the multinomial regression model of Section 3.3. We fit the model using the `polr()` function.

```
> library(package = MASS)
> levels(diet$bloat)
[1] "none"   "low"    "medium" "high"
> mod.fit.ord <- polr(formula = bloat ~ fiber, weights = count,
   data = diet, method = "logistic")
> summary(mod.fit.ord)

Re-fitting to get Hessian

Call: polr(formula = bloat ~ fiber, data = diet, weights =
   count, method = "logistic")

Coefficients:
             Value Std. Error t value
fiberbran -0.3859     0.7813  -0.494
fibergum   2.4426     0.8433   2.896
fiberboth  1.4235     0.7687   1.852

Intercepts:
             Value  Std. Error t value
none|low      0.0218 0.5522     0.0395
low|medium    1.6573 0.6138     2.7002
medium|high   3.0113 0.7249     4.1539

Residual Deviance: 112.2242
AIC: 124.2242
```

The estimated model is

$$\text{logit}(\widehat{P}(Y \leq j)) = \hat{\beta}_{j0} + 0.3859\texttt{bran} - 2.4426\texttt{gum} - 1.4235\texttt{both},$$

where $\hat{\beta}_{10} = 0.0218$, $\hat{\beta}_{20} = 1.6573$, and $\hat{\beta}_{30} = 3.0113$.

An LRT for independence tests $H_0 : \beta_2 = \beta_3 = \beta_4 = 0$ vs. H_a : any $\beta_i \neq 0$, $i = 2, 3, 4$. This produces the following results.

```
> library(package = car)
> Anova(mod.fit.ord)
Analysis of Deviance Table (Type II tests)

Response: bloat
      LR Chisq Df Pr(>Chisq)
fiber   15.048  3   0.001776 **
```

Because $-2\log(\Lambda) = 15.048$ is large relative to a χ^2_3 distribution (p-value = 0.0018), there is strong evidence that association exists in the form of a trend among the log-odds for the cumulative probabilities. Remember that with the multinomial regression model of Section 3.3 and the tests for independence given in Section 3.2, there was only marginal evidence of association. By narrowing the type of association considered by our test, we now have more power to reject the null hypothesis.

The estimated odds ratios for comparing each fiber source to using no fiber are $\widehat{OR} = \exp(\hat{\beta}_r)$ for $r = 2, 3,$ and 4. Below are these odds ratios and the corresponding profile LR intervals.

```
> round(exp(-coefficients(mod.fit.ord)), 2)
fiberbran  fibergum fiberboth
     1.47      0.09      0.24

> conf.beta <- confint(object = mod.fit.ord, level = 0.95)
Waiting for profiling to be done...

Re-fitting to get Hessian

> ci <- exp(-conf.beta)
> round(data.frame(low = ci[,2], up = ci[,1]), 2)
           low   up
fiberbran 0.32 7.06
fibergum  0.02 0.43
fiberboth 0.05 1.05
> round(data.frame(low = 1/ci[,1], up = 1/ci[,2]), 2)
           low    up
fiberbran 0.14  3.15
fibergum  2.32 65.01
fiberboth 0.95 19.82
```

Because the profile LR interval for gum is less than 1, there is sufficient evidence to indicate that using gum as a fiber source increases bloating severity; however, there is not sufficient evidence that bran increases bloating severity because its confidence interval contains 1. When combining both sources of fiber, there is marginal evidence of an effect, because the lower bound of the interval is close to 1.

Formally, we can write an interpretation for the gum and no fiber source confidence interval as:

> With 95% confidence, the odds of bloating severity being below a particular level are between 2.32 and 65.01 times as large when there is no fiber source as when using gum as a fiber source.

Equivalently, this interval can be interpreted as

> With 95% confidence, the odds of bloating severity being *above* a particular level are between 2.32 and 65.01 times as large *when using gum as a fiber source as when using no fiber.*

To compare gum directly to bran, the simplest way is to use the `relevel()` or `factor()` functions to change the base level of the fiber source variable. Similar code to that presented above can be used then to refit the model and calculate the profile LR intervals (code is within the program). We find that

> With 95% confidence, the odds of bloating severity being above a particular level are between 3.29 and 101.26 times as large when using gum as a fiber source than when using bran.

Thus, this indicates that gum leads to more bloating than bran.

Overall, if given a choice between the two fiber sources and their combination, bran would be preferred if the only criterion was bloating severity.

3.4.3 Non-proportional odds model

A proportional odds model is a preferred way to account for an ordered multinomial response, because slope regression parameters are constant over the response categories. While this results in a simplified model, it imposes the assumption that association affects the log-odds of cumulative probabilities in the same way for all $j = 1, \ldots, J-1$. This may not be true in all situations. An alternative model that relaxes this assumption is

$$\text{logit}(P(Y \leq j)) = \beta_{j0} + \beta_{j1}x_1 + \cdots + \beta_{jp}x_p, \tag{3.16}$$

where $j = 1, \ldots J-1$. Notice that all the regression parameters are now allowed to vary across the levels of Y. Equation 3.16 is referred to as the *non-proportional odds model.*

Because the proportional odds model is a special case of Equation 3.16, we can test the proportional odds assumption through the hypotheses $H_0 : \beta_{1r} = \cdots = \beta_{J-1,r}$ for each $r = 1, \ldots, p$ vs. H_a : Not all equal for at least one r. The test is conducted as an LRT where the degrees of freedom for the reference χ^2 distribution are found from the difference in the number of parameters for the two models, $(p+1)(J-1) - (p+J-1) = p(J-2)$.

Rejecting the proportional odds assumption suggests that the non-proportional odds model may be preferred. Estimated probabilities and odds ratios have a different form because of the extra parameters, and Exercise 7 discusses these formulations. However, the proportional odds model may still be preferred because of its smaller number of parameters. Each parameter estimated adds variability to the subsequent estimation of probabilities and odds ratios, so one may get odds ratio estimates that are closer to the truth on average by using a smaller model with a minor defect than by using a much larger model without the defect. Also, a very large sample size could result in a rejection of the null hypothesis even though the data deviate only slightly from the proportional odds assumption.

Failing to reject the proportional odds hypothesis is not proof that it is true. However, it does offer some assurance that a proportional odds model provides a reasonable approximation to true relationships between Y and the explanatory variables. We will examine additional ways to evaluate a model's fit in Chapter 5.

Example: Fiber-enriched crackers (Fiber.R, Fiber.csv)

Unfortunately, the `polr()` function does not provide a way to test the proportional odds assumption. We instead use the `vglm()` function to help perform the test. First, we fit the proportional odds model and examine a summary of the results.

```
> library(package = VGAM)
> mod.fit.po <- vglm(formula = bloat ~ fiber, family =
    cumulative(parallel = TRUE), weights = count, data =
    diet[diet$count != 0,])
> summary(mod.fit.po)

Call:
vglm(formula = bloat ~ fiber, family = cumulative(parallel =
    TRUE), data = diet[diet$count != 0, ], weights = count)

Coefficients:
               Estimate Std. Error z value Pr(>|z|)
(Intercept):1  0.02182     0.55654   0.039  0.96872
(Intercept):2  1.65734     0.62533   2.650  0.00804 **
(Intercept):3  3.01128     0.72766   4.138  3.5e-05 ***
```

```
fiberbran      0.38593   0.79412   0.486  0.62697
fibergum      -2.44258   0.82591  -2.957  0.00310 **
fiberboth     -1.42348   0.78090  -1.823  0.06832 .
---
Signif. codes:  0 '***' 0.001 '**' 0.01 '*' 0.05 '.' 0.1 ' ' 1

Names of linear predictors: logitlink(P[Y<=1]),
   logitlink(P[Y<=2]),  logitlink(P[Y<=3])

Residual deviance: 112.2242 on 36 degrees of freedom

Log-likelihood: -56.1121 on 36 degrees of freedom

Number of Fisher scoring iterations: 6

No Hauck-Donner effect found in any of the estimates

Exponentiated coefficients:
 fiberbran   fibergum  fiberboth
1.47098886 0.08693587 0.24087418
Warning message: In eval(slot(family, "initialize")) :
  response should be ordinal---see ordered()
```

The estimated model is practically the same as found earlier with `polr()`. The `weights` argument within `vglm()` is used to denote the number of individuals that each row of the data frame represents. As described in Section 3.3.2, this function does not work with 0 counts, so these corresponding rows of the data frame are removed. If the data set had been alternatively formulated so that each row represented an individual observation (i.e., a person testing a cracker), the `weights` argument would not have been needed.

Next, we fit the non-proportional odds model

$$\text{logit}(P(Y \leq j)) = \beta_{j0} + \beta_{j2}\texttt{bran} + \beta_{j3}\texttt{gum} + \beta_{j4}\texttt{both}$$

using `vglm()` with the `parallel = FALSE` argument value.

```
> mod.fit.npo <- vglm(formula = bloat ~ fiber, family =
   cumulative(parallel = FALSE), weights = count, data =
   diet[diet$count != 0,])

Call:
vglm(formula = bloat ~ fiber, family = cumulative(parallel =
   FALSE), data = diet[diet$count != 0, ], weights = count)

Coefficients:
                 Estimate Std. Error z value Pr(>|z|)
(Intercept):1  -1.680e-15  5.774e-01   0.000   1.0000
(Intercept):2   1.609e+00  7.746e-01   2.078   0.0377 *
(Intercept):3   1.802e+01  1.433e+03      NA       NA
fiberbran:1     3.365e-01  8.223e-01   0.409   0.6824
fiberbran:2     7.885e-01  1.300e+00   0.606   0.5443
fiberbran:3     4.083e-09  2.027e+03   0.000   1.0000
fibergum:1     -1.609e+00  9.661e-01  -1.666   0.0957 .
```

```
fibergum:2      -2.303e+00  9.874e-01   -2.332    0.0197 *
fibergum:3      -1.768e+01  1.433e+03   -0.012    0.9902
fiberboth:1     -1.609e+00  9.661e-01   -1.666    0.0957 .
fiberboth:2     -1.273e+00  9.710e-01   -1.311    0.1899
fiberboth:3     -1.641e+01  1.433e+03   -0.011    0.9909
---
Signif. codes:   0 '***' 0.001 '**' 0.01 '*' 0.05 '.' 0.1 ' ' 1

Names of linear predictors: logitlink(P[Y<=1]),
logitlink(P[Y<=2]), logitlink(P[Y<=3])

Residual deviance: 108.3914 on 30 degrees of freedom

Log-likelihood: -54.1957 on 30 degrees of freedom

Number of Fisher scoring iterations: 16

Warning: Hauck-Donner effect detected in the following
   estimate(s): '(Intercept):3'

Exponentiated coefficients:
 fiberbran:1  fiberbran:2  fiberbran:3    fibergum:1
1.400000e+00 2.200000e+00 1.000000e+00 2.000000e-01
  fibergum:2   fibergum:3  fiberboth:1  fiberboth:2
1.000000e-01 2.089257e-08 2.000000e-01 2.800000e-01
 fiberboth:3
7.461632e-08
Warning messages:
1: In eval(slot(family, "initialize")) :
  response should be ordinal---see ordered()
2: In eval(slot(family, "initialize")) :
  response should be ordinal---see ordered()
```

Notice that there are now $p(J-2) = 6$ additional parameters estimated when compared to the proportional odds model. The hypotheses for the test of the proportional odds assumption are $H_0 : \beta_{1i} = \beta_{2i} = \beta_{3i}$ for each $i = 2, 3, 4$ vs. H_a : Not all equal for at least one i. The `anova()` function provides the needed calculations for the test.

```
> anova(mod.fit.po, mod.fit.npo, type = "I")
Analysis of Deviance Table

Model 1: bloat ~ fiber
Model 2: bloat ~ fiber
  Resid. Df Resid. Dev Df Deviance Pr(>Chi)
1        36     112.22
2        30     108.39  6   3.8328   0.6993
```

The LRT results in a statistic of $-2\log(\Lambda) = 3.83$ and a p-value of 0.6993. Thus, there is not sufficient evidence to indicate the proportional odds assumption has been violated. Notice that the `anova()` function is used here in a similar manner as introduced in Section 2.2.2 when including two objects containing model-fit information.

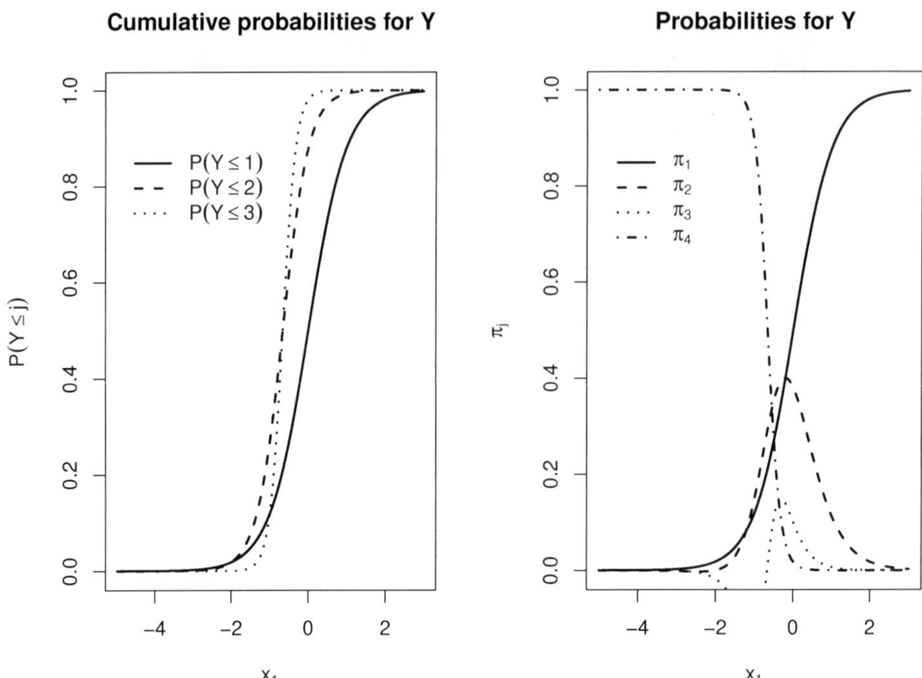

Figure 3.6: Non-proportional odds model for $\beta_{10} = 0$, $\beta_{20} = 2$, $\beta_{30} = 4$, $\beta_{11} = 2$, $\beta_{21} = 3$, and $\beta_{31} = 6$.

A requirement for the corresponding method function from `VGAM` is that the `type = "I"` argument value is included to represent a sequential test.

A problem with using non-proportional odds models for general use is that the model does not adequately constrain the parameters to prevent $P(Y \leq j) < P(Y \leq j')$ for $j > j'$. Thus, the cumulative probabilities can decrease at some point causing an individual category probability to be less than 0. This violation of probability rules occurs because the effect an explanatory variable has on $\text{logit}(P(Y \leq j))$ can change for each j; i.e., the $\beta_{j1}, \ldots, \beta_{jp}$ parameters can vary freely over the levels of Y. For this reason, caution needs to be used with these models to make sure that nonsensical probabilities do not occur. The next example illustrates a case where a non-proportional odds model is inappropriate.

Example: Non-proportional odds model plots (CumulativeLogitModelPlot.R)

Consider the model $\text{logit}(P(Y \leq j)) = \beta_{j0} + \beta_{j1}x_1$, where $\beta_{10} = 0, \beta_{20} = 2, \beta_{30} = 4, \beta_{11} = 2, \beta_{21} = 3, \beta_{31} = 6$, and $J = 4$. Compared to the model used for Figure 3.6, the $j = 1$ case is the same here, but the $j = 2$ and 3 cases have different "slope" parameters ($\beta_{21} \neq \beta_{11}$ and $\beta_{31} \neq \beta_{11}$). Figure 3.6 shows plots of the cumulative and individual category probabilities. Notice that the cumulative probability curve for $P(Y \leq 3)$ crosses the other two curves. Thus, the model allows for $P(Y \leq 3) < P(Y \leq 2)$ and $P(Y \leq 3) < P(Y \leq 1)$ for some values of x_1. As a result, π_3 is less than 0 for some values of x_1. Please see our program for the code to create these plots.

3.5 Additional regression models

Sections 3.3 and 3.4 give the two most common ways to model multinomial responses. There are many other approaches. For example, we are not limited to only using a logit transformation of an odds. The probit and complementary log-log transformations shown in Section 2.3 can be used instead. The `polr()` function fits these models for ordinal responses by using the appropriate `method` argument value within the function call. The `vglm()` function fits these models as well by including a `link` argument within `cumulative()`. Examples are provided in the program for the wheat kernels data.

Models exist between proportional and non-proportional odds formulations. These models allow some explanatory variables to have the same effect on all log-odds for cumulative probabilities while allowing others to have effects that may vary. For example, the model

$$\text{logit}(P(Y \leq j)) = \beta_{j0} + \beta_1 x_1 + \beta_{j2} x_2$$

assumes that x_1 has the same effect on all log-odds, while x_2 has a different effect for each one. This model can be fit using `vglm()` with the argument `family = cumulative(parallel = FALSE ~ x2)`, where `x2` represents x_2.

Another model for multicategory responses is the *adjacent-categories model*:

$$\log(\pi_{j+1}/\pi_j) = \beta_{j0} + \beta_{j1} x_1 + \cdots + \beta_{jp} x_p,$$

where $j = 1, \ldots, J-1$. This model produces the same probability estimates as the multinomial regression model, but simplifies the process of estimating odds ratios comparing consecutive categories, as one might want to do in an ordinal problem. Also, we can use a simplified version of the model,

$$\log(\pi_{j+1}/\pi_j) = \beta_{j0} + \beta_1 x_1 + \cdots + \beta_p x_p,$$

where the regression slope parameters are forced to be the same for all j. This model assumes that an explanatory variable has the same effect on the odds for every successive category comparison in much the same way the proportional odds regression model does with odds based on cumulative probabilities. Both of these adjacent-categories models can be estimated in R using the `vglm()` function. We explore these models further in Exercise 19.

3.6 Exercises

1. For the case of $I = J = 2$, show that $\sum_{i=1}^{I}\sum_{j=1}^{J}\frac{(n_{ij} - n_{i+}n_{+j}/n)^2}{n_{i+}n_{+j}/n}$ is equivalent to X^2 as given in Chapter 1. We recommend writing out each of the four terms in the sum to see this equivalence.

2. Show that $-2\log(\Lambda)$ simplifies to $-2\sum_{i=1}^{I}\sum_{j=1}^{J} n_{ij}\log\left(\frac{n_{ij}}{n_{i+}n_{+j}/n}\right)$ when testing for independence between two variables.

3. Consider a 2×2 contingency table structure as a special case of Table 3.1.

   (a) Discuss why the marginal distribution of n_{1+} is binomial with parameter π_{1+}.

(b) The conditional joint distribution of n_{11}, n_{12}, and n_{21} given n_{1+} can be shown to be $f(n_{11}, n_{12}, n_{21}|n_{1+}) = f(n_{11}, n_{12}, n_{21})/f(n_{1+})$, where $f(n_{11}, n_{12}, n_{21})$ is the joint distribution of n_{11}, n_{12}, n_{21} and $f(n_{1+})$ is the marginal distribution of n_{1+}. Note that we do not need to state n_{22} within $f(n_{11}, n_{12}, n_{21})$ because $n = n_{11} + n_{12} + n_{21} + n_{22}$ and n is known. Similarly, we do not need to state n_{2+} within $f(n_{1+})$. Using Equation 3.2 for the joint distribution of $f(n_{11}, n_{12}, n_{21})$ and the binomial distribution for $f(n_{1+})$, find $f(n_{11}, n_{12}, n_{21}|n_{1+})$ and show it is equivalent to Equation 3.3 when $I = J = 2$.

(c) In general for $I \times J$ contingency table structures, extend the result in (b) to show that Equation 3.2 leads to Equation 3.3 when conditioning on the row marginal counts.

4. Suppose that we have one continuous explanatory variable x that is to be represented in a regression model by both a linear and quadratic term.

   (a) For a c-unit increase in x, derive the odds ratio that compares a category j ($j \neq 1$) response to a category 1 response for a multinomial regression model. Show the form of the variance that would be used in a Wald confidence interval.

   (b) For a c-unit increase in x, derive the odds ratio that compares cumulative probabilities for a proportional odds regression model. Show the form of the variance that would be used in a Wald confidence interval.

5. Suppose a regression model has two continuous explanatory variables x_1 and x_2, and they are represented in the model by their linear and interaction terms.

   (a) For a c-unit increase in x_1, derive the odds ratio that compares a category j response to a category 1 response for a multinomial regression model. Derive the variance that would be used in a Wald confidence interval.

   (b) For a c-unit increase in x_1, derive the odds ratio that compares cumulative probabilities for a proportional odds regression model. Show the form of the variance that would be used in a Wald confidence interval.

6. With the help of the CumulativeLogitModelPlot.R program discussed in Section 3.4, complete the following parts.

   (a) Construct plots similar to Figure 3.4 using different values of the parameters. What happens to the plots when β_1 is increased or decreased while β_{10}, β_{20}, and β_{30} remain constant? What happens to the plots when β_{10}, β_{20}, and β_{30} have more or less separation among them while β_1 remains constant?

   (b) For a proportional odds regression model, find a set of regression parameter values that lead to some categories never having the largest probability at any x_1. State a mathematical reason relative to the model for why this occurs.

   (c) Construct similar plots to Figure 3.6 using different values of the parameters. Find a set of parameters where nonsensical probabilities do not occur (i.e., the parameters satisfy $P(Y \leq j) < P(Y \leq j')$ for all $j < j'$).

7. Consider the non-proportional odds regression model $\text{logit}(P(Y \leq j)) = \beta_{j0} + \beta_{j1}x_1$, where $j = 1, \ldots J - 1$.

   (a) Show that $Odds_{x_1+c}(Y \leq j)/Odds_{x_1}(Y \leq j) = \exp(c\beta_{j1})$. Provide a formal interpretation of the odds ratio.

(b) Find π_j for $j = 1, \ldots, J$. Why can a π_j be less than 0 with this model?

8. Exercise 19 of Chapter 2 examined the prevalence of hepatitis C among healthcare workers in the Glasgow area of Scotland. Convert the data for this exercise to a contingency table structure where occupational groups are located on the rows and the presence and absence of hepatitis are located on the columns. Perform Pearson chi-square and LR tests for independence using these data. Are your results here similar to what was found for the data analysis in Chapter 2? Explain.

9. Classification systems are often used by medical doctors to help make sense of patient symptoms and to guide them toward applying appropriate treatments. For this purpose, Cognard et al. (1995) propose a new classification system with respect to dural arteriovenous fistulas (DAVFs). DAVFs are abnormal connections between arteries and veins in a person's head. If left untreated, they can lead to death because of stroke or other complications. Table 3.3 provides the data on 205 cases included in their study. The columns of the table provide the classification of each individual, where the classifications generally increase in DAVF severity from left to right. Each individual is given only one classification. The rows represent different symptoms experienced by the individuals. Assume that each individual also has only one symptom.

   (a) What would independence and dependence between the rows and columns mean in the context of DAVFs? Which conclusion do you think the researchers would prefer to justify their classification system? Explain.

   (b) A Pearson chi-square test for independence does not provide information about the type of dependence that may be present. Are there potential types of dependence that may be undesirable in the context of developing a classification system? Explain.

   (c) Cognard et al. (1995) state that "the contingency table shows a statistically significant difference (P = .0001)" without mentioning the particular statistical test conducted. Perform a Pearson chi-square test for independence to produce a similar significance result.

   (d) When performing the test in (c) with `chisq.test()`, R states `Chi-squared approximation may be incorrect`. Examine if the $\chi^2_{(I-1)(J-1)}$ distribution approximation is appropriate using Monte Carlo simulation. Compute the p-value using the simulated distribution.

   (e) It is unlikely that patients with a DAVF will exhibit only one symptom. While not stated in this paper, the authors may have summarized the patient's "primary" or "most extreme" symptom, so that each patient contributed to the count of only one cell in the table. However, if this was not the case, a patient could contribute to none, one, or more than one cell (e.g., a patient could have intracranial hypertension and seizures while having a level 3 classification). In fact, the total counts for the table could be greater or less than the number of patients. Would this alternative scenario invalidate the use of the Pearson chi-square test in this situation? Explain.[11]

[11]Section 6.4 examines situations where individuals could contribute to more than one row and/or column response combination.

Table 3.3: Classification of DAVFs by symptoms. Data source is Cognard et al. (1995).

		Classification						
		1	2a	2b	2a and 2b	3	4	5
Symptoms	Hemorrhage	0	0	2	1	10	19	5
	Intracranial hypertension	1	8	1	2	0	4	1
	Focal neurologic deficit	0	0	0	6	8	2	0
	Seizures	0	1	0	2	1	3	0
	Cardiac deficiency	0	1	0	1	0	0	0
	Myelopathy	0	0	0	0	0	0	6
	Non-aggressive symptoms	83	17	7	6	6	1	0

Table 3.4: Sex of a child and whether the mother was exposed to antibiotics during pregnancy. Data source is Uldbjerg et al. (2021).

	Female	Male
Exposed	2,755	2,767
Not exposed	13,634	13,495

10. Continuing Exercise 32 of Chapter 1 and Exercise 32 of Chapter 2, Uldbjerg et al. (2021) provide 11 two-variable contingency tables in their Table 1. Each of these tables examines the relationship between antibiotic exposure and another variable using a Pearson chi-square test for independence. Complete the following parts with regard to these tables.

    (a) Table 3.4 provides one of these 11 tables. Perform a Pearson chi-square test for independence with the data. Interpret the results.

    (b) If you have access to the paper, answer the questions below.

        i. For the remaining contingency tables given by Uldbjerg et al. (2021), compute the p-values for the Pearson chi-square tests of independence. Verify your results match those of the paper.

        ii. The purpose of the paper is to examine the relationship between antibiotic use and childhood asthma. Why do you think the authors examine these other contingency tables?

11. Section 3.3.2 discusses how the fiber-enriched crackers example can be analyzed by separating the fiber-source variable into two separate binary variables for bran and gum. This type of analysis may be preferred to assess separately the effects of these two sources on bloating. Use this form of the data to complete the parts below.

    (a) Fit the proportional odds regression model involving bran, gum, and their interaction. Is there sufficient evidence to indicate an interaction is important?

    (b) Evaluate the proportional odds assumption for the model in (a) by estimating the corresponding non-proportional odds model and then performing an LRT.

    (c) With the most appropriate model found through the investigations in (a) and (b), use odds ratios to interpret the effect of bran and gum on increasing bloating severity.

12. For the wheat kernel data, complete the following parts.

    (a) Interpret the estimated odds ratios and corresponding confidence intervals computed for the other explanatory variables not specifically discussed in the examples of Sections 3.3.1 and 3.4.1. Relate your results back to Figure 3.2.

(b) For the multinomial regression model, change the base level of `type` to `"Scab"` and refit the model with `multinom()`. Compute odds ratio estimates comparing sprout vs. scab for all of the explanatory variables. Compute the corresponding 95% profile LR confidence intervals and interpret these intervals in the context of the data.

13. To maximize sales, items within grocery stores are strategically placed to draw customer attention. This exercise examines one type of item—breakfast cereal. Typically, in large grocery stores, boxes of cereal are placed on shelves located on one side of the aisle. By placing particular boxes of cereals on specific shelves, grocery stores may better attract customers. To investigate this further, a random sample of size 10 was taken from each of four shelves at a Dillons grocery store in Manhattan, KS. These data are given in the CerealDillons.csv file. The response variable is the shelf number, which is numbered from bottom (1) to top (4), and the explanatory variables are the sugar, fat, and sodium content of the cereals.

    (a) The explanatory variables need to be reformatted to take into account the different serving sizes for each cereal. Divide each explanatory variable by its serving size to account for the different serving sizes among the cereals. Below is code that can be used to reformat the data after the data file has been read into an object named `cereal`.

    ```
    cereal.reformat <- data.frame(Shelf =
        as.factor(cereal$Shelf), sugar =
        cereal$sugar_g/cereal$size_g, fat =
        cereal$fat_g/cereal$size_g, sodium =
        cereal$sodium_mg/cereal$size_g)
    ```

    This code also converts the shelf response to a factor type. Some method functions used with results from `vglm()` will not work properly unless the response variable is a factor.

    (b) Construct side-by-side box plots with dot plots overlaid for each of the explanatory variables. Below is code that can be used for plots involving sugar.

    ```
    boxplot(formula = sugar ~ Shelf, data = cereal.reformat,
        ylab = "Sugar", xlab = "Shelf", pars = list(outpch=NA),
        col = NULL)
    stripchart(x = cereal.reformat$sugar ~
        cereal.reformat$Shelf, lwd = 2, col = "red", method =
        "jitter", vertical = TRUE, pch = 1, add = TRUE)
    ```

    Also, construct a parallel coordinates plot for the explanatory variables and the shelf number. Discuss whether possible content differences exist among the shelves.

    (c) The response has values of $1, 2, 3,$ and 4. Under what setting would it be desirable to take into account ordinality? Do you think this occurs here?

    (d) Fit a multinomial regression model with linear forms of the sugar, fat, and sodium variables. Use the `vglm()` function for the estimation. Perform LRTs to examine the importance of each explanatory variable.

(e) Refit the multinomial regression model including an interaction for each pair of variables and the three-way interaction. Use a single test to show that there are no significant interactions among the explanatory variables

(f) Kellogg's Apple Jacks is a cereal marketed toward children. For a serving size of 28 grams, its sugar content is 12 grams, fat content is 0.5 grams, and sodium content is 130 milligrams. Estimate the shelf probabilities for Apple Jacks using the model from part (d).

(g) Construct a plot similar to Figure 3.3 where the estimated probability for a shelf is on the y-axis and the sugar content is on the x-axis. Use the model from part (d) and hold the fat and sodium variables at their mean values. Comment on how sugar content affects the four probabilities of shelf placement.

(h) Estimate odds ratios and calculate corresponding profile LR intervals for each explanatory variable. Compute these quantities for each pair of shelves using the model from part (d). Relate your interpretations back to the plots constructed for this exercise.

14. Section 3.3 discusses that convergence for multinomial regression models can be slow with `multinom()`. The purpose of this exercise is to explore these convergence issues and make comparisons to the results obtained from `vglm()` for the data examined in Exercise 13. Fit a multinomial regression model with linear forms of the sugar, fat, and sodium variables again, but now use the `multinom()` function. Show that the model does not converge when using the default convergence criteria.

   (a) Allow for a larger number of iterations when estimating the model from part (a) to show that convergence can be obtained. Compare the parameter estimates of `multinom()` and `vglm()`. Which of these functions produced an estimated model with a larger log likelihood? The log-likelihood function evaluated at the parameter estimates can be obtained using the `logLik()` function.

   (b) Allow for a larger number of iterations when estimating the model from part (a) and decrease the convergence criterion to `reltol = 1.0e-11`. Compare the parameter estimates and the log-likelihood function evaluated at these estimates to what was obtained by `vglm()`.

   (c) The author of `multinom()` suggests that rescaling each variable to be within 0 and 1 can help with the convergence of parameter estimates. Below is code that we used to perform this rescaling and the initial reformating.

```
stand01 <- function(x) { (x - min(x))/(max(x) - min(x)) }
cereal2 <- data.frame(Shelf = cereal$Shelf, sugar =
   stand01(x = cereal$sugar_g/cereal$size_g), fat =
   stand01(x = cereal$fat_g/cereal$size_g), sodium =
   stand01(x = cereal$sodium_mg/cereal$size_g))
```

   Fit the model again from (a) using the default number of iterations, convergence criterion, and `cereal2`. Does convergence occur and does it occur more quickly than previously? What is a disadvantage of using rescaled data like this?

   (d) Compute 95% Wald intervals for Apple Jack's shelf probabilities. Discuss any problems associated with these intervals.

15. Similar to Exercise 13, the CerealSupersaver.csv file contains the sugar, fat, and sodium content for cereals placed on shelves at a Super Saver grocery store in Lincoln,

NE. Analyze these data to better understand the shelf placement of the cereals at this grocery store.

16. An example from Section 4.2.5 examines data from the 1991 U.S. General Social Survey that cross-classifies people according to

    - Political ideology: Very liberal (`VL`), Slightly liberal (`SL`), Moderate (`M`), Slightly conservative (`SC`), and Very conservative (`VC`)
    - Political party: Democrat (`D`) or Republican (`R`)
    - Gender: Female (`F`) or Male (`M`); no other genders reported

    Consider political ideology to be an ordinal response variable, and political party and gender to be explanatory variables. The data are available in the file PolIdeolData.csv.

    (a) Use the `factor()` function with the ideology variable to make sure that R places the levels of the ideology variable in the correct order.

    (b) Because the two explanatory variables are categorical, we can view the entire data set in a three-dimensional contingency table structure. The `xtabs()` and `ftable()` functions are useful for this purpose. Below is how these functions can be used after the data file has been read into R (the data frame `set1` contains the original data).

    ```
    > c.table <- xtabs(formula = count ~ party + ideol + gender,
       data = set1)
    > ftable(x = c.table, row.vars = c("gender", "party"),
       col.vars = "ideol")
    ```

    Run this code and report the results. This approach for viewing the data may be helpful for part (c).

    (c) Using multinomial and proportional odds regression models that include party, gender, and their interaction, complete the following parts.

        i. Fit the models and perform LRTs to test the importance of each explanatory variable.

        ii. Compute the estimated probabilities for each ideology level given all possible combinations of the party and gender levels.

        iii. Construct a contingency table with estimated counts from the model. These estimated counts are found by taking the estimated probability for each ideology level multiplied by their corresponding number of observations for a party and gender combination. For example, there are 264 observations for `gender = "F"` and `party = "D"`. Because the multinomial regression model results in $\hat{\pi}_{\text{VL}} = 0.1667$, this model's estimated count is $0.1667 \times 264 = 44$.

        iv. Are the estimated counts the same as the observed? Explain.

        v. Use odds ratios computed from the estimated models to help understand relationships between the explanatory variables and the response.

    (d) Compare the results for the two models. Discuss which model may be more appropriate for this setting.

17. Continuing Exercise 16, consider again the proportional odds model and the contingency table containing the estimated counts from it. Calculate the estimated odds of

a very liberal response for female democrats using the model. Show that this same odds can be calculated with the estimated counts from the contingency table. Discuss how these equivalences can be shown for other estimated odds.

18. Researchers at Penn State University performed a study to determine the optimal fat content for consumer preferences of ice cream. Details of the study are available at `https://online.stat.psu.edu/stat504/lesson/case-study-ice-cream-study-penn-state`. In summary, 493 individuals were asked to taste and then rate a particular type of ice cream on a 9-point scale (1 to 9 with 1 equating to not liking and 9 equating to really liking). The ice cream given to the individuals had fat proportion levels of 0, 0.04, ..., or 0.28. The data are available in IceCream.csv, where the `count` column represents the number of individuals in a particular fat group giving a particular rating.

    (a) Display the data in an 8×9 contingency table structure.

    (b) Within each fat content level, find the observed proportion of responses for each rating score. Construct a single plot of these observed proportions vs. fat content with a separate line for each rating score. Are there any trends in the plot reflecting a relationship between fat content and rating? What does the plot suggest as values for the optimal fat content?

    (c) Discuss why Section 3.2's Pearson chi-square and LR tests for independence are not ideal analysis choices for these data.

    (d) Fit a proportional odds model using fat content to predict rating. Investigate whether or not a quadratic term is helpful. Note that the rating variable needs to be converted to a `factor` class type if `polr()` is used to fit the model.

    (e) Perform an LRT to assess the proportional odds assumption. Include the linear and quadratic terms in the proportional and non-proportional odds models for the test.

    (f) Use odds ratios and plots of the estimated probabilities to interpret the resulting model. Recommend a fat content.[12]

19. For the fiber-enriched crackers data, an adjacent-category model can be written as $\log(\pi_{j+1}/\pi_j) = \beta_{j0} + \beta_{j1}\texttt{bran} + \beta_{j2}\texttt{gum} + \beta_{j3}\texttt{both}$ for $j = 1, 2,$ and 3.

    (a) Why is the adjacent categories model equivalent to the multinomial regression model from Section 3.3?

    (b) Fit the adjacent-category model using the following as a continuation of the code used in Fiber.R.

```
> mod.fit.full <- vglm(formula = bloat ~ fiber, weights =
    count, family = acat(parallel = FALSE), data =
    diet[diet$count != 0,])
> summary(mod.fit.full)
```

[12]If `polr()` and the **emmeans** package are used to calculate odds ratios and the corresponding Wald intervals, we recommend `rating` be a factor in the data frame used with `polr()`. The odds ratio calculations will be performed as shown in Section 3.4.1. If the `factor()` function is applied instead within the `formula` argument for `polr()`, a quirk of the **emmeans** package leads to a calculation method that is not as desirable (the delta method is applied directly to the odds ratio rather than the log of the odds ratio).

The `acat()` function used in the `family` argument specifies an adjacent category model. Show that similar convergence problems occur here as those that occurred for the example involving this data in Section 3.3.2.

(c) To better take advantage of bloating severity's ordinal response, we can fit the model as $\log(\pi_{j+1}/\pi_j) = \beta_{j0} + \beta_1$`bran` $+ \beta_2$`gum` $+ \beta_3$`both` where parameters are shared across the response categories for the fiber sources. Estimate this model by using `acat(parallel = TRUE)` in the `family` argument of `vglm()`.

(d) Examine odds ratios using the fits of both models in (b) and (c). Compare their values.

(e) State in words what is tested by $H_0 : \beta_{11} = \beta_{21} = \beta_{31}, \beta_{12} = \beta_{22} = \beta_{32}, \beta_{13} = \beta_{23} = \beta_{33}$ vs. H_a : At least one inequality. Perform the test using an LRT.

20. For the wheat kernel data, consider a model to estimate the kernel condition using the density explanatory variable as a linear term.

(a) Write an R function that computes the log-likelihood function for the multinomial regression model. Evaluate the function at the parameter estimates produced by `multinom()` or `vglm()`, and verify that your computed value is the same as that produced by `logLik()` (use the object saved from `multinom()` or `vglm()` within this function).

(b) Maximize the log-likelihood function using `optim()` to obtain the MLEs and the estimated covariance matrix. Compare your answers to what is obtained by `multinom()` or `vglm()`. Note that to obtain starting values for `optim()`, one approach is to estimate separate logistic regression models for $\log(\pi_2/\pi_1)$ and $\log(\pi_3/\pi_1)$. These models are estimated only for those observations that have the corresponding responses (e.g., a $Y = 1$ or $Y = 2$ for $\log(\pi_2/\pi_1)$).

(c) Repeat (a) and (b), but now for a proportional odds regression model. Compare your answers to what is obtained by using `polr()` or `vglm()`. Note that to obtain starting values for `optim()`, one approach is to estimate separate logistic regression models for $\text{logit}(P(Y \leq 1))$ and $\text{logit}(P(Y \leq 2))$. The intercepts are used as initial estimates for β_{10} and β_{20}, and the two slopes are averaged to obtain an initial estimate of β_1.

21. For the wheat kernel data, consider again the estimated multinomial and proportional odds models from Sections 3.3 and 3.4 that included all of the explanatory variables as linear terms.

(a) For each model, construct a plot of the estimated probability on the y-axis and `density` on the x-axis. Because `class` is binary, use separate plots for its levels. For the remaining explanatory variables, use their mean values. Interpret the plots and compare the models.

(b) How do the plots from (a) change when the remaining numerical explanatory variables are set to values other than their means?

(c) Construct additional plots as in (a), but now for explanatory variables other than `density` on the x-axis.

22. Changes to habitat affect different species of wildlife differently. Birds that hunt for insects on the ground, for example, may be less affected by clearing rainforest than those that feed on plant nectars, like hummingbirds, or that extract insects from trees, like woodpeckers. Root (1967) defines a *guild* as "a group of species that exploit

Table 3.5: Cross-tabulation of mist-netted Ecuadorean birds by habitat degradation and food-source guild.

Guild	Degradation		
	High	Moderate	Low
Air-Arth	8	39	48
Leaf-Arth	69	139	131
Wood-Arth	2	39	50
Ground-Arth	25	57	115
Fruit	20	48	16
Seeds	7	20	19
Nectar	100	177	190

the same class of environmental resources in a similar way." The mist-netted birds described in an example from Section 4.2.3 were classified into guilds according to their feeding habits. The guilds are labeled by food source as Air-Arth, Leaf-Arth, Wood-Arth, Ground-Arth, Carnivore, Fruit, Seeds, and Nectar, where "Arth" means arthropod (insect and spider). We wish to know whether the relative abundance of birds from different guilds is related to the amount of habitat "degradation" at the six locations. Degradation is defined here as a factor with three levels representing how much the habitat has changed from its original forested state. The two forest locations are "Low," the forest edge and forest fragment are "Moderate," and the two pasture locations are "High." The data are given in Table 3.5. Perform an analysis to determine which guilds are affected more or which are affected less by habitat degradation.

23. *Pneumoconiosis* (black lung disease) "is a lung disease that results from breathing in dust from coal, graphite, or man-made carbon over a long period of time" (`https://www.nlm.nih.gov/medlineplus/ency/article/000130.htm`). The data frame `pneumo` in the `VGAM` package contains counts of coal miners aggregated into eight groups based on their time working in the mine (in years) and their severity of *pneumoconiosis* (three levels: normal, mild, and severe). Analyze these data with the specific goals of (1) understanding the relationship between time in the mines and resulting severity of the disease (an ordinal categorical variable), and (2) predicting the disease status of miners after 5, 10, 15, 20, or 25 years of working in a mine.

24. Many individuals have a snack after exercising. Gustafson et al. (2018) examine if the timing of the snack decision has an effect on whether a healthy or unhealthy one is chosen. The authors sampled 256 individuals at the University of Nebraska-Lincoln's student recreation center that were preparing to start exercising for their workout. Each individual was offered a food choice—brownie, apple, or none—as a reward to be received at the end of their workout. Some of these individuals were asked to make the food choice before their workout (still to be received at the end of their workout), while others were asked to make the choice after their workout. Additional information was collected on each individual as well. The full data set is available in the file FoodChoice.csv. Below is a partial listing of the data and a description of each variable.

```
> set1 <- read.csv(file = "C:\\data\\FoodChoice.csv",
   stringsAsFactors = TRUE)
> head(set1, n = 5)
```

```
   Choice Condition       BMI Age Gender Duration Food
1 brownie     After 21.42627  19      F       78    0
2   apple     After 26.21011  19      F       77    0
3   apple     After 23.17496  25      F       35    0
4 brownie     After 27.37311  19      M       92    0
5 brownie     After 24.84756  23      F       85    0
  Aerobic Resistance Other
1       1          1     0
2       1          0     0
3       0          1     0
4       1          1     0
5       0          1     0
```

- `Choice`: Food selected by study participant with levels of `apple`, `brownie`, or `neither`
- `Condition`: Food choice was offered `After` or `Before` the workout
- `BMI`: Body mass index
- `Age`: Age in years
- `Gender`: Male (`M`) or Female (`F`); no other genders reported
- `Duration`: Number of minutes for the workout
- `Food`: Food consumed (other than water) during the workout with levels of 1 (yes) and 0 (no)
- `Aerobic`: Aerobic exercise during workout with levels of 1 (yes) and 0 (no)
- `Resistance`: Resistance exercise during workout with levels of 1 (yes) and 0 (no)
- `Other`: Other types of exercise during workout with levels of 1 (yes) and 0 (no)

Schrage (2019) provides additional information about the research, including an interview with the authors. Both Gustafson et al. (2018) and Schrage (2019) are open access to readers. Using these two citations, complete the following with regards to the study design. Exercise 25 examines the data using the analysis methods described in Chapter 3.

(a) Specifically describe the sample used for the study. What is the population that the authors can make inferences to? What do you think is a potential larger population that the authors would like to make inferences to?

(b) Why do the authors think that new research is needed in this area?

(c) Page 3 of Gustafson et al. (2018) says that "the study was approved by the University of Nebraska-Lincoln Institutional Review Board." These types of review boards are common in the United States. Perform a brief investigation into what these boards represent and why research often needs to be approved by these boards.

(d) Page 4 of Gustafson et al. (2018) says that "individuals who participated in the study multiple times were eliminated from the data set such that only the first study visit was included in the analysis." Why were these observations removed from the data set? Answer this question relative to assumptions needed for the methods in Chapter 3.

25. Continuing Exercise 24, complete the following parts. All page references are with respect to Gustafson et al. (2018).

(a) Construct a contingency table with the levels of `Condition` as the rows and the levels of `Choice` as the columns. Perform a Pearson chi-square test for independence between these two variables using this table. This test is performed on p. 5, but the numerical values given are incorrect.

(b) The authors are primarily interested in the relationship between the food choice timing and the food chosen, but they also collect information on additional variables, like body mass index and workout duration. They present an alternative analysis to what was done in part (a), where `Condition` and these additional variables are included in a multinomial regression model to estimate the probability of choosing a particular food. What reason did the authors give for including these additional variables?

(c) Fit a multinomial regression model for `Choice` using `Condition` and the additional variables in the data set as explanatory variables. Do not include any interactions among variables. Transform the `BMI` variable to be equal to 1 if the body mass index is above 25 and 0 if it is 25 or below. This transformation is performed by the authors because of the current guidelines for being overweight (index greater than 25). No other transformations should be made upon the variables.

(d) Using the model from part (c), complete the odds ratio calculations and interpretations as given below.

   i. Estimate the odds ratios to compare the food choices for each of the explanatory variables. Remember to compare each food choice level to the other two levels (each explanatory variable will have three odds ratios associated with it).

   ii. Calculate the corresponding 95% confidence intervals for i.

   iii. Compare your numerical values from i. and ii. to those given in Table 2 of the paper. This table provides the odds ratios for comparing `apple` to `neither` and `brownie` to `neither`. Note that the authors represent `Condition` in their model with a 1 for the `After` level and a 0 for the `Before` level, which is the opposite of the default indicator variable coding in R.

   iv. Interpret `Condition` using its odds ratios and corresponding confidence intervals. Do your interpretations coincide with those given by the authors on the bottom of p. 5? Explain.

   v. Why does it not make sense to include the "Intercept" row of Table 2 in the paper?

(e) Using the model from part (c), estimate the probabilities of choosing a particular food (`apple`, `brownie`, `neither`) when the choice is given before the workout and when the choice is given after the workout. For the other variables in the model, use a BMI of 23, 23 years old, female, 60-minute workout, no food during the workout, aerobic exercise, and resistance exercise. Specifically for brownie chosen before the workout, show how these calculations are performed by writing out the actual equations and verifying the numerical value using R.

(f) Most statistical analysis can be approached in more than one way. If you had performed the analysis without seeing this paper first, what is one thing that you would have done for the analysis that was not mentioned in the paper? Explain.

(g) Page 3 states a "primary hypothesis" for the research. Given your results for the previous parts, is there evidence to support this hypothesis? Explain.

26. Plinko is a game played on *The Price is Right* television show. Contestants drop a circular disk, referred to as a chip, from the top of a rectangular pegboard. The chip travels down the board and bumps into pegs that can change its direction. Eventually, the chip will reach the bottom of the board and land in one of nine slots. These slots are labeled with \$10,000 in the middle (5th of the 9 slots) and then \$0, \$1,000, \$500, and \$100 successively from the middle outward in each direction. A contestant wins the dollar amount corresponding to where the chip landed. Contestants can choose to drop the chip starting from above any of the nine slots, which may affect winnings. For example, perhaps dropping the chip above the \$10,000 slot may maximize the probability of winning this dollar amount. The purpose of this problem is to determine if the drop location matters, and, if so, develop a strategy to play the game.

    The Plinko.csv file contains information on every chip drop over a one-year period of time. Below is a description of each variable.

    - `drop`: Slot above which the chip was dropped at the top of the board, with 1 = \$100, 2 = \$500, 3 = \$1,000, 4 = \$0, and 5 = \$10,000
    - `win`: Slot the chip landed in at the bottom of the board with 1 = \$100, 2 = \$500, 3 = \$1,000, 4 = \$0, and 5 = \$10,000
    - `count`: Frequency of occurrences during the year

    The `drop` and `win` variables are coded with five levels, even though there are nine slots, because the pegboard is symmetric. Complete the following parts using this data.

    (a) Why is a proportional odds regression model potentially better than a multinomial regression model for this data?

    (b) Fit a proportional odds regression model using the win category as the response and drop location as the explanatory variable. For the win category, order its factor values by the numerical values given in the data ("1" is the lowest value and "5" is the highest value) rather than ordering by the dollar amounts. For the drop location, represent it by one linear term in the model. Use this model for the remainder of the problem.

    (c) Estimate the probabilities of winning each possible dollar amount for each drop location.

    (d) Plot the estimated model. Include a legend to differentiate among the different lines on the plot. Interpret the plot.

    (e) Estimate the odds ratio for the `drop` variable and calculate the corresponding 95% confidence interval. Interpret the interval.

    (f) Perform a likelihood ratio test to evaluate the importance of the drop variable.

    (g) If the goal is to maximize the probability of winning \$10,000, where should the chip be dropped? Explain.

27. This problem continues Exercise 26, but now using multinomial regression.

    (a) Fit a multinomial regression model using the same response and explanatory variable. Use winning \$100 (slot 1) as the base level.

    (b) Complete parts (c) through (f) from Exercise 26 again, but now use the multinomial regression model. Because the structure of this model, there will be more than one odds ratio.

(c) Contestants tend to focus on winning \$10,000. Therefore, it may be best to use this amount as the base level for the model. How can the base level be changed to \$10,000?

(d) Compare the results for the multinomial and proportional odds regression models.

28. Citrus greasy spot is a citrus tree disease caused by a fungus. Infection is characterized by spots appearing on leaves. These leaves can fall prematurely from the tree, leading to lower citrus yields. Aguilera-Cogley and Vicent (2020) investigate the effectiveness of the fungicide fenbuconazole on grapefruit and orange trees to reduce the severity of the infection. In their open access paper, the authors use proportional odds regression models to relate infection severity to fungicide application in two separate experiments repeated over multiple years. Read this paper and use it to complete the parts below.

    (a) What is the population of interest?

    (b) Infection severity is quantified as a four-category ordinal response representing the number of spots on leaves. A binary variable for whether the fungicide is applied is the only explanatory variable. These variables are used in separate models for each year of both experiments. For the third year of the first experiment, complete the following:

        i. Formally write out the estimated model in terms of the response variable, the parameter estimates, and the explanatory variable. Make sure to define all quantities in the model.

        ii. What approximately is the p-value for the hypothesis test evaluating the importance of the explanatory variable.

        iii. What is the estimated odds ratio and the corresponding 95% confidence interval that measures the fungicide effectiveness. Formally interpret the odds ratio.

        iv. Relate the findings in ii. and iii. to the box and dot plots of Figure 2 in the paper.

    (c) The authors state the proportional odds assumption is tested (results are not stated in the paper). Why is it important to test this assumption?

    (d) There are 95% confidence intervals given for odds ratios resulting from the models. Unfortunately, the type of confidence interval used is not given. Why would it be of interest to know the type? Explain.

    (e) Is there sufficient evidence to indicate the fungicide reduces the number of spots on the leaves? Explain using the statistical evidence presented in the paper.

    (f) A four-category ordinal response is chosen by the authors to quantify the severity of infection. Through a personal communication, the authors indicate the reason for using this categorical scale, rather than using the number of spots, is due to the difficulty in evaluating the leaves in practice. If this difficulty did not occur, suggest an alternative analysis approach for using the number of spots as the response variable without this categorization system.

29. Section 2.2.8 examines properties of logistic regression through Monte Carlo simulation. The purpose of this exercise is to examine many of these same properties now for the multinomial regression of

$$\log(\pi_2/\pi_1) = 1 + 2x \text{ and } \log(\pi_3/\pi_1) = -1 + 3x.$$

(a) Run the code below to simulate one data set (`set1`) corresponding to the multinomial regression model.

```
m <- 100

# A semicolon can be used to end one complete statement
beta20 <- 1; beta21 <- 2; beta30 <- -1; beta31 <- 3

set.seed(8934)
x <- runif(n = m, min = -3, max = 3)
den <- 1 + exp(beta20 + beta21*x) + exp(beta30 + beta31*x)
pi1 <- 1 / den
pi2 <- exp(beta20 + beta21*x) / den
pi3 <- exp(beta30 + beta31*x) / den

save.set <- matrix(data = NA, nrow = m, ncol = 2)
for (i in 1:m) {
  vec <- rmultinom(n = 1, size = 1, prob = c(pi1[i], pi2[i],
      pi3[i]))
  y <- which(vec == 1)
  save.set[i,] <- c(y, x[i])
}

set1 <- data.frame(y = save.set[,1], x = save.set[,2])
```

Describe what each line of the code does relative to the data simulation.

(b) Fit the multinomial regression model to the data simulated in part (a). Compare the regression parameter estimates to the true parameter values. Include a plot of the estimated and true models in these comparisons.

(c) Perform the data set simulation process of part (a) 10,000 separate times to simulate 10,000 different data sets of size 100. Fit the multinomial regression model to each simulated data set. Construct one plot similar to Figure 2.9 showing the estimated models for the first 50 data sets. Using this plot and numerical summaries of all regression parameter estimates, discuss how well the estimated models perform relative to the true model.

(d) Repeat part (c) with samples of size 1,000. Compare your results to those obtained for part (c).

(e) Examine how well a normal distribution approximates the sampling distribution of the regression parameter estimates for samples of size 100 and 1,000.

30. Repeat Exercise 29 for a proportional odds regression model. For this exercise, use one explanatory variable and regression parameters $\beta_{10} = -3.5$, $\beta_{20} = 3.5$, and $\beta_1 = 3$. Simulate the explanatory variable in the same manner as in Exercise 29. We recommend using `vglm()` for model estimation because `polr()` occasionally encounters problems with its iterative numerical procedure for some simulated data sets.

Chapter 4

Analyzing a count response

In Chapter 1, we presented the binomial distribution to model count data arising from observing a binary response on n independent Bernoulli trials. Counts observed in this manner are constrained between 0 and n, with variability that decreases as the probability of success approaches 0 or 1. Count data can arise from other mechanisms that have nothing to do with Bernoulli trials. The number of cars crossing a certain bridge in a given day, the number of weeds in a plot of crop land, the number of people standing in line at a service counter, and the number of moles on a person's body are examples of counts that are not constrained by any specific number of trials. In this chapter, we use the Poisson distribution to model counts of this type. We start by exploring the model for a single count, and then extend this framework to Poisson regression models that can handle any number of explanatory variables of any type. Finally, we consider some special cases of models that are used to describe situations that do not fit cleanly into the standard Poisson regression framework. We show in all cases how inference is conducted and how R can be used to aid in the analysis.

4.1 Poisson model for count data

4.1.1 Poisson distribution

Consider a situation in which some kind of events happen over a fixed period of time, such as cars passing over a bridge. It is easy to imagine that the *rate* or *intensity* at which cars cross the bridge might vary according to time of day (e.g., more cars during rush hour, fewer at 3:00 a.m.) or day of the week (more on a workday, fewer on weekends and holidays). Suppose we count cars for exactly one hour at the same time every week, say Wednesday from 7:00–8:00 a.m. Even if there are no practical differences in the population intensity from week to week, we would not expect the exact same number of cars to cross the bridge each week. Some individuals who normally cross the bridge during this hour may carpool or take transit; they may be sick, on vacation, at a meeting, or tele-commuting; or they may simply be earlier or later than usual and miss the cutoffs for the hour of counting. Thus, we expect the actual count of cars during this time to vary randomly, even if the population rate underlying the process is not changing.

This is the nature of the Poisson distribution for counts. If we can assume that (1) all counts taken on some process have the same underlying intensity, and (2) the period of observation is constant for each count, then the counts follow a Poisson distribution. If Y is a Poisson random variable, then the PMF of Y is

$$P(Y = y) = \frac{\exp(-\mu)\mu^y}{y!},\ y = 0, 1, 2, \ldots,$$

DOI: 10.1201/9781003093091-4

where $\mu > 0$ is a parameter. We can abbreviate this distribution by writing $Y \sim Po(\mu)$. Adaptations of this model will be given later for cases where the intensity is not constant for all counts (Section 4.2) and for when the observation period is not constant for all counts (Section 4.3).

Both the binomial distribution and the Poisson distribution serve as useful models for counts. One may wonder how to decide when one model is more appropriate than the other. The key is that binomial counts must be driven by counting successes and failures from a series of Bernoulli trials, whereas Poisson counts are not. That is, there is a natural denominator associated with a binomial count—the number of trials—and therefore it is natural to express the counts in terms of proportions. Poisson counts do not naturally make proportions. Where one might view multiple observations of counts as being analogous to trials in a binomial, the Poisson count is not restricted to be either 0 or 1 for each observation.

Like all models, the Poisson distribution does not have to fit a particular count phenomenon perfectly to be useful. For example, the Poisson distribution does put probability greater than zero on literally every positive integer value, whereas in most situations where counts are recorded, there is a practical limit on how high the count can be. For example, bridge capacity and local population size may prevent, say, 1 billion cars from crossing the bridge in an hour. The Poisson PMF typically puts infinitesimal probability on counts that are impractically large, thereby maintaining a close approximation to reality.

Properties of the Poisson distribution

The Poisson distribution has a number of appealing properties that make it a convenient model from which to work. First of all, the form of the distribution is fairly simple. There is only one parameter, μ, which represents both the mean and variance of the distribution:

$$E(Y) = Var(Y) = \mu.$$

Notice that, unlike the normal distribution, the variance of the Poisson distribution changes as the mean changes. This makes sense because counts are bounded below by 0. For example, suppose means for two different groups of counts are 5 and 50. The counts in the first group might be mostly in the 0–10 range, while those in the second group might naturally range from, say, 30–70. There is more "room" for variability with a larger mean, so we expect the variance to be increasing with the mean. The Poisson model is more specific than this: it requires that the variance *equals* the mean. We shall see in Section 5.3 that this requirement is sometimes too restrictive.

Another useful property is that sums of Poisson random variables are also Poisson random variables: If $Y_1, Y_2, \ldots, Y_n$ are independent with $Y_i \sim Po(\mu_i)$ for $i = 1, \ldots, n$, then $\sum_{i=1}^{n} Y_i \sim Po(\sum_{i=1}^{n} \mu_i)$. Thus, totals of several counts can be modeled with Poisson distributions as long as the constituent counts can.

The shape of the Poisson distribution approaches that of the normal distribution as μ grows. In fact, the normal distribution is sometimes used as a model for counts, even though it is not a discrete distribution. Exercise 23 demonstrates when the normal distribution is a reasonable approximation to the Poisson.

Example: Cars at an intersection (Stoplight.R, Stoplight.csv)

The intersection of 33rd and Holdrege streets in Lincoln, Nebraska, is a typical north-south/east-west, four-way intersection, except that there is a fire station located approximately 45 meters (150 feet) north of the intersection on the west side of the street. A back-up of vehicles waiting to go south at the intersection could block the fire station's driveway, which would prevent emergency vehicles from exiting the station. To determine the probability that this could happen, 40 consecutive stoplight cycles were watched, and the number of vehicles stopped on the north side of the intersection was recorded for each cycle. Note that there were no vehicles remaining at the intersection for more than one stoplight cycle.

To choose a model for these counts, we note that there are 40 observed counts, and that each count arises from one cycle of a traffic light. If we were to treat the cycles as "trials," a binomial model would require that there are only two possible outcomes for each cycle. However, that is not the case here—the number of cars in a cycle is not restricted to be either 0 or 1. Therefore, the binomial model is not appropriate for these data.

To use the Poisson model, we must assume that the observation period at the intersection was short enough that the intensity with which cars arrive at the intersection is not changing considerably throughout the period. This would be violated, for example, if a nearby school let students out during the observation period, so that we might expect an increase of traffic through the intersection for a particular portion of the time.

Below, we read the data into R and calculate the mean and variance for the 40 counts.

```
> stoplight <- read.csv(file  = "C:\\data\\Stoplight.csv")
> head(stoplight)
  Observation vehicles
1           1        4
2           2        6
3           3        1
4           4        2
5           5        3
6           6        3

> # Summary statistics
> mean(stoplight$vehicles)
[1] 3.875
> var(stoplight$vehicles)
[1] 4.317308
```

The sample mean is $\bar{y} = 3.9$. The sample variance, $s^2 = 4.3$, is fairly close to the mean, as is expected if the car count random variable follows a Poisson distribution. A plot of a Poisson probability distribution using $\mu = 3.875$ and the relative frequency distribution of the data are shown in Figure 4.1.

```
> # Frequencies
> table(stoplight$vehicles)  # Note that y = 0, 1, ..., 8 all
   have positive counts
0 1 2 3 4 5 6 7 8
1 5 7 3 8 7 5 2 2
> rel.freq <- table(stoplight$vehicles) /
   length(stoplight$vehicles)
> rel.freq2 <- c(rel.freq, rep(0, times = 7))

> # Poisson calculations
```

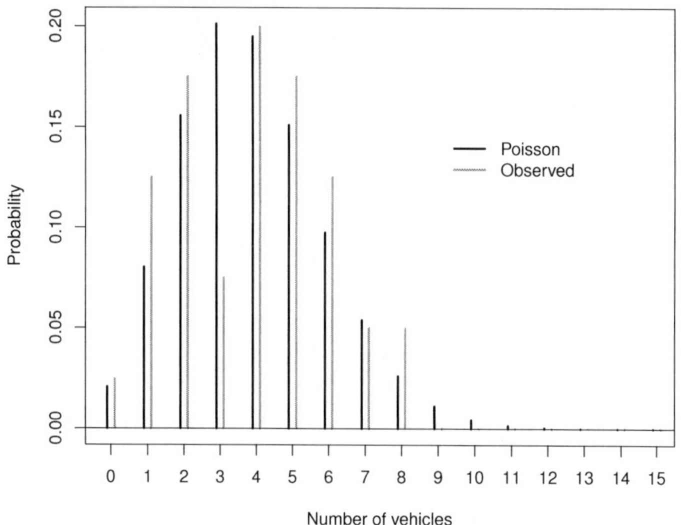

Figure 4.1: Comparison of the observed distribution of car counts and a Poisson distribution with the same mean.

```
> y <- 0:15
> prob <- round(dpois(x = y, lambda = mean(stoplight$vehicles)),
   4)

> # Observed and Poisson
> data.frame(y, prob, rel.freq = rel.freq2)
    y   prob rel.freq
1   0 0.0208    0.025
2   1 0.0804    0.125
3   2 0.1558    0.175
4   3 0.2013    0.075
5   4 0.1950    0.200
6   5 0.1511    0.175
7   6 0.0976    0.125
8   7 0.0540    0.050
9   8 0.0262    0.050
10  9 0.0113    0.000
11 10 0.0044    0.000
12 11 0.0015    0.000
13 12 0.0005    0.000
14 13 0.0001    0.000
15 14 0.0000    0.000
16 15 0.0000    0.000

> plot(x = y-0.1, y = prob, type = "h", ylab = "Probability",
   xlab = "Number of vehicles", lwd = 2, xaxt = "n")
> axis(side = 1, at = 0:15)
> lines(x = y+0.1, y = rel.freq2, type = "h", lwd = 2, lty =
   "solid", col = "gray70")
> abline(h = 0)
> legend(x = 9, y = 0.15, legend = c("Poisson", "Observed"), lty
```

```
= c("solid", "solid"), lwd = c(2,2), col = c("black",
"gray70"), bty = "n")
```

The `dpois()` function computes the PMF for the Poisson distribution with mean given by `lambda`. The code creates side-by-side vertical lines by using the `type = "h"` argument value in both `plot()` and `lines()`, and by plotting at points on the x-axis that are slightly above and below the actual integer values. The plot shows generally good agreement with the Poisson PMF that assumes $\mu = \bar{y} = 3.875$. The only large deviation appears at $y = 3$. We do not believe that there is any physical reason that would prevent groups of three cars from lining up at the intersection, so we assume that this deviation is just a result of chance. Overall, it appears the Poisson distribution provides a satisfactory approximation to the distribution of cars stopped at the stoplight. The use of this approximation to estimate the probability of blocking the fire station appears in Exercise 4.

4.1.2 Poisson likelihood and inference

The parameter μ in the Poisson distribution is estimated using maximum likelihood. Given a random sample of counts, $y_1 \ldots, y_n$, from a Poisson distribution with mean μ, the likelihood function is

$$L(\mu; y_1 \ldots, y_n) = \prod_{i=1}^{n} \frac{\exp(-\mu)\mu^{y_i}}{y_i!}.$$

The MLE is shown in Appendix B.3.1 to be the sample mean, $\hat{\mu} = \bar{y}$. Using the second derivative of the log likelihood as described in Appendix B.3.4, the variance of $\hat{\mu}$ is shown to be estimated by $\widehat{Var}(\hat{\mu}) = \hat{\mu}/n$. Notice that this differs from our usual estimate for the variance of $\bar{y}$, s^2/n, because the mean and variance are the same in the Poisson distribution. In large samples from a Poisson distribution, the sample mean and variance will be nearly identical, but natural variability from sampling can make them appear quite different in small samples.

As with the probability-of-success parameter π in a binomial distribution, there are many ways to perform hypothesis tests and form confidence intervals for the Poisson mean parameter μ. See Swift (2009) for a discussion of the properties of some of these methods for confidence intervals. We discuss here the Wald, likelihood ratio, score, and exact methods to develop a confidence interval for μ.

The $100(1-\alpha)\%$ Wald confidence interval is

$$\hat{\mu} \pm Z_{1-\alpha/2}\sqrt{\hat{\mu}/n}.$$

The score interval is derived from the hypothesis tests of $H_0 : \mu = \mu_0$, which can be conducted using the score statistic

$$Z_S = \frac{\hat{\mu} - \mu_0}{\sqrt{\mu_0/n}},$$

comparing Z_S to appropriate quantiles of the standard normal distribution. Inverting this test results in the confidence interval,

$$\left(\hat{\mu} + \frac{Z^2_{1-\alpha/2}}{2n}\right) \pm Z_{1-\alpha/2}\sqrt{\frac{\hat{\mu} + Z^2_{1-\alpha/2}/4n}{n}}. \quad (4.1)$$

Table 4.1: Properties of possible estimates of variance for a Poisson distribution.

n	Average sample mean	Average sample variance	Variance of sample means	Variance of sample variances	Variance ratio
2	9.98	10.29	4.37	221.64	50.76
5	9.95	9.60	1.88	45.61	24.28
10	10.02	10.14	1.02	23.48	23.00
20	9.99	9.97	0.54	10.82	19.95
50	10.01	10.03	0.19	4.36	22.81
100	9.98	10.02	0.10	1.89	18.33
500	10.00	10.00	0.02	0.43	20.57
1000	10.00	10.01	0.01	0.21	21.78

The LR interval is also based on an inverted test and consists of the set of values for μ_0 that satisfy $-2\log[L(\mu_0|y_1 \ldots, y_n)/L(\hat{\mu}|y_1 \ldots, y_n)] \leq \chi^2_{1,1-\alpha}$. This does not have a closed-form solution and must be found using iterative numerical procedures. It is therefore not generally used for this simple problem. The exact interval is developed using logic similar to what was used for the Clopper-Pearson interval for the binomial probability from Section 1.1.2. Its form is

$$\chi^2_{2n\hat{\mu},\alpha/2}/2n < \mu < \chi^2_{2(n\hat{\mu}+1),1-\alpha/2}/2n.$$

Some of these methods are similar to tests and confidence intervals used for inference on means from the normal distribution. The main advantage of these Poisson-based methods is their use of the Poisson likelihood function to obtain $\widehat{Var}(\hat{\mu}) = \hat{\mu}/n$. The statistic $\hat{\mu}$ is estimated more precisely than s^2. This allows shorter confidence intervals and more powerful tests to be computed, especially in smaller samples.

Example: Comparison of sample mean and sample variance as estimates of variance in the Poisson distribution (PoissonVarianceEstimationSim.R)

We use a small Monte Carlo simulation to demonstrate the gain in precision from using the sample mean $\hat{\mu}$ instead of the sample variance s^2 as an estimate of the variance for a Poisson distribution. For a given sample size n, we simulate 1000 data sets of that size from $Po(\mu)$ where $\mu = 10$. For each data set, we calculate both the sample mean and the sample variance. Then, we look at how precisely the two statistics estimate the true Poisson variance, μ, by computing the variance of each statistic across the 1,000 data sets. We repeat this for several different sample sizes. Details are in the program for this example. The results are shown in Table 4.1.

For sample sizes ranging from 2 to 1,000, both the sample mean and sample variance provide average estimates very close to the population value, 10. However, the sample mean varies much less across the 1,000 data sets than the sample variance does. The final column is simply the ratio of the previous two, showing that sample variances are around 20 times as variable than sample means are, with a higher ratio in smaller samples.

These results are specific to $\mu = 10$. It can be shown that the ratio of variances for the two statistics is actually $2\mu(n/(n-1))$. Exercise 44 explores this concept further.

These results show the advantage of using a Poisson model when data are truly generated from a Poisson distribution. However, the use of the Poisson distribution as a model can also be a disadvantage, because it is common that the processes that generate counts have intensity that does not remain rigidly constant. This causes the counts to behave as if they have more variability than the Poisson model expects.[1] In these cases, $\widehat{Var}(\hat{\mu}) = \hat{\mu}/n$ often underestimates the true variability of the mean count. Although s^2 is a less precise estimate of $Var(Y)$ when the Poisson distribution holds exactly, it estimates the variability present from *all* sources, including any variations in process intensity. Therefore, ordinary t-distribution tests and confidence intervals for a population mean are more robust to deviations from the Poisson assumption and are sometimes used with count data drawn from a single population. However, notice that the confidence intervals developed specifically for the Poisson distribution can all be used even when $n = 1$, while the t-based interval cannot.

Example: Cars at an intersection (Stoplight.R, Stoplight.csv)

We next find confidence intervals for the mean number of cars stopped at the intersection.[2]

```
> alpha <- 0.05
> n <- length(stoplight$vehicles)
> mu.hat <- mean(stoplight$vehicles)

> # Wald
> mu.hat + qnorm(p = c(alpha/2, 1-alpha/2))*sqrt(mu.hat/n)
[1] 3.264966 4.485034

> # Score
> (mu.hat + qnorm(p = c(alpha/2, 1 - alpha/2))^2/(2*n)) +
   qnorm(p = c(alpha/2, 1 - alpha/2)) * sqrt((mu.hat + qnorm(p =
   1 - alpha/2)^2/(4*n))/n)
[1] 3.311097 4.534939

> # Exact
> qchisq(p = c(alpha/2, 1-alpha/2), df = c(2*n*mu.hat,
   2*(n*mu.hat+1)))/(2*n)
[1] 3.288979 4.535323

> # Usual t-distribution based interval
> t.test(x = stoplight$vehicles, conf.level = 0.95)
        One Sample t-test

data:  stoplight$vehicles
t = 11.7949, df = 39, p-value = 1.955e-14
alternative hypothesis: true mean is not equal to 0
95 percent confidence interval:
 3.210483 4.539517
sample estimates:
mean of x
    3.875
```

[1]This is called *overdispersion* and is covered in Section 5.3.

[2]The `poisson.test()` function in the `stats` package can be used to find an exact interval. The package `epitools` also contains functions that can compute confidence intervals for the Poisson mean parameter. In particular, `pois.exact()` produces the exact interval and `pois.approx()` produces the Wald interval. Examples of their use are included in the program for this example.

The 95% score confidence interval is $3.3 < \mu < 4.5$; there is a 95% chance that intervals produced in this way contain the true mean number of cars at the stoplight. Other confidence intervals are fairly similar and have a similar interpretation.

Our recommendations regarding which intervals to use for estimating the mean parameter of the Poisson distribution mirror those for the probability of success parameter from the binomial. The Wald interval is not good, because its true confidence level can be very low and rarely reaches its stated level. Conversely, the exact interval can be excessively conservative and hence wider than necessary. The score interval is a good compromise in that its true confidence level is generally better than the Wald, and it is generally shorter than the exact interval.

Swift (2009) compares numerous intervals (excluding Wald) and finds other intervals that he prefers to the score interval. For example, there are improvements to the exact interval, similar to those described for confidence intervals for the binomial probability parameter in Section 1.1.2, that can make it shorter while retaining the same true confidence level. The differences among the various intervals are often slight, however, as the next example shows. For more details on different possible confidence intervals and their properties, see Swift (2009).

Example: Comparison of confidence intervals for Poisson mean (PoiConfLevels.R)

In this example, we compare the true confidence levels and mean lengths of the Wald, score, exact, and t-based confidence intervals for the population mean when data are sampled from a Poisson distribution. The algorithm for finding the true confidence levels for the Wald and score intervals follows that used in Section 1.1.2 for a binomial probability of success. We find the probability that a sample of size n has an observed sum equal to $0, 1, \ldots, 40n$ when the sample is drawn from $Po(\mu)$ for $\mu = 0.01, 0.02, \ldots, 20$, respectively. To do this, we use the fact the sum has a $Po(n\mu)$ distribution.[3]

Because the t-based interval depends on both the estimated mean and variance, it is much more difficult to find its true confidence level exactly. We would need to identify every possible combination of sample mean and variance for a given sample size and find the Poisson probability for each of the resulting confidence intervals. Instead, we estimate the true confidence level by using Monte Carlo simulation as in Section 1.1.3. We simulate data sets of count responses from $Po(\mu)$, use these data sets to find confidence intervals, and estimate the true confidence level with the observed proportion of these intervals that contain the μ that was used to generate the data. Estimated mean lengths are just the average length of all t-based intervals calculated from each μ.

From Figure 4.2 and investigations for other sample sizes, it is obvious that the true confidence level for the exact interval is always at least 95%; for small μ it is considerably higher. If strict maintenance of the confidence level is required (as in some regulatory situations), then this interval or one of its variants should be used. The confidence level for the score interval is well maintained near 95% at any sample size. The score interval only becomes erratic for very small μ (e.g., $\mu < 1$), but it is

[3]Technically, there are infinitely many possible values of the sum for any given μ, because the $Po(\mu)$ distribution places non-zero probability on responses $y = 0, 1, 2, \ldots$. All except a few happen with probabilities so small that they are ignorable. The program computes probabilities for the relatively small number of values of $\bar{y}$ that have non-negligible probabilities.

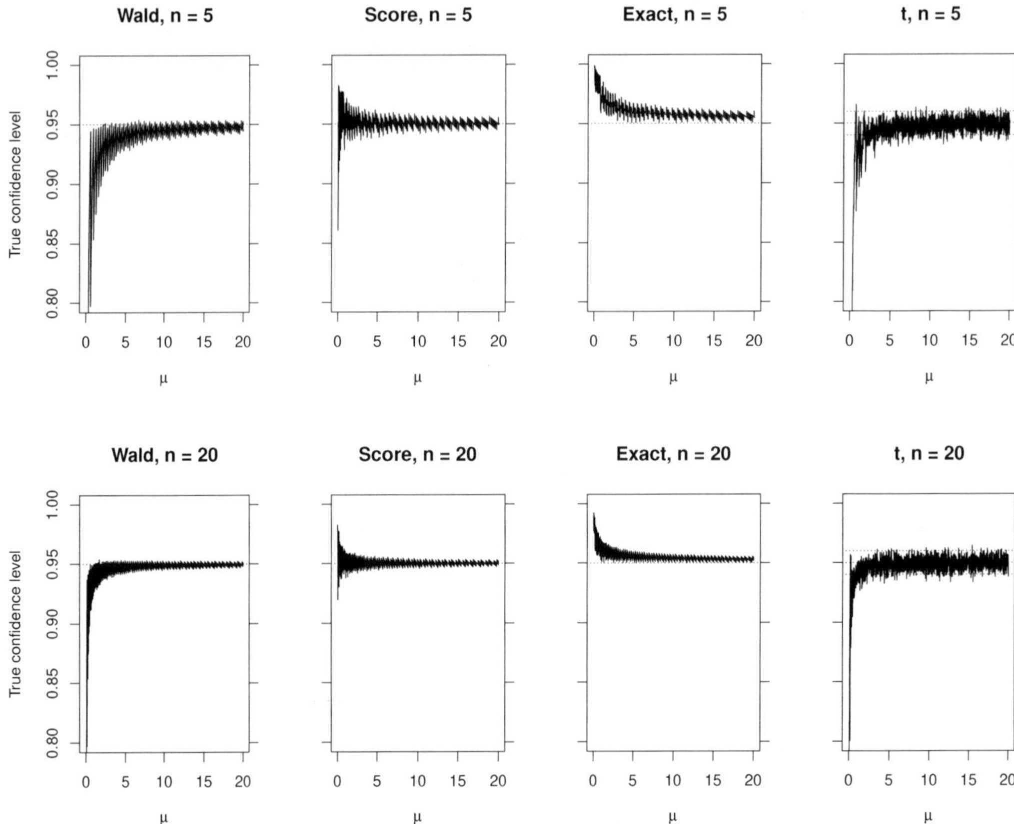

Figure 4.2: True confidence level for Wald, score, and exact 95% confidence intervals and estimated true confidence level for 95% t-based interval. The dotted line at 0.95 represents the stated level; additional horizontal lines in t-based interval plot are limits within which the estimated true confidence levels should fall 95% of the time if the true confidence level is 0.95.

still the only interval among the three approximate intervals whose true confidence level does not drop to 0 as μ decreases toward 0. Therefore, this interval can be recommended for general use as long as the Poisson distribution is expected to hold. The t-based interval's estimated true confidence level seems fine—mostly within its bounds—for $n = 5$ as long as μ is large enough that the distribution of Y is reasonably approximated by a normal ($\mu > 7$ or so). It is a viable alternative to the score interval in these cases, although it creates longer intervals, as Figure 4.3 shows. However, at $n = 2$ (not shown), the t-based interval breaks down and no longer achieves 95% for any $\mu < 20$. The score interval maintains its stated level even for $n = 1$.

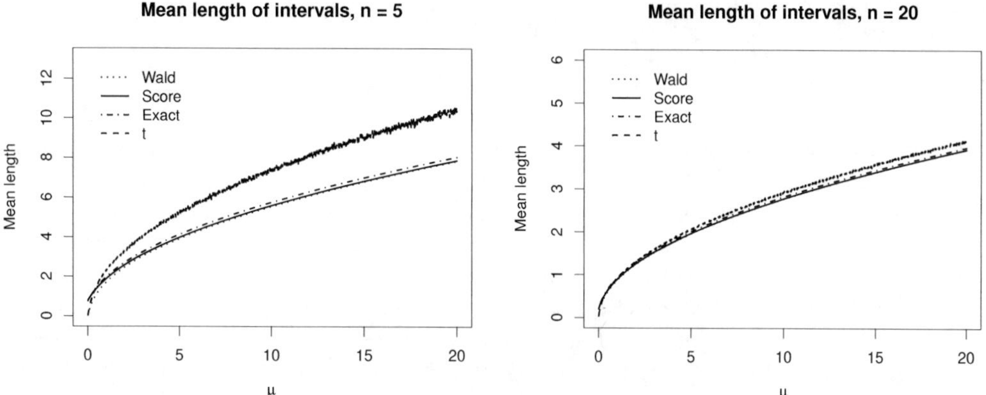

Figure 4.3: Mean 95% confidence interval length for four confidence interval methods. Note that the Wald and score intervals' lengths are nearly identical beyond about $\mu = 2$.

4.2 Poisson regression models for count responses

As we saw in Chapter 2, binary regression models provide a powerful framework within which a wide variety of problems can be addressed. Regression models for count data have also been developed based on the Poisson distribution. These are generically called *Poisson regression models*, and they are closely related to logistic regression models by the fact that both families are classes of generalized linear models that are described in Section 2.3. A special type of Poisson regression model, called a *loglinear model*, not only replicates the classical analysis of contingency tables described in Sections 1.2.3, 2.2.6, and 3.2, but can extend these analyses to any number of variables and can accommodate both continuous and ordinal variables. Poisson regression models can even be used to replicate logistic regression analyses, although their use in this manner is somewhat cumbersome.

The modeling steps with a Poisson regression are very much the same as those undertaken in any other regression modeling process. These steps include (1) specifying the model—the distribution, the parameters to be modeled (e.g., probabilities or means), and their relationship to the explanatory variables; (2) selecting explanatory variables; (3) estimating the parameters in the model; (4) assessing the model fit; and (5) performing inferences on model parameters and other quantities of interest. We assume for now that the explanatory variables to be used in the model are already known or chosen, leaving until Chapter 5 any discussion of the process by which they might have been selected. We also leave assessing the fit of the model until Chapter 5.

4.2.1 Model for mean: log link

We have n observations of a response random variable Y, and $p \geq 1$ fixed explanatory variables, $x_1, \ldots, x_p$. We assume that for observation $i = 1, \ldots, n$,

$$Y_i \sim Po(\mu_i),$$

where

$$\mu_i = \exp(\beta_0 + \beta_1 x_{i1} + \ldots + \beta_p x_{ip}).$$

Thus, our generalized linear model has a Poisson random component, a linear systematic component $\beta_0 + \beta_1 x_1 + \ldots + \beta_p x_p$, and a log link,

$$\log(\mu) = \beta_0 + \beta_1 x_1 + \ldots + \beta_p x_p. \tag{4.2}$$

Each of these three specifications is an assumption that is open to question. For example, the Poisson distribution may provide a poor fit to the data. We will see later in this chapter and in Chapter 5 that there are different distributions that can serve as random components for count data in this case. Also, the linear predictor could conceivably be replaced with something that relates the explanatory variables to the parameters in a nonlinear fashion, at the cost of interpretative and computational complexity. Finally, the link could take an alternative form. In particular, the identity link, $\mu = \beta_0 + \beta_1 x_1 + \ldots + \beta_p x_p$, might seem a simpler form to use. However, the Poisson distribution requires that $\mu > 0$, and the identity link can lead to a non-positive value of μ for particular values of the explanatory variables. The log link function guarantees $\mu > 0$, so it is almost universally used.

The regression parameters in the Poisson regression model have analogous interpretations to those in linear or logistic regression, adapted to the log link rather than the identity or logit. The "intercept" β_0 is the log mean of Y when $x_j = 0$ for all $j = 1, \ldots, p$. Each "slope" parameter β_j is the change in log mean of Y when x_j increases by 1 unit, holding all other variables constant.

A consequence of this is that the explanatory variables affect the response mean multiplicatively. Consider a Poisson regression model with one explanatory variable: $\mu(x) = \exp(\beta_0 + \beta_1 x)$, where our notation emphasizes that μ changes as a function of x. When we increase the explanatory variable by c units, the result is $\mu(x + c) = \exp(\beta_0 + \beta_1(x + c)) = \mu(x)\exp(c\beta_1)$. Thus, the ratio of the means at $x + c$ and at x is

$$\frac{\mu(x+c)}{\mu(x)} = \frac{\exp(\beta_0 + \beta_1(x+c))}{\exp(\beta_0 + \beta_1 x)} = \exp(c\beta_1).$$

This leads to a convenient way to interpret the effect of x:

> The percentage change in the mean response that results from a c-unit change in x is $PC = 100(\exp(c\beta_1) - 1)\%$.

If there were additional variables in the model, the same interpretation would apply, provided we hold the additional explanatory variables constant. When there are interaction terms or transformations involving the explanatory variable of interest, the ratio of means is more complicated, but can be derived in a similar manner as shown for odds ratios in Section 2.2.5. These derivations are the subject of Exercise 6.

Example: Poisson regression means (PoRegLinks.R)

Figure 4.4 shows what the mean function $\mu(x) = \exp(\beta_0 + \beta_1 x)$ looks like for a Poisson regression model when $\beta_1 > 0$ or $\beta_1 < 0$. It also shows the regions likely to contain most of the data. Notice in particular that because the variance and mean are equal, these regions expand as the mean grows.

4.2.2 Parameter estimation and inference

The Poisson regression model assumes that observations $y_1, y_2, \ldots, y_n$ are independent (for example, they are not serially correlated, nor do they form clusters), and hence the

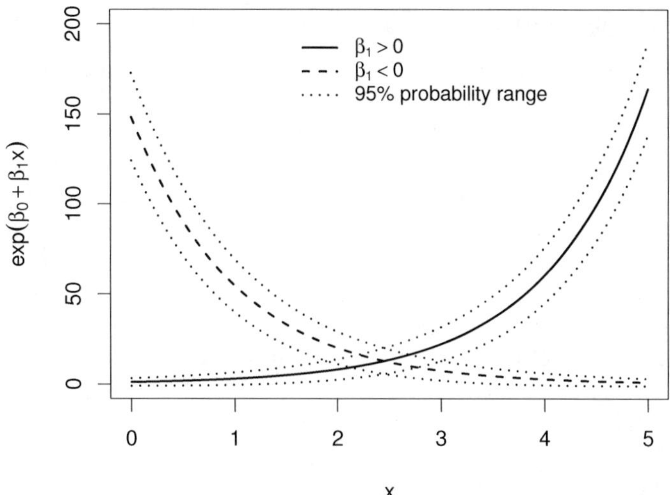

Figure 4.4: Mean functions using log links for $\beta_0 = 0.1, \beta_1 = 1$ (solid line) and $\beta_0 = 5, \beta_1 = -1$ (dashed line). Dotted lines represent approximate regions within which 95% of observations should fall according to the Poisson regression model (± 2 standard deviations).

likelihood is formed by the product of individual PMFs. Following the same steps as laid out in Equations 2.4–2.5, this leads to the log likelihood

$$\log[L(\beta_0, \ldots, \beta_p | y_1, \ldots, y_n)] = \sum_{i=1}^{n}[-\exp(\beta_0 + \beta_1 x_{i1} + \ldots + \beta_p x_{ip}) + y_i(\beta_0 + \beta_1 x_{i1} + \ldots + \beta_p x_{ip}) - \log(y_i!)]. \quad (4.3)$$

As usual, we differentiate the log-likelihood with respect to each parameter β_j, set each of the resulting $p+1$ equations equal to 0, and solve the system of equations to find the MLEs $\hat{\beta}_0, \hat{\beta}_1, \ldots, \hat{\beta}_p$. As was the case with logistic regression in Section 2.2.1, these equations do not typically yield closed-form solutions, so the solutions must be found using iterative numerical procedures.

Once we have MLEs for the regression coefficients, we can compute MLEs for any function of the coefficients by taking the same function of the MLEs. For example, the *fitted value* or *prediction* is $\hat{\mu}_i = \exp(\hat{\beta}_0 + \hat{\beta}_1 x_{i1} + \ldots + \hat{\beta}_p x_{ip})$, and the estimated percentage change in the mean for a c-unit change in x_j is $\widehat{PC} = 100(\exp(c\hat{\beta}_1) - 1)$, holding the other explanatory variables constant.

Inference

The standard approaches to inference in maximum likelihood estimation are used here. We can derive Wald confidence intervals and tests for regression parameters, and various functions of them, by appealing to the large-sample normal approximation for the distribution of the MLEs $\hat{\beta}_0, \hat{\beta}_1, \ldots, \hat{\beta}_p$. While Wald methods are relatively easy to carry out, they do not always perform well. LR methods are better choices when the computational routines are available. On the other hand, programming the parameter coefficients needed for LR methods via `mcprofile` takes more time and can be more prone to mistakes than using `emmeans` for Wald inference. Therefore, we tend to use LR methods for estimating simpler quantities and for smaller sample sizes settings, while using Wald methods for more complex quantities and for larger sample size settings.

Example: Alcohol consumption (AlcoholPoRegs.R, DeHartSaturday.csv)

DeHart et al. (2008) describe a study in which "moderate to heavy drinkers" (at least 12 alcoholic drinks/week for women, 15 for men) were recruited to keep a daily record of each drink that they consumed over a 30-day study period. Participants also completed a variety of rating scales covering daily events in their lives and items related to self-esteem. Among the researchers' hypotheses was that negative events—particularly those involving romantic relationships—might be related to amount of alcohol consumed, especially among those with low self-esteem.

The repeated-measures aspect of this study (the 30 days of measurements on each participant, see Section 6.5) is beyond our current scope. We instead focus on each participant's first Saturday in the study, because Saturday is a day in which alcohol consumption might normally be high. We will model the number of drinks consumed—our count response—as a function of the variables measuring total positive and negative events. Each of these explanatory variables is a combination of several items scored on a 0–3 scale. High values of positive or negative events represent larger numbers of events and/or more extreme events.

The data contain 13 variables, measured over the first Saturday (coded as "6" in the variable `dayweek`) for each study participant.[4] Not all of the variables are needed in our current analysis, so our first step is to create a data frame that contains the required subset of these data. The variables are `id` (participant identifier), `numall` (number of alcoholic beverages, or "drinks", consumed in one day), `negevent` (an index for combining the total number and intensity of negative events experienced during the day), and `posevent` (same as `negevent`, except with positive events).

```
> dehart <- read.csv("C:\\Data\\DeHartSaturday.csv")
> head(dehart)
  id studyday dayweek numall      nrel      prel  negevent
1  1        1       6      9 1.0000000 0.0000000 0.4000000
2  2        4       6      4 5.8333333 0.8333333 2.3766667
3  4        4       6      1 0.3333333 4.0000000 0.2333333
4  5        3       6      0 0.0000000 0.0000000 0.2000000
5  7        7       6      2 0.0000000 2.3333333 0.0000000
6  9        4       6      7 1.0000000 3.0000000 0.5500000
   posevent gender rosn      age  desired    state
1 0.5250000      2  3.3 39.48528 5.666667 4.000000
2 0.9241667      2  3.9 38.00137 5.666667 4.111111
3 1.3464286      2  3.7 30.04791 5.000000 4.111111
4 1.5000000      2  3.0 27.60849 1.666667 4.222222
5 1.6333333      2  3.3 40.35044 4.000000 4.444444
6 0.6250000      2  3.5 33.04586 7.333333 4.222222

> # Reduce data to what is needed for examples
> saturday <- dehart[, c(1,4,7,8)]
> head(round(x = saturday, digits = 3))
   id numall negevent posevent
1   1      9    0.400    0.525
11  2      4    2.377    0.924
18  4      1    0.233    1.346
```

[4]Data kindly provided by Dr. Steve Armeli, School of Psychology, Fairleigh Dickinson University.

```
24  5        0    0.200      1.500
35  7        2    0.000      1.633
39  9        7    0.550      0.625

> dim(saturday)
[1] 89  4
```

The `dim()` function reports the dimensions of a data frame, array, or matrix object. Notice that our data frame, `saturday`, contains data on 89 participants and 4 variables.

Poisson regression models are fit using `glm()` with essentially the same syntax as shown in Section 2.2.1. The main adaptation for Poisson regression is that we must now specify `family = poisson(link = "log")`. Follow-up analysis uses the same generic functions as in logistic regression—`summary()`, `confint()`, `anova()`, `Anova()`, `predict()`, and so forth—to supply parameter estimates, confidence intervals, and tests.

The first model we fit is a simple one, $Y_i \sim Po(\mu_i)$ with $\log(\mu_i) = \beta_0 + \beta_1 \texttt{negevent}_i$, where Y_i is the number of drinks consumed (`numall`) for person $i = 1, \ldots, 89$. The code below fits the model, estimates the percentage change associated with a 1-unit increase in negative events, and tests the significance of the relationship using an LRT.

```
> mod.neg <- glm(formula = numall ~ negevent, family =
    poisson(link = "log"), data = saturday)
> summary(mod.neg)

Call:
glm(formula = numall ~ negevent, family = poisson(link = "log"),
    data = saturday)

Deviance Residuals:
    Min       1Q   Median       3Q      Max
-2.9855  -1.3563  -0.2745   0.4736   5.8854

Coefficients:
            Estimate Std. Error z value Pr(>|z|)
(Intercept)  1.52049    0.07524  20.208   <2e-16 ***
negevent    -0.26118    0.13597  -1.921   0.0547 .

(Dispersion parameter for poisson family taken to be 1)

    Null deviance: 250.34  on 88  degrees of freedom
Residual deviance: 246.39  on 87  degrees of freedom
AIC: 505.79

Number of Fisher Scoring iterations: 5

> 100*(exp(mod.neg$coefficients[2]) - 1)
 negevent
-22.98564

> beta1.int <- confint(mod.neg, parm = "negevent", level = 0.95)
Waiting for profiling to be done...
> 100*(exp(beta1.int) - 1)
```

```
       2.5 %       97.5 %
-41.5294841   -0.3479521

> library(package = car)
> Anova(mod.neg)
Analysis of Deviance Table (Type II tests)
Response: numall
         LR Chisq Df Pr(>Chisq)
negevent   3.9495  1    0.04688 *

> # Matching confidence level with p-value of LRT to demonstrate
   equivalence
> confint(mod.neg, parm = "negevent", level = 1-0.04688)
Waiting for profiling to be done...
         2.3 %          97.7 %
-5.406268e-01  -5.932462e-06
```

The parameter estimates are $\hat{\beta}_0 = 1.52$ and $\hat{\beta}_1 = -0.26$. The negative "slope" parameter indicates that the number of drinks is decreasing as the negative events increase. We calculate $\widehat{PC} = -22.99$, which means that a 1-unit increase in negative events leads to an estimated 23.0% decrease in number of alcoholic beverages with 95% profile LR confidence interval $-41.5\% < PC < -0.3\%$. This change is marginally significant according to a Wald test (p-value = 0.0547) and LRT (p-value = 0.047). For illustration purposes, we reran the confidence interval calculations setting α to the p-value from the test, to show the equivalence between the profile LR confidence interval and the LRT. As expected, this yields a confidence interval for β_1 with an endpoint essentially at zero, confirming that the test and confidence interval are based on the same calculations.

Figure 4.5 shows a plot of the data with the estimated mean curve, $\hat{\mu} = \exp(1.52 - 0.26\texttt{negevent}) = 4.57(0.77)^{\texttt{negevent}}$, using code given in the program for this example. It is apparent from the plot that the mean decreases as the negative events increase, but the change is somewhat small, decreasing from about 4.5 drinks to about 2.5 drinks across the range of negative events. The plot also shows some other interesting features. Very few of these "moderate-to-heavy" drinkers consumed no alcohol on their first Saturday in the study. One person reported consuming 21 drinks (we do not recommend this).

The inverse relationship that was observed between negative events and alcohol consumption is somewhat surprising. Considering that we are using a very simple model and analyzing only one day of the responses from each study participant, we would view these results as preliminary and in need of further verification. Finally, a legitimate question that we do not address here is whether these people behaved differently because they were participating in a study (e.g., did recording their consumption cause them to change their drinking behavior?). If so, then this might explain why none of the people who experienced the most extreme negative events were among the heavier alcohol consumers on this day. Fuller investigation of the data (e.g., comparing the first Saturday with later Saturdays) might help to resolve this question.

Next, we consider an extended model that also takes into account the effects of positive events on drinking behavior. We now model the mean as $\log(\mu) = \beta_0 + \beta_1\texttt{negevent} + \beta_2\texttt{posevent} + \beta_3\texttt{negevent} \times \texttt{posevent}$. The parameter estimates and other analyses are given below.

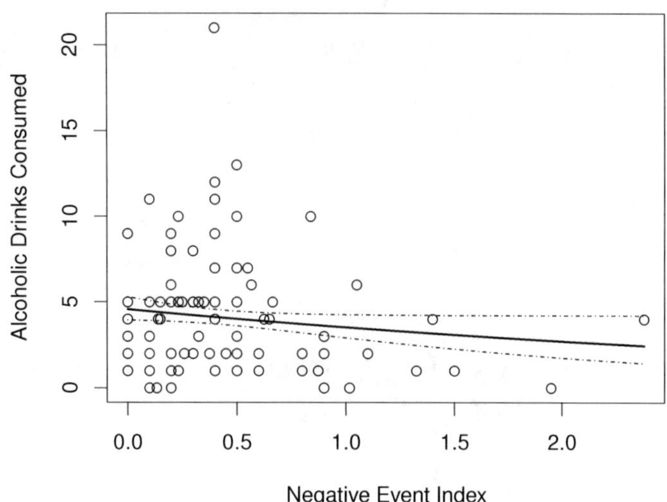

Figure 4.5: Plot of data and estimated mean (with 95% confidence interval) from Poisson regression of number of drinks consumed on total negative events for the first Saturday night in the alcohol consumption study.

```
> mod.negpos <- glm(formula = numall ~ negevent*posevent, family
   = poisson(link = "log"), data = saturday)
> summary(mod.negpos)

Call:
glm(formula = numall ~ negevent * posevent, family =
   poisson(link = "log"),
    data = saturday)

Deviance Residuals:
    Min       1Q   Median       3Q      Max
-3.0208  -1.2673  -0.5198   0.5172   5.9333

Coefficients:
                  Estimate Std. Error z value Pr(>|z|)
(Intercept)         1.7571     0.1565  11.228  < 2e-16 ***
negevent           -1.1988     0.3231  -3.710 0.000207 ***
posevent           -0.1996     0.1199  -1.665 0.095941 .
negevent:posevent   0.7525     0.2253   3.341 0.000836 ***

(Dispersion parameter for poisson family taken to be 1)

    Null deviance: 250.34  on 88  degrees of freedom
Residual deviance: 233.07  on 85  degrees of freedom
AIC: 496.47

Number of Fisher Scoring iterations: 5

> confint(mod.negpos)
Waiting for profiling to be done...
```

```
                        2.5 %      97.5 %
(Intercept)          1.4490794  2.06217629
negevent            -1.8492097 -0.58240139
posevent            -0.4388871  0.03049091
negevent:posevent    0.3137587  1.19645560

> Anova(mod.negpos)
Analysis of Deviance Table (Type II tests)

Response: numall
                  LR Chisq Df Pr(>Chisq)
negevent            4.0153  1  0.0450894 *
posevent            1.9081  1  0.1671724
negevent:posevent  11.4129  1  0.0007293 ***
```

The estimated model is $\log(\hat{\mu}) = 1.76 - 1.20\texttt{negevent} - 0.20\texttt{posevent} + 0.75\texttt{negevent} \times \texttt{posevent}$. An interesting feature here is the strength of the interaction term. The profile LR confidence interval for β_3 from `confint()` is $0.31 < \beta_3 < 1.20$, quite clearly excluding 0. The LRT from `Anova()` places the p-value for testing $H_0 : \beta_3 = 0$ at 0.0007. The left panel of Figure 4.6 shows a plot of the data with respect to both explanatory variables, where points are shaded according to number of drinks, with darker shades indicating more drinks. Contours of the estimated mean from the model are overlaid. We see that for small values of the positive event index (e.g., < 1), the shades seem to get lighter for *larger* values of negative events. However, for larger values of the positive event index (e.g., > 2), the shades tend to be lighter for *smaller* values of negative events. This clearly depicts the nature of the interaction. The contours show this too: they are increasing from left to right in the upper half of the plot and decreasing in the lower half. Without interaction, the contours would have curved in the same direction throughout the plot. Notice also that the estimated means in the upper-right corner of the plot, where we have no data, are obviously absurd. This serves as a reminder of the dangers of extrapolating a regression function beyond the range of the data.

The right panel shows the estimated mean curves for drinks against negative events at the three quartiles of positive events, 0.68, 1.03, and 1.43 (a technique suggested by Milliken and Johnson (2001) for examining interactions involving continuous variables). The three curves show that when there are low positive events, there is a more acute decrease in drinks as negative events increase than when there are higher positive events.[5] We can also calculate the ratio of means corresponding to a 1-unit increase in negative events at each of the three positive-event quartiles. For a fixed value of positive events, say a, the ratio of means is

$$\frac{\mu(\texttt{negevent} + 1, a)}{\mu(\texttt{negevent}, a)} = \exp(\beta_1 + a\beta_3).$$

These quantities are computed below, along with the corresponding percent changes, $100(\exp(\beta_1 + a\beta_3) - 1)$, and their confidence intervals.[6]

[5]The code for these plots is in the accompanying program for this example. Also in the program is code to produce an interactive three-dimensional plot of the same points and surface. We suggest rotating this surface in different directions to see how it fits the points.

[6]The computational process used by `mcprofile()` can sometimes produce slightly different confidence intervals in different runs of the same program. The differences are generally negligible, but can be eliminated by using `set.seed()` before `mcprofile()`.

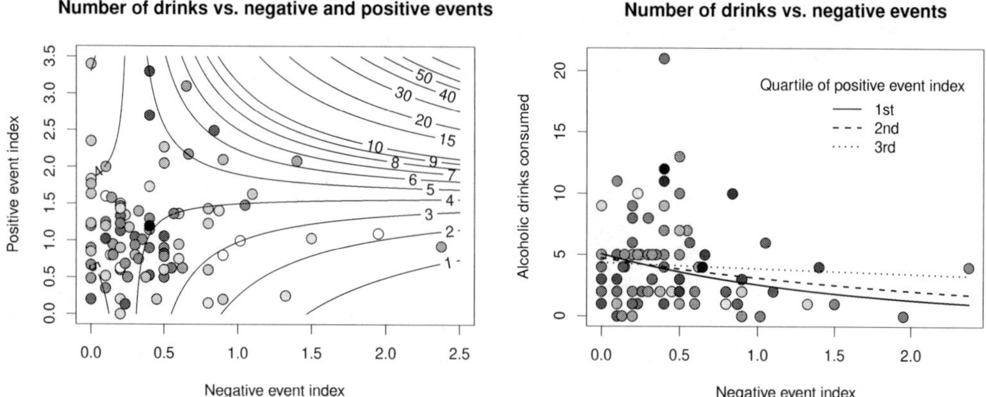

Figure 4.6: (Left) Plot of positive events vs. negative events, with the response variable, number of drinks, represented as increasingly dark shading with increasing drinks. Overlaid contours are from the estimated Poisson regression function for the number of drinks against both variables and their interaction. (Right) Plot for the number of drinks against negative events with the other explanatory variable, positive events, represented as increasingly dark shading with increasing positive events. Overlaid curves are from the same regression, evaluated at the three quartiles of positive events.

```
> posev.quart <- summary(saturday$posevent)
> posev.quart
  Min. 1st Qu.  Median    Mean 3rd Qu.    Max.
0.0000  0.6833  1.0330  1.1580  1.4330  3.4000

> mean.ratio <- exp(mod.negpos$coefficients[2] +
   posev.quart[c(2,3,5)]*mod.negpos$coefficients[4])
> mean.ratio
  1st Qu.    Median   3rd Qu.
0.5042493 0.6560323 0.8864268
> 100*(mean.ratio - 1)
  1st Qu.    Median   3rd Qu.
-49.57507 -34.39677 -11.35732

> library(package = mcprofile)
> K <- matrix(data = c(0, 1, 0, 1*posev.quart[2],
                       0, 1, 0, 1*posev.quart[3],
                       0, 1, 0, 1*posev.quart[5]), nrow = 3,
                         ncol = 4, byrow = TRUE)
> linear.combo <- mcprofile(object = mod.negpos, CM = K)
> ci.beta <- confint(object = linear.combo, level = 0.95)
> 100*(exp(ci.beta$estimate) - 1)  # Verifies same answer
   Estimate
C1 -49.57507
C2 -34.39677
C3 -11.35732
> 100*(exp(ci.beta$confint) - 1)
       lower      upper
```

```
C1 -68.15792 -23.150640
C2 -54.18113  -8.890579
C3 -36.66921  21.936802
```

We see that estimated alcohol consumption decreases at the first quartile of positive events by nearly 50% per unit increase in negative events. The corresponding confidence interval clearly excludes 0 ($-68\% < PC < -23\%$). At the third quartile of positive events, the increase is just 11% per unit, and the confidence interval does not exclude 0 ($-37\% < PC < 22\%$).

Similar results are found from Wald-based inference methods using `emmeans`. The `emmeans()` function creates estimated linear predictors (log-means) for the combinations of variables listed in `specs` at the fixed values listed in `at`. Here, we compute log-means for a 1-unit increase in negative events (taking the difference between `negevent` at 0 and 1) for each quartile of `posevent`. The main code and results are shown below.

```
> library(package = emmeans)
> emm.negpos <- emmeans(object = mod.negpos, specs = ~negevent +
    posevent, at = list(negevent = c(0,1), posevent =
    c(posev.quart[c(2,3,5)])))
> con.emm.negpos <- contrast(object = emm.negpos, method =
    "consec", by = "posevent", reverse = FALSE)
> ci.emm.negpos <- confint(object = con.emm.negpos, type =
    "response")
> 100*(ci.emm.negpos[, c("ratio", "asymp.LCL", "asymp.UCL")] - 1)
                ratio asymp.LCL asymp.UCL
1st Qu. -49.57381 -65.73472 -25.79076
Median  -34.38032 -51.45449 -11.30090
3rd Qu. -11.33508 -33.46937  18.16313
```

The next step in this analysis would be an assessment of the model fit. This is covered in Section 5.2. For the time being, we can explore one possible concern. We question whether the observation with 21 drinks in one evening is a legitimate measurement. We do not have access to the person who recorded this value, but we can check whether it has an inordinate influence on the analysis results. We leave this to Exercise 45.

4.2.3 Categorical explanatory variables

Categorical explanatory variables can be included in a Poisson regression model in a similar manner to regression models for binary or multicategory responses. A categorical explanatory variable X with I levels is converted into $I-1$ indicator variables, each indicating one of the levels of X. For later notational convenience, and to emulate the way that R parameterizes its indicator variables, we denote these variables as $x_2, x_3, \ldots, x_I$. Suppose for the moment that X has four levels, and that it is the only explanatory variable in the model. Then, we can write the model for the mean as

$$\log(\mu) = \beta_0 + \beta_2 x_2 + \beta_3 x_3 + \beta_4 x_4.$$

This specifies a structure analogous to a 1-way ANOVA model as described by Table 4.2.

Table 4.2 also demonstrates that we can write the model as $\log(\mu_i) = \beta_0 + \beta_i x_i$ for level $i = 2, \ldots, I$ of X. This leads immediately to the interpretation of the parameters and makes

Table 4.2: Indicator variables for a 4-level categorical explanatory variable.

Level	x_2	x_3	x_4	Log-mean
1	0	0	0	$\log(\mu_1) = \beta_0$
2	1	0	0	$\log(\mu_2) = \beta_0 + \beta_2$
3	0	1	0	$\log(\mu_3) = \beta_0 + \beta_3$
4	0	0	1	$\log(\mu_4) = \beta_0 + \beta_4$

clear our choice of index labels: β_0 is the log-mean for level 1 of X; β_2 is the difference in log-means between levels 2 and 1; and other parameters are interpreted similarly as β_2. The regression parameters $\beta_2, \ldots, \beta_I$ are often called *effect parameters* for the categorical variable X. Notice that $\beta_2 = \log(\mu_2) - \log(\mu_1)$, which implies that $\exp(\beta_2) = \mu_2/\mu_1$, with a similar interpretation for β_3 and β_4. Thus, the effect parameters are very useful for comparing means at different levels against the mean at the first level. Furthermore, the difference between any two effect parameters also measures a ratio between two means. For example, $\log(\mu_2) - \log(\mu_3) = \beta_2 - \beta_3$, so that $\mu_2/\mu_3 = \exp(\beta_2 - \beta_3)$. If we choose to write $\beta_1 = 0$ for convenience, then all ratios between pairs of means are of the form $\mu_i/\mu_{i'} = \exp(\beta_i - \beta_{i'})$.

Inferences typically focus on estimating the means for each level of X and comparing those means in some chosen way. Means are estimated as $\hat{\mu}_i = \exp(\hat{\beta}_0 + \hat{\beta}_i)$, $i = 1, \ldots, I$. Confidence intervals for individual means are found by exponentiating confidence intervals for corresponding sums of regression parameters, $\beta_0 + \beta_i$.

It is common to test $H_0 : \mu_1 = \mu_2 = \ldots = \mu_I$, which is equivalent to $H_0 : \beta_2 = \ldots = \beta_I = 0$. This hypothesis is easily tested using an LRT. Further comparisons among the means are performed using tests or confidence intervals on combinations of parameters. For example, as noted above, $\mu_i/\mu_{i'} = \exp(\beta_i - \beta_{i'})$, so a confidence interval for the difference between two effect parameters can be exponentiated into a confidence interval for the ratio of their means.

Profile LR methods are preferred for confidence intervals. The `confint.glm()` method function can find confidence intervals for individual parameters, thereby allowing comparisons directly between μ_1 and any other mean. To compare other means, one can reorder levels of the categorical explanatory variable, refit the model, and use `confint()` again (see Section 2.2.6). More generally, LR confidence intervals for linear combinations of parameters are available from the `mcprofile` package, while Wald intervals are available from the `emmeans` package.

Tests for individual means or ratios are carried out using the same approach as for confidence intervals by creating linear combinations of the corresponding model parameters. Various packages in R have the capability to produce these tests (e.g., the `deltaMethod()` function in `car`, the `glht()` function in `multcomp`, the `test()` function in `emmeans`, or the `summary()` function in `mcprofile`). Tests can also be carried out informally by "inverting" the confidence interval (e.g., determining whether a confidence interval for $\beta_2 - \beta_3$ contains 0), although this does not produce a p-value.

Notice that we continually work with ratios of means rather than differences. This is because the log link leads very naturally to computing differences in the log scale, which leads to ratios of the means. The regression parameter estimators are approximately normally distributed and have variances that are relatively easy to estimate as described in Appendix B.3.4. Linear combinations of normal random variables are also normal random variables, so that linear combinations of parameter estimators are also approximately normally distributed. Therefore, confidence intervals for certain functions of linear combinations of parameters, such as $\exp(\beta_i - \beta_{i'})$, are easy to develop and compute.

Table 4.3: Data on bird counts at 6 locations in Ecuador.

Location	Counts
Forest A	155, 84
Forest B	77, 57, 38, 40, 51, 67
Edge	53, 49, 44, 47
Fragment	35, 50, 66, 51, 49, 75
Pasture A	40, 39
Pasture B	33, 31, 38, 50

Although less commonly used, linear combinations of means that can not be reduced to linear combinations of parameters, such as $(\mu_2 + \mu_3)/2 = (\exp(\beta_0 + \beta_2) + \exp(\beta_0 + \beta_3))/2$ and $\mu_2 - \mu_3 = \exp(\beta_0 + \beta_2) - \exp(\beta_0 + \beta_3)$, can sometimes be of interest. The relevant estimates and corresponding inferences are easily created by adding the `regrid = "response"` argument to `emmeans()`. This causes all calculations to be performed in the mean scale rather than on the linear predictors. Manual calculation is also available by starting with the `deltaMethod()` function in `car`, which can be further extended to other functions of the means than linear combinations. Both approaches apply the delta method to find an approximate standard error for the function (see Appendix B.4.2).

Example: Ecuadorean bird counts (BirdCountPoReg.R, BirdCounts.csv)

Clearing of forests for farmland usually results in a change in the composition of plant and animal life in the area. In a study of bird species inhabiting cloud forest regions of Ecuador, researchers used mist nets to capture and identify birds at six different locations. Two locations were within undisturbed forest, one was in a fragment of forest surrounded by cleared land, one was on the edge between the forest and cleared land, and two were in pasture that had previously been forested. Several netting visits were made to each location over three years. For each bird captured, researchers recorded the species, gender, and various other measurements. See Becker et al. (2008) for details on the study. The total counts of all birds from each visit are listed in Table 4.3.[7]

We are interested in comparing overall abundance of birds across different types of habitat. To do this, we fit a Poisson regression model to estimate the mean count of bird captures at each location as an index of the overall abundance, and compare these means across different types of habitats. Thus, we use `glm()` to estimate the six means using five indicator variables indicating the last five locations in alphabetical order. The data are entered and these indicator variables are displayed below. Notice that the results from `contrasts()` show a form identical to Table 4.2.

```
alldata <- read.csv(file = "C:\\data\\BirdCounts.csv",
   stringsAsFactors = TRUE)
> head(alldata)
   Loc Birds
1 ForA   155
2 ForA    84
3 ForB    77
4 ForB    57
```

[7]Data kindly provided by Dr. Dusti Becker, International Conservation Program Coordinator at Life Net Nature (`http://www.lifenetnature.org`).

```
5 ForB      38
6 ForB      40

> contrasts(alldata$Loc)
     ForA ForB Frag PasA PasB
Edge    0    0    0    0    0
ForA    1    0    0    0    0
ForB    0    1    0    0    0
Frag    0    0    1    0    0
PasA    0    0    0    1    0
PasB    0    0    0    0    1
```

Next, the model is fit and the summary is obtained.

```
> M1 <- glm(formula = Birds ~ Loc, family = poisson(link =
    "log"), data = alldata)
> summary(M1)

Call:
glm(formula = Birds ~ Loc, family = poisson(link = "log"), data
    = alldata)

Deviance Residuals:
    Min        1Q    Median        3Q       Max
-3.4322   -0.7594   -0.1302    0.8874    3.1038

Coefficients:
            Estimate Std. Error z value Pr(>|z|)
(Intercept)  3.87640    0.07198  53.853   <2e-16 ***
LocForA      0.90692    0.09678   9.371   <2e-16 ***
LocForB      0.13094    0.09062   1.445   0.1485
LocFrag      0.11874    0.09082   1.307   0.1911
LocPasA     -0.20010    0.13356  -1.498   0.1341
LocPasB     -0.23881    0.10844  -2.202   0.0277 *
---
Signif. codes:
0 '***' 0.001 '**' 0.01 '*' 0.05 '.' 0.1 ' ' 1

(Dispersion parameter for poisson family taken to be 1)

    Null deviance: 216.944  on 23  degrees of freedom
Residual deviance:  67.224  on 18  degrees of freedom
AIC: 217.87

Number of Fisher Scoring iterations: 4
```

Because $\hat{\beta}_0 = 3.88$, the estimated mean count for the edge habitat is $\hat{\mu}_1 = \exp(3.88) = 48.3$. The remaining $\hat{\beta}_i$'s, $i = 2, \ldots, 6$, measure the estimated difference in log-means between the given location and edge, $\log(\hat{\mu}_i) - \log(\hat{\mu}_1)$, $i = 2, \ldots, 6$. For example, the difference in log-means between fragment and edge is represented by $\hat{\beta}_4 = 0.11874$. Thus, the ratio of these two means is $\hat{\mu}_4/\hat{\mu}_1 = \exp(0.11874) = 1.126$.

We give code in the program for this example to show that all estimated means from this model are the same as their respective sample means. This happens because we are allowing the model to estimate a separate mean for each habitat, and the MLEs in this case can be shown to be the respective sample means. Thus, it seems that we are going to a great deal of extra effort to do something that should be easier to do. The advantage of the model-based approach lies in the numerous inference procedures that accompany the model fit. We show several of these next.

To test whether all means are equal,

$$H_0 : \mu_1 = \mu_2 = \ldots = \mu_6$$
$$H_a : \text{Not all means are equal,}$$

we use anova() to perform the LRT (Anova() could be used as well).

```
> anova(M1, test = "Chisq")
Analysis of Deviance Table
Model: poisson, link: log
Response: Birds
Terms added sequentially (first to last)
     Df Deviance Resid. Df Resid. Dev P(>|Chi|)
NULL                    23    216.944
Loc   5   149.72        18     67.224 < 2.2e-16 ***
```

This null hypothesis is strongly rejected ($-2\log(\Lambda) = 67.2$ and p-value is less than 2.2×10^{-16}), so clearly not all locations have the same mean. To understand the cause of this result, a plot of the estimated means and corresponding confidence intervals can help.

We show next how to compute estimated means and corresponding confidence intervals using the mcprofile and emmeans packages. For LR intervals, we start by specifying the coefficient matrix K for estimating each $\log(\hat{\mu}_i) = \hat{\beta}_0 + \hat{\beta}_i$. The rows are arranged in order of decreasing forest habitat: forests, fragment, edge, pastures. We add row and column labels to the matrix using the rownames() and colnames() functions, respectively. The Wald intervals can be found more easily using the emmeans package because it will find the K matrix on its own. The results are very similar between the two methods.

```
> # Profile LR intervals using mcprofile
> K <- matrix(data = c(1, 1, 0, 0, 0, 0,
                       1, 0, 1, 0, 0, 0,
                       1, 0, 0, 1, 0, 0,
                       1, 0, 0, 0, 0, 0,
                       1, 0, 0, 0, 1, 0,
                       1, 0, 0, 0, 0, 1), nrow = 6, ncol = 6,
                byrow = TRUE)
> colnames(K) <- c("beta0", "beta2", "beta3", "beta4", "beta5",
   "beta6")
> rownames(K) <- c("ForA", "ForB", "Frag", "Edge", "PasA",
   "PasB")
> K
     beta0 beta2 beta3 beta4 beta5 beta6
ForA     1     1     0     0     0     0
ForB     1     0     1     0     0     0
Frag     1     0     0     1     0     0
Edge     1     0     0     0     0     0
```

```
PasA      1     0     0     0     1     0
PasB      1     0     0     0     0     1

> linear.combo <- mcprofile(object = M1, CM = K)
> ci.log.mu <- confint(object = linear.combo, level = 0.95,
    adjust = "none")
> exp(ci.log.mu)
   mcprofile - Confidence Intervals

level:              0.95
adjustment:         none

     Estimate lower upper
ForA    119.5 105.0 135.3
ForB     55.0  49.3  61.1
Frag     54.3  48.6  60.4
Edge     48.3  41.8  55.4
PasA     39.5  31.4  48.9
PasB     38.0  32.3  44.4

> # Wald intervals using emmeans
> library(package = emmeans)
> emm1 <- emmeans(object = M1, specs = ~ Loc)
> confint(object = emm1, adjust = "none", type =
    "response")[c(2,3,4,1,5,6),]
 Loc   rate   SE  df asymp.LCL asymp.UCL
 ForA 119.5 7.73 Inf     105.3     135.7
 ForB  55.0 3.03 Inf      49.4      61.3
 Frag  54.3 3.01 Inf      48.7      60.6
 Edge  48.2 3.47 Inf      41.9      55.6
 PasA  39.5 4.44 Inf      31.7      49.2
 PasB  38.0 3.08 Inf      32.4      44.5

Confidence level used: 0.95
Intervals are back-transformed from the log scale
```

The plot in Figure 4.7 shows that there is a general increasing trend in the mean counts as the amount of forestation increases. It is also apparent that one forest location has a mean that is drastically different from the other locations. Forests were expected to have higher mean counts, and those means were not necessarily expected to be identical in all forests. This shows that the variability from one location to the next can be very large.

Comparing the two forests or the two pastures to each other is actually not of interest. Instead, we want to combine the means for each pair of locations of the same habitat before comparing them to other habitats. It would be inappropriate to combine their data and estimate a single mean for the two locations of the same habitat, because the data from pairs of locations do not necessarily come from the same Poisson distribution. Instead, we estimate their means separately, and then combine the means to make comparisons. To carry out inferences, we need to develop a matrix of coefficients for the linear combinations of parameters that represent the comparisons being made. For example, to compare the two forest habitats against the fragment, we can compare the ratio for the geometric mean of the two forest means

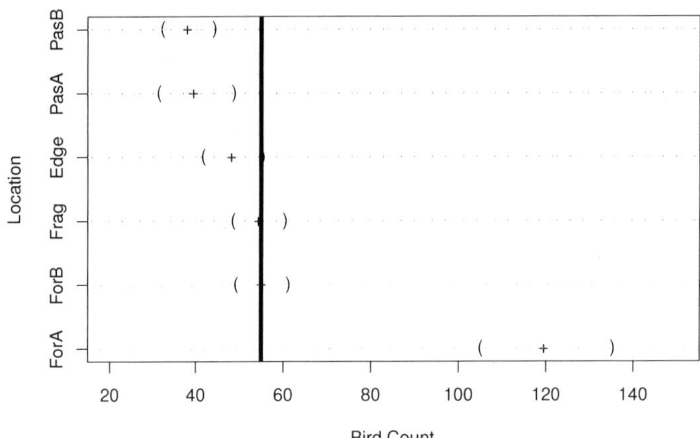

Figure 4.7: Plot of MLEs for mean bird counts, denoted by "+", and LR confidence intervals, denoted by parentheses, for mist-netted bird captures at six locations in Ecuador. The vertical line represents the overall mean of all counts.

against the fragment mean, i.e., $(\mu_2\mu_3)^{1/2}/\mu_4$.[8]

The log of this quantity is $(\log(\mu_2) + \log(\mu_3))/2 - \log(\mu_4)$. Because the `emmeans()` function directly computes estimates of these log-means by default, it is easy to create the required linear combination in the `contrast()` function and return results in terms of the ratio of interest. To compute the LR confidence intervals using `mcprofile()`, we need to further re-express the comparison of log-means in terms of the regression parameters:

$$\begin{aligned}(\log(\mu_2) + \log(\mu_3))/2 - \log(\mu_4) &= (\beta_0 + \beta_2 + \beta_0 + \beta_3)/2 - (\beta_0 + \beta_4) \\ &= (\beta_2 + \beta_3)/2 - \beta_4.\end{aligned}$$

In both cases, the coefficients for this ratio are (0, 0.5, 0.5, −1, 0, 0). For `emmeans()`, the coefficients are entered into `contrast()` as a named member of a `list` for the value of the `method` argument (see Section 3.3.1 for another example of this use). The confidence intervals for each method are shown below.

```
> # Wald intervals using emmeans
> emm.lincon <- contrast(object = emm1, method = list(For.Frag =
   c(0, 0.5, 0.5, -1, 0, 0)), type = "response", adjust = "none")
> emm.lincon
 contrast ratio    SE  df z.ratio p.value
 For.Frag   1.49 0.104 Inf   5.734  <.0001

Tests are performed on the log scale
> confint(object = emm.lincon, level = 0.95)
 contrast ratio    SE  df asymp.LCL asymp.UCL
 For.Frag   1.49 0.104 Inf       1.3      1.71

Confidence level used: 0.95
```

[8]The geometric mean of $a_1, \ldots, a_n$ is $\left(\prod_{i=1}^n a_i\right)^{1/n}$, so that the log of the geometric mean is $n^{-1}\sum_{i=1}^n \log(a_i)$.

```
Intervals are back-transformed from the log scale

> # Profile LR intervals using mcprofile
> K2 <- matrix(data = c(0, 0.5, 0.5, -1, 0, 0), nrow = 1, ncol =
   6, byrow = TRUE)
> rownames(K2) <- "For-Frag"
> colnames(K2) <- colnames(K)
> linear.combo <- mcprofile(object = M1, CM = K2)
> summary(linear.combo)

   mcprofile - Multiple Testing

Adjustment:        single-step
Margin:              0
Alternative:      two.sided

                Estimate Statistic Pr(>|z|)
For-Frag == 0        0.4      5.81   <2e-16 ***
---
Signif. codes:  0 '***' 0.001 '**' 0.01 '*' 0.05 '.' 0.1 ' ' 1

> exp(confint(linear.combo)$confint)
     lower    upper
1 1.302285 1.712259
```

Similar sets of coefficients can be found for comparing each pair of habitats. Completing this process is left as Exercise 46. The program for this example shows several other ways to calculate confidence intervals comparing the means in this example.

4.2.4 Poisson regression for contingency tables: loglinear models

A special use for categorical explanatory variables in a Poisson regression model is to model counts from contingency tables. Practically any analysis that one might normally conduct using traditional methods for contingency tables—such as those described in Sections 1.2 and 3.2—can be carried out using Poisson regression. Tools that are available within the generalized linear model framework to perform additional analyses make it an appealing alternative to traditional table-based methods. Because Poisson regression on contingency tables is so common, it has an alternative name—the *loglinear model*—which stems from the use of the log link for the mean. Note that the loglinear model can be shown to be largely equivalent to the multinomial regression models discussed in Section 3.3.2. We will compare advantages and disadvantages of the two approaches at the end of the next section.

Using Poisson regression models for a two-dimensional contingency table structure requires nothing more than to treat both the row variable and the column variable as categorical explanatory variables. Poisson regression is then applied, using the resulting sets of indicator variables together in one model. This process is very similar to ANOVA modeling of continuous responses. However, some explanation of the details is needed to see how the model reproduces traditional statistics from contingency tables.

Let X represent the row variable with I levels and Z represent the column variable with J levels. We use the indicator variables $x_2, \ldots, x_I$ for X and $z_2, \ldots, z_J$ for Z within a

Table 4.4: Indicator variables and log-means from a loglinear model for a 2×2 contingency table.

X	Z	x_2	z_2	Log-mean
1	1	0	0	$\log(\mu_{11}) = \beta_0$
1	2	0	1	$\log(\mu_{12}) = \beta_0 + \beta_2^Z$
2	1	1	0	$\log(\mu_{21}) = \beta_0 + \beta_2^X$
2	2	1	1	$\log(\mu_{22}) = \beta_0 + \beta_2^X + \beta_2^Z$

basic Poisson regression model:

$$\log(\mu) = \beta_0 + \beta_2^X x_2 + \beta_3^X x_3 + \ldots + \beta_I^X x_I + \beta_2^Z z_2 + \beta_3^Z z_3 + \ldots + \beta_J^Z z_J, \quad (4.4)$$

where β_i^X, $i = 2, \ldots, I$, and β_j^Z, $j = 2, \ldots, J$ are ordinary regression effect parameters with the superscript added for convenience. Thus, the model for the mean count in cell (i, j) of the contingency table, μ_{ij}, can be written in a manner similar to ANOVA models as

$$\log(\mu_{ij}) = \beta_0 + \beta_i^X + \beta_j^Z,\ i = 1, \ldots, I,\ j = 1, \ldots, J, \quad (4.5)$$

where is it implicit that $\beta_1^X = \beta_1^Z = 0$.

To see what these parameters measure, we evaluate Equations 4.4 and 4.5 for a 2×2 contingency table in Table 4.4. We see that

$$\log(\mu_{21}) - \log(\mu_{11}) = \log(\mu_{22}) - \log(\mu_{12}) = \beta_2^X,$$

so β_2^X measures the difference between log-means from rows 2 and 1 within the same column. Thus, $\exp(\beta_2^X)$ measures the ratio between any two such means. Similarly,

$$\log(\mu_{12}) - \log(\mu_{11}) = \log(\mu_{22}) - \log(\mu_{21}) = \beta_2^Z,$$

so β_2^Z measures the difference in log-mean between the two counts in the same row from columns 2 and 1, and $\exp(\beta_2^Z)$ measures the ratio of those means. In larger tables, β_i^X measures the difference in log-means (i.e., the log of the ratio of means) between counts in rows i and 1 in the same column. An analogous interpretation holds for β_j^Z. It can therefore be shown that these parameters also measure differences in log-means, or log-ratios of means, in the *margins* of the table: $\beta_i^X = \log(\mu_{i+}/\mu_{1+})$ (see Exercise 8).

There are no parameters in Equation 4.4 that allow individual cells in an $I \times J$ table to deviate from what their row and column margins dictate. This corresponds to *independence* between X and Z (see Section 3.2), because the effects of one variable are the same regardless of the level of the other variable. Another way to see this is to consider the odds ratio between any two rows and columns. For example, suppose we have a 2×2 table with cell probabilities π_{ij}, $i = 1, 2$; $j = 1, 2$, as in Section 3.2.1. Using the equivalence between the Poisson and multinomial models discussed in Section 4.2.5, we can denote the Poisson regression model's mean count in cell (i, j) by $\mu_{ij} = n\pi_{ij}$. It is then fairly easy to show that

$$OR = \frac{\mu_{11}/\mu_{12}}{\mu_{21}/\mu_{22}} = \frac{\mu_{11}\mu_{22}}{\mu_{21}\mu_{12}}.$$

Equation 4.4 implies that

$$\begin{aligned}\log(OR) &= \log(\mu_{11}) - \log(\mu_{12}) - \log(\mu_{21}) + \log(\mu_{22}) \\ &= \beta_0 - (\beta_0 + \beta_2^Z) - (\beta_0 + \beta_2^X) + (\beta_0 + \beta_2^X + \beta_2^Z) \\ &= 0\end{aligned}$$

Thus, $OR = 1$. The same result occurs in models for larger tables regardless of which two rows or columns are chosen. That is, the loglinear model represented by Equation 4.4 represents independence between X and Z, because the conditional distribution across categories of X is the same for each level of Z and vice versa. The parameters in this model serve only to make the table of estimated mean counts have the same margins as the table of observed counts. In fact, the estimated mean counts from Equation 4.4 are exactly the same as the expected counts we computed in Section 3.2 for testing independence in a contingency table.

Because inference on margins is not typically our main concern in a contingency table, we need to augment Equation 4.4 to allow for association between X and Z. This can be accomplished by adding to the model cross product terms of the form $\beta_{ij}^{XZ}x_iz_j$, $i = 2,\ldots,I$; $j = 2,\ldots,J$, to create an interaction between X and Z. Because there are $I-1$ indicators x_i and $J-1$ indicators z_j, there are $(I-1)(J-1)$ new terms to be added to the model. This creates a saturated model, because the total number of parameters is equal to the number of cell means being estimated. To see this, note first that the table consists of IJ Poisson-distributed counts sampled independently according to their respective means. The intercept β_0 accounts for 1 degree of freedom (df). The effect parameters for X, $\beta_2^X,\ldots,\beta_I^X$, account for $I-1$ df, while those for Z account for an additional $J-1$. With the interaction terms, the new model consumes $1+(I-1)+(J-1)+(I-1)(J-1) = IJ$ df.

The new model can be written as

$$\log(\mu_{ij}) = \beta_0 + \beta_i^X + \beta_j^Z + \beta_{ij}^{XZ}\ i = 1,\ldots,I,\ j = 1,\ldots,J. \tag{4.6}$$

The effect parameters β_i^X and β_j^Z again operate on the marginal counts, but now the XZ-interaction parameters β_{ij}^{XZ}, $i = 1,\ldots,I$, $j = 1,\ldots,J$, allow the ratios of means between cells in two rows to change across columns and vice versa. To see this, consider again the 2×2 table, and recall that we set $\beta_{ij}^{XZ} = 0$ when either $i = 1$ or $j = 1$. Then

$$\begin{aligned}\log(OR) &= \log(\mu_{11}) + \log(\mu_{22}) - \log(\mu_{21}) - \log(\mu_{12})\\ &= (\beta_0 + \beta_1^X + \beta_1^Z + \beta_{11}^{XZ}) + (\beta_0 + \beta_2^X + \beta_2^Z + \beta_{22}^{XZ})\\ &\quad - (\beta_0 + \beta_2^X + \beta_1^Z + \beta_{21}^{XZ}) - (\beta_0 + \beta_1^X + \beta_2^Z + \beta_{12}^{XZ})\\ &= \beta_{11}^{XZ} + \beta_{22}^{XZ} - \beta_{21}^{XZ} - \beta_{12}^{XZ}\\ &= \beta_{22}^{XZ}.\end{aligned}$$

Thus, the odds ratio in a 2×2 table is just $\exp(\beta_{22}^{XZ})$. More generally, the odds ratio between rows i and i' and columns j and j' is

$$OR_{ii',jj'} = \exp(\beta_{ij'}^{XZ} + \beta_{i'j'}^{XZ} - \beta_{i'j}^{XZ} - \beta_{ij'}^{XZ}), \tag{4.7}$$

where the extra subscripts remind us that there is more than one possible odds ratio. Because Equation 4.6 is a saturated model, its estimated means $\hat{\mu}_{ij}$ match their corresponding observed counts y_{ij} exactly. Therefore, odds ratios estimated from Equation 4.7 are exactly the same as those based on the corresponding counts in the observed table.

Tests for independence in contingency tables that were described in Section 3.2.3 can be replicated exactly from the Poisson regression model through model comparison tests. The null hypothesis of independence implies Equation 4.4, while the alternative implies the unrestricted model in Equation 4.6. Fitting both models to the tabulated counts y_{ij}, $i = 1,\ldots,I$, $j = 1,\ldots,J$, and comparing residual deviances provides the LRT statistic. In this case, the statistic is also just the residual deviance from the null model, because the alternative model is saturated, so its residual deviance is zero. This statistic can be shown to be exactly the same as the LRT statistic presented in Section 3.2 for testing independence. The degrees of freedom for the χ^2 sampling distribution, $(I-1)(J-1)$, come from the number of additional parameters in the full model compared to the null model (not counting those set to zero). Odds ratio estimates and confidence intervals are found by exponentiating estimates and confidence intervals for linear combinations of the XZ interaction parameters as determined by Equation 4.7.

It should be noted that the Pearson statistic presented in Section 3.2.3 is generally preferred to the LR statistic for testing independence in an $I \times J$ contingency table (see also the discussion in 1.2.3). However, the LR statistic is easier to adapt for tests in other models, as we discuss in Sections 4.2.5 and 4.2.6.

Example: COVID-19 vaccine clinical trial (COVID19vaccineLoglinear.R)

This example re-examines the clinical trial data for the Pfizer-BioNTech COVID-19 vaccine (Section 1.2.4) using a loglinear model. We will estimate the odds ratio of a positive COVID-19 diagnosis for the vaccine group relative to the placebo group. We show that the odds ratio estimate and Wald confidence interval are exactly the same as those given in Section 1.2.5. We further provide a LR confidence interval for the odds ratio.

Corresponding to Table 1.3, there are two variables, `Treatment`, with levels "Vaccine" and "Placebo," and `Outcome`, with levels "Positive" and "Negative." For this example, we make use of the object `c.table` created in Chapter 1, and convert the object into a data frame. Note that this changes the ordering of factor levels from the default alphanumeric ordering to the order used in the `dimnames` argument of `array`. The results are shown below.

```
> c.table <- array(data = c(8, 162, 17403, 17349), dim = c(2,2),
    dimnames = list(Treatment = c("Vaccine", "Placebo"), Outcome
    = c("Positive", "Negative")))
> c.table
         Outcome
Treatment Positive Negative
  Vaccine        8    17403
  Placebo      162    17349

> # Convert data to data frame for glm()
> all.data <- as.data.frame(as.table(c.table))
> all.data
  Treatment  Outcome  Freq
1   Vaccine Positive     8
2   Placebo Positive   162
3   Vaccine Negative 17403
4   Placebo Negative 17349
> levels(all.data$Treatment)
[1] "Vaccine" "Placebo"
> levels(all.data$Outcome)
[1] "Positive" "Negative"
```

The counts are contained in the variable `Freq`. We fit the saturated model $\log(\mu_{ij}) = \beta_0 + \beta_i^T + \beta_j^O + \beta_{ij}^{TO}$, where μ_{ij} is the mean count in cell (i,j), $i = 1, 2$, $j = 1, 2$, and superscript index T stands for "Treatment" and O for "Outcome." The model is estimated by using the `formula` argument value `Freq ~ Treatment*Outcome` within glm(). Given the ordering of levels for our factors, β_0 is the log-mean count of participants with a positive COVID-19 outcome and vaccine treatment; β_2^T is the log-ratio of mean counts for participants receiving placebo to participants receiving vaccine among those with positive COVID-19; and β_2^O is the log-ratio of mean counts for negative versus positive COVID-19 outcome among those in the vaccine group. The interaction parameter β_{22}^{TO} is the log of the ratio of [the odds of *negative COVID-19 outcome* given that they received the placebo] to [the odds of *negative COVID-19 outcome* given that they received the vaccine]. The summary of the model fit is below.

```
> M1 <- glm(formula = Freq ~ Treatment*Outcome, family =
    poisson(link = "log"), data = all.data)
> summary(M1)

Call: glm(formula = Freq ~ Treatment * Outcome, family =
    poisson(link = "log"), data = all.data)

Deviance Residuals:
[1]  0  0  0  0

Coefficients:
                                    Estimate Std. Error
(Intercept)                           2.0794     0.3536
TreatmentPlacebo                      3.0082     0.3622
OutcomeNegative                       7.6850     0.3536
TreatmentPlacebo:OutcomeNegative     -3.0113     0.3623
                                    z value Pr(>|z|)
(Intercept)                           5.882 4.06e-09 ***
TreatmentPlacebo                      8.306  < 2e-16 ***
OutcomeNegative                      21.731  < 2e-16 ***
TreatmentPlacebo:OutcomeNegative     -8.311  < 2e-16 ***
---
Signif. codes:  0 '***' 0.001 '**' 0.01 '*' 0.05 '.' 0.1 ' ' 1

(Dispersion parameter for poisson family taken to be 1)

    Null deviance:  4.6434e+04  on 3  degrees of freedom
Residual deviance: -7.0921e-13  on 0  degrees of freedom
AIC: 42.066

> predict(object = M1, type = "response")
    1     2     3     4
    8   162 17403 17349
```

The estimated parameters are $\hat{\beta}_0 = 2.08, \hat{\beta}_2^T = 3.01, \hat{\beta}_2^O = 7.69$, and $\hat{\beta}_{22}^{TO} = -3.01$. Using Equation 4.6, we can estimate each mean count according to the model parameters. Because the model is saturated, the estimated means match the observed counts exactly. For example, $\hat{\mu}_{21} = \exp(\hat{\beta}_0 + \hat{\beta}_2^T + \hat{\beta}_1^O + \hat{\beta}_{21}^{TO}) = \exp(2.08 + 3.01 + 0 + 0) = 16$ is exactly the same as the observed count for the number of participants in the placebo group who contracted COVID-19.

The odds ratio in this 2×2 table is controlled by β_{22}^{TO}. We follow up with the usual analysis functions to estimate the odds ratio and find confidence intervals.

```
> round(exp(M1$coefficients[4]), digits = 3)
TreatmentPlacebo:OutcomeNegative
                           0.049
> # LR CI
> inter <- confint(object = M1, parm =
    "TreatmentPlacebo:OutcomeNegative")
Waiting for profiling to be done...
> round(exp(inter), digits = 3)
 2.5 % 97.5 %
 0.022  0.094
> # Wald CI
> round(exp(confint.default(M1, parm =
    "TreatmentPlacebo:OutcomeNegative")), digits = 3)
                                 2.5 % 97.5 %
TreatmentPlacebo:OutcomeNegative 0.024    0.1
```

The estimated odds ratio is $\widehat{OR} = 0.049$, so the estimated odds of contracting COVID-19 after receiving a vaccine are only about 5% as high as they are after receiving a placebo. This implies that the vaccine provides an estimated 95% reduction in odds of COVID-19 compared to no vaccine. The profile LR confidence interval from exponentiating the confidence interval on the interaction parameter is $0.022 < OR < 0.094$, so with 95% confidence, the odds of COVID-19 are between 2.2% and 9.4% as large following vaccine as they are under placebo. For comparison purposes, the Wald confidence interval calculated by `confint.default()` is $0.024 < OR < 0.10$, which matches what was given in Section 1.2.5.

The Wald confidence interval can instead be found using the `emmeans` package. Its `emmeans()` function produces estimated log-means for each combination of `Treatment` and `Outcome`. Thereafter, pairwise interaction contrasts compute the log-OR as differences of differences of log-means; e.g., as $[\log(\mu_{22}) - \log(\mu_{21})] - [\log(\mu_{12}) - \log(\mu_{11})]$. The results for this analysis are shown below.

```
> library(package = emmeans)
> emm1 <- emmeans(object = M1, specs = ~ Treatment + Outcome)
> emm1 # log-means for each combination
 Treatment Outcome  emmean       SE  df asymp.LCL asymp.UCL
 Vaccine   Positive  2.079 0.353553 Inf     1.386     2.772
 Placebo   Positive  5.088 0.078567 Inf     4.934     5.242
 Vaccine   Negative  9.764 0.007580 Inf     9.750     9.779
 Placebo   Negative  9.761 0.007592 Inf     9.746     9.776

Results are given on the log (not the response) scale.
Confidence level used: 0.95

> # "pairwise" takes difference of log-means for pairs of
> #      levels within a variable
> # "interaction" repeats difference on differences
> #     (i.e., interaction term)
> con1 <- contrast(object = emm1, interaction = list("pairwise"))
> confint(object = con1, type = "response")
Treatment_pairwise Outcome_pairwise      ratio     SE  df
Vaccine / Placebo  Positive / Negative 0.0492 0.0178 Inf
```

```
asymp.LCL asymp.UCL
   0.0242       0.1

Confidence level used: 0.95
Intervals are back-transformed from the log scale
```

The estimate and Wald confidence intervals match those calculated before.

Finally, the LRT for $H_0 : \beta_{22}^{GC} = 0$, $H_a : \beta_{22}^{GC} \neq 0$, which is equivalent to $H_0 : OR = 1$, $H_a : OR \neq 1$, is provided using the `Anova()` function.

```
> library(package = car)
> Anova(M1)
Analysis of Deviance Table (Type II tests)

Response: Freq
                  LR Chisq Df Pr(>Chisq)
Treatment                0  1     0.5926
Outcome              46262  1     <2e-16 ***
Treatment:Outcome      171  1     <2e-16 ***
---
Signif. codes:  0 '***' 0.001 '**' 0.01 '*' 0.05 '.' 0.1 ' ' 1
```

The test statistic is $-2\log(\Lambda) = 171$ with a tiny p-value, indicating that the odds (and hence probability) of COVID-19 after vaccination are significantly different than after placebo.

4.2.5 Large loglinear models

Loglinear models can be extended to model counts cross-classified by arbitrarily many categorical variables. A table of counts corresponding to the cross-classification of p categorical variables is called a *p-way table*. For example, consider the three-way table consisting of three categorical variables, X, Z, and W with I, J, and K levels, respectively. We can extend our loglinear model by adding a subscript for variable W to the means and observed counts: y_{ijk} has mean μ_{ijk}, $i = 1, \ldots, I$; $j = 1, \ldots, J$; $k = 1, \ldots, K$. The counts in a three-way table can be thought of as being laid out in a cube with dimension $I \times J \times K$. In other words, we can imagine a stack of K separate $I \times J$ tables, each summarizing the counts for X and Z at one level of W.

To analyze this structure, we first consider expanding the independence model in Equation 4.4. We add indicator variables for W—$w_2, w_3, \ldots, w_K$—defined analogously to those for X and Z. Their corresponding regression parameters, $\beta_2^W, \ldots, \beta_K^W$, affect the W-marginal counts of a fitted three-way table—summed across all levels of both X and Z— so that ratios of means from levels k and k' of W are constant. Thus, $\log(\mu_{ijk}/\mu_{ijk'}) = \exp(\beta_k^W - \beta_{k'}^W)$, across all levels of X and Z. This model also represents independence among all three variables, because any odds ratio between two levels of one variable and two levels of another variable is 1 at a fixed level of the third. Because of this, it is sometimes called the *mutual independence* model.

Models that allow for associations among the three variables are developed depending on what associations are of interest or expected to be present. Some of these are summarized in Table 4.5. There are now three pairwise interactions to consider: XZ, XW, and ZW. Each

Table 4.5: Common models and their associated odds ratios for a three-way loglinear model with variables $X, Z,$ and W.

Model	Notation	XZ odds ratio at level k of W
Mutual independence	X, Z, W	1 for all k
Homogeneous association	XZ, XW, ZW	$\exp(\beta_{ij}^{XZ} + \beta_{i'j'}^{XZ} - \beta_{i'j}^{XZ} - \beta_{ij'}^{XZ})$ for all k
Saturated	XZW	$\exp[(\beta_{ij}^{XZ} + \beta_{i'j'}^{XZ} - \beta_{i'j}^{XZ} - \beta_{ij'}^{XZ})$
(Heterogeneous association)		$+ (\beta_{ijk}^{XZW} + \beta_{i'j'k}^{XZW} - \beta_{i'jk}^{XZW} - \beta_{ij'k}^{XZW})]$

is formed by adding products of the indicator variables to the regression model as described on p. 269. For example, to create the ZW interaction, we add the terms $z_j w_k$, $j = 2, \ldots, J$; $k = 2, \ldots, K$ with corresponding parameters β_{jk}^{ZW} to the model containing all of the main-effect indicator variables for X, Z, and W. Each set of interaction parameters allows odds ratios for the corresponding pair of variables to be something other than 1. If all three pairwise interactions are included in the model, we have

$$\log(\mu_{ijk}) = \beta_0 + \beta_i^X + \beta_j^Z + \beta_k^W + \beta_{ij}^{XZ} + \beta_{ik}^{XW} + \beta_{jk}^{ZW}. \quad (4.8)$$

To see the odds ratios for this model, consider computing OR for levels i, i' of X and j, j' of Z at level k of W:

$$OR_{ii',jj'(k)} = \frac{\mu_{ijk}\mu_{i'j'k}}{\mu_{i'jk}\mu_{ij'k}}.$$

Applying Equation 4.8, we have

$$\begin{aligned} \log(OR_{ii',jj'(k)}) &= \log(\mu_{ijk}) + \log(\mu_{i'j'k}) - \log(\mu_{i'jk}) - \log(\mu_{ij'k}) \\ &= \beta_{ij}^{XZ} + \beta_{i'j'}^{XZ} - \beta_{i'j}^{XZ} - \beta_{ij'}^{XZ} \end{aligned} \quad (4.9)$$

All other parameters cancel except those involved in the interaction from which the odds ratio of interest is computed. The same happens with any pair of categorical variables at any pairs of levels. Therefore, this model allows all odds ratios corresponding to any pair of variables to vary freely across levels of those two variables. Notice, however, that a given odds ratio is forced to be constant across levels of the third variable. For example, the odds ratio above is $\exp(\beta_{ij}^{XZ} + \beta_{i'j'}^{XZ} - \beta_{i'j}^{XZ} - \beta_{ij'}^{XZ})$, regardless of the level k of W. Thus, this model is referred to as the *homogeneous association* model. Versions of this model can be considered in which only one or two of the interactions are present, allowing associations to be modeled between some pairs of variables and assuming independence between others. In all cases, tests and estimates are obtained using the same tools as before.

Finally, interactions of higher order are possible. The three-variable interaction allows associations between any pair of variables to vary across levels of the third. Terms of the form β_{ijk}^{XZW} are added to Equation 4.8, representing parameters attached to three-way cross products of indicators, $x_i z_j w_k$. These terms enter into any odds ratios. For example, the odds ratio for levels i and i' of X and levels j and j' of Z at level k of W becomes

$$\begin{aligned} \log(OR_{ii',jj'(k)}) &= \log(\mu_{ijk}) + \log(\mu_{i'j'k}) - \log(\mu_{i'jk}) - \log(\mu_{ij'k}) \\ &= \beta_{ij}^{XZ} + \beta_{i'j'}^{XZ} - \beta_{i'j}^{XZ} - \beta_{ij'}^{XZ} \\ &\quad + \beta_{ijk}^{XZW} + \beta_{i'j'k}^{XZW} - \beta_{i'jk}^{XZW} - \beta_{ij'k}^{XZW}, \end{aligned} \quad (4.10)$$

which shows that the XZ odds ratios have the potential to vary depending on the level k of W. While this form may seem complicated, it is predictable and not difficult to estimate in R.

In models with more than three variables, terms of even higher order are technically possible. Sets of interaction parameters, indexed by the names and levels of the variables, can be created to correspond to any combination of variables. However, higher-order interactions are difficult to interpret. A blanket interpretation can be developed recursively as follows: a three-variable interaction allows two-variable interactions (i.e., odds ratios) to vary depending on the level of the third variable; a four-variable interaction allows three-variable interactions to vary depending on the level of the fourth variable; and so on. Understanding what these statements imply in the context of a particular problem is a different matter. Careful tabulation or graphing of odds ratios across levels of other variables can help to extract the underlying meaning from a higher-order interaction.

Often, the goal of an analysis is to determine which associations, if any, are important, and to subsequently estimate odds ratios for these associations. Generally, it is accepted that the main-effect terms that control the marginal counts should remain in the model when two-variable and higher-order interaction terms are the focus of study.[9] Furthermore, the principle of *marginality* dictates that all lower-order terms that are subsets of any higher-order terms should also remain in the model. For instance, the XZW interaction contains XZ, XW, and ZW, as well as X, Z, and W. If a model contains XZW, then it also contains these other six terms. Following this convention, we can use abbreviated notation that identifies any model by listing only the highest-order terms in which each variable appears in the model. Our current example is denoted as "Model XZW." Writing models in this way also corresponds to the terms that need to be specified in the `formula` argument of `glm()` in order to fit the model. For example, `formula = count ~ X*Z*W` would fit this model, where `count` denotes a cell count. As another example, "Model X, ZW" would consist of terms for X and ZW, and by marginality, also Z and W. This model specifies independence between X and W and between X and Z, but allows for association between Z and W that is homogeneous across levels of X. This model can be fit using `formula = count ~ X + Z*W`.

In most cases, the exact nature of the association structure is not known in advance, so model terms for associations need to be selected through statistical analysis. How to select these terms is the subject of some debate. Procedures analogous to forward selection or backward elimination methods that were once popular in regression analysis (Kutner et al., 2004) are sometimes recommended (e.g., Agresti, 2019). While significance tests can tell us whether a particular hypothesis should or should not be rejected, multiple and sequential use of significance tests to repeatedly select or eliminate associations according to their estimated size violates the probability foundations behind the definition of type I error (Burnham and Anderson, 2002). This procedure should therefore be used as a last resort when no better methods are available, and the level of α used should not be interpreted literally as the probability of any particular type I error. We discuss methods of model selection in Chapter 5.

[9]Leaving main effects out of the model in `glm()` when the corresponding interaction remains in the model results in a completely different interpretation of the interaction term. To see this, try running `M2 <- glm(formula = Freq ~ Treatment + Treatment:Outcome, family = poisson(link = "log"), data = all.data)` with the COVID-19 Pfizer vaccine example and examine the results. In particular, look at `summary(M2)`, `model.matrix(M2)`, and `predict(M2)`.

Example: Political ideology (PolIdeolNominal.R, PolIdeolData.csv)

Agresti (2007) presents a summary from the 1991 U.S. General Social Survey[10] cross-classifying people according to

- Political ideology, a five-level variable with levels 1: "Very Liberal" ("VL"), 2: "Slightly Liberal" ("SL"), 3: "Moderate" ("M"), 4: "Slightly Conservative" ("SC"), and 5: "Very Conservative" ("VC")
- Political party, a two-level variable with levels "Republican" ("R") and "Democrat" ("D")
- Gender, a two-level variable[11] with levels "Female" ("F") and "Male" ("M"); no other genders reported.

We fit to these data various models that differ in their interpretation of which variables are associated and whether the associations are homogeneous. Notice that the ideology variable is ordinal. In the next section, we explore how that structure can be exploited to provide a more parsimonious description of the association. For now, we treat ideology as nominal. The data are shown below, where we can see that there is a party-by-ideology table for each level of gender.

```
> alldata <- read.csv("C:\\Data\\PolIdeolData.csv",
    stringsAsFactors=TRUE)
> head(alldata)
  gender party ideol count
1      F     D    VL    44
2      F     D    SL    47
3      F     D     M   118
4      F     D    SC    23
5      F     D    VC    32
6      F     R    VL    18

> levels(alldata$ideol)
[1] "M"  "SC" "SL" "VC" "VL"
> #Reorder levels
> alldata$ideol <- factor(alldata$ideol, levels =
    levels(alldata$ideol)[c(5,3,1,2,4)])
> levels(alldata$ideol)
[1] "VL" "SL" "M"  "SC" "VC"

> # Show tables of the data
> tab.gpi <- xtabs(formula = count ~ party + ideol + gender,
    data = alldata)
> tab.gpi
, , gender = F
```

[10]The source of these data is a large-scale survey with a complex probability-based sampling design. Failing to account for the survey design in the analysis could lead to biased parameter estimates and misrepresentation of the true relationships among the variables. Proper techniques for analyzing data from this type of survey are presented in Section 6.3 and should be applied here. The analyses we present here are used solely as an illustration of the present analysis techniques and as a comparison to previously published work.

[11]In 1991, most surveys assumed a binary representation of gender.

```
      ideol
party  VL  SL   M  SC  VC
    D  44  47 118  23  32
    R  18  28  86  39  48

, , gender = M

      ideol
party  VL  SL   M  SC  VC
    D  36  34  53  18  23
    R  12  18  62  45  51
```

The models we consider are all Poisson regressions with the variables gender (G), party (P), and ideol (I) present as main effects. The models differ in how the interactions are handled. The saturated model (GPI) is

$$\log(\mu_{ijk}) = \beta_0 + \beta_i^G + \beta_j^P + \beta_k^I + \beta_{ij}^{GP} + \beta_{ik}^{GI} + \beta_{jk}^{PI} + \beta_{ijk}^{GPI},$$

where $i = F, M$, $j = D, R$, and $k = 1, \ldots, 5$. The homogeneous association model (GP, GI, PI) is given by

$$\log(\mu_{ijk}) = \beta_0 + \beta_i^G + \beta_j^P + \beta_k^I + \beta_{ij}^{GP} + \beta_{ik}^{GI} + \beta_{jk}^{PI}.$$

We consider also the model that assumes mutual independence among all variables (G, P, I),

$$\log(\mu_{ijk}) = \beta_0 + \beta_i^G + \beta_j^P + \beta_k^I.$$

One more model is considered that is intermediate to the last two. We feel fairly certain that a PI association should exist because differing ideologies is part of what defines the two parties. Therefore, we expect Model G, P, I to provide a poor fit, so we consider an augmentation of it to allow an association for PI (Model G, PI). Comparing this model to Model GP, GI, PI will allow us to test whether the other two associations, whose importance is uncertain, can simultaneously be dropped from the model.

We fit all models using glm() and compare models using anova().

```
> # Saturated Model: GPI
> mod.sat <- glm(formula = count ~ gender*party*ideol, family =
    poisson(link = "log"), data = alldata)

> # Homogeneous association model in all 3 associations: GP,GI,PI
> mod.homo <- glm(formula = count ~ (gender + party + ideol)^2,
    family = poisson(link = "log"), data = alldata)
> anova(mod.homo, mod.sat, test = "Chisq")
Analysis of Deviance Table
Model 1: count ~ (gender + party + ideol)^2
Model 2: count ~ gender * party * ideol
  Resid. Df Resid. Dev Df Deviance P(>|Chi|)
1         4     3.2454
2         0     0.0000  4   3.2454    0.5176

> # Model assuming only PI association: G,PI
> mod.homo.PI <- glm(formula = count ~ gender + party*ideol,
    family = poisson(link = "log"), data = alldata)
```

Table 4.6: Deviance analysis of party-ideology association. The test in each row uses that row's model as the null hypothesis and the model directly above it as the alternative.

Model	Residual deviance	Residual df	LRT statistic $-2\log(\Lambda)$	LRT df	p-value
GPI	0	0			
GP, GI, PI	3.25	4	3.25	4	0.52
G, PI	17.52	9	14.27	5	0.01
G, P, I	79.85	13	62.33	4	≈ 0

```
> anova(mod.homo.PI, mod.homo, test = "Chisq")
Analysis of Deviance Table
Model 1: count ~ gender + party * ideol
Model 2: count ~ (gender + party + ideol)^2
  Resid. Df Resid. Dev Df Deviance P(>|Chi|)
1         9    17.5186
2         4     3.2454  5   14.273   0.01396 *

> # Model assuming mutual independence: G,P,I
> mod.indep <- glm(formula = count ~ gender + party + ideol,
    family = poisson(link = "log"), data = alldata)
> anova(mod.indep, mod.homo.PI, test = "Chisq")
Analysis of Deviance Table
Model 1: count ~ gender + party + ideol
Model 2: count ~ gender + party * ideol
  Resid. Df Resid. Dev Df Deviance P(>|Chi|)
1        13     79.851
2         9     17.519  4   62.333 9.376e-13 ***
```

The model comparisons are summarized in Table 4.6. For example, the comparison of the last two models tests $H_0 : \beta_{jk}^{PI} = 0, j = D, R, k = 1, \ldots, 5$ against H_a : Not all $\beta_{jk}^{PI} = 0$. Model GP, GI, PI appears to fit reasonably well relative to the saturated model ($-2\log(\Lambda) = 3.25$ and p-value $= 0.52$), indicating that all two-variable associations seem to be reasonably consistent across levels of the third variable. The GP and GI associations seem to contribute something to the fit of the model, as indicated by the somewhat poor fit of G, PI relative to GP, GI, PI ($-2\log(\Lambda) = 14.27$ and p-value $= 0.01$). Clearly, the independence model G, P, I does not fit well, as expected ($-2\log(\Lambda) = 62.33$ and p-value ≈ 0), showing the importance of the PI association. In view of these results, we consider GP, GI, PI as the working model from which to examine these associations further.

Our next step is to obtain a summary of the model parameter estimates so that we can see the order in which they are listed.[12] We also perform tests of each term in the model using `Anova()`.

```
> round(summary(mod.homo)$coefficients, digits = 3)
            Estimate Std. Error z value Pr(>|z|)
(Intercept)    3.841      0.139  27.673    0.000
```

[12]Although we can predict this ordering based on knowing the factor levels and the order in which they appear in the fitted model, examining them can help with forming the linear combinations that are needed for odds ratio estimation.

```
genderM              -0.332        0.197   -1.686      0.092
partyR               -1.105        0.225   -4.909      0.000
ideolSL               0.067        0.190    0.352      0.725
ideolM                0.899        0.162    5.534      0.000
ideolSC              -0.740        0.226   -3.279      0.001
ideolVC              -0.407        0.207   -1.970      0.049
genderM:partyR        0.276        0.147    1.878      0.060
genderM:ideolSL      -0.136        0.265   -0.513      0.608
genderM:ideolM       -0.372        0.227   -1.636      0.102
genderM:ideolSC       0.163        0.269    0.604      0.546
genderM:ideolVC       0.076        0.257    0.297      0.767
partyR:ideolSL        0.424        0.283    1.497      0.134
partyR:ideolM         0.861        0.243    3.548      0.000
partyR:ideolSC        1.687        0.287    5.875      0.000
partyR:ideolVC        1.563        0.273    5.731      0.000

> library(package = car)
> Anova(mod.homo)
Analysis of Deviance Table (Type II tests)

Response: count
             LR Chisq Df Pr(>Chisq)
gender         20.637  1  5.551e-06 ***
party           0.528  1    0.46736
ideol         154.131  4  < 2.2e-16 ***
gender:party    3.530  1    0.06026 .
gender:ideol    8.965  4    0.06198 .
party:ideol    60.555  4  2.218e-12 ***
```

The LRTs for model terms show that the PI association is highly significant, but there is only marginal evidence for the existence of the other two (both p-values are 0.06).[13]

With the order of parameter estimates now clear, we can use them to estimate odds ratios for the three associations, GP, GI, and PI. Because associations are homogeneous, we need not consider a third variable when estimating an odds ratio between the other two. Each odds ratio is related to the model parameters according to Equation 4.9. For GP, we construct the ratio of the odds of being Republican for males vs. females:

$$
\begin{aligned}
\log(OR) &= \log\left(\frac{\mu_{MR}/\mu_{MD}}{\mu_{FR}/\mu_{FD}}\right) \\
&= \beta_{MR}^{GP} - \beta_{MD}^{GP} - \beta_{FR}^{GP} + \beta_{FD}^{GP} \\
&= \beta_{MR}^{GP}
\end{aligned}
$$

The last equality results from the fact that $\beta_{MD}^{GP} = \beta_{FR}^{GP} = \beta_{FD}^{GP} = 0$ because "F" represents level 1 of `gender` and "D" represents level 1 of `party`. Thus, the estimated odds ratio is $\exp(\hat{\beta}_{MR}^{GP}) = \exp(0.276) = 1.3$. For GI, we will construct the ratio of odds of being in a more liberal ideology relative to a less liberal ideology for males vs. females. As an example, we compare the odds of being Very Liberal vs. Slightly

[13]It turns out that both terms *are* significant at $\alpha = 0.05$ if the other term is first removed from the model. This is an example of the difficulty with using hypothesis tests as a model-selection tool. See Chapter 5 for alternative methods of model selection.

Liberal for males, against the same ideology odds for females. For this example,

$$\log(OR) = (\beta^{GI}_{M1} - \beta^{GI}_{M2}) - (\beta^{GI}_{F1} - \beta^{GI}_{F2}) = -\beta^{GI}_{M2},$$

where, as a reminder, 1=VL and 2=SL. We use a similar treatment of PI, comparing the odds of being in a more liberal ideology for Republicans vs. Democrats. Here, we expect that Republicans are more conservative, so we expect the log of the odds ratio to be negative and the magnitude to increase with increasing difference between the two ideologies being compared.

We can form Wald intervals for the odds ratios using the `emmeans` package. The `emmeans()` function creates estimates of linear predictors (log-means) for all combinations of variables listed in the `specs` argument. These estimates automatically incorporate information about any interactions in the model, so the user does not need to manually identify which model parameters are required in each estimate. Below is partial output from the table it produces when we ask for linear predictors from all combinations of all three variables.

```
> library(package = emmeans)
> emm1 <- emmeans(object = mod.homo, specs = ~ gender + party +
    ideol)
> head(summary(emm1))
 gender party ideol emmean    SE  df asymp.LCL asymp.UCL
 F      D     VL      3.84 0.139 Inf      3.57      4.11
 M      D     VL      3.51 0.160 Inf      3.20      3.82
 F      R     VL      2.74 0.212 Inf      2.32      3.15
 M      R     VL      2.68 0.215 Inf      2.26      3.10
 F      D     SL      3.91 0.133 Inf      3.65      4.17
 M      D     SL      3.44 0.161 Inf      3.13      3.76

Results are given on the log (not the response) scale.
Confidence level used: 0.95
```

To obtain odds ratios and confidence intervals for two factors averaged across all levels of other factors, we list only those two factors in `specs`. This makes sense when there is no three-way interaction so that these odds ratios are the same at all levels of the third factor. The specific odds ratio we want to compute here compares the odds of Republican to Democrat for males, divided by the same ratio for females, which can be expressed symbolically as "R/D for M/F." Therefore, in the `specs` argument formula, we should enter `party` before `gender`. Then in the `contrast()` function, requesting `interaction = "pairwise"` or `"revpairwise"` provides all possible comparisons of log-odds involving levels of the two factors, either in the order in which they are stored or in the opposite order, respectively. Because the order for `party` is `D,R` and for `gender` is `F,M`, we want to use reverse pairwise comparisons on both variables. The results are shown below.

```
> emm.GP1 <- emmeans(object = mod.homo, specs = ~ party + gender)
> # Linear predictors for each combination with unadjusted 95%
    CIs.
> summary(emm.GP1)
 party gender emmean     SE  df asymp.LCL asymp.UCL
 D     F        3.80 0.0692 Inf      3.67      3.94
 R     F        3.61 0.0768 Inf      3.46      3.76
```

```
D      M         3.42 0.0817 Inf       3.26        3.58
R      M         3.50 0.0804 Inf       3.34        3.65

Results are averaged over the levels of: ideol
Results are given on the log (not the response) scale.
Confidence level used: 0.95

> aa.GP1 <- contrast(object = emm.GP1, interaction =
   list("revpairwise", "revpairwise"))
> confint(object = aa.GP1, type = "response")
 party_revpairwise gender_revpairwise ratio    SE  df
 R / D                  M / F           1.32 0.193 Inf
 asymp.LCL asymp.UCL
     0.988      1.76

Results are averaged over the levels of: ideol
Confidence level used: 0.95
Intervals are back-transformed from the log scale
```

Notice that the output from `confint()` lists the variable levels the same way we expressed them symbolically above, R/D for M/F. The estimated odds ratio is 1.32, indicating that the R/D ratio is estimated to be 32% higher for males than for females. However, the 95% Wald interval for the odds ratio is 0.99 to 1.76. Because this interval barely includes 1, there is just marginal evidence of a difference between the odds of Republican among males and the odds of Republican among females.

To get LR confidence intervals for specific odds ratios, we use Equation 4.9 to identify the four log-means that make up each log odds ratio. To use `mcprofile()`, we need to create vectors of contrast coefficients whose entries are are 0, 1, or -1 as determined by the $\log(OR)$ calculations. There is one odds ratio for GP, 10 for GI, and 10 for PI. This leads to 21 sets of coefficients for linear combinations on the 16 parameters listed in the output above. The first seven parameters do not contribute to the odds ratios, so their coefficients are all 0. This leads to the contrast matrices, such as `contr.mat.GI`, shown in the program accompanying this example. With these matrices, we can get profile LR confidence intervals using `mcprofile()`. In particular, we put coefficients of $+1$ on log-means from the numerator, -1 on log-means from the denominator, and 0 on all other log-means. We then perform the algebraic additions and subtractions to determine the coefficients on individual parameters that form one row of the coefficient matrix. For example, to get the ratio of the odds of being Republican for males vs. females for very liberal people,

$$\log(OR_{MF,RD,1}) = \log\left(\frac{\mu_{MR1}\mu_{FD1}}{\mu_{FR1}\mu_{MD1}}\right),$$

we need a $+1$ on the `gender = "M", party = "R", ideol = "VL"` and `gender = "F", party = "D", ideol = "VL"` log-means, and a -1 on the `gender = "F", party = "R", ideol = "VL"` and `gender = "M", party = "D", ideol = "VL"` log-means. The same process can be repeated for each level of ideology, resulting in five separate estimates and confidence intervals. However, because we are working with the homogeneous association model, the GP odds ratio is the same at all levels of I. Thus, all five estimates and confidence intervals will be identical. The program for this example confirms this fact.

Both the `mcprofile` and `emmeans` packages allow for computations with or without adjustment for simultaneous confidence levels. For simultaneous inference, we need to define a "family" of intervals about which we want our confidence statements to apply. While we might like to have 95% confidence that all confidence intervals in a large analysis will contain their parameters, the drawback of a large family size is that the intervals become wider. For example, if we chose all 21 intervals to be one family here, then the confidence level for each interval using the Bonferroni adjustment would be $1-\alpha' = 1-0.0024 = 0.9976$ (hence $Z_{1-\alpha'} = 3.04$). Instead, it seems sensible to make a separate confidence statement about the odds ratios for each of the three interaction effects. Thus, the GP family contains 1 interval, while the GI and PI families each contain 10. A Bonferroni adjustment in the latter families would use $1-\alpha' = 1-0.005 = 0.995$ (hence $Z_{1-\alpha'} = 2.81$) as the confidence level instead, leading to somewhat narrower intervals. Other more complex methods to control the familywise confidence level can achieve the same goal with narrower intervals. We next calculate the Wald intervals for the GI family with a 95% familywise confidence level using the multivariate t method adjustment, which is the same as the single-step method used by `mcprofile()` (see Section 2.2.5).

```
> emm.GI <- emmeans(object = mod.homo, specs = ~ ideol + gender)
> aa.GI <- contrast(object = emm.GI, interaction =
   list("revpairwise", "revpairwise"))
> ci.GI <- confint(object = aa.GI, type = "response", adjust =
   "mvt")
> # Making column names shorter for printing
> colnames(ci.GI) <- c("ideol", "gender", colnames(ci.GI)[3:7])
> cbind(ci.GI[1:2],round(ci.GI[c(3,4,6,7)],2))
     ideol gender ratio   SE asymp.LCL asymp.UCL
1  SL / VL  M / F  0.87 0.23      0.43      1.79
2   M / VL  M / F  0.69 0.16      0.37      1.28
3   M / SL  M / F  0.79 0.17      0.44      1.42
4  SC / VL  M / F  1.18 0.32      0.57      2.45
5  SC / SL  M / F  1.35 0.35      0.67      2.72
6   SC / M  M / F  1.71 0.37      0.95      3.07
7  VC / VL  M / F  1.08 0.28      0.54      2.17
8  VC / SL  M / F  1.24 0.30      0.63      2.41
9   VC / M  M / F  1.57 0.31      0.91      2.70
10 VC / SC  M / F  0.92 0.22      0.48      1.77
```

All odds ratios compare a more conservative ideology to a less conservative one for males vs. females. We see that the estimated odds ratios are all relatively close to 1, with adjusted confidence intervals that all contain 1, so we do not find reason to believe that males are systematically more or less conservative than females. Furthermore, we are 95% confident that this procedure produces a set of intervals that *all* contain their respective parameters.

Loglinear vs. logistic and multinomial models

We have seen that the loglinear model and the multinomial regression model in Equation 3.10 can both be applied to contingency tables. Is there a clear preference for one or the other?

In many cases the answer is no. The two approaches for modeling counts in a contingency table give identical answers because of a mathematical relationship between the Poisson and

Table 4.7: Table and parameter structures for logistic regression (top) and loglinear model (bottom) when X is a 2-level response variable and A and B are each 2-level explanatory variables.

		$B=1$				$B=2$		
		$X=0$	$X=1$			$X=0$	$X=1$	
A	1	π_{11}	$1-\pi_{11}$	1	1	π_{12}	$1-\pi_{12}$	1
	2	π_{21}	$1-\pi_{21}$	1	2	π_{22}	$1-\pi_{22}$	1

		$B=1$				$B=2$		
		$X=0$	$X=1$			$X=0$	$X=1$	
A	1	μ_{111}	μ_{112}	μ_{11+}	1	μ_{121}	μ_{122}	μ_{12+}
	2	μ_{211}	μ_{212}	μ_{21+}	2	μ_{221}	μ_{222}	μ_{22+}

multinomial distributions. On the one hand, the multinomial distribution assumes that the total count in the entire table was fixed by design (e.g., by specifying a total sample size prior to data collection). On the other hand, the Poisson model assumes that the total is a random quantity, because it is a sum of independent random variables. However, it turns out that the joint probability distribution for any set of independent Poisson random variables, conditioned on their sum, is a multinomial distribution. That is, if we treat the total count from a contingency table as a fixed quantity, the two models are identical. Furthermore, there is never any harm in treating this sum as fixed, because it is not a quantity that we attempt to model. Therefore, the two models can be used interchangeably in most settings.

A unique feature of loglinear models is that they can be used not only in cases where there is one particular variable that is considered a "response" variable, but also in cases where there are multiple response variables or none at all. Logistic regression and multinomial models require one response variable whose probabilities are to be modeled. In a loglinear model analysis with one response variable, the main interest is in modeling associations between that variable and the others. It is typical in that case to include in any model all main effects and interactions involving the explanatory variables and not explore these effects any further. For example, if the response variable is X and the explanatory variables are A, B, C, then the minimal model to be fit would be Model ABC, X.

To see why this is needed, consider the case where the response variable X has two levels, and there are two explanatory variables, A and B, each with two levels. The loglinear model assumes that a count at level i of A, level j of B, and level k of X is distributed as $Y_{ijk} \sim Po(\mu_{ijk})$. On the other hand, a logistic regression model assumes that these counts accrue through observing repeated Bernoulli trials of X at each level of A and B where each trial has probability of success π_{ij}. The table structures and parameters used by the two models are laid out in Table 4.7. The saturated loglinear model is

$$\log(\mu_{ijk}) = \beta_0 + \beta_i^A + \beta_j^B + \beta_{ij}^{AB} + \beta_k^X + \beta_{ik}^{AX} + \beta_{jk}^{BX} + \beta_{ijk}^{ABX},\ i=1,2;\ j=1,2,\ k=1,2,$$

while the saturated logistic regression model written in an ANOVA format specifies

$$\log\left(\frac{\pi_{ij}}{1-\pi_{ij}}\right) = \lambda + \lambda_i^A + \lambda_j^B + \lambda_{ij}^{AB},\ i=1,2,\ j=1,2,$$

where we use λ's to represent the regression parameters here to avoid confusion. The parameters from these models have some common interpretations:

- Recall that main effects in the logistic regression model (λ_i^A and λ_j^B) represent associations between that variable and the response variable X. In the loglinear model, two-variable interactions with X (β_{ik}^{AX} and β_{jk}^{BX}) represent the same associations. Odds ratios in the logistic model are found from exponentiated differences in main effect parameters. In the loglinear model, the same odds ratios are found from exponentiated $(1,-1,-1,1)$ contrasts among the X-interaction parameters. It can be shown that *odds ratios calculated from these two models are identical.*

- In the logistic regression model, the λ_{ij}^{AB} parameters control the heterogeneity of association between X and each other variable. In the loglinear model, the β_{ijk}^{ABX} parameters serve the same purpose. In both models, estimated odds ratios between either A and X or B and X computed separately for each level of the other explanatory variable are the same as the corresponding observed odds ratios in the tables. Also, in both models, absence of the highest-order interaction results in a model with homogeneous association.

The end result is that any logistic regression model for categorical explanatory variables can be written as an equivalent loglinear model with (a) all main effects and interactions among the explanatory variables, (b) a main effect for the response variable, and (c) an interaction with the response for each term in the logistic model (e.g., an A effect in the logistic model produces an AX effect in the loglinear model). If this is done, then the loglinear association parameters β_{ik}^{AX} and β_{jk}^{BX}, $i=1,\ldots,I$; $j=1,\ldots,J$, $k=1,2$ are *identical* to the logistic regression parameters λ_i^A and λ_j^B, respectively, and the heterogeneity parameters β_{ijk}^{ABX} are identical to the interaction parameters λ_{ij}^{AB}.[14]

To fit a Poisson-based model as an alternative to a multinomial regression, each observed set of J multinomial counts is converted into J separate observed counts, each with the same values for the explanatory variables as the original multinomial count had. Then, a new categorical explanatory variable is created consisting of the levels of the response categories. The response variable for each observation is the observed count for that observation's explanatory variables and response category. The model then contains all of the main effects of the explanatory variables, all interactions for these explanatory variables, and the new categorical variable for the response category. The main interests lie in determining which interactions to include between the response and explanatory variables.

Because the translation to odds ratios and probabilities is a little bit simpler in the logistic regression model, we tend to prefer it over the loglinear model when there is a two-level response variable. Multinomial regression models are not as easy to use and interpret as logistic regression, so the choice between loglinear and multinomial models when a response variable has more than two levels is not as clear. The advantage of the loglinear formulation is that much simpler model creation, diagnostic, and correction tools can be used, as discussed in Chapter 5. Also, when there is an ordinal response variable, the loglinear model is more flexible than the multinomial regression model in terms of the kinds of associations that can be modeled, as is described next.

4.2.6 Ordinal categorical variables

Recall that our political ideology has an explanatory variable, `ideol`, that has a natural ordering to its levels. In Section 3.4, we saw one approach to modeling counts when a response variable is ordinal and arises from a multinomial distribution. Here, we provide

[14]For example, the code `logit.homo <- glm(formula = party ~ gender + ideol, family = binomial(link = "logit"), weights = count, data = alldata)` estimates the same model as `mod.homo` in the Political ideology example.

an alternative approach using a Poisson regression model. This model also works when one or more explanatory variables are ordinal or when there is no designated response variable.

There are two general approaches to analyzing counts at different levels of an ordinal variable: ignore the ordinality or assign numerical scores to represent the ordered levels. In the former case, the analyses of the previous sections are applied, but information relating the mean counts and odds ratios to the ordering of the levels is lost. This is not an efficient or informative type of analysis if the mean counts or odds ratios vary in some systematic manner with the ordering. Indeed, in the example in the previous section, we saw that analyzing the associations between the ordinal ideology variable and the nominal variables was somewhat cumbersome. We therefore try to incorporate the ordinal structure of variables into an analysis whenever we can. In a Poisson regression model, assigning numerical scores to represent levels of an ordinal variable allows us to treat the ordinal variable like a numerical variable. Once again, methods from the previous sections are used for the analysis.

Thus, the presence of ordinal variables represents no difficulty for Poisson regression modeling. We therefore focus of the selection of scores and the interpretation of the results. We also describe a procedure to check whether the choice of scores has failed to uncover some important aspect of the relationship between the ordinal variable and the counts.

Choosing Scores

Assigning scores is subjective, primarily because there is typically no single scoring system that is obviously better than all others. For example, ordered categorical data often arises from questionnaires that ask a respondent to rate something on a scale like "Excellent", "Very Good", "Good", "Fair", and "Poor." If one assigns numerical scores to represent these levels, what should the scores be? Should we choose 1-2-3-4-5, or 5-4-3-2-1, or some other sequence? Does it matter? The answer varies depending on the problem. Different scores can sometimes heavily influence the results of the analysis. Other times, all reasonable choices of scores lead to essentially the same results.

Wherever possible, scores should be chosen in consultation with the subject-matter specialist whose questions are being answered. The important feature in any set of scores is the relative sizes of the spacings, or *gaps*, between scores. The more similar two categories are, the narrower the gap should be between their scores. For the previous example, if we think that the notion behind "Fair" is closer to "Good" than to "Poor", then the gap between "Good" and "Fair" should be smaller than the gap between "Fair" and "Poor." How much closer is a matter of judgment. Whether those three scores should be 4-3-1 (gaps of 1 and 2), 5-4-1 (gaps of 1 and 3), 6-4-1 (gaps of 2 and 3), or something else entirely is often a matter of guesswork.

For choosing scores, the *relative* sizes of the different gaps matters, not the actual numbers assigned. For example, scores 1-2-3 provide the same analysis results as do 10-20-30 or 3-2-1. Of course, estimates of ratios of means will depend on the numerical values that define a "1-unit change" in score. For example, with 1-2-3 scores, a 1-unit change equals moving to an adjacent category; with 3-2-1, the movement is to an adjacent category in the opposite direction; and with 10-20-30, a 10-unit change is required to change categories, so slope parameters relating to a 1-unit change may need to be multiplied by 10 to be meaningful. We therefore generally arrange scores so that a 1-unit increase in scores relates to some interpretable concept, such as the change between two important categories. This is convenient, but not strictly necessary, because test results are identical for any sets of scores where the relative gap sizes are in the same ratios.

When there is some uncertainty in the definition of scores, it is often good practice to duplicate an analysis on several sets of reasonable scores. If the results are very similar, then the choice of scores does not matter much and conclusions can be drawn accordingly. If the results *do* change appreciably, then more exploration may be necessary to understand the cause of the difference. In any case, unless one set of scores was strongly preferred prior to seeing the analysis results, both sets of results should be reported with roughly equal consideration.

Example: Political ideology (PolIdeolOrd.R, PolIdeolData.csv)

We saw that the ideology variable in this example is ordinal, so we will attempt to use this structure to construct a more meaningful and parsimonious description of the association. As this example arises from "interesting data" rather than a problem brought to us by a political scientist, we are acting as subject-matter specialists ourselves. We choose to address two different sets of questions by using the ordering of the categories for the ideology variable in different ways. The first set of questions makes use of the fact that the five categories are naturally ordered from more liberal to less liberal (equivalently, from less conservative to more conservative). As a default, we choose 1-2-3-4-5 to represent this progression, where higher scores represent more conservative ideologies. Then a 1-unit increase represents a 1-category move toward a more conservative ideology.

However, we suspect that the "slightly" liberal and conservative groups might each be closer to the middle than to the extremes of their respective ideologies. We therefore consider scores with wider gaps between "very" and "slightly" than between "slightly" and "moderate." Somewhat arbitrarily, we choose to double the gap sizes, creating 0-2-3-4-6 scores. We certainly expect either set of scores to be associated with the political parties, as each party deliberately seeks to appeal to a different ideology. It would be interesting to see whether there is an association between gender and ideology (are men more liberal or conservative than women?) or whether the association between party and ideology is different for men and women.

The second set of questions relates to whether one party or gender holds more *extreme* views than the other. In other words, does one party or gender tend to have more people with "very" strong views as opposed to "slightly" strong or central views? To examine this, we create scores to measure extremeness of ideology: 2-1-0-1-2. This essentially creates three categories measuring strength of ideology, low, medium, and high, with 0-1-2 scores on those levels. The scores are created and added to the data as shown below.

```
> lin.score1 <- c(rep(x = c(1,2,3,4,5), times = 4))
> lin.score2 <- c(rep(x = c(0,2,3,4,6), times = 4))
> extrm.score <- c(rep(x = c(2,1,0,1,2), times = 4))
> alldata <- data.frame(alldata, lin.score1, lin.score2,
    extrm.score)
> head(alldata)
  gender party ideol count lin.score1 lin.score2 extrm.score
1      F     D    VL    44          1          0           2
2      F     D    SL    47          2          2           1
3      F     D     M   118          3          3           0
4      F     D    SC    23          4          4           1
5      F     D    VC    32          5          6           2
6      F     R    VL    18          1          0           2
```

Note that the `ifelse()` function could have been used instead to create the scores.

Models and interpretation

For a single ordered categorical variable X with I levels, let s_i^X, $i = 1, \ldots, I$ represent the set of numerical scores assigned to the levels. The ordinal Poisson regression model treats the potential count Y_i as $Po(\mu_i)$, with $\log(\mu_i) = \beta_0 + \beta_1 s_i^X$. Interpretation of the regression parameters is the same as with any numerical explanatory variable in a Poisson regression as described in Section 4.2.1. However, the "units" applied to scores usually have no physical meaning, except in those cases where the ordered categories represent intervals of a continuum. For example, if we replace categories like "Good", "Fair", and "Poor" with 3-2-1 scores, then a 1-unit change in score is merely a change in category. However, if we replace age groups "<25", "25-45", and ">45" with scores like 20-35-60, then a 1-unit change in score is approximately equivalent to a 1-year change in age.

Examining ratios of means when s changes by 1 unit may not adequately address all interests if some of the gaps between scores are more than 1. Therefore, we will often wish to examine changes for a c-unit change in s, where c represents the difference in scores between any two categories of interest. As described in Section 4.2.1, this simply means analyzing $c\beta$.

When there are two or more categorical explanatory variables, at least one of which is ordinal, then how the scores enter into the model depends on the goal of the analysis; i.e., whether the focus is on modeling and predicting means or on modeling associations between variables. If the problem is more like a typical regression and means are the primary focus, then any score sets are treated as numerical variables and the usual analyses are performed as described in Section 4.2.1. Adding interaction terms to the model is optional and depends on whether the mean ratios created by changing scores in one variable are expected to be different across levels of the other variable(s).

When an analysis focuses on testing and characterizing associations, special models are needed. Associations between categorical variables are measured by odds ratios, so we need to consider the assumptions that different models make regarding how the odds ratios are structured.

To start, consider a problem with one nominal variable, X, and one ordinal variable, Z, with scores s_j^Z, $j = 1, \ldots, J$, such as a contingency table with X as the row variable and Z as the column variable. Further, assume that the goal is to understand the associations between X and Z by testing the interaction and estimating mean ratios and odds ratios. Consider first the usual two-variable loglinear nominal model with nominal main effects and interaction, $\log(\mu_{ij}) = \beta_0 + \beta_i^X + \beta_j^Z + \beta_{ij}^{XZ}$. This saturated model has $1 + (I-1) + (J-1)$ parameters for intercept and main effects to ensure that estimated marginal totals match observed marginal counts in each row and column. The $(I-1)(J-1)$ interaction parameters assume no particular structure to the odds ratios between X and Z. In particular, as we have seen, odds ratios for two levels of each variable are modeled as $OR = \exp(\beta_{ij}^{XY} + \beta_{i'j'}^{XY} - \beta_{i'j}^{XY} - \beta_{ij'}^{XY})$. Also, notice that ignoring the ordinality with this model implies that ratios between means (i.e., the odds) at different levels of Z at a fixed level of X can be written as

$$\frac{\mu_{ij}}{\mu_{ij'}} = \exp([\beta_j^Z - \beta_{j'}^Z] + [\beta_{ij}^{XZ} - \beta_{ij'}^{XZ}]).$$

Here, the first term in the exponent is simply the mean ratio in row $i = 1$, $\exp([\beta_j^Z - \beta_{j'}^Z])$, while the second term is an adjustment to this ratio for the row i in which the comparison takes place.

To model associations using the ordinality in Z, we want a model that continues to match marginal counts, while allowing mean ratios and odds ratios to depend on differences in scores. This is achieved using the model

$$\log(\mu_{ij}) = \beta_0 + \beta_i^X + \beta_j^Z + \beta_i^{XZ} s_j^Z. \tag{4.11}$$

This model has the same intercept and main-effect parameters, but uses only $I - 1$ parameters to model the association, with a separate β_i^{XZ} for each row, $2, \ldots, I$.

To see the structure that this model imposes on associations, consider how it models the ratio of means between levels j and j' of Z at level i of X:

$$\frac{\mu_{ij}}{\mu_{ij'}} = \exp([\beta_j^Z - \beta_{j'}^Z] + \beta_i^{XZ}[s_j^Z - s_{j'}^Z]). \tag{4.12}$$

We see the same first term as in the nominal model, but now find that the second term makes structured adjustments in row i according to the difference in scores between the two levels of Z. In particular, any two levels of Z that have the same gap between their scores produce identical mean ratios in a given row. For example, if consecutive integer scores are used, then the mean ratio between each pair of consecutive levels of Z is the same.

The effect of the ordinal association on the odds ratio follows directly from Equation (4.12), because an odds ratio is simply the ratio of two mean ratios, e.g., at levels i and i' of X. The result is

$$\begin{aligned}\log(OR_{ii',jj'}) &= \log(\mu_{ij}) - \log(\mu_{ij'}) - \log(\mu_{i'j}) + \log(\mu_{i'j'}) \\ &= \beta_i^{XZ} s_j^Z + \beta_i^{XZ} s_{j'}^Z - \beta_{i'}^{XZ} s_j^Z - \beta_{i'}^{XZ} s_{j'}^Z \\ &= (\beta_i^{XZ} - \beta_{i'}^{XZ})(s_j^Z - s_{j'}^Z).\end{aligned} \tag{4.13}$$

Given the two levels of X, the log-odds ratio changes linearly as the ordinal scores on Z change. For example, suppose $J = 3$ and scores are 1-2-3, and we want to estimate odds ratios between two levels of Z across levels 1 and 2 of X. Then we have

$$\begin{aligned}\log(OR_{12,12}) &= -1(\beta_1^{XZ} - \beta_2^{XZ}) \\ \log(OR_{12,13}) &= -2(\beta_1^{XZ} - \beta_2^{XZ}) \\ \log(OR_{12,23}) &= -1(\beta_1^{XZ} - \beta_2^{XZ}).\end{aligned}$$

This model is sometimes called a *linear association* model, although the linearity is technically for the *log* of the odds ratios.

The model in Equation 4.11 allows us to model data with fewer parameters, which is always desirable, but at a cost of more assumptions about the structure of associations in the data. This is a good trade-off when the assumptions are at least approximately true, but can result in an incorrect summary of the association structure if the linear log-odds assumption is not true. Later in this section, we will see how we can test whether the added assumptions about association structure fit the data well.

Notice also that the scores are *not* used in defining the effects of Z on the marginal counts: the main effect β_j^Z is retained instead of $\beta^Z s_j^Z$. This is because we are examining whether the *associations* are affected by X and Z, and the main effect terms do not enter into the odds ratios. We therefore maintain a nominal model for each of the marginal counts. Then any difference in the overall fit of the model due to the use of linear, rather than nominal, association is caused directly by the change in the association structure, and not also by a change in the fitting of marginal means (see Exercise 9).

A test for the interaction term ($H_0 : \beta_i^{XZ} = 0$ for $i = 1, \ldots I$) for this model tests the null hypothesis of independence against the alternative of a linear change in $\log(OR)$. Failure to reject H_0 here does not mean that there is *no* association between X and Z. It means only that there is not sufficient evidence of a linear trend for $\log(OR)$. An association pattern could exist such that the odds ratios are not continually getting more extreme as the difference between their scores increases. Similarly, rejecting the null hypothesis does not imply that the association is *entirely* linear for $\log(OR)$. There can be a general increasing or decreasing trend with increasing distance between scores, but the trend does not have to be linear. Testing for goodness of fit of a model as described after the next example can help to assess the possibility for alternative patterns.

Example: Political ideology (PolIdeolOrd.R, PolIdeolData.csv)

Our work with nominal models has suggested that the three-variable interaction is not needed, so we start with a homogeneous association model and consider adding scores. To begin, we consider the 1-2-3-4-5 scores contained in `lin.score.1`, where higher scores mean greater conservativism. We apply the scores to the PI term, because we believe that the odds ratios from this association should increase as the distance between ideologies increases. We leave everything else as it was before, in particular, allowing the association between gender and ideology to remain nominal, because we do not yet have reason to believe that this association operates ordinally.

Thus, the model we fit is

$$\log(\mu_{ijk}) = \beta_0 + \beta_i^G + \beta_j^P + \beta_k^I + \beta_{ij}^{GP} + \beta_{ik}^{GI} + \beta_j^{PI} s_k^I, \tag{4.14}$$

which assumes linear association between party and ideology that is homogeneous across genders. The model assumes that a log-odds ratio comparing Republicans to Democrats at ideology levels k and k' has the form

$$\log(OR_{RD,kk'}^{PI}) = (\beta_R^{PI} - \beta_D^{PI})(s_k^I - s_{k'}^I)$$

for both genders, where the value of $s_k^I - s_{k'}^I$ is the difference in scores between the two ideologies being compared.

Notice that using consecutive integer scores (1-2-3-4-5) implies that the odds ratios involving any two consecutive ideologies are the same, since the difference is always 1 or -1, depending on the direction in which ratios are formed. More generally, odds ratios involving any two ideologies separated by the same score give the same result. For example, the odds ratio comparing the odds of Republican for very conservative vs. slightly liberal ideologies is the same as the one comparing the odds of Republican between slightly conservative and very liberal (both score differences are 3). We fit and analyze this linear PI-association model in detail below. More models are fit in the accompanying program for this example.

The terms in Equation 4.14 are specified concisely in `glm()` by using only the interaction terms as shown below.

```
> mod.homo.lin1.PI <- glm(formula = count ~ gender*party +
    gender*ideol + party*lin.score1, family = poisson(link =
    "log"), data = alldata)
> summary(mod.homo.lin1.PI)

Call: glm(formula = count ~ gender * party + gender * ideol +
    party * lin.score1, family = poisson(link = "log"), data =
    alldata)

Deviance Residuals:
     Min         1Q     Median         3Q        Max
```

```
 -1.24363   -0.54155    0.02178    0.52057    0.90547

Coefficients: (1 not defined because of singularities)
                    Estimate Std. Error z value Pr(>|z|)
(Intercept)          4.72852    0.08169  57.881  < 2e-16 ***
genderM             -0.71139    0.13656  -5.209 1.90e-07 ***
partyR              -1.51082    0.20862  -7.242 4.42e-13 ***
ideolSC             -1.40587    0.14897  -9.437  < 2e-16 ***
ideolSL             -0.83121    0.13685  -6.074 1.25e-09 ***
ideolVC             -1.41136    0.15424  -9.150  < 2e-16 ***
ideolVL             -0.89370    0.14906  -5.996 2.03e-09 ***
lin.score1                NA         NA      NA       NA
genderM:partyR       0.28723    0.14555   1.973   0.0484 *
genderM:ideolSC      0.55859    0.21413   2.609   0.0091 **
genderM:ideolSL      0.23706    0.21543   1.100   0.2711
genderM:ideolVC      0.43621    0.20135   2.166   0.0303 *
genderM:ideolVL      0.37450    0.22682   1.651   0.0987 .
partyR:lin.score1    0.43058    0.06005   7.170 7.48e-13 ***
---
Signif. codes:  0 '***' 0.001 '**' 0.01 '*' 0.05 '.' 0.1 ' ' 1

(Dispersion parameter for poisson family taken to be 1)

    Null deviance: 255.1481  on 19  degrees of freedom
Residual deviance:   8.3986  on  7  degrees of freedom
AIC: 142.83

Number of Fisher Scoring iterations: 4
```

Adding `lin.score1` into the model only through the interaction with `party` causes some difficulty for `glm()`. When we use `party*lin.score1`, `glm()` also creates the main effect for `lin.score1`. But this variable represents a duplication of `ideol` that appears earlier in the model formula, because both have the same distinct values (just coded differently) for each row of the data set. This duplication leads to the note about "singularities" in the listing of estimated model coefficients and the "`NA`" results for the `lin.score.1` main effect.[15] This redundancy and resulting lack of estimate creates no difficulty for interpretation, because we are not trying to understand the main effects of ideology score. Note that we could add only the interaction term without the main effect by using `party:lin.score1`. However, this formulation creates two interaction parameters, causing another singularity, and it estimates the interaction differently from how we expect. Therefore, we use the model with the superfluous main effect and address the issue when we compute odds ratios.

We start with tests for effects using `Anova()`.

```
> library(package = car)
```

[15]The exact cause of the "singularity" is beyond the scope of this book. Readers who are familiar with *model matrices*—for example the matrix $\mathbf{X}$ in the linear model specification $\mathbf{Y} = \mathbf{X}\boldsymbol{\beta} + \epsilon$—may appreciate the fact that the `lin.score1` variable can be written as a linear combination of `(Intercept)` and the four indicators representing the `ideol` variable. As a result, `lin.score1` contains no new information that hasn't already been accounted for by the `ideol` term, so the columns of the matrix representing this model are not linearly independent.

```
> Anova(mod.homo.lin1.PI)
Analysis of Deviance Table (Type II tests)
Response: count
                  LR Chisq Df Pr(>Chisq)
gender              20.637  1  5.551e-06 ***
party                0.528  1    0.46736
ideol              163.007  3  < 2.2e-16 ***
lin.score1                  0
gender:party         3.901  1    0.04826 *
gender:ideol         9.336  4    0.05323 .
party:lin.score1    55.402  1  9.825e-14 ***
```

We see that there is a very strong linear association between party and ideology, which we already knew, but now the LRT has just 1 degree of freedom. Tests for the other two associations yield results similar to those in the nominal case, although if one uses a rigid 0.05 significance level, GP is now significant. Practically speaking, the two tests are telling us the same thing: there is marginal evidence to suggest that GP and GI interactions are present.

The usual R functions for estimating odds ratios—`mcprofile()` and `emmeans()`—are all adversely impacted by the missing parameter estimate for `lin.score.1`. While relatively simple adjustments allow the two Wald methods to work properly, we have found no way to coerce `mcprofile()` to create estimates for this problem. We therefore focus on Wald intervals using `emmeans()`. The `emmeans()` function automatically performs all computations at the mean value of any numeric variables, so it will attempt to use the average score for all of the log-mean computations. However, the model connects this score value to at most one level of the ordinal variable—for example, the average of our scores is 3, which is the score for `ideol = "M"`—and therefore `emmeans()` will not compute linear predictors for any levels other than the one for which the average applies. See the program for an example of the results. To fix this, we use the `ref_grid()` function in the `emmeans` package. We link the numeric scores in `lin.score1` to their respective levels in the nominal variable `ideol` using the `cov.reduce` argument.[16] Using the output of this function in the `object` argument of `emmeans()` allows the remaining programming for `emmeans()` and associated functions to proceed as usual.

Because GP and GI are entirely nominal, their odds ratios and confidence intervals are calculated as shown in the previous example. For PI, `emmeans()` uses these two variables in the `specs` argument and produces log-mean estimates for combinations of these variables that automatically incorporate the information about the scores. The results are shown below.

```
> rg1 <- ref_grid(object = mod.homo.lin1.PI, cov.reduce =
    lin.score1 ~ ideol)
> emm.PI1 <- emmeans(object = rg1, specs = ~ ideol + party)
> emm.PI1
 ideol party emmean     SE  df asymp.LCL asymp.UCL
 VL    D       3.67 0.1048 Inf      3.46      3.87
 SL    D       3.66 0.0978 Inf      3.47      3.85
```

[16]The `cov.reduce` argument name refers to "covariate reduce" because it reduces the number of distinct explanatory variables—i.e., covariates—to be considered.

```
M      D      4.37 0.0684 Inf      4.24      4.51
SC     D      3.25 0.1041 Inf      3.04      3.45
VC     D      3.18 0.1226 Inf      2.94      3.42
VL     R      2.73 0.1424 Inf      2.45      3.01
SL     R      3.15 0.1092 Inf      2.94      3.37
M      R      4.30 0.0694 Inf      4.16      4.43
SC     R      3.60 0.0967 Inf      3.41      3.79
VC     R      3.97 0.0909 Inf      3.79      4.14
```

The effects of the scores on these estimates are not obvious yet, but will be apparent when we study the odds ratios.

To estimate odds ratios, we need to choose two levels of each variable. We start by comparing mean ratios at consecutive levels of `ideol` for Repubicans divided by the same mean ratio for Democrats. For example, we compute $OR_{kk'.RD}$, for combinations of (k, k') consisting of (SL,VL), (M,SL), (SC,M), and (VC,SC). We know that these comparisons all compare ideologies with a score difference of 1, so they should all give the same result. The `interaction` argument in the `contrast()` function therefore uses values `consec` for `ideol` and `revpairwise` to place `R` in the numerator and `D` in the denominator for `party`. This yields the results below.

```
> aa.PI <- contrast(object = emm.PI1, interaction =
    list("consec", "revpairwise"))
> ci.PI <- confint(object = aa.PI, type = "response")
> # Column names are too long to fit output on one line
> colnames(ci.PI) <- c("ideol", "party", colnames(ci.PI)[3:7])
> ci.PI
 ideol   party ratio     SE  df asymp.LCL asymp.UCL
 SL / VL R / D  1.54 0.0924 Inf      1.37      1.73
 M / SL  R / D  1.54 0.0924 Inf      1.37      1.73
 SC / M  R / D  1.54 0.0924 Inf      1.37      1.73
 VC / SC R / D  1.54 0.0924 Inf      1.37      1.73

Results are averaged over the levels of: gender
Confidence level used: 0.95
Intervals (GP, GI, PI)are back-transformed from the log scale
```

Indeed, all estimated odds ratios are 1.54, indicating that each one-category shift toward a more conservative ideology increases the odds of being a Republican by 54%. Alternatively, for any two consecutive ideological categories, Republicans have 54% higher estimated odds of choosing the more conservative ideology. The 95% confidence interval is $1.37 < OR < 1.73$, which is clearly distinct from 1.

With a small change, we can estimate odds ratios for other interesting comparisons of ideology, representing one-, two-, three-, or four-category changes. These can be found quickly in `emmeans()` by replacing the `consec` value in the `interaction` argument with `trt.vs.ctrl1`, which compares each level of `ideol` with the first level as shown below.

```
> aa.PI2 <- contrast(object = emm.PI1, interaction =
    list("trt.vs.ctrl1", "revpairwise"))
> ci.PI2 <- confint(object = aa.PI2, type = "response")
> # Column names are too long to fit output on one line
```

```
> colnames(ci.PI2) <- c("ideol","party",colnames(ci.PI2)[3:7])
> ci.PI2
 ideol    party ratio     SE  df asymp.LCL asymp.UCL
 SL / VL R / D   1.54 0.0924 Inf      1.37      1.73
 M / VL  R / D   2.37 0.2841 Inf      1.87      2.99
 SC / VL R / D   3.64 0.6556 Inf      2.56      5.18
 VC / VL R / D   5.60 1.3445 Inf      3.50      8.96

Results are averaged over the levels of: gender
Confidence level used: 0.95
Intervals are back-transformed from the log scale
```

The four odds ratios given above estimate how the odds of being Republican change as ideologies differ by progressively more categories. For a two-category difference, the estimated odds of being Republican increase to 2.37 times as large in the more conservative ideology ($1.87 < OR < 2.99$); for three categories the estimated odds ratio is 3.64 ($2.56 < OR < 5.18$); and finally the estimated odds of being Republican are 5.6 times as large ($3.50 < OR < 8.96$) for very conservative ideology as for very liberal ideology.

Notice that both the estimation and the interpretation of the PI association are much easier when the ordinality is used. This highlights the appeal of using ordinality when it is available and when the association can reasonably be described by the linear trend across the log odds ratios. In the homogeneous association model with nominal association, (GP, GI, PI), the estimates based on consecutive ideologies were 1.53, 1.55, 2.28, and 0.88 (code available in PolIdeolNominal.R). Our estimate of 1.54 lies in the middle of these values. However, the confidence intervals for the odds ratio involving `VC/SC` and `SC/M` in the nominal analysis do not have much overlap with those in the ordinal analysis here. This suggests that these two intervals are estimating odds ratios that may not both have the same value. Thus, it remains to be seen whether the linear pattern for $\log(OR)$ really fits these data.

To further compare odds ratios from different models, Table 4.8 shows a summary for some of the odds ratios from the saturated model, the homogeneous association model, and this linear association model. Our program contains the corresponding code for these computations. Comparing odds ratio estimates and confidence intervals from different models allows us to understand the models' properties better.

There is very close agreement for GP odds ratios across all models. The estimates from the two models that assume homogeneous associations are in the middle of those from the saturated model, (GPI). The confidence intervals are narrowest when we use all data to estimate a single odds ratio, rather than try to estimate them separately for each ideology. A similar result holds for the two representative GI odds ratios shown in Table 4.8. In this case, the increase in precision is less apparent because the saturated model contains only versions of each odds ratio, one for each party. The PI odds ratio estimates are also somewhat similar across models for each comparison, but the confidence intervals from the linear association model are considerably narrower than those from the homogeneous association model with nominal association. This demonstrates the other big advantage of using ordinality explicitly: confidence intervals for comparable odds ratios are generally much shorter in the ordinal analysis than in the nominal analysis. The ability to represent the PI association using fewer parameters allows those parameters to be estimated more precisely.

Table 4.8: Estimates of various odds ratios and 95% Wald confidence intervals from different loglinear models.

Model	Odds ratio estimates (confidence intervals)		
	GP	$GI:VC/VL$	$GI:SL/VL$
(GPI)	$VL:0.8$ (0.3, 1.9)	$D:0.9$ (0.4, 1.8)	$D:0.9$ (0.5. 1.6)
	$SL:0.9$ (0.4, 1.9)	$R:1.6$ (0.7, 3.7)	$R:1.0$ (0.4, 2.5)
	$M:1.6$ (1.0, 2.5)		
	$SC:1.5$ (0.7, 3.1)		
	$VC:1.5$ (0.8, 2.9)		
(GP, GI, PI)	1.3 (1.0, 1.8)	1.1 (0.5, 2.2)	0.9 (0.4, 1.8)
(GP, GI, PI) with linear PI	1.3 (1.0, 1.8)	1.1 (0.6, 1.8)	0.9 (0.5, 1.5)

Model	Odds ratio estimates (confidence intervals)	
	$PI:VC/VL$	$PI:SL/VL$
(GPI)	$F:3.7$ (1.8, 7.4)	$F:1.5$ (0.7, 3.0)
	$M:6.7$ (2.9, 15.1)	$M:1.6$ (0.7, 3.8)
(GP, GI, PI)	4.8 (2.3, 10.0)	1.5 (0.7, 3.3)
(GP, GI, PI) with linear PI	5.6 (3.5, 9.0)	1.5 (1.4, 1.7)

When both variables in a two-variable model are ordinal, we can define scores $s_i^X, i = 1, \ldots, I$ for X and $s_j^Z, j = 1, \ldots, J$ for Z. The $I-1$ association parameters from Equation 4.11 are replaced with a single parameter by using the model

$$\log(\mu_{ij}) = \beta_0 + \beta_i^X + \beta_j^Z + \beta^{XZ} s_i^X s_j^Z.$$

It is easy to see that for this model $\log(OR_{ii',jj'}) = \beta^{XZ}(s_i^X - s_{i'}^X)(s_j^Z - s_{j'}^Z)$ (Exercise 11). This structure is sometimes called the *linear-by-linear association.* The reduction of association parameters again comes at the cost of more-structured assumptions regarding the nature of the log-odds ratios. Therefore, when a test for $H_0 : \beta^{XZ} = 0$ is rejected, it does not imply that the association between X and Z is entirely linear; there may be additional structure not captured by β^{XZ}. Similarly, failing to reject $H_0 : \beta^{XZ} = 0$ means only that there is insufficient evidence of a *linear-by-linear* association, not that there is no association at all.

With more than two variables, associations continue to be defined by two-variable and higher-order interactions. Nominal, linear, and linear-by-linear association terms for different pairs of variables can appear together in any combination—indeed, the political ideology example contained two nominal association terms and one linear association term. Placing linear structures on *three*-variable interactions assumes that the $\log(OR)$'s between two variables change linearly across the scores of the third variable. Modeling interactions involving more than three variables can be done. The main difficulty then is interpreting the resulting models. Carrying out assessments as shown in Equations 4.10 and 4.13 provides the key to interpreting models.

Checking goodness of fit for a score set

For a single ordered categorical variable X with I levels, let s_i^X represent the set of scores assigned to the levels. As in Section 4.2.3, let $x_2, \ldots, x_I$ be the indicator variables that would be used to represent the levels of X if the ordering is not considered. We can consider the nominal model $\log(\mu_i) = \beta_0 + \beta_2 x_2 + \ldots + \beta_I x_I$ and the ordinal model $\log(\mu_i) = \gamma_0 + \gamma_1 s_i^X$ as two alternative explanations for the changes in mean count. The nominal model creates a separate parameter for each mean, and thus represents a saturated model, so that the residual deviance is zero when there is one observation per level of X. The ordinal model assumes that the population means change log-linearly with the scores. When this is true, the data ought to fit the model reasonably closely, and hence the residual deviance should be relatively small. If the ordinal model is wrong, then the means it estimates may lie far from the observed means, and the residual deviance may be large. A comparison of the residual deviances of the nominal and ordinal models represents a test for the assumption that the means follow a linear pattern with the chosen scores.

Formally, this is an LRT. The test statistic $-2\log(\Lambda)$ is the difference between the nominal- and ordinal-model deviances. The degrees of freedom for the test are derived from the difference in the number of parameters in the two models. In the nominal model, the variable X uses $I-1$ degrees of freedom (the parameters $\beta_2, \ldots, \beta_I$), while in the ordinal model, X uses only 1 degree of freedom for the parameter γ_1. Therefore, the test statistic is compared to a χ^2_{I-2} distribution. Notice that using different sets of scores can affect the residual deviance of the ordinal model, and hence the test statistic and conclusion. Therefore, this test is often used to help determine whether a particular set of scores is appropriate. This particular type of test is referred to as a *goodness-of-fit* test.

The hypotheses for the goodness-of-fit test are $H_0 : \log(\mu_i) = \gamma_0 + \gamma_1 s_i^X$ vs. $H_a : \log(\mu_i) = \beta_0 + \beta_2 x_2 + \ldots + \beta_I x_I$. In words, we are testing the null hypothesis that the scores provide a reasonable explanation for the pattern of mean counts against the alternative that the pattern of means does *not* fit the shape implied by the scores. Failing to reject H_0 implies that the scores seem to do a reasonably good job of capturing the trend in means across levels of X. Note that this is not the same as saying that the scores are *correct*. It is quite possible that models based on several different sets of scores can provide "adequate" fits based on the lack of significance in the test. Note also that this test is different from the LRT for the significance of the score variable s^X in the ordinal model, for which the hypotheses are $H_0 : \log(\mu_i) = \gamma_0$ vs. $H_a : \log(\mu_i) = \gamma_0 + \gamma_1 s_i^X$. These two questions are different from each other; their null hypotheses can be true or false in any combination.

This procedure extends to comparisons of linear- and nominal-association models for two or more variables. The null hypothesis for the goodness-of-fit test is always represented by the model that uses scores, while the alternative is the model in which one or more sets of scores have been replaced with indicator variables. Rejecting H_0 means that the linear structure imposed by the scores is not an adequate description of the association for which they are used.

When there are multiple ordinal variables, it is possible to create a series of models by successively loosening the assumptions in a given association. For example, if X and Z are both ordinal, then, in each variable, we have a choice of whether or not to use scores in a linear structure. Suppose we use a subscript L to denote that a variable appears in an interaction in its linear-score form. In the interaction term, we could use scores for both variables ($X_L Z_L$), scores for X only ($X_L Z$), scores for Z only ($X Z_L$), or leave both variables nominal (XZ). The middle two models, which use scores for only one variable, are both generalizations of the linear-by-linear model and are both restricted versions of the fully nominal model. Thus, model-comparison tests can be used in several different ways, comparing $H_0 : X_L Z_L$ to one of several possible alternatives, $H_a : X_L Z$ or $X Z_L$ or XZ. We can also test $H_0 : X_L Z$ or $H_0 : X Z_L$ against $H_a : XZ$. Only $X_L Z$ and $X Z_L$ are not directly comparable using LRTs.

Testing pairs of models at a time can end with ambiguous results. It could happen, for example, that $X_L Z_L$ is rejected as a reduced model for $X_L Z$ and/or $X Z_L$ but not as a reduced model for XZ. Then one is left wondering whether $X_L Z_L$ is an adequate model. This is a problem whenever we use a series of tests to try to select a single model. An alternative is to use information criteria or some other measure to assess the fit of each model individually, rather than in comparison to some other model. These measures are discussed in greater detail in Chapter 5.

Example: Political ideology (PolIdeolOrd.R, PolIdeolData.csv)

We test the fit of the model that assumes a linear association in the PI interaction against the one that assumes a nominal association. The null hypothesis is H_0 : *The PI association is adequately described by the 1-2-3-4-5 scores*, and the alternative is simply that H_0 is not true. The test is shown below.

```
> anova(mod.homo.lin1.PI, mod.homo, test = "Chisq")
Analysis of Deviance Table

Model 1: count ~ gender * party + gender * ideol + party *
    lin.score1
Model 2: count ~ (gender + party + ideol)^2
  Resid. Df Resid. Dev Df Deviance P(>|Chi|)
1         7     8.3986
2         4     3.2454  3   5.1532    0.1609
```

The test statistic is $-2\log(\Lambda) = 5.15$ and the p-value $= 0.16$, so using any common level of α we would not reject H_0. We can conclude that there is not sufficient evidence that the extra parameters given by the nominal variable version of the model are needed. Less formally, this means that the assumption of loglinear odds ratios using the 1-2-3-4-5 score set provides a reasonable explanation for the association between party and ideology.

Further simplifications of this model can be considered. We can try using the same scores to describe the GI association; that is, we can specify that the odds formed from two ideologies must have a ratio for males vs. females that changes in a loglinear fashion with the distance between the ideologies (as measured by the scores). We can try the alternative 0-2-3-4-6 scores for either or both of the ordinal associations. And finally, and perhaps most interestingly, we can use the "extremeness" scores, 2-1-0-1-2, in the associations with party, gender, or both. In fact, we can even use these scores *in addition to* the ones that measure ideologies linearly. We leave these pursuits as exercises.

4.3 Poisson rate regression

Consider the following claim: "Most automobile accidents occur within five miles of home." If this claim is true, it would seem to imply that the most dangerous roads are close to your home—and close to our homes, and close to everyone else's home! A narrative that one might manufacture for this claim is that we become so familiar with the roads near our own homes that we relax too much when we drive near home and, thus, open ourselves up to grave danger.

A much simpler explanation for this result is that perhaps most *driving* occurs within five miles of home! If this is so, then the *rate* at which we collide with others may be no different, or even lower, near home than elsewhere, but the *exposure* to potential accidents is much greater. In this context, counting the *number* of accidents at different distances from home may be a less meaningful measure of risk than estimating the *rate* of accidents *per mile driven.*

This highlights an important point that was made at the start of this chapter: an assumption behind the use of the Poisson model for counts is that both the intensity (or rate) of event occurrence and the opportunity (or exposure) for counting are constant for all available observations. In Section 4.2, we investigated models that allow a mean count to change. Implicitly, we were assuming that the exposure was constant for all observations. For example, in counting the drinks consumed on a Saturday, the length of observation ("Saturday") was constant for all participants, so that only their rate or intensity of drink consumption varied. We then allowed mean counts (i.e., rate of consumption) to vary by using a Poisson regression model to relate them to explanatory variables.

In problems where exposure is not constant among all observations, modeling the counts directly leaves interpretations unclear. For example, if we measured some participants in the alcohol study for one hour and others for one day and others for a week or a month, we should expect that the number of drinks consumed among participants would differ simply because of the way we conducted our measurements. This might obscure any effects that the explanatory variables might have. Therefore, we would need to take the exposure effect out of consideration, so that we may get to the heart of the issue relating the explanatory variables to the *rate* of consumption.

Very generally, "rate" is defined as mean count per unit exposure (e.g., drinks per day, animals trapped per visit, accidents per mile driven, trees per hectare, and so forth). That is, $R = \mu/t$, where R is the rate, t is the exposure, and μ is the mean count over an exposure duration of t. When our observed counts all have the same exposure, modeling the mean count μ as a function of explanatory variables $x_1, \ldots, x_p$ is the same as modeling the rate. When exposures vary, we can still use a Poisson regression model for the means, but we need to account for the exposure in the model. This is done using a *Poisson rate regression* model by writing the model as:

$$\log(R_i) = \log(\mu_i/t_i) = \beta_0 + \beta_1 x_{1i} + \ldots + \beta_p x_{pi}$$

or

$$\log(\mu_i) = \log(t_i) + \beta_0 + \beta_1 x_{1i} + \ldots + \beta_p x_{pi},\ i = 1, \ldots, n. \tag{4.15}$$

Notice that the $\log(t_i)$ term is known and is not multiplied by a parameter. It simply adjusts the mean for each different observation to account for its exposure. The term $\log(t)$ is called an *offset*, so this model is sometimes called *Poisson regression with offsets.* A direct consequence of Equation 4.15 is that $\mu = t \times \exp(\beta_0 + \beta_1 x_1 + \ldots + \beta_p x_p)$, which shows the direct influence of exposure on the mean count.

There is a similarity between rates and binomial proportions that sometimes causes confusion. Both are expressed as $\{\text{counts of events}\}/\{\text{opportunities for events}\}$. One might incorrectly use the number of trials in a binomial model as an exposure and apply a Poisson rate regression, or conversely, might incorrectly think of the exposure in a Poisson rate problem as a number of trials and apply a logistic regression. However, the two situations are different, and the distinction is easily made by asking the following question: Can each unit of exposure conceivably produce more than one event, or must each unit of exposure produce exactly 0 or 1 event? If more than one event is possible in a single unit of exposure, then this violates the binomial model assumptions and the Poisson rate model is more appropriate. For example, there is nothing that prevents multiple accidents from happening in a single mile driven, and indeed, a very intoxicated or careless driver may experience this misfortune. A binomial model would be inappropriate in this case. On the other hand, each placekick or free throw must be either a success or a failure, and one cannot count more than one successful event per trial. For these problems, the logistic regression model should be used.

Parameters $\beta_0, \ldots, \beta_p$ from Equation 4.15 are estimated using maximum likelihood estimation as before, but their interpretation is in terms of the *rate* rather than the mean. For example, β_0 is the log of the true event rate when all explanatory variables are set to 0, and β_1 is the change in the log rate of occurrence per unit increase in x_1. Note that the units of measurement for the rate are count-units per exposure-unit (e.g., accidents per mile driven).

It is sometimes helpful to convert the exposure to other units to create a more interpretable rate. For example, if we are analyzing accidents, and exposure is given as miles driven, the average number of accidents in one mile should be extremely small. If we prefer, we can choose to rewrite the rate as a more interpretable quantity like "accidents per 1000 miles" by dividing the exposure miles by 1000 before applying the model.

Given a set of estimates $\hat{\beta}_0, \ldots, \hat{\beta}_p$, estimated rates for given values $x_1, \ldots, x_p$ are found from $\hat{R} = \exp(\hat{\beta}_0 + \hat{\beta}_1 x_1 + \ldots + \hat{\beta}_p x_p)$. An estimated mean count for a given exposure t is found simply by multiplying the rate by the exposure, $\hat{\mu} = t \times \hat{R}$.

Example: Beetle egg-laying response to crowding (BeetleEggCrowding.R, BeetleEggCrowding.txt)[17]

Tauber et al. (1996) describe an experiment examining the effects of crowding on reproductive properties of a certain species of leaf beetle, *Galerucella nymphaeae*. Cages of a fixed size containing either one female and one male or five females and five males were kept in temperature chambers at either 21°C or 24°C. Among the measurements taken on each cage is the number of egg masses produced by the females in the cage over the course of the study. We use Poisson regression to examine whether crowding and/or temperature influence the number of egg masses produced by each female. A complicating feature is that in cages with five females, there is no easy way to identify which females laid a given egg mass, so we must consider "egg masses per female" as the *rate* response of interest.

The data are entered using the code below. The temperature variable `Temp` takes values 21 and 24; `TRT`, which is the crowding treatment variable, takes values `"I"` for cages with individual females and `"G"` for cages with groups of 5; `Unit` identifies the cage within a combination of temperature and crowding level (there are 35 individual cages and 7 group cages at each temperature); `NumEggs` counts the egg masses laid in the box over the course of the study. We create a variable, "`females`", to represent the number of females in the cage that will serve as our exposure variable.

```
> eggdata <- read.csv(file = "C:\\Data\\BeetleEggCrowding.csv",
    stringsAsFactors = TRUE)
> eggdata$females <- ifelse(test = eggdata$TRT == "I", yes = 1,
    no = 5)
> head(eggdata)
  Temp TRT Unit NumEggs females
1   21   I    1       8       1
2   21   I    2       0       1
```

[17] Data kindly provided by Dr. Jim Nechols, Department of Entomology, Kansas State University.

```
3   21   I   3     3     1
4   21   I   4     5     1
5   21   I   5     1     1
6   21   I   6     3     1
```

As an initial summary, the sample mean numbers of eggs masses per female for each group are calculated within the program, yielding the results 3.54 for `TRT` = `"I"`, `Temp` = 21, 4.34 for `TRT` = `"I"`, `Temp` = 24, 0.51 for `TRT` = `"G"`, `Temp` = 21, and 1.46 for `TRT` = `"G"`, `Temp` = 24.

The model that we will fit is $\log(\mu) = \log(t) + \beta_0 + \beta_1\texttt{Temp} + \beta_2\texttt{TRTI} + \beta_3\texttt{Temp} \times \texttt{TRTI}$, where `TRTI` is the indicator variable for `TRT` = `"I"`, and t is the number of females in the cage (1 or 5). In `glm()`, the offset is incorporated into the model using the `offset` argument as shown below.

```
> eggmod1 <- glm(formula = NumEggs ~ Temp*TRT, family =
    poisson(link = "log"), offset = log(females), data = eggdata)
> summary(eggmod1)

Call:
glm(formula = NumEggs ~ Temp * TRT, family = poisson(link =
    "log"), data = eggdata, offset = log(females))

Deviance Residuals:
   Min       1Q   Median       3Q      Max
-3.817   -2.662   -1.076    1.575    5.174

Coefficients:
            Estimate Std. Error z value Pr(>|z|)
(Intercept) -7.95515    2.12517  -3.743 0.000182 ***
Temp         0.34715    0.09139   3.799 0.000145 ***
TRTI         7.79489    2.31405   3.369 0.000756 ***
Temp:TRTI   -0.27928    0.09989  -2.796 0.005176 **
---
Signif. codes: 0 '***' 0.001 '**' 0.01 '*' 0.05 '.' 0.1 ' ' 1

(Dispersion parameter for poisson family taken to be 1)

    Null deviance: 637.85  on 83  degrees of freedom
Residual deviance: 485.57  on 80  degrees of freedom
AIC: 673.43

Number of Fisher Scoring iterations: 6
```

From the positive coefficients on the main effect terms, it is apparent that the egg-laying rate is higher at the higher temperature and in the individual cages. The negative interaction coefficient indicates that the ratio of rates between individual and group cages is higher at the lower temperature. Rate estimates can be calculated from these coefficients. For example, the estimated rate for `TRT` = `I`, `Temp` = 21 is $\exp[-7.955 + 0.347(21) + 7.795(1) - 0.279(21)(1)] = 3.54$, matching the previous estimate.

An easy way to get estimated *mean counts* for each cage is to use `predict()` on the model object. This produces one estimate for each cage, which, in this example, repeats the same estimates numerous times. Instead, estimated means and rates can be found once for each combination of TRT and Temp by specifying these levels in a separate data frame as shown below.

```
> newdata <- data.frame(TRT = c("G", "I", "G", "I"), Temp =
    c(21, 21, 24, 24), females = c(5, 1, 5, 1))
> mu.hat <- round(predict(object = eggmod1, newdata = newdata,
    type = "response"), 2)
> # Convert means to rates
> data.frame(newdata, mu.hat, rate = mu.hat/newdata$females)
  TRT Temp females mu.hat  rate
1   G   21       5   2.57 0.514
2   I   21       1   3.54 3.540
3   G   24       5   7.29 1.458
4   I   24       1   4.34 4.340
```

Estimated mean *rates* for each combination of variable levels are also computed using `emmeans()` as shown below. By default, the function produces means that can be scaled to rates per N units using `offset = log(N)`. For example, we create rates on a per unit ($N = 1$) basis using `offset = log(1)` below. Also, in our case, exposures are constant within each TRT level, so the function produces a warning about a nesting structure unless we add `nesting = NULL`.

```
> emm1o <- emmeans(object = eggmod1, specs = ~ TRT + Temp,
    offset = log(1), nesting = NULL)
> confint(object = emm1o, type = "response")
 TRT Temp  rate    SE  df asymp.LCL asymp.UCL
 G     21 0.514 0.121 Inf     0.324     0.816
 I     21 3.543 0.318 Inf     2.971     4.225
 G     24 1.457 0.204 Inf     1.107     1.917
 I     24 4.343 0.352 Inf     3.705     5.091

Results are averaged over the levels of: .offset.
Confidence level used: 0.95
Intervals are back-transformed from the log scale
```

The reason that the estimated egg-laying rates (rate) are identical to our sample estimates is that the model contains as many parameters as there are unique rates to be estimated. Thus, even though the model is not saturated—the residual deviance is not zero—the four observed rates are fully explained by the 4 degrees of freedom accounted for by β_0, β_1, β_2, and β_3. This does not happen in every problem.

Below is a summary of the LRTs for the model terms.

```
> library(package = car)
> Anova(eggmod1)
Analysis of Deviance Table (Type II tests)
Response: NumEggs
         LR Chisq Df Pr(>Chisq)
Temp       10.842  1   0.000992 ***
TRT       132.994  1  < 2.2e-16 ***
```

```
Temp:TRT      8.450  1   0.003650 **
```

The `Temp:TRT` interaction term is significant, indicating that the ratios of rates between the two crowding treatments are not the same at both temperatures ($-2\log(\Lambda) = 8.4, \text{df} = 1$, p-value $= 0.004$). Because of this interaction, we compare rates of eggs per female in groups vs. alone separately for each temperature. This is done with Wald intervals using the `contrast()` function from `emmeans` as shown below.

```
> mean.rat1 <- contrast(object = emm1o, method = "pairwise",
    simple = list("TRT"))
> confint(object = mean.rat1, type = "response")
$`simple contrasts for TRT`
Temp = 21:
 contrast ratio     SE  df asymp.LCL asymp.UCL
 G / I    0.145 0.0366 Inf    0.0885     0.238

Temp = 24:
 contrast ratio     SE  df asymp.LCL asymp.UCL
 G / I    0.336 0.0543 Inf    0.2443     0.461

Confidence level used: 0.95
Intervals are back-transformed from the log scale
```

At 21°C, females in grouped cages produced an estimated 15% as many eggs per female as when in individual cages, with 95% confidence interval (9%, 24%). At 24°C, the egg-laying rates were more similar; females in groups produced an estimated 34% as many eggs per female as when in individual cages, with 95% confidence interval (24%, 46%). In both cases, the grouping condition clearly resulted in reduced egg-laying rates. Additional comparisons are made in the program accompanying this example, including profile LR confidence intervals. Estimated rates and Wald confidence intervals are depicted in Figure 4.8. The effect of crowding is apparent: fewer egg masses are laid per female in the cages with the crowded condition. The interaction effect is explored in more detail in Exercise 21

Rates sometimes depend on the amount of exposure as well as other variables. For example, crime rates (e.g., crimes per 10,000 population) may sometimes be higher in cities with larger populations. This was also implicit in the beetle egg example above, where the crowding treatment related to the number of females in each box, which was also the exposure for the rate of eggs per female. As is clear from the example, it is certainly possible to use the exposure variable as another explanatory variable in the regression, in addition to using it as an offset. That is, we can define $x_{p+1} = t$ (or $\log(t)$ if preferred) and add $\beta_{p+1}x_{p+1}$ to Equation 4.15. The dual use of the exposure causes no problems for the model parameter estimation like those seen in the political ideology example of Section 4.2.6 because no parameter is estimated for the offset (in essence, the parameter is fixed at 1).

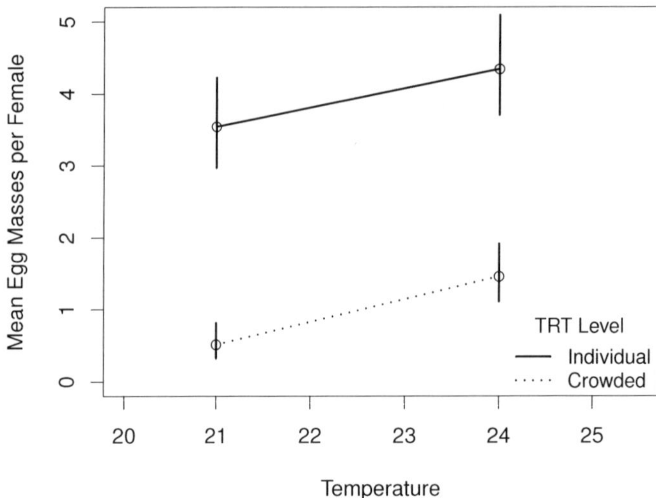

Figure 4.8: Estimated rates of egg mass production and confidence intervals for all treatment-temperature combinations in the beetle egg-laying example. Individual cages have one female and male; crowded cages have five females and males. The vertical lines represent 95% Wald confidence intervals for the rate.

4.4 Zero inflation

Many populations consist of subgroups of different types of units. A particularly common situation is that a population consists of two classes: units that have the potential to respond with a count and units that can only respond with a count of zero. For example, magnetic resonance imaging (MRI) machines are large and very expensive pieces of medical equipment. Some, but not all, hospitals have an MRI machine. If a hospital has one, it surely does not sit idle for long periods of time. Suppose that hospital administrators are asked in a survey to report the number of MRI scans they perform in a month. Hospitals with no MRI machine are not performing any MRI scans, while the rest will report some count greater than 0.

As a different example, suppose that we record the number of fish caught on various lakes in four-hour fishing trips in Minnesota. Some lakes in Minnesota are too shallow for fish to survive the winter, so fishing on those lakes will yield no catch. On the other hand, even on a lake where fish are plentiful, we may or may not catch any fish because of the conditions or our own competence. Thus, the number of fish caught will be zero if the lake doesn't support fish, and will be zero, one, or more if it does.

These situations are examples where the responses come from *mixtures* of distributions, and, in particular, are cases where one of the distributions places all of its probability mass on zero. That is, the subpopulation of "susceptibles" are capable of returning a count according to some distribution, while the subpopulation of "immunes" all return a zero.

Data that result from this type of mixture have the property that there are more zero counts than would be expected by a single distributional model. This is referred to as *zero inflation*. In general, let π be the proportion of immunes in the population, and let the counts for susceptibles have a $Po(\mu)$ distribution. This mixture is called the *zero-inflated Poisson* (ZIP) model. Figure 4.9 shows histograms of the PMFs from ZIP distributions with

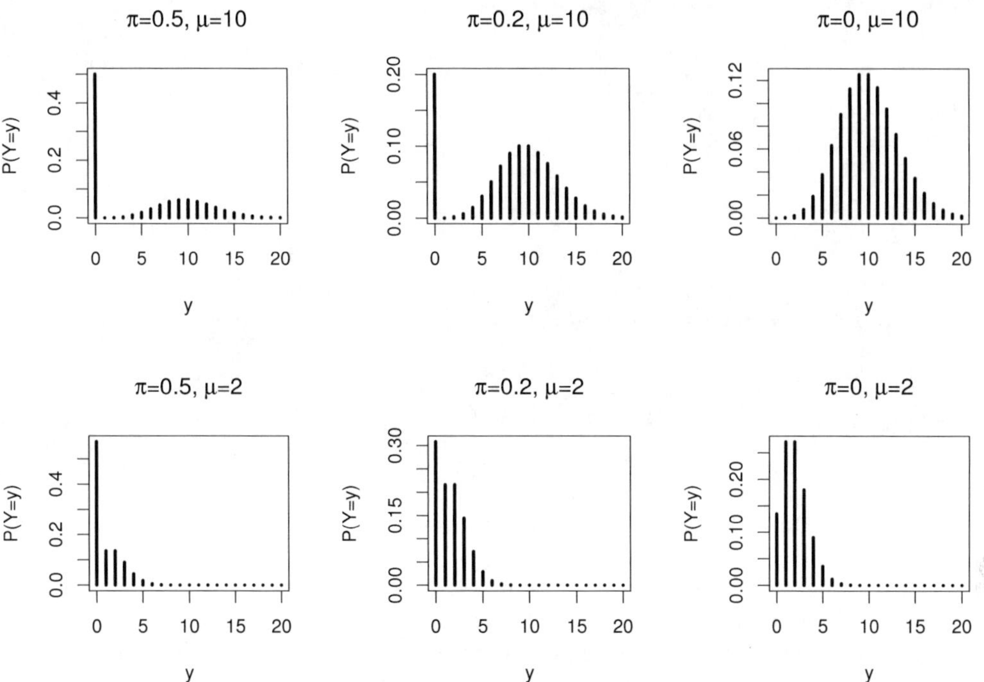

Figure 4.9: Probability mass histograms from zero-inflated Poisson models with varying means and levels of zero inflation. These plots are created by ZeroInflPoiDistPlots.R.

varying values of π and μ (note that $\pi = 0$ is the ordinary Poisson distribution). The zero inflation is obvious when the mean is far from zero as is the case in the first two panels in the top row. One would hardly be tempted to use a single Poisson distribution as a model for data with histograms like these. However, when the mean count among susceptibles can place appreciable mass on zero or one (bottom row), then the presence of zero inflation can go unnoticed if the nature of the problem does not suggest it in advance, especially if π is not large. The presence of explanatory variables can also blur the distinction between the classes, because then mean counts for susceptibles and probabilities of immunity can vary greatly among members of the population. In practice, zero inflation is often discovered only upon fitting a model and finding the fit poor using techniques as described in Chapter 5.

Corrective models

When a population is known or believed to consist of two subgroups as described above, or when an excess of zero counts appears unexpectedly, then ordinary Poisson models will likely yield poor fits to the data. Several models have been proposed to account for the excess zeroes. The main model is the ZIP model of Lambert (1992). The ZIP model specifies that immunity is a binary random variable whose probability may depend on explanatory variables collectively denoted by z, while the counts among susceptibles follow a Poisson distribution whose mean may depend on explanatory variables collectively denoted by x. Thus,

$$\begin{array}{lll} Y = & 0 & \text{with probability} \quad \pi(z) \\ Y \sim & Po(\mu(x)) & \text{with probability } 1 - \pi(z) \end{array}$$

Typically, a logistic model is assumed for $\pi(z)$,

$$\text{logit}(\pi(z)) = \gamma_0 + \gamma_1 z_1 + \ldots + \gamma_q z_q.$$

while the usual log link is used for $\mu(x)$,

$$\log(\mu(x)) = \beta_0 + \beta_1 x_1 + \ldots + \beta_p x_p,$$

where $\gamma_0, \gamma_1, \ldots, \gamma_q$ and $\beta_0, \beta_1, \ldots, \beta_p$ are unknown regression parameters. It is permissible that $x = z$ so that the same explanatory variables are used both to distinguish the classes and to model the mean for susceptibles, but this is not required. Generally, x and z can consist of arbitrarily similar or different sets of variables. One can show that

1. $P(Y = 0|x, z) = \pi(z) + (1 - \pi(z))e^{-\mu(x)}$,
2. $E(Y|x, z) = \mu(x)(1 - \pi(z))$, and
3. $Var(Y|x, z) = \mu(x)(1 - \pi(z))(1 + \mu(x)\pi(z))$.

Thus, the mean of Y when $\pi(z) > 0$ is always less than $\mu(x)$, but the variance of Y is greater than its mean by a factor of $1 + \mu(x)\pi(z)$.

A relative of the ZIP model, called the *hurdle model* (Mullahy, 1986), differs by assuming that a count is always greater than zero whenever some real or conceptual hurdle is crossed. The MRI example that started this section is of this type because if a hospital has gone to the expense of installing an MRI machine (the "hurdle"), then they surely use it. In this case, *all* zeroes correspond to immunes (those units that have not crossed the hurdle of installing a machine), while the count model for susceptibles is assumed to take values starting at 1 rather than 0. We therefore need to employ a count model that excludes zero. This is done with a Poisson model by imposing 0 probability at $Y = 0$ and then increasing all other probabilities proportionately so that they once again sum to 1. The resulting distribution is called a "left-truncated Poisson," because the left tail of the distribution ($Y = 0$) is cut off from the usual Poisson distribution. The logistic model for probability of immune and the log link for the mean of the left-truncated Poisson distribution of counts for susceptibles are used in the hurdle model as in the ZIP model.

Both models can be estimated by maximum likelihood using functions in the `pscl` package. The syntax for fitting both models is very similar and is described in detail by Zeileis et al. (2008). The model fitting can be easier with the hurdle model, because it is immediately clear that all zeroes are immunes. Then, the modeling of the mean function is based on the subset of data with $Y > 0$ and is independent of the probability of immunes. However, the ZIP model seems to be preferred in most areas. Our preference is generally determined by which model for the immunes better agrees with the source of the data. If a zero count *defines* an immune, as with the MRI example, a hurdle makes better sense. If susceptibles can sometimes provide zero counts, as with the fishing example, then this points toward a ZIP model. If it is not clear whether susceptibles can produce zero counts, then empirical considerations such as measures of fit for the two models can preside (see Chapter 5 for details).

Example: Beetle egg-laying response to crowding (BeetleEggCrowdingZIP.R, BeetleEggCrowding.csv)

One of the goals of the experiment in Tauber et al. (1996) was to observe a particular response called *diapause* in the females. Diapause occurs in insects when they temporarily cease biological development in some way as a response to environmental conditions. In this case, the researchers suspected that females' reproductive cycles

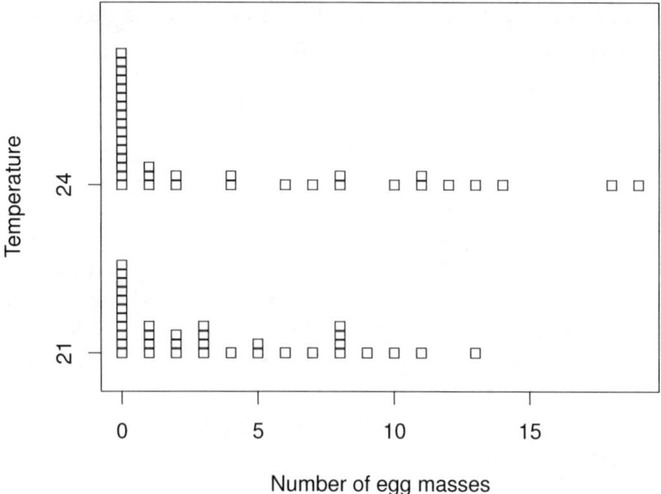

Figure 4.10: Histograms of beetle egg masses per female, showing zero inflation.

might stop in response to temperature or perceived crowding. Therefore, they expected that females might enter diapause and produce no eggs with higher probability for some conditions than for others.

The fact that some females in the group cages might enter diapause while others do not creates a complication because those cages of mixed females will still produce eggs, but at a reduced rate. Thus, estimating the probability that a particular female produces zero egg masses in a group cage for the entire study duration is an example of a group testing problem (see Section 6.7 for an introduction to group testing). This is beyond our current scope, so for this example we will limit ourselves to the individual cages. We will use a ZIP model to determine whether temperature affects either the probability of diapause or the mean number of eggs laid by females not in diapause. We leave the fitting of a hurdle model as Exercise 15.

First we examine the extent of the zero inflation. We use the `stripchart()` function to plot these data in Figure 4.10.

```
> eggdata2 <- eggdata[eggdata$TRT == "I",]
> stripchart(NumEggs ~ Temp, method = "stack", data = eggdata2,
   ylab = "Temperature", xlab = "Number of egg masses")
```

In our previous example, the estimated means were 3.54 for `TRT=I`, `Temp`=21 and 4.34 for `TRT=I`, `Temp`=24. The plots show that there are considerably more zeroes than would be expected from Poisson distributions with these means (the estimated probabilities are $P(Y = 0) = e^{-\hat{\mu}} = 0.029$ and 0.013, respectively). Furthermore, there are a few more zeroes at `Temp`=24 than at `Temp`=21 (16 vs. 11), and the means of the non-zero counts appear to differ between temperatures. Overall, the ZIP model is preferred for this setting because a female not in diapause could nonetheless produce a zero count (e.g., because of some other biological process such as infertility or early death).

Our ZIP model uses a logistic regression to describe the probability of "immunity" to egg-laying (i.e., diapause) and a Poisson regression on the counts for all females who are not "immune." In both models, we will allow the probabilities and the mean counts to vary according to temperature. That is, we are fitting

$$\log\left(\pi/(1+\pi)\right) = \gamma_0 + \gamma_1 \texttt{Temp}$$

and

$$\log(\mu) = \beta_0 + \beta_1 \texttt{Temp}.$$

The ZIP model is fit using the function `zeroinfl()` from the `pscl` package. The main argument required for `zeroinfl()` is the model formula. In general, the syntax `formula = y ~ x1 + x2 +...|z1 + z2 +...` fits the count model to the x-variables and the probability-of-immune model to the z-variables. If the same variables are to be used in both models, as is the case for the beetles, the shortcut `y ~ x1 + x2 ...` can be used.

We first fit the model in which both the probability of immune and the means are allowed to vary by temperature.

```
> library(package = pscl)
> zip.mod.tt <- zeroinfl(formula = NumEggs ~ Temp | Temp, dist =
    "poisson", data = eggdata2)
> summary(zip.mod.tt)

Call: zeroinfl(formula = NumEggs ~ Temp | Temp, data = eggdata2,
    dist = "poisson")

Pearson residuals:
    Min      1Q  Median      3Q     Max
-1.1688 -0.9659 -0.5211  0.8134  3.2601

Count model coefficients (poisson with log link):
            Estimate Std. Error z value Pr(>|z|)
(Intercept)  -1.4631     0.9249  -1.582 0.113672
Temp          0.1476     0.0407   3.626 0.000288 ***

Zero-inflation model coefficients (binomial with logit link):
            Estimate Std. Error z value Pr(>|z|)
(Intercept)  -5.1843     3.7897  -1.368    0.171
Temp          0.2088     0.1671   1.249    0.212

Number of iterations in BFGS optimization: 18
Log-likelihood: -187.2 on 4 Df
```

The output from the `zeroinfl()` call gives the parameter estimates for the Poisson mean model first, followed by those for the logistic probability model. The estimated mean function is $\hat{\mu} = \exp(-1.46{+}0.15\ \texttt{Temp})$, and the estimated probability of immune is $\hat{\pi} = \exp(-5.18 + 0.21\ \texttt{Temp})/(1 + \exp(-5.18 + 0.21\ \texttt{Temp}))$. The Wald tests for temperature effects suggest significance in the mean model but not in the probability model. We explore this later.

There is no `anova()` method function for objects produced by `zeroinfl()`, so we must instead use other tools for LRTs. The package `lmtest` provides a variety of testing and diagnostic tools for linear models, some of which have been extended to handle `zeroinfl`-class objects. In particular, the `lrtest()` function from the `lmtest` package can produce LRTs for comparing nested models. We use it here to compare the full model given in `zip.mod.tt` to the fits of two reduced models: (1) no temperature effect on π and (2) no temperature effect on μ.

```
> # Fit ZIP Models: Temp in mean only
> zip.mod.t0 <- zeroinfl(formula = NumEggs ~ Temp | 1, dist =
    "poisson", data = eggdata2)
> # Fit ZIP Models: Temp in probability only
> zip.mod.0t <- zeroinfl(formula = NumEggs ~ 1 | Temp, dist =
    "poisson", data = eggdata2)

> # LRTs of each reduced model against largest model
> library(package = lmtest)
> lrtest(zip.mod.t0, zip.mod.tt)
Likelihood ratio test

Model 1: NumEggs ~ Temp | 1
Model 2: NumEggs ~ Temp | Temp
  #Df  LogLik Df  Chisq Pr(>Chisq)
1   3 -187.97
2   4 -187.18  1 1.5866     0.2078

> lrtest(zip.mod.0t, zip.mod.tt)
Likelihood ratio test

Model 1: NumEggs ~ 1 | Temp
Model 2: NumEggs ~ Temp | Temp
  #Df  LogLik Df  Chisq Pr(>Chisq)
1   3 -193.82
2   4 -187.18  1 13.294  0.0002662 ***
```

The first LRT shows that that reducing from the full model `zip.mod.tt`, with temperature effects on both mean and probability, to `zip.mod.t0`, with temperature effects only on the mean, does not cause a significant drop in the observed value of the likelihood function ($-2\log(\Lambda) = 1.59$, df $= 1$, p-value $= 0.21$). However, removing the temperature effect from the mean does result in a significantly worse fit ($-2\log(\Lambda) = 13.29$, df $= 1$, p-value $= 0.0002$). Therefore, we would consider `zip.mod.t0` to be a good working model, pending diagnostic evaluation as discussed in Chapter 5.

We summarize this model in several ways using code given in the program for this example. First, we estimate the effect of temperature as the ratio of mean egg-mass counts for the 3°C increase in temperature used in the experiment. We also find a 95% confidence interval for this quantity using `confint()`, which produces Wald intervals using the method function `confint.default` (there is no specific `confint()` method function for models with class `zeroinfl`). The estimated ratio is $\exp(3 \times 0.1476) = 1.55$, and the confidence interval is (1.22, 1.97). Thus, with 95% confidence, we estimate that the mean number of egg masses laid by females not in diapause increases between 22% and 97% for this 3°C increase in temperature.

Next, we use emmeans to perform Wald inferences on the probability model, the count model, and on the expected counts, $E(Y|x,z) = \mu(x)(1-\pi(z))$.[18] For ZIP models, the emmeans() function has an additional argument, mode, which can take values including "zero" for inference on estimates from the logistic model, "count" for inference on estimates from the count model (i.e., mean counts conditional on susceptible status), and "response" for inference on estimates on the expected counts (i.e., mean counts including zeroes from immunes). For "zero" or "count" modes, adding lin.pred = TRUE performs calculations in the logit or log scale, respectively, so that comparisons of probabilities can be expressed as odds ratios and comparisons of means can be expressed as ratios. For the "response" mode, calculations are carried out on the count scale. The code for these analyses is applied to the original full model with temperature in both parts of the model, so that we can demonstrate the analysis better. The same analysis is applied to zip.mod.t0 in Exercise 48.

```
> # Using mode = "zero" for inference on binomial model
> emm.tt.z <- emmeans(object = zip.mod.tt, specs = ~ Temp, mode
    = "zero", lin.pred = TRUE)
> confint(object = emm.tt.z, type = "response")
 Temp  prob     SE  df asymp.LCL asymp.UCL
   21 0.310 0.0790 Inf     0.179     0.481
   24 0.457 0.0842 Inf     0.302     0.621

Confidence level used: 0.95
Intervals are back-transformed from the logit scale

> emm.ci.z <- contrast(object = emm.tt.z, method = "revpairwise")
> confint(object = emm.ci.z, type = "response")
 contrast        odds.ratio    SE  df asymp.LCL asymp.UCL
 Temp24 / Temp21       1.87 0.938 Inf       0.7         5

Confidence level used: 0.95
Intervals are back-transformed from the log odds ratio scale

> # Using mode = "count" for inference on Poisson model
> emm.tt.c <- emmeans(object = zip.mod.tt, specs = ~ Temp, mode
    = "count", lin.pred = TRUE)
> confint(object = emm.tt.c, type = "response")
 Temp count    SE  df asymp.LCL asymp.UCL
   21  5.14 0.469 Inf      4.30      6.14
   24  8.00 0.649 Inf      6.82      9.38

Confidence level used: 0.95
Intervals are back-transformed from the log scale

> emm.ci.c <- contrast(object = emm.tt.c, method = "revpairwise")
> confint(object = emm.ci.c, type = "response")
 contrast        ratio    SE  df asymp.LCL asymp.UCL
```

[18]Note that the default in emmeans() is to treat numeric variables with only two levels, like temp in this example, as if they were two-level factors. Thus, calculations are performed separately at each level rather than at the mean value of the numeric variable, which is the usual behavior of the function. This can be overridden by using the at argument to specify levels at which calculations are performed.

```
 Temp24 / Temp21   1.56 0.19 Inf        1.23        1.98

Confidence level used: 0.95
Intervals are back-transformed from the log scale

> # Using mode = "response" for inference on estimated means
> emm.tt.m <- emmeans(object = zip.mod.tt, specs = ~ Temp, mode
   = "response")
> confint(object = emm.tt.m, type = "response")
 Temp emmean    SE  df asymp.LCL asymp.UCL
   21   3.54 0.512 Inf      2.54      4.55
   24   4.34 0.760 Inf      2.85      5.83

Confidence level used: 0.95

> emm.ci.m <- contrast(object = emm.tt.m, method = "revpairwise")
> confint(object = emm.ci.m, type = "response")
 contrast        estimate    SE  df asymp.LCL asymp.UCL
 Temp24 - Temp21      0.8 0.917 Inf    -0.997       2.6

Confidence level used: 0.95
```

The output first shows estimates and 95% confidence intervals for the probability of immune (i.e., a female in diapause) at each temperature. At 21°C, $\hat{\pi} = 0.31$ $(0.18 < \pi < 0.48)$, while at 24°C, $\hat{\pi} = 0.46$ $(0.30 < \pi < 0.62)$. The resulting estimated odds ratio comparing the odds of diapause at 24°C to 21°C are 1.87 $(0.70 < OR < 5.0)$. Thus, we find that there is not sufficient evidence that the odds of diapause are different at these two temperatures, which agrees with our previous LRT for the temperature parameter.

Next, we see that, among susceptibles (i.e., females not in diapause), the mean number of eggs laid is $\hat{\mu} = 5.14$ $(4.30 < \mu < 6.14)$ at 21°C and $\hat{\mu} = 8.00$ $(6.82 < \mu < 9.38)$ at 24°C. The ratio of estimated mean count at 24°C vs. 21°C is 1.56, indicating that the mean number of eggs is about 56% larger at 24°C than at 21°C (95% confidence interval is (1.23, 1.98)). Thus, we do have sufficient evidence to conclude that the mean number of eggs laid by females not in diapause is different at these two temperatures.

Finally, looking at expected counts including the zeroes from females in diapause, we have the estimates $\hat{E}(Y|x,z) = 3.54$ at 21°C $(2.54 < E(Y|x,z) < 4.55)$, and $\hat{E}(Y|x,z) = 4.34$ at 24°C $(2.85 < E(Y|x,z) < 5.83)$. Here, there is not an option to compute these estimates in a different scale, so the comparison is based on a difference. The estimated difference between expected counts at 24°C and 21°C is 0.80, with a confidence interval from -1.0 to 2.6. Thus, we find that there is not sufficient evidence of a difference in the expected number of eggs among all females at 21°C and 24°C. This result seems to contradict the previous finding that the mean egg counts were different for females not in diapause at 21°C and 24°C. However, at 24°C, there is a higher estimated probability of diapause and, hence, more zero counts than occur at 21°C.

To compare the handling of zeroes in these ZIP models with the Poisson model fit in Section 4.3, we estimated the expected number of zero counts by the three ZIP models above, plus one more model that assumes that both the probability of immune and the mean count for susceptibles are the same at the two temperatures. We also included the Poisson model for comparison. In the ZIP models, the expected number of zeroes is calculated by computing the estimated probability of a zero response for each observation based on its explanatory variables, and then summing these probabilities across the full data set. The code is given in the program for this example. The observed number of zeroes was 27. Each ZIP model produces an estimated count that rounds to 27.0. The Poisson model that does not account for zero inflation predicts only 1.47 zeroes. Clearly, the Poisson model is not adequate for these data. However, diagnostics done in Exercise 31 from Chapter 5 reveal that the final ZIP model given in `zip.mod.t0` has problems as well!

A potential difficulty with the ZIP model is the fact that zero counts can arise from two separate sources: immune members of the population and susceptible members who happen to record a zero. Individual observations of zero generally do not provide any direct indication as to whether they are from immune or susceptible members. Therefore, it is difficult to separate the two sources precisely using only data on observed counts. This is particularly true when the mean for susceptibles is close to zero, so that an appreciable fraction of the susceptible population yields counts of zero.

In the next example, we use a Monte Carlo simulation to explore properties of the estimates of π and μ under various conditions when their true values are known.

Example: Simulation comparing ZIP and Poisson models (ZIPSim.R)

In this example, we generate a large number of data sets from a ZIP model with fixed values for π and μ. We then use the ZIP model to estimate these parameters and find Wald confidence intervals for them. We calculate the average value of the parameter estimates so that we can check for bias in the estimates. We estimate the true confidence level of the Wald intervals and compute the median endpoints to see what a "typical" interval might look like.

The data generation is performed as follows. Given the sample size n for a single data set and the number of data sets we wish to create, say B, we generate nB independent observations of two independent random variables. The first is a Bernoulli random variable, say S, that creates the immune and susceptible members of the sample. We set $S = 1$ for susceptible for reasons that will be apparent momentarily. Then, $P(S = 1) = 1 - \pi$. The second random variable is a count variable, say C, from $Po(\mu)$. Finally, we create ZIP observations by taking $Y = S \times C$. This way, immunes are generated with probability π and their counts are set to 0 because $S = 0$, while susceptibles are generated with probability $1 - \pi$ and their counts are distributed as $C \sim Po(\mu)$. We form a data matrix of these counts and apply the ZIP analysis function to them, one set of n observations at a time. We then compute confidence intervals for both μ and π.

For demonstration, we chose $n = 35$ because this was the sample size in the beetle egg-laying example. We set $\mu = 1$ as a case where the count distribution will naturally produce a fairly large number of zeroes ($P(C = 0) = \exp(-1) \approx 0.37$) and explore values of $\pi = 0, 0.1, 0.2, 0.5, 0.8, 0.9$. Finally, we use $B = 1000$ so that we can obtain reasonably good estimates for the distributions of $\hat{\pi}$ and $\hat{\mu}$ and for the true level of the corresponding confidence intervals.

Pilot runs highlighted a problem with the use of large values of π: some data sets were generated that contained 35 counts of 0. This created problems that are described in the program for this example. Therefore, we added code to omit these cases from the calculation of the summary statistics. The results from the simulations are presented in Table 4.9 and Figure 4.11.

Table 4.9: Simulation results of ZIP parameter estimation from 1,000 simulated data sets of $n = 35$ generated from a ZIP model. True confidence level was 95% and $\mu = 1$.

	$\hat{\pi}$				$\hat{\mu}$			
		Conf.	Median CI limit			Conf.	Median CI limit	
π	Mean	Level	Lower	Upper	Mean	Level	Lower	Upper
0	0.06	0	0^b	1	1.07	94.2	0.72	1.59
0.1	0.13	88.5	0.00	0.91	1.05	92.4	0.65	1.70
0.2	0.19	93.0	0.02	0.77	1.01	91.7	0.58	1.76
0.5	0.44	96.5	0.19	0.81	0.99	90.4	0.47	2.01
0.8^a	0.64	98.4	0.45	0.95	0.96	80.1	0.30	3.08
0.9^a	0.58	89.1	0.47	0.99	0.91	57.0	0.14	3.96

[a]Summaries based on 991 ($\pi = 0.8$) and 899 ($\pi = 0.9$) data sets.

[b]Rounded. Actual median is 7.2×10^{-99}. No lower confidence limits contain 0.

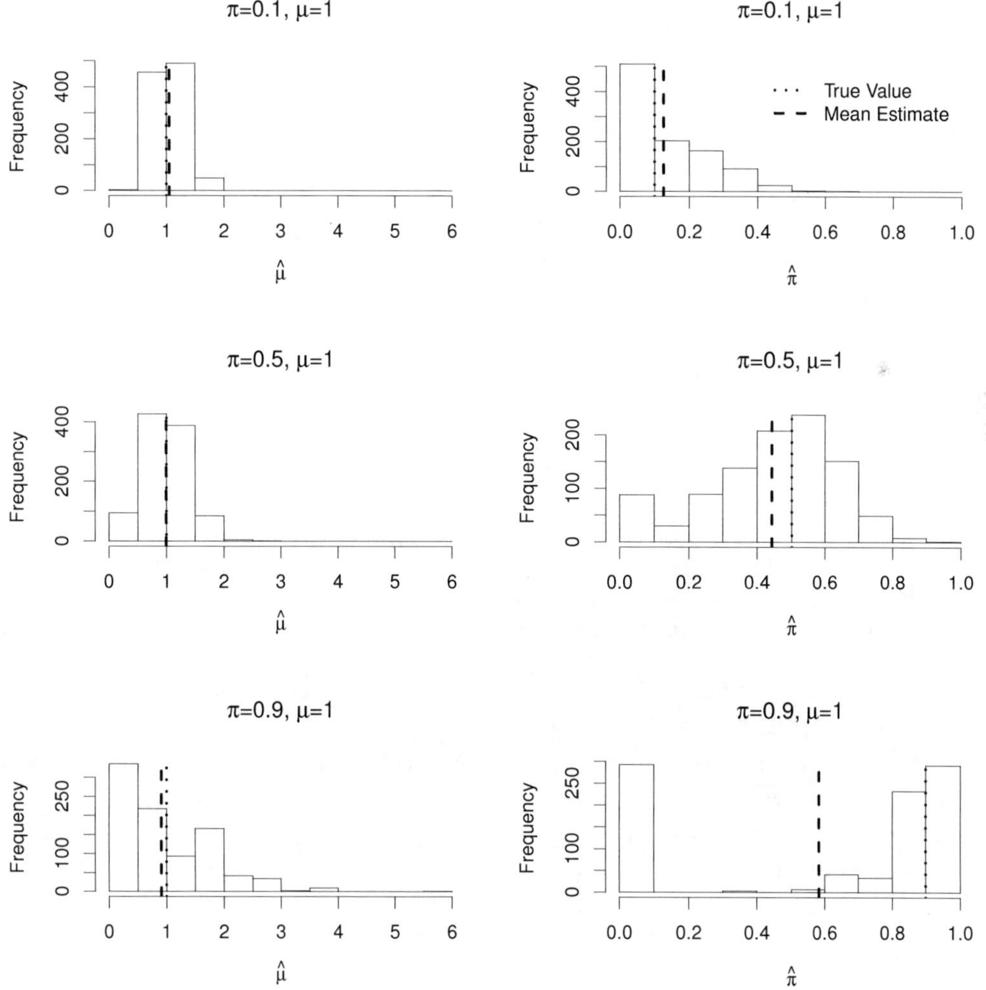

Figure 4.11: Simulated sampling distributions of estimates from ZIP model with various parameter values.

It is evident from the results that it can sometimes be difficult to estimate the parameters from a ZIP model when μ is not far from 0. When the probability of immune is small, there is great uncertainty in the estimates of π, as indicated by the excessively wide confidence intervals. The estimated true confidence levels of the corresponding confidence intervals are too low, despite their vast width. With $\mu = 1$, the probability of a zero response from a susceptible is 0.37. When the immunes add very few additional zeroes, the model estimates a near-zero probability of immune, but with an immense standard error for logit($\hat{\pi}$). Thus, the interval endpoints are transformed to nearly (but not quite) 0 and 1.

On the other hand, with large values of π, there are too few non-zero counts to provide reliable estimates of μ. The estimates $\hat{\mu}$ and $\hat{\pi}$ are mostly too small, indicating that the procedure tends to prefer attributing the many zeroes to a small μ rather than a large π. Most of this difficulty is associated with the small sample size. There is not enough information with which to reliably estimate a small mean and a large probability of immune with $n = 35$. Additional simulations using $n = 100$ and $n = 500$ with $\pi = 0.9$ provide mean estimates of 0.99 and 0.98 for μ and 0.83 and 0.90 for π, respectively.

From this example, we can conclude that some caution is needed in using the ZIP models with small sample sizes, especially when there are very few non-zero observations. In Exercise 22, the simulations can be repeated for the hurdle model.

4.5 Exercises

1. There are many simple data-collection exercises that can be performed with the goal of estimating a population mean count for a particular problem. We encourage you to perform your own. The parts below discuss some possible mean counts that one could estimate. For each part, outline how to take a sample of counts in such a way that it might be reasonable to assume that the data represent a random sample from a single Poisson distribution. In particular, discuss how your sampling plan maintains independence of individual counts and the assumption of a constant mean. If possible, collect the data and examine the appropriateness of the Poisson distribution by comparing the sample mean to the sample variance and comparing observed relative frequencies and the corresponding probabilities from a $Po(\hat{\mu})$ distribution.

   (a) Determine the mean number of people standing in line at once to check out at a particular grocery store.

   (b) Determine the mean number of Ph.D. graduates from a certain academic department at a particular university.

   (c) Determine the mean number of cars passing through a particular intersection of streets within a five-minute period of time.

2. Suppose you count 225 cars crossing a particular bridge between 8:00–9:00 a.m. on a "typical" Monday.

   (a) Assume that a Poisson distribution is a reasonable model for the process of cars crossing the bridge between 8:00 and 9:00 a.m. on "typical" Mondays. Find a confidence interval for the population mean car count for this time period.

(b) If the intensity of cars crossing this bridge is variable from week to week, what impact will this have on your calculation? What impact would it have on the interval's true confidence level or even your interpretation?

3. The Great Plains region of the United States usually has a number of tornadoes during the non-winter months. Groups of scientists, called *storm chasers*, travel throughout the region when thunderstorms are expected in the hopes of encountering a tornado and taking various measurements. The file Stormchaser.csv contains the number of tornadoes encountered on 69, 12-hour long storm chases during 1999–2004 for one particular storm chaser.

   (a) Compute the sample mean of the number of tornadoes in a storm.

   (b) Make a plot comparing the observed counts of tornadoes against the Poisson PMF with mean estimated in (a).

   (c) Comment on the fit of the Poisson distribution. In particular, do any tornado counts occur with a frequency that is much higher or lower than the Poisson distribution expects?

4. Refer to the "Cars at an intersection" example in Section 4.1.1 to answer the questions below.

   (a) The data for this problem were collected between 3:25 and 4:05 p.m. on a non-holiday weekday. Ideally, it would also be important to understand the traffic patterns for other time periods. The number of cars stopped at the intersection is likely to vary because of factors such as time of day, day of the week, whether school is in session, and so forth. Discuss how these factors could have been accounted for in designing this study.

   (b) Are the observations truly independent? If not, discuss what assumptions must be made to use the Poisson distribution for this problem.

   (c) The lengths of vehicles and the distances between stopped vehicles vary. For the purpose of this problem, suppose all vehicles are 14 feet long with a distance between cars of 4 feet when stopped at the intersection. This suggests that nine vehicles ($150/18 = 8.3$) or more will at least partially block the fire station's driveway. Using a Poisson distribution, estimate the probability this will happen for one stoplight cycle. Considering part (a), what caveats need to be placed on the interpretation of this probability?

   (d) Using the probability from part (c), estimate the probability that the fire station's driveway is at least partially blocked one or more times over 60 cycles of the light (roughly one hour). Use the binomial distribution to help answer this problem.

5. Once each week for a year, you drive around a county and observe the number of houses being built. You record the number each week. Is it reasonable to think of these 52 counts as a random sample from a Poisson distribution with mean μ? Why or why not? (Hint: It can take several months, or even longer, to build a house. Is construction cyclic?)

6. Consider the Poisson regression models below involving numerical variables x_1, x_2, and x_3:

   - $\mu = \exp(\beta_0 + \beta_1 x_1 + \beta_2 x_2 + \beta_3 x_3)$

- $\mu = \exp(\beta_0 + \beta_1 x_1 + \beta_2 x_2 + \beta_3 x_1 x_2 + \beta_4 x_3)$
- $\mu = \exp(\beta_0 + \beta_1 x_1 + \beta_2 x_2 + \beta_3 x_2^2 + \beta_4 x_3)$

For each model, complete the following parts for x_2.

(a) Derive the percent change in the mean response that results from a c-unit change in the explanatory variable.

(b) Provide a formal interpretation for the expression derived in (a).

(c) State the Wald confidence interval for the expression derived in (a). Derive the necessary variance expression needed for the interval so that the variance is a simple function of variances and/or covariances of individual regression parameter estimators.

7. For a Poisson regression with $\mu_i = \exp(\beta_0 + \beta_1 x_{i1} + \ldots + \beta_p x_{ip})$, let $x_{i1}, x_{i2}, \ldots, x_{ip}$ be the explanatory variables corresponding to response y_i, $i = 1, \ldots, n$. Derive the log-likelihood in Equation 4.3.

8. In Equation 4.5, show that $\beta_i^X = \log(\mu_{i+}/\mu_{1+})$.

9. Consider a two-way contingency table with counts y_{ij}, $i = 1, \ldots, I$, $j = 1, \ldots, J$. Following Section 4.2.4, use Equation 4.4 to model the mean counts.

   (a) Write out the log-likelihood $\log[L(\beta_0, \beta_2^X, \ldots, \beta_I^X, \beta_2^Z, \ldots, \beta_J^Z | y_{11}, \ldots, y_{IJ})]$ analogously to Equation 4.3.

   (b) Differentiate the log-likelihood with respect to β_2^X, and set the result equal to zero. Show that this equation implies that $\hat{\mu}_{2+} = \sum_{j=1}^J y_{ij}$; i.e., the MLE for the row-marginal total is the sum of counts in that row. (This is also true for $i = 3, \ldots, I$.)

   (c) Repeat this calculation for β_2^Z and hence infer that the MLE for a column total is the observed column total.

   (d) Now consider adding interactions as shown on p. 269. How does this affect the results of parts (b) and (c)?

   (e) Suppose that ordinal forms of interaction were added instead, but the main effects remained nominal. Would this affect the results from (b) and (c)? Explain.

10. For a $2 \times 2 \times 2$ contingency table, assume that the model is given by Equation 4.8.

    (a) Write out the corresponding regression model in terms of indicator variables x_2, z_2, and w_2.

    (b) Write out the log-means in a format like Table 4.4.

11. For the linear-by-linear association model given on p. 294, show that $\log(OR_{ii',jj'}) = \beta^{XZ}(s_i^X - s_{i'}^X)(s_j^Z - s_{j'}^Z)$.

12. The purpose of this problem is to further investigate the political ideology data of Section 4.2.6.

    (a) Refit the model in Equation 4.14 by trying different score sets for I in the PI interaction as suggested in the example (0-2-3-4-6 and 2-1-0-1-2). Do any of them fit as well as 1-2-3-4-5?

(b) Fix the PI scores at 1-2-3-4-5. Now consider the ordinality of I in the GI interaction.

   i. Starting from Equation 4.14, replace the GI term with an ordinal term using 1-2-3-4-5 scores and fit the model. Fit the model two more times, replacing GI scores with 0-2-3-4-6 and 2-1-0-1-2. Compare deviances for these three models to the one for Equation 4.14 to test whether the reduced number of parameters in each ordinal model still provides an adequate explanation for the association.

   ii. Is the ordinal GI association significant in each case?

(c) Use the GI linear association model with the smallest residual deviance from part (b) as the final model.

   i. Estimate the GI odds ratio for a 1-unit difference in scores. Interpret what this measures and find the Wald confidence interval for the estimate.

   ii. Compare the results of the ordinally estimated odds ratio to the nominal estimates obtained in the example on p. 293. In particular, compare the confidence intervals for each odds ratio. Do they give similar interpretations?

(d) Write a brief conclusion for the results of this analysis. Explain the nature of any associations that were found. Do not use mathematical symbols in this discussion.

13. In the downtown areas of very large cities, it is common for Starbucks locations to be within a block of one another. Why does Starbucks decide to put their locations so close together? One would expect that it has something to do with how busy a current location is. If an order line is long, a potential customer may not even get into line and, instead, leave without making a purchase, which is lost business for the store.

Using this as motivation, a Starbucks location in downtown Lincoln, Nebraska was visited between 8:00 a.m. and 8:30 a.m. every weekday for five weeks. The number of customers waiting in line was counted at the start of each visit. The collected data are stored within the file Starbucks.csv, where `Count` (number of customers) is the response variable and `Day` (day of week) is the explanatory variable.

(a) What is the population of inference? In other words, define the setting to which one would like to extend inferences based on this sample.

(b) Construct side-by-side dot plots of the data where the y-axis gives the number of customers and the x-axis is for the day of the week. Describe what information this plot provides regarding the mean number of customers per day. In particular, does it seem plausible that the population mean count is constant across the days? (We recommend putting the factor values given within `Day` in their chronological order using the `factor()` function before completing this plot.)

(c) Using a Poisson regression model that allows different mean counts on different days, complete the following:

   i. Fit the model.

   ii. Perform an LRT to determine if there is evidence that day of the week affects the number of customers waiting in line.

   iii. Estimate the ratio of means comparing each pair of the days of the week and compute 95% confidence intervals for these same comparisons. Do this both with and without control for the familywise confidence level for the family of intervals. Interpret the results.

iv. Compute the estimated mean number of customers for each day of the week using the model. Compare these estimates to the observed means. Also, compute 95% confidence intervals for the mean number of customers for each day of the week.

(d) The hypotheses for the LRT in part (c) can be written as $H_0 : \beta_2 = \beta_3 = \beta_4 = \beta_5 = 0$ vs. H_a : At least one $\beta_r \neq 0$. These hypotheses can be equivalently expressed as $H_0 : \mu_{\text{Monday}} = \mu_{\text{Tuesday}} = \mu_{\text{Wednesday}} = \mu_{\text{Thursday}} = \mu_{\text{Friday}}$ vs. H_a : At least one pair of means is unequal, where μ_{i} represents the mean number of customers in line on day i. Discuss why these two ways of writing the hypotheses are equivalent. Write out the proper forms of the Poisson regression model to support your result.

14. Often the word "rate" is used informally to refer to something that would better be described as a proportion. For each of the situations below, indicate whether a logistic regression or a Poisson rate regression would be more appropriate.

 (a) Crime rate: in different cities, measure counts of violent crimes relative to population sizes.
 (b) Product defect rate for a particular product: on different days of production, measure the number of defective products produced (or sampled) relative to total number produced (or sampled).
 (c) Egg-laying rate: on different plants, measure the number of insect eggs laid relative to the number of adult females placed on the plants.
 (d) Alcohol consumption rate: at different soccer games, measure the number of beers sold relative to number of fans.
 (e) Illegal parking rate: in different parking lots, measure the number of cars illegally parked relative to the total number of cars in the lot.
 (f) Medical resource usage rate: for different patients, measure the number of doctor visits since last heart attack relative to time since last heart attack.
 (g) Caesarean delivery rate: at different hospitals, measure the number of babies delivered by Caesarean section relative to total number of babies delivered at the hospital.

15. Use the `hurdle()` function in `pscl` to repeat the analysis of the beetle egg example from Section 4.4 with a hurdle model. Compare the results to those from the ZIP model. Are there any differences that lead you to prefer one model over the other? In terms of the beetles, what is the difference in assumptions between the two models regarding the source of zero counts? (Note that the coefficients in the probability model estimate $P(Y > 0|z)$, but `predict(..., type = "prob")` returns $P(Y = 0|z)$.)

16. Deb and Trivedi (1997) and Zeileis et al. (2008) examine the relationship between the number of physician office visits for a person (`ofp`) and a set of explanatory variables for individuals on Medicare. Their data are contained in the file DT.csv. The explanatory variables are number of hospital stays (`hosp`), number of chronic conditions (`numchron`), gender (`gender`; male = 1, female = 0, no other genders reported), number of years of education (`school`), and private insurance (`privins`; yes = 1, no = 0). Two additional explanatory variables given by the authors are denoted as `health_excellent` and `health_poor` in the data file. These are self-perceived health status indicators that take on a value of yes = 1 or no = 0, and they cannot both be 1 (both equal to 0 indicates "average" health).

(a) Fit the Poisson regression model to predict the number of physician office visits. Use all of the explanatory variables in a linear form without any transformations.

(b) Interpret the effect that each explanatory variable has on the number of physician office visits.

(c) Compare the number of zero-visit counts in the data to the number expected from the model and comment. Can you think of a possible explanation for why there are so many zeroes in the data? See Exercise 17 for an alternative analysis that accounts for these excess zeroes.

17. Refer to Exercise 16 on counts of hospital visits. It was discovered that there are more zeroes in the response counts than the fitted Poisson regression model expects.

    (a) Fit the ZIP regression model to predict the number of physician office visits. Use all of the explanatory variables in a linear form without any transformations for the $\log(\mu_i)$ part of the model and no explanatory variables in the π_i part of the model. Interpret the model-fit results.

    (b) Complete part (a) again, but now use all of the explanatory variables in a linear form in both parts of the model.

        i. Compare this model to the previous ZIP model using an LRT.
        ii. Use and interpret LRTs for each variable in the mean portion of this model.
        iii. Which variables have the strongest association with the probability that a person is "immune to" (has no need for) hospital visits? Are they the same variables that relate to the number of visits for people who might still need them?

    (c) Examine how well each model estimates the number of 0 counts.

18. Explore the effectiveness of ML estimation in the ZIP model using ZIPSim.R from Section 4.4. Below are ideas for completing this exercise.

    (a) Examine how well it estimates the probability of diapause and mean number egg masses for the beetle egg-laying example at 21°C. This can be done using the function call `save.beetle <- sim.zip(n = 35, pi = 0.38, mu = 5.14, sets = 1000, seed = 1200201)`. Include the following in your examination.

        i. Recall that the Wald interval is based on the large-sample normality of $\log(\hat{\mu})$. Make a histogram and a normal quantile (QQ) plot (`qqnorm()`) of the estimated log-mean (the estimated means are stored in the first row of `save.beetle`) and indicate how much faith you have in the use of a normal approximation for the distribution of $\log(\hat{\mu})$. What does this suggest about the general quality of the Wald intervals for μ here?
        ii. Repeat for $\log(\hat{\pi}/(1-\hat{\pi}))$ (from the fourth row of `save.beetle`).

    (b) Rerun simulations as in the example from Section 4.4, but this time hold π constant at 0.8 and let μ vary across 0.5, 1, 2, 5, and 10. Summarize the results and draw conclusions about the effect of different population means on the quality of estimation of π when it is relatively large and $n = 35$. As an example, `save.05 <- sim.zip(n = 35, pi = 0.8, mu = 0.5, sets = 1000, seed = 29950288)` will generate the first results.

    (c) Explore the effect of sample size on these conclusions. Repeat the simulations from this exercise using $n = 100$. How do the conclusions change?

Table 4.10: Salamander count data.

Years after burn	12	12	32	20	20	27	23	19	23	26
Salamanders	3	4	8	6	10	5	4	7	2	8
Years after burn	21	3	8	35	2	19	8	25	33	35
Salamanders	6	0	2	6	1	5	1	5	4	10

19. The data in Table 4.10 are from a survey of salamanders counts taken on plots of fixed size sampled from prairie regions in Manitoba, Canada.[19] The prairies are subject to periodic controlled burning to reduce the chance of uncontrolled wildfires. The research question being investigated is how salamander populations react to these controlled burns. Use a Poisson regression model to address this question. Also comment on the apparent fit of the model.

20. The file BladderCancer.csv contains data concerning recurrence of bladder cancer tumors in patients who had received previous treatment to remove a primary tumor (Seeber, 2005). Patients were observed for different periods of time post-surgery, which were measured in months, and have different sizes of primary tumor, classified as 0 if the tumor was small (< 3 cm) and 1 if large (> 3 cm). The question to be answered is whether the size of the initial primary tumor relates to the number of tumors (`Tumors` in the data file) discovered in follow-up. The different lengths of follow-up time represent different exposures to potential tumor development. Analyze these data.

21. Use the data from the egg-laying example from Section 4.3 to find confidence intervals for the ratio of egg-laying rates between individual and group cages separately at 21°C and 24°C. Use these results to explain the interaction. Interpret exactly what the interaction coefficient, -0.279, measures in terms of egg-laying rates.

22. Repeat the simulations from Section 4.4 for the hurdle model. How does its performance compare to the ZIP model? Draw some conclusions about the relative performance of the two models, making sure to indicate the limitations of your conclusions. (Considering that the data were actually generated from a ZIP model, this is a demonstration that sometimes one gets better answers from an approximate model than from the correct one!)

23. Count data are often modeled using the usual linear regression and ANOVA models that assume normally distributed responses. While such models ignore the discrete nature of count data and the potential for unequal variances, they are supported by the fact that the Poisson distribution is increasingly well approximated by the normal distribution $N(\mu, \mu)$ as μ increases. The program PoissonNormalDistPlots.R generates a probability mass histogram for a Poisson distribution with $\mu = 1$ and a probability density curve for the corresponding normal. Try different values of μ with this program and observe the relationship between the two distributions. Above what value of μ do you feel that the normal approximation seems excellent? Below what value does it seem poor?

24. McCullagh and Nelder (1989, p. 205) provide data on the number of reported incidents of wave damage incurred by ships of five different types (`ShipType`), four different

[19]Data kindly provided by Dr. Carl Schwarz, Department of Statistics and Actuarial Science, Simon Fraser University.

construction periods (ConstYear), and two different operation periods. An amended version of these data, consisting of just one of the two periods of operation, is in the file Ships.csv. There are different numbers of ships of each type, and different years of construction and, hence, different potential for damage incidence in each combination of year and ship type. There is an additional measure, "Aggregate months of service" (Months), that represents the exposure.

(a) Use a Poisson rate regression with variables ShipType and ConstYear to test whether each affects the rate of damage incidents per month. Interpret all tests using $\alpha = 0.05$ and draw conclusions. Note that there is one observation for which there was no exposure! This will cause an error if left in the data. It can be removed by adding a condition to the data in the glm() function call, such as dataframename[Months>0,].

(b) Use a 95% confidence interval to estimate the geometric mean incident rate per 1,000 months of service for each ship type across all years of construction. Also perform pairwise comparisons at $\alpha = 0.05$. Which ships appear to be least or most prone to damage? Make a plot to support your conclusions.

(c) Now include in the model an interaction between ship type and construction period.

   i. Explain what such an interaction term is attempting to measure.

   ii. Does there appear to be a significant interaction between ship type and construction year? Explain.

(d) Draw conclusions about the problem without referring to mathematical symbols.

25. In Section 4.2, we analyzed alcohol consumption as a function of the number of positive and negative events for a sample of moderate-to-heavy drinkers during their first Saturday on the study. Repeat the analysis for Friday using the data in DeHartFriday.csv and draw conclusions. Are the results similar to those for Saturday?

26. The researchers in the alcohol consumption study proposed the following hypothesis (DeHart et al., 2008, p. 529):

> We hypothesized that negative interactions with romantic partners would be associated with alcohol consumption (and an increased desire to drink). We predicted that people with low trait self-esteem would drink more on days they experienced more negative relationship interactions compared with days during which they experienced fewer negative relationship interactions. The relation between drinking and negative relationship interactions should not be evident for individuals with high trait self-esteem.

In DeHartSaturday.csv, trait self-esteem (a long-term view of self-worth) is measured by the variable rosn, while the measure of negative relationship interactions is nrel. Conduct an analysis to address this hypothesis, using the data for the first Saturday in the study.

27. Consider an expanded regression model for the alcohol consumption study from Section 4.2. We will use a model that regresses the number of drinks consumed (numall) against positive romantic-relationship events (prel), negative romantic-relationship events (nrel), age (age), trait (long-term) self-esteem (rosn), state (short-term) self-esteem (state), and two other variables that we will create below. We will again use Saturday data only.

The `negevent` variable is the average of the ratings across 10 different types of "life events", one of which is romantic relationships. We want to isolate the relationship events from other events, so create a new variable, `negother`, as 10*`negevent` – `nrel`. Do the same with positive events to create the variable `posother`.

(a) Construct plots of the number of drinks consumed against the explanatory variables `prel`, `nrel`, `posother`, `negother`, `age`, `rosn`, and `state`. Comment on the results: which variables seem to have any relationship with the response?

(b) Fit the full model with each of the variables in a linear form. Report the regression parameter estimates, standard errors, and confidence intervals. Do these estimates make sense considering the plots from part (a)?

(c) Conduct LRTs on the regression parameters to determine which corresponding variables make a significant contribution to the model. State the hypotheses, test statistic, p-value, and use the results to draw conclusions regarding the contributions of each variable to the model.

(d) Determine whether any variables except the two negative events variables are needed in the model. To do this, refit the model with only `nrel` and `negother`. Perform an LRT comparing the full model above with this reduced model. State the hypotheses, test statistic and df, p-value, and conclusions.

28. Suppose you take a random sample, $y_{11}, y_{12}, \ldots, y_{1n_1}$, from a $Po(\mu_1)$ distribution and an independent random sample, $y_{21}, y_{22}, \ldots, y_{2n_2}$, from a $Po(\mu_2)$ distribution.

(a) Write out a single likelihood function for both parameters using both samples.

(b) Find the MLEs for μ_1 and μ_2 from this likelihood. Do the estimates make sense?

(c) Now consider the test $H_0 : \mu_1 = \mu_2$. Under H_0, both samples are drawn from the same Poisson distribution, say $Po(\mu_0)$. Show that the MLE for μ_0 is $(n_1\bar{y}_1 + n_2\bar{y}_2)/(n_1 + n_2)$, where $\bar{y}_j = (1/n_j)\sum_{i=1}^{n_j} y_{ji}$.

(d) Derive the LRT statistic for this test.

29. Prove the three properties of the ZIP model listed on p. 304.

30. Refer to Table 1.7 and Exercise 19 in Chapter 1. Reanalyze these data using a Poisson loglinear model.

(a) Fit the saturated loglinear model and report the parameter estimates.

(b) Explain what independence of the two variables means in the context of the problem.

(c) Use an LRT to test for independence between strategy and field goal outcome. Report all parts of the test and draw conclusions.

(d) Explain what the population odds ratio measures in the context of this problem.

(e) Estimate the odds ratio between strategy and field goal outcome. Include a 95% confidence interval for the population odds ratio.

(f) Draw conclusions from the analysis: Does calling a time out appear to reduce the kicker's field goal success rate?

31. Use Poisson regression methods to perform the analysis suggested in Exercise 22 of Chapter 3.

32. Recall the fiber-enriched cracker example from Table 3.2 in Section 3.2.

    (a) Consider at least three different possible sets of scores for the ordinal response variable, bloating severity. Explain why each might be relevant.

    (b) Reanalyze these data using ordinal loglinear models. Draw conclusions about the effect of the fiber source on bloating.

33. The data file Pregnancy.csv gives results on a study of *pre-eclampsia* (also known as *toxaemia*), a medical condition affecting women in pregnancy. It is indicated by the presence of both high blood pressure, or *hypertension* (HT), and protein in the urine, or *proteinurea* (PU). The cause of pre-eclampsia is not immediately known, and other medical conditions may cause either of these symptoms to appear without the other. It is therefore of interest to understand the factors that relate to HT, PU, and their association with each other. Brown et al. (1983) report on a study in which women in their first pregnancy were classified according to smoking status (`Smoke`, with levels 0, 1–19, 20+, coded as 1, 2, 3), social class (`Social`, with levels I, II, III, IV, V, coded as 1, 2, 3, 4, 5), and presence of symptoms `HT` (yes/no) and `PU` (yes/no). The goal was to see whether smoking and social class relate to either symptom's frequency and/or whether they are associated with the symptoms' interaction.

    (a) Examine the data using the steps below.

        i. Convert the explanatory variables `Smoke` and `Social` to variables `Smokef` and `Socialf` using `as.factor()`.

        ii. Use these new factors to create a cross tabulation of the data (e.g., `xtabs(formula = Count ~ HT + PU + Smokef + Socialf, ...)`).

        iii. Use `ftable(..., col.vars = c("HT", "PU"))` to create a nicer-looking table and print out both the counts and the proportions (use `prop.table()` for the latter).

        iv. Comment on these table summaries: What effects do the explanatory variables have on the two symptoms?

    (b) Treat social class and smoking status levels as nominal for now. Consider a loglinear model using `formula = Count ~ (Smokef + Socialf + HT + PU)^3`.

        i. Assign the letters S, C, H, P to represent the four variables in this model, respectively. Use the notation given in Section 4.2.5 to describe the model being fit and identify all of its terms.

        ii. Write out the corresponding regression model for the log-mean count of women in each factor-level combination using indicators for factor levels.

        iii. Considering the study's objectives, why is this model a reasonable starting point?

        iv. Which terms in the model address whether the explanatory variables relate to the symptoms' frequency?

        v. Which terms address whether they relate to the symptoms' association?

    (c) Fit the model and perform the following analyses.

        i. Use the residual deviance to test whether the excluded four-variable interaction is needed in the model. Report results and draw conclusions.

        ii. Use `Anova()` from the `car` package to test the significance of terms identified in part (b) as being important for the analysis. Draw conclusions.

(d) Fit the simpler model that omits all three-variable interactions and perform the following analyses.

i. Test whether the simpler model provides an adequate fit relative to the model that includes all three-variable interactions. State the hypotheses, test statistic, and p-value. Draw conclusions.

ii. According to these results, are smoking or social class related to the association between PU and HT? Explain.

iii. Use odds ratios based on the SH term to examine the relationship between smoking and HT. Compare all pairs of smoking levels. List the estimated odds ratios and 95% confidence intervals and draw conclusions.

iv. Repeat this for the relationship between social class and PU.

v. Analyze the HP odds ratio in the same way. Draw conclusions.

34. Using the data described in Exercise 33, consider treating smoking as an ordinal variable.

(a) Discuss possible choices of scores.

(b) Starting from the model with all two-variable interactions, use 1-2-3 scores to analyze smoking effects on HT and PU. (Leave the smoking main effect nominal as well as its interaction with social class.)

(c) Compare the fit of the ordinal model relative to the nominal model from part (d) of Exercise 33. Draw conclusions about the ordinal model.

35. The `datasets` package in R contains a `table` object called `HairEyeColor` consisting of a cross-classification of 592 people by hair color (black, brown, red, blond) and eye color (brown, blue, hazel, green). According to Snee (1974), the sample was taken as a class project by students in a Statistics class at the University of Delaware. It is not clear how the sampling was conducted.

(a) The data were later artificially split out by sex by Friendly (1992), so the `HairEyeColor` table is actually a three-way table with variables `Sex`, `Hair`, and `Eye`. We first need to remove the artificial sex-split using `margin.table(x = HairEyeColor, margin = c(1,2))`. The result should be a 4×4 table. Print out the resulting table and comment on the results. Do there appear to be any associations between the hair color and eye color?

(b) Fit the model in Equation 4.6 to the data. Obtain the summary of the fit. Explain what the symbols in Equation 4.6 mean in terms of the hair and eye colors.

(c) Write out the log-odds ratio comparing the odds of blue vs. brown eyes for someone with blond hair vs. brown hair as in Equation 4.7. Estimate the odds ratio with a 95% confidence interval and interpret the result.

36. Continuing Exercise 35, consider the hypothesis that brown and black hair have the same eye-color distribution and that red and blond hair have the same eye-color distribution, but that these two distributions differ.

(a) Create a new variable (say `Hair2`) to represent these two groups ("dark" vs. "light" hair), and refit the model using this new variable instead of the variable `Hair`. Perform an LRT of the proposed hypothesis. Provide the test statistic and p-value and draw conclusions about the hypothesis.

(b) A second hypothesis is that eye color can be reduced to two groups: "dark"=brown and "light"=(blue, hazel, green). Starting from the model on the full 4×4 hair-color data, repeat the LRT analysis on the eye-color hypothesis.

37. Long (1990) gathered data comparing the publication success of recent male and female Ph.D. recipients in biochemistry. The data are available in the data frame `bioChemists` in the `pscl` package. The response is `art`, the number of publications produced in the three-year period ranging from one year before the year in which the Ph.D. was granted to one year after. Explanatory variables available include `fem`, an indicator for female gender of the subject; `mar`, an indicator for married status in the year of the Ph.D.; `kid5`, indicating whether the subject had children aged 5 or younger; `phd`, a numerical index of the Ph.D. department's prestige; and `ment`, the number of journal articles published by the student's mentor over the same period. The main research hypothesis is that the various factors, other than gender, have a more serious effect on females' publication counts than on males'. That is, gender likely interacts with the other factors in such a way that they have more of an effect on females, especially in reducing publications, than on males.

    (a) Make a histogram of `art` and note the large number of zeroes. Report this number.

    (b) Fit a Poisson regression model using `art` as the response and all other variables in their linear forms as explanatory. Use the model to estimate the expected number of zero counts in the data. Is this number close to the actual number?

    (c) Repeat the analysis with a ZIP regression, using linear terms for all variables in both parts of the model.[20]

        i. Examine the expected number of zeroes produced. Does this seem closer to the actual number?

        ii. Use an LRT to compare the ZIP model to the Poisson model. Does adding the model for probability of zero improve the fit of the model?

        iii. Create a summary of the model fit and explain what effects female gender have on the probability of zero articles and the mean count.

    (d) Add interaction effects between `fem` and all other explanatory variables to the count portion of the model.

        i. Create a summary of the model and identify which variables, if any, interact with female gender in their effect on the number of articles published. Use the corresponding parameter estimates to describe the nature of the interactions.

        ii. Use `emmeans` to study the effect of gender on the total expected number of articles separately for people from schools with prestige levels of 1, 2, 3, and 4.

38. Air traffic controllers (ATCs) are required to retire at age 56 in the United States. This relatively early age is used because of concerns that an individual's ability to perform this job may diminish as one gets older. Broach and Schroeder (2006) explore this age requirement further to determine if there was actual data-based evidence to validate it. The authors collected information on experience and age of ATCs and analyzed

[20] The model produces warnings indicating that log(1-phi) produces NaNs, suggesting that estimates of $\pi(z)$ are approaching 1. An examination of the model-estimated probabilities of zero using `predict()` with `response = "zero"` shows that does not appear to be an issue.

the aggregated data available in ATC.csv. The `experience` variable has values of 1 for 3 or fewer years of experience, 2 for 4–9 years, 3 for 10–14 years, 4 for 15–19 years, 5 for 20–24 years, and 6 for 25 or more years. The `age` variable is equal to 1 for 55 years old or less and a 2 for 56 years old or more. Despite the age requirement, some ATCs obviously continued to work past their 56th birthday. This is partially because some ATCs were hired prior to when the age requirement was put into place. For a specific experience and age combination, the `errors` variable in the data set represents the number of errors by ATCs and the `population` variable is the number of ATCs.

(a) Plot the data with error rate (`errors` / `population`) on the y-axis and experience group on the x-axis. Differentiate the plotting points appropriately so that one can distinguish between the two age groups. Discuss trends shown in the plot.

(b) Why is it important to use the `population` variable as part of an offset in a Poisson rate regression model to estimate the mean number of errors?

(c) Fit and state the Poisson rate regression model for the mean number of errors using `experience` and `age` as explanatory variables (no interaction) and `population` in the offset. Treat `experience` as ordinal with scores equal to the group number.

(d) Add the estimated model to the plot in part (a) (use $t = 1$ because the error rate is on the y-axis). Comment on how well the model follows the trend among the data points in the plot.

(e) Refit and state the Poisson rate regression model, but now also include a quadratic term for `experience`. Again, plot the data and the estimated model. Comment on how well the model follows the trend among the data points in the plot.

(f) The final conclusions reached by the authors are "the results reported from this study cast doubt on the explicit characterization of the mandatory early retirement of controllers as 'primarily a safety measure'." Do you reach the same conclusions? Explain your answer using the appropriate statistical inference methods.

(g) The `age` variable is binary with values of 1 and 2. Why would the conclusions about the variable remain the same if an alternative coding of 0 and 1 was used?

39. All research should attempt to use correct statistical analyses and be reproducible. To explore this relative to Broach and Schroeder (2006) in Exercise 38, we continue examining this paper further.

(a) The goal of the paper was to investigate if older age was an indicator of an increased number of errors. While the authors of the paper used Poisson regression, it does not appear that they included `population` in their model. Why could omission of the offset for `population` lead to an invalid conclusion?

(b) Page 370 of the paper states:

> Overall, the Poisson regression model fit the data poorly, likelihood ratio $\chi^2(5, N = 3,231) = 283.81$, $p < .001$. The parameter estimate for the main effect of age (3.50) was significantly different from 0 (95% confidence interval = 3.29 to 3.70), as were the parameter estimates for [experience].

Table 4.11: Star Trek crew and casualities. Data source is Barsalou (2013).

Shirt color	Casualties	Crew members
Blue	7	136
Gold	9	55
Red	24	239

Unfortunately, the authors do not explicitly state what model was used. Through trying a number of possible Poisson regression models, we discovered the model had `errors` as the response variable and `experience` and `age` as explanatory variables without an interaction term. No offset term was included in the model. Complete the following:

i. What does "significantly different from 0" mean in the above quote relative to age? Does this match the authors' conclusion as given in part (f) of the previous exercise?

ii. The authors treat `experience` as a nominal categorical variable. What potential information could be lost by making this choice rather than treating the variable as ordinal?

iii. Show that the 3.5 given in the quote should actually be −3.5. How could this change the interpretation of age in the model?

iv. The $\chi^2(5, N = 3,231)$ in the above quote is in reference to an LRT for testing the model with both explanatory variables versus a saturated model. Show that the "283.81" in the quote should be 238.81. Section 5.2.2 discusses why this measure assesses how well the model fits the data.

40. On the *Star Trek: The Original Series* television show, crew members of the starship Enterprise wore blue, gold, or red shirts as part of their uniform to designate their roles on the ship. Many Star Trek fans believe that crew members with red shirts were more likely to be casualties than those wearing other shirt colors. Table 4.11 provides casualty data for the series, where the "Crew members" column reflects the typical complement for the Enterprise. The specific individuals on the starship varied over the course of the show because of factors including death, reassignment, or resignation. We assume that they would be replaced and that the number of crew members in each role remained constant.

    (a) Why can one not simply conclude that the red uniform crew were more likely to be a casualty because they have a larger number of casualties in the data set?

    (b) Fit a Poisson rate regression model using casualties as the response variable, color as an explanatory variable, and crew members as the exposure.

    (c) Does uniform color matter with respect to casualty rate? Perform an LRT to answer this question.

    (d) Find the ratio of the mean number of casualties for the red and gold shirts in terms of the proposed model (do not include parameter estimates).

    (e) Compute 95% profile LR intervals for *PC* comparing red to gold, red to blue, and gold to blue. Interpret the intervals.

    (f) What is your overall conclusion regarding the red shirts relative to Star Trek fan perceptions?

41. Continuing Exercise 26 of Chapter 1, the purpose here is to show how the analysis of the Olestra data can be performed using a loglinear model.

(a) Convert the data from the contingency table format to the appropriate format needed for `glm()`. Use R code for this conversion rather than retyping the data into a data frame.

(b) Fit a saturated Poisson regression model using both the row and column variables as explanatory variables in the model for the counts in the table. When stating the estimated model, use indicator variables to represent the model's terms.

(c) State the estimated model in a format like what is shown for Equation 4.6.

(d) Show that the estimated mean counts for the model in (b) are equal to the observed counts as should occur with a saturated model.

(e) Perform an LRT for independence between the row and column variables using the model in (b). Compare your results here to that which can also be obtained using the `assocstats()` function from the `vcd` package.

42. Derive the score confidence interval for the Poisson mean parameter as given in Equation 4.1.

43. Refer to Exercise 3.

(a) What different assumptions do the hurdle and ZIP models make regarding the cause of zero counts for tornadoes?

(b) There are many different kinds of thunderstorms, but storm chasers specifically seek those that seem to have the potential for tornadic development. With this in mind,

i. Describe what the parameter π would represent in a zero-inflated model.

ii. Describe what the parameter π would represent in a hurdle model.

(c) Fit both the ZIP and hurdle models, using only an intercept for both the mean and probability portions of the models. Report all model parameter estimates.

(d) For both models, estimate the respective probabilities and mean counts with 95% confidence intervals and interpret the results.

44. Refer to the example on p. 247, where we calculated the variances of the sample mean and sample variance for estimating the variance of a Poisson distribution.

(a) Show that the ratio of variances between the sample variance and the sample mean is $2\mu n/(n-1)+1$.

(b) Rerun the simulation for $\mu = 2$ and $\mu = 50$ and report the results. Do they support the mathematical derivation that the ratio of variances is $2\mu n/(n-1)$?

45. Refer to the example on alcohol consumption in Section 4.2.2. We question whether the observation of 21 drinks in one evening is genuine. Remove this observation and rerun the regression analysis on positive and negative events and their interaction.

(a) Compare the confidence intervals for the regression parameters to the original results. Do they seem very different?

(b) Compare the LRT for each parameter to the original results. Has it changed enough to result in different conclusions about the model?

(c) Compare the plots to those in Figure 4.6. Do they tell the same basic story?

(d) Is the model fit heavily influenced by the questionable value? Provide your overall conclusions.

46. Refer to the example on Ecuadorean bird counts in Section 4.2.3. In that example, we wanted to compare mean bird counts at each habitat type to each other type using ratios, where the two forest locations were considered to be the same habitat, and the two pasture locations were considered to be the same habitat. In the example, we demonstrated how to derive the required coefficients to compute the ratio of forest mean to edge mean, expressing the linear combination both in terms of log-means (for `emmeans`) and in terms of regression parameters (for `mcprofile`).

    (a) Write out the remaining five ratios in terms of the means, $\mu_1 \ldots, \mu_6$: forest to edge, forest to pasture, edge to fragment, edge to pasture, and fragment to pasture.

    (b) Write out each linear combination of log means. Then express each linear combination as six coefficients for use with `emmeans`.

    (c) Write out each linear combination in terms of regression parameters. Then express each linear combination as six coefficients for use with `mcprofile`.

    (d) Compute and present all six estimated ratios and Wald confidence intervals using `emmeans`, without adjusting for multiple confidence intervals. Which habitat types appear to have larger or smaller means than others?

    (e) Repeat part (d) with LR confidence intervals using `mcprofile`. Which habitat types appear to have larger or smaller means than others?

47. Repeat part (e) in Exercise 46 using the default adjustment for multiple inferences. Do the conclusions change compared to the previous part?

48. Refer to the example on p. 308 where we fit four different ZIP models. In the example, we identified `zip.mod.t0` as the model we preferred, but we performed the analysis using `zip.mod.tt`. Now repeat that analysis using `zip.mod.t0`. Specifically:

    (a) Find and interpret 95% profile LR confidence intervals for all model parameters.

    (b) Estimate and compute 95% confidence intervals for the mean number of eggs laid by a female not in diapause at each temperature.

    (c) Estimate and compute 95% confidence intervals for the ratio of the mean egg count at 21°C to the mean egg count at 24°C.

    (d) Estimate and compute the 95% confidence interval for the probability that a female is in diapause.

    (e) Estimate and compute 95% confidence intervals for the expected average number of eggs laid by all females at each temperature, and for the difference in these means between 21°C and 24°C.

49. Refer to the example on p. 282 where we fit the homogeneous association model to the political ideology data. In that example, we did an analysis of the odds ratios relating to the *GI* interaction. Repeat that analysis on the *PI* interaction.

50. In an early paper on the relationship between smoking and early death, Hammond and Horn (1954) study past smoking status and death rates of 187,766 men over a 20-month period. The data are given in the file JAMA1954Smoking.csv. The variables are `Smoker` (`0` if very little or no smoking in their lifetimes, `1` otherwise), `Agecat` (`1` for men of ages 50–54 at the start of the study, `2` for 55–59, `3` for 60–64, and `4` for ages 65–69), `Death` (`1` if the subject died during the 20-month follow-up period, `0` otherwise), and `Count` of the number of men in each combination for the other

three variables. The main question is whether smoking and death are associated. A secondary interest is in whether the association between smoking and death changes across age groups. If it does, then how is it different?

(a) Fit a saturated loglinear model to these data. Use `summary()` to ensure that the correct number of parameters is estimated for each term.

(b) Perform LRTs on all terms in the model. Report test statistics and p-values.

(c) Interpret the results of the test for the three-way interaction. Specifically, what question does this test answer for the researchers?

(d) Since there is a significant three-way interaction, we explore the relationship between smoking and death separately for each age group.

i. Compute the ratio of the odds of death for smokers to the odds of death for nonsmokers for each age category.

ii. Report 95% Wald confidence intervals for each of these odds ratios and interpret. Specifically, does smoking appear to be associated with increased risk of death in each age group?

iii. Repeat using LR confidence intervals, also reporting the coefficients used for each of the log-odds ratio calculations.

(e) Because the three-way interaction is significant, it suggests that not all age categories have the odds ratio for death by smoking status. We now perform pairwise comparisons of these odds ratios by forming ratios of two odds ratios at a time.

i. Compute ratio of odds ratios for death by smoking status for each pair of age categories, with the older category in the numerator.

ii. Use Wald methods to compute 95% confidence intervals for these ratios of odds ratios. Use these to identify which age categories have higher or lower risk of death associated with smoking.

iii. Repeat using LR methods, also reporting the coefficients used for each of the comparisons.

51. Refer to Exercise 50 on the relationship between smoking and death in 187,766 men. Note that the `Agecat` variable, which groups ages into five-year intervals, is an ordinal variable. Consider an ordinal version of `Agecat` using 1-2-3-4 scores.

(a) What does a 1-unit increase in score represent with respect to age?

(b) Suppose we use an ordinal version of `Agecat` in the three-factor interaction with `Smoker` and `Death` and a nominal version elsewhere. What assumption does this form of interaction make about the relationship between age and the odds ratio for death by smoking status?

(c) Fit the model using an ordinal version of `Agecat` in the three-factor interaction with `Smoker` and `Death`.

i. Is the new three-factor interaction significant? What does this result suggest about the relationship between the odds ratio for death by smoking status and age?

ii. Compare this model to the saturated model using an LRT. Is the relationship between the odds ratio for death by smoking status and age adequately described by this new assumption?

52. Flaherty et al. (2012) examine red squirrel habitats in Scotland. For 52 equally sized plots of land, information on the following variables was collected:

- `plot`: Plot number
- `cones`: Number of pine cones stripped
- `ntrees`: Total number of trees
- `dbh`: Mean diameter at breast height of trees in meters
- `height`: Mean tree height in meters
- `cover`: Mean percentage of canopy closure

The data are available in Squirrels.csv. The number of stripped cones for a plot is an indicator of habitat preference by squirrels (the more stripped cones, the greater the preference). Therefore, the authors chose to estimate the number of stripped cones as a function of the variables available for each plot.

(a) The authors focused on a Poisson regression model that uses canopy cover, number of trees, and mean tree height as linear terms without interactions. Estimate this model. For readers comparing their estimated model to what is presented in the paper, the standard errors given in the paper are adjusted from those obtained from a regular Poisson regression model. This adjustment will be discussed in Section 5.3 and Exercise 38 of Chapter 5. For this part, do not adjust the standard errors.

(b) For each explanatory variable in the model from part (a), complete the following items.

   i. Perform LRTs to evaluate the importance of the explanatory variable.

   ii. Estimate PC for each explanatory variable. Use a one standard deviation increase for the explanatory variable in your calculation.

   iii. Compute and interpret a 95% profile LR interval for each PC.

(c) Estimate the mean number of stripped cones using the model from part (a) at the mean values for the three explanatory variables. Compute the corresponding 95% confidence interval. Interpret the interval.

(d) Construct a plot of the estimated model from part (a) with canopy cover on the x-axis and the number of trees and tree height fixed at their mean values. Include a 95% confidence interval band for the mean number of stripped cones. Interpret the plot.

(e) The ultimate goal of the paper was to determine what forest characteristics lead to more red squirrels. Based on the results for this exercise, develop overall conclusions relative to the goals of the paper. Because the authors used adjusted standard errors for their analysis, conclusions reached in the paper may be different than those for this exercise.

53. Continuing Exercise 52, complete the following parts.

(a) Suppose the plots of land were not equally sized. Could this affect the analysis? How could one take into account size differences using methods discussed in this chapter?

(b) Many authors do not examine interactions within models. What are the potential problems with this omission? Should interaction terms be included in the model for the data in this exercise? Answer the question by examining one LRT for all possible interaction terms.

(c) The paper is open access to readers. Using the paper, answer the questions below.

i. The second column of p. 441 states a model. While the estimated regression parameters are the same as those one should obtain for part (a) of Exercise 52, the model is not stated correctly. List the problems with the model statement.

ii. The Figure 2 caption of the paper states the Poisson regression model written in terms of the regression parameters. Again, the model is not stated correctly. List the problems with the model statement.

iii. The asterisk note with Table 1 of the paper is important with regard to why a Poisson regression model is used rather than a logistic regression model. Examine this note and provide the reason why a logistic regression model could not be used.

Chapter 5

Model selection and evaluation

In Chapters 1-4, we developed a variety of probability models for analyzing different types of categorical data. In each case, we started with a fixed set of explanatory variables and explored techniques of model fitting and inference assuming that the model and the chosen variables were correct. However, each of these probability models is an *assumption* that may or may not be satisfied by the data from a particular problem. Also, in practice, there is often uncertainty regarding which explanatory variables are needed in a model. Indeed, the goal of many categorical regression analyses is to identify which variables from a large number of candidates are associated with a response or with one another, and which are not.

"Model selection" consists of both (1) identifying an appropriate probability model for a problem and (2) identifying an appropriate set of explanatory variables to be used in this model. In this chapter, we first present techniques that can be used to select an appropriate set of explanatory variables from among a larger pool of candidate variables. We show both classical techniques and more recent developments that have distinct advantages over their older counterparts. Once variable selection is addressed, we then explore methods for assessing whether the assumptions that surround our probability model are satisfied. We introduce residuals and show how they can be used in plots and tests to identify model assumptions that may be violated. In particular, we identify a common model violation—*overdispersion*—and discuss its causes, its impacts on inferences, and adjustments that can be made to a model that can reduce its adverse effects. We include in this chapter two full examples, based on the placekicking data introduced in Chapter 2 and the alcohol consumption data introduced in Chapter 4, that go through the entire model selection and assessment process.

5.1 Variable selection

Variable selection refers to the process of reducing the size of the model from a potentially large number of explanatory variables to a more manageable and interpretable set. There are many approaches to selecting a subset of variables from a larger pool. All have strengths and weaknesses, and new approaches are being developed each year. We will first present the methods that are historically the most often used, and indeed are still in common use today. However, these methods are no longer recommended by experts on variable selection. We will present critiques of these methods and offer alternatives that are generally more reliable.

DOI: 10.1201/9781003093091-5

5.1.1 Overview of variable selection

Why do variable selection at all? Why not simply use the model with all available variables in it? There are three answers to these questions. First, researchers often want to know which explanatory variables have the greatest association with a response. The entire goal of the analysis may be to identify these variables to aid in understanding a phenomenon. For example, Korall et al. (2018) use methods described in our Section 5.1.6 to identify staff/management attributes that relate to encouraging residents to use protective hip pads in long-term care facilities.

Second, in many modern applications, the number of available variables may be too large to form an interpretable model. In areas like genetics and medicine, the number of available variables may even be larger than the number of observations, a problem referred to as "$p > n$." In these applications, the full model cannot actually be fit, which means that some sort of variable selection is needed.

The last answer has to do with the *bias-variance tradeoff*. Errors made by a model's predictions are a result of two sources: bias and variance. Bias results from the fact that "all models are wrong" (Box, 1976), meaning that real problems are more complex and involve more factors than we can typically identify. When we formulate a model, we propose an overly simple structure as an educated guess about the correct relationship between the response and the explanatory variables in the model. We may be wrong in many ways, such as fitting a straight-line model to a curved relationship, or omitting important variables, perhaps because we did not know we needed to measure them.

Using all available variables would seem to be a remedy for the last of these problems, and indeed minimizes the potential bias caused by variable selection. However, there is a drawback to fitting large models. The process of estimating parameters adds variability to any predicted values, because the parameter estimate does not usually match the true parameter value exactly. To see this, consider fitting a model to, say, 10 variables when only one is actually important. Although the true parameter values should be 0 for the other nine variables, they will likely be estimated by some random non-zero amount. Even if we were to guess the one important variable's parameter *exactly*, the random noise in the other parameter estimates would cause predicted values to vary from their true means. The more unimportant variables we add to a model, the more random noise is added to the predictions.

It can even be the case that a variable's true effect on the response is not zero, but is so small that the increase in bias caused by leaving it out of a model is smaller than the variability added from having to estimate its parameter. When the main goal of an analysis is predicting new observations, it may be worthwhile to use a model that has *too few* variables in it rather than one that has all of the right ones. This is known as *sparsity* or *parsimony* in the variable-selection literature. Of course, in problems where understanding the relationships between the response and the explanatory variables is the main goal, we may not want to overlook small contributions. Both goals are worthwhile in different contexts. Exercise 1 explores the bias-variance tradeoff in the context of logistic regression.

Variable selection is therefore a common and important part of many statistical analyses. See Heinze et al. (2018) for an overview of variable selection with a wealth of references for deeper study. All of the variable-selection techniques we discuss apply to any of the probability models from the previous chapters. We therefore begin by assuming that we have selected a probability model that we believe is appropriate for our problem (e.g., Poisson, binomial, or multinomial), and that there is a "pool" of P explanatory variables, $x_1, \ldots, x_P$, that are candidates for selection into a regression model. In accordance with generalized linear model notation, we assume that the parameters of the model (e.g., the mean μ for a Poisson model or the probability π for a binomial) are linked to the regression parameters $\beta_0, \beta_1, \ldots, \beta_P$ via an appropriate link function $g(\cdot)$ (e.g., log or logit; see Section 2.3). We therefore express all of our models in terms of the linear predictor $g(\cdot) = \beta_0 + \beta_1 x_1 + \ldots + \beta_P x_P$.

Obviously, we want the final model to contain all of the important variables and none of the unimportant ones. If knowledge or some theory about the problem tell us that certain variables are definitely important, then these should be included in the model at the start and are no longer considered for selection. Similarly, if any variables are somehow known to be unrelated to the response, then these can be immediately excluded from the model.

The P remaining variables form the pool upon which the selection process should focus. This pool of variables may include transformations of other variables in the pool or interactions among variables. We need to somehow create models out of these variables, compare them using some criterion, and select variables or models that give the best results. We first discuss popular criteria for comparing models. Then, we describe algorithms for creating models that can be compared using these (and other) criteria. We finally explain how the selected models should (and should not) be used.

5.1.2 Model comparison criteria

Comparison of two models can be done using a hypothesis test as long as one of the models is *nested* within the other. That is, it must be possible to obtain the smaller model by setting some parameters in the larger model to specific values. For example, model $g_1(\cdot) = \beta_0 + \beta_1 x_1$ can be obtained from model $g_2(\cdot) = \beta_0 + \beta_1 x_1 + \beta_2 x_2$ by setting $\beta_2 = 0$. On the other hand, there is no way to obtain model $g_1(\cdot)$ by fixing values for parameters in $g_3(\cdot) = \beta_0 + \beta_2 x_2$. This severely limits the use of hypothesis tests for variable selection. Instead, a criterion is needed that can assess how well *any* model explains the data.

When models are fit using ML estimation, the residual deviance provides an aggregate measure of how far the model's predictions lie from the observed data, with smaller values indicating a closer fit. It might seem that deviance could be used as a variable-selection criterion in the sense that a model with a smaller deviance would be preferred over one with a larger deviance. However, like the sum of squared errors in linear regression, residual deviance cannot increase when a new variable is added to a model. The model with all available variables always has the smallest deviance, or equivalently the largest value for its evaluated likelihood function.

The residual deviance or log-likelihood value needs to be adjusted somehow so that adding a variable does not automatically improve the measure. *Information criteria* (IC) are measures based on the log likelihood that include a "penalty" for each parameter estimated by the model (see, e.g., Burnham and Anderson, 2002). Adding variables to a model improves the likelihood but also increases the penalty, and the combination can result in either a better or a worse value of the criterion.

Numerous versions of information criteria have been proposed that use different penalties for the size of the model (i.e., the number of regression parameters and any other parameters in the model, such as the variance in normal linear regression). Let n be the sample size, r be the the size of the model, and $LL = \log(L(\hat{\boldsymbol{\beta}}|y_1, \ldots.y_n))$ be the log likelihood of an estimated model evaluated at the MLEs for the parameters, where $\hat{\boldsymbol{\beta}}$ is a vector consisting of the estimated regression parameters. The general form of most information criteria is

$$IC(k) = -2LL + kr,$$

where k is a penalty coefficient that is chosen in advance and used for all models. For a given k, models with smaller values of $IC(k)$ have larger log-likelihood relative to their penalty, and hence are better models according to the criterion.

The three most common information criteria are

1. Akaike's Information Criterion (AIC; Akaike, 1974):
$$AIC = IC(2) = -2LL + 2r$$

2. Corrected AIC (AIC$_c$; Sugiura, 1978):
$$AIC_c = IC(2n/(n-r-1)) = -2LL + \frac{2n}{n-r-1}r = AIC + \frac{2r(r+1)}{n-r-1}$$

3. Bayesian Information Criterion (BIC; Schwarz, 1978):
$$BIC = IC(\log(n)) = -2LL + \log(n)r$$

The penalty coefficient k for AIC$_c$ is always larger than that for AIC. It increases as the model size increases, but is similar to the AIC penalty when the sample size is large relative to the model size. The BIC penalty becomes more severe as n grows. It is larger than the AIC penalty for all $n \geq 8$, and larger than the AIC$_c$ penalty whenever $n-1-(2n/\log(n)) > r$, which is the case in many problems.

Information criteria are used for variable selection as follows. Suppose we have a group of models that differ in their explanatory variables. We select a penalty k and compute $IC(k)$ on all models. The model with the smallest $IC(k)$ value is the one preferred by that information criterion. Notice that this does not require the nesting of models, which gives information criteria a distinct advantage over hypothesis tests for variable selection.

It is possible that many models have values of $IC(k)$ that are near that of the best model. Roughly, models whose $IC(k)$ values are within about two units of the best are considered to have similar fit to the best model (Burnham and Anderson, 2002, p. 70; see also Section 5.1.6). The more models that are near the best, the less definitive the selection of the best model is, and the greater the chance that a small change in the data could result in a different model being selected. For this reason, *model averaging* has been developed for variable selection and related inference. We discuss this topic in Section 5.1.6.

Notice that using larger values of k results in a criterion that favors smaller models. Thus, AIC tends to choose larger models than BIC and AIC$_c$, while BIC will often choose smaller models than AIC$_c$. Opinions vary as to which among these is best, as there are theoretical results that support the use of each of these. BIC has a property called *consistency*, which means that as n increases, it chooses the "right" model (i.e., the one with the same explanatory variables as the model from which the data arises) with probability approaching 1, assuming that the true model is among those being examined. AIC enjoys a property called *efficiency*, meaning that, as n increases, it is increasingly likely to select models that minimize the mean squared prediction error, which accounts for both the error in estimating the parameters and the variability of the data (Burnham and Anderson, 2002; Hastie et al., 2009). AIC$_c$ was developed for linear models to extend the efficiency property to smaller samples. It is used in generalized linear models to approximately achieve the same goal. A more rigorously justified correction has been developed by Imori et al. (2014), but its form is rather complex and depends on the data. It does not appear to be in regular use.

Given that parsimonious models are generally preferred over models with too many variables, we generally prefer to use AIC$_c$ when the goal of an analysis is ultimately to predict future responses or BIC when the goal is to identify the most important variables. Both AIC and BIC are generally easy to compute in R using generic functions, like `AIC()`, which can be applied to any model-fit object that produces a log-likelihood accessible through the logLik() generic function. Objects obtained from glm() fits possess this property. Setting the argument value k = 2 in AIC() gives AIC, while k = log(n) gives BIC, where n is an object containing the sample size of the data set. There is no automatic way to calculate AIC$_c$, but it is typically not hard to calculate once AIC is calculated.

Different R functions may give different numerical values of an information criterion, even if the same model object is used for the calculation. The reason is that some functions exclude computing constants that are part of the likelihood function but do not change across different models. For example, the log-likelihood function for Poisson regression given in Equation 4.3 contains a term, $\sum_{i=1}^{n}\log(y_i!)$, that relates only to the data and not to the fitted model's parameters. Some R functions may leave this term out of their computation, which affects the numerical values of information criteria but not the differences among them for different models. To prevent misleading results, we strongly recommend that information criteria for a particular comparison of models all be computed using the same function in R.

5.1.3 All-subsets regression

Given a criterion for comparing and selecting models, the next step is deciding which models to compare. Perhaps the most obvious thing to consider is to make every possible combination of variables and select the one that "fits best." Indeed, this approach—called *all-subsets regression*—is popular where it can be used, so we describe it in more detail.

When there are P candidate explanatory variables, including any transformations and interactions that might be of interest, then there are 2^P different models that can be formed. We refer to this set of all possible models as the *model space*. All-subsets regression for generalized linear models is limited to problems in which P is not too large, because each model must be fitted using iterative numerical techniques. While modern computers can be fast, the sheer scope of the problem can be overwhelming. For example, if $P = 10$, then there are just over 1,000 models to be fit, and this can be done fairly quickly in most cases. If $P = 20$, the number of models is over 1 million, while if $P = 30$, the number is over 1 billion. Time or memory requirements may make it infeasible to assess every model in these large model spaces.

When an exhaustive search of the entire model space *is* possible, an information criterion $IC(k)$ is computed for each of the 2^P models. The model with the smallest $IC(k)$ is considered "best," although it is quite common that there are numerous models with similar values of $IC(k)$. This is especially likely when P is large, so that there are many possible variables whose importance is borderline.

When an exhaustive search is not possible, an alternative search algorithm can be used to explore the model space without evaluating every model. These algorithms can be either deterministic (for a given data set, they explore a specific set of models) or stochastic (they can explore different sets of models every time they are run). Several deterministic algorithms are described in Sections 5.1.4 and 5.1.5. An example of a stochastic search algorithm is the *genetic algorithm* (Michalewicz, 1996). This algorithm makes a first "generation" of models out of randomly selected combinations of variables, and then makes a new generation of models by splicing together parts of those models from the previous generation that have the lowest $IC(k)$ values. This step is iterated many times with occasional random additions and deletions of variables (called *mutations*). The combining structure allows the algorithm to identify and focus on combinations of variables that work well together, while the mutations allow it to explore the model space more broadly. The algorithm eventually converges to a "best" model according to $IC(k)$. This model is not guaranteed to be the one with the smallest $IC(k)$ among all possible models, but in many problems, it does find the best model or one very close to it. Because it is a random search process and the optimal selection is not guaranteed, we recommend running the algorithm a few times to see whether better models can be found.

Several functions can perform all-subsets regression for GLMs, including `bic.glm()` in the `BMA` package, `bestglm()` in the `bestglm` package, `dredge()` in the `MuMIn` package, and `glmulti()` in the `glmulti` package. Each has different features and limitations:

- `bic.glm()` in `BMA` can handle up to 50 variables, the most among functions that perform an exhaustive search of the model space. It is fast and appears to be applicable to GLMs of any family fit using the `glm()` function, although the documentation is slightly vague about this. It can perform model averaging, which we describe in Section 5.1.6. However, it can only use BIC for model comparisons. We use it for large models when we are satisfied with using BIC.

- `bestglm()` in `bestglm` is fast and efficient and allows the user to select AIC, BIC, or two additional information criteria not described here. However, it is limited to models of no more than 15 variables and cannot compute AIC_c.

- `dredge()` in `MuMIn` has the greatest built-in flexibility among the functions described here. It allows selection using AIC, BIC, and AIC_c and can be extended to a very wide range of models, including all `glm()` families and most of the models used in this book that are fit using functions other than `glm()`. It performs model averaging and can handle up to 30 variables. It is also the only one of these function whose results can be fed directly into `emmeans`. Its main drawback is that the algorithm can be slower than the previous two functions and may not be feasible for model sizes near its limit.

- `glmulti()` in `glmulti` allows selection using AIC, BIC, and AIC_c, and it does model averaging. It is the best function for handling interactions, because it has built-in capacity for imposing marginality (described below). Importantly, it is the only function that uses the genetic algorithm, so it can be used with models of larger sizes than the other functions. It has the capacity to be extended to models outside of `glm()` families, although doing so requires writing several additional functions for each new model that is used. The biggest difficulty with this function is that it requires a Java installation using the same architecture (32- or 64-bit) as R on the computer where it is run, which is a limitation for many users. A somewhat smaller difficulty with using the function is its exact results from the genetic algorithm are not reproducible. Setting a seed number prior to running the function has no effect, because its random aspect uses a Java random number generator, and no argument is provided to control it.

In summary, any of these functions can handle models fit using `glm()` with up to 15 variables. Only `dredge()` and `glmulti()` can use all three information criteria described in Section 5.1.2 and can handle other models, which `dredge()` does much more easily. We therefore suggest `dredge()` for all-subsets regression or model averaging on most data sets, using `glmulti()` when the number of variables is larger than 30, or large enough that `dredge()` takes too long.

Example: Placekicking (AllSubsetsPlacekick.R, Placekick.csv)

We implement all-subsets regression on the placekick data, selecting from all available explanatory variables using AIC. We demonstrate both exhaustive search using `dredge()` and genetic algorithm using `glmulti()`, starting from a `glm()` model fit called `mod.fit` using all explanatory variables in linear form (we use `formula = good`

~ . as a shortcut to request that the model be fitted to all remaining variables from the data = placekick). In the program for this example, we also show exhaustive search with BIC using the other three functions.

The dredge() function receives the data, variables, and model information from the existing model-fit object entered as the value for the global.model argument. Its rank argument specifies the selection criterion and includes values "AIC", "BIC", and "AICc". Other optional arguments can control largest and smallest subset sizes to consider, which variables to retain in all models, and numerous other model, output, and implementation details. Applying subset() to the object produced by dredge() selects only a subset of models for output or further computations, the details of which can be specified in various ways. Below, the delta argument is used to select all models whose AIC values are within 1 of the best.

```
> library(package = MuMIn)
> allsub.aic <- dredge(global.model = mod.fit, rank = "AIC")
Fixed term is "(Intercept)"
> subset(x = allsub.aic, subset = delta < 1)
Global model call: glm(formula = good ~ ., family =
    binomial(link = "logit"),
data = placekick, na.action = na.fail)
---
Model selection table
    (Intrc)    chang     dstnc   PAT      week     wind df
148   4.752 -0.3351 -0.08724 1.230              -0.5234  5
212   4.990 -0.3478 -0.08723 1.228 -0.02468 -0.4765  6
84    4.950 -0.3538 -0.08653 1.243 -0.02773           5
20    4.676 -0.3402 -0.08646 1.245                    4
147   4.599         -0.08676 1.319              -0.5326  4
      logLik    AIC delta weight
148 -378.343 766.7  0.00  0.256
212 -377.537 767.1  0.39  0.211
84  -378.629 767.3  0.57  0.192
20  -379.666 767.3  0.65  0.185
147 -379.831 767.7  0.98  0.157
Models ranked by AIC(x)
```

The output from subset() shows five models within 1 AIC-unit of the best model, listed in increasing order of AIC. The parameter estimates from each model are given underneath their respective variable names, with a blank when the variable is not in that particular model. The degrees of freedom column df shows the number of variables in each model, *including* the intercept. The log-likelihood and AIC values are shown next, followed by delta, showing the AIC difference from the best model. The weight column is explained Section 5.1.6.

The best model according to AIC includes distance, change, PAT, and wind. Among the top five models, we see some patterns to the variables selected: distance and PAT are always in the top models, while elap30, type, and field are never selected (no columns are present for variables that do not appear in the selected subset of models). There is some uncertainty regarding the importance of week, change, and wind. In Section 5.1.6, we will describe a more formal approach to this assessment.

Adding pairwise interactions to the set of variables would be desirable. However, this hugely increases the number of possible models, from $2^8 = 256$ to $2^{36} = 68.7$ billion, which is and is more variables than dredge() can handle. This is where the genetic search algorithm from `glmulti()` is useful. The argument `y` in `glmulti()` can contain a formula, as shown below, or an object like a previous `glm()` fit of the full model. The `fitfunction` argument takes the name of any R function that can evaluate a log likelihood from a `formula`-class object, including `"glm"`, `"lm"`, or similar user-defined functions. Other important arguments include `level` (1 for main effects only, 2 to include two-way interactions) and `method` (`"h"` for exhaustive search, `"g"` for genetic algorithm). When `level = 2` is chosen, adding `marginality = TRUE` requires that when a model includes any interactions, it also includes the main effects from those interactions. The `print()` generic function is used instead of `summary()` to give a summary of the results, while `weightable()` lists the models in order of their $IC(k)$ values. By default, a status report is given after every 10 generations.

```
> search.gmarg.aic <- glmulti(y = good ~ ., data = placekick,
    fitfunction = "glm", level = 2, marginality = TRUE, method =
    "g", crit = "aic", family = binomial(link = "logit"))
Initialization...
TASK: Genetic algorithm in the candidate set.
Initialization...
Algorithm started...

After 10 generations:
Best model:
good ~ 1 + week + distance + change + elap30 + PAT + type +
    field + wind + change:distance + PAT:distance + PAT:elap30 +
    type:distance + type:change + type:elap30 + field:PAT +
    field:type + wind:distance + wind:type + wind:field
Crit= 774.419815118267
Mean crit= 785.409187040214
Change in best IC: -9225.58018488173 / Change in mean IC:
    -9214.59081295979

<OUTPUT EDITED>

> print(search.gmarg.aic)
glmulti.analysis
Method: g / Fitting: glm / IC used: aic
Level: 2 / Marginality: TRUE
From 100 models:
Best IC: 765.199563838363
Best model:
[1] "good ~ 1 + week + distance + change + PAT + type + field +
    wind + "
[2] "    field:type + wind:distance + wind:type + wind:field"
[1] "good ~ 1 + week + distance + change + PAT + type + field +
    wind + "
[2] "    PAT:distance + field:type + wind:distance + wind:field"
[1] "good ~ 1 + week + distance + change + PAT + type + field +
    wind + "
[2] "    PAT:distance + field:type + wind:distance + wind:type +
    wind:field"
[1] "good ~ 1 + week + distance + change + PAT + type + field +
    wind + "
[2] "    PAT:distance + wind:distance + wind:type + wind:field"
```

```
Evidence weight: 0.0476361121172935
Worst IC: 777.574811706961
25 models within 2 IC units.
40 models to reach 95% of evidence weight.
Convergence after 370 generations.
Time elapsed: 3.18166240056356 minutes.

> head(weightable(search.gmarg.aic))

model
1                                good ~ 1 + week + distance + change
   + PAT + type + field + wind + field:type + wind:distance +
   wind:type + wind:field
2                             good ~ 1 + week + distance + change +
   PAT + type + field + wind + PAT:distance + field:type +
   wind:distance + wind:field
3            good ~ 1 + week + distance + change + PAT + type +
   field + wind + PAT:distance + field:type + wind:distance +
   wind:type + wind:field
4                              good ~ 1 + week + distance + change +
   PAT + type + field + wind + PAT:distance + wind:distance +
   wind:type + wind:field
5                       good ~ 1 + distance + change + PAT + type +
   field + wind + PAT:distance + field:type + wind:distance +
   wind:type + wind:field
6 good ~ 1 + week + distance + change + PAT + type + field +
   wind + PAT:distance + field:type + wind:week + wind:distance
   + wind:type + wind:field
       aic    weights
1 765.1996 0.04763611
2 765.1996 0.04763611
3 765.1996 0.04763611
4 765.1996 0.04763611
5 765.2632 0.04614369
6 765.3531 0.04411553
```

The algorithm starts with a randomly generated set of models. After 10 generations, the best model has many variables and interactions resulting in an AIC of 777.4. Many more updates are printed before the algorithm converges to a final set of models. The `print()` function shows that the algorithm concluded after approximately 3.2 minutes and converged after 370 generations, meaning that it did not find any better models to add to its top 100 for 50 consecutive generations. Interestingly, there are four models with exactly the same AIC values, which is quite unexpected under usual circumstances. However, there are certain combinations of variables that cannot occur in these data: PATs are taken only at distances of 20 yards, and indoor stadiums cannot be windy or have natural turf fields. Thus, the `PAT:distance`, `wind:type` and `field:type` interactions do not create any new columns in the matrix $\mathbf{X}$ from which the parameters are estimated, as described in Section 2.2.1. The resulting models with and without any of these interactions provide equivalent fits to the data. The model without these terms is the one we would choose, although this particular genetic algorithm run did not find this model. All of these equivalent models have an AIC of 765.2, but there are 25 models within 2 IC units of the best model. This indicates that the declaration of this model as "best" is not very definitive. The worst of the top 100 models had an AIC of 777.6.

The top five models are shown next along with their AIC values. It is meaningful to note that all of these top models include `wind:distance` and `wind:field`, confirming our belief that wind interactions may have an important role to play in the probability of success for a placekick.

Finally, we ran the function three more times to see how consistent the results were. It turns out that this first run did not find the model with the smallest AIC! A model with $AIC = 763.8$ was discovered in one run that had terms similar to the best one above but without all of the `type` terms. While the improvement in IC is modest, it does affirm the potential variability in results from different runs of a genetic algorithm, as well as the need to rerun it several times.

Burnham and Anderson (2002) strongly urge caution in using automated variable-selection routines because, in the presence of so many candidate models, finding the true "best" model becomes a needle-in-a-haystack problem. Many models may have similar $IC(k)$ values simply by luck. For example, imagine that a world-champion darts player throws one dart at a dart board and, then, a novice randomly scatters another 64 billion darts around the board. The chances are extremely good that some of the random darts will land closer to the center of the board than the champion's, no matter how good the champion is. The list of candidate explanatory variables should therefore be made as small as possible using subject-matter knowledge before beginning all-subsets regression. Indeed, the wisdom of routinely adding pairwise interactions and transformations to a variable search is questionable in this light. Instead, it often makes better sense to include only certain interactions that are of specific interest and to use subject-matter knowledge and diagnostic tools from Section 5.2 to aid in selecting appropriate models.

5.1.4 Stepwise variable selection

As noted in the previous section, unless the number of variables, P, is fairly small—typically less than 50—it is not feasible to complete an exhaustive search of the entire model space to locate the "best" model. In cases where P is too large, we must use some kind of algorithm to explore models that can be quickly identified as "promising." *Stepwise search algorithms* are simple and fast methods for selecting a relatively small number of potentially good models from the model space. These algorithms have shortfalls that are not widely known among practitioners, and so while we describe them below, we do not generally recommend them unless no better approach is available.

Stepwise variable selection uses a structured algorithm to examine models in only a small portion of the model space. The key to the algorithm is to evaluate the models in a specific sequence, where each model differs from the previous model by only a single variable. Once a sequence of models is chosen, an information criterion is used to compare them and select a "best" model. There are several different versions of the algorithm, as described shortly. In each version, let x_j, $j = 1, \ldots, P$, represent either a single variable or a set of variables that must be included or excluded together (e.g., a categorical explanatory variable that is represented by a group of indicator variables that are included in the model or excluded from it as a group).

Forward Selection

Forward selection starts from a model with no variables in it and then adds one variable at a time until either all of the variables are in the model or, optionally, some stopping rule has been reached. Specifically:

1. Select a k for the information criterion.

2. Start with an empty model, $g(\cdot) = \beta_0$, and compute $IC(k)$ on this model. All variables are in the "selection pool."

3. Add variables to the model one at a time as follows:

   (a) Compute $IC(k)$ for each model that can be formed by adding a single variable from the selection pool into the current model. In other words, we consider models that contain all of the variables in the current model plus one of the remaining variables.

   (b) Choose the variable whose model has the smallest $IC(k)$ and add it to the model.

   (c) (Optional) If the model with the added variable has a larger $IC(k)$ than the previous model, stop. Report the model with the smaller $IC(k)$ value as the final model.

4. Repeat step 3 until the termination rule in part (c) is used or until all variables have been added into the model.

5. If no stopping rule is used, then a sequence of $P + 1$ models is created, each with a different number of variables. Report the model with the smallest $IC(k)$ as the final model.

Part (c) of step 3 is most often implemented in practice. This can allow the algorithm to stop before reaching a model with P variables, saving time, but it risks missing a model in which a combination of additional variables might reduce $IC(k)$ although none of them achieves this alone.

Notice that the model size at each step increases by at most one variable. The logic here is that, at each step, the variable added is the one that improves the model most. Therefore, including it into the model yields the best model of its size, given the previous model. The next example illustrates this algorithm.

Example: Placekicking (StepwisePlacekick.R, Placekick.csv)

Stepwise variable selection methods are implemented in several R functions. We use the `step()` function because it is part of the `stats` package and it is fairly easy to use. To begin, the smallest and largest models to be considered are first fit using `glm()` and saved as objects. The smallest model becomes the value for the `object` argument in `step()`, and the largest model is specified in the `scope` argument. If interactions are to be considered, they must be included in the largest model, either by listing them explicitly or by using a "`^`2" around the main-effects portion of the model syntax. The `anova()` function summarizes the selection process.

Forward selection is denoted by `direction = "forward"` in `step()`. We start from a logistic regression model with no variables (`empty.mod`) for the placekicking data and add one variable at a time until no variable can be added that improves the criterion or the model with all variables (`full.mod`) is reached. Therefore, we include part (c) of step 3 in our algorithm implementation. Unfortunately, `step()` requires a constant penalty coefficient for all models, so AIC$_c$ is not available. Instead, we base the selection on BIC, $IC(\log(n))$, by specifying the argument value `k = log(nrow(placekick))`.

```
> empty.mod <- glm(formula = good ~ 1, family = binomial(link =
   logit), data = placekick)
> full.mod <- glm(formula = good ~ .,  family = binomial(link =
   logit), data = placekick)
> forw.sel <- step(object = empty.mod, scope = list(upper =
   full.mod), direction = "forward", k = log(nrow(placekick)),
   trace = TRUE)
Start:  AIC=1020.69
good ~ 1

           Df Deviance     AIC
+ distance  1   775.75  790.27
+ PAT       1   834.41  848.93
+ change    1   989.15 1003.67
<none>         1013.43 1020.69
+ elap30    1  1007.71 1022.23
+ wind      1  1010.59 1025.11
+ week      1  1011.24 1025.76
+ type      1  1011.39 1025.92
+ field     1  1012.98 1027.50

Step:  AIC=790.27
good ~ distance

         Df Deviance    AIC
+ PAT     1   762.41 784.20
<none>        775.75 790.27
+ change  1   770.50 792.29
+ wind    1   772.53 794.32
+ week    1   773.86 795.64
+ type    1   775.67 797.45
+ elap30  1   775.68 797.47
+ field   1   775.74 797.53

Step:  AIC=784.2
good ~ distance + PAT

         Df Deviance    AIC
<none>        762.41 784.20
+ change  1   759.33 788.38
+ wind    1   759.66 788.71
+ week    1   760.57 789.62
+ type    1   762.25 791.30
+ elap30  1   762.31 791.36
+ field   1   762.41 791.46

> anova(forw.sel)
Analysis of Deviance Table
Model: binomial, link: logit
Response: good
Terms added sequentially (first to last)
         Df Deviance Resid. Df Resid. Dev
NULL                      1424    1013.43
```

```
distance   1   237.681        1423      775.75
PAT        1    13.335        1422      762.41
```

The `trace = TRUE` argument value allows one to see the full forward selection process at each step. In the first step, the algorithm starts with an intercept only and gives the $IC(k)$ values—listed as `AIC`, regardless of the value specified for `k`—for each model that arises based on adding one of the eight explanatory variables. The models with each added variable and the model with no added variables ("`<none>`") are sorted by their $IC(k)$ values. We see that `distance` has the smallest BIC and, hence, is the first variable entered. The next step begins with `distance` in the model and finds that adding `PAT` results in a smaller BIC (notice that BIC is different now than it was for `PAT` in the first step because `distance` is now in the model). Thus, `PAT` is added to the model and the next step begins. Because no additional variable improves BIC further, the algorithm terminates and declares that our final model uses `distance` and `PAT`. Exercise 42 repeats this process using AIC.

Backward elimination

A different stepwise algorithm is sometimes used that approaches the problem from the opposite direction. Backward elimination starts from a model that has *all* available variables in it (assuming that this model can be fit) and then removes one variable at a time as long as doing so improves $IC(k)$ for the model. Specifically,

1. Select a k for the information criterion.
2. Start with a full model, $g(\cdot) = \beta_0 + \beta_1 x_1 + \ldots + \beta_P x_P$, and compute $IC(k)$ on this model.
3. Remove variables from the model one at a time as follows:
   (a) Compute $IC(k)$ for each model that can be formed by deleting a single variable from the current model. In other words, we consider models that contain all but one of the variables in the current model.
   (b) Choose the variable whose deletion results in the smallest $IC(k)$ and remove it from the model.
   (c) (Optional) If the new model with the variable deleted has a larger $IC(k)$ than the previous model, stop. Report the model with the smaller $IC(k)$ value as the final model.
4. Repeat step 3 until the termination rule in part (c) is used or until all variables have been removed from the model.
5. If no stopping rule is used, then a sequence of $P+1$ models is created, each with a different number of variables. Report the model with the smallest $IC(k)$ as the final model.

Notice that at each step the current model is compared against all models that are exactly one variable smaller. Thus, model size decreases by at most one variable. The logic here is that, at each step, the variable removed is the one contributing the least to the current model.

Example: Placekicking (StepwisePlacekick.R, Placekick.csv)

Only minor modifications to the code from the previous example are needed to perform backward elimination. The starting model in `object` is now the full model, the ending model is listed in `scope`, and we set `direction = "backward"`. The output for this example is long, so we show an abridged version of it.

```
> back.sel <- step(object = full.mod, scope = list(lower =
    empty.mod), direction = "backward", k = log(nrow(placekick)),
    trace = TRUE)
Start:  AIC=818.87
good ~ week + distance + change + elap30 + PAT + type + field +
    wind

           Df Deviance    AIC
- elap30    1   753.72 811.82
- field     1   754.25 812.35
- type      1   754.76 812.86
- week      1   755.13 813.23
- change    1   756.69 814.79
- wind      1   756.74 814.84
<none>          753.52 818.87
- PAT       1   764.63 822.73
- distance  1   822.64 880.73

Step:  AIC=811.82
good ~ week + distance + change + PAT + type + field + wind

           Df Deviance    AIC
- field     1   754.49 805.32
- type      1   755.04 805.87
- week      1   755.32 806.16
- change    1   756.78 807.61
- wind      1   756.96 807.79
<none>          753.72 811.82
- PAT       1   764.81 815.64
- distance  1   824.69 875.53

<OUTPUT EDITED>

Step:  AIC=788.38
good ~ distance + change + PAT

           Df Deviance    AIC
- change    1   762.41 784.20
<none>          759.33 788.38
- PAT       1   770.50 792.29
- distance  1   831.75 853.53

Step:  AIC=784.2
good ~ distance + PAT

           Df Deviance    AIC
<none>          762.41 784.20
```

```
- PAT          1    775.75 790.27
- distance     1    834.41 848.93
```

Starting from a model with all eight variables, removing `elap30` results in the most improvement in BIC. At the next step, `field` is removed. Steps continue to remove variables until only `PAT` and `distance` remain. Neither of these can be removed from the model without increasing BIC, so this is the final model.

Alternating stepwise selection

One of the criticisms of the previous two algorithms is that once a variable is added to (or removed from) the model, this decision can never be reversed. But in all forms of regression, relationships among the explanatory variables can cause the perceived importance of a variable to change depending on which other variables are in the same model. For example, a person's average daily calorie intake may seem important by itself in a model for the probability of high blood pressure, but that relationship might disappear once we also account for their weight and exercise level. Therefore, it can be useful in some cases to have an algorithm that allows past decisions to be reconsidered as the model structure changes. The standard such algorithm is based on forward selection, but with the amendment that each time a variable is added, a round of backward elimination is carried out so that variables that have been rendered unimportant by the new addition can be removed from the model.

Example: Placekicking (StepwisePlacekick.R, Placekick.csv)

Because this algorithm is the default in `step()`, the forward-selection algorithm code can be modified to run stepwise selection simply by deleting the `direction` argument (`"both"` is its default value). We leave running the code and interpreting the results as Exercise 7. In this case, the results are identical to forward selection, because there are no steps at which any previously added variables can be removed.

Stepwise procedures have historically been applied using hypothesis tests to determine the sequence of models to be considered. However, these tests are not actually valid hypothesis tests. The theoretical foundations on which hypothesis tests are based are ruined when the same data are used to both select and test the hypotheses (Miller, 1984; see also Section 5.5). Therefore, it is becoming more common to apply stepwise methods using an information criterion as we have shown above. The program corresponding to the previous example provides code showing how to manipulate the `step()` function to use hypothesis tests for variable addition and deletion. It is important to note that both hypothesis tests and information criteria order the variables the same way at each step and, thus, produce exactly the same sequence of models if allowed to run until all variables are added or removed. The only possible difference between using a test and an information criterion is the selection of a final model from the sequence.

Although the forward, backward, and stepwise procedures all pointed to the same "best" model in the placekicking example, this does not generally have to happen. It is quite possible that the three procedures select three different models. In fact, Miller (1984) discusses an example with eleven explanatory variables in which the first variable entered into the model by forward selection is also the first one removed by backward elimination. See Miller (1984), Grechanovsky (1987), Shtatland et al. (2003), and Kutner et al. (2004) for more details and critiques of stepwise procedures.

5.1.5 Modern variable selection methods

Variable selection is an active research area that continues to produce new developments both for general use and for specialized problems. The most popular among the general-use methods is the *least absolute shrinkage and selection operator* (LASSO) proposed by Tibshirani (1996). The LASSO has been refined, improved, and redeveloped in many different ways, and research into expanding its capabilities for inference continues. We next describe the original LASSO and mention several improvements to it.

The LASSO

A variable selection procedure is more likely to select a particular variable when random chance in the data causes the variable to seem more important than it really is, as compared to those times when random chance makes it seem less important than it really is. As a result, parameter estimates for the selected variables tend to be biased: they are frequently estimated farther away from zero than they should be. This is the main motivation for the LASSO, which attempts to simultaneously select variables and shrink their estimates back toward zero to counteract this bias.

The LASSO obtains parameter estimates using a procedure that adds a penalty to the log-likelihood to prevent the estimates from being too large. Specifically, for a model with p explanatory variables, the LASSO parameter estimates $\hat{\beta}_{0,\text{LASSO}}, \hat{\beta}_{1,\text{LASSO}}, \ldots, \hat{\beta}_{p,\text{LASSO}}$ maximize

$$\log(L(\beta_0, \beta_1, \ldots, \beta_p | y, \ldots, y_n) - \lambda \sum_{j=1}^{p} |\beta_j|, \tag{5.1}$$

where λ is a *tuning parameter*[1] that needs to be determined. For a given value of λ, the penalty term $\sum_{j=1}^{p} |\beta_j|$ in the likelihood criterion has two effects. It causes some of the regression parameter estimates to remain at zero, effectively excluding their corresponding variables from the model. Also, it discourages other parameter estimates from becoming too large unless increasing their magnitudes provides a sufficient increase in the likelihood to overcome the additional penalty. Thus, the procedure simultaneously selects variables and shrinks their estimates toward zero, which can help to overcome the bias noted above.

As the penalty parameter λ grows larger, more shrinkage occurs and smaller models are chosen. This parameter is usually chosen by *cross-validation* (CV), which randomly splits the data into several groups and predicts responses within each group based on models fit to the rest of the data. A measure of prediction error is computed that compares the predicted and observed responses in some way (e.g., sum of squares or deviance). The prediction error is re-evaluated for a sequence of different values for λ, and the λ that has the smallest estimated prediction error is a possible choice for the penalty. Note that CV is a randomized procedure. It is likely that slightly different prediction errors would result from repeated runs of the CV algorithm, leading to different chosen values for λ and resulting in different models and parameter estimates. Furthermore, there are often many values of λ that lead to similar values of prediction error. In practice it is common to use the smallest model whose prediction error is within 1 standard error of the lowest CV error, where the standard error refers to the variability of the CV-prediction-error estimates and can be calculated in several ways. See Hastie et al. (2009) for details.

[1]Tuning parameters are unknown quantities that affect the performance of an algorithm by making it more sensitive or less sensitive to the variations in a data set. They are different from model parameters in that they do not directly relate to a mean, probability, or other population quantity, but rather affect how the algorithm estimates these quantities.

Although the LASSO estimates often result in better prediction than the ordinary MLEs, subsequent inference procedures have not yet been fully developed. For example, confidence intervals for regression parameters or for predicted values do not yet exist. Therefore the LASSO is used primarily as a variable selection tool or for making predictions where interval estimates are not required. An alternative procedure that can provide interval estimates is considered in the next subsection.

Example: Placekicking (LASSOPlacekick.R, Placekick.csv)

The `glmnet` package includes functions that can compute LASSO parameter estimates and predicted values for binomial (logit link), Poisson (log link), and multinomial (nominal response) categorical regression models. The `glmnet()` function computes LASSO estimates for a sequence of up to 100 values of λ. The `cv.glmnet()` function uses CV to choose a value of λ. Both `glmnet()` and `cv.glmnet()` require that the observations for the explanatory variables and response be separated into numeric `matrix`-class objects and entered as argument values for `x` and `y`, respectively. While both of these functions have various additional control arguments, the default levels serve fairly well for a relatively easy analysis. The objects created by these functions can be accessed using `coef()`, `predict()`, and `plot()` generic functions. By default, `coef()` produces coefficient estimates and `predict()` produces estimated logits, probabilities, or means for every value of λ for objects from `glmnet()`; alternatively, values of λ can be specified using the `s` argument.

We apply these functions to the placekicking data below. After creating the required matrices (stored as objects `yy` and `xx`), we run the LASSO algorithm and examine the results.

```
> yy <- as.matrix(placekick[,9])
> xx <- as.matrix(placekick[,1:8])
> lasso.fit <- glmnet(y = yy, x = xx, family = "binomial")
> round(coef(lasso.fit), digits = 3)
9 x 63 sparse Matrix of class "dgCMatrix"
   [[ suppressing 63 column names 's0', 's1', 's2' ... ]]

(Intercept) 2.047  2.359  2.632  2.875  3.095  3.297  3.482
week        .      .      .      .      .      .      .
distance    .     -0.011 -0.021 -0.029 -0.036 -0.042 -0.048
change      .      .      .      .      .      .      .
elap30      .      .      .      .      .      .      .
PAT         .      .      .      .      .      .      .
type        .      .      .      .      .      .      .
field       .      .      .      .      .      .      .
wind        .      .      .      .      .      .      .

<OUTPUT EDITED>

(Intercept)  4.789  4.788  4.788  4.787  4.786  4.786  4.785
week        -0.023 -0.023 -0.023 -0.023 -0.023 -0.023 -0.024
distance    -0.086 -0.086 -0.086 -0.086 -0.086 -0.086 -0.086
change      -0.337 -0.338 -0.339 -0.340 -0.341 -0.342 -0.342
elap30       0.004  0.004  0.004  0.004  0.004  0.004  0.004
PAT          1.195  1.199  1.203  1.207  1.210  1.214  1.216
type         0.238  0.245  0.250  0.256  0.260  0.265  0.269
```

```
field          -0.151 -0.156 -0.162 -0.166 -0.171 -0.175 -0.178
wind           -0.566 -0.572 -0.577 -0.581 -0.586 -0.590 -0.593
```

The algorithm produced parameter estimates that maximize Equation 5.1 at 63 values of λ, of which we display only the first and last few. Each column in the resulting matrix corresponds to a particular λ. The estimates are listed in order of *decreasing* λ, so that models progress from intercept-only to a nearly unpenalized fit. In the program for this example, we provide additional coding to get the λ values printed in each column along with the estimates. For example, this additional code helps one see that the last column corresponds to the parameter estimates when $\lambda = 0.0004$.

We next use `cv.glmnet()` to select a value for λ using the defaults for the CV process and then plot and print the results. We set a seed so that the CV procedure is reproducible.

```
> set.seed(27498272)
> cv.lasso.fit <- cv.glmnet(y = yy, x = xx, family = "binomial")

> # Default plot method for cv.lambda() produces CV errors +/- 1
   SE  at each lambda.
> plot(cv.lasso.fit)
> # Print out coefficients at optimal lambda
> coef(cv.lasso.fit)
9 x 1 sparse Matrix of class "dgCMatrix"
                      s1
(Intercept)  4.09748542
week         .
distance    -0.06697266
change       .
elap30       .
PAT          .
type         .
field        .
wind         .
> predict.las <- predict(cv.lasso.fit, newx = xx, type =
   "response")
> head(cbind(placekick$distance, round(predict.las, digits = 3)))
         lambda.1se
[1,] 21      0.936
[2,] 21      0.936
[3,] 20      0.940
[4,] 28      0.902
[5,] 20      0.940
[6,] 25      0.919
```

The CV estimates of error at each λ are shown in Figure 5.1; note that the binomial residual deviance is the function evaluated to compare the predicted and observed data in CV. The minimum CV estimate of deviance occurs with a six-variable model, but there are many models with practically the same deviance. The smallest model whose deviance is within 1 standard error of the minimum has only one variable. The print of the coefficients for this model shows that the variable is `distance`, with a parameter estimate of -0.067. Comparing this to the MLE, -0.115, we see that some

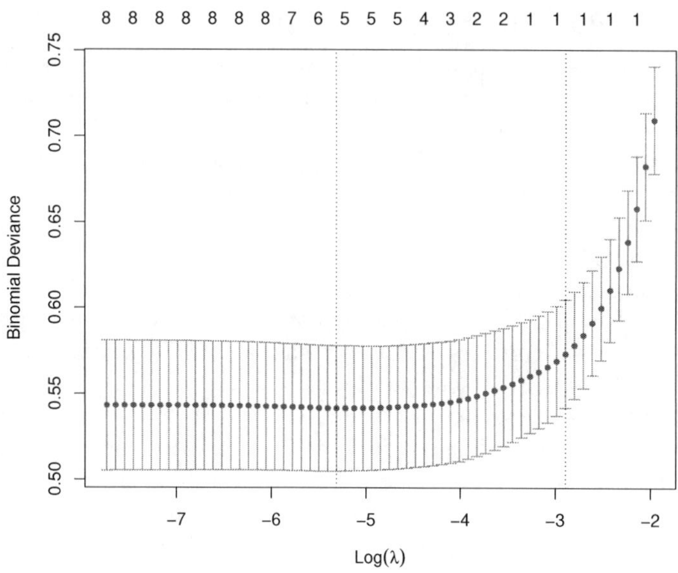

Figure 5.1: Cross-validation estimates of prediction error (binomial residual deviance) ± 1 standard error for each value of λ. Numbers at the top of the plot are the number of variables at the corresponding λ. The first dotted vertical line is where the smallest CV error occurs; the second is the smallest model where the CV error is within 1 standard error of the smallest.

shrinkage has taken place. The predicted probabilities of a successful field goal are printed out for the first few observations in the data set.

The LASSO has been improved in several ways since its introduction. The original formulation for the LASSO does not handle categorical explanatory variables, so a revised version, called the *grouped LASSO* was developed to correct this (Yuan and Lin, 2006). The grouped LASSO is available for logistic and Poisson regression in the `grplasso` package. Also, it has been noted in several places that the LASSO may perform poorly when faced with a problem in which there are many variables in the pool but few are important (see, e.g., Meinshausen, 2007). It has a tendency to include many more variables than are required, albeit with very small parameter estimates. Improvements such as the *relaxed LASSO* (Meinshausen, 2007) and the *adaptive LASSO* (Zou, 2006) both correct this tendency by separating the task of shrinkage from the task of variable selection. An approximation to the relaxed LASSO is available through `cv.glmnet()` using `relax = TRUE`. We demonstrate this in our `LASSOplacekick.R` program. Interestingly, the optimal relaxed LASSO chooses a model with `distance` as the only variable and estimates the model's parameters with no shrinkage; in other words, it uses the MLEs from this model. It is often the case when there is a dominant variable in a data set, like `distance` here, that the ordinary LASSO has difficulty balancing between shrinking parameter estimates for unneeded variables all the way to zero and maintaining a sufficiently large parameter estimate for the most important variable. By separating the tasks of variable selection and shrinkage, the relaxed LASSO avoids this difficulty.

5.1.6 Model averaging

As noted earlier in this section, it is often the case that there are many models with values of $IC(k)$ very close to the smallest value. This is an indication that there is some uncertainty regarding which model is truly best. In such cases, changing the data slightly could result in selecting a different model as best. In Exercise 2, we illustrate this uncertainty by splitting the placekick data approximately in half and showing that there are considerable differences in the top five models selected by the two halves. This uncertainty is difficult to measure if variable selection is applied using the traditional approach of selecting a single model and using it for all further inference. In particular, if a model is selected and then fit to the data, any variables not in that model implicitly have their regression parameters estimated as zero, *with a standard error of zero*. That is, we act as if we *know* that these variables do not belong in the model, when in fact the data cannot determine this for certain. It would be useful to be able to better estimate the uncertainty in each of our parameter estimates.

When all-subsets variable selection is feasible, then *model averaging* can account for model-selection uncertainty in subsequent inferences (see, e.g., Burnham and Anderson, 2002). In particular, *Bayesian model averaging* (BMA; Hoeting et al., 1999) uses Bayesian theory to compute the posterior probability that each possible model is the correct model (assuming that one of them is, indeed, correct; see Section 6.6 for a more detailed description of posterior distributions). It turns out that these probabilities can be approximated by relatively simple functions of the BIC value for each model. Suppose that a total of M models are fit and denote the BIC for model m by BIC_m, $m = 1, \ldots, M$. Denote the smallest value of BIC among all models by BIC_0, and define $\Delta_m = BIC_m - BIC_0$, which is greater than or equal to 0 for all models. Assuming that all models were considered equally likely before the data analysis began, then the estimated posterior probability that model m is correct, τ_m, is approximately

$$\hat{\tau}_m = \frac{\exp(-\frac{1}{2}\Delta_m)}{\sum_{a=1}^{M}\exp(-\frac{1}{2}\Delta_a)}. \tag{5.2}$$

The model with the smallest BIC has $\Delta_m = 0$, so $\exp(-\Delta_m/2) = 1$; for all other models, $\exp(-\Delta_m/2) < 1$, and this number shrinks very quickly as Δ_m grows.

Notice that when $\Delta_m = 2$, $\exp(-\Delta_m/2) = 0.37$, so that the posterior probability of this model is about one-third as large as that for the best-fitting model. This is not a big difference—if they were the only two models being considered, then the less-likely model would have a 25% chance of being correct vs. 75% for the more likely model—and suggests that we should not feel overly confident that the data have unambiguously found the "correct" model. It is useful to know how many other models are also supported by the data and to have a measure of the comparative evidence for each one. On the other hand, variable-selection procedures that select a single model essentially declare one model to have probability 1 and the rest to have probability 0. This seems a questionable strategy, considering that there were 46 models within 2 IC units of the best in our placekick example of Section 5.1.3.

Once we have estimated posterior probabilities for each model, we can use these to combine information from several models into a single estimate of any parameter. For now, let θ be any quantity that we want to estimate from the models (such as a regression parameter, an odds ratio, or a predicted value). Denote its estimate in model m by $\hat{\theta}_m$ with corresponding variance estimated by $\widehat{Var}(\hat{\theta}_m)$. Then the model-averaged estimate of the parameter is

$$\hat{\theta}_{\text{MA}} = \sum_{m=1}^{M} \hat{\tau}_m \hat{\theta}_m, \tag{5.3}$$

and the variance of this estimate is estimated by

$$\widehat{Var}(\hat{\theta}_{\text{MA}}) = \sum_{m=1}^{M} \hat{\tau}_m[(\hat{\theta}_m - \hat{\theta}_{\text{MA}})^2 + \widehat{Var}(\hat{\theta}_m)]. \tag{5.4}$$

If one model is clearly better than all others, then its $\hat{\tau}_m \approx 1$. In this case, the model-averaged estimate, $\hat{\theta}_{\text{MA}}$, is not much different from this model's estimate, and the variance of $\hat{\theta}_{\text{MA}}$ is essentially just the estimated variance of $\hat{\theta}_m$ from its model. On the other hand, if many models have comparable probabilities, then the variance of the estimate $\hat{\theta}_{\text{MA}}$ depends both on how variable the parameter estimates are from model to model through $(\hat{\theta}_m - \hat{\theta}_{\text{MA}})^2$ and on the variance of each model's parameter estimates.

This process can be applied to any number of parameters associated with the model. As an example, consider the case of a single regression parameter, $\theta = \beta_r$. There is an estimate, $\hat{\beta}_{r,m}$, in each model. The model-averaged estimate of β_r is $\hat{\beta}_{r,\text{MA}} = \sum_{m=1}^{M} \hat{\tau}_m \hat{\beta}_{r,m}$. Notice here that half of the models in all-subsets selection exclude X_r, so $\hat{\beta}_{r,m} = 0$ for these models. In models where X_r does appear, and that also have high posterior probability attached to them, β_r is likely to have its magnitude overestimated. These facts have several implications:

1. Model averaging generally results in a non-zero estimate for *all* regression parameters.

2. Variables that are truly unimportant are unlikely to appear in the models that have high posterior probability. Thus, their model-averaged estimates are close to zero, because the probabilities for the models in which they appear are all small, and there is little variability in these estimates. These variables have little impact on model-averaged predictions. From a variable-selection perspective, there is negligible evidence that they are needed in the model, but the BMA procedure itself does not remove them.

3. On the other hand, variables that appear in some, but not all, of the models with the highest posterior probability typically have their estimates shrunk by being averaged with zeroes from those models in which they do not appear. This reduces the post-selection bias noted in Section 5.1.5.

There is some controversy surrounding the use of BMA estimates for regression parameters. It is well understood that the interpretation of any regression model parameter depends on what other variables are included in the model. Specifically, regression parameters represent changes in a linear predictor, *holding all other variables in the model fixed* (e.g., see Sections 2.2.3 and 4.2). However, variables not in the model are not constrained in this way. Hence, BMA parameter estimates are amalgamations of estimates of parameters that have inherently different meanings. This matter is discussed by Heinze et al. (2018) in the general context of variable selection. The problem seems to be at least partly resolved if one starts from the assumption that all other variables in the pool are held constant when interpreting a particular parameter's estimate, regardless of the immediate model's context.

Model averaging as a variable-importance measure

As noted in Section 5.1.3, it is common in an exhaustive search that there are many models with $IC(k)$ values very similar to the best model. These models may contain many of the same explanatory variables, thereby providing evidence regarding which variables are most important. We can formalize this evidence using model averaging.

If a variable is important, then the models where that variable is present—i.e., where its $\beta_r \neq 0$—tend to be among those with the highest probabilities, and models without that variable—where its $\beta_r = 0$—all have very small probabilities. Thus, the models that contain the variable have probabilities that add to nearly 1. Similarly, an unimportant variable tends not to be among the top models, so its probabilities sum to something fairly small. Therefore, we can measure a variable's importance in the model by summing the posterior probabilities for all models in which the variable appears.

Raftery (1995) suggests guidelines for interpreting the posterior probability that a variable belongs in the model. First, any variable with posterior probability below 0.5 is more likely not to belong in the model than to belong. Evidence that a variable belongs in the model is "weak" when its posterior probability is between 0.50 and 0.75, "positive" when it lies between 0.75 and 0.95, "strong" when it lies between 0.95 and 0.99, and "very strong" when its posterior probability is above 0.99. Of course, these are not the only interpretations possible.

Example: Placekicking (BMAPlacekick.R, Placekick.csv)

As described in Section 5.1.3, there are several R packages that can perform all-subsets regression. Among these `BMA`, `MuMin`, and `glmulti` all can perform model averaging. We use `MuMIn` here because it is relatively easy and can be applied to models fitted using many different R functions. The program for this example also performs the analysis using the other packages.

Below, we start from the model-fit object and use the `dredge()` function to evaluate the IC, here BIC, on each possible model. The `model.avg()` function computes the posterior probabilities for each model and each variable. An optional argument `subset` allows one to specify a subset of models on which the calculations are to be performed, which can reduce computational burden when the full set contains billions of models, most of which have negligible probability. The object created by `model.avg()` contains several useful components that can be displayed using `summary()` or accessed separately for more control. We show the 10 models with the smallest BIC and, hence, largest posterior probabilities, in the `msTable` component of the saved object. The `sw` component of this same object gives the posterior probabilities for each variable. The `coefmat.full` component from `summary()` gives the tables of model-averaged estimates, standard errors, and standard normal test results for each parameter. Confidence intervals for each parameter are available from `confint()`.

```
> mod.fit <- glm(good ~ ., family = binomial(link = "logit"),
    data = placekick, na.action = na.fail)
> try1 <- dredge(global.model = mod.fit, rank = "BIC")
Fixed term is "(Intercept)"

> # Model averaging with model.avg
> ma1 <- model.avg(object = try1)
> # Models and posterior probabilities ("weight") are in msTable
> ma1$msTable[1:10,]
     df    logLik      BIC    delta     weight
25    3 -381.2049 784.1956 0.000000 0.65688254
125   4 -379.6662 788.3802 4.184578 0.08106215
258   4 -379.8309 788.7094 4.513828 0.06875786
257   4 -380.2855 789.6186 5.423004 0.04364126
2     2 -387.8725 790.2689 6.073279 0.03152768
256   4 -381.1263 791.3004 7.104740 0.01882405
235   4 -381.1557 791.3591 7.163511 0.01827894
```

```
245    4 -381.2037 791.4550 7.259429 0.01742299
12     3 -385.2498 792.2853 8.089662 0.01150376
1258   5 -378.3433 792.9961 8.800516 0.00806269

> # Variable weights
> ma1$sw
                     distance PAT  change wind week type elap30
Sum of weights:      1.00     0.94 0.12   0.09 0.06 0.03 0.03
N containing models:  128      128  128    128  128  128  128
                     field
Sum of weights:      0.03
N containing models:  128

> # Model-averaged coefficients in table
> round(summary(ma1)$coefmat.full, digits = 3)
            Estimate Std. Error Adjusted SE z value Pr(>|z|)
(Intercept)    4.626      0.551       0.551   8.390    0.000
distance      -0.088      0.013       0.013   6.882    0.000
PAT            1.252      0.480       0.481   2.605    0.009
change        -0.042      0.133       0.133   0.317    0.752
wind          -0.051      0.184       0.184   0.276    0.783
week          -0.002      0.008       0.008   0.204    0.838
type           0.003      0.038       0.038   0.068    0.946
elap30         0.000      0.002       0.002   0.052    0.958
field          0.000      0.031       0.031   0.012    0.990

> # Confidence intervals for parameters
> round(cbind(coef(ma1), confint(ma1)), digits = 3)
                    2.5 % 97.5 %
(Intercept)  4.626  3.546  5.707
distance    -0.088 -0.113 -0.063
PAT          1.326  0.578  2.074
change      -0.354 -0.740  0.031
wind        -0.535 -1.150  0.081
week        -0.026 -0.064  0.012
type         0.091 -0.315  0.497
elap30       0.003 -0.017  0.024
field       -0.014 -0.388  0.359

> # Exponentiating into ORs, removing intercept
> round(exp(cbind(coef(ma1), confint(ma1))[-1,]), digits = 2)
              2.5 % 97.5 %
distance 0.92  0.89   0.94
PAT      3.77  1.78   7.96
change   0.70  0.48   1.03
wind     0.59  0.32   1.08
week     0.97  0.94   1.01
type     1.10  0.73   1.64
elap30   1.00  0.98   1.02
field    0.99  0.68   1.43
```

The msTable codes variables in each model according to their alphanumeric order. In this example, 1 is change, 2 is distance, and so forth. Thus, the first model listed is given as 25, which includes distance and PAT. The output gives the degrees of freedom (number of parameters in the model, including intercept), the BIC value, the

Δ_m value, and the estimated posterior probability (`weight`) for each listed model. We see that the model with `distance` and `PAT` has the smallest BIC and largest estimated posterior probability, 0.66. Any model with a posterior probability more than 0.5 is believed by the process to be more likely the correct model than not. Adding `change` to the model increases BIC by 4.18 units and decreases the posterior probability to 0.08. Notably, `distance` is in all of these top models, and `PAT` is in most of them. This is reflected in the estimated posterior probabilities for the variables, where `distance` has a probability that rounds to 1.00, while `PAT` has probability 0.94. No other variables have posterior probabilities greater than 0.5, which suggests that there is less evidence in favor of their inclusion in the correct model than against it.

The model-averaged parameter estimates are rounded to three digits for easier reading. They have been rearranged from highest to lowest posterior probability. There are two columns for standard errors, where the `Adjusted SE` version is based on equation (5.4), and the other is based on a different formula from Burnham and Anderson (2002) that gives essentially the same results in this case. For `distance`, $\hat{\beta}_{\text{MA}} = 0.088$, with standard error 0.013, while for `PAT`, $\hat{\beta}_{\text{MA}} = 1.25$, with standard error 0.48. The p-values from the standard normal tests in the last column in the table provide evidence that both `distance` and `PAT` have parameters that are not zero, while failing to provide such evidence for the remaining variables' regression parameters. These tests are not based on any post-selection process, but rather on the combined results from many fitted models, and hence are valid subject to the usual concerns about the small-sample performance of tests that use large-sample approximations to their distributions. Finally, the 95% confidence intervals for regression parameters and corresponding odds ratios corroborate the previous observations about the importance of `distance` and `PAT`.

Although the theory for BMA applies only to BIC, the methods of this section are often used with other $IC(k)$ as well. Burnham and Anderson (2002) refer to the model and variable probabilities calculated in this section as *evidence weights* rather than posterior probabilities when alternative $IC(k)$ are used. This explains why the posterior probabilities are referred to as `weights` in the output from `MuMIn`.

5.2 Tools to assess model fit

When we fit a generalized linear model (GLM), we assume that the following three things are true about the population from which the data are drawn: (1) the distribution, or random component, is the one we specify (e.g., binomial or Poisson); (2) the mean of that distribution, say ξ, is linked to the explanatory variables by the function g that we specify (e.g., $g(\xi) = \beta_0 + \beta_1 x$); and (3) the link relates to the specified explanatory variables in a linear fashion. Any of these assumptions could be violated; indeed, we expect that a model is only an approximation to the truth. In this section, we present tools that can help us to identify when a model is not a good approximation.

The methods we use are largely analogous to those used in normal linear regression, particularly in their reliance on *residuals*. For a refresher on the use of residuals in diagnosing problems with linear models with normally distributed errors, see a book on linear regression such as Kutner et al. (2004). In this section, we show how to compute and interpret residuals that are appropriate for diagnosing problems in distributions where the mean and variance are related and the data are not necessarily from normal distributions. We also offer some methods for testing the fit of a model. These techniques should always be used as part of an iterative model-building process. A model is proposed, fit, and checked. If it is found lacking, a new model is proposed, fit, and checked, and this process is repeated until a satisfactory model is found. The final test of any model should be on data that have been held out from this iterative process, if possible, to ensure that the model does not "overfit" the data (see Section 5.5).

This section discusses tools that are used primarily with models for a single count response per observation, such as binomial and Poisson models. Historically, more effort has been spent developing diagnostic tools for binomial regression models—particularly, logistic regression—than for other models. Also, binomial models require special consideration because the counts that are modeled may be based on different numbers of trials for different subjects. Therefore, we focus our discussion on binomial models, and generalize to other models for counts as appropriate. Diagnostics for the multicategory regression models of Sections 3.3 and 3.4, which model multiple responses simultaneously, are less well developed and require special attention. We discuss them briefly at the end of the section.

5.2.1 Residuals

Generally, a residual is a comparison between an observed and a predicted value. With count data in regression models, the observed data are counts, and the predicted values are estimated mean counts produced by the model. For the Poisson model, this is straightforward, because the mean count is modeled and predicted directly. However, in some other cases, such as binomial, multinomial, and Poisson rate regressions, models are written in terms of parameters that are not necessarily on the count scale, such as the probability of success or the rate. These parameters need to be converted into counts before computing residuals. For a Poisson rate regression, this means simply multiplying the rate by the exposure for each observation. For binomial regression, this requires a little more care.

Consider a binomial regression problem in which there are M unique combinations of the explanatory variable(s). We refer to these combinations as *explanatory variable patterns* (EVPs). For EVP $m = 1, \ldots, M$, let n_m be the number of trials observed for variable(s) $x_{m1}, \ldots, x_{mp}$. For example, in the context of the placekicking example with distance as the only explanatory variable (x_{m1}), there are 3 placekicks at a distance of 18 yards, 7 at 19 yards, 789 at 20 yards, and so forth, leading to $n_1 = 3, n_2 = 7, n_3 = 789, \ldots$. We have a choice between fitting the model to the individual binary responses or first aggregating the trials and fitting the model to the summarized counts of successes, w_m, as shown in Section 2.2.1. The parameter estimates are the same either way and lead to the same estimated probability of success, $\hat{\pi}_m$. But modeling the binary trial responses individually produces n_m separate residuals for each EVP. Because the observed responses are binary, w_m of these residuals are $1 - \hat{\pi}_m$ and the remaining $n_m - w_m$ are $-\hat{\pi}_m$, regardless of whether $\hat{\pi}_m$ is at all close to the observed proportion of success, w_m/n_m. On the other hand, modeling the aggregated counts results in a single residual at each EVP that compares the observed and predicted counts directly as described below. Therefore, prior to computing residuals or any of the quantities that we later derive from them, it is important to aggregate all trials for each EVP into a single count as shown in the example on p. 85.

Henceforth, we assume that all binomial data have been aggregated in this way into counts w_m, $m = 1, \ldots, M$. In models where no aggregation is needed, including Poisson and related models, then w_m is simply the observed response. We can denote the corresponding random variable as W_m and its population mean as μ_m, $m = 1, \ldots, M$, where in the binomial case $\mu_m = n_m\pi_m$. The estimated mean count is $\hat{w}_m$, which is either directly estimated by the model for mean counts or computed from the rate or probability.

Often the fits of several possible models are to be compared, for example through likelihood ratio tests or some other comparison of residual deviance. Likelihood functions based on different EVP structures behave as though they were calculated on different data sets, so they are not comparable. Aggregation of binomial counts should, therefore, be done at the level of the most complex model being compared before fitting all models of interest.

Counts from the distributional models discussed in Chapters 2–4 all have variances that depend on their respective means. Therefore, the "raw residual" that is commonly used in normal linear regression, $w_m - \hat{w}_m$, $m = 1, \ldots, M$, is of limited use because a difference of a given size may be either negligible or extraordinary depending on the size of the μ_m. We must instead use quantities that adjust for the variability of the count. The basic residual used with count data is the *Pearson residual,*

$$e_m = \frac{w_m - \hat{w}_m}{\sqrt{\widehat{Var}(W_m)}}, \; m = 1, \ldots, M$$

where $\widehat{Var}(W_m)$ is the estimated variance of the count W_m based on the model (e.g. $\widehat{Var}(W_m) = \hat{w}_m$ for Poisson, $\widehat{Var}(W_m) = n_m\hat{\pi}_m(1 - \hat{\pi}_m)$ for binomial). Pearson residuals behave approximately like a sample from a standard normal distribution, particularly when the $\hat{w}_m$ (and $n_m - \hat{w}_m$ in the binomial case) are large for all m. However, the denominator overestimates the standard deviation of $w_m - \hat{w}_m$, so that the variance of e_m is smaller than 1. To correct this, we compute the *standardized Pearson residual,*

$$r_m = \frac{w_m - \hat{w}_m}{\sqrt{\widehat{Var}(W_m - \hat{W}_m)}} = \frac{w_m - \hat{w}_m}{\sqrt{\widehat{Var}(W_m)(1 - h_m)}},$$

where h_m is the mth diagonal element of the hat matrix described in Sections 2.2.7 and 5.2.3.

A different kind of residual is sometimes used that is based on the model's residual deviance. As first discussed in Section 2.2.2, the residual deviance is a measure of how far the model lies from the data and can be written in the form $\sum_{m=1}^{M} d_m$, where the exact nature of d_m depends on the model for the distribution. For example, Equation 2.10 shows that $d_m = -2y_m\log[\hat{w}_m/w_m] - 2(n_m - w_m)\log[(n_m - w_m)/(n_m - w_m)]$ in the binomial case. The quantity $e_m^D = \text{sign}(w_m - \hat{w}_m)\sqrt{d_m}$ is called the *deviance residual*, which is commonly displayed in `summary()` output from `glm()` fits and has an approximate standard normal distribution in large samples.. The *standardized deviance residual* $r_m^D = e_m^D/\sqrt{(1 - h_m)}$ has a more stable variance and is used more often for model assessment. Among the different types of residuals, we prefer to use the standardized Pearson residuals when they are available. They are usually easy to interpret, and they are also not difficult to compute from many common fitted model objects.

Standardized residuals can be interpreted approximately as observations from a standard normal distribution. We use convenient thresholds based on the standard normal distribution as rough guidelines for identifying unusual residuals. For example, values beyond ± 2 should happen only for about 5% of the sample when the model is correct, values beyond ± 3 are extremely rare, and values beyond ± 4 are not expected at all.

However, for EVPs where only a small number of possible responses are likely to occur, the standard normal approximation is poor, and these guidelines should be viewed as tentative. For example, binary responses ($n_m = 1$) can produce only two possible raw residual values, $0 - \hat{\pi}_m$ and $1 - \hat{\pi}_m$ and, hence, also only two possible Pearson, deviance, or standardized residual values. This discrete situation is not at all well-approximated by a normal distribution, which is for continuous random variables. More generally, there are

$n_m + 1$ possible residual values for an EVP with n_m trials, so with small n_m there are not enough different possible residual values to allow meaningful interpretation. The normal approximation can also be poor for larger n_m when $\hat{\pi}_m$ is near 0 or 1. Again, these cases tend to generate only a few possible responses unless the number of trials is large. As a consequence, very rarely are standardized residuals for such EVPs found beyond the rough guidelines even when the fitted model is very different from the true structure. When they do appear, it is often merely because there are many EVPs with individual trials and relatively extreme values of $\hat{\pi}_m$, so that *any* response in the opposite direction has standardized residual with large magnitude.

A similar phenomenon can happen with Poisson models where the estimated means are very small, for example $\hat{w}_m < 0.5$. Then most observed w_m are either 0 or 1, with only occasional larger values, and there are very few likely residual values. However, those larger values that have small probability do occasionally occur, and residuals can help to identify if they happen more often than the model expects. See Exercise 16.

In addition to the interpretation using a standard normal distribution approximation, we suggest computing the exact tail probability of residuals, assuming that the model is correct. For each observation, the tail probability of the residual is the probability that the respective count would be so far away from its mean relative to its estimated probability distribution. A small tail probability implies that the observation is somewhat unusual compared to where the model expects it to be. Conversely, a tail probability that is not small suggests that the observation is not particularly unusual.

In the binomial case, we find the minimum of $P(W \geq w_m)$ and $P(W \leq w_m)$ for W that has a binomial distribution with n_m trials and $\hat{\pi}_m$ as the estimated probability of success as calculated from the model. The Poisson case follows in a similar manner using a Poisson distribution with $\hat{w}_m$ as the estimated mean. An appropriate threshold on the tail probability such as 0.025 or 0.005 is used to identify potentially unusual residual values for a model, which suggests where the model may not fit the data well. We typically use normal-based thresholds mainly for constructing guidelines on residual plots that aid in rapid assessment of model fit, as shown below.

Computing residuals in R

Many model-fitting functions have method functions associated with them that can compute various kinds of residuals for models shown in Chapters 2–4. Most notably, the generic function `residuals()` has method functions that work on model-fit objects produced by `glm()`, `vglm()`, `zeroinfl()`, and `hurdle()`. All of these functions use the `type` argument to select the types of residuals computed. Pearson or raw residuals are produced with the argument values `"pearson"` or `"response"`, respectively. The `residuals.glm()` method also has a `"deviance"` argument value to produce deviance residuals. For models fit with `logistf()`, there are currently no functions available that produce useful residuals automatically. For these objects, relatively simple manual calculation can create Pearson or deviance residuals.

Standardized Pearson and deviance residuals are available for `glm`-class objects using `rstandard()`. Computing standardizing residuals for fits from other models entails calculating the hat matrix manually. Finding the diagonal elements of the hat matrix is aided by the code for computing $(\mathbf{X}'\mathbf{V}\mathbf{X})^{-1}$ shown in the example on p. 80, although not all model-fit objects readily provide all of the required elements to complete these calculations easily.

Graphical assessment of residuals

The information content in a set of residuals is usually best understood in graphical displays. Most of these displays mirror the residual plots used in linear regression (see, e.g., Kutner et al., 2004). However, their interpretation in generalized linear models can be slightly different. In particular, whereas residual plots in linear regression mainly diagnose problems with the mean model and possible unusual observations, in generalized linear models the plots also can diagnose when other model assumptions do not fit the data well, including the choice of distributional family. Of course, interpretation of any residual plot for a count-data model is subject to the caveats noted previously when there are limited numbers of possible responses.

A plot of the standardized residuals against each explanatory variable, x_r, $r = 1, \ldots, p$, can show whether the form for that explanatory variable is appropriate. The plot should show roughly the same variance throughout the range of the explanatory variable of interest, x_r, and should show no serious fluctuations in the mean value. Curvature in the plot suggests that a transformation or additional polynomial terms are needed for that variable, or possibly that the link needs to be changed. We recommend adding a smooth curve to the plot, such as a *loess* smoother[2], representing the average residual over the range of the x-axis. This can make finding changes in the mean residuals easier. The curve should wiggle randomly around zero when the model is a reasonable fit. Clear patterns of curvature, such as a "smile" or a "frown," are indications of a problem. In binomial models, the loess curve can be weighted by n_m, so that EVPs based on larger numbers of trials contribute relatively more to the curve placement than those with relatively few. Note that loess curves are highly variable where data are sparse or near the extreme values of the variable on the x-axis. Care should be taken not to "overinterpret" apparent changes in the curve at the edges of the plot.

Similarly, a plot of standardized residuals against the fitted values $\hat{w}$ ($\hat{\pi}$ in binomial models) is useful for assessing when the link function is not fitting well. The points should again have roughly constant variance and show no clear curvature. Alternatively, a plot of these residuals against the estimated linear predictor, $g(\hat{w})$, may show patterns of change in the residuals more clearly and can help to diagnose how the link $g(\cdot)$ should be changed to better fit the data.

Any of these plots can be read to check for extreme residuals. Where appropriate, thresholds based on the normal distribution can be added to the plot to identify unusual values. As discussed earlier, only about 5% of standardized residuals should be beyond ± 2, about $3/1000$ should lie beyond ± 3, and usually none beyond ± 4 when the normal distribution approximation is appropriate. Presence of a very small number of very extreme residuals could indicate *outliers*—unusual observations that should be checked for possible data entry errors or some other underlying cause. However, outliers are rare by definition. If there are numerous residuals whose magnitudes are larger than expected, and these values are randomly scattered across the range of the plot's x-axis, then this may be a sign of *overdispersion*, meaning that there is more variability to the counts than what the model assumes there should be. This is an indication that there may be important explanatory variables missing from the model, or that a different distribution may be needed to model the data. See Section 5.3 for details regarding overdispersion.

Unusual patterns may develop in residual plots when w can take on only a few possible values with appreciable probability, for example in Poisson models with very small means,

[2]Loess (sometimes LOWESS) is short for "locally weighted scatterplot smoother." For a given value of the explanatory variable x, say x_0, a response is estimated using a regression where points with values of x near x_0 are given higher weight than distant points.

binomial models with small n_m for each m, or binomial models with probabilities very close to 0 or 1. Hosmer and Lemeshow (2000) show that the dispersion of standardized residuals is greater in these regions than in regions where these values are more moderate. There are two main reasons for this. First, hat values h_m tend to be larger for more extreme values of the explanatory variables, for which the probability-of-success estimates may also be extreme. Second, the raw residual values with the largest magnitudes are $0 - n_m\hat{\pi}_m$ or $n_m - n_m\hat{\pi}_m$ when $\hat{\pi}_m$ is near 1 or 0, respectively. Calculating the actual tail probabilities for these residuals can aid in their interpretation, rather than relying on normal distribution thresholds.

Also, when the response counts take on relatively few unique values, residual plots often show odd-looking bands of points. As the most extreme example, consider a plot of raw residuals against $\hat{\pi}$ in a logistic regression with binary responses. As noted above, for each value of $\hat{\pi}$ there are only two possible residuals. The plot will therefore show just two linear bands of points: one at $-\hat{\pi}$ and the other at $1 - \hat{\pi}$. This lack of "randomness" in the plot is an indication that the response counts on which the model is based are rather small and limited in range. In this case, residual plots may not be very helpful for detecting violations of model assumptions.

Example: Placekicking (PlacekickDiagnostics.R, Placekick.csv)

We demonstrate the interpretation of residuals and residual plots for logistic regression using the placekicking data with a model containing only the `distance` variable. Aggregation of these data into EVP form for this model was shown in Section 2.2.1. The model fit and functions calling the residuals are shown below, along with the code for computing the binomial tail probability for the extremeness of each residual.

```
> mod.fit.bin <- glm(formula = success/trials ~ distance,
    weights = trials, family = binomial(link = logit), data = w.n)

> pi.hat <- predict(mod.fit.bin, type = "response")
> s.res <- rstandard(mod.fit.bin, type = "pearson")
> w.n <- data.frame(w.n, pi.hat, s.res)
> round(head(w.n), digits = 3)
  distance success trials  prop pi.hat  s.res
1       18       2      3 0.667  0.977 -3.575
2       19       7      7 1.000  0.974  0.433
3       20     776    789 0.984  0.971  3.628
4       21      19     20 0.950  0.968 -0.448
5       22      12     14 0.857  0.964 -2.149
6       23      26     27 0.963  0.960  0.091

> # Compute exact lower and upper cumulative probability of each
> #  observed response (How extreme is the residual?)
> # P(W <= w_m)
> prob.smaller <- pbinom(q = w.n$success, size = w.n$trials,
    prob = w.n$pi.hat, lower.tail = TRUE)
> # P(W >= w_m)
> prob.higher <- pbinom(q = w.n$success, size = w.n$trials, prob
    = w.n$pi.hat, lower.tail = FALSE) + dbinom(x = w.n$success,
    size = w.n$trials, prob = w.n$pi.hat)
> # Mininum of P(W <= w_m) and P(W >= w_m)
> w.n$tail.prob <- apply(X = cbind(prob.smaller, prob.higher),
    MARGIN = 1, FUN = min)
```

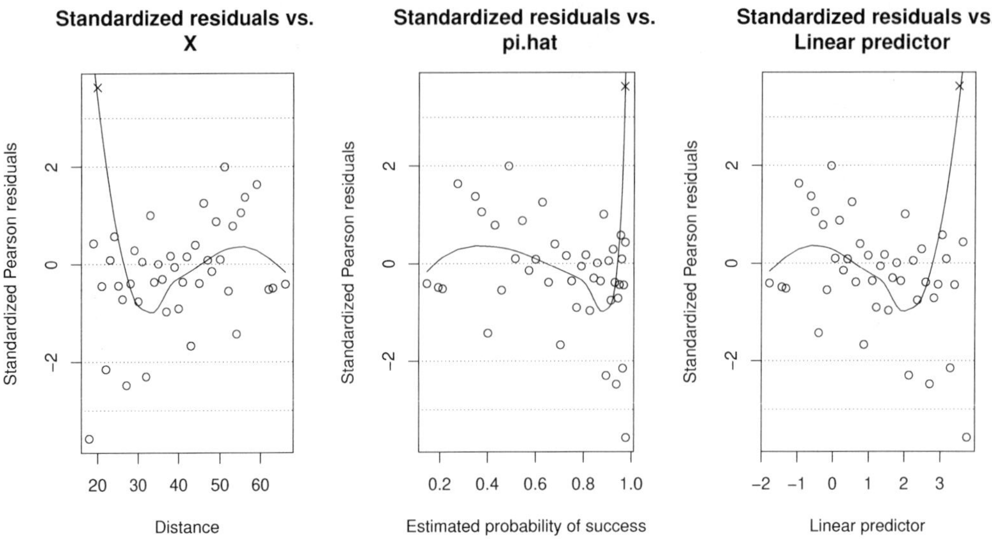

Figure 5.2: Residual plots for modeling of the probability of success vs. distance.

```
> round(head(w.n), digits = 3)
  distance success trials  prop pi.hat  s.res tail.prob
1       18       2      3 0.667  0.977 -3.575     0.068
2       19       7      7 1.000  0.974  0.433     0.832
3       20     776    789 0.984  0.971  3.628     0.017
4       21      19     20 0.950  0.968 -0.448     0.483
5       22      12     14 0.857  0.964 -2.149     0.089
6       23      26     27 0.963  0.960  0.091     0.672

> round(w.n[w.n$tail.prob < 0.025,], digits = 3)
  distance success trials  prop pi.hat s.res tail.prob
3       20     776    789 0.984  0.971 3.628     0.017
> round(w.n[abs(w.n$s.res) > 1.96,], digits = 3)
   distance success trials  prop pi.hat  s.res tail.prob
1        18       2      3 0.667  0.977 -3.575     0.068
3        20     776    789 0.984  0.971  3.628     0.017
5        22      12     14 0.857  0.964 -2.149     0.089
10       27      24     29 0.828  0.937 -2.476     0.032
15       32      23     30 0.767  0.894 -2.299     0.034
34       51      11     15 0.733  0.486  1.995     0.048
```

Printing out the residuals with tail probability less than 0.025 suggests that only the EVP at 20 yards has an extreme residual, with tail probability 0.017. Using the normal approximation suggests that there are five additional EVPs with extreme residuals, although their tail probabilities are generally not as extreme as the normal approximation suggests.

The code below creates the plot of standardized Pearson residuals against the explanatory variable, yielding the first plot in Figure 5.2. The other two plots in Figure 5.2 use very similar code to plot residuals against estimated probabilities and linear predictors. Notice that the character symbol is changed to an "×" for residuals with tail probability below 0.025. Note also the use of `loess()` to create an object whose predicted values add the loess smooth trend for the mean residual. The `weights = trials` argument value causes `loess()` to use the number of trials in the EVP—denoted in the `w.n` data frame as `trials`—to weight the mean trend. Thus, EVPs with many trials influence the trend curve more than those with fewer.

```
> chars <- ifelse(test = w.n$tail.prob < 0.025, yes = 4, no = 1)

> # Standardized Pearson residual vs. X plot
> plot(x = w.n$distance, y = w.n$s.res, xlab = "Distance", ylab
   = "Standardized Pearson residuals", main = "Standardized
   residuals vs. \n X", pch = chars)
> abline(h = c(3, 2, 0, -2, -3), lty = "dotted", col = "blue")

> smooth.stand <- loess(formula = s.res ~ distance, data = w.n,
   weights = trials)
> # Make sure that loess estimates are ordered by "X" for the
> #  plots, so that they are displayed properly
> order.dist <- order(w.n$distance)
> lines(x = w.n$distance[order.dist], y =
   predict(smooth.stand)[order.dist], lty = "solid", col =
   "red", lwd = 1)
```

Because `distance` is the only variable in the model and it has a negative coefficient, the plot of the residuals against the linear predictor is just a mirror image of the plot against `distance`. All three plots show the same general features. First, note the general trend of increasing variability as $\hat{\pi}$ approaches 1, which is most evident in the center plot. This is expected to some degree even when the model fits well. However, the extent to which this phenomenon manifests here is a bit unusual. The residuals with the two largest magnitudes, both of which are beyond 3, lie near the boundary of each plot. These correspond to distances of 18 and 20 yards. The magnitude of the residual at 18 yards is large because there were only 2 successes in 3 trials when the probability of success is estimated to be 0.977. The normal approximation for the standardized residuals is not expected to hold well in this situation, so the bounds at ± 2 and ± 3 shown on the plot are not very accurate. In the previous output, we computed that the tail probability of a residual this extreme was actually 0.068.

Only the residual at 20 yards was identified as unusual by the binomial tail probability, and this observation should be investigated further. It is interesting to note that this EVP contains more than half of the trials in the data because it combines both attempted field goals and PATs. Therefore, it is likely to have high influence on both the fitted logistic regression model and the loess curves on the plots. Indeed, if one focuses only on the remaining points in the plots, it appears that there may be a general decreasing or increasing trend to them, depending on the plot. Unlike ordinary field goals, PATs are placed exactly in front of the center of the goalposts, creating the largest possible angle for a successful kick, and they are worth only one point rather than three. It might be possible that these kicks have a higher probability of success than would be expected for an ordinary field goal at a comparable distance, and therefore the observation at 20 yards pulls the logistic regression curve higher than the neighboring points suggest. Perhaps the model could be improved by adding a variable that allows PATs to have a different probability of success than other placekicks. We explore this possibility in a further example in Section 5.4.1.

5.2.2 Goodness of fit

Examining residual plots is essential for understanding problems with the fit of any generalized linear model, but interpretation of these plots is somewhat subjective. Goodness-of-fit (GOF) statistics are often computed as more objective measures of the overall fit for a model. They attempt to summarize how well certain model assumptions are supported by the data. While these statistics are potentially useful, they provide little information regarding the cause of any poor fit that they might detect. We strongly recommend that these statistics be used as an adjunct to interpreting residual plots, rather than as a replacement for them.

In linear regression, people sometimes use the coefficient of determination, widely denoted as R^2, as a measure of how well a model fits. However, R^2 does not technically measure goodness of fit. Specifically, R^2 measures what fraction of the total variability in a set of data can be explained by the fitted model. It says nothing about whether the model is a good approximation to the true mean structure for the population from which the data were drawn. Therefore, while there do exist variants of R^2 designed for GLMs fitted by ML estimation rather than least squares, we do not consider them to be GOF statistics.

Instead, statistical assessment of overall model fit typically begins with looking at the residual deviance and Pearson statistics. The residual deviance, which we can denote here by D, is a comparison of the model-estimated counts to the observed counts and has a form that depends on the distribution used for the model (see Appendix B.5.3). The Pearson statistic has a more generic form but can also be calculated in several different ways. For distributions with a univariate response, such as the Poisson or binomial, the Pearson statistic is calculated as

$$X^2 = \sum_{m=1}^{M} \frac{(w_m - \hat{w}_m)^2}{\widehat{Var}(W_m)}.$$

For example, $\widehat{Var}(W_m) = \hat{w}_m$ for a Poisson model and $\widehat{Var}(W_m) = n_m\hat{\pi}_m(1 - \hat{\pi}_m)$ for a binomial model. For binomial models, it is sometimes more convenient to use an equivalent form of the statistic based on both the success and failure counts,

$$X^2 = \sum_{m=1}^{M} \frac{(w_m - \hat{w}_m)^2}{\hat{w}_m} + \sum_{m=1}^{M} \frac{((n_m - w_m) - (n_m - \hat{w}_m))^2}{n_m - \hat{w}_m}. \quad (5.5)$$

The D and X^2 statistics are summary measures of the distance between the model-based estimates and the observed data. As such, they are often used in formal tests of goodness of fit, which means testing the null hypothesis that the model is correct against the alternative that it is not. Rejecting the null hypothesis suggests that there is something wrong with the model, but the alternative is very general and provides no direct evidence for what the problem might be. Of course, failure to reject the null hypothesis does not mean that the model is "correct," but rather that the data do not provide sufficient evidence against the model.

Both statistics have large-sample $\chi^2_{M-(p+1)}$ distributions, where $p+1$ is the total number of regression parameters estimated in the model. However, this result is only valid under the assumption that the set of EVPs in the data is fixed and would not change with additional sampling. This is true if, for example, all explanatory variables are categorical, and all possible categories have already been observed, or if numerical values are fixed by the design of an experiment. However, in most cases with continuous explanatory variables, additional sampling would surely yield new EVPs, and so the distributions of D and X^2 may not be approximated well by a $\chi^2_{M-(p+1)}$ distribution even if the model is correct. Furthermore, the $\chi^2_{M-(p+1)}$ approximation holds reasonably well only if all predicted counts $\hat{w}_m$—and

$n_m - \hat{w}_m$ in binomial models—are large enough (e.g., all are at least 1 and many are at least 5). Thus, the models in which these statistics can be used in formal tests are limited primarily to those with only categorical predictors, and not too many of them, so that all combinations have reasonably high estimated counts.

As an informal alternative to formal hypothesis testing, it is common to compute the ratio of residual deviance to residual degrees of freedom, which we denote in the text as "deviance/df," or symbolically by $D/(M-(p+1))$. Because the expected value of a $\chi^2_{M-(p+1)}$ random variable is $M-(p+1)$, we can expect that this ratio ought not be too far from 1 when the model is correct. There are no clear guidelines for how far is "too far," owing in part to the uncertainty surrounding whether the $\chi^2_{M-(p+1)}$ distribution is a reasonable approximation to the distribution of D. *Very* rough guidelines can be derived assuming that the approximation is reasonable. In this case, when the number of degrees of freedom is large, values of D that are more than, say, two or three standard deviations above their expected values would be unusual when the model is correct. Because the variance of a chi-square random variable is twice its degrees of freedom, then this suggests a rough guideline of $D/(M-(p+1)) > 1 + 2\sqrt{2/(M-(p+1))}$ to indicate a potential problem, and $D/(M-(p+1)) > 1 + 3\sqrt{2/(M-(p+1))}$ to indicate a poor fit. We do not intend these thresholds to be taken as firm rules, but rather to provide the user with some way to reduce the subjectivity in interpreting this statistic.

Goodness-of-fit tests

There are several alternative GOF tests that can be used even when the number of EVPs may increase with additional sampling. These are especially common in logistic regression, but variants of them can be used in other regression models as well. Chapter 5 of Hosmer and Lemeshow (2000) provides a nice discussion of these tests for logistic regression. We summarize their discussion and provide our own insight.

The best-known GOF test for EVPs with continuous variables is the Hosmer and Lemeshow (HL) test (Hosmer and Lemeshow, 1980). The idea is to aggregate similar observations into groups that have large enough samples so that a Pearson statistic computed on the observed and predicted counts from the groups has approximately a chi-square distribution. As an example, consider a single explanatory variable, such as distance in the placekicking data. We might expect that placekicks from similar distances have similar enough success probabilities that their observed successes and failures could be grouped into a single pair of counts with *roughly* constant probability of success. This could be done before fitting a model (i.e., "binning" the data into intervals), but doing so can sometimes create a poor-looking fit out of a good model. Instead, Hosmer and Lemeshow suggested fitting the model to the data first, and then forming g groups of "similar" EVPs according to their estimated probabilities. The aggregated counts of successes and failures within each group are then combined with the aggregated model-estimated counts within the same group into a Pearson statistic as shown in Equation 5.5, summing over both successes and failures for the g groups. The test statistic, X^2_{HL}, is compared to a χ^2_{g-2} distribution.

Failing to reject the null hypothesis that the model is correct does not mean that the model is, indeed, a good fit for the data. The power of the test can be affected by sample size and clustering of EVPs, and the process of aggregation can sometimes mask potential problems. For example, a large positive outlier and a large negative outlier in the same group will somewhat "cancel" each other out, resulting in a smaller value of X^2_{HL} than some other grouping might have created. This points to another difficulty with the test: there are numerous ways to form the groups of similar EVPs, and the outcome of the test can depend on the chosen grouping pattern. In the most common implementation of the test, EVPs are ordered by their values of $\hat{\pi}$ and divided into $g = 10$ groups of approximately equal size. However, there are several other ways to perform the grouping—for example, making the number of EVPs, trials, or expected counts in each group approximately constant—and these can give different groupings. Test results may vary with the grouping, and no one algorithm is clearly better than others. Thus, it is a good idea to try several groupings, for example by setting g to be several different values (sample size permitting), and to ensure that the test results from the multiple groupings are in substantive agreement before drawing conclusions from this test. In general, the more groups that are used, the less chance that very dissimilar EVPs are grouped together, so the test can be more sensitive to poor fits. At the same time, it is important that the estimated expected counts—i.e., aggregated predicted successes and failures from the model—are large enough to justify using the large-sample chi-square approximation.

The HL test has other properties that call its reliability into question. For example, adding variables to a logistic regression model does not have a predictable impact on the HL test. The test may indicate that the larger model looks better or worse than the smaller one, and may do so without regard to the importance of the added terms. In general, the test maintains power to detect poor fits with moderately large models only when each EVP has $n_m = 1$ (Surjanovic and Loughin, 2023).

Hosmer et al. (1997) compare several GOF tests for logistic regression and recommend two other tests that can be used in addition to the HL test. The first of these is the Osius-Rojek test, which is based on the original model's Pearson GOF statistic. Osius and Rojek (1992) derived the large-sample mean and variance for X^2 in such a way that allows the number of sampled categories to grow as the sample size grows, enabling it to be used as a test statistic even when there are continuous explanatory variables. Then a standardized version of X^2 is compared against a standard normal distribution, with extreme values representing evidence that the fitted model is not correct. A second test is the Stukel test (Stukel, 1988), which explores whether the logit link is adequate in particular ways. The test uses a second logistic regression model that includes two additional explanatory variables calculated from the squared estimated logits from the original model fit. A test of the null hypothesis that the regression coefficients for these variables are both zero indicates whether the original model shape is adequate—which is implied by the null hypothesis—or whether a different link or a transformation of continuous variables might be needed. We forgo the calculation details here and direct interested readers to refer to Hosmer et al. (1997) and to our R code described in the next example.

Our recommendation with GOF tests is to use several of them in conjunction with residual plots and seek a consensus. The Osius-Rojek, Stukel, and HL tests can all be applied, the last with several different values of g, in an attempt to quantify any apparent concerns from the residual plots. If there is discrepancy among the test results, then approach the conclusion with uncertainty. Furthermore, Hosmer and Hjort (2002) recommend examining tables of Pearson residuals comparing the aggregated observed and model-predicted counts from the HL test, seeking patterns that may help to explain any evidence against the model.

Example: Placekicking (PlacekickDiagnostics.R, AllGOFTests.R, Placekick.csv)

In the placekicking example, we have a numeric variable, `distance`. While `distance` is discrete, with a somewhat limited set of values that could possibly be observed, we have not observed each of these values in the data. And if we were to add additional variables to the model, then our ability to observe every possible EVP would diminish. Therefore, we cannot treat the residual deviance from this model as a test statistic with an approximate $\chi^2_{m-(p+1)}$ distribution. We must instead use other numeric measures and tests to judge the fit of the model.

We start with the simple deviance/df measure. The statistic and its rough guidelines are simple to compute from the `deviance` and `df.residual` components taken from the fitted model.

```
> rdev <- mod.fit.bin$deviance
> dfr <- mod.fit.bin$df.residual
> ddf <- rdev/dfr
> thresh2 <- 1 + 2*sqrt(2/dfr)
> thresh3 <- 1 + 3*sqrt(2/dfr)
> c(rdev, dfr, ddf, thresh2, thresh3)
[1] 44.499448 41.000000  1.085352  1.441726  1.662589
```

The residual deviance, 44.5, is not much larger than the residual degrees of freedom, 41. The ratio, 1.09, is not larger than the lower threshold 1.44, so this rough check suggests no serious issues with the model fit.

None of the HL, Osius-Rojek, or Stukel tests are included in the default distribution of R, so we have written functions to perform them: `HLTest()`, `o.r.test()`, and `stukel.test()`, respectively. All three are contained in the file AllGOFTests.R and not presented here. For all three functions, the `obj` argument is a `glm` object fit using `family = binomial`. The `HLTest()` function includes an additional argument for g, which is set to `g = 10` by default. The function creates g groups with approximately equal numbers of EVPs in each group. Note that when EVPs have very different numbers of trials—as happens in the placekick data—the total numbers of trials in different groups may be very different as well. Each function's R code needs to be run once before carrying out the tests (e.g., using `source("AllGOFTests.R")`). For each function, the test statistic and p-value are produced. The `HLTest()` output also includes the observed and expected counts in each group and the corresponding unstandardized Pearson residuals that are used to calculate the test statistic. We run this test here using $g = 10$. In Exercise 29, the test is repeated for $g = 5, 6, \ldots, 12$.

```
> HL <- HLTest(obj = mod.fit.bin, g = 10)
Warning message:
In HLTest(obj = mod.fit.bin, g = 10) :
  Some expected counts are less than 5. Use smaller number of
     groups
> cbind(HL$observed, round(HL$expect, digits = 1))
              Y0  Y1 Y0hat Y1hat
[0.144,0.353]  3   2   3.8   1.2
(0.353,0.469] 19  13  18.3  13.7
(0.469,0.589] 25  39  30.1  33.9
(0.589,0.699] 24  49  26.4  46.6
(0.699,0.79]  32  74  26.6  79.4
(0.79,0.859]  18  75  15.9  77.1
(0.859,0.908] 12  69   9.2  71.8
(0.908,0.941] 10  72   5.9  76.1
(0.941,0.963]  3  53   2.6  53.4
(0.963,0.977] 17 816  24.3 808.7

> HL
        Hosmer and Lemeshow goodness-of-fit test with 10 bins
data:  mod.fit.bin
```

```
X2 = 11.028, df = 8, p-value = 0.2001

> round(HL$pear, digits = 1)  # Pearson residuals for each group
interval            Y0   Y1
  [0.144,0.353] -0.4  0.8
  (0.353,0.469]  0.2 -0.2
  (0.469,0.589] -0.9  0.9
  (0.589,0.699] -0.5  0.4
  (0.699,0.79]   1.0 -0.6
  (0.79,0.859]   0.5 -0.2
  (0.859,0.908]  0.9 -0.3
  (0.908,0.941]  1.7 -0.5
  (0.941,0.963]  0.3 -0.1
  (0.963,0.977] -1.5  0.3

> o.r.test(obj = mod.fit.bin)
z = 1.563042 with p-value = 0.1180426

> stukel.test(obj = mod.fit.bin)
Stukel Test Stat = 6.977164 with p-value = 0.03054416
```

The HL test function first reports a warning that some estimated expected counts are less than 5 in the 10×2 table displaying these counts for the groups. We print out the observed successes and failures (labeled `Y1` and `Y0`, respectively) and estimated expected number of successes and failures (`Y1hat` and `Y0hat`, respectively) in each group. For example, among all EVPs with $\hat{\pi}$ in the interval [0.144, 0.353][3] there are only 5 trials, and their estimated expected counts are split as 3.8 failures and 1.2 successes according to the fitted model. Also, the interval (0.941, 0.963] has only 2.6 estimated expected failures. These counts are not drastically low (e.g., below 1), and there are not too many of them (only 15% of cells), so the sample size is large enough to potentially support the χ^2_8 approximation for the test statistic. Also, the model has only one variable, so concerns about loss of power with larger models do not apply.

The test statistic is $X^2_{HL} = 11.0$ with a p-value of 0.20, which does not provide evidence of a poor fit. P-values for other values of g range from 0.09 to 0.23 (Exercise 29), which is in general agreement with this conclusion. However, the Pearson residuals for the grouped counts show a potential pattern in the signs. In each row, there must be a positive and a negative residual unless both are zero, and these residuals should be equally likely to occur with either success or failure category. Notice that five intervals in a row have the same pattern of positive residual for the $Y = 0$ response. Although none of these values is large (e.g., none larger than 2), the pattern suggests potential for a specific problem of overestimating the true success probability when the probability is large. We have already seen evidence of this in the residual plots of Figure 5.2.

The Osius-Rojek test yields a test statistic of 1.56 with a p-value of 0.12, which offers no serious evidence of a poor fit. However, the Stukel test provides a p-value of 0.03, which provides some evidence that the model is not fitting the data well. In particular, because this test targets deviations from the model separately for large and

[3]Square brackets, "[" or "]", indicate that the corresponding interval endpoint is included in the interval, while round parentheses, "(" or ")", indicate that the corresponding interval endpoint is *not* in that interval. For example, the interval (0.353,0.469] includes all values of $\hat{\pi}$ such that $0.353 < \hat{\pi} \leq 0.469$. Using this notation, there is no ambiguity regarding which interval contains the endpoints.

small probabilities, it may be more sensitive to the possible model flaw suggested by Figure 5.2 and the Pearson residuals from the HL grouping. We should investigate further to determine what may be causing this problem. We revisit this issue in Section 5.4.1.

For response variables from other distributions, there is no formal GOF test that is recognized as standard when continuous explanatory variables are involved. For models with a univariate response, a generalized version of the HL test can be conducted, wherein groupings are formed based on fitted values from the model, such as predicted counts from a Poisson model or rates from a Poisson rate model (Surjanovic et al., 2023). A package to perform this test is available at `https://github.com/nikola-sur/goodGLM`.

Note that the p-values from any GOF test are valid only if the model being tested could have been chosen before beginning the data analysis. The reasons are the same as those discussed in Section 5.1.4. In particular, using the data to select the variables to include in a model has the effect of including the variables that explain *that sample* the best. The selected model can sometimes appear to fit the data much better than it would fit a different sample from the same population. Applying a GOF test to the selected model using the same data tends to result in larger p-values than would be found if the fit were assessed more fairly on an independent data set.

This leads to two important points. First, whenever possible, models should be tested on independent data before they are adopted for some broader purpose than describing the data on which they are based (see Section 5.5). Second, the p-value from a GOF test needs to be interpreted carefully when the test is done on a model that results from a variable-selection process. If the p-value is "small," then this is evidence that the model does not fit well. However, it is not clear what interpretation can be applied when the p-value is not small.

Finally, all GOF tests are subject to a strange phenomenon: they may become *less* useful when the sample size grows extremely large. Because models are merely mathematically convenient approximations to more complicated phenomena, it should be expected that a GOF test should eventually be able to detect flaws in any model when given enough data. Thus, while a GOF test may yield a large p-value when applied with little data to a poor model, it may produce tiny p-values for seemingly excellent approximations just because the sample size is huge. Box (1976) argues that "the scientist must be alert to what is importantly wrong. It is inappropriate to be concerned about mice when there are tigers abroad." That is, we must decide whether flaws in a model are serious enough to warrant changing the model, or whether the model remains useful despite inevitable imperfection. The question of whether the model is useful is more an issue for a subject-matter specialist than a statistical analyst, but there are things that the statistician can do to "correct" overly aggressive GOF tests. See Nattino et al. (2020) and Paul et al. (2013).

5.2.3 Influence

The fit of a regression model is influenced by each of the observations in the sample. Removing or changing any observation (or equivalently removing an EVP in a binomial regression) can result in a change to the parameter estimates, predicted values, and GOF statistics. In variable selection, removing an observation can even influence whether or not a particular variable is selected. An observation is considered to be *influential* if the regression results change a lot when it is temporarily removed from the data set. For example, we speculated in Section 5.2.1 that the EVP at 20 yards might have high influence on the fitted regression, and that the relationship might change somewhat if the PATs were removed. Obviously, we do not want our conclusions to depend too strongly on a single observation, especially if it causes the model to fit less well for other observations, so it is wise to check for influential observations. *Influence analysis* consists of computing and interpreting statistics that measure different ways in which a model changes when each observation is changed or removed.

Removing observations and refitting models to measure how they change can be cumbersome. In linear regression the quantities most commonly computed to measure influence all have "shortcut" formulas that allow them to be computed from a single model fit. The same quantities are not so easily computed from a GLM model fit, but they can be approximated using formulas analogous to those used in linear regression. We present an approximate formula for two useful statistics here, and we discuss another quantity, leverage, that is useful for identifying potentially influential observations. For more details on influence analysis for linear models, see Belsley et al. (1980) or Kutner et al. (2004). Pregibon (1981) and Fox (2008) show how to apply some of these measures to GLMs.

An observation might be considered to be influential if the model fit changes considerably when the model is refitted to data without that observation. It is challenging to summarize in one statistic all of the changes that take place in the fit of a model when an observation is removed, but there are several statistics that can measure some aspects of this concept. For example, a large change in GOF statistics can indicate a highly influential observation. It turns out to be quite easy to approximate the change in Pearson or deviance GOF statistics when m is removed. The change in the Pearson statistic X^2 caused by the deletion of observation m is measured approximately by

$$\Delta X^2_m = r^2_m,$$

while the change in the residual deviance statistic is approximately

$$\Delta D_m = (e^D_m)^2 + h_m r^2_m.$$

The latter approximation is especially excellent, in many cases replicating the actual change in the residual deviance almost perfectly. For both statistics, we use thresholds of 4 and 9 to suggest observations that have moderate or strong influence on their respective fit statistics. These numbers are derived from squaring the ± 2 and ± 3 suggested thresholds that we use for the residuals upon which they are based.

Alternatively, we can measure the aggregate change in *all* of the parameter estimates simultaneously when observation m is temporarily deleted. Cook's distance does this for linear regression models. An approximate version of Cook's distance for GLMs is found as

$$CD_m = \frac{r^2_m h_m}{(p+1)(1-h_m)^2}, \; m = 1, \ldots, M.$$

In general, points with values of CD_m that stand out above the rest are candidates for further investigation. Alternatively, observations with $CD_m > 1$ might be considered to have high influence on the regression parameter estimates, while those with $CD_m > 4/M$ have moderately high influence.

In linear regression, *leverage* values are often computed as a way to measure the potential for an observation to influence the regression model through its extremeness relative to other observations. In Section 2.2.7, we defined the *hat matrix* in a logistic regression as $\mathbf{H} = \mathbf{V}^{1/2}\mathbf{X}(\mathbf{X}'\mathbf{V}\mathbf{X})^{-1}\mathbf{X}'\mathbf{V}^{1/2}$, where $\mathbf{X}$ is the $M \times (p+1)$ matrix whose rows $\mathbf{x}'_m$, $m = 1, \ldots, M$, contain the values for the p explanatory variables and a 1 in the first position (see p. 80), and $\mathbf{V}$ is a diagonal matrix whose mth diagonal element is $v_m = \widehat{Var}(W_m)$, the variance of the response variable calculated using the regression estimate of the mean or probability at $\mathbf{x}'_m$. This definition comes about through an analog to weighted least squares estimation used for normal linear regression models and applies to all GLMs (Pregibon, 1981). The diagonal elements of $\mathbf{H}$—h_m, $m = 1, \ldots, M$—are the leverage values for the GLM. Because the average value of h_m in any model is p/M, observations with $h_m > 2p/M$ are considered to have moderately high leverage, and those with $h_m > 3p/M$ to have high leverage.

Pregibon (1981) and Hosmer and Lemeshow (2000) show that h_m can sometimes measure the *potential* for observation m to have large influence on the model fit. However, the dependence of h_m on $\widehat{Var}(W_m)$ can create generally higher leverage values in high-variance locations and lower values in low-variance places, which clouds its interpretation. Leverage values should not be used by themselves to definitively identify observations with potential for high influence. Whether influence is actually exerted depends on whether an extreme *response* occurs, and this can be measured by CD_m, ΔX^2_m, and ΔD^2_m.

Plots

Plotting the four measures—$h_m, CD_m, \Delta X^2_m$, and ΔD_m—against the fitted values is recommended by Hosmer and Lemeshow (2000) for helping to locate influential observations for logistic regression models. These plots can be used for other GLMs as well. Hosmer and Lemeshow (2000) also recommend plotting the latter three statistics against the leverage values, with appropriate threshold lines on each measure, as a way to simultaneously judge potential influence. However, all three of these statistics are direct functions of h_m, so large leverage can affect their values. Also, all of them are functions of residuals, so they may be large because of large residuals associated with observations that are not particularly influential. This is especially true for ΔX^2_m and ΔD_m.

Once potentially influential observations have been identified, we are left to decide what to do about them. We discuss this after the next example.

Example: Placekicking (PlacekickDiagnostics.R, glmDiagnostics.R, Placekick.csv)

For `glm`-class model objects, the influence measures are all easy to calculate: `hatvalues()` produces h_m; `cooks.distance()` produces CD_m; `residuals()` produces e^D_m with a `type = "deviance"` argument value; and `rstandard()` produces r_m with a `type = "pearson"` argument value. The leave-one-out measures of influence can be subsequently calculated using r_m, e^D_m , and h_m. For convenience, we have created a function, `glmInflDiag()`, to produce the plots and form a list of all observations that are identified as potentially influential by at least one statistic. This function expects a `glm`-class object for its first argument, but will also work on any other object for which there are `residuals()` and `hatvalues()` method functions and for which `summary()$dispersion` is available.

The output from this function is given below. We include the `print.output` and `which.plots` arguments in the function to control the output to be printed and sets of plots to be constructed; both are set here to their default values to maximize the information given. The observations printed by `glmInflDiag()` correspond to those where $h_m > 3p/M, CD_m > 4/M, \Delta X^2_m > 9$, or $\Delta D_m > 9$, and those with both $\Delta X^2_m > 4$ and $h_m > 2p/M$ or $\Delta D_m > 4$ and $h_m > 2p/M$. Figures 5.3 and 5.4 show the plots.

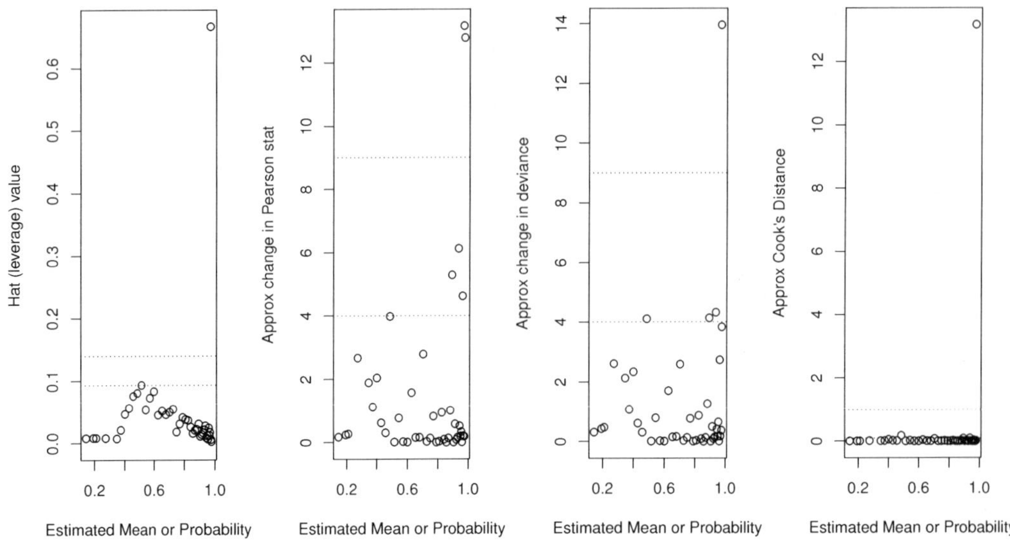

Figure 5.3: Plots of influence measures vs. estimated probabilities of success for the placekicking data. Dotted lines represent thresholds for potentially high influence.

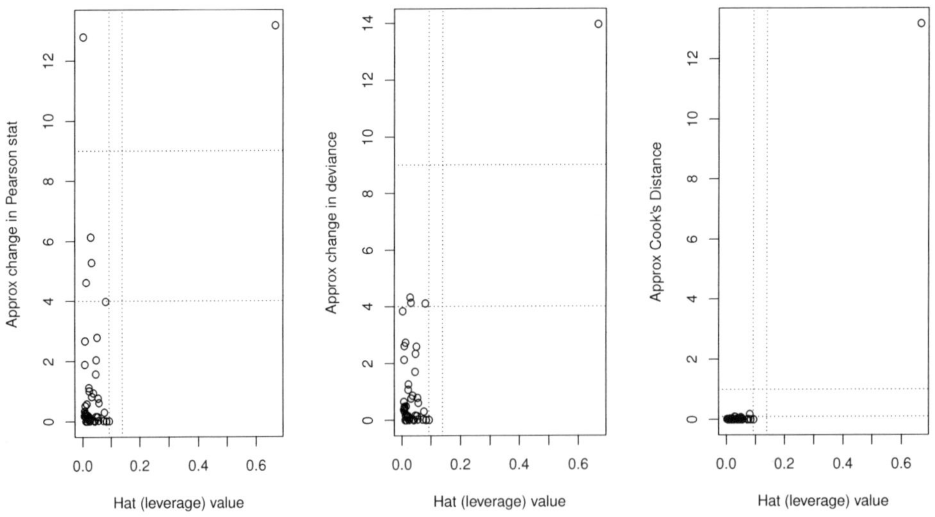

Figure 5.4: Plots of influence measures vs. leverage values for the placekicking data. Dotted lines represent thresholds for potentially high influence.

```
> source("glmDiagnostics.R")
> save.diag <- glmInflDiag(mod.fit = mod.fit.bin, print.output =
    TRUE, which.plots = 1:2)
Potentially influential observations by any measures
      h hflag Del.X2 xflag Del.dev dflag Cooks.D cflag
1  0.00        12.78    **    3.84          0.015
3  0.67    **  13.16    **   13.95    **   13.163    **
34 0.08         3.98          4.11     *    0.174     *
```

```
Data for potentially influential observations
    distance success trials    prop  yhat
1         18       2       3 0.6667 0.977
3         20     776     789 0.9835 0.971
34        51      11      15 0.7333 0.486
```

The function identifies three potentially influential observations: #1, #3, and #34. Observation #1 has large ΔX_1^2 without a large h_1 value, indicating that the large influence measure value is due to the size of its standardized residual r_1. This case was addressed in Section 5.2.1, where we saw that the large residual was merely a result of observing 1 failure out of 3 trials when the estimated probability of success is very large. There is no cause for alarm with this observation. Observation #3 has large ΔX_3^2, ΔD_3, and CD_3 values along with a large h_3. This observation was also identified as having an extreme residual in Section 5.2.1. Its high leverage is undoubtedly due to the fact that it accounts for more than half of the total 1,425 trials in the data set. Thus, we expect this observation to be influential, although its high influence is not necessarily a bad thing. However, because it also has an extreme Cook's distance, its influence is clearly affecting the parameter estimates and the overall fit of the model. We will find out in Section 5.4.1 that simply adding the `PAT` explanatory variable to the model eliminates the extreme standardized residual, but still leaves an influential observation corresponding to this same distance.

Observation #34 is also potentially influential, but the evidence is mild. Both CD_{34} and ΔD_{34} are only slightly above their respective thresholds. It has a somewhat smaller estimated probability of success than the observed proportion. A quick calculation shows that the probability of 11 or more successes out of 15 trials when $\hat{\pi}_{34} = 0.486$ is 0.047, so this observation is slightly, but not extremely, atypical and hence is not really of great concern.

Model adjustments for influence and outliers

How one handles outliers and influential values depends on the cause of the extreme results and the goals of the analysis. In general, any flagged observations should be investigated for other features that might cause their outlying and/or influential status. If there is reason to believe that an error has been made in measuring or recording their data, then this should obviously be fixed if possible, and the analysis rerun. If the observations are fundamentally different in some way, such as a goat in a study of sheep, or a professional athlete in a study of regular people, then it is reasonable to exclude the observation(s), provided that the reason for doing so is made clear and the results of the analysis are explicitly applied to this narrower population (e.g., people who are not professional athletes). In the context of our placekicking data, there are differences between PATs and other field goals that might cause their probabilities of success at the same distance to be different. We could choose to exclude the PATs from the data and rerun the analysis, in which case our inferences would apply only to field goals and not to all placekicks.

When there is no clear reason for the influence or extremeness of an observation, then removing it from the study should not be automatic. When there are not too many flagged values to consider, it is possible to rerun the analysis both with and without the extreme values. If the main inferences from the analysis do not change in any substantive way, then there is no worry that these values are interfering with the analysis, and there is no problem with presenting the original results. On the other hand, if the inferences do change meaningfully, then there is uncertainty regarding what conclusions should be drawn.

Reporting of the results should reflect this uncertainty by clearly indicating the differences and explaining that it is not clear which results are closer to the truth.

5.2.4 Diagnostics for multicategory response models

As noted in the introduction to this section, diagnostic tools for multicategory response models are less well developed than for models with single responses. To begin the discussion of what is available or can be created, we assume that the data have been aggregated into EVP form as in a binomial regression model. Then, there are M EVPs, where EVP m consists of n_m trials for $m = 1, \ldots, M$. If there are J categories in the response variable, then each response is a list of J counts rather than the single count of successes that the binomial models use. Represent the counts as y_{mj}, $m = 1, \ldots, M$; $j = 1, \ldots, J$.

Model fitting provides estimated probabilities $\hat{\pi}_{mj}$ for all J categories for each EVP, and hence estimated expected counts $\hat{y}_{mj} = n_m\hat{\pi}_{mj}$. Then there are J residuals, $y_{mj} - \hat{y}_{mj}$, $j = 1, \ldots, J$ per EVP, although these are not independent, because both the observed counts and the expected counts must sum to n_m. A set of Pearson residuals can be defined for each observed count as

$$e_{mj} = \frac{y_{mj} - \hat{y}_{mj}}{\sqrt{\hat{y}_{mj}}}, \; j = 1, \ldots .J.$$

Each EVP's contribution to the model-fit Pearson statistic is the sum of its squared Pearson residuals across categories, $X^2_m = \sum_{j=1}^{J} e^2_{mj}$. Each EVP's contribution to the residual deviance can be computed similarly.

From the Pearson residuals, a number of further quantities can be computed similar to those in Sections 5.2.1 and 5.2.3. Defining the hat matrix, and hence the leverage values h_m, requires some care because of the multivariate nature of the response. Mathematical details are given in Lesaffre and Albert (1989), who extend the measures from Pregibon (1981) to the multicategory case.

We have written R functions that calculate statistics and create plots that can help to identify outlying or influential EVPs. These are contained in the R program multinomDiagnostics.R for models fit using `multinom()`. The statistics calculated on each EVP include Pearson residuals, the contributions to the Pearson and residual deviance GOF statistics, leverage, Cook's distance, and the two case-deletion statistics, ΔD_m and ΔX^2_m.

Formal GOF tests are available for multinomial models. All of them are based on grouping EVPs in a manner similar to the HL test. Fagerland et al. (2008) create groupings based on fitted probabilities for whichever category was selected as the base level. Hence the statistic could change if a different category is used as the base level, even though all model-estimated probabilities would be the same. For ordinal models, several tests are available as reviewed by Fagerland and Hosmer (2016). All of these tests use grouped EVPs with different algorithms for grouping and different test statistics. Functions to perform these tests are available in the `generalhoslem` package.

As an alternative to performing specialized diagnostics for multinomial models, the problem can be broken down into a series of ordinary logistic regressions and each of these can be assessed for fit separately. For example, the multinomial regression model for a nominal response,

$$\log(\pi_j/\pi_1) = \beta_{j0} + \beta_{j1}x_1 + \cdots + \beta_{jp}x_p,$$

uses $J-1$ logits, each comparing response j to response 1, $j = 2, \ldots, J$. A series of $J-1$ logistic regressions can be fit, each considering only the data from category 1 and category j, and the fit for each of these models can be assessed using standard tools described earlier in this section. The appeal of this approach is that the diagnostic tools are easily accessed and well understood. However, each model is only a portion of the total model.

5.3 Overdispersion

For linear regression problems, we assume that errors are normally distributed with constant variance. Under this assumption, the variance of the data is not affected by the model we pose for the mean because the mean and variance are controlled independently by separate parameters. That is, suppose we temporarily write the mean response in a conditional expectation form as $E(Y|x) = \mu(x)$ to emphasize the relationship between the mean and the explanatory variable(s) represented by x. We use analogous notation for $Var(Y|x)$. In the normal model, we have $Var(Y|x) = \sigma^2$, a constant that does not depend on x.

In most other GLM settings, such as logistic and Poisson regressions, the mean and variance are related. For Poisson regression with $E(Y|x) = \mu(x)$, we have $Var(Y|x) = \mu(x)$. In a logistic regression, an EVP with n_m trials has $E(Y|x) = \mu(x) = n_m\pi(x)$, so that $Var(Y|x) = n_m\pi(x)(1 - \pi(x))$. Thus, with most GLMs, when we specify a model for the relationship between the mean and x, we are implicitly imposing a model on the relationship between the variance and x.

In linear regression, it is somewhat common that the variance does not follow its assumed model; that is, the variance is *not* constant for all values of x. This situation is known as *heteroscedasticity*, or more simply *non-constant variance*, and it is often detected through analysis of the residuals (see, e.g., Kutner et al., 2004). In other GLMs, the variance may similarly not follow the structure that is imposed on it by the distributional model. In particular, it is fairly common to have counts or proportions that exhibit more variability than what the models indicate there should be. This indicates a bad fit of the model, even when portion of the model representing the estimated mean or probability seems to fit the data well. This phenomenon is called *overdispersion*. The next subsections explain what causes overdispersion, how to detect it, and how to correct models to account for it.

5.3.1 Causes and implications

Overdispersion is a failure of the *model*, not a failure of the *data*. It is a symptom of another problem rather than a problem by itself.

In linear regression, when an important variable is added to the model, the variability that it explains is taken out of the error sum of squares and put into the model sum of squares. Looking at this another way, when an important variable is *left out* of the model, the error sum of squares is much larger than it should be. As a result, the mean square error becomes inflated and overestimates the true variance σ^2. In a sense, this is an example of overdispersion, except that the parameter for this variance (also called *dispersion*) is estimated separately from the mean and disguises the model's failure. Omitting an important variable still adds bias to the estimated regression parameters. However, the inflated mean square error makes test statistics less extreme and confidence intervals wider. While this does not necessarily make inferences correct, the added uncertainty helps to compensate somewhat for using the wrong model.

Models such as the Poisson and binomial have no separate variance parameter to allow them to adapt to the absence of important variables. When an important variable is left out of the model, any observations that have the same mean according to the model may actually have different means if they have different values of the missing variable. Thus, the responses may exhibit more variability—overdispersion—around their estimated means than what the model expects.

Inferences assume that the model-based estimates of variance are correct, when in fact they are too small. Thus, confidence intervals are too narrow—that is, they do not achieve their stated confidence levels—and p-values are smaller than they should be, meaning that type I error rates are higher than stated. The end result is that inferences tend to detect effects that are not really there.

The next example demonstrates how the overdispersion phenomenon occurs with data from Poisson distributions with variable means that are not properly modeled in a GLM. The estimated response mean from a fitted Poisson regression model underestimates the amount of variability in the data around that mean, and inferences performed at a level α actually have much higher error rates. Exactly the same phenomenon occurs with other distributional models that lack a separate parameter for variance, such as the binomial.

Example: Simulation of overdispersion (OverdispSim.R)

In this example, we simulate data under a Poisson model with and without overdispersion. First, we simulate data directly from a Poisson distribution with mean 100, and compute the sample mean and variance of the data (called `mean0` and `var0`, respectively). Then, we repeat the simulation, except that the mean is allowed to vary by simulating it from a normal distribution with mean 100 and standard deviation 20. The sample mean and variance is again computed (called `mean20` and `var20`, respectively). Both simulations use sample sizes of 50, and the simulations are repeated 10 times so that the patterns are apparent. The program and results are given below.

```
> set.seed(389201892)
> poi0 <- matrix(data = rpois(n = 500, lambda = rep(x = 100,
   times = 500)), nrow = 50, ncol = 10)
> poi20 <- matrix(data = rpois(n = 500, lambda = rnorm(n = 500,
   mean = 100, sd = 20)), nrow = 50, ncol = 10)

> mean0 <- apply(X = poi0, MARGIN = 2, FUN = mean)
> var0 <- apply(X = poi0, MARGIN = 2, FUN = var)
> mean20 <- apply(X = poi20, MARGIN = 2, FUN = mean)
> var20 <- apply(X = poi20, MARGIN = 2, FUN = var)

> all <- cbind(mean0, var0, mean20, var20)
> round(all, digits = 1)
      mean0  var0 mean20 var20
 [1,]  99.9  85.9  104.3 456.0
 [2,] 101.5  91.7   97.0 533.5
 [3,]  99.1  83.3  104.5 534.0
 [4,]  99.9 121.1  100.4 452.2
 [5,]  99.5 102.4  101.0 452.3
 [6,]  99.7  73.8  104.0 408.8
 [7,] 101.0  94.5   96.9 509.0
 [8,] 100.4 110.9  102.1 292.8
 [9,]  98.0  92.4   98.1 303.8
[10,]  99.8  96.1  104.3 524.6
> round(apply(X = all, MARGIN = 2, FUN = mean), digits = 1)
 mean0   var0 mean20  var20
  99.9   95.2  101.3  446.7
```

The sample means for both cases vary randomly around the true value of 100, although there is much more variability for the case where the true means are allowed to vary by

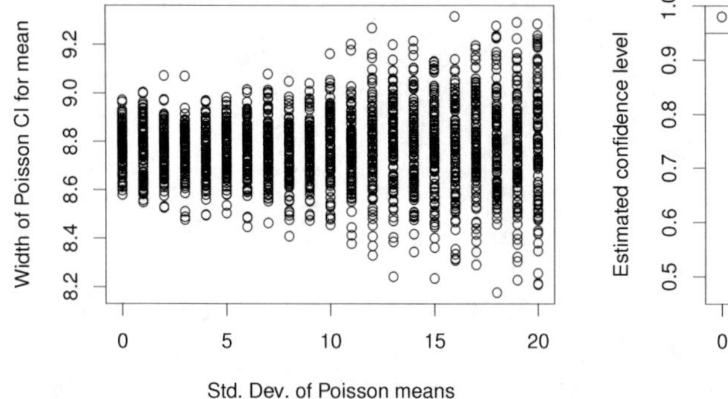

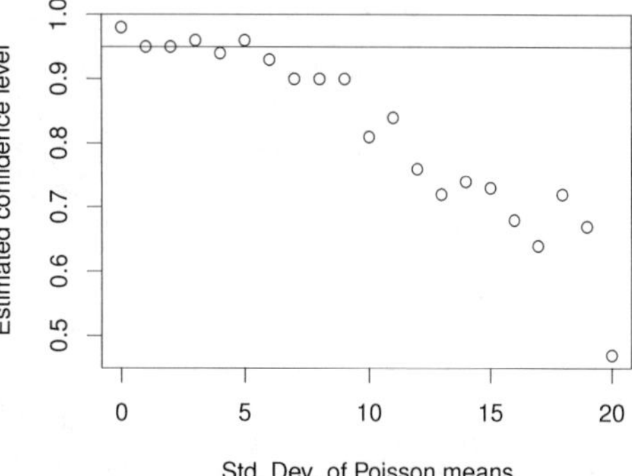

Figure 5.5: Likelihood ratio confidence interval results from simulations using overdispersed Poisson data. Confidence interval widths (left) and estimated confidence levels (right) are plotted against the standard deviation of the means used to generate the Poisson data. The horizontal line in the confidence level plot shows the stated confidence level of 95%.

a normal distribution.[4] Thus, it is clear that generating data from a situation where the means may vary causes the resulting data to exhibit much more variability than is expected under a Poisson model.

The program for this example also contains code to simulate data sets from the same model used above, except that the amount of overdispersion is controlled more finely by setting the standard deviation on the random means to 0, 1, ..., 20. For each standard deviation, we generated 100 data sets of size 20. These data sets are then analyzed by a Poisson regression model using only an intercept, ignoring the added variablility from the random normal draws. Model-based LR confidence intervals for μ are computed, and the average confidence interval widths and estimated true confidence levels—the fraction of data sets whose confidence interval correctly contains μ—are calculated. These widths and estimated confidence levels are plotted against the standard deviations used to generate the Poisson means in Figure 5.5.

The plot of confidence interval widths shows that the average width is not affected much by the increased levels of overdispersion. There does appear to be a little more variability to the widths as overdispersion increases, which reflects the observation made in the first simulation that the means from the overdispersed case have more variability than those from the standard Poisson. The major problem we see is in the estimated confidence levels. These levels decrease as the overdispersion increases, and drop substantially below the stated 95% level for larger standard deviations. This is an indication that inferences are unreliable when ordinary Poisson regression models are fit to overdispersed data.

Other causes for overdispersion have been given in the literature. For example, McCullagh and Nelder (1989), among others, point out that the standard binomial and Poisson models assume that each trial or observation is independent of all others. However, sometimes the sampling process is such that this assumption is not satisfied. A common example is sampling data in *clusters*, meaning that groups of observations are sampled together. Typically,

[4]It can be shown that the true mean and variance of the response Y under the second model are 100 and 500, respectively. See Exercise 30.

observations within a cluster respond more similarly to one another than observations in different clusters. For example, married couples may hold opinions that are more similar to each other's opinions than to those of some other random person. Cattle in the same pen or field tend to have more similar health issues than cattle selected from different locations. In the Salk vaccine trial described in Chapter 2's Exercise 33, children from the same school who participated in the trial might all be more likely to test positive for polio if another child in the same school does. If sampling is done on couples, pens of cattle, or schools—i.e., in clusters—then the resulting individuals will tend to be positively correlated within their clusters. The positive correlation within the clusters causes means or probabilities to have more variability than their respective models expect. Thus, treating units that were gathered in clusters as if they were independent is likely to lead to overdispersion. This phenomenon is discussed in more detail in Section 6.5.

5.3.2 Detection

The main symptom of overdispersion is a poor model fit without any obvious cause. In particular, large GOF statistics described in Section 5.2.2, such as the deviance/df statistic or formal GOF tests, may indicate a problem with the model. However, this is not evidence of overdispersion by itself. Examination of standardized residual plots is required to rule out other issues. The plots must not show a poor fit of the mean model, nor identify one or perhaps two particular outliers that might be causing the excessively large deviance and/or Pearson statistic. If these other problems *are* present, then they must be addressed before considering whether there is overdispersion. Typically, overdispersion causes too many standardized Pearson residuals to lie near or beyond the expected boundaries for "extreme" values—considerably more than 5% of them beyond ± 2 and often several beyond ± 3. These extreme residuals generally occur fairly uniformly across all values of the linear predictor, the estimated mean or probability, or any explanatory variable.

Example: Ecuadorean bird counts (BirdOverdisp.R, BirdCounts.csv)

In this example, we revisit the Ecuadorean bird count data originally analyzed in Section 4.2.3. We refit the Poisson model used for these data and examine the results for signs of overdispersion. This is done using the deviance/df statistic and a plot of the standardized residuals against the estimated means. The relevant results are given below.

```
> Mpoi <- glm(formula = Birds ~ Loc, family = poisson(link =
    "log"), data = alldata)
> summary(Mpoi)

<OUTPUT EDITED>

(Dispersion parameter for poisson family taken to be 1)

    Null deviance: 216.944  on 23  degrees of freedom
Residual deviance:  67.224  on 18  degrees of freedom
AIC: 217.87

Number of Fisher Scoring iterations: 4

> pred <- predict(Mpoi, type = "response")
```

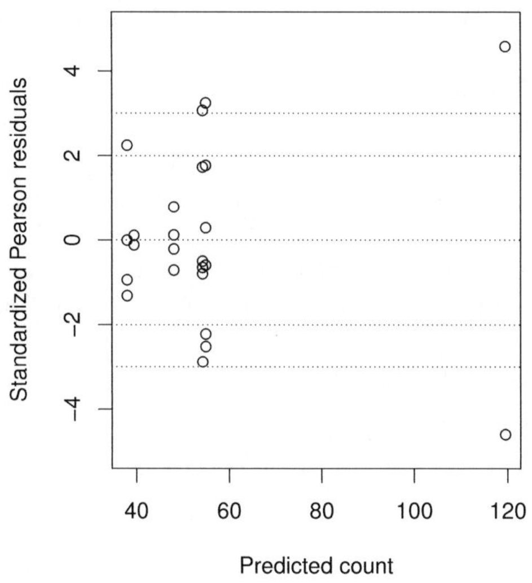

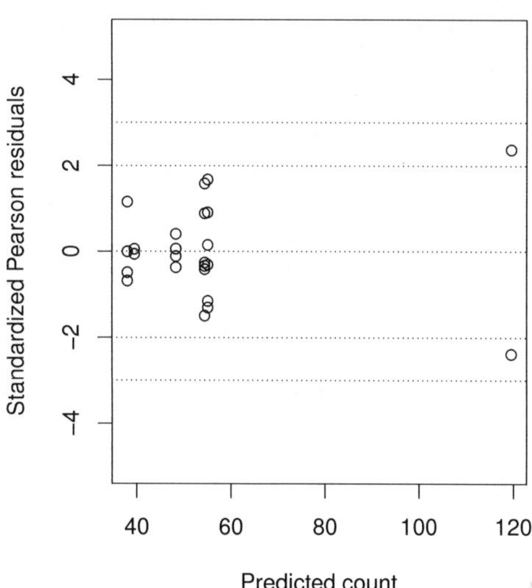

Figure 5.6: Standardized residuals plotted against the mean for the Ecuadorean bird count data. Left: Residuals from the fit of the original Poisson model. Right: Residuals from the fit of a quasi-Poisson model. In both cases, upper and lower dotted lines are given at ±2 and ±3 to aid interpretation of the plots.

```
> # Standardized Pearson residuals
> stand.resid <- rstandard(model = Mpoi, type = "pearson")
> plot(x = pred, y = stand.resid, xlab = "Predicted count", ylab
   = "Standardized Pearson residuals", main = "Residuals from
   regular likelihood", ylim = c(-5,5))
> abline(h = c(qnorm(0.995), 0, qnorm(0.005)), lty = "dotted",
   col = "red")
```

The residual deviance/df ratio is $67.2/18 = 3.73$, which is much larger than the suggested guidelines in Section 5.2.2: $1 + 3\sqrt{2/18} = 2.0$. The residual plot on the left in Figure 5.6 shows that 4 out of 24 residuals lie beyond ±3, and 2 even lie beyond ±4. Because the explanatory variable is categorical, there is no concern with nonlinearity in this plot. Furthermore, there are no other variables to consider adding to this model, and there is no clustering to the sampling process. Therefore, we conclude that the poor model fit is likely caused by overdispersion. We examine solutions to the problem in the next subsection.

Note that overdispersion occurs when the apparent variance of response counts is larger than the model suggests. In binomial (or multinomial) regressions where all EVPs have $n_m = 1$, overdispersion cannot be detected, even when it might be expected to occur (e.g., if there are clusters of highly correlated data that are mistakenly being handled as independent observations). This is because when $n_m = 1$, the responses can take on only values 0 or 1 and cannot express extra variability, whereas when $n_m > 1$, counts can be distributed farther from $n_m\hat{\pi}_m$ than expected.

5.3.3 Solutions

Because overdispersion is a result of an inadequacy in the model, the obvious way to fix the overdispersion problem is to fix the model. How one does this depends on the source of the overdispersion.

The easiest case to deal with is when there are additional explanatory variables that have been measured but not included in the model. Then adding one or more of these into the model will create predictions that are closer to the observed counts, especially if the added variables are strongly related to the response. Residuals from the new model will be generally smaller than those from the original model, and the residual deviance will be reduced. If augmenting the model in this way is successful in removing the symptoms of overdispersion, then inferences can be carried out in the usual manner using the new model.

Example: Placekicking (OverdispAddVariable.R, Placekick.csv)

We have seen in previous examples that the distance at which a placekick is attempted is the most important predictor of the kick's probability of success. This result makes sense given an understanding of placekicking in football. In many other problems, however, there may be insufficient intuition as to which explanatory variables will turn out to be important. In this example, we demonstrate what can happen when an important explanatory variable—represented by `distance` in the placekicking data—is left out of the model.

We fit a logistic regression model using only `wind` as a predictor and assess the fit of this model. Then, we add `distance` to the model, refit, and show the revised assessment of fit. Prior to the model fitting and assessment, the data are aggregated into EVP form corresponding to both `wind` and `distance`. This allows us to compare fit statistics between models with and without the distance variable.

```
> w <- aggregate(x = good ~ distance + wind, data = placekick,
    FUN = sum)
> n <- aggregate(x = good ~ distance + wind, data = placekick,
    FUN = length)
> w.n <- data.frame(success = w$good, trials = n$good, distance
    = w$distance, wind = w$wind)

> mod.fit.nodist <- glm(formula = success/trials ~ wind, weights
    = trials, family = binomial(link = logit), data = w.n)
> summary(mod.fit.nodist)

<OUTPUT EDITED>

    Null deviance: 317.35  on 70  degrees of freedom
Residual deviance: 314.51  on 69  degrees of freedom
AIC: 423.81

> pred <- predict(mod.fit.nodist)
> stand.resid <- rstandard(model = mod.fit.nodist, type =
    "pearson")

> par(mfrow = c(1,2))
> plot(x = pred, y = stand.resid, xlab = "Estimated
    logit(success)", ylab = "Standardized Pearson residuals",
```

```
  main = "Standardized residuals vs. Estimated logit", ylim =
  c(-6, 13))
> abline(h = c(qnorm(0.995), 0, qnorm(0.005)), lty = "dotted",
  col = "red")

> plot(w.n$distance, y = stand.resid, xlab = "Distance", ylab =
  "Standardized Pearson residuals", main = "Standardized
  residuals vs. Distance", ylim = c(-6, 13))
> abline(h = c(qnorm(0.995), 0, qnorm(0.005)), lty = "dotted",
  col = "red")
> dist.ord <- order(w.n$distance)
> lines(x = w.n$distance[dist.ord], y =
  predict(loess(stand.resid ~ w.n$distance))[dist.ord], col =
  "blue")
```

The deviance/df ratio for the model fit without `distance` is $314.51/69 = 4.5$, well above the expected upper limit for this statistic, $1 + 3\sqrt{(2/69)} = 1.5$. This result leaves no doubt that the model does not fit well, but does not explain why. Plotting the standardized residuals against the linear predictor (Figure 5.7, left plot) shows a generally overdispersed pattern, with numerous residuals lying well outside the expected range. This plot does nothing to explain the cause of the extreme residuals, and might generically be thought to indicate overdispersion. However, the plot on the right shows that the residuals have a very clear decreasing pattern as the omitted variable `distance` increases. This is because the model assumes a constant probability of success for all distances at a given level of `wind`. Therefore, observed success probabilities are higher than the model expects for short placekicks, leading to generally positive residuals, and the observed success probabilities are lower than the model expects for long placekicks, leading to negative residuals.

Next, we add `distance` to the model and repeat the analysis.

```
> mod.fit.dist <- glm(formula = success/trials ~ distance +
  wind, weights = trials, family = binomial(link = logit), data
  = w.n)
> summary(mod.fit.dist)

<OUTPUT EDITED>

    Null deviance: 317.345  on 70  degrees of freedom
Residual deviance:  76.453  on 68  degrees of freedom
AIC: 187.76
```

The added variable reduces the deviance/df ratio to $76.453/68 = 1.1$, which is well within the lower threshold of $1 + 2\sqrt{(2/68)} = 1.3$. The residuals are shown in Figure 5.8 on the same y-axis scale as the previous plots to make comparison easier. The weighted loess mean-trend curve shows a pattern very similar to that observed in Figure 5.2, and the interpretation offered for that plot in the corresponding example applies here. Although there are still a few points outside the ± 2 and ± 3 thresholds, none are as extreme as when `distance` was left out of the model.

Measured variables with a large enough effect to create noteworthy overdispersion are rarely omitted from a model when an appropriate variable-selection procedure has been

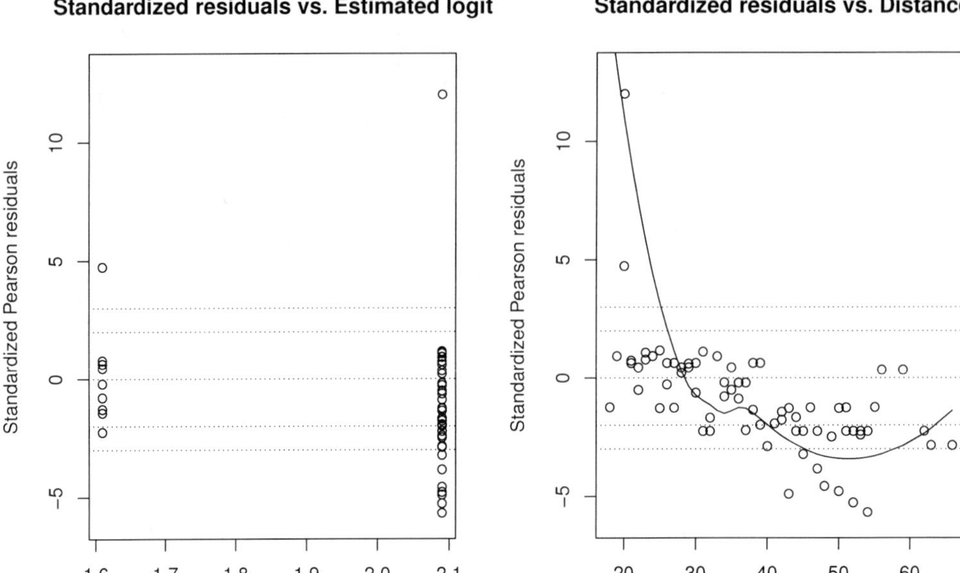

Figure 5.7: Standardized residual plots for placekick data when the logistic regression model contains only wind. Lines at ± 2 and ± 3 are approximate thresholds for standardized residuals. Left: Residuals versus the logit of the estimated probability of success. Right: Residuals versus the omitted variable `distance`, with the curve from a weighted loess smoother added to accentuate the mean trend.

applied to the data. However, variable selection often does not consider interactions until a poor model fit suggests that they might be needed. Thus, if overdispersion appears and no other variables are available that have not already been considered, it is possible that adding one or more interactions to the model may alleviate the problem.

Other causes of overdispersion require different model amendments. For example, if the apparent overdispersion is caused by analyzing clustered data as if they were independent, the problem can often be addressed by adding a random-effect term to the model with a different level for each cluster. Then a separate parameter is fit which accounts for the extra variability that is created by the within-cluster correlation. The result is a *generalized linear mixed model*, meaning a model containing both fixed- and random-effect variables. Details on how to formulate and account for random effects are given in Section 6.5.

Quasi-likelihood models

When there is no apparent cause for the overdispersion, then it may be due to the omission of important variables that are unknown and not measured, sometimes called *lurking variables* (Moore and Notz, 2009). Adding these variables to the model is obviously not an option. Instead, the solution is to change the distributional family for the response variable to one with an that allows dispersion to be estimated separately from the mean. There are several options, each of which provides a more flexible model for the variance than the Poisson or binomial families allow. *We strongly recommend against the routine use of these methods unless all other avenues for model improvement have been exhausted.*

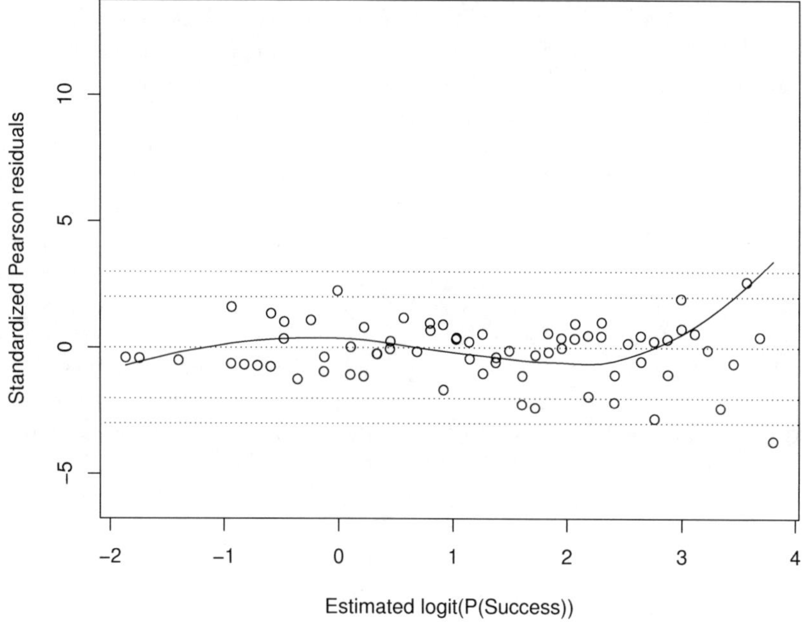

Figure 5.8: Standardized residuals against linear predictor for placekick data when the logistic regression model for success contains wind speed and distance. Upper and lower lines at ± 2 and ± 3 are approximate thresholds for standardized residuals. The solid curve is from a weighted loess smoother to highlight any mean trend.

These methods are often just a patch that covers up a symptom of a problem, rather than a cure for the problem.

One approach for doing this, called *quasi-likelihood*, can be used for both binomial and Poisson analyses. Quasi-likelihood is rigorously described in Wedderburn (1974) and McCullagh and Nelder (1989). In the current context, a quasi-likelihood function is a likelihood function that has extra parameters added to it that are not a part of the distribution on which the likelihood is based. As an example, suppose we model counts as Poisson with mean $\mu(x)$, where x represents one or more explanatory variables. Suppose that the overdispersion stems from a constant inflation of the variance; that is, suppose that $Var(Y|x) = \gamma\mu(x)$, where $\gamma \geq 1$ is an unknown constant called a *dispersion parameter*. A quasi-likelihood can be created to account for this by dividing the usual Poisson log-likelihood by γ. This has no effect on the regression-parameter estimates, but after they are computed, an estimate of γ is found by dividing the Pearson GOF statistic for the model by its residual degrees of freedom, $M - \tilde{p}$, where $\tilde{p}$ is the number of parameters estimated in the model. See Wedderburn (1974) for details.

Parameter estimates and test statistics computed from a quasi-likelihood function have the same properties as those computed from a regular likelihood, so inferences are carried out using the same basic tools, with minor changes. First, the dispersion parameter estimate, $\hat{\gamma}$, which is greater than 1 when overdispersion exists, is used to adjust standard errors and test statistics for subsequent inferences. Second, the sampling distributions used for tests and confidence intervals need to reflect the fact that the dispersion is being estimated with a statistic, based on the model Pearson statistic, that has approximately a chi-square distribution with $M - \tilde{p}$ degrees of freedom in large samples.

Specifically, consider an LRT statistic for a model effect that would normally be compared to a chi-square distribution with q df. The statistic is divided by $\hat{\gamma}$, which not only makes it smaller, but also changes its distribution. Further dividing the statistic by q creates a new statistic that, in large samples, can be compared to an F distribution with $(q, M - \tilde{p})$ degrees of freedom. Similarly, standard errors for regression parameters are multiplied by $\sqrt{\hat{\gamma}}$, which changes the large-sample distribution on which confidence intervals are based from normal to a t distribution with $M - \tilde{p}$ degrees of freedom ($t_{M-\tilde{p}}$). Diagnostic statistics like standardized residuals and Cook's distance are also appropriately adjusted by $\hat{\gamma}$.

Finally, it is often recommended that $\hat{\gamma}$ be computed from a "maximal model," meaning one that contains all available variables, even when the final model to which quasi-likelihood is applied is a smaller model. This is done to ensure that there can be no artificial inflation of the dispersion parameter estimate caused by missing explanatory variables. However, in modern studies, there are often huge numbers of explanatory variables available, many of which are expected to be unimportant. In these cases, basing $\hat{\gamma}$ on a maximal model can be excessively conservative, if not impossible to calculate. If the residual degrees of freedom from the full model are much smaller than the residual degrees of freedom from a model containing just the important variables, power for tests and widths of confidence intervals may be adversely affected. Estimating γ from a full model is "safe," but if one can be fairly confident that no important measured variables have been left out of a particular smaller model, then that model should be able to provide a reasonable estimate of γ that can result in more powerful inferences.[5]

When quasi-likelihood is applied to a Poisson model as described above, the result is sometimes referred to as a *quasi-Poisson* model. The same quasi-likelihood approach can be applied to a binomial model, in which case we model $Var(Y|x_m) = \gamma n_m \pi(x_m)(1 - \pi(x_m))$, where $\pi(x_m)$ is the probability of success and n_m is the number of trials for EVP m. The result is sometimes called a *quasi-binomial* model. This model is only relevant when at least some $n_m > 1$; otherwise, the Pearson statistic cannot be used to measure overdispersion.

Example: Ecuadorean bird counts (BirdOverdisp.R, BirdCounts.csv)

In this example, we refit the Ecuadorean bird count data using a quasi-Poisson model. Quasi-Poisson and quasi-binomial model fitting are available in `glm()` using `family = quasipoisson(link = "log")` or `quasibinomial(link = "logit")`, respectively. Alternative links can be specified. Here, we apply a quasi-Poisson model to the Ecuadorean bird counts.

```
> Mqp <- glm(formula = Birds ~ Loc, family = quasipoisson(link =
    "log"), data = alldata)
> (sumq <- summary(Mqp))

Call:
glm(formula = Birds ~ Loc, family = quasipoisson(link = "log"),
    data = alldata)

Deviance Residuals:
    Min        1Q    Median        3Q       Max
-3.4322   -0.7594   -0.1302    0.8874    3.1038
```

[5]There still may be concern about potential post-model-selection inference issues. See Section 5.5.

```
Coefficients:
            Estimate Std. Error t value Pr(>|t|)
(Intercept)   3.8764     0.1391  27.877 2.93e-16 ***
LocForA       0.9069     0.1870   4.851 0.000128 ***
LocForB       0.1309     0.1751   0.748 0.464139
LocFrag       0.1187     0.1755   0.677 0.507153
LocPasA      -0.2001     0.2580  -0.775 0.448115
LocPasB      -0.2388     0.2095  -1.140 0.269257
---
Signif. codes:  0 '***' 0.001 '**' 0.01 '*' 0.05 '.' 0.1 ' ' 1

(Dispersion parameter for quasipoisson family taken to be
   3.731881)

    Null deviance: 216.944  on 23  degrees of freedom
Residual deviance:  67.224  on 18  degrees of freedom
AIC: NA

Number of Fisher Scoring iterations: 4


> # Demonstrate calculation of dispersion parameter
> pearson <- residuals(Mpoi, type = "pearson")
> sum(pearson^2)/Mpoi$df.residual
[1] 3.731881
```

Notice several things about these results. First, the parameter estimates and residual deviance are identical to those from the Poisson model fit. This is an indication that the portion of the likelihood function that is used to estimate the regression parameters has not been changed. The dispersion parameter is now estimated to be 3.73, and we show code demonstrating that this is the Pearson statistic divided by its degrees of freedom. A plot of the standardized residuals is given in the plot on the right in Figure 5.6 (code is given in the program for this example). All residuals are now contained within ± 3 and only two remain outside ± 2.

To compare inferences from the Poisson and quasi-Poisson models, we compute LRTs for the effect of locations on the means using the `anova()` function. The F-test is not the default option in `anova()` and must be specified with the argument `test = "F"`. Similarly, the `Anova()` function requires `test.statistic = "F"`. Following the tests, we compute Wald confidence intervals for the mean bird counts using `emmeans`. Profile LR intervals can be found using code given in the program that causes `glm()` to estimate means directly instead of requiring use of `predict()`.

```
> anova(Mpoi, test = "Chisq")
Analysis of Deviance Table
Model: poisson, link: log
Response: Birds
Terms added sequentially (first to last)
     Df Deviance Resid. Df Resid. Dev  Pr(>Chi)
NULL                    23    216.944
Loc   5   149.72        18     67.224 < 2.2e-16 ***
---
Signif. codes:  0 '***' 0.001 '**' 0.01 '*' 0.05 '.' 0.1 ' ' 1
```

```
> anova(Mqp, test = "F")
Analysis of Deviance Table
Model: quasipoisson, link: log
Response: Birds
Terms added sequentially (first to last)
     Df Deviance Resid. Df Resid. Dev      F    Pr(>F)
NULL                   23    216.944
Loc   5   149.72       18     67.224 8.0238 0.0003964 ***
---
Signif. codes:  0 '***' 0.001 '**' 0.01 '*' 0.05 '.' 0.1 ' ' 1

> emmeans(object = Mpoi, specs = ~ Loc, type = "response", level
   = 0.95)
 Loc   rate   SE  df asymp.LCL asymp.UCL
 Edge  48.2 3.47 Inf      41.9      55.6
 ForA 119.5 7.73 Inf     105.3     135.7
 ForB  55.0 3.03 Inf      49.4      61.3
 Frag  54.3 3.01 Inf      48.7      60.6
 PasA  39.5 4.44 Inf      31.7      49.2
 PasB  38.0 3.08 Inf      32.4      44.5

Confidence level used: 0.95
Intervals are back-transformed from the log scale
> emmeans(object = Mqp, specs = ~ Loc, type = "response", level
   = 0.95)
 Loc   rate    SE  df asymp.LCL asymp.UCL
 Edge  48.2  6.71 Inf      36.7      63.4
 ForA 119.5 14.93 Inf      93.5     152.7
 ForB  55.0  5.85 Inf      44.7      67.7
 Frag  54.3  5.81 Inf      44.1      67.0
 PasA  39.5  8.59 Inf      25.8      60.5
 PasB  38.0  5.95 Inf      28.0      51.7

Confidence level used: 0.95
Intervals are back-transformed from the log scale
```

The test for $H_0 : \mu_1 = \mu_2 = \ldots = \mu_6$ rejects this null hypothesis in both models, although overdispersion correction does alter the extremeness of the test statistic, as measured by the p-values. The LRT statistic for the Poisson model is $-2\log(\Lambda) = 149.72$, which has a very small p-value using a χ^2_5 approximation. On the other hand, the quasi-Poisson model uses $F = (149.72/5)/(67.22/18) = 8.33$, which has a p-value of 0.0004 using a $F_{5,18}$ distribution approximation.

The means estimated by the two models are identical as expected, because both models use the same likelihood to estimate the regression parameters. However, the confidence intervals for the quasi-Poisson model are wider than those for the Poisson. Again, the main conclusion is not changed: Forest A has a considerably higher mean than all other locations. However, with the quasi-Poisson there is more overlap between the confidence intervals for the two pastures and those for the edge, fragment, and Forest B.

Notice that $\hat{\gamma}$ is not found by maximum likelihood—the computation is performed *after* the MLEs are found for the regression parameters. Thus, the value of $\hat{\gamma}$ does not impact the value of the log-likelihood. In particular, this means that the standard information criteria described in Section 5.1.2 cannot be computed on quasi-likelihood estimation results. There is instead a series of quasi-information criteria, which we can denote by $QIC(k, \gamma)$, that can be used to compare different regression models within the same quasi-likelihood family. These are obtained simply by dividing the log-likelihood $\log(L(\hat{\beta}|y_1, ...y_n))$ by $\hat{\gamma}$ in the formulas for any $IC(k)$, and adding 1 to the number of parameters. For example, $QAIC = -2\log(L(\hat{\beta}|y_1, ...y_n))/\hat{\gamma}+2(k+1)$, where k is the number of regression parameters estimated, including the intercept; $QAIC_c$ and $QBIC$ are similarly defined.

Using quasi-information criteria to compare different regression models must be done carefully. Because $\hat{\gamma}$ is computed externally to the MLE calculations, the effect of changing its value from model to model is not properly accounted for in the calculation of $QIC(k, \gamma)$. It is therefore important that all regression models be compared using the same value of $\hat{\gamma}$. An appropriate strategy is to estimate γ from the largest model being considered, and then use the same estimate in the computations of $QIC(k, \gamma)$ for all models being compared. As noted before in the bird count example, this also ensures that the Pearson GOF statistic on which γ is based is least likely to be inflated due to omitted variables. For details regarding computing $QIC(k)$ values in R, see Bolker (2009).

Negative binomial and beta-binomial models

There are other models that can be used as alternatives to the standard models or their quasi-counterparts. The *negative binomial* distribution[6] is most often used when an alternative to the Poisson distribution is needed. The negative binomial distribution results from allowing the mean μ in a Poisson distribution to be a random variable with a gamma distribution. This is very similar to the problem we studied in our simulation example on p. 374, except that we used a normal distribution for the mean rather than a gamma distribution. When a gamma distribution is used instead, then a specific relationship arises between the variance of the count random variable Y and its mean μ:

$$Var(Y|x) = \mu(x) + \theta\mu(x)^2,$$

where $\theta \geq 0$ is an unknown parameter. Notice that $\theta = 0$ returns the same mean-variance relationship as in the Poisson model. Also notice that this relationship is different from the one assumed by the quasi-Poisson model, $Var(Y) = \gamma\mu$. Therefore, these two models are distinct and there can be instances where one model is preferred over the other.

Ver Hoef and Boveng (2007) provide some guidance for helping to decide between these two models. In particular, a plot of $(y_m - \hat{\mu}_m)^2$ vs. $\hat{\mu}_m$ from the Poisson model can help to identify the variance-mean relationship. If a smoothed version of this plot follows a mostly linear trend, then a quasi-Poisson model is appropriate. If the trend is more of an increasing quadratic, then the negative binomial is preferred. Ver Hoef and Boveng (2007) suggest that the trend in the plot can be made clearer by first grouping the data according to similar values of $\hat{\mu}_m$ (similar to what is done in the HL test discussed in Section 5.2.2), then plotting the average squared residual in each group against the average $\hat{\mu}_m$ in the group.

Unlike the quasi-Poisson model, all parameters of the negative binomial model, including θ, are estimated using MLEs. Therefore, information criteria can be used without further adjustment to compare different negative binomial regression models. The value of θ counts as one additional parameter in the penalty calculations of $IC(k)$. Also, because these are

[6]The negative binomial was originally derived from properties of Bernoulli trials. It is the distribution of the number of successes that are observed before the rth failure, which gives rise to the name "negative binomial." See p. 95 of Casella and Berger (2002).

the usual information criteria, they can be used to compare models across the negative binomial and Poisson families. However, no direct comparison to a quasi-Poisson model is possible, because $QIC(k)$ is a different criterion.

Example: Ecuadorean bird counts (BirdOverdisp.R, BirdCounts.csv)

We now refit the Ecuadorean bird count data using a negative binomial distribution. The `MASS` package has a function `glm.nb()` that fits this model and estimates θ. Other packages, including `VGAM`, also have functions that can fit negative binomial models. A log link is the default in `glm.nb()`, but alternative links can be specified using a `link` argument. We use this function below to fit the model and examine the results.

```
> library(package = MASS)

> M.nb <- glm.nb(formula = Birds ~ Loc, data = alldata)
> summary(M.nb)
Call: glm.nb(formula = Birds ~ Loc, data = alldata, init.theta =
    33.38468149, link = log)
Deviance Residuals:
     Min         1Q     Median        3Q       Max
-1.80758   -0.48773   -0.08495   0.55106   1.65702

Coefficients:
            Estimate Std. Error z value Pr(>|z|)
(Intercept)   3.8764     0.1126  34.438  < 2e-16 ***
LocForA       0.9069     0.1784   5.083 3.71e-07 ***
LocForB       0.1309     0.1438   0.910    0.363
LocFrag       0.1187     0.1440   0.825    0.410
LocPasA      -0.2001     0.2008  -0.997    0.319
LocPasB      -0.2388     0.1635  -1.460    0.144
---
Signif. codes:  0 '***' 0.001 '**' 0.01 '*' 0.05 '.' 0.1 ' ' 1

(Dispersion parameter for Negative Binomial(33.3847) family
    taken to be 1)

    Null deviance: 73.370  on 23  degrees of freedom
Residual deviance: 22.705  on 18  degrees of freedom
AIC: 198.06

Number of Fisher Scoring iterations: 1

              Theta:  33.4
          Std. Err.:  14.9

2 x log-likelihood:  -184.061

> anova(M.nb, test = "Chisq")
Analysis of Deviance Table

Model: Negative Binomial(33.3847), link: log

Response: Birds
```

```
Terms added sequentially (first to last)
     Df Deviance Resid. Df Resid. Dev  Pr(>Chi)
NULL                   23     73.370
Loc   5   50.665       18     22.705 1.013e-09 ***
---
Signif. codes:  0 '***' 0.001 '**' 0.01 '*' 0.05 '.' 0.1 ' ' 1
Warning message:
In anova.negbin(M.nb, test = "Chisq") :
  tests made without re-estimating 'theta'

> # Wald CIs for means
> emmeans(object = M.nb, specs = ~ Loc, type = "response", level
   = 0.95)
 Loc  response    SE  df asymp.LCL asymp.UCL
 Edge     48.2  5.43 Inf      38.7      60.2
 ForA    119.5 16.54 Inf      91.1     156.7
 ForB     55.0  4.93 Inf      46.1      65.6
 Frag     54.3  4.88 Inf      45.6      64.8
 PasA     39.5  6.57 Inf      28.5      54.7
 PasB     38.0  4.51 Inf      30.1      47.9

Confidence level used: 0.95
Intervals are back-transformed from the log scale
```

The goodness-of-fit results from `summary()` show that the negative binomial model fits somewhat better than the Poisson model: the residual deviance is only 22.7 on 18 df. The extra dispersion parameter θ is estimated to be 33.4 with a standard error of 14.9. This estimate is more than two standard errors from 0, indicating that the overdispersion correction is important. The LRT for equality of the means at the six locations is given in the `anova()` results. The deviance of 50.7 is highly significant, although less extreme than in the uncorrected Poisson model. The warning about not re-estimating θ is not a problem. It is an artifact from using `anova()` on a single model, as opposed to using it to compare two or more `glm.nb()` fits that have different estimates of θ.

The estimated means are the same in this model as in the previous models (this does not have to be so, but happens here because we are using only a nominal explanatory variable). The confidence intervals are similar to those from the quasi-Poisson model, except wider for larger means and narrower for smaller means. This is consistent with the variance-mean relationship of the negative binomial model.

Which of the two models to correct for overdispersion is better for these data? To answer this, we make a plot of $(y_i - \hat{\mu}_i)^2$ vs. $\hat{\mu}_i$. We plot a linear and a quadratic fit to examine trends and see which one fits better.

```
> res.sq <- residuals(object = Mpoi, type = "response")^2
> set1 <- data.frame(res.sq, mu.hat = Mpoi$fitted.values)

> fit.lin <- lm(formula = res.sq ~ mu.hat, data = set1)
> fit.quad <- lm(formula = res.sq ~ mu.hat + I(mu.hat^2), data =
   set1)
> summary(fit.quad)
```

```
Call: lm(formula = res.sq ~ mu.hat + I(mu.hat^2), data = set1)

Residuals:
    Min      1Q  Median      3Q     Max
-159.49  -95.80  -19.95   41.66  320.51

Coefficients:
              Estimate Std. Error t value Pr(>|t|)
(Intercept) -108.88160  331.02260  -0.329    0.745
mu.hat        -0.61133    9.69505  -0.063    0.950
I(mu.hat^2)    0.10115    0.05993   1.688    0.106

Residual standard error: 137.5 on 21 degrees of freedom
Multiple R-squared: 0.8638,     Adjusted R-squared: 0.8508
F-statistic:  66.6 on 2 and 21 DF,  p-value: 8.1e-10

> plot(x = set1$mu.hat, y = set1$res.sq, xlab = "Predicted
   count", ylab = "Squared Residual")
> curve(expr = predict(object = fit.lin, newdata =
   data.frame(mu.hat = x), type = "response"), col = "blue", add
   = TRUE, lty = "solid")
> curve(expr = predict(object = fit.quad, newdata =
   data.frame(mu.hat = x), type = "response"), col = "red", add
   = TRUE, lty = "dashed")
> legend(x = 50, y = 1000, legend = c("Linear", "Quadratic"),
   col = c("red", "blue"), lty = c("solid", "dashed"), bty = "n")
```

The results are given in Figure 5.9. The quadratic trend seems to describe the relationship slightly better than the straight line, although the improvement is not vast, and neither is an excellent fit. The estimated regression parameters for the quadratic fit of the squared residuals against the predicted mean are shown in the output. The quadratic coefficient has a p-value of 0.1, which is not significant but not very large. The results in this case are somewhat inconclusive, so there is justification for either model.

Similar to the negative binomial model, the *beta-binomial* model (Williams, 1975) can serve as an alternative to quasi-binomial as a model for overdispersed counts from fixed numbers of trials. Suppose that we observe binomial counts from M EVPs, $w_1, w_2, \ldots, w_M$, with w_m resulting from n_m trials with probability π_m, $m = 1, \ldots, M$. The beta-binomial model assumes that each π_m follows a beta distribution, the mean of which may depend on the explanatory variables. Reparameterizing the beta distribution for this purpose leads to the result that the expected count for EVP m is $E(W_m) = n_m\pi_m$, while the variance is $Var(W_m) = n_m\pi_m(1-\pi_m)(1+\phi(n_m-1))$, where $0 \leq \phi < 1$ is an unknown dispersion parameter. Notice that the mean of this distribution is the same as in the binomial distribution. The variance is the same too when either $\phi = 0$ or $n_m = 1$. When $\phi > 0$ and $n_m > 1$, the variance of the count is strictly greater than the usual variance specified by the binomial distribution, thereby allowing for overdispersion. When only individual binary trial responses are observed, each with its own explanatory variables, then $n_m = 1$ for all $m = 1, \ldots, M$, and the beta-binomial model cannot be fit. Finally, notice that when all n_m are equal, the factor $(1 + \phi(n_m - 1))$ is the same for all observations, so that the beta-binomial assumes the same mean-variance relationship as the quasi-binomial model.

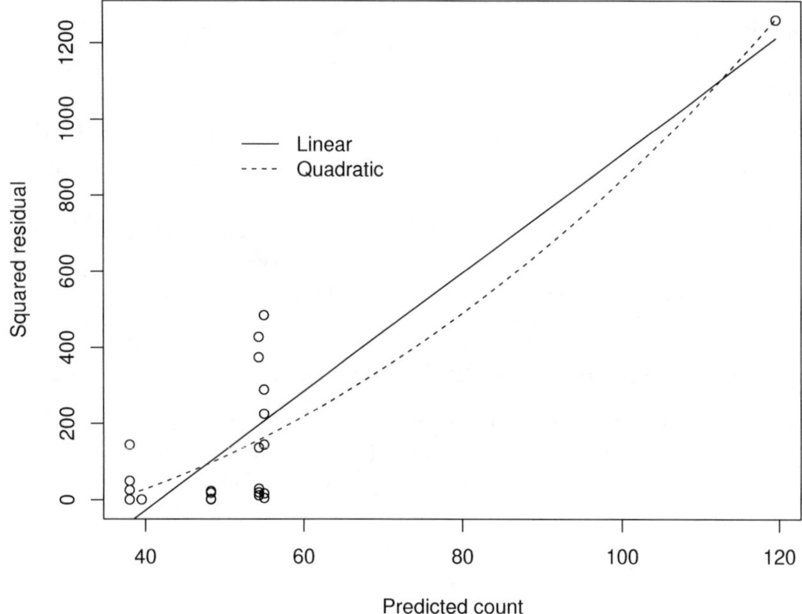

Figure 5.9: Plot of squared residuals vs. predicted values for choosing between quasi-Poisson and negative binomial models. Solid line is the straight line fit for quasi-Poisson model; dashed line is quadratic fit for negative binomial.

Parameters of the beta-binomial model are estimated using MLEs, so that standard inference procedures can be used. Also, information criteria can be computed. This allows this model to be compared to an ordinary binomial regression model to see whether the overdispersion is severe enough to warrant the fitting of a model with an extra parameter. The model fitting can be done by several R functions, including the `betabin()` function of the `aod` (short for "analysis of overdispersed data") package and the `vglm()` function of the `VGAM` package (`family = betabinomial`). Both functions have capacity to further allow ϕ to depend on explanatory variables. An example of `betabin()` is given in Exercise 35d.

Overdispersion can affect multicategory regression models like Equation 3.4. Unfortunately, the complexity of a multicategory regression model makes detecting and resolving overdispersion somewhat more complicated than in single-response models. Generally, it can present itself as an inflation of GOF statistics that cannot explained by other diagnostic investigations. The simplest solution is a quasi-likelihood-type formulation, in which the variances and covariances among the counts for different categories are all multiplied by a constant $\gamma > 1$. This process is discussed in McCullagh and Nelder (1989, p. 174). Alternative methods of formulating and fitting the multinomial model have been proposed by Mebane and Sekhon (2004) and Landsman et al. (2019). An extension of the beta-binomial model to multivariate response, called the Dirichlet-multinomial model, can also be used as a way to introduce extra variability into the multinomial regression model. Capacity for fitting this model is available in the `MGLM` package.

As a simpler alternative to fitting complex multinomial regression models, a Poisson-based formulation for the multivariate counts could be used as discussed in Section 4.2.5. This allows the use of simpler tools for model diagnostics and for correcting overdispersion.

5.4 Examples

In this section we reanalyze two examples that have thus far been presented only in pieces throughout the book: the placekicking data using logistic regression and the alcohol consumption data using Poisson regression. Our goal is to demonstrate the process of undertaking a complete analysis, starting with variable selection, continuing through model fitting and assessment, and concluding with inferences. We present R code and graphics liberally and explain our rationale for each decision made during the analyses.

5.4.1 Logistic regression: placekicking data set

We now examine the placekicking data set more closely to find models for the probability of a successful placekick that fit the data well. The data used here include 13 additional observations that were not part of any previous example. The reason these additional observations are included (and then eventually excluded) is made clear from our analysis. The data used here also include these additional variables:

- `altitude`: Official altitude in feet for the city where the placekick is attempted (not necessarily the exact altitude of the stadium)
- `home`: Binary variable denoting placekicks attempted at the placekicker's home (1) vs. away (0) stadium
- `precip`: Binary variable indicating whether precipitation is falling at game time (1) vs. no precipitation (0)
- `temp72`: Fahrenheit temperature at game time, where 72° is assigned to placekicks attempted within domed stadiums

The data are stored in the PlacekickMB.csv file, and we include all of our R code in PlacekickFindBestModel.R.

Variable selection

Our first goal is to identify a plausible model for the probability of success for all placekicks in our data. It is the subject-matter specialist's responsibility to limit the pool of variables to those that are considered relevant to the modeling problem. Because we are serving as the "experts" in this example, we start by identifying the variables that we consider plausible. In particular, all variables in the data set are considered in linear form. We believe that it may be important to include several interactions among variables in the data, particularly interactions with `distance`. Blindly including all 66 interactions into our pool of potential variables would create an unwieldy variable-selection task that only the genetic algorithm of `glmulti()` could handle. We instead limit the interactions to those that make sense in the context of the problem. Based on our experience as fans and analysts, we believe that the following two-way interactions are plausible:

- `distance` with `altitude`, `precip`, `wind`, `change`, `elap30`, `PAT`, `field`, and `temp72`; distance clearly has the largest effect on the probability of success, and all of these other factors could conceivably have a greater effect on longer kicks than on shorter ones.
- `home` with `wind`; a placekicker on his home field may be more accustomed to local wind effects than a visiting kicker.

- `precip` with `type`, `field`, and `temp72`; precipitation cannot directly affect a game in a domed stadium, can have a bigger effect on grass than on artificial turf, and can make players miserable if the temperature is cold enough.

Variable selection with interactions presents a unique challenge relating to the principle of marginality first described in Section 2.2.5. A model respects marginality if it includes all lower-order terms that comprise any interactions in the model. Models that include interactions without including the corresponding main effects can have undesirable properties. For example, in a linear regression where x and xz are included in a model without z, the slope of Y against z is forced to be exactly 0 when $x = 0$, which may not be a realistic restriction in many applications. We therefore want to consider only models that respect marginality by including interactions into a model only if the two linear terms are in it as well.

Unfortunately, the options for incorporating marginality into variable selection are limited. If we were to consider all interactions as plausible rather than just these 12, we could do all-subsets selection with `glmulti()`, which has an argument that limits the search to marginal models. It also has an `exclude` argument that can supposedly remove unwanted interactions from consideration, but the argument does not work well and failed in our example.

As an alternative, we take a two-step approach to identify a reasonable model. We first identify the best main-effects model by all-subsets selection using the `dredge()` function from the `MuMIn` package with the AIC criterion. Starting from the selected model, we then add interactions or excluded main effects in a forward stepwise manner using the `step()` function. This function respects marginality by default, because at each step it evaluates interactions only if the two main effects are already in the model. However, it has the limitation that it can only add one variable at a time. This could cause it to miss an important interaction because its main effects must first be added. The main effect might worsen the AIC criterion resulting in premature termination of the algorithm, even if adding the interaction later would have improved the criterion. To circumvent this possibility, we also show code in our program where we first run `step()` without an IC penalty. This causes the function to keep adding the best variable to the model until all variables are included, creating a sequence of "best" models of each size. We then compute the AIC on each of these models using `extractAIC()`, and select as our final model the one with the smallest AIC. In this instance, both approaches select the final same model, so we present only the simpler code and results below.

Similar to the results from Section 5.1.2, the model with `distance`, `wind`, `change`, and `PAT` has the smallest AIC among models with main effects only. We fit that model and save it as `mod.best.lin`. Then we fit the model with all main effects and the chosen interactions, save it as `mod.best.upper`, and specify these two models as the `object` and `scope` arguments in `step()`. The resulting model with added interactions is below.

```
> mod.best.lin <- glm(formula = good ~ distance + wind + change +
    PAT, family = binomial(link = logit), data = placekick.mb)
> mod.best.upper <- glm(formula = good ~ distance*altitude +
    distance*precip + distance*wind + distance*change +
    distance*elap30 + distance*PAT + distance*field + distance*temp72
    + home*wind + precip*type + precip*field + precip*temp72, family
    = binomial(link = logit), data = placekick.mb)
> step.inter <- step(object = mod.best.lin, scope = list(upper =
    mod.best.upper), k = 2, trace = FALSE)
> summary(step.inter)
```

```
Call:
glm(formula = good ~ distance + wind + change + PAT + distance:wind
    + distance:PAT, family = binomial(link = logit), data =
    placekick.mb)

Coefficients:
              Estimate Std. Error z value Pr(>|z|)
(Intercept)    4.49640    0.48142   9.340  < 2e-16 ***
distance      -0.08069    0.01143  -7.062 1.64e-12 ***
wind           2.92477    1.78448   1.639  0.10121
change        -0.33205    0.19454  -1.707  0.08785 .
PAT            6.71190    2.11356   3.176  0.00150 **
distance:wind -0.09183    0.04568  -2.010  0.04443 *
distance:PAT  -0.27171    0.09799  -2.773  0.00556 **
---
Signif. codes:  0 '***' 0.001 '**' 0.01 '*' 0.05 '.' 0.1 ' ' 1

(Dispersion parameter for binomial family taken to be 1)

    Null deviance: 1024.77  on 1437  degrees of freedom
Residual deviance:  759.72  on 1431  degrees of freedom
AIC: 773.72

Number of Fisher Scoring iterations: 6
```

Forward selection runs for two steps and finds that the "best" model adds `distance:wind` and `distance:PAT` to the base model. Therefore, a preliminary model for our data includes `distance`, `PAT`, `wind`, `change`, `distance:wind`, and `distance:PAT`.

This is not the only possible approach to variable selection in this slightly complicated setting. Alternative methods make different assumptions and have different limitations, while not always leading to the same chosen model. For comparison purposes, we also used all-subsets selection with the main effects of `distance`, `PAT`, `wind`, and `change` along with the pairwise interactions among only these variables. This identified the same model as with our previous forward selection.

Assessing the model fit: Preliminary model

To begin the model assessment process, we must convert the data into EVP form corresponding to the four variables in the model, and refit the model to these data. The code for this is below.

```
> # Convert data to EVP form; interactions are not needed in
    aggregate() because they do not change the number of unique
    combinations of explanatory variables.
> w <- aggregate(x = good ~ distance + wind + change + PAT, data =
    placekick.mb, FUN = sum)
> n <- aggregate(x = good ~ distance + wind + change + PAT, data =
    placekick.mb, FUN = length)
> w.n <- data.frame(w, trials = n$good, prop = round(w$good/n$good,
    4))
> head(w.n)
  distance wind change PAT good trials prop
```

```
1        18    0        0    0      1        1 1.00
2        19    0        0    0      3        3 1.00
3        20    0        0    0     15       15 1.00
4        21    0        0    0     11       12 0.92
5        22    0        0    0      7        8 0.88
6        23    0        0    0     15       15 1.00
> nrow(w.n)  # Number of EVPs (M)
[1] 124
>  sum(w.n$trials)  # Number of observations
[1] 1438

> mod.prelim1 <- glm(formula = good/trials ~ distance + wind +
   change + PAT + distance:wind + distance:PAT, family =
   binomial(link = logit), data = w.n, weights = trials)
```

We first explore residual plots from our selected model. Because `PAT`, `wind`, and `change` are binary explanatory variables, residual plots against these variables would yield minimal insight into possible model flaws. Therefore, we plot standardized Pearson residuals only against `distance`.

```
> stand.resid <- rstandard(model = mod.prelim1, type = "pearson")
> plot(x = w.n$distance, y = stand.resid, ylim = c(min(-3,
   stand.resid), max(3, stand.resid)), ylab = "Standardized Pearson
   residuals", xlab = "Distance")
> abline(h = c(3, 2, 0, -2, -3), lty = "dotted", col = "blue")
> smooth.stand <- loess(formula = stand.resid ~ distance, data =
   w.n, weights = trials)
> ord.dist <- order(w.n$distance)
> lines(x = w.n$distance[ord.dist], y =
   predict(smooth.stand)[ord.dist], lty = "solid", col = "red")
```

Figure 5.10 shows the plot along with a loess curve. The curve undulates gently around 0, showing no apparent "smile" or "frown" shapes that would suggest a nonlinear relationship between the log-odds of success and `distance`. For illustration purposes, we also temporarily added a quadratic term for `distance` to the model. This increased the AIC, which agrees with our findings that adding nonlinear terms for distance is unnecessary.

To expedite the process of further assessing the fit of the preliminary model, we have written a function that implements most of the methods outlined in Section 5.2 in a simplified manner specifically for logistic regression. This function, called `examine.logistic.reg()`, is available from the textbook's website in the program Examine.logistic.reg.R. The only required argument is `mod.fit.obj`, which corresponds to the `glm`-class object containing the model whose fit is to be assessed. Optional arguments include `identify.points` that allows users to interactively identify points on a plot, `scale.n` and `scale.cookd` which allow users to rescale numerical values used as the circle size in bubble plots, and `pearson.dev` to denote whether standardized Pearson or standardized deviance residuals are plotted (default is Pearson). These arguments are explained in more detail shortly. The function's use is demonstrated below, and the resulting plots are in Figure 5.11.

```
> # Used with examine.logistic.reg() to rescale numerical values
> one.fourth.root <- function(x) {
    x^0.25
  }
```

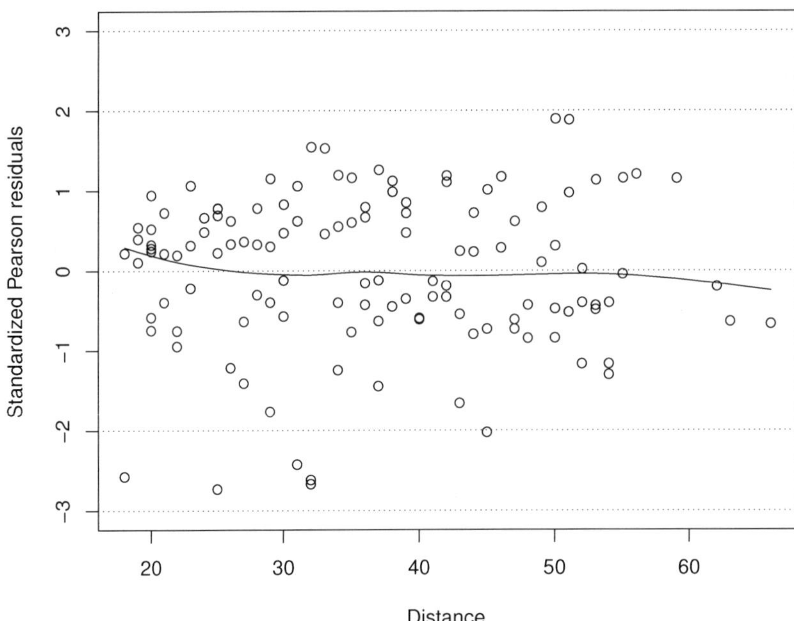

Figure 5.10: Standardized Pearson residuals vs. distance of the placekick.

```
> one.fourth.root(16)  # Example
[1] 2

> # Read in file containing examine.logistic.reg() and run function
> source(file = "C:\\Rprograms\\Examine.logistic.model.R")
> save.info1 <- examine.logistic.reg(mod.fit.obj = mod.prelim1,
   identify.points = TRUE, scale.n = one.fourth.root,  scale.cookd =
   sqrt)
> names(save.info1)
 [1] "pearson"         "stand.resid"      "stand.dev.resid"
 [4] "deltaXsq"        "deltaD"           "cookd"
 [7] "pear.stat"       "dev"              "dev.df"
[10] "gof.threshold"   "pi.hat"           "h"
```

After each plot is created, R prompts the user to identify points in that plot. This identification is done by left-clicking on or near the points of interest, and these are subsequently labeled by the "row name" in `placekick.mb`.[7] This is how row numbers were placed on the plots in Figure 5.11. To end the identification process for a particular plot, a user can right-click the mouse and select Stop from the resulting window.[8] By default, the

[7]Every row of a data frame has a "name." These row names are shown in the left-most column of a printed data frame. Typically, the row number of the data frame is the row name. However, these row names can be changed to be more descriptive. For example, if `set1` is a two-row data frame, then `row.names(set1) <- c("row 1", "row 2")` replaces the default row names for `set1` with the names `row 1` and `row 2`.

[8]The exact process may differ because of the operating system and/or integrated development environment for which R is being used within. What we describe here is the process for a Windows operating system using R directly. For readers using RStudio, its default graphics window cannot handle the interactive identification process. A simple fix is to run `dev.new()` which will open a regular R graphics window outside of RStudio's graphical user interface. Until this window is closed, all plots will be sent to it.

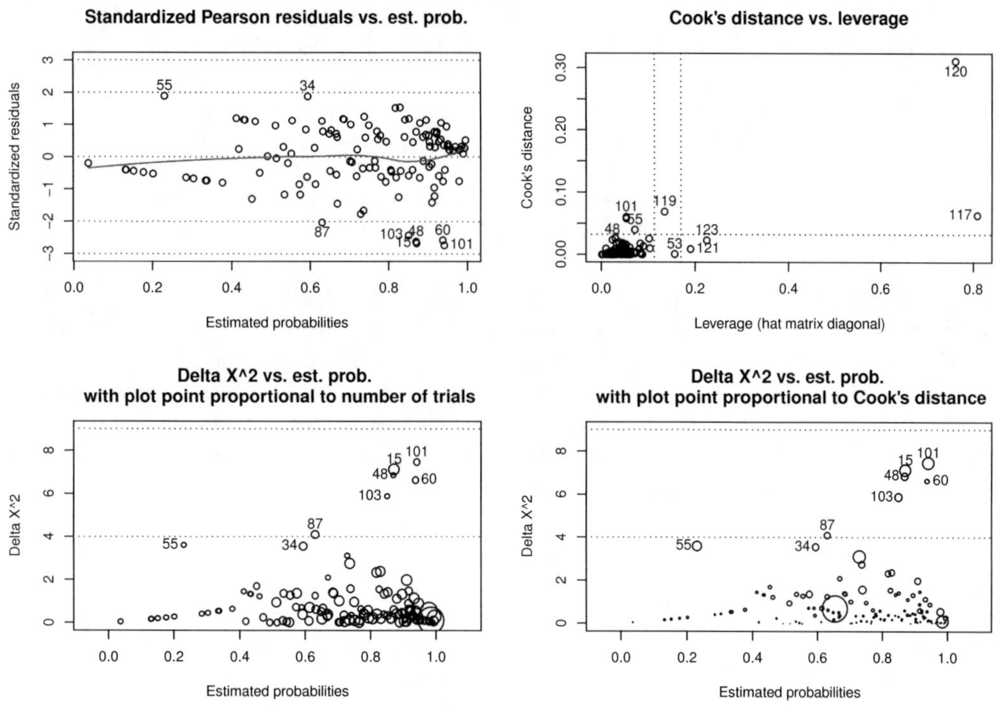

Deviance/df = 1.01; GOF thresholds: 2 SD = 1.26, 3 SD = 1.39

Figure 5.11: Diagnostic plots for the placekicking logistic regression model; all observations are included in the data set.

`identify.points` argument is set to `TRUE`, but can be changed to `FALSE` in the function call if identification is not of interest.

The two plots in the top row are similar to those shown in Section 5.2 for the standardized Pearson residuals, Cook's distances, and leverages. There are a few EVPs outside or close to ± 2 that we may want to investigate further, but none beyond ± 3, which is a good sign. A loess curve included on the top-left plot shows no strong relationship between the residuals and estimated probabilities.

The top-right plot shows that EVP 120 has a very large value of Cook's distance in comparison to the other EVPs—beyond the $4/M$ guideline but not beyond 1—so we will want to investigate this EVP further. There are a few other EVPs with Cook's distances that stand out from the rest that we may want to examine more closely as well. We further see that that both EVP 117 and 120 have very high leverage, far beyond the vertical line at $3p/M$. Two other EVPs, 121 and 123, also have leverage values beyond $3p/M$, indicating the potential for influence.

The plots on the bottom row both give bubble plots of ΔX_m^2 against the model-estimated probability of success, using number of trials and Cook's distance, respectively, to determine the bubble sizes. This makes it easier to simultaneously assess multiple measures of influence in a single plot. After some experimenting, we used the `scale.n` argument to rescale the bubble sizes to the fourth root of the number of trials, and the `scale.cookd` argument to rescale Cook's D by a square root. These scales made large bubble sizes that were informatively different.

The bottom-left bubble plot shows that a few EVPs with large estimated probabilities have somewhat large ΔX_m^2 values too. However, these corresponding EVPs have a small number of trials as shown by the size of their plotting point (except perhaps EVP 15), so these EVPs may not necessarily be truly unusual. Still, we should examine these EVPs more closely before making a final judgment. An interesting feature of the bottom-right plot is that the very large Cook's distance value (EVP 120, the largest circle on plot) has a relatively small squared standardized residual. This could indicate a very influential EVP that is "pulling" the model toward it to better its fit.

Below the plots, the deviance/df statistic is shown to be $D/(M-(p+1)) = 1.01$. The rough thresholds of $1+2\sqrt{2/(M-(p+1))} = 1.26$ and $1+3\sqrt{2/(M-(p+1))} = 1.39$ are also given at the bottom of the plot. Because the deviance/df is well below these thresholds, it does not offer evidence of any problems with the overall fit of the model.

We also applied the HL test with $g = 8$, 10, and 12 (output not shown here). All three tests gave p-values larger than 0.8, indicating again there is insufficient evidence of a problem with the model's overall fit. On the other hand, the Stukel test identified a potentially poor model fit, with a p-value of 0.002. It appears that there is some aspect of the model's fit that is too subtle for very general statistics and tests to identify, but that can be found by a test that targets a particular deviation between the data and the model. This test also indicated concerns with the simpler model studied in Section 5.2.2.

We investigate further by examining potentially influential EVPs. Our preferred method for this is to examine all EVPs with their model-estimated probabilities, standardized Pearson residuals, binomial tail probabilities for the residuals, Cook's distances, and leverage values together in one data set. With only 124 EVPs, this is manageable to do; however, to save space here and to illustrate what can be done when the number of EVPs is much larger, we print only those EVPs with standardized Pearson residuals greater than 2 in absolute value, binomial tail probabilities below 0.025, Cook's distances greater than $4/124 = 0.0323$, or leverage values greater than $3 \times 7/124 = 0.17$. The code below illustrates the process of extracting these EVPs from the data frame w.n.

```
> w.n.diag1 <- data.frame(w.n, pi.hat = round(save.info1$pi.hat, 2),
   std.res = round(save.info1$stand.resid, 2),  tail.prob =
   round(save.info1$tail.prob, 3), cookd = round(save.info1$cookd,
   2), h = round(save.info1$h,2))

> # Potential EVPs to examine further
> p <- length(mod.prelim1$coefficients)
> ck.out <- abs(w.n.diag1$std.res) > 2 | w.n.diag1$cookd >
   4/nrow(w.n) | w.n.diag1$h > 3*p/nrow(w.n) | w.n.diag1$tail.prob <
   0.025
> # Extract EVPs
> extract.EVPs <- w.n.diag1[ck.out,]
> # Order by distance
> extract.EVPs[order(extract.EVPs$distance),]
    distance wind change PAT good trials prop pi.hat
60        18    0      1   0    1      2 0.50   0.94
117       20    0      0   1  605    614 0.99   0.98
121       20    1      0   1   42     42 1.00   0.99
123       20    0      1   1   94     97 0.97   0.98
101       25    1      1   0    1      2 0.50   0.94
119       29    0      0   1    0      1 0.00   0.73
120       30    0      0   1    3      4 0.75   0.65
```

```
103         31    1      1   0    0      1 0.00   0.85
15          32    0      0   0   12     18 0.67   0.87
48          32    1      0   0    0      1 0.00   0.87
87          45    0      1   0    1      5 0.20   0.63
55          50    1      0   0    1      1 1.00   0.23
     std.res tail.prob cookd    h
60     -2.57     0.121  0.01 0.01
117     0.32     0.530  0.06 0.81
121     0.52     0.802  0.01 0.19
123    -0.75     0.341  0.02 0.23
101    -2.73     0.114  0.06 0.05
119    -1.76     0.271  0.07 0.13
120     0.83     0.570  0.31 0.76
103    -2.43     0.149  0.03 0.03
15     -2.67     0.021  0.06 0.05
48     -2.62     0.130  0.02 0.02
87     -2.03     0.066  0.02 0.03
55      1.90     0.231  0.04 0.07
```

Many of these EVPs have standardized Pearson residuals beyond ± 2. Most of these EVPs are not truly of concern because they consist of a very small number of trials. As detailed in Section 5.2.1, thresholds of 2 and 3 based on a standard normal distribution approximation are not very useful in this situation, and the relatively large residual tail probabilities (`tail.prob`) confirm this. The slightly large leverage (`h`) values for EVPs 121 and 123 are likely because of their relatively large numbers of trials, which automatically gives them the potential for influence, but they show no other concerning features.

There are three EVPs that require further discussion. EVP 117 has one of the larger Cook's distances and the largest leverage. However, this EVP consists of 614 out of the 1,438 total trials in the data set. We would expect one EVP with such a large percentage of the total observations to be influential. As mentioned in Section 5.2.3, there may be some concern that such an influential observation could cause a poor model fit for other nearby EVPs or could unduly influence the variable selection, but we do not see this occurring here. There are a few slightly extreme negative standardized residuals near this EVP that may be related to its influence, but they are not especially concerning. Overall, the model does not appear to be detrimentally affected by this EVP.

EVP 15 consists of 32-yard field goals with low wind and no lead change. This appears simply to have an unusually low number of successes compared to its estimated probability of success. The result is a somewhat extreme standardized Pearson residual (tail probability is 0.02) and Cook's distance. Neighboring EVPs between 29–35 yards all have sample proportions above this EVP's model estimate of 0.87, suggesting that this low number of successes is just "bad luck," and not a model flaw. In any case, EVP 15 is the only one with a tail probability below 0.025, which is not unexpected for a data set with 124 EVPs.

Finally, EVP 120 has the largest Cook's distance by far and the second largest leverage. This suggests that it is likely quite influential. Its observed explanatory variable values correspond to a very unusual type of placekick—a 30-yard PAT, rather than the customary 20-yard PAT that occurred in the NFL when the data was collected. Placekicks like this happen only as a result of a penalty on an initial placekick attempt. In fact, we note that EVP 119—which has the second-highest Cook's distance and slightly elevated leverage—is a single PAT also taken at a non-standard distance. These EVPs, which are the only ones taken beyond 25 yards, combine for only three successes in five attempts. For reference, the model estimates that field goals taken under the same conditions should have success probabilities of 0.90 and 0.89 at 29 and 30 yards, respectively. This, too, could have been "bad luck", but we nonetheless wonder whether these unusual types of placekicks are somehow different than the more typical types of placekicks and perhaps need separate attention.

To examine EVPs 119 and 120 and their effect on the model more closely, we temporarily remove them from the data set and refit the model.

```
> mod.prelim1.wo119.120 <- glm(formula = good/trials ~ distance +
    wind + change + PAT + distance:wind + distance:PAT, family =
    binomial(link = logit), data = w.n[-c(119, 120),], weights =
    trials)
> round(summary(mod.prelim1.wo119.120)$coefficients, digits = 4)
              Estimate Std. Error z value Pr(>|z|)
(Intercept)     4.4985     0.4816  9.3400   0.0000
distance       -0.0807     0.0114 -7.0640   0.0000
wind            2.8770     1.7866  1.6103   0.1073
change         -0.3308     0.1945 -1.7010   0.0889
PAT           -12.0703    49.2169 -0.2452   0.8063
distance:wind  -0.0907     0.0457 -1.9851   0.0471
distance:PAT    0.6666     2.4607  0.2709   0.7865
```

The output shows a dramatic change in the estimate corresponding to `distance:PAT`. When EVP 119 and 120 are in the data set, the estimate is −0.2717 with a Wald test p-value of 0.0056. Now, the estimate is 0.6666 with a Wald test p-value of 0.7865. Therefore, it appears the presence of this interaction was solely due to the five placekicks corresponding to these EVPs.

There are five EVPs containing a total of 13 observations which are non-20-yard PATs in the data set, and 11 of these placekicks are successes (the two failures are included in EVPs 119 and 120). Because of the small number of placekicks of this type, it is difficult to determine if what we observe here is a real trend or an anomaly in the data. We are particularly concerned that two missed placekicks out of a sample of 1,438 created a need for a separate term in the model to describe them, which seems to be a rather extreme and disproportionate reaction. This suggests two approaches to handle the problem:

1. Return EVPs 119 and 120 to the data set and acknowledge that they are responsible for making the `distance:PAT` interaction appear important.

2. Reduce the population of inference to focus on only certain types of placekicks. We see two options here:

    (a) Remove all PATs from the data set and find separate models for field goals and PATs.

    (b) Remove all non-20-yard PATs from the data set.

We decided that option #2b was a slightly better choice than the other two. We are *very* hesitant to remove data from an analysis, as this is a dangerous practice in general and has potential to create biased results. Also, the huge number of 20-yard PATs provides valuable information about the effects of other variables—at least for shorter placekicks—so we would like to keep them in the data set. But non-20-yard PATs are somewhat rare events and always follow penalties. If they truly have different probabilities of success, it may be because of the after-effects of penalties rather than, or in addition to, the increase in distance. Penalties undoubtedly occurred prior to a small fraction of other placekicks, but we do not have a record of them. We have no way to model this effect and are uncomfortable with creating an additional model term solely for these few observations. Therefore, we choose to remove these kicks from further consideration so that we can focus on field goals in general and PATs of 20 yards. A new, larger data set might contain enough of these placekicks to offer a more complete assessment of any possible differences in their success probabilities.

Assessing the model fit: Revised model

Below is the code to remove the non-20-yard PATs.

```
> # Remove non-20 yard PATs - "!" negates and "&" means "and"
> placekick.mb2 <- placekick.mb[!(placekick.mb$distance!=20 &
    placekick.mb$PAT==1),]
> nrow(placekick.mb2)  # Number of observations after 13 were removed
[1] 1425
```

We now return to the variable selection step with this revised data set containing 1,425 total placekicks. Details are given in our program, and the resulting model terms are the same as before, except that `distance:PAT` is no longer available to be selected. Below is the output from fitting the new model.

```
> # w.n2 contains 119 EVPs
> mod.prelim2 <- glm(formula = good/trials ~ distance + wind +
    change + PAT + distance:wind, family = binomial(link = logit),
    data = w.n2, weights = trials)
> summary(mod.prelim2)

Call:
glm(formula = good/trials ~ distance + wind + change + PAT +
    distance:wind, family = binomial(link = logit), data = w.n2,
    weights = trials)

Deviance Residuals:
    Min       1Q   Median       3Q      Max
-2.2386  -0.5836   0.1965   0.8736   2.2822

Coefficients:
              Estimate Std. Error z value Pr(>|z|)
(Intercept)    4.49835    0.48163   9.340  < 2e-16 ***
distance      -0.08074    0.01143  -7.064 1.62e-12 ***
wind           2.87783    1.78643   1.611  0.10719
change        -0.33056    0.19445  -1.700  0.08914 .
PAT            1.25916    0.38714   3.252  0.00114 **
distance:wind -0.09074    0.04570  -1.986  0.04706 *
---
Signif. codes:  0 '***' 0.001 '**' 0.01 '*' 0.05 '.' 0.1 ' ' 1

(Dispersion parameter for binomial family taken to be 1)

    Null deviance: 376.01  on 118  degrees of freedom
Residual deviance: 113.86  on 113  degrees of freedom
```

```
AIC: 260.69

Number of Fisher Scoring iterations: 5
```

We again use `examine.logistic.reg()` and GOF tests to evaluate the fit of the model. The test results are very similar to those seen with the preliminary model—all three HL tests are not significant with p-values > 0.8, but the Stukel test remains significant, albeit with a larger p-value, 0.02. The corresponding plots are not given here because they essentially show the same results as before, excluding those EVPs that we removed. EVP 116 (formerly EVP 117) has the largest Cook's distance now. This EVP contains the 614 PATs at 20 yards, so we would again expect this to be influential. The other EVPs with slightly elevated Cook's distances all occur under somewhat different conditions where the observed proportion of successes was a bit different from the model's estimated probability. We took the extra step of temporarily removing a few of these EVPs one at a time and refitting the model to determine if the regression parameter estimates changed substantially. These estimates did not, so we make no further changes to the data or the model. We conclude that there may remain some relatively minor flaw in the model, but it does not appear to be a serious issue nor have an obvious cause. We might investigate further by eliminating all PATs and focusing on the shape of the logistic regression curve more closely; we leave that to the reader.

Our final model is

$$\text{logit}(\hat{\pi}) = 4.4983 - 0.08074\texttt{distance} + 2.8778\texttt{wind} - 0.3306\texttt{change} + 1.2592\texttt{PAT}$$
$$-0.09074\texttt{distance} \times \texttt{wind}.$$

Interpreting the model

We calculate odds ratios and corresponding profile LR confidence intervals to interpret the explanatory variables in the model. Although we include confidence intervals in our calculations, these intervals should be interpreted with care because they are computed using the same data on which the model was selected. As described in Section 5.5, the stated confidence levels are likely overestimates of the true confidence levels, because they are based on the same data that was used for selecting the model. In light of this, we use 99% confidence levels to widen the intervals, but we remain uncertain about the exact confidence level for these intervals.

```
> library(package = mcprofile)

> OR.name <- c("Change", "PAT", "Distance, 10-yard decrease, windy",
   "Distance, 10-yard decrease, not windy", "Wind, distance = 20",
   "Wind, distance = 30", "Wind, distance = 40", "Wind, distance =
   50", "Wind, distance = 60")
> var.name <- c("int", "distance", "wind", "change", "PAT",
   "distance:wind")
> K <- matrix(data = c(0,    0, 0, 1, 0,   0,
                       0,    0, 0, 0, 1,   0,
                       0,  -10, 0, 0, 0, -10,
                       0,  -10, 0, 0, 0,   0,
                       0,    0, 1, 0, 0,  20,
                       0,    0, 1, 0, 0,  30,
                       0,    0, 1, 0, 0,  40,
```

```
                     0,    0, 1, 0, 0,   50,
                     0,    0, 1, 0, 0,   60), nrow = 9, ncol = 6,
                        byrow = TRUE, dimnames = list(OR.name,
                        var.name))
> # K # Check matrix - excluded to save space
> linear.combo <- mcprofile(object = mod.prelim2, CM = K)
> ci.log.OR <- confint(object = linear.combo, level = 0.99, adjust =
    "none")
> exp(ci.log.OR)

   mcprofile - Confidence Intervals

level:          0.99
adjustment:     none

                                       Estimate   lower   upper
Change                                   0.7185 0.43617   1.190
PAT                                      3.5225 1.33291  10.054
Distance, 10-yard decrease, windy        5.5557 2.11131  23.526
Distance, 10-yard decrease, not windy    2.2421 1.68481   3.041
Wind, distance = 20                      2.8950 0.41977  54.261
Wind, distance = 30                      1.1683 0.37033   6.024
Wind, distance = 40                      0.4715 0.17738   1.240
Wind, distance = 50                      0.1903 0.02794   0.867
Wind, distance = 60                      0.0768 0.00297   0.845
```

The change and PAT explanatory variable interpretations are the simplest because they are not involved in any interactions in the model. From the model, we estimate that higher-pressure placekicks have odds of success that are 0.72 times as large as the odds under normal pressure, when holding the other variables constant. The corresponding confidence interval is (0.43, 1.19). It may seem surprising that the chosen model includes a variable whose confidence interval contains 1. There are reasons for this result. First, the large confidence level widens the interval. In this case, a 90% confidence interval is (0.55, 0.99), which excludes 1. Second, the AIC was used to select variables, rather than hypothesis testing. This can lead to including variables in a model that do not achieve statistical significance at some particular level. Overall, our results indicate that any pressure effect for placekicks may lower the probability of success under higher pressure, but the evidence is not substantial. With respect to the PAT variable, the estimated odds of a success are 3.5 times as large for 20-yard PATs as for field goals of the same distance, when holding the other variables constant. The corresponding 99% confidence interval is (1.3, 10.1).

The distance and wind explanatory variable interpretations need to take into account their interaction. For a 10-yard decrease in distance, the estimated odds of success increase by 5.6 times (interval is (2.1, 23.5)) when there are windy conditions and increase by 2.2 times (interval is (1.7, 3.0)) when there are not windy conditions. This means that decreasing the distance of a placekick is even more important when conditions are windy than when they are not. We also calculated the estimated odds ratios and corresponding 99% confidence intervals for windy vs. not windy conditions for placekicks of 20, 30, 40, 50, and 60 yards. Again, we see that the effects of windy conditions are minimal for short placekicks, but pronounced for longer placekicks.

The estimated probability of success along with profile LR intervals can be calculated for specific combinations of explanatory variables. For example, we examine the probabilities of success for PATs and field goals at a distance of 20 yards with wind = 0 and change = 0 as follows.

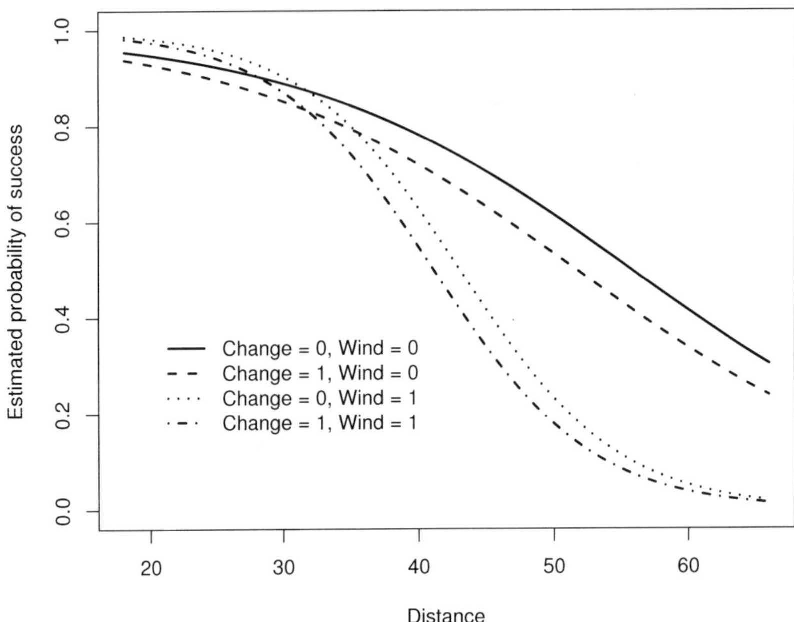

Figure 5.12: Estimated probability of success vs. the field goal (PAT = 0) distance for four combinations of change and wind.

```
> K <- matrix(data = c(1, 20, 0, 0, 1, 0,
                       1, 20, 0, 0, 0, 0), nrow = 2, ncol = 6, byrow
                         = TRUE, dimnames = list(c("PAT", "FG"),
                         var.name))
> linear.combo <- mcprofile(object = mod.prelim2, CM = K)
> ci.lin.pred <- confint(object = linear.combo, level = 0.99, adjust
    = "none")
> # as.matrix() is needed to get the proper class for plogis()
> round(plogis(q = as.matrix(ci.lin.pred$estimate)), digits = 3)
    Estimate
PAT    0.984
FG     0.947
> round(plogis(q = as.matrix(ci.lin.pred$confint)), digits = 3)
     lower upper
[1,] 0.970 0.993
[2,] 0.902 0.974
```

The results here show that the probability of success is larger for PATs than for field goals, as the odds ratios demonstrated earlier.

Figure 5.12 plots the estimated model as a function of `distance` (the code is included in the program). With the addition of "risk" factors, windy conditions and lead-change attempts, the estimated probability of success generally decreases.

5.4.2 Poisson regression: alcohol consumption data set

We return to the Poisson regression analysis of the alcohol consumption data introduced in Section 4.2.2. The focus again is on the subjects' consumption during their first Saturday in the study. We now perform a more complete analysis of the data, including initial examination of the data, variable selection, assessment of fit, correction for overdispersion, and analysis and interpretation of the results. The code for the analysis is contained in AlcoholDetailed.R, and the data are in DeHartSaturday.csv.

Examining the data

First, we introduce new variables that have not been used before. The example in Section 4.2.2 regressed the number of drinks consumed (`numall`) against an index for the positive events (`posevent`) and negative events (`negevent`) experienced by the subject each day. In Exercise 27 of Chapter 4, we considered additional explanatory variables for positive romantic-relationship events (`prel`), negative romantic-relationship events (`nrel`), age (`age`), trait (long-term) self-esteem (`rosn`), and state (short-term) self-esteem (`state`). In this example we add two more variables: `gender` (1=male, 2=female; no other genders are reported in the data) and `desired`, which measures a subject's desire to drink each day, with a higher score meaning greater desire. Note that trait self-esteem was measured once at the start of the study, while state self-esteem was measured daily. Also note that, although `gender` is a numerical representation of a categorical variable, there is no harm in treating it as numerical because there are only two levels. The "slope" coefficient is exactly the same as the "difference" parameter for a two-level categorical variable when it is coded with numbers that are 1 unit apart.

We start with a scatterplot matrix of the nine numerical variables using the function `spm()` from the `car` package. Figure 5.13 gives this plot. The diagonal of the matrix is an estimate of the density function for each variable. This helps to identify variables with skewness or outlying data that may lead to observations with high influence. The scatterplots help to show whether any pairs of variables are highly correlated, which could cause difficulties with separating their effects for variable selection and parameter estimation. They also give a preliminary impression regarding which explanatory variables might be related to the response.

The density estimates show some marked right skewness in `nrel` and `negevent`, which may be a potential source of high influence in a regression. We also see somewhat milder skewness in their positive counterparts, and a bit of left skewness in `rosn`. In the response variable `numall`, the extreme drink count of 21 that has been mentioned in previous examples stands out. This represents a possible outlier that may have a large residual, or it could be influential by forcing variables into the model to try to explain it or by altering regression parameter estimates on some variables.

The scatterplot matrix reveals no substantial correlations among the explanatory variables. The most linear-looking relationship seems to be `prel` vs. `posevent`. However, there is considerable variability to the relationship, so there is not likely to be substantial difficulty separating their effects during variable selection or parameter estimation. Of course, problems can occur because of correlations among combinations of more than two variables, and this plot cannot depict such relationships.

Among associations between response and explanatory variables, the strongest positive relationship seems to be with `desired`, which suggests that a strong desire to drink alcohol may be followed by increased consumption. It is also interesting to note that the subject with the extreme `numall` value reported the maximum possible desire to drink (`desired` = 8, tied with a few others) and a zero value for negative relationships (tied with many others), but was otherwise not extreme in any other measurement.

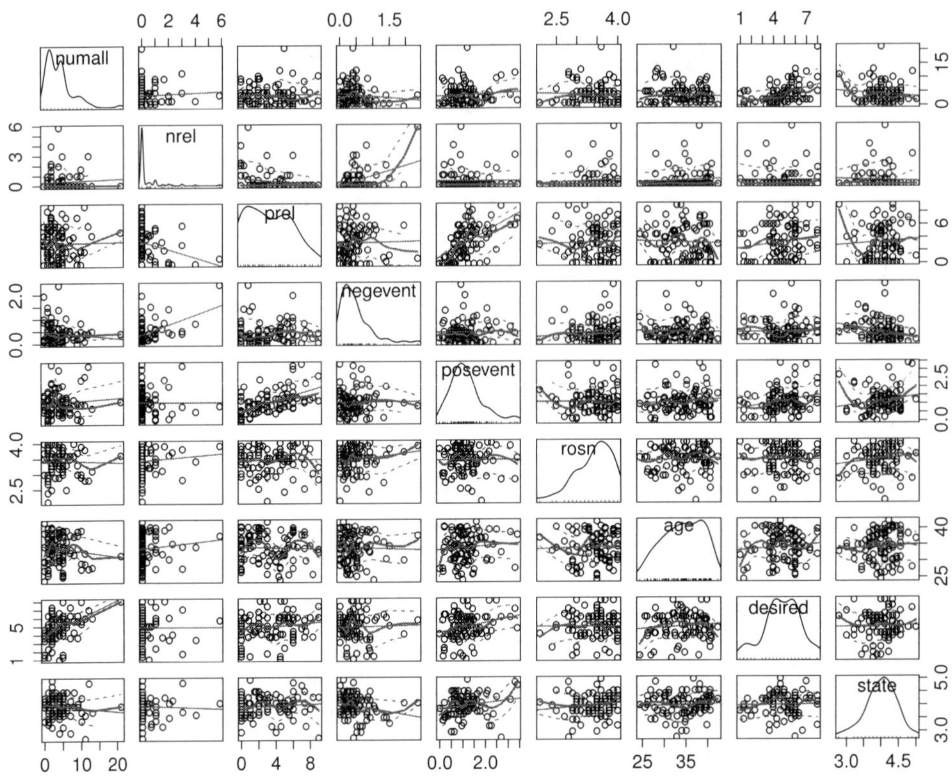

Figure 5.13: Scatterplot matrix of response (`numall`) and explanatory variables for alcohol consumption data.

Selecting variables

We use model averaging (Section 5.1.6) in several steps to identify important variables and suggest a model. There are nine explanatory variables, so an exhaustive search of the main effects is possible as a first step to see which variables seem most strongly related to the drink count. We use AIC_c as the criterion for assessing each model. The code and results from the `dredge()` function of the `MuMIn` package are below.

```
> library(package = MuMIn)
> mod.fit <- glm(numall ~ ., family = poisson(link = "log"), data =
    saturday, na.action = na.fail)
> try1 <- dredge(global.model = mod.fit, rank = "AICc")
Fixed term is "(Intercept)"
> ma1 <- model.avg(object = try1)
> # Models and posterior probabilities ("weight") are in msTable
> ma1$msTable[1:10,]
        df    logLik     AICc     delta     weight
12459    6 -208.5428 430.1099 0.000000 0.25638311
124589   7 -208.3400 432.0628 1.952904 0.09656543
124579   7 -208.4283 432.2393 2.129395 0.08840914
```

```
124569    7 -208.5338 432.4503 2.340403 0.07955682
123459    7 -208.5409 432.4645 2.354609 0.07899371
1245789   8 -208.2427 434.2853 4.175397 0.03178436
1245689   8 -208.3350 434.4701 4.360162 0.02897960
1234589   8 -208.3400 434.4800 4.370065 0.02883645
1245679   8 -208.4124 434.6247 4.514799 0.02682337
1234579   8 -208.4256 434.6513 4.541364 0.02646943
> # Variable weights
> ma1$sw
                     desired negevent state nrel age  rosn
Sum of weights:      1.00    0.99     0.93  0.93 0.92 0.26
N containing models:  256     256      256   256  256  256
                     prel posevent gender
Sum of weights:      0.26 0.24     0.24
N containing models:  256  256      256
```

There are $2^9 = 512$ models in the search. The `msTable` element from the `model.avg` output in `ma1` contains the information criteria and evidence weights for all 512 models, so we print out only the top 10. This shows that the top model contains five variables and has approximately 0.25 weight, so support for the top model is not overwhelmingly large. Note that the variables are coded according to their alphabetical order, so that "`12459`" corresponds to the variables `age`, `desired`, `negevent`, `nrel`, and `state`, respectively. The next-best models all have weights between 0.05 and 0.10, so there is at least a 2.5-to-1 preference for the top model relative to any other single model. Notice that all of these other models contain the five variables from the top model along with an assortment of other variables. The `sw` component shows the variable weights for all variables, which clearly indicate that these five variables are important, with weights of at least 0.9. The remaining variables have weights between 0.2 and 0.3, so they are not completely useless, but none are well supported by the data.

DeHart et al. (2008) were interested in possible interactions among certain variables. As described in the placekicking case study, extending the search to consider only those interactions is difficult to do using all-subsets selection. Rather than repeat the analysis from that example here, we demonstrate the process considering interactions more broadly between all pairs of variables. To do this using AIC_c, we rely on the capabilities of `glmulti()`.

Among the nine main effects, there are $9 \times 8/2 = 36$ possible pairwise interactions that can be formed, leading to a potential variable-selection problem consisting of 45 variables. This is much too large to be carried out exhaustively. Even imposing marginality on the search—i.e., requiring main effects to be included in any model that contains their interactions—still leaves too large a set of possible models to complete an exhaustive search.[9] One way around this problem is to consider interactions only among the main effects that the previous search has chosen as important. Then there are only $5 \times 4/2 = 10$ interactions, yielding a set of "only" $2^{5+10} = 32{,}768$ models. Imposing marginality reduces this set further. Running this selection takes just a few seconds and results in a best model with the five main effects, plus `state:negevent` and `age:desired`. The AIC_c for this model is 421.2, compared to 430.1 for the best main-effects-only model, so including these interactions does improve the model.

[9]Running `glmulti()` with `method = "d"` calculates the number of possible models in the search without evaluating them. This is useful for checking whether a particular problem can be computed feasibly. In this case, it runs several minutes and returns, "`Your candidate set contains more than 1 billion (1e9) models.`").

An alternative approach is to not limit the variables considered, and instead use a genetic algorithm to explore the model space randomly but intelligently. Because this is a randomized sampling process, it is not guaranteed to find the best model every time. It is possible for a given run of the algorithm to become "stuck" searching around slightly inferior models and never find the ones that perform the best. To increase the chances of finding the best models, we run the algorithm search several times. We choose to run the search four times, which is a compromise between increasing the chance of finding the best models and decreasing execution time. We retain the top 100 models from each run. These four sets of 100 models are not identical, but contain considerable overlap. Many models appear in more than one search result, but not all of the best results appear in any one search. Combining results of the multiple runs of the genetic algorithm gives a "consensus" in which the best models from each run are brought together into a final set of 100 models. Calculations of evidence weights are carried out from this consensus set. The process completed in around 4.5 minutes on a 6-core, 3.2 GHz processor with 32 GB of RAM. Abridged results are shown below and depicted in Figure 5.14. Unfortunately, these results change with each run because `glmulti()` does not recognize seeds set in R. However, by combining the results from several runs of the genetic algorithm, we hope that the combined results are reasonably stable.

```
> search.2g0.aicc <- glmulti(y = numall ~ ., data =
   saturday[,-c(1:3)], fitfunction = "glm", level = 2, marginality =
   TRUE, method = "g", crit = "aicc", family = poisson(link = "log"))
After 730 generations:
Best model: Crit= 407.129916150943
Mean crit= 425.066862696667
Algorithm is declared to have converged.

> search.2g1.aicc <- glmulti(y = numall ~ ., data =
   saturday[,-c(1:3)], fitfunction = "glm", level = 2, marginality =
   TRUE, method = "g", crit = "aicc", family = poisson(link = "log"))
After 590 generations:
Best model: Crit= 407.129916150943
Mean crit= 423.983364416536
Algorithm is declared to have converged.

> search.2g2.aicc <- glmulti(y = numall ~ ., data =
   saturday[,-c(1:3)], fitfunction = "glm", level = 2, marginality =
   TRUE, method = "g", crit = "aicc", family = poisson(link = "log"))
After 610 generations:
Best model: Crit= 406.09300880825
Mean crit= 420.313095864784
Algorithm is declared to have converged.

> search.2g3.aicc <- glmulti(y = numall ~ ., data =
   saturday[,-c(1:3)], fitfunction = "glm", level = 2, marginality =
   TRUE, method = "g", crit = "aicc", family = poisson(link = "log"))
After 760 generations:
Best model: Crit= 406.09300880825
Mean crit= 420.043673540324
Algorithm is declared to have converged.

> search.2allg.aicc <- consensus(xs = list(search.2g0.aicc,
```

```
   search.2g1.aicc, search.2g2.aicc, search.2g3.aicc), confsetsize =
   100)
> print(search.2allg.aicc)
consensus of 4-glmulti.analysis
Method: g / Fitting: glm / IC used: aicc
Level: 2 / Marginality: TRUE
From 100 models:
Best IC: 406.09300880825
Best model:
[1] "numall ~ 1 + prel + negevent + gender + rosn + age + "
[2] "desired + state + rosn:prel + age:rosn + desired:gender + "
[3] "desired:age + state:negevent"
Evidence weight: 0.0664972544357641
Worst IC: 414.234599845948
6 models within 2 IC units.
74 models to reach 95% of evidence weight.
```

It is perhaps disconcerting to note that the best model is not always the same in the four runs of the algorithm, as indicated by the different AIC$_c$ values (407.1, 407.1, 406.1, 406.1). Based on this variability, we cannot be certain that the best among these, 406.1, is truly the best value among all models. This might lead to a lack of faith in the genetic algorithm. However, it is important to remember that variable selection is inherently a very uncertain process unless the problem has one clear model that stands out above the rest. Finding the "correct" model is usually a hopeless goal; finding a good model is often possible. Further, from a practical point of view, all four runs had best models that were substantially better than what was found in the first analysis using a reduced set of variables. In fact, the *worst* of the 100 models in the consensus top 100 has AIC$_c$ = 414.2, which is still much better than the best model from the search on the restricted set of variables. This indicates clearly that (1) interactions among variables may be important, even when their main effects appear not to be, and (2) the genetic algorithm is a useful, although not perfect, tool for finding these interactions.

The variables preferred in this unrestricted search include all of those from the exhaustive search on the restricted set of variables, except `nrel`. They also include interactions `desired:gender`, `prel:rosn`, and `age:rosn`, as well as the additional main effects required by these interactions, `rosn`, `gender`, and `prel`. All of these variables have evidence weights of at least 0.94, whereas the top omitted variable, `posevent`, has evidence weight of 0.45. These variables also combine to form the top model in the consensus results from the parallel algorithm runs. Therefore, we use this model as an initial working model.

Assessing the model

To assess the fit of the chosen model, we first fit the Poisson regression using `glm()` and then use the tools from Section 5.2 on the model object. As an initial check, we look at the deviance/df GOF measure.

```
> mod.fit <- glm(formula = numall ~ prel + negevent + gender + rosn
   + age + desired + state + rosn:prel + age:rosn + desired:gender +
   desired:age + state:negevent, family = poisson(link = "log"),
   data = saturday)
> deviance(mod.fit)/mod.fit$df.residual
[1] 1.576825
```

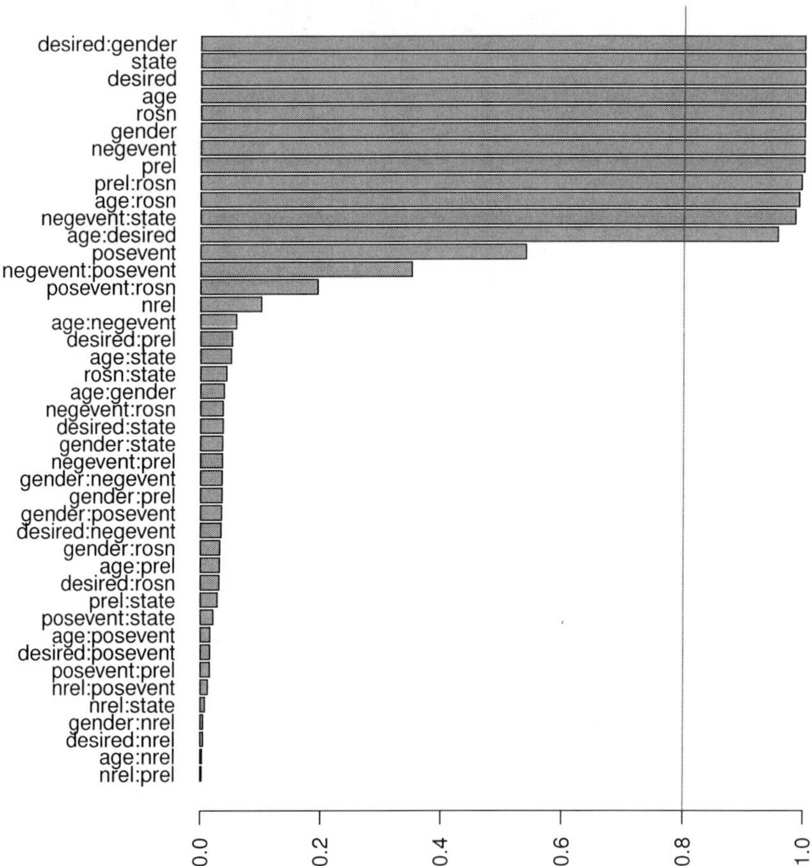

Figure 5.14: Results of genetic algorithm variable search for the alcohol consumption example using model averaging with evidence weights on all variables and two-way interactions.

```
> 1 + 2*sqrt(2/mod.fit$df.residual)
[1] 1.324443
> 1 + 3*sqrt(2/mod.fit$df.residual)
[1] 1.486664
```

The model's deviance/df ratio is 1.57, which is somewhat larger than the rough thresholds. We therefore conclude that this model *may* have some problems. To explore this further, we first check for zero inflation; that is, we check whether there are more subjects consuming no alcoholic beverages than the model expects. We compute the observed number of zero counts, n_0, along with the expected number based on the model. For a subject with estimated expected value $\hat{\mu}_i$, the probability that they consume zero drinks according to the Poisson model is $\exp(-\hat{\mu}_i)$. The expected number of zeroes according to the model is therefore $\hat{n}_0 = \sum_{i=1}^{M} \exp(-\hat{\mu}_i)$.

```
> # Estimated expected number of zeroes for each observation
> mu.poi <- exp(predict(mod.fit))
> # Estimated expected number of zeroes for Poisson model
> zero.poi <- sum(exp(-mu.poi))
> # Total zeroes in data
> zero.obs <- sum(saturday$numall == 0)
> c(zero.poi,zero.obs)
[1] 7.075871 7.000000
```

There were 7 zero counts in the data, while the model estimates that there should be about 7.08. These numbers are so close that there is clearly no evidence to suggest a problem with zero inflation. We could further confirm this by fitting the ZIP model and comparing its AIC$_c$ to that of the Poisson fit. We do not pursue this here.

Next, we check residuals and influence for individual observations. Standardized Pearson residuals are plotted against each explanatory variable's main effect and against the estimated means and linear predictors using code given in the program for this example. The results are shown in Figure 5.15. Keeping in mind that loess smoothing is unreliable near the edges of the range of the x-axis, the plots do not show any clear problems with poor choice of link or clear curvature in any explanatory variables. The loess curves generally wiggle around the horizontal line plotted at 0. The plot for `negevent` may show some curvature, but the loess curve in the right half of the plot is estimated using very few data points. It may be distorted by these points that attract the curve toward their residual values. There is one very large standardized residual (> 4) and a small number between 2 and 3. In all, 6 out of 89 lie beyond ± 2. This number is perhaps a few more than what might be expected on average to happen by chance (5% of $M = 89$ is 4.45), and there are several negative residuals just inside the threshold at -2. While it is possible that there is some mild overdispersion here, what we see in these plots overall is not necessarily unusual for a Poisson regression model fit.

We further investigate each observation with residuals beyond ± 2 by computing its exact two-tailed cumulative probability assuming a Poisson distribution with mean as estimated from the model. The code and results from this are below.

```
> saturday$mu.hat <- predict(mod.fit, type = "response")
> saturday$p.res <- residuals(mod.fit, type = "pearson")
> saturday$s.res <- rstandard(mod.fit, type = "pearson")
> saturday$lin.pred <- mod.fit$linear.predictors

> sres.prob <-  cbind(ppois(q = saturday$numall, lambda =
    saturday$mu.hat, lower.tail = TRUE), ppois(q = saturday$numall,
    lambda = saturday$mu.hat, lower.tail = FALSE) + dpois(x =
    saturday$numall, lambda = saturday$mu.hat))
> saturday$tail.prob <- apply(X = sres.prob, MARGIN = 1, FUN = min)
> saturday[abs(saturday$s.res)>2,]
    numall nrel     prel negevent posevent gender rosn
1        9  1.0 0.000000     0.40 0.525000      2  3.3
52       4  0.5 4.500000     0.40 1.150000      2  4.0
57       8  0.0 1.500000     0.30 1.050000      1  4.0
58      21  0.0 5.000000     0.40 1.200000      1  3.6
122      6  0.0 6.000000     1.05 1.483333      1  3.3
137      9  0.0 7.333333     0.20 1.133333      2  3.7
          age  desired    state    mu.hat     p.res     s.res
```

```
1   39.48528 5.666667 4.000000  4.222784 2.324746 2.416673
52  31.68241 1.000000 3.777778  1.354016 2.273920 2.403968
57  37.97399 4.333333 4.555556  3.206604 2.676829 2.852893
58  31.99179 8.000000 4.000000 14.635833 1.663538 2.173019
122 39.77276 5.666667 3.888889  2.517945 2.194384 2.281660
137 38.25051 3.666667 4.111111  2.532968 4.063405 4.228319
     lin.pred   tail.prob
1   1.4404946 0.028761147
52  0.3030751 0.048675978
57  1.1652124 0.017014077
58  2.6834729 0.068660436
122 0.9234429 0.043230517
137 0.9293918 0.001246447
```

The probabilities are all somewhat small, corroborating the interpretation using an approximate standard normal distribution for the standardized Pearson residuals. In particular, the one standardized Pearson residual beyond 4 corresponds to a person who consumed 9 drinks when the model predicted a mean of about 2.5. We cannot easily determine why the model missed this observation so badly, because the residual plots do not indicate any obvious model flaws in the neighborhood of this extreme residual. However, it likely contributed substantially to the large deviance/df statistic we observed earlier.

Finally, we perform influence diagnostics using our function `glmInfDiag()` in the program glmDiagnostics.R. The results are plotted in Figure 5.16 and shown below.

```
> source("glmDiagnostics.R")
> save.info <- glmInflDiag(mod.fit = mod.fit)
Potentially influential observations by any measures
       h hflag Del.X2 xflag Del.dev dflag Cooks.D cflag
2   0.46    **   0.01          0.01          0.000
5   0.22         2.26          4.01     *    0.050     *
39  0.14         3.83          5.11     *    0.047     *
52  0.11         5.78     *    3.98          0.052     *
57  0.12         8.14     *    6.01     *    0.085     *
58  0.41     *   4.72     *    4.39     *    0.257     *
61  0.35     *   3.02          2.73          0.123     *
108 0.32     *   1.92          2.16          0.069     *
113 0.56    **   0.18          0.19          0.018
118 0.66    **   0.12          0.12          0.018
129 0.44    **   0.05          0.05          0.003
137 0.08        17.88    **   11.25    **    0.114     *

<OUTPUT EDITED>
```

From the plots and the listings, six points stand out: four have large leverage values (observations 2, 113, 118, and 129); one has large ΔX^2_m and ΔD^2_m and moderately large Cook's distance, but low leverage (observation 137); and one has a combination of moderately large values of all measures, including a Cook's distance that stands out from the rest (observation 58). Further examination of the first four reveals that observation 2 has the largest value of `negevent` but with only a moderate estimated mean; observation 113 has a large `prel`, an odd combination of very low trait self-esteem but very high state self-esteem, and a very high estimated mean; observation 118 has the largest `prel`, the smallest `state`, and a high estimated mean; and observation 129 has moderate values of all variables but a high

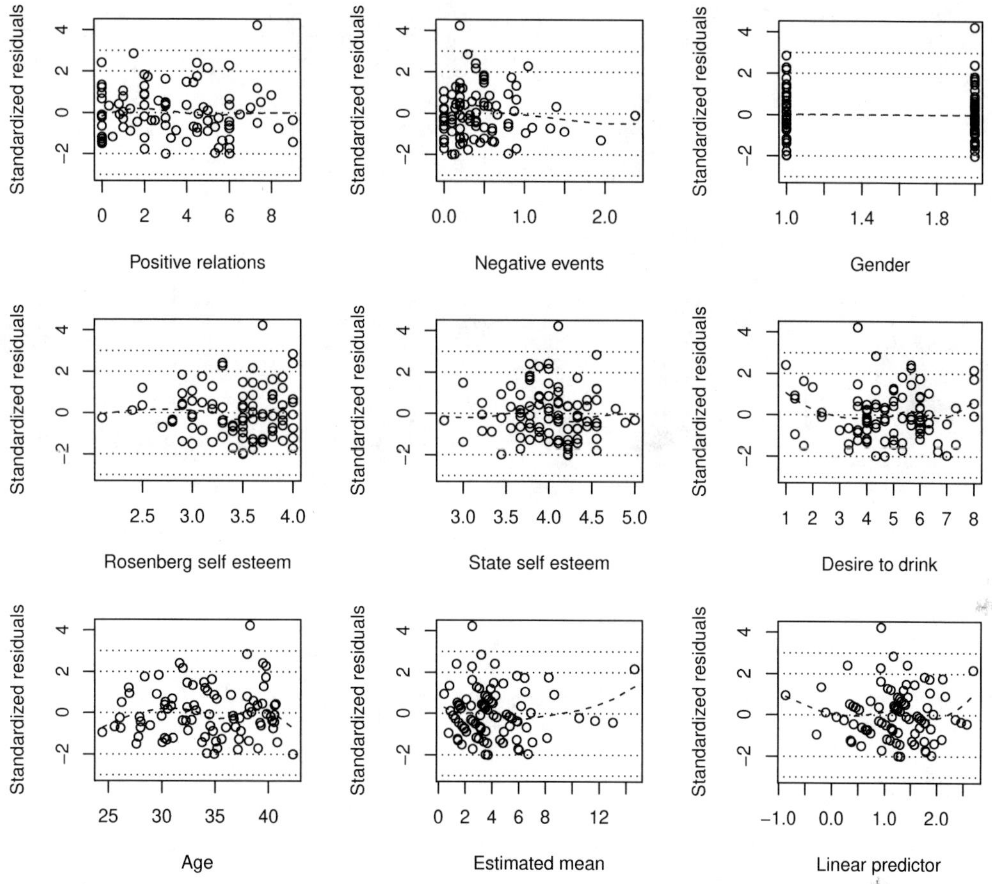

Figure 5.15: Residual plots from the working Poisson regression model for the alcohol consumption data.

estimated mean. In the cases of the high estimates, the actual count is very close to the estimated mean and the actual influence on the model parameters is not large according to Cook's distance. Therefore, we are not especially worried about these four subjects. The subject with the very large ΔX_m^2 and ΔD_m^2 is the one we examined above with the extreme residual. There is a somewhat large `prel` and a somewhat low `negevent` and `desired`. This subject appears to influence the regression mainly by having an unexpectedly high count. This person's data could be checked for accuracy. The same is true for subject 58, whose 21 drinks is by far the highest observed. This subject has the maximum values of `prel` and `desired`, and the largest estimated mean at 14.6.

To highlight these findings, we created a parallel coordinates plots (see Section 3.3) showing each observation's position relative to the rest of the data. An example for observation 58 is shown in Figure 5.17. The code to make these plots is in the program for this example.

Because we do not have access to the original log-books or to the subjects themselves, we cannot check the veracity of their data. We therefore have two options: (1) ignore the problems and proceed, or (2) rerun the analysis without one or both of the questionable observations and compare the results to the present analysis. Following the second option, we reran the genetic algorithm without observation 58 and found that the `gender` and `desired:gender` interaction were no longer very important. Their evidence weight was

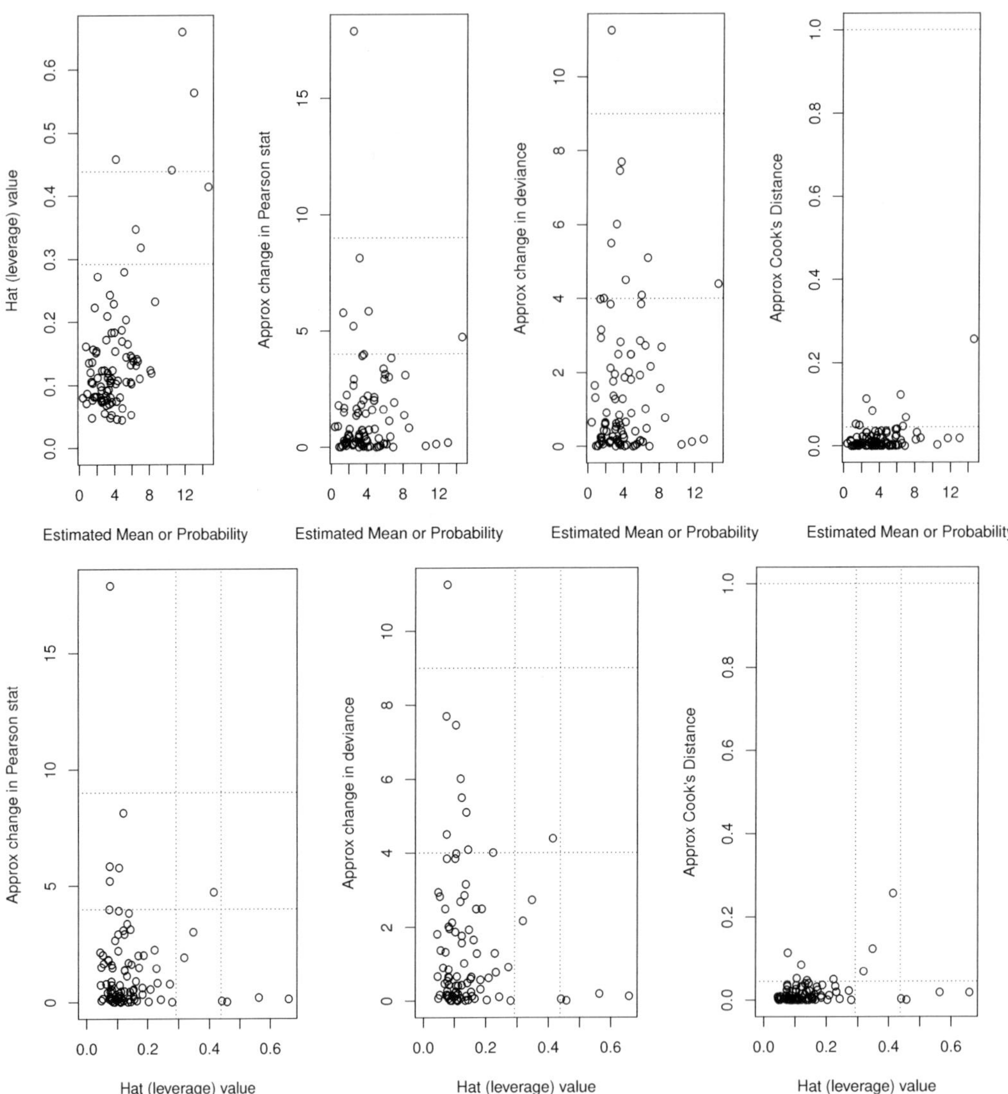

Figure 5.16: Influence diagnostic plots for alcohol consumption data.

reduced from nearly 1 to about 0.4. All other variables remained important and no new ones rose to be considered. We therefore conclude that this one observation has considerable influence on the importance of the `desired:gender` interaction. We now have additional uncertainty regarding the choice of model that would ideally be resolved in conversation with the researchers. For the remainder of this example, we assume that observation 58 is a legitimate value. We note there were recorded drink counts of 18, 15, 14, and 13 on other days not included in this data, so the 21 drinks consumed by this person on this Saturday is not as extreme relative to the larger data set.

Addressing overdispersion

Our investigation of the model fit has provided no concrete evidence of any specific flaw in our model. But the deviance/df indicator still suggests a potential problem with the

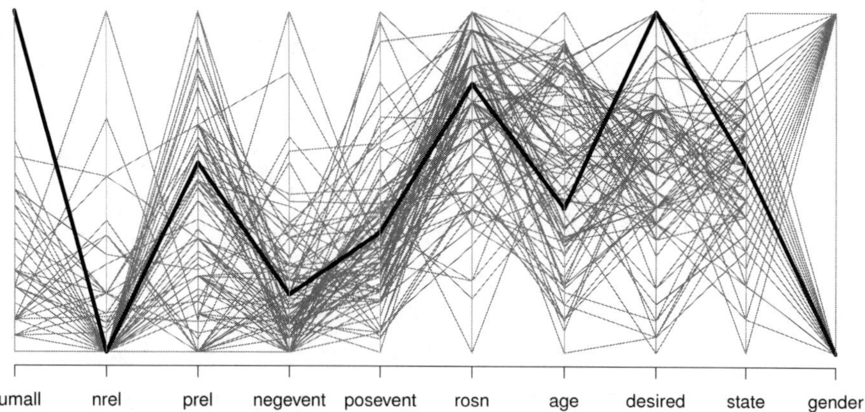

Figure 5.17: Parallel coordinate plot highlighting observation 58's position relative to the rest of the data.

model fit. Lacking any other explanation, we conclude that there may be overdispersion for some reason. Perhaps there are additional important variables that have not been measured, or maybe three-way or higher interactions could be useful. One hundred twenty different three-way interactions can be formed from 10 explanatory variables. We do not have enough data to fit them all, and we lack guidance from the researchers regarding which ones might be important. We could consider a forward-selection exercise like in the placekicking example of the previous section, although it would be somewhat complicated, since we should also reconsider all two-way interactions as well to maintain marginality. Lacking better directions to pursue, we instead try adjusting the inferences to account for the overdispersion. Even though we cannot be sure that the Poisson model is truly a poor fit for the data, making the inferences more conservative (e.g., making standard errors a little larger) helps to reflect our uncertainty that something seems amiss.

First, we assess whether a quasi-Poisson or negative binomial model is preferred. Applying the analysis from Section 5.3.3, we fit a regression of the squared residuals against the fitted means and look for curvature in the relationship.

```
> res.sq <- residuals(object = mod.fit, type = "response")^2
> set1 <- data.frame(res.sq, mu.hat = saturday$mu.hat)
> fit.quad <- lm(formula = res.sq ~ mu.hat + I(mu.hat^2), data =
    set1)
> anova(fit.quad)
Analysis of Variance Table

Response: res.sq
            Df Sum Sq Mean Sq F value    Pr(>F)
mu.hat       1  681.9  681.93 11.9460 0.0008532 ***
I(mu.hat^2)  1   20.1   20.10  0.3521 0.5544753
Residuals   86 4909.2   57.08
```

The quadratic term is clearly not important, indicating that a quasi-Poisson model is preferred. *Our next step should be to return to the variable selection stage and repeat the model-averaging assessment of variables using a quasi-Poisson model.* We have done this in the program for this example, and found that the same variables are considered important as before, although with somewhat smaller evidence weight in some cases. We therefore proceed to the analysis of the chosen quasi-Poisson model.

Analyzing and interpreting parameter estimates

Our final step is to quantify the explanatory variables' effects on the number of drinks consumed. This analysis is complicated by the fact that every variable we consider is involved in at least one interaction. We therefore need to estimate the effects of a given explanatory variable separately at different levels of its interacting variables. This involves computing certain linear combinations of parameter estimates and computing their variances accordingly.

```
> mq <- glm(formula = numall ~ prel + negevent + gender + rosn + age
    + desired + state + rosn:prel + age:rosn + desired:gender +
    desired:age + state:negevent, family = quasipoisson(link =
    "log"), data = saturday)
> round(summary(mq)$coefficients, 3)
               Estimate Std. Error t value Pr(>|t|)
(Intercept)       8.016      5.548   1.445    0.153
prel              0.474      0.210   2.255    0.027
negevent         -4.961      2.189  -2.266    0.026
gender            1.645      0.593   2.775    0.007
rosn             -2.869      1.462  -1.962    0.053
age              -0.260      0.160  -1.624    0.109
desired           1.482      0.375   3.951    0.000
state            -0.925      0.266  -3.473    0.001
prel:rosn        -0.156      0.063  -2.489    0.015
rosn:age          0.102      0.043   2.379    0.020
gender:desired   -0.294      0.106  -2.775    0.007
age:desired      -0.021      0.011  -1.802    0.075
negevent:state    1.143      0.550   2.079    0.041
```

The estimated model is $\log(\hat{\mu}) = 8.016 + 0.474\texttt{prel} - 4.961\texttt{negevent} + \ldots - 0.021(\texttt{age} \times \texttt{desired}) + 1.143(\texttt{negevent} \times \texttt{state})$. Seeing the signs and magnitudes of the coefficients helps us to understand their relationships between the response and the explanatory variables. For example, `prel` has a coefficient of 0.474, while that for `prel:rosn` is −0.156. Thus, the effect of `prel` on the number of drinks is estimated to be

$$\frac{\hat{\mu}(\texttt{prel}+1, \texttt{rosn})}{\hat{\mu}(\texttt{prel}, \texttt{rosn})} = \exp(0.474 - 0.156\texttt{rosn}).$$

Furthermore, from Figure 5.13, we see that the range of `rosn` is 2–4. Thus, when we hold all other explanatory variables constant, drinking increases as the number of positive relationship events increases for lower values of trait self-esteem (e.g., $\exp(0.474 - 0.156 \times 2) = 1.18$). However, when trait self-esteem is high, the effect of positive relationship events on drinking is reversed (e.g., $\exp(0.474 - 0.156 \times 4) = 0.86$). Code in the program for this example shows how to make a three-dimensional plot of this surface.

To further this analysis, we compute the ratio of estimated means corresponding to a 1-unit increase for an explanatory variable at each of the quartiles of the other variables (in the case of `desired:gender`, at the two levels of `gender`). For example, to further assess the effects of `prel`, we first compute the three quartiles of `rosn`, then compute $\exp(0.474 - 0.156\texttt{rosn})$. The code for doing this manually is shown below.

```
> bhat <- mq$coefficients
> rosn.quart <- summary(saturday$rosn)[c(2,3,5)]
> rosn.quart
1st Qu.  Median 3rd Qu.
    3.2     3.5     3.8
> mprel.rosn <- exp(bhat["prel"] + rosn.quart * bhat["prel:rosn"])
> mprel.rosn
  1st Qu.    Median   3rd Qu.
0.9753995 0.9308308 0.8882985
> 100*(mprel.rosn - 1)
  1st Qu.     Median    3rd Qu.
-2.460046  -6.916922 -11.170151
```

For convenience, we specified the element of the coefficient vector by its name, as in `bhat["prel"]`. At the first quartile of `rosn`, the effect of `prel` is already negative, reducing the mean number of drinks consumed by about $100(1 - 0.975) = 2.5\%$ for each 1-unit increase in `prel`. The decrease in number of drinks reaches about 11% per 1-unit increase in `prel` by the third quartile of `rosn`.

We can use these quantities to obtain profile (quasi-)likelihood confidence intervals for the true parameter values using the `mcprofile` package. See Sections 2.2.4 and 4.2.2 for previous examples using this package. The matrix of coefficients contains a row for each estimate and a column for each parameter from the model. An entry of 1 is placed in the position for the main effect for the variable whose slope we are estimating. The value of the quartile of the interacting variable goes into the position for the interaction. All other positions are set to coefficients of 0. For the confidence interval calculation, we have not specified a method for multiplicity adjustment in the code below (no `adjust` argument), so the default single-step method is used (see the description in Section 2.2.5). Finally, we re-express the parameter estimates and confidence intervals as percentage changes.

```
> library(package = mcprofile)
> # Create coefficient matrices as
> #   1*target coefficient + quartile*interaction(s)
> # Look at bhat for ordering of coefficients
> # CI for prel, controlling for rosn
> K.prel <- matrix(data =
    c(0, 1, 0, 0, 0, 0, 0, 0, rosn.quart[1], 0, 0, 0, 0,
      0, 1, 0, 0, 0, 0, 0, 0, rosn.quart[2], 0, 0, 0, 0,
      0, 1, 0, 0, 0, 0, 0, 0, rosn.quart[3], 0, 0, 0, 0),
                   nrow = 3, byrow = TRUE)

> profile.prel <- mcprofile(object = mq, CM = K.prel)
> ci.prel <- confint(object = profile.prel, level = 0.95)

> 100*(exp(ci.prel$estimate) - 1)  # Verifies same answer
     Estimate
C1  -2.460046
C2  -6.916922
C3 -11.170151
> 100*(exp(ci.prel$confint) - 1)
      lower     upper
1  -8.491358  3.843081
```

```
2 -12.856058 -0.717001
3 -18.980590 -2.814959
```

The estimates match with what we previously calculated manually. The confidence interval at the first quartile of `rosn` contains zero, so the effect of positive relationships on drinking is not clear. However, the association is decreasing for the higher quartiles of `rosn`. The code for finding confidence intervals for the other variables' effects is given in the program for this example, including how to find the confidence intervals for variables that are involved in two interactions.

Note that we have used the same data to estimate the parameters and means as we used to select the model. As the next section explains, the resulting estimates are subject to selection bias, and their standard errors may be underestimated. Confidence levels should also be interpreted with care. If model averaging could be used for parameter estimation and interpretation, we could avoid some of these biases. Unfortunately, variances for linear combinations of parameters cannot be computed accurately using the model-averaged parameter estimates, because `glmulti()` does not provide the full variance-covariance matrix of the parameter estimates. We could compute the model-averaged mean estimates and ratios, but would have no reliable estimate of their variability. For illustration purposes, we show the model-averaged analysis of individual parameters in the program for this example.

5.5 Inference after model selection

Variable selection is an extremely common and useful part of many statistical analyses, but care must be taken in performing inferences with the resulting models. This process, called *post-selection inference*, has subtle properties that researchers have only recently begun to understand and address seriously. In particular, hypothesis tests can have larger type I error rates and confidence intervals have lower confidence levels than stated when they are computed on the same data that was used to select the model (Leeb and Poetscher, 2005). The example below offers a demonstration as to why this occurs.

Example: Type I error rate after model selection (PostSelect.R)

The purpose here is to show an extreme example of why inference after model selection behaves differently from how it is designed to behave. We use a Poisson regression model with $n = 100$ and $p = 20$ explanatory variables that are randomly generated from a standard normal distribution. The response variable has a $Po(10)$ distribution and is independent of all explanatory variables. Thus, the true model is $\log(\mu) = \beta_0$. We simulate 20 such data sets. All of these settings can be adjusted in the program if desired.

For each data set, we implement all-subsets regression using BIC. If this process allows any variables into the model, we refit the selected model using `glm()` and extract the p-value from the Wald test for $H_0 : \beta_r = 0$ vs. $H_a : \beta_r \neq 0$ for each variable that was selected into the model. Below is a summary of the results:

- Across 20 data sets with 20 explanatory variables each, there are 400 opportunities for a variable to be selected by all-subsets. BIC selects only 14 (3.5%) of these—less than one per data set, on average—demonstrating its resistance to "false positive" identification of important variables.

- BIC correctly identified the model with no variables in it 8 times, leaving 12 data sets where one or more explanatory variables were included in the model. In 10 of these, only one variable was selected, while the other 2 had two.

- The p-values ranged from 0.002 to 0.031 for the 14 Wald tests for the selected variables. Thus, all 14 tests incorrectly rejected the null hypothesis that the corresponding parameter was zero at the $\alpha = 0.05$ level. This is an estimated type I error rate of 1.0. Using $\alpha = 0.01$, 6 of the 14 (42.9%) Wald tests still rejected the null hypothesis.

- Because of the equivalence between Wald tests and confidence intervals, all 14 95% confidence intervals fail to contain the true value of 0. Therefore, we have an estimated 0% confidence level when finding 95% confidence intervals for these parameters.

Other variable-selection methods and inference procedures might yield slightly different results from these. However, it is nonetheless clear that we cannot expect the usual statistical properties of inference procedures to hold on the subset of variables that are selected by some technique using the same data.

One might argue that the procedure, as a whole, had only a 3.5% type I error rate, because the remaining 96.5% of variables were correctly excluded from further consideration. However, we approach this from the perspective of how inference would be typically done in practice *after* the final model was selected. In that case, the same data would be used again to answer questions about the importance of particular variables. Testing $H_0 : \beta_r = 0$ vs. $H_a : \beta_r \neq 0$ on selected variables results in a biased process that leads to rejection with high probability, regardless of whether the underlying parameter is actually zero. Nothing new is learned from the test that the selection process has not already told us.

A statistical procedure that is used to identify potentially important relationships among data may return both real relationships and false positives. The false positives are relationships that look strong because of the unfortunate alignment of the random noise in the responses with the values of the corresponding explanatory variable. The more variables that are compared to any given set of responses, the more false positives we expect to be identified. Inferences on these relationships using the same data—with the same accidentally strong relationships between response and the incorrectly selected explanatory variables—are likely to agree with the false finding of a strong relationship. This represents a bias in the inference procedure. It happens whenever we use the same set of data to both ask and answer a question using standard inference methods.

What can be done? Statistical research is ongoing in this area; see Kuchibhotla et al. (2022) for a review. Some work looks at using simultaneous inference methods while other work looks at using conditional inference methods. These methods try to use the data both for model selection and post-selection inference by making adjustments to one or both processes, but not enough work has been done for standards of practice to be developed. An alternative approach is to collect new data. The chosen model is fit to this new data and inference can be performed without worrying about reduced levels of confidence. However, it is typically not possible to collect more data, because of time, money, or feasibility constraints.

A second alternative approach is to split the data set into two separate parts. One part is used to choose the model, and the second part is used for inference. Thus, it is like collecting new data for inference purposes without actually collecting new data.

The problem now is a much smaller sample size is used to choose the model and then to perform inferences. This could lead to missing important explanatory variables because of a reduction in power. Also, this could lead to less precision (e.g., wider conference intervals) when stating an explanatory variable's relationship with the response. While these problems are disappointing, their effects can be lessened by starting with a very large overall sample size.

Data splitting

Because splitting the data is a viable option for some data analyses, we examine it further for the remainder of this section. The two separate data sets are known by a variety of names in practice. The set on which the model is chosen is referred to as the training, modeling, or development set. The set on which the inferences are conducted is referred to as the validation, test, or hold-out set.

The validation set can be used in a number of ways to help confirm whether the results from the training set generalize well to the population. Typically, the model chosen from the training set is refit to the validation set, and inferences are carried out using the new parameter estimates. Tests and confidence intervals can be conducted, and the results can be compared to the training set to determine whether the model-selection process has potentially over-reacted to random deviations in the data that may align to appear like an actual pattern. Assuming that the model fits from the two data sets broadly agree that a model is satisfactory, a final model fit using the combined, full data set gives more precise estimates of the model parameters than either set separately (Faraway, 2016).

There are no agreed-upon recommendations on what proportion of the data to put in the training or validation sets. If the selection process uses too little data, the ability to identify real associations will be hindered, and important explanatory variables may be missed. Therefore, the training set should be relatively large—some recommendations suggest that variable selection is not effective unless $n > 6P$ or even $n > 10P$, where P again represents the number of candidate explanatory variables. On the other hand, if the validation set is too small, then hypothesis tests will have low power, and they may fail to reject null hypotheses for variables that are actually important. Similarly, confidence intervals will be wide if the validation set is too small. Thus, there is a balance needed with data splitting to ensure that both sets are large enough to achieve their goals. Our recommendation for most data analyses is to use a 50/50 split—half the data is put into the training set, and the other half is put into the validation set. This essentially ensures the same amount of precision is used with both data sets.

There are additional considerations for choosing data splits. As has been pointed out in previous chapters, the notion of a "large sample" for validity of some procedures for tests and confidence intervals depends not only on the numbers of observations, but also on how close the parameters lie to the boundary value of 0 (or 1 for a probability). For example, suppose there were a sample of size 10,000. Generally, this might seem an adequate sample size for data splitting. However, if there were only 10 failures out of 10,000 observations, there would be very little information in the data regarding how probabilities of success change with the explanatory variables—most probabilities would be very near 1. Splitting the data in such cases would invariably lead to insufficient power to detect signals in either half. Indeed, there would be a non-negligible chance that one set might contain no failures. The same issue can occur for multicategory responses (e.g., one category is rarely observed) and for count responses (e.g., a response with a very small mean where most observed counts are zero).

How to split the data depends somewhat on the nature of the data. In situations where data are collected over time, it is common to split temporally—earlier data are the training set and later data are the validation set. This mimics the way in which models are typically used on future data after being trained on available data. Of course, this assumes that no substantive changes have taken place during the interval in which the data were gathered, which might make later data different somehow from earlier data. In that case, modeling should carefully account for these changes. In other situations where data have no specific temporal index or time is not considered a relevant factor, data should be split randomly. This usually ensures that all important features and variable values are roughly equally distributed between the two sets and prevents any data selection bias from being introduced.

Example: Placekicking and inference after variable selection (Placekick2.R, Placekick2.csv)

This example shows how to split a data set into training and validation components. We also explore how one can choose a model from the training set and use this model for inferences with the validation set. Our investigations are done in a familiar setting for readers of the book: placekicking. Rather than use our previous placekicking data introduced in Chapter 2, we use a data set from the 2015 season that includes all of the same variables and also the location (`city`) of the placekick.[10] While we might ordinarily use these data as a validation set for the model we found previously, a very important rule change happened prior to the 2015 season that changed the distance of most PATs from 20 yards to 33 yards. Also, as with most sports, how football is played (strategy, player conditioning, and so forth) changes over time. The impact might not be noticeable over a small number of consecutive seasons, but in our case more than a decade passed between the two data sets. We therefore focus only on this data set and use random splitting to create two sets of roughly equal size.

There are a number of ways to perform the random split in R. Here, we use the `sample.int()` function to select $n/2$ integers between 1 and n. The chosen numbers (`index.sample`) represent the observation numbers from the original data set that are used to create the validation set. The remaining observations form the training set.

```
> # Original data set
> set1 <- read.csv(file = "c:\\data\\Placekick2.csv")
> head(set1, n = 3)
  week distance altitude home type precip wind change
1    1       44    255.9    0    1      1    0      1
2    1       33    255.9    1    1      1    0      0
3    1       46    255.9    0    1      1    0      0
   elap30 PAT field good temp72 city
1 24.7333   0     0    0     65   NE
2 11.1833   1     0    1     65   NE
3  8.7167   0     0    0     65   NE

> n <- nrow(set1)
> # Select random half of numbers between 1 and n
> set.seed(31010742)
> index.sample <- sample.int(n = n, size = round(n/2))
> # Form the two data sets
> set.train <- set1[-index.sample,]
> set.valid <- set1[index.sample,]
```

[10] Data kindly provided by Robert Fitzsimmons, University of Pennsylvania.

```
> n
[1] 2203
> nrow(set.train)
[1] 1101
> nrow(set.valid)
[1] 1102
> head(sort(index.sample))
[1]  2  4  5  7  9 10
> head(index.sample)
[1] 1222  583 2079 1664  644  100

> head(set.train, n = 3)
  week distance altitude home type precip wind change
1    1       44    255.9    0    1      1    0      1
3    1       46    255.9    0    1      1    0      0
6    1       33    255.9    1    1      1    0      0
   elap30 PAT field good temp72 city
1 24.7333   0     0    0     65   NE
3  8.7167   0     0    0     65   NE
6 25.6167   1     0    1     65   NE

> head(set.valid, n = 3)
     week distance altitude home type precip wind change
1222   10       23    723.7    0    1      0    0      1
583     5       31     36.3    0    1      0    1      0
2079   17       19      8.6    1    1      0    0      0
      elap30 PAT field good temp72 city
1222 23.1500   0     1    1     60  PIT
583  25.1167   0     1    1     85   TB
2079  1.9333   0     1    1     77  MIA
```

The training set `set.train` is the original set `set1` with observation numbers in `index.sample` removed. The validation set `set.valid` contains the observations in `set1` corresponding to the observation numbers in `index.sample`.

We follow the same process as described in Section 5.4.1 to find the best model with the training set, using AIC as the criterion for variable selection. The details are provided in the corresponding program. Below is the output from the chosen model.

```
> mod.fit.train <- glm(formula = good ~ distance + altitude,
    family = binomial(link = logit), data = set.train)
> summary(mod.fit.train)

Call:
glm(formula = good ~ distance + altitude, family = binomial(link
    = logit), data = set.train)

Coefficients:
              Estimate Std. Error z value Pr(>|z|)
(Intercept)  6.5186787  0.5105483  12.768   <2e-16 ***
distance    -0.1172706  0.0124135  -9.447   <2e-16 ***
altitude     0.0001947  0.0001378   1.413    0.158
---
```

```
Signif. codes:  0 '***' 0.001 '**' 0.01 '*' 0.05 '.' 0.1 ' ' 1

(Dispersion parameter for binomial family taken to be 1)

    Null deviance: 741.41  on 1100  degrees of freedom
Residual deviance: 648.90  on 1098  degrees of freedom
AIC: 654.9
Number of Fisher Scoring iterations: 5

> library(car)
> Anova(mod.fit.train)
Analysis of Deviance Table (Type II tests)

Response: good
       LR Chisq Df Pr(>Chisq)
distance   91.796  1     <2e-16 ***
altitude    2.377  1     0.1232
---
Signif. codes:  0 '***' 0.001 '**' 0.01 '*' 0.05 '.' 0.1 ' ' 1

> confint(mod.fit.train)
Waiting for profiling to be done...
                    2.5 %        97.5 %
(Intercept)  5.537973e+00  7.5425600627
distance    -1.419330e-01 -0.0931870158
altitude    -4.767481e-05  0.0005023618
```

The model includes only `distance` and `altitude`. The LRT p-value from the hypothesis test for the regression parameter corresponding to `distance` is very small, leaving little doubt about the importance of this variable. The 95% LR confidence interval for this parameter is $(-0.142, -0.093)$. The LRT for `altitude` results in a p-value of 0.12, which is a little larger than what would normally be deemed marginally significant. This can occur when AIC is used for variable selection, because it is a more permissive criterion (has a smaller penalty for additional variables) than other ICs. The 95% LR confidence interval is $(-0.00005, 0.00050)$, which is mostly positive with a lower limit slightly below 0.

We fit the same model to the validation set.

```
> mod.fit.valid <- glm(formula = good ~ distance + altitude,
    family = binomial(link = logit), data = set.valid)
> summary(mod.fit.valid)

Call:
glm(formula = good ~ distance + altitude, family = binomial(link
    = logit), data = set.valid)

Coefficients:
             Estimate Std. Error z value Pr(>|z|)
(Intercept)  6.696e+00  5.157e-01  12.984   <2e-16 ***
distance    -1.182e-01  1.246e-02  -9.483   <2e-16 ***
altitude     1.583e-05  1.076e-04   0.147    0.883
---
Signif. codes:  0 '***' 0.001 '**' 0.01 '*' 0.05 '.' 0.1 ' ' 1
```

```
(Dispersion parameter for binomial family taken to be 1)

    Null deviance: 702.32  on 1101  degrees of freedom
Residual deviance: 609.06  on 1099  degrees of freedom
AIC: 615.06

Number of Fisher Scoring iterations: 5

> Anova(mod.fit.valid) Analysis of Deviance Table (Type II tests)

Response: good
         LR Chisq Df Pr(>Chisq)
distance   92.878  1     <2e-16 ***
altitude    0.022  1     0.8823
---
Signif. codes:  0 '***' 0.001 '**' 0.01 '*' 0.05 '.' 0.1 ' ' 1

> confint(mod.fit.valid)
Waiting for profiling to be done...
                    2.5 %        97.5 %
(Intercept)  5.7081932803  7.7332123461
distance    -0.1429722847 -0.0940263004
altitude    -0.0001803199  0.0002471711

> data.frame(train = round(mod.fit.train$coefficients, 6), valid
   = round(mod.fit.valid$coefficients, 6))
                train     valid
(Intercept)  6.518679  6.696119
distance    -0.117271 -0.118195
altitude     0.000195  0.000016
```

The estimated regression parameter, standard error, test statistic, and confidence interval for `distance` are very similar for both the training and validation sets. This is typical for variables whose importance is so strong that they need no "good luck" from the data to be found using variable-selection techniques. The same cannot be said though for `altitude`. In the validation set, the LRT statistic is much smaller than it was in the training set; the p-value is much larger; and the confidence interval, $(-0.00018, 0.00024)$, is much more evenly spread around 0 than in the training set. Thus, it is apparent that the selection of `altitude` into the model may have been partly influenced by a fortunate random alignment of its values with the successes and failures in the training set.

Because the training and validation set do not agree on `altitude`, we are very hesitant to combine the two data sets to obtain one final model and perform inferences on important parameters. Instead, we would focus on interpreting distance with the validation set by examining odds ratios and corresponding confidence intervals.

This example further highlights the considerable amount of uncertainty that accompanies variable selection. If we wanted to gain greater certainty about which variables truly ought to be included into the model, we could repeat the data-splitting process several times and get a composite impression of which variables are always, sometimes, or never included. We would still be left with a decision about which of the "sometimes" variables to include in the model. Complex methods using multiple levels of splitting have been used to determine these thresholds (Freeman et al., 2009).

However, because this process uses all of the data across different splits, there are no data left for valid inference. In these cases, if the size of the data set permits, a validation set specifically for inferences—primarily for valid confidence intervals—can be held out prior to beginning variable selection on the remaining data.

Exercise 43 examines this placekicking data further. In particular, the exercise uses the more restrictive BIC for variable selection instead of AIC. This results in a model that only includes `distance`. Also, in the data split, the roles of the training and validation sets are reversed, demonstrating some of the uncertainty with variable selection described above.

5.6 Exercises

1. The bias-variance tradeoff described on p. 332 is sometimes difficult to picture. The program LogisticRegBiasVarianceSim.R contains a simulation designed to demonstrate that it can sometimes be better to use a model that is too small rather than one that is too large, or even the correct model! The program simulates binary data, $Y_1, \ldots, Y_n$, with probability of success π_i defined by $\text{logit}(\pi_i) = \beta_0 + \beta_1 x_{i1} + \beta_2 x_{i2} + \beta_3 x_{i3}$, where each x_{ij} is selected from a uniform distribution between 0 and 1. The values of the regression parameters are initially set to $\beta_0 = -1, \beta_1 = 1, \beta_2 = 1, \beta_3 = 0$, so that the correct model has just x_1 and x_2. The program fits three models with variable sets $\{x_1\}, \{x_1, x_2\}$, and$\{x_1, x_2, x_3\}$, and compares the models' estimated probabilities to the true probabilities over a grid of (x_1, x_2, x_3) combinations. The average bias, variance, and mean squared error (MSE = bias^2 + variance) for the predictions are calculated at each (x_1, x_2, x_3) combination over 200 simulated data sets. The closer to 0 that each of these three summary quantities is, the better the model.

   (a) Run the model using the default settings, which includes $n = 20$. Examine the plots of bias, variance, and MSE as well as the averages for these quantities printed out at the end of the program. Which models perform best and worst with respect to each quantity? Is the correct model ("Model 2" in the program) always the best? Explain.

   (b) Increase the sample size to 100 and repeat the analysis. How do the bias, variance, and MSE change for the three models? Do the best and worst models change?

   (c) Increase the sample size to 200 and repeat the analysis again. Answer the same questions as given in (b).

   (d) Using $n = 20$, try increasing and decreasing the values of β_1 and β_2, maintaining $\beta_0 = -(\beta_1 + \beta_2)/2$ so that probabilities remain centered at 0.5 (the `logit` summary results should remain symmetric around zero). Can you draw conclusions about how each parameter's size affects bias, variance, and MSE?

2. Split the placekick data set, Placekick.csv, approximately in half and repeat all-subsets variable selection using AIC$_c$ on each half. Determine whether results are similar in the two halves. Partial code for doing this is shown below (assuming that data are contained in the data frame called `placekick`). Note that `get.models()` extracts

fitted models from the object returned by dredge(), and subset specifies the rank(s) of the model(s) requested.

```
set.seed(32876582)
n <- nrow(placekick)
# Select random half of numbers between 1 and n
set1 <- sample.int(n = n, size = round(n/2))

# Create named data frames for each half because dredge()
#   requires named data sets without subsetting
placekick1 <- placekick[set1,]
# Remove set1 to create the other half
placekick2 <- placekick[-set1,]

library(package = MuMIn)

mod.fit1 <- glm(good ~ ., family = binomial(link = "logit"),
   data = placekick1, na.action = na.fail)
try1 <- dredge(global.model = mod.fit1, rank = "AICc")

# Get best model
get.models(object = try1, subset = 1)
```

3. Refer to the definitions of AIC, AIC$_c$, and BIC in Section 5.1.2. Construct a plot comparing the penalty coefficients (k) against number of variables (p) for the three criteria with $n = 10$ and $p = 1, 2, \ldots, n/2$. Considering that a larger penalty coefficient causes an information criterion to prefer a smaller model, what do these results imply for models selected by these three criteria? Repeat for $n = 100$ and $n = 1000$. Do the conclusions change?

4. Refer to Exercise 7 in Chapter 2 for the placekicking data collected by Berry and Wood (2004). Using Distance, Weather, Wind15, Temperature, Grass, Pressure, and Ice as explanatory variables and Good as the response, perform model selection in the following ways.

   (a) Compare which variables are selected under forward selection, backward elimination, and stepwise selection using BIC.

   (b) Perform BMA using BIC. Which variables seem important and which seem clearly unimportant?

   (c) Estimate the regression parameters for the stepwise analyses. Compare these estimates to those of BMA. How are they different?

   (d) Explain why the regression parameter estimate for Grass using BMA is somewhat closer to 0 than it is using stepwise.

5. Continuing Exercise 4 in Chapter 2, consider the model using temperature as the only explanatory variable.

   (a) Refit the model using a probit and a complementary log-log link instead of a logit link. Plot all three curves with the data, using a temperature range of 31° to 81° on the x-axis. Does one appear to fit better than the others? Do they provide similar probabilities of failure at 31°?

(b) Compare BIC values for the three different links. Which link appears to fit best?

(c) Compute the BMA posterior probabilities for the models using these three links. Comment on the uncertainty of the model selection among these three links.

6. Refer to the example and program on all-subsets regression using AIC from Section 5.1.3 with the placekicking data set. Run the genetic algorithm using AIC_c and include pairwise interactions in the model with `marginality = TRUE` (no pairwise interactions between variables are considered unless both variables are in the model as well). Note that the function will take a few minutes to run.

    (a) Report the best model and its AIC_c. Compare this to the AIC_c from the best model found without interactions. Do the interactions improve the model?

    (b) How many models are within 2 AIC_c units of the best model? What does this imply about our confidence in having found the best model?

    (c) How many generations did the algorithm produce?

    (d) Rerun the algorithm. Re-examine the same items as requested in (a)–(c) and compare to what you obtained initially.

    (e) Rerun the algorithm, increasing the mutation probability to 0.01 (add the argument value `mutrate = 0.01`; the default is 0.001). Report the results as requested in parts (a)–(c). Are the results any different from those in (a)–(c)? Note that in this context, increasing the mutation probability increases the probability that a variable is randomly added to or removed from a model. In general, increasing the mutation probability can make the algorithm better at finding the top models, but can increase run time.

7. Section 5.1.4 includes an example involving the placekicking data set and alternating stepwise selection of its variables for a logistic regression model. Run the alternating stepwise selection program referred to in the example and report the results. For the final model, compute how much BIC would increase with each possible change that could be made to the model.

8. Refer again to the example referenced in Exercise 7 and its program using alternating stepwise selection. Run the alternating stepwise selection code using `k = 0`.

    (a) Notice that all variables are in the final model. Why must this happen?

    (b) In the final model, the variables are in order relative to how much $IC(0)$ would change if the variable were dropped. Notice that this is not exactly the same as the order in which the variables were entered. Explain why this can happen.

9. Refer to the hospital-visit data described in Exercise 16 of Chapter 4. Consider the problem of trying to predict whether a person has private insurance based on their pattern of health-care usage. This suggests a logistic regression model for `privins`, a binary variable. Use the LASSO to identify which of the remaining variables relate to the probability that a person has private insurance. Interpret the results; in particular, estimate the effect of each explanatory variable on the odds of private insurance.

10. A criminologist studying capital punishment was interested in identifying whether certain social, economic, and political attributes of a country related to its use of the death penalty. She gathered data from public sources on 194 countries, recording the following variables:

- `COUNTRY`: The name of the country
- `STATUS`: The country's laws on death penalties, coded as `a` = Public executions; `b` = Private executions; `c` = Executions allowed but none carried out in last 10 years; `d` = No death penalty
- `LEGAL`: The basis for the country's legal system, coded as `a` = Islamic; `b` = Civil; `c` = Common; `d` = Civil/Common; `e` = Socialist; `f` = Other
- `HDI`: Human Development Index, a numerical measure ranging from 0 (very low human development) to 1 (very high human development)
- `GINI`: GINI Index of income inequality, a numerical measure ranging from 0 (perfect equality) to 1 (perfect inequality)
- `GNI`: Gross National Income per capita (US$)
- `LITERACY`: Literacy rate (% of adults aged 15 and above who are literate)
- `URBAN`: Percentage of total population living in urban areas
- `POL`: Political Instability Index for the level of threat posed to governments by social protest, a numerical measure ranging from 0 (no risk) to 10 (high risk)
- `CONFLICT`: Level of conflict experienced in the country, coded as `a` = No conflict, `b` = latent (nonviolent) conflict, `c` = manifest conflict, `d` = crisis, `e` = severe crisis, `f` = war

The data for the 141 countries with complete records are available in the DeathPenalty.csv file.[11] The response variable is the country's death penalty status, while all other variables except `COUNTRY` are explanatory. Note that the large difference in scale between `GNI` and other variables can cause problems with some model fitting functions. Change the scale to thousands of dollars (`capital$GNI <- capital$GNI/1000`) to resolve the issue.

(a) Fit a multinomial (nominal response) regression model to these data using all available explanatory variables as linear terms. Compute the AIC and the BIC for this model. Also compute AIC_c by using the formula on p. 334. Note that the model degrees of freedom that are needed for computing AIC_c can be found from the `extractAIC()` function.

(b) Explain why `STATUS` can be viewed as an ordinal variable.

(c) Fit a proportional odds regression model (see Section 3.4) to these data using all available explanatory variables as linear terms. Compute AIC, BIC, and AIC_c as in the multinomial regression model.

(d) Note that the proportional odds model assumes both the ordinality of the response and equal coefficients across different logits. Based on the information criteria, is there evidence to suggest that these added assumptions lead to a worse model fit than treating `STATUS` as nominal? Explain.

11. Continuing Exercise 10, the researcher's main interest was to determine which, if any, of the explanatory variables are associated with the country's use of the death penalty. Complete the items below with this in mind.

(a) Explain why a model-averaging approach would be more appropriate than a stepwise approach given the context of this problem.

[11] Data courtesy of Diana Peel, Department of Criminology, Simon Fraser University.

(b) Treating the response as nominal, carry out the model-averaging analysis using a multinomial model with main effects only and AIC_c. Report the results and draw conclusions.

12. Repeat the analysis from Exercise 11 using a proportional odds regression model.

    (a) For each variable, report the posterior probability that it belongs in the model and interpret the results.
    (b) Obtain model-averaged estimates of the odds ratios corresponding to the effect of each explanatory variable. Interpret these values.
    (c) Compute 95% confidence intervals for each odds ratio and interpret. In particular, which variables appear to be most important in describing countries' use of the death penalty?
    (d) Repeat part (c) on the odds ratios from a single proportional odds model using all explanatory variables. Compare the results to the model-averaged values.

13. Refer to the hospital-visit data described in Exercise 16 of Chapter 4. Use BMA with a BIC criterion on a ZIP model to identify which of the explanatory variables are related to the number of physician office visits for a person (`ofp`) and the probability of zero visits. Note that `health_excellent` and `health_poor` are really just two levels of a three-level factor. Combine them into a single new variable with three levels, 1, 0, and -1, by taking `health = as.factor(health_excellent - health_poor)`. Include this new variable in the search instead of the separate two indicator variables.

    For each variable, report:

    (a) The estimated probability that it belongs in the model
    (b) Its parameter estimates and confidence intervals in both parts of the model

    Draw conclusions from your investigation.

14. Repeat Exercise 13 including pairwise interactions into the list of variables, maintaining marginality of the model. Note that this will require using the genetic algorithm. We suggest turning off the plotting (`plotty = FALSE`) because it produces an enormous amount of graphical output that may crash some systems. Note that this may take several hours to run.

15. Using the placekick data (Placekick.csv) from the examples in Section 5.1, fit a logistic regression model with variables `distance` and `PAT`. Save the results as an object, say `mod.fit`.

    (a) Execute `summary(mod.fit)` and notice that there is a value called `AIC` in the output. Report this value. Report also the residual deviance.
    (b) Obtain the AIC and residual deviance by extracting them from the model fit object using `mod.fit$aic` and `mod.fit$deviance`. Extract the number of parameters in the model using `mod.fit$rank`. Using the resulting values, compute the AIC and the AIC_c.
    (c) The functions `AIC()` and `extractAIC()` can compute values of $IC(k)$ for given values of k. Use these functions to compute the model deviance, the AIC, the BIC, and the AIC_c. Note that for AIC_c you will need to use the value of `mod.fit$rank` from the previous part in the computation of the penalty parameter `k`.

16. Refer to the discussion of residuals for Poisson models on p. 357. Demonstrate how well the standard normal approximation holds for large residuals from a Poisson model with different means. Specifically, for counts `y <- c(0:10)`, and for mean `yhat <- 2`:

    (a) Compute Pearson residuals `pear <- (y-yhat)/sqrt(yhat)`.

    (b) Compute both the Poisson and the standard normal cumulative probabilities for a *larger* value than the Pearson residual, `pp <- 1-ppois(y,yhat)` and `pn <- 1-pnorm(pear)`, respectively.

    (c) Use `cbind(y,pear,pp,pn)` to print the results. Compare these probabilities for any Pearson residual above 2, 3, and 4, the guidelines given for interpreting "large" residuals. Are the probabilities similar? Are there any other Pearson residuals below 2 that have probability < 0.05? Are there any other Pearson residuals above 2 that have probability > 0.05?

    (d) From parts (a)–(c), what can you conclude about interpreting Pearson residuals from models where $\hat{w} = 2$?

    (e) Repeat these steps for $\hat{w} = 1, 0.5, 0.25, 0.1$. Comment on whether you are comfortable using the 2, 3, and 4 guidelines to identify large Pearson residuals in these cases.

17. Continuing Exercise 4, use the variables `Distance` and `Grass` as suggested by the BMA. Include the interaction of these two variables in the model and perform diagnostics as follows:

    (a) Plot standardized Pearson residuals against distance, the estimated probabilities, and the linear predictor. Interpret the results.

    (b) Perform the HL test using 10 groups. State hypotheses, test statistic, p-value, and conclusions. Also, examine the Pearson residuals from the groupings and indicate whether they show any particular pattern.

    (c) Perform the Osius-Rojek and Stukel tests and draw conclusions.

    (d) Conduct an influence analysis. Identify any influential observations and indicate how they are influential. Interpret the results.

    (e) Write a brief summary of the results of the diagnostics and conclude whether the model is reasonably good for the data.

18. Margolin et al. (1981) provide data showing the number of revertant colonies of *Salmonella* in response to varying doses of quinoline. There are six different dose levels, and three independent observations per dose. The data are provided in the data frame `salmonella` from the `aod` package.

    (a) Notice that the spacings among the dose levels are approximately constant on a log scale. Plot the response (`y`) against both dose and base-10 log-dose (`log10(y+1)`) and comment on any apparent trends. Note that we use base 10 here because the dose levels 10, 100, and 1,000 are simply 1, 2, and 3 in base-10 log. The +1 is used in the log here because $\log_{10} 0$ is not defined, but $\log_{10} 1 = 0$. In this problem all logs should be taken this way.

    (b) Estimate three separate Poisson regression models for `y` using different interpretations of dose as the explanatory variable: one with dose as a linear term, one with log-dose as linear, and one with a categorical version of dose (treat dose as a six-level categorical factor by using `factor(dose)` in the `formula`). Test the effect of dose in each model. What conclusions can you draw about the relationship between the number of revertant colonies and dose of quinoline based on these tests?

(c) Examine the deviance/df statistics for each model. Do the model fits appear to be good? Explain.

(d) Plot the standardized residuals against fitted means for the two models that use dose in a numerical manner. Comment on what these plots tell you.

(e) Plot the standardized residuals against fitted means for the model treating dose as categorical. There is no concern about the linearity of the response or the appropriateness of the link function here. What does this plot suggest?

19. Refer to the Challenger data from Exercise 4 in Chapter 2.

    (a) Fit a logistic regression model including only temperature as an explanatory variable, *without first aggregating the counts and trials by temperature.* Plot standardized Pearson residuals against $\hat{\pi}$. On the plot, separately identify all the points corresponding to $w = 0, 1$, or 2 successes. What do you notice about these three sets of points?

    (b) Aggregate the data by temperature and refit the model for the remainder of the exercise. Examine the deviance/df statistic.

    (c) Conduct a graphical assessment of the residuals. Discuss the fit of the model.

    (d) Examine the goodness of fit tests for the model and draw conclusions.

    (e) Conduct an influence analysis and draw conclusions.

        i. Explain why h_1 is so large.

        ii. Refit the logistic regression without observation 1 in the data set. For both the model with and without this observation, compare confidence intervals for the parameters, and plot the estimated probability of O-ring failure against temperature on one plot. Do the results change in a meaningful way?

20. Refer to the death penalty data from Exercise 10. Consider a binary interpretation of the response for presence or absence of death penalty, "Allows Death Penalty" (`STATUS` is `a`, `b`, or `c`) versus "No Death Penalty" (`STATUS` is `d`). Use a logistic regression model consisting of the variables `LEGAL`, `HDI`, and `GNI` without interactions. Assess the model fit and identify any concerns.

21. Refer to Exercise 12 in Chapter 4. Fit the nominal homogeneous association model and each of the three models from part (a) of that exercise. Plot each model's standardized residuals vs. political ideology, scored as 1-2-3-4-5 on the x-axis for the plots. Which of these models fit the data reasonably well?

22. Agresti (2019) provides data on the social behavior of horseshoe crabs. These data are contained in the HorseshoeCrabs.csv file available on our website. Each observation corresponds to one female crab. The response variable is `Sat`, the number of "satellite" males in her vicinity. Physical measurements of the female—`Color` (4-level ordinal), `Spine` (3-level ordinal), `Width` (cm), and `Weight` (kg)—are explanatory variables.

    (a) Fit a Poisson regression model with a log link using `Width` and `Weight` as linear terms and `Color` and `Spine` as nominal categorical variables. Test their significance and summarize results.

(b) Compute deviance/df and interpret its value.

(c) Plot standardized Pearson residuals against estimated means and against each explanatory variable. Using `jitter(as.numeric(...))` on the factor variables in the plots can help with seeing individual points that would otherwise completely overlap on the respective plots. This function adds a small amount of noise to the plotted points so that they are separated on the plot. Identify any potential problems with the model.

23. Continuing Exercise 22, conduct an influence analysis. Interpret the results.

24. Continuing Exercise 22, notice that there is one crab with a weight that is substantially different from the rest. This can be seen, for example, in a histogram of the weights. Remove this crab from the data and repeat the steps from Exercise 22. Has this fixed any problems with the model? Are there any other problems with the model, and what could be done to solve these problems?

25. Refer to Exercise 19 from Chapter 4 on the salamander count data.

    (a) Compute deviance/df. What does this value suggest about the model?

    (b) Examine the standardized residuals and perform an influence analysis. Interpret the results.

26. Refer to Exercise 27 from Chapter 4 on the alcohol consumption data, in which a Poisson regression model was fit using first Saturday drink consumption as the response and `prel`, `nrel`, `posother`, `negother`, `age`, `rosn`, and `state` as the explanatory variables. Examine the fit of this model and draw conclusions.

27. Refer to the wheat kernels example from Section 3.3. Examine the fit of the model with the six explanatory variables. Use the `multinomDiag()` function contained in our R program multinomDiagnostics.R to help with your calculations.

28. Refer to Exercise 16 from Chapter 4. Fit the ZIP model with all listed variables as linear terms.

    (a) Plot the Pearson residuals against each explanatory variable and interpret the results.

    (b) Add a quadratic term for `numchron` to the model. Is this term significant? How well does the model fit now? Explain.

    (c) Suggest how the model might be improved further.

29. Refer to the example on p. 365, where we performed the HL test on the placekick data with `distance` as the explanatory variable. Repeat the test using $g = 5, 6, \ldots, 12$ groups. Report the results and comment on the sensitivity of the test to the number of groups.

30. Refer to the example on p. 374 regarding simulating data with overdispersion. In general, consider a model in which $Y \sim Po(\mu)$, where μ has a normal distribution with mean τ and variance σ^2. Using the fact that $E(Y) = E_N[E_P(Y|\mu)]$ and $Var(Y) = E_N[Var_P(Y|\mu)] + Var_N[E_P(Y|\mu)]$, where E_N, Var_N are taken with respect to the normal distribution for the mean μ and E_P, Var_P are taken with respect to the Poisson distribution for the response Y, show that $E(Y) = \tau$ and $Var(Y) = \tau + \sigma^2$.

31. Refer to the example on "Beetle egg-laying response to crowding" in Section 4.4.

(a) Plot standardized residuals against temperature for the final model used in the example (contained in `zip.mod.t0`) to check for overdispersion and comment on the results. When plotting the Pearson residuals versus temperature, the `jitter()` function will be useful. This function adds a small amount of random noise to the temperature values, which then helps to distinguish between similar Pearson residuals at the same temperature.

(b) The zero-inflated negative binomial model is the analog to the ZIP regression model for overdispersed counts. Fit a zero-inflated negative binomial model (`dist = "negbin"` in `zeroinfl()`). Compare the parameter estimates and confidence intervals to those from the ZIP model. Check this model for signs of overdispersion.

32. Continuing Exercise 13 from Chapter 4, complete the following parts.

(a) Find the sample mean and variance for each day of the week. Is there evidence of overdispersion? Explain.

(b) Create a plot of $(y - \hat{\mu})^2$ vs. $\hat{\mu}$ as suggested in Section 5.3.3 to examine whether a quasi-Poisson or negative binomial model would be most helpful to account for overdispersion. According to this plot, which model would be an appropriate alternative to the Poisson regression model?

(c) Regardless of the outcome in (b), fit both the quasi-Poisson and negative binomial regression models and complete all of the parts given in Exercise 13c from Chapter 4.

(d) Consider the ANOVA model $Y_{ij} = \mu + \alpha_i + \epsilon_{ij}$ where ϵ_{ij} are independent, normally distributed errors with mean 0 and variance $\sigma^2 > 0$. This is the usual linear model found in a one-way analysis of variance. Below is the code that can be used to find the ANOVA table and perform multiple comparisons, where `starbucks` is the name of our data frame.

```
mod.fit.anova <- aov(formula = Count ~ Day, data = starbucks)
summary(mod.fit.anova)

library(package = emmeans)
calc.est <- emmeans(mod.fit.anova, specs = ~ Day)
test.info <- contrast(object = calc.est, method = "pairwise")
confint(object = test.info, adjust = "none", level = 0.95)
confint(object = test.info, adjust = "tukey", level = 0.95)
```

Compare these results to those obtained using the Poisson, quasi-Poisson, and negative binomial regression models.

(e) Compute AIC on each model examined for this exercise. Which one seems best according to this criterion?

33. For the Salmonella data from Exercise 18, there may be signs of overdispersion. Consider only the model that treats dose as a categorical explanatory variable.

(a) Repeat the analysis from the example on p. 387 to determine whether a quasi-Poisson or a negative binomial is a better option. Specifically, create a plot like Figure 5.9 and compute the quadratic regression as shown in the example. What do you conclude about the two model options?

(b) Fit the quasi-Poisson model to the data and test the significance of the dose effect. How do the results compare to those from Exercise 18?

(c) Plot standardized residuals from the quasi-Poisson model against the estimated means. Comment on the results.

34. For the horseshoe crab data from Exercise 22, there may be signs of overdispersion.

(a) Repeat the analysis from the example on p. 387 to determine whether a quasi-Poisson or a negative binomial is a better option. Specifically, create a plot like Figure 5.9 and compute the quadratic regression as shown in the example. What do you conclude about the two model options?

(b) Fit the quasi-Poisson model to the data and test the model effects again. How do the results compare to those from Exercise 22?

(c) Plot standardized residuals from the quasi-Poisson model against the estimated means. Comment on the results.

35. Kupper and Haseman (1978) vaguely describe an experiment in which pregnant female mice were assigned to one of two groups, labeled "treatment" and "control." The data are available in the `mice` data frame of the `aod` package. From each female, the number of pups born in the litter (`n`) and the count of those that were affected in some way (`y`) were recorded.

(a) Fit a logistic regression model to these data. Test the significance of the treatment effect and find a confidence interval for the odds ratio regarding the effect of "treatment" relative to "control."

(b) Note that the number of explanatory variable levels is fixed, so it would not increase if the sample size were to grow. This means that we can use the residual deviance to test the fit of the model formally. Perform this test. State the hypotheses, the test statistic, the p-value, and the conclusions.

(c) Fit a quasi-binomial regression model and repeat the analysis from part (a). Are the results substantially different?

(d) Use a beta-binomial model to repeat the analysis from part (a). This can be done using the `betabin()` function of the package `aod`. Are the results substantially different?

36. Continuing Exercise 38 of Chapter 4, fit a sequence of Poisson rate regression models for different combinations of variables as follows:

(a) `experience` as a nominal categorical variable and `age`

(b) same as in part (a), but also include the interaction

(c) `experience` as a numerical variable and `age` (model used for Exercise 38c)

(d) same as in part (c), but also include the interaction

(e) same as in part (c), but also include a quadratic term for `experience` as well (model used for Exercise 38e)

(f) same as in part (e), but also include an interaction term between `experience` and `age`

(g) same as in part (f), but also include an interaction term between $\texttt{experience}^2$ and `age`

Which of these is the best model using AIC? Which is best using BIC?

37. Exercises 26 and 27 of Chapter 3 examined proportional odds and multinomial regression models for the Plinko game on *The Price is Right.* Use information criteria to choose between these two models.

38. Exercise 52 in Chapter 4 examined a Poisson regression model to understand the habitat preferences of red squirrels in Scotland. Complete the following parts using the data from that exercise.

    (a) Fit the Poisson regression model for the mean number of stripped cones using canopy cover, number of trees, and mean tree height as linear terms without interactions. Construct a plot of the standardized Pearson residuals vs. estimated mean response. Interpret the plot.

    (b) Fit the same model as in (a) but now use quasi-likelihood methods. Verify that the regression parameter estimates and corresponding standard errors match those given by Flaherty et al. (2012).

    (c) Construct a plot of the standardized Pearson residuals vs. estimated mean response using the model from (b). Interpret the plot and compare it to the plot constructed in (a).

    (d) Complete (b)–(d) of Exercise 52 in Chapter 4 using the quasi-Poisson regression model and the appropriate adjustments for inferences.

    (e) Describe the effect that accounting for overdisperson has on interpretations.

    (f) The ultimate goal of the paper was to determine what forest characteristics lead to more red squirrels. Based on the results for this exercise, develop overall conclusions relative to the goals of the paper. Compare your conclusions to those stated in the paper.

39. Continuing Exercise 38, fit the same model but now use negative binomial modeling methods. Compare the results to those obtained by the quasi-Poisson regression model. Which model should be used?

40. Exercise 52 in Chapter 4 examined a Poisson regression model to understand the habitat preferences of red squirrels in Scotland. Complete the following using the data from that exercise.

    (a) Perform all-subsets selection using AIC on a Poisson regression model fitting the mean of `cones` against `ntrees`, `dbh`, `height`, and `cover`.

        i. Which variables are selected in the best model?

        ii. How much difference is there between the AIC for this best model and the next-best one? What does this suggest about the strength of evidence favoring the best model?

    (b) Repeat all-subsets selection, but include two-way interactions among variables while maintaining marginality.

        i. Which variables are in the new best model?

        ii. How much difference is there between the AIC for the best model with interactions and the best one without them? What does this tell you about the strength of evidence favoring the model with interactions?

41. Refer to the example and program on all-subsets regression from Section 5.1.3 with the placekicking data set. Run all-subsets selection on the linear (main-effect) terms using BIC instead of AIC. Report the best model and its BIC. How does this compare this to the best model selected by AIC in the example on p. 392?

42. Section 5.1.4 shows forward and backward stepwise selection on the placekick data using BIC. Repeat these procedures using AIC and compare the results to the respective models selected by BIC.

43. Section 5.5 examines post-selection inference for a new placekicking data set (Placekick2.csv). This exercise examines alternative investigations of the data.

    (a) Use the BIC with the training set to choose a model. Apply this model to the validation set. Do the parameter estimates, standard errors, and inferences from the validation set agree with those obtained from the training set?

    (b) From this same data split, treat the training set as the validation set and vice versa. Use AIC with the new training set to choose a model, and apply the chosen model to the validation set. Do the parameter estimates, standard errors, and inferences from the validation set agree with those obtained from the training set?

    (c) Change the seed prior to data splitting and rerun the variable selection using both AIC and BIC. Repeat this process of changing the seed and selecting variables for a total of 10 times. For each variable, identify how often it is selected. Are there differences in which variables are selected among the splits?

Chapter 6

Additional topics

6.1 Binary responses and testing error

Measurement processes are often imperfect. Instruments may be improperly calibrated (e.g., bathroom scales), local conditions may not all be identical (e.g., temperatures in a region), or a target may be difficult to measure (e.g., homeless persons in a census). Even binary responses are sometimes measured with error: a "success" may actually be a mismeasured "failure" and vice versa. For example, a referee may mistakenly declare a placekick successful when it was not.

We focus in this section on those binary responses resulting from diagnostic tests for infectious diseases. For example, tests for infections such as HIV, COVID-19, and chlamydia usually are not 100% accurate. The reasons for testing error can include (1) the amount of a pathogen in a specimen is below a detectable level (often the case for new infections), (2) inhibitors within a specimen block a positive outcome from being detected, and (3) laboratory errors. Fortunately, diagnostic tests usually have high accuracy levels, as we saw in Exercise 6 of Chapter 1 for the Aptima Combo 2 Assay. However, the possibility of testing error still needs to be taken into account for a statistical analysis to enable correct inferences. The purpose of this section is to discuss basic techniques to account for testing error that can be used in situations examined in Chapters 1 and 2.

6.1.1 Estimating the probability of success

Define Y as a Bernoulli random variable that denotes a diagnostic test's *measured* result, either positive (1) or negative (0). When Y is measured with the possibility of testing error, we need to define another Bernoulli random variable $\tilde{Y}$ that denotes the *true* status, positive (1) or negative (0). This leads to the following measures of accuracy:

- Sensitivity: $S_e = P(Y = 1|\tilde{Y} = 1)$
- Specificity: $S_p = P(Y = 0|\tilde{Y} = 0)$

Thus, sensitivity is the probability that a measured result is positive given that the underlying condition is truly positive, and specificity is the probability that a measured result is negative given that the underlying condition is truly negative. Ideally, we would like each of these conditional probabilities to equal 1, which means a perfect diagnostic test. Unfortunately, this is often not the case. Thus, the testing error rate is $1 - S_e = P(Y = 0|\tilde{Y} = 1)$ for truly positive individuals and $1 - S_p = P(Y = 1|\tilde{Y} = 0)$ for truly negative individuals.

In the evaluation of a diagnostic test, values for S_e and S_p are usually determined by performing tests on known true positive and negative specimens. For example, infectious disease assays often go through a clinical trial in this manner, and the resulting outcomes are given as part of a product insert for the assay. In fact, the data used for Exercise 6 of Chapter 1 were obtained from the assay's product insert. Interestingly, the estimates of

DOI: 10.1201/9781003093091-6

sensitivity and specificity from these trials are typically taken to be the true values without the acknowledgment of variability that inevitably occurs. We will do the same here, but we address alternative analysis methods in Section 6.1.3.

Often we wish to estimate the population prevalence for a disease, $\tilde{\pi} = P(\tilde{Y} = 1)$. To do this, we must assume that we have a random sample of individuals from the population within which the disease occurs. In particular, if individuals are more likely to be tested when they are positive than when they are negative, as is the case when testing occurs primarily among symptomatic individuals, then the process that follows leads to biased and inappropriate estimates of disease prevalence.

We estimate prevalence by using the marginal and conditional probabilities relating to test results, given true conditions. Specifically, the probability of a positive test response is

$$\begin{aligned} P(Y = 1) &= P(Y = 1 \text{ and } \tilde{Y} = 1) + P(Y = 1 \text{ and } \tilde{Y} = 0) \\ &= P(Y = 1|\tilde{Y} = 1)P(\tilde{Y} = 1) + P(Y = 1|\tilde{Y} = 0)P(\tilde{Y} = 0) \\ &= S_e P(\tilde{Y} = 1) + (1 - S_p)P(\tilde{Y} = 0). \end{aligned}$$

By letting $P(Y = 1) = \pi$ and $P(\tilde{Y} = 1) = \tilde{\pi}$ and then solving for $\tilde{\pi}$ in the above expression, we can compactly write the prevalence as

$$\tilde{\pi} = \frac{\pi + S_p - 1}{S_e + S_p - 1}. \tag{6.1}$$

To estimate $\tilde{\pi}$, suppose that a random sample of n individuals is tested for some disease, yielding responses $y_1, \ldots, y_n$. For example, $y_1, \ldots, y_n$ may represent the measured positive and negative HIV test results for blood donations submitted to a blood bank. The observed responses can be modeled using a Bernoulli distribution with probability π. From this model, the maximum likelihood estimate of $\tilde{\pi}$, say $\hat{\tilde{\pi}}$, can be found by the usual methods. The likelihood function is

$$\begin{aligned} L(\tilde{\pi}|y_1, \ldots, y_n) &= \pi^w (1 - \pi)^{n-w} \\ &= [S_e\tilde{\pi} + (1 - S_p)(1 - \tilde{\pi})]^w \, [1 - S_e\tilde{\pi} - (1 - S_p)(1 - \tilde{\pi})]^{n-w} \end{aligned} \tag{6.2}$$

where $w = \sum_{i=1}^{n} y_i$ is the observed number of positive test responses in a sample of size n. Maximizing Equation 6.2 results in the MLE of $\tilde{\pi}$:

$$\hat{\tilde{\pi}} = \frac{\hat{\pi} + S_p - 1}{S_e + S_p - 1}, \tag{6.3}$$

where $\hat{\pi} = w/n$ (see Exercise 4). Alternatively, one can use the invariance property of maximum likelihood estimators (see Appendix B.4) and simply replace π with $\hat{\pi}$ in Equation 6.1. Notice that when there is no testing error ($S_e = S_p = 1$), $\hat{\tilde{\pi}} = \hat{\pi}$ as would be expected.

There is one unfortunate problem with Equation 6.3. Typically, S_p is a large number close to 1. However, some diseases have such low infection rates and, hence, small values of $\tilde{\pi}$, that the value of S_p may not be large enough to prevent the numerator from being negative. Because the denominator is positive in all realistic applications, this leads to a negative value for $\hat{\tilde{\pi}}$, which does not make sense for a probability. A simple solution, although not necessarily ideal, is to set the estimate of $\tilde{\pi}$ to 0 if the MLE is negative. On the other hand, there may be a larger issue to solve when the MLE is negative; that is, perhaps the diagnostic test is just not accurate enough to obtain a meaningful estimate of $\tilde{\pi}$.

The estimated variance of $\hat{\tilde{\pi}}$ can also be found in a similar manner as in Section 1.1.2. Exercise 4 shows that the variance is

$$\widehat{Var}(\hat{\tilde{\pi}}) = \frac{\hat{\pi}(1 - \hat{\pi})}{n(S_e + S_p - 1)^2}.$$

This variance is the same as given in Equation 1.3 for the case of no testing error, except for $(S_e + S_p - 1)^2$ in the denominator. Thus, because $1 < S_e + S_p < 2$ in all realistic applications, the presence of testing error essentially reduces the effective sample size from n to $n(S_e + S_p - 1)^2$. The end result is larger variability for the estimator when there is testing error present. This result is intuitive because variability is a measurement of uncertainty. If we are uncertain about a true response, this should be reflected by a larger variance!

The $(1 - \alpha)100\%$ Wald confidence interval for $\tilde{\pi}$ is

$$\hat{\tilde{\pi}} - Z_{1-\alpha/2}\sqrt{\frac{\hat{\pi}(1-\hat{\pi})}{n(S_e + S_p - 1)^2}} < \tilde{\pi} < \hat{\tilde{\pi}} + Z_{1-\alpha/2}\sqrt{\frac{\hat{\pi}(1-\hat{\pi})}{n(S_e + S_p - 1)^2}}.$$

The performance of the interval is similar to that for the Wald interval for π. Exercise 3 explores this behavior further. A likelihood ratio interval for $\tilde{\pi}$ is presented in Exercise 6 as an alternative to the Wald interval.

Example: Hepatitis C prevalence among blood donors when testing error is included (HepCPrevSeSp.R)

Section 1.1.2 gives an example where 42 out of 1,875 blood donors tested positive for hepatitis C in Xuzhou City, China. While no values of S_e and S_p are given in the corresponding paper by Liu et al. (1997), it is common for diagnostic tests to not be perfectly accurate in similar settings. For example, Wilkins et al. (2010) list values of $S_e = 0.96$ and $S_p = 0.99$ for assays that detect the viral RNA. These authors also list both higher and lower values for S_e and S_p with other types of hepatitis C diagnostic tests.

The proportion of individuals testing positive for hepatitis C was $\hat{\pi} = 42/1875 = 0.0224$. Assuming accuracy levels of $S_e = 0.96$ and $S_p = 0.99$, the MLE for the overall prevalence becomes

$$\hat{\tilde{\pi}} = \frac{0.0224 + 0.99 - 1}{0.96 + 0.99 - 1} = 0.0131.$$

The estimated variance of $\hat{\pi}$ is $0.0224(1 - 0.0224)/1875 = 0.00001168$, and the estimated variance of $\hat{\tilde{\pi}}$ is $0.0224(1 - 0.0224)/[1875(0.96 + 0.99 - 1)^2] = 0.00001294$. When testing error is included in the analysis, we see that the estimated prevalence decreases while the estimated variance increases. The 95% Wald intervals for π and $\tilde{\pi}$ are $0.01570 < \pi < 0.02910$ and $0.006002 < \tilde{\pi} < 0.02010$, respectively.

To examine the effects of testing error further, additional calculations are shown in Table 6.1 for values of S_e and S_p given in Wilkins et al. (2010) for other diagnostic tests. The rows of the table are ordered by the overall uncertainty in the accuracy of the test. We see that as the overall uncertainty increases, the estimated variance increases as well. Also, nonsensical estimates of $\tilde{\pi}$ occur for smaller values of S_p, including an estimate of -0.301 in the last row, which uses the S_e and S_p from a recombinant immunoblot assay. Such nonsensical estimates indicate that these data are unlikely to have arisen from a test with these properties. In particular, such a low specificity would lead to a large number of false positives, which would increase $\hat{\pi}$. Still, these comparisons serve the purpose of showing that increasing uncertainty leads to increasing variability and perhaps even to nonsensical estimates.

Table 6.1: Prevalence estimates for various values of S_e and S_p.

S_e	S_p	$\hat{\tilde{\pi}}$	$\widehat{Var}(\hat{\tilde{\pi}})$	95% Wald confidence interval
1.00	0.98	0.002	1.22×10^{-5}	(−0.004, 0.009)
0.94	0.97	−0.008	1.41×10^{-5}	(−0.016, −0.001)
0.87	0.99	0.014	1.58×10^{-5}	(0.007, 0.022)
0.79	0.80	−0.301	3.36×10^{-5}	(−0.312, −0.290)

6.1.2 Binary regression models

The presence of testing error can be incorporated into the estimation of a logistic regression model (or other binary regression models) in a similar manner as was shown in Section 6.1.1. The general approach is to model the true probability based on responses $y_1, ..., y_n$ that include possible testing error. Then, we apply Equation 6.1 to work back and forth between the two different probabilities.

The logistic regression model now has the form

$$\text{logit}(\tilde{\pi}_i) = \beta_0 + \beta_1 x_{i1} + \cdots + \beta_p x_{ip},$$

where $\tilde{\pi}_i$ is the true probability of success for observation $i = 1, \ldots, n$. The model uses a likelihood function in a familiar form to what was given in Section 2.2.1:

$$\begin{aligned} L(\boldsymbol{\beta}|\mathbf{y}) &= \prod_{i=1}^{n} \pi_i^{y_i}(1-\pi_i)^{1-y_i} \\ &= \prod_{i=1}^{n} \left[S_e\tilde{\pi}_i + (1-S_p)(1-\tilde{\pi}_i)\right]^{y_i} \left[1 - (S_e\tilde{\pi}_i - (1-S_p)(1-\tilde{\pi}_i))\right]^{1-y_i}, \end{aligned}$$

where π_i is the probability of a positive test for observation y_i, $\boldsymbol{\beta} = (\beta_0, \ldots, \beta_p)$, and $\mathbf{y} = (y_1, ..., y_n)$. Iterative numerical procedures again need to be used to estimate the regression parameters. The estimated covariance matrix can be found by inverting the negative of the Hessian matrix, as discussed in Appendix B.3.4 and in Section 2.2.1.

Example: Prenatal infectious disease screening (HIVKenya.R, HIVKenya.csv)

Verstraeten et al. (1998) screened a sample of pregnant women in Kenya for HIV with the purpose of evaluating a new screening method. While not discussed in this paper, subsequent work by Vansteelandt et al. (2000) and others examined the data in the context of binary regression models. A portion of their data is available in the data set for this example. The response variable `hiv` is a 0 for a negative test and a 1 for a positive test. The explanatory variables are `parity` (number of times the woman has previously given birth), `age` (in years), `marital.status` (1 = single, 2 = married polygamous, 3 = married monogamous, 4 = divorced, 5 = widow), and `education` (1 = none, 2 = primary, 3 = secondary, 4 = higher). We use these data to model prevalence of HIV among pregnant women in Kenya. Focus is on the `age` explanatory variable here, while Exercise 5 examines the other explanatory variables. Additional variables in the data set are examined in Section 6.7. Below are the results from fitting a logistic regression model to the data assuming no testing error.

```
> set1 <- read.csv(file = "C:\\data\\HIVKenya.csv")
> head(set1)
  parity age marital.status education hiv gnum groupres retest
1      1  16              3         2   0    1        0     NA
2      0  17              3         1   0    1        0     NA
3      2  26              3         2   0    1        0     NA
4      3  20              3         2   0    1        0     NA
5      1  18              3         1   0    1        0     NA
6      5  35              3         2   0    2        1      0

> mod.fit <- glm(formula = hiv ~ age, data = set1, family =
    binomial(link = logit))
> round(summary(mod.fit)$coefficients, 4)
            Estimate Std. Error z value Pr(>|z|)
(Intercept)  -1.8618     0.6917 -2.6917   0.0071
age          -0.0273     0.0287 -0.9528   0.3407
```

The actual testing involved a series of up to three tests per individual to reduce the probability of a misdiagnosis. Unfortunately, the data from each test is not available, and only the final test diagnoses are given by `hiv` in the data set. For illustration purposes, we examine next how to fit the model assuming that the final diagnosis has properties $S_e = S_p = 0.98$.

We present two ways to fit the model using R. First, we follow a similar process as illustrated in Section 2.2.1 when we used our own function to evaluate the log-likelihood and then `optim()` to maximize it. Below is our code and output.

```
> Se <- 0.98
> Sp <- 0.98
> X <- model.matrix(mod.fit)
> logL <- function(beta, X, Y, Se, Sp) {
    pi.tilde <- plogis(X%*%beta)  # exp(X*beta)/(1 + exp(X*beta))
    pi <- Se*pi.tilde + (1 - Sp)*(1 - pi.tilde)
    sum(Y*log(pi) + (1-Y)*log(1-pi))
  }

> mod.fit.opt <- optim(par = mod.fit$coefficients, fn = logL,
    hessian = TRUE, X = X, Y = set1$hiv, control = list(fnscale =
    -1), Se = Se, Sp = Sp, method = "BFGS")
> mod.fit.opt$par  # beta.hats
(Intercept)         age
 -2.0068169  -0.0334271
> mod.fit.opt$value  # log(L)
[1] -177.27
> mod.fit.opt$convergence  # 0 means converged
[1] 0

> # Estimated covariance matrix (multiply by -1 because of
    fnscale)
> cov.mat <- -solve(mod.fit.opt$hessian)
> cov.mat
            (Intercept)          age
(Intercept)   0.7884776 -0.032263002
```

```
age              -0.0322630   0.001388222
> sqrt(diag(cov.mat))   # SEs
(Intercept)         age
0.88796260   0.03725886
> z <- mod.fit.opt$par[2]/sqrt(diag(cov.mat))[2]   # Wald stat
> 2*(1-pnorm(q = abs(z)))   # p-value
      age
0.3696344
```

The `logL()` function evaluates the log-likelihood function using matrix algebra to calculate $\tilde{\pi}_i$ for $i = 1, \ldots, n$. This matrix algebra representation allows the code to be easily generalized to more than one explanatory variable. A non-matrix-algebra alternative for one explanatory variable is given in our program. The estimated logistic regression model is

$$\text{logit}(\hat{\tilde{\pi}}) = -2.0068 - 0.03343\text{age}.$$

A numerical estimate of the Hessian matrix is produced by specifying `hessian = TRUE` in `optim()` and leads to the estimated covariance matrix for $\hat{\beta}_0$ and $\hat{\beta}_1$ as given by Equation B.6. A Wald test of $H_0 : \beta_1 = 0$ vs. $H_a : \beta_1 \neq 0$ has a p-value of 0.3696, which indicates little evidence that a linear age term is needed.

The model can also be estimated by writing a new link function for the `family` argument of `glm()`. Specifically, the link function form can be seen by rewriting our model as

$$\text{logit}\left(\frac{\pi + S_p - 1}{S_e + S_p - 1}\right) = \beta_0 + \beta_1\texttt{age},$$

where Equation 6.1 is used to express the model in terms of π. Solving for π on the left side results in the inverse link function

$$\pi = \frac{S_e\exp(\beta_0 + \beta_1\texttt{age}) - S_p + 1}{1 + \exp(\beta_0 + \beta_1\texttt{age})}.$$

We code these equations into our own function below.

```
> my.link <- function(Se, Sp) {
    # mu = E(Y) = pi
    linkfun <- function(mu) {
      pi.tilde <- (mu + Sp - 1)/(Se + Sp - 1)
      log(pi.tilde/(1-pi.tilde))
    }
    linkinv <- function(eta) {
      (exp(eta)*Se - Sp + 1)/(1 + exp(eta))
    }
    mu.eta <- function(eta) {
      exp(eta)*(Se + Sp - 1)/(1 + exp(eta))^2
    }
    save.it <- list(linkfun = linkfun, linkinv = linkinv, mu.eta
       = mu.eta)
    class(save.it) <- "link-glm"
    save.it
  }
```

```
> mod.fit2 <- glm(formula = hiv ~ age, data = set1, family =
    binomial(link = my.link(Se, Sp)))
> round(summary(mod.fit2)$coefficients, 4)
             Estimate Std. Error z value Pr(>|z|)
(Intercept)   -2.0097     0.9398 -2.1383   0.0325
age           -0.0333     0.0397 -0.8396   0.4011
> vcov(mod.fit2)
             (Intercept)          age
(Intercept)   0.88329971 -0.036451992
age          -0.03645199  0.001573203
```

The `mu.eta()` function within `my.link()` gives the partial derivative of π with respect to the systematic component, $\eta = \beta_0 + \beta_1$`age`:

$$\frac{\partial \pi}{\partial \eta} = \frac{\exp(\eta)(S_e + S_p - 1)}{[1 + \exp(\eta)]^2}$$

(see Section 2.3). The `glm()` function is used then in the same manner as with logistic regression except that `my.link()` is the value for the `link` argument.[1] The regression parameter estimates are practically the same as those obtained from using `optim()`. The small differences between the estimated covariance matrices from the two methods arise because `glm()` uses Equation B.5 in its calculation, whereas `optim()` uses Equation B.6.

The square root of the estimated variance for $\hat{\beta}_1$ is 0.0287 when there is no testing error and 0.0397 when there is testing error. This increase occurs because there is more uncertainty in the response when testing error is present, which then is exhibited in the variances for the regression parameter estimates. Exercise 7 examines this behavior further.

6.1.3 Other methods

The methods described in Section 6.1.2 are meant to be used in situations when the values of S_e and S_p are known or when the variability from estimating S_e and S_p cannot be easily obtained. However, sampling variability from estimating S_e and S_p adds uncertainty to the estimation of $\tilde{\pi}$ that should be accounted for when possible. In other situations, there are more than two response categories of interest for a given problem, as we saw in Chapter 3. These additional response categories subsequently lead to more types of misclassification that need to be addressed. Chapter 2 of the book by Buonaccorsi (2010) discusses both of the above situations, and we refer the reader to this book for detailed discussion.

Küchenhoff et al. (2006) provide an innovative way to account for testing error in categorical response and/or explanatory variable settings. Their "misclassification, simulation, and extrapolation" (MC-SIMEX) method is implemented by the `simex` package of R (Lederer and Küchenhoff, 2006), and we show how to use this package in Exercise 8. In situations when only a binary response variable is subject to testing error, simulations in Küchenhoff et al. (2006) show that the maximum likelihood methodology we described in Sections 6.1.1 and 6.1.2 provides nearly unbiased estimators, while MC-SIMEX produces somewhat biased

[1]The `mis()` function of the `brglm2` package can be used instead of `my.link()`. We illustrate its use in our program.

estimators. Therefore, our presented analysis methods generally would be preferred when S_e and S_p are known.

The analysis of data measured with error, either in response or explanatory variables, is a large and active research area. Books such as Buonaccorsi (2010), Carroll et al. (2006), Gustafson (2004), and Yi (2017) provide thorough accounts.

6.2 Exact inference

Most of the inference procedures examined so far rely on a statement similar to the following:

> As the sample size goes to infinity, the statistic's distribution approaches a χ^2 (or normal) distribution.

Unfortunately, infinity is the one sample size that we can never have. There are many situations where a parametric distribution, such as normal or chi-square, serves as a good approximation to the actual distribution for the statistic. However, there are some situations where a parametric distribution does not, particularly when the sample size is small. The purpose of this section is to develop inference procedures that do not rely on large-sample approximations. These inference procedures are often referred to as *exact* in the sense that they use the actual distribution for the statistic of interest under minimal assumptions about the structure of the data. These exact inference procedures often also provide us ways to assess whether or not a large-sample distributional approximation for a statistic is appropriate.

Exact inference methods were first discussed in Section 1.1.2 with respect to the Clopper-Pearson interval for the probability-of-success parameter in a binomial distribution. We showed that the interval always had a true confidence level at least at the stated level, but it could be much more. This led us to labeling the interval as conservative. Unfortunately, most other exact inference methods are conservative as well. For hypothesis testing, exact tests tend to reject the null hypothesis less often than the stated type I error level when the null hypothesis is true (it will never reject at a higher rate). This conservative tendency persists when the null hypothesis is false, so that exact tests may have lower power than other tests. Still, a conservative test is generally better than a liberal test when error-rate control is important, and especially in settings with small sample sizes when non-exact tests may become excessively liberal.

We do not have space to cover all possible exact inference procedures, so we begin in Section 6.2.1 with one of the most commonly used exact procedures, Fisher's exact test. We generalize the ideas of exact inference in Section 6.2.2, where permutation methods allow us to test for independence. In Section 6.2.3, we introduce exact inference in the context of logistic regression. Finally, we conclude with a survey of other exact inference procedures in Section 6.2.4 for those readers who would like to consider them further.

6.2.1 Fisher's exact test for independence

We begin by examining a test for independence in a 2×2 contingency table, a structure studied previously in Section 1.2. Let two groups be represented by the rows of the table and the responses "success" and "failure" by the columns. Recall that an odds ratio (OR) with a value of 1 means that the odds of success for group 1 are the same as the odds of

Table 6.2: Two representations of a 2×2 table. The representation on the left uses the notation from Section 1.2.1, and the representation on the right uses the notation from the hypergeometric distribution in the context of the urn example.

		Response					Urn		
		1	2				Drawn out	Remaining	
Group	1	w_1	$n_1 - w_1$	n_1	Color	Red	m	$a - m$	a
	2	w_2	$n_2 - w_2$	n_2		Blue	$k - m$	$b - k + m$	b
		w_+	$n_+ - w_+$	n_+			k	$n - k$	n

success for group 2; i.e., the odds are independent of group designation. Because of this result, we will simply write $H_0 : OR = 1$ vs. $H_a : OR \neq 1$ as our hypotheses in a test for independence between the column responses and the row responses.

Hypergeometric distribution

The hypergeometric probability distribution plays an important role in constructing an exact test for independence, so we provide a brief review here. The distribution is typically introduced using an example such as the following:

> Suppose an urn has a red balls and b blue balls with $n = a + b$. Suppose $k \leq n$ balls are randomly drawn from the urn without replacement. Let M be the number of red balls drawn out.

The random variable M has a hypergeometric distribution. The PMF is

$$P(M = m) = \frac{\binom{a}{m}\binom{b}{k-m}}{\binom{n}{k}}$$

for $m = 0, \ldots, k$ subject to $m \leq a$ and $k - m \leq b$. Note that $a, b, n,$ and k are all fixed quantities. The only variable on the right side is m.

The hypergeometric distribution can be used to find the probability of observing a particular 2×2 table under independence. Table 6.2 shows a 2×2 table from two perspectives. The left side shows the notation for a 2×2 table under the independent binomial model that was given in Section 1.2.1. The right side shows the same table written in the notation for a hypergeometric distribution. The big difference between the two is that all of the marginal counts are fixed (i.e., known) before any sampling takes place for the hypergeometric distribution, whereas in the binomial model the column margins—representing the total number of observed successes and failures—are considered random. However, under certain assumptions, models for the two tables can be made equivalent, as we show next.

In the independent binomial model, we have two random variables, W_1 and W_2. We use the observed values w_1 and w_2 to compare the respective probabilities of success, π_1 and π_2. The total number of successes, w_+, tells us nothing about how different π_1 and π_2 might be and, hence, it is not important to the comparison. Thus, there is no harm in "conditioning on" w_+ (i.e., assuming it is known instead of random). In fact, there are mathematical benefits to making this assumption. When $H_0 : \pi_1 = \pi_2$ is true, then it can be shown that conditioning on w_+ in the independent binomial model leads to the hypergeometric model. See Exercise 1 for details of this derivation.

Because all of the marginal totals are known in the hypergeometric model, once we observe one cell count, the remaining three counts are obtained by subtraction and the whole table is known. Thus, the hypergeometric distribution gives us the probability of observing a

Table 6.3: Hypothetical outcome of Fisher's experiment.

		Lady's response		
		Milk	Tea	
Actual	Milk	3	1	4
	Tea	1	3	4
		4	4	8

Table 6.4: Probability of possible outcomes for Fisher's experiment. Note that the last odds ratio is $4 \times 4/(0 \times 0)$, which is undefined. This has been labeled as ∞ to indicate that the estimated odds ratio would become larger as the denominator decreases toward 0.

M	$P(M = m)$	$\widehat{OR}$	X^2
0	0.0143	0	8
1	0.2286	1/9	2
2	0.5143	1	0
3	0.2286	9	2
4	0.0143	∞	8

particular contingency table under independence. This is useful for constructing a test for independence because it provides a method for calculating the exact p-value for a test. If there is low probability of a table as extreme as the one observed, this casts doubt on the independence assumption. Fisher's exact test, which we discuss next, is based on this idea.

Fisher's exact test

Salsburg (2001) describes one of the most frequently discussed events in statistical history as follows.[2,3] One day in Cambridge, England, during the late 1920s, a number of individuals gathered together for afternoon tea. A lady in the group said that she could differentiate between whether the milk or the tea was poured first into a cup. Sir Ronald Fisher, who contributed immensely to the early foundations of statistics, was in the group and proposed a simple experiment to test the lady's claim. He suggested to pour tea first into four cups and to pour milk first into four cups. The lady was blinded to which cups had milk or tea poured first, but she knew there were four of each. Thus, it would be expected that she would actually choose four milk and four tea, leading to fixed values for both the row and column totals within a 2×2 contingency table summarizing her choices. Table 6.3 shows a hypothetical outcome of the experiment where two cups were identified incorrectly.

If the lady could not actually differentiate which was added to a cup first, then her selection of milk responses would in reality be four random cups, and the odds of a milk response would be the same regardless of whether milk or tea was truly added first (odds for row 1 equal to odds for row 2). This is the same sampling structure as in the "ball-and-urn" example from earlier, where milk and tea now represent the two colors of the balls. We can therefore calculate the probabilities of all possible 2×2 tables under this scenario using the hypergeometric distribution. These probabilities are shown in Table 6.4 where m is the count in the (1,1) cell of the contingency table. We see that the closer to 1 the estimated odds ratio is, the more likely a particular contingency table is observed by random selection. The opposite is true as the estimated odds ratio becomes farther from 1.

[2]The television show *Nova* presents a re-enactment in their episode "Prediction by the Numbers" that aired in 2018 (Season 45, Episode 6); see `https://youtu.be/9OIe15NUG7Q?t=1346`.

[3]Richardson (2021) provides additional details and alternative accounts.

Suppose the lady responded as shown in Table 6.3. The estimated odds ratio is 9, indicating that the estimated odds of a milk response are 9 times as large when milk is actually poured first as when tea is poured first. Thus, it might appear that the lady is able to detect what was poured into the cup first. The probabilities given in Table 6.4 measure how likely this or other choices are, if she did not really know the difference. We can use these probabilities then in the context of a standard hypothesis test, where we calculate a p-value as the probability of an event at least as extreme as the one observed. The probability of randomly guessing three or more of the milk cups correctly is $P(M \geq 3) = 0.2286 + 0.0143 = 0.2429$. This probability is fairly large, so we would conclude that the lady's response would not be unusual for someone who was guessing at random. Suppose instead that she had correctly identified all four milk-first cups. Then, we would begin to believe that this was a skill on her part, because the probability of doing this by guessing is only $P(M \geq 4) = 0.0143$.

The scenarios described in the last paragraph show how to calculate a p-value for Fisher's exact test of $H_0 : OR \leq 1$ vs. $H_a : OR > 1$. A one-sided alternative hypothesis makes more sense here than a two-sided because only $OR > 1$ means that the lady can correctly detect what was poured into the cup first. In general for a one-sided test, the p-value is calculated as the sum of the table probabilities only in the direction represented by the alternative hypothesis.

For other situations, a two-sided test may be of interest. The p-value for a two-sided test is calculated as the probability of observing an M at least as extreme as the one observed (i.e., observing a contingency table with probability less than or equal to $P(M = m)$). For demonstration purposes, a two-sided test p-value is $0.0143 + 0.2286 + 0.2286 + 0.0143 = 0.4858$ if Table 6.3 is observed.[4]

We remarked earlier that exact inference procedures are often conservative in rejecting the null hypothesis. This is the case for Fisher's exact test as well, due to the highly discrete nature of the hypergeometric distribution. It rarely happens that there is a configuration of table probabilities under H_0 that leads to exactly the stated significance level α. For the one-sided version of the lady tasting tea, there is only a 0.0143 probability of observing a p-value below $\alpha = 0.05$ under independence, so a test at this level actually has a type I error rate of 0.0143, and no test with exactly a 0.05 type I error rate is possible.

Example: Tea tasting (Tea.R)

Salsburg (2001) indicates that the lady did correctly respond for all eight cups, leading to a p-value of 0.0143. To duplicate these calculations in R, we can use the `dhyper()` function to find the probabilities of the hypergeometric distribution or use `fisher.test()` to compute the p-value. Below are the calculations.

```
> M <- 0:4
> # Syntax for dhyper(m, a, b, k)
> data.frame(M, prob = round(dhyper(M, 4, 4, 4), 4))
```

[4]An alternative method sometimes used for calculating two-sided test p-values is simply to take twice the minimum of the two one-sided p-values. For the data in Table 6.3, we calculate

$$\begin{aligned} 2\min\{0.5, P(M \leq 3), P(M \geq 3)\} &= 2\min\{0.5, 0.9857, 0.2429\} \\ &= 0.4858, \end{aligned}$$

where 0.5 is included in the brackets to make sure the calculated p-value does not exceed 1. The two presented p-value calculation methods are equivalent when the corresponding distribution is symmetric.

```
  M prob
1 0 0.0143
2 1 0.2286
3 2 0.5143
4 3 0.2286
5 4 0.0143

> c.table <- array(data = c(4, 0, 0, 4), dim = c(2,2), dimnames
   = list(Actual = c("Milk", "Tea"), Response = c("Milk",
   "Tea")))
> c.table
      Response
Actual Milk Tea
  Milk    4   0
  Tea     0   4

> fisher.test(x = c.table, alternative = "greater")

        Fisher's Exact Test for Count Data

data:  c.table
 p-value = 0.01429
alternative hypothesis: true odds ratio is greater than 1
95 percent confidence interval:
 2.003768 Inf
sample estimates:
odds ratio
       Inf
```

We use the `alternative = "greater"` argument value in `fisher.test()` to specify $H_a : OR > 1$. The default value for the argument is `"two.sided"`. The p-value given matches what we calculated before using the hypergeometric distribution. Note that there is also an estimate of the odds ratio along with a confidence interval. The confidence interval endpoints are calculated using a non-central hypergeometric distribution that does not assume independence (details are available on p. 606 of Agresti, 2013).

Multiple hypergeometric distribution for $I \times J$ tables

Fisher's exact test can be extended to tables larger than 2×2 by using the multiple hypergeometric distribution. The row and column totals are still fixed in this setting, but now there are $(I-1)(J-1)$ different random variables. Probabilities for each possible contingency table can be calculated leading to a test for independence in the same manner as before. The probability of observing a specific set of cell counts, n_{ij}, $i = 1, \ldots, I-1$; $j = 1, \ldots, J-1$ is

$$\frac{\left(\prod_{i=1}^{I} n_{i+}!\right)\left(\prod_{j=1}^{J} n_{+j}!\right)}{n!\left(\prod_{i=1}^{I}\prod_{j=1}^{J} n_{ij}!\right)},$$

where we use the notation of Section 3.2 to describe the table counts. Because there may be a very large number of contingency tables, efficient algorithms have been developed to perform the calculations. A related algorithm is discussed in Section 6.2.2.

Example: Fiber-enriched crackers (FiberExact.R, Fiber.csv)

An example in Section 3.2 examined testing for independence between fiber source for crackers and bloating severity in subjects eating the crackers. A Pearson chi-square and LR test for independence resulted in p-values of 0.0496 and 0.0262, respectively. There were a number of very low expected cell counts, so we were somewhat concerned that the large-sample χ^2_9 approximation used with the test statistics might not be close to their actual distributions. Applying Fisher's exact test allows us to avoid this concern. Below is code continuing from that example.

```
> fisher.test(x = diet.table)

        Fisher's Exact Test for Count Data

data:  diet.table
p-value = 0.06636
alternative hypothesis: two.sided
```

The p-value is 0.0664, which again indicates marginal evidence against independence. It is interesting to note that this p-value is a little larger than the p-values for the Pearson chi-square and LR tests. This could be due to the conservative nature of the Fisher's exact test and/or the χ^2_9 approximation not working as well as we would like. We will perform a more detailed examination of this approximation in the next section.

6.2.2 Permutation test for independence

The actual probability distribution for the Pearson chi-square test statistic X^2 is not exactly $\chi^2_{(I-1)(J-1)}$ in a test for independence. Rather, it is closely related to the hypergeometric (or multiple hypergeometric) distribution when the row and column totals for a contingency table are fixed. Table 6.4 gives each possible value of X^2 that could be observed for the lady-tasting-tea experiment. By summing probabilities over duplicate values of X^2, we obtain the exact PMF under independence as $P(X^2 = 0) = 0.5143$, $P(X^2 = 2) = 0.4571$, and $P(X^2 = 8) = 0.0286$. Figure 6.1 plots the exact cumulative distribution function (CDF) for X^2 along with a χ^2_1 CDF approximation (please see Tea.R for the code). While the χ^2_1 distribution roughly goes through the middle of the exact function, we see there are definitely some differences between the two. For example, the p-value for $X^2 = 2$ is $P(X^2 \geq 2) = 0.4857$ using the exact distribution, while the χ^2_1 approximation yields 0.1572. For $X^2 = 8$, the exact p-value is 0.0286, while the approximate is 0.0047.

Using the hypergeometric distribution is easy for 2×2 contingency tables; however, when either the sample size or the size of the table is large, the number of tables whose probabilities comprise the p-value may be so great that it is very difficult to compute the p-value exactly. Instead, we can use a general procedure known as a *permutation test* to estimate the exact distribution of a test statistic. A permutation test randomly permutes (reorders) the observed data a large number of times. Each permutation is done in a way that allows the null hypothesis to be true and the marginal counts in the contingency

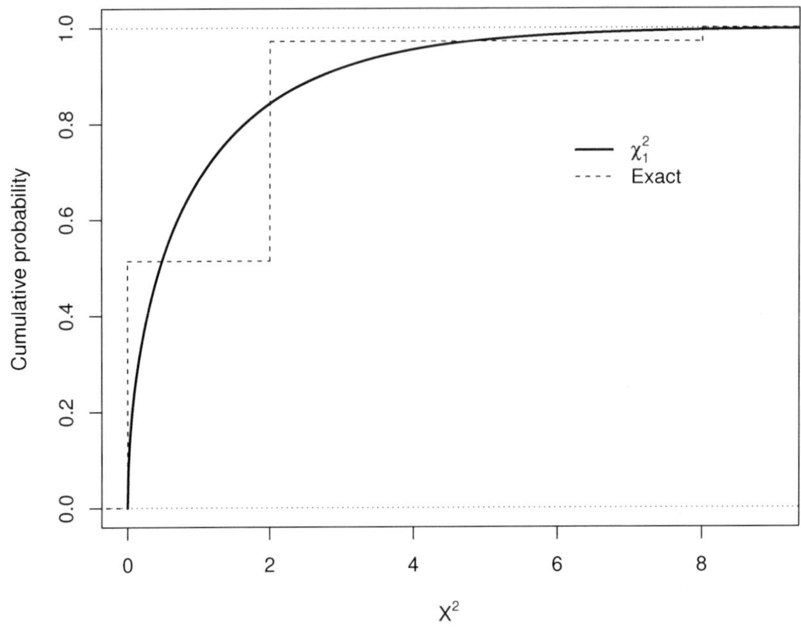

Figure 6.1: Exact and chi-square CDFs for tea-tasting example.

table to remain unchanged.[5] For each permutation, the statistic of interest is calculated. The estimated PMF under the null hypothesis is calculated from the collection of these statistics by simply forming a relative frequency table or histogram. P-values are then calculated directly from this PMF.

To test for independence with X^2, we can first decompose the contingency table into its raw data form (see Section 1.2.1); i.e, we create a data set where each row represents an observation and two variables represent the row category and column category of an observation in the contingency table. Table 6.5 (left) shows this format for the data in Table 6.3. We identify each column response in Table 6.5 with its original observation number using z_j, $j = 1, \ldots, 8$. This allows us to see that many distinct permutations of the column responses may lead to the same contingency table. For example, the original data and the center permutation in Table 6.5 lead to the same contingency table, for which $X^2 = 2$. The permutation on the right causes a different contingency table to be formed, for which $X^2 = 0$.

Under independence between the row and column variables and conditional on the observed marginal row and column counts, any random permutation of the column observations is equally likely to be observed with the given row observations. There are $8! = 40,320$ different possible permutations of the original data in Table 6.5 with each having a probability of $1/40,320$ of occurring under independence. For this setting, one can show mathematically that there are 20,736 permutations that lead to $X^2 = 0$, 18,432 permutations that lead to $X^2 = 2$, and 1,152 permutations that lead to $X^2 = 8$. Using these permutations,

[5]Remember that for hypothesis testing in general, we assume the null hypothesis to be true and find a probability distribution for the test statistic of interest.

Table 6.5: Raw form of the data in Table 6.3. The left table contains the original data. The center and right tables show possible permutations.

Row	Column	Row	Column	Row	Column
1	$z_1 = 1$	1	$z_2 = 1$	1	$z_1 = 1$
1	$z_2 = 1$	1	$z_1 = 1$	1	$z_2 = 1$
1	$z_3 = 1$	1	$z_3 = 1$	1	$z_7 = 2$
1	$z_4 = 2$	1	$z_4 = 2$	1	$z_4 = 2$
2	$z_5 = 1$	2	$z_5 = 1$	2	$z_5 = 1$
2	$z_6 = 2$	2	$z_6 = 2$	2	$z_8 = 2$
2	$z_7 = 2$	2	$z_7 = 2$	2	$z_3 = 1$
2	$z_8 = 2$	2	$z_8 = 2$	2	$z_6 = 2$

we can calculate the exact distribution for X^2 as $P(X^2 = 0) = 20,736/40,320 = 0.5143$, $P(X^2 = 2) = 18,432/40,320 = 0.4571$, and $P(X^2 = 8) = 1,152/40,320 = 0.0286$. These probabilities are the same as those provided by the hypergeometric distribution!

Frequently, the number of permutations and the number of possible X^2 values are so large that it is difficult to calculate the PMF by evaluating every possible permutation. Instead, we can estimate the p-value similar to what was done in Section 3.2.3 for the multinomial distribution: randomly select a large number of permutations, say B, calculate X^2 on each permutation, and use the empirical distribution of these calculated values as an estimate of the exact distribution. This estimate is often referred to as the *permutation distribution* of the statistic X^2. Using this distribution to do a hypothesis test is referred to as a permutation test.

Below is an algorithm for finding the permutation estimate of the p-value using Monte Carlo simulation:

1. Randomly permute the column observations while keeping the row observations unchanged.[6]
2. Calculate the Pearson chi-square statistic on the newly formed data. Denote this statistic by X^{2*} to distinguish it from the value calculated for the original sample.
3. Repeat steps 1 and 2 B times, where B is a large number (e.g., 1,000 or more).
4. (Optional) Plot a histogram of the X^{2*} values. This serves as a visual estimate of the exact distribution of X^2.
5. Calculate the p-value as the proportion of the X^{2*} values greater than or equal to the observed X^2; i.e., calculate (Number of $X^{2*} \geq X^2)/B$.

Note that the p-value may change from one implementation of these steps to the next. However, the p-values will be very similar as long as a large B is used. Also, note that we could replace the Pearson chi-square statistic with the LRT statistic to perform a different permutation test for independence.

Example: Fiber-enriched crackers (FiberExact.R, Fiber.csv)

The `chisq.test()` function performs a permutation test using the X^2 statistic when the `simulate.p.value = TRUE` argument value is specified.

[6]Permuting the row observations in addition to the column observations is not necessary. Only one set needs to be permuted to achieve an equal probability for any possible set of pairings.

```
> set.seed(8912)
> chisq.test(x = diet.table, correct = FALSE, simulate.p.value =
   TRUE, B = 1000)

        Pearson's Chi-squared test with simulated p-value
        (based on 1000 replicates)

data:  diet.table
X-squared = 16.943, df = NA, p-value = 0.03896
```

We use the `set.seed()` function at the beginning so that we can reproduce the same results whenever we run the code. The `B = 1000` argument value in `chisq.test()` specifies the number of permuted data sets. We obtain a p-value of 0.03896, which indicates marginal evidence against independence.[7]

Running the previous code again, but without `set.seed(8912)`, will likely result in a slightly different p-value. Because this p-value is a sample proportion, we can use a confidence interval procedure from Section 1.1.3 to estimate the true probability that would result from using all possible permutations. A 95% Wilson confidence interval is (0.029, 0.053), which does not change our original conclusion about independence. A much larger number of permutations can be used provided that it does not take too long to complete the calculations. Using $B = 100,000$ permutations and the same seed number as before, we obtain a p-value of 0.0455 with a 95% Wilson confidence interval (0.0442, 0.0468).

Unfortunately, the `chisq.test()` function does not provide any information about the permutation distribution other than the relative position of X^2. To estimate the entire permutation distribution and determine whether a χ^2_9 approximation is really appropriate for X^2, we need to perform some calculations ourselves. We begin doing this by arranging the contingency table counts into the raw form of the data. This is accomplished by first converting `diet.table` to a data frame. Next, we repeat each row of the data frame, using the `rep()` function, corresponding to the number of times a particular `fiber` and `bloat` combination is observed. The raw data are contained in the `set2` data frame as shown below.

```
> set1 <- as.data.frame(as.table(diet.table))
> tail(set1)  # Notice 2 obs. for fiber = both and bloat = high
   fiber  bloat Freq
11   gum medium    3
12  both medium    3
13  none   high    0
14  bran   high    0
15   gum   high    5
16  both   high    2

> set2 <- set1[rep(1:nrow(set1), times = set1$Freq), -3]
> tail(set2)  # Notice 2 rows for fiber = both and bloat = high
```

[7]Note that `chisq.test()` calculates its p-value as (1 + Number of $X^{2*} \geq X^2)/(1 + B)$, which is slightly different than our p-value formula. This alternative formula treats the original sample as one of the random permutations and ensures that no p-value is ever exactly 0, which otherwise can happen if $X^{2*} < X^2$ for all permutations. Both formulas are widely used to calculate p-values.

```
     fiber bloat
15.1   gum  high
15.2   gum  high
15.3   gum  high
15.4   gum  high
16    both  high
16.1  both  high

> # Verify data is correct
> xtabs(formula = ~ set2[,1] + set2[,2])
         set2[, 2]
set2[, 1] none low medium high
     none    6   4      2    0
     bran    7   4      1    0
     gum     2   2      3    5
     both    2   5      3    2

> X.sq <- chisq.test(set2[,1], set2[,2], correct = FALSE)
Warning message:
In chisq.test(set2[, 1], set2[, 2], correct = FALSE) :
  Chi-squared approximation may be incorrect
> X.sq$statistic
X-squared
 16.94267
```

The xtabs() and chisq.test() functions confirm that the raw data were created correctly and the same X^2 value as before was obtained.

To illustrate one permutation of the data, we use the following code.

```
> set.seed(4088)
> set2.star <- data.frame(row = set2[,1], column =
   sample(set2[,2], replace = FALSE))
> xtabs(formula = ~ set2.star[,1] + set2.star[,2])
              set2.star[, 2]
set2.star[, 1] none low medium high
          none    3   6      2    1
          bran    6   4      1    1
          gum     5   4      3    0
          both    3   1      3    5
> X.sq.star <- chisq.test(set2.star[,1], set2.star[,2],  correct
   = FALSE)
Warning message:
In chisq.test(set2.star[, 1], set2.star[, 2], correct = FALSE) :
  Chi-squared approximation may be incorrect
> X.sq.star$statistic
X-squared
 14.63903
```

The sample() function randomly permutes the column observations, and data.frame() combines the row observations with these permuted column observations to form a permuted data set. Notice that $X^{2\star} = 14.64$. This process is repeated $B = 1,000$ times using the for() function.

```
> B <- 1000
> X.sq.star.save <- matrix(data = NA, nrow = B, ncol = 1)

> set.seed(1938)
> for(i in 1:B) {
    set2.star <- data.frame(row = set2[,1], column =
        sample(set2[,2], replace = FALSE))
    X.sq.star <- chisq.test(set2.star[,1], set2.star[,2],
        correct = FALSE)
    X.sq.star.save[i,1] <- X.sq.star$statistic
  }
There were 50 or more warnings (use warnings() to see the first
   50)

> mean(X.sq.star.save >= X.sq$statistic)
[1] 0.039
> summarize(result.set = X.sq.star.save, statistic =
   X.sq$statistic, df = (nrow(diet.table) - 1) *
   (ncol(diet.table) - 1), B = B)
[1] 0.039
```

The p-value is 0.039 indicating marginal evidence against independence, which agrees with our results from before. Note that the warning messages given by R are not a cause for concern here because these are generated when `chisq.test()` encounters small expected cell counts, which would be an issue only if we were comparing the computed X^{2*} values to the large-sample χ^2_9 distribution.[8] We have written a function named `summarize()` (see our program) to produce the plots in Figure 6.2 and calculate the p-value. We see that the χ^2_9 distribution approximates the permutation distribution very well, leaving little reason to doubt the χ^2_9 approximation for X^2 under independence. For example, the QQ-plot gives the quantiles of a χ^2_9 distribution plotted against corresponding ordered $X^{2\star}$ values. These plotted values are generally very similar as evidenced by the vast majority of points lying on a 1-to-1 line drawn out from the origin (for example, $\chi^2_{9,0.5005} = 8.348$ is plotted against the 501st ordered $X^{2\star}$ value of 8.762).

There are other ways to obtain the permutation distribution. For example, the `boot()` function in the `boot` package provides a convenient way to perform permutation tests in general, where the `sim = "permutation"` argument value needs to be specified in the function call. We provide an example of how to use `boot()` in our program.

6.2.3 Exact logistic regression

Exact inference methods for logistic regression provide an alternative to the large-sample methods discussed in Chapter 2. In that chapter, we made inferences about a regression parameter β_j using an approximate normal distribution for the parameter's corresponding estimator $\hat{\beta}_j$. The use of the normal distribution was justified by the fact that MLEs are approximately normally distributed in large samples. Exact methods are useful in

[8] If desired, using `options(warn = -1)` before the `for()` function will prevent the warnings from being printed. This should be done only when one is sure that the warnings are not of concern. Changing the `warn` argument value to `0` allows the warnings to be printed again.

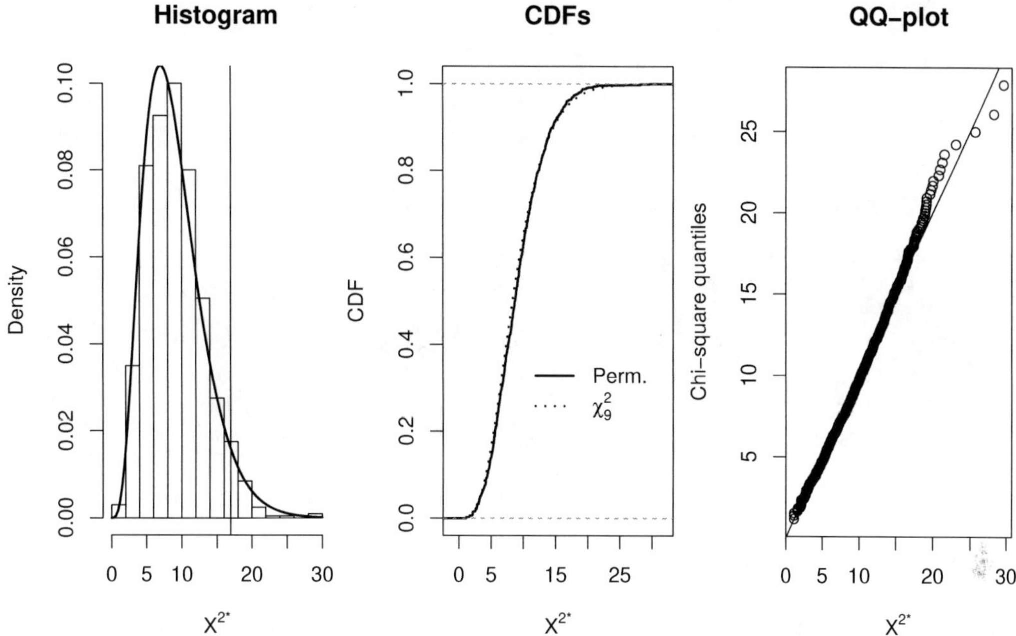

Figure 6.2: Permutation and chi-square distributions. The vertical line in the histogram is drawn at $X^2 = 16.94$.

situations when small sample sizes may not be large enough for such approximations to work well. Furthermore, exact methods allow estimation and inference when the convergence of maximum likelihood estimates may not occur, such as in complete separation situations (see Section 2.2.7).

The general approach to exact logistic regression is similar to that used in Fisher's exact test. For each parameter of interest, we first identify which statistics do or do not contribute to that parameter's estimation. Those statistics that do contribute are called the sufficient statistics for the parameter.[9] As shown in the previous section, there are ways to permute the observed data that leave the values of certain statistics unchanged. We apply this approach to inference on regression parameters by holding the explanatory variables fixed and permuting the observed responses. We then focus on the subset of these permutations in which the values of certain sufficient statistics are held constant and look at the distribution of other statistics within that subset. This is referred to as the exact conditional distribution of the latter statistics.

To formalize these ideas, write the joint probability distribution (also the likelihood function) for logistic regression as

$$\begin{aligned} P(Y_1 = y_1, \ldots, Y_n = y_n) &= \prod_{i=1}^{n} P(Y_i = y_i) \\ &= \prod_{i=1}^{n} \pi_i^{y_i}(1 - \pi_i)^{1-y_i}. \end{aligned}$$

[9]Further details on sufficiency are available in Section 6.2 of Casella and Berger (2002).

Because

$$\pi_i = \frac{\exp(\beta_0 + \beta_1 x_{i1} + \cdots + \beta_p x_{ip})}{1 + \exp(\beta_0 + \beta_1 x_{i1} + \cdots + \beta_p x_{ip})}$$

and

$$1 - \pi_i = \frac{1}{1 + \exp(\beta_0 + \beta_1 x_{i1} + \cdots + \beta_p x_{ip})},$$

we can write the joint probability compactly as

$$P(Y_1 = y_1, \ldots, Y_n = y_n) = \frac{\exp[\sum_{i=1}^n y_i(\beta_0 + \beta_1 x_{i1} + \cdots + \beta_p x_{ip})]}{\prod_{i=1}^n [1 + \exp(\beta_0 + \beta_1 x_{i1} + \cdots + \beta_p x_{ip})]}. \tag{6.4}$$

Applying Theorem 6.2.10 of Casella and Berger (2002) to Equation 6.4, the sufficient statistic for the regression parameter β_j is $\sum_{i=1}^n y_i x_{ij}$, $j = 0, ..., p$, where $x_{i0} = 1$ for $i = 1, \ldots, n$ corresponds to the intercept parameter β_0.

Without loss of generality, suppose our interest is in β_p. Inference on β_p is carried out using the exact conditional distribution of its sufficient statistic, say $T = \sum_{i=1}^n Y_i x_{ip}$, holding constant the values of the other sufficient statistics at their observed values. Let $t = \sum_{i=1}^n y_i x_{ip}$ be the observed value of T and let I be a vector denoting values of the sufficient statistics for all other parameters, $I = \{\sum_{i=1}^n y_i x_{ij}$ for $j = 0, \ldots, p-1\}$. Denote a permutation of $y_1, \ldots, y_n$ by $y_1^*, \ldots, y_n^*$. The joint probability for $Y_1, \ldots, Y_n$ conditional on I is shown in Exercise 10 to be

$$P(Y_1 = y_1, \ldots, Y_n = y_n | I) = \frac{\exp(\beta_p \sum_{i=1}^n y_i x_{ip})}{\sum_R \exp(\beta_p \sum_{i=1}^n y_i^* x_{ip})}, \tag{6.5}$$

where R is the set of all permutations of $y_1, \ldots, y_n$ such that the values in I remain unchanged.

Next, define U as the number of possible values that $\sum_{i=1}^n y_i^* x_{ip}$ can take on as formed by different permutations of $y_1, \ldots, y_n$ in R. Let these distinct values be denoted as $t_1, t_2, \ldots, t_U$. Also, define $c(t_u)$ as a count for the number of permutations in R such that $\sum_{i=1}^n y_i^* x_{ip} = t_u$. Then, by using Equation 6.5, Exercise 10 also shows that the exact conditional PMF of T given I is

$$P(T = t_u | I) = \frac{c(t_u)\exp(\beta_p t_u)}{\sum_{v=1}^U c(t_v)\exp[\beta_p t_v]}, \tag{6.6}$$

for $u = 1, \ldots, U$.

Because finding the set R is not trivial, specific algorithms have been developed to identify the permutations that comprise R (e.g., see Hirji et al., 1987) and to compute the PMF. Once found, this PMF is used in a similar manner as permutation distributions were used in Section 6.2.2. Simply, we can calculate a p-value based on how extreme the observed value t is relative to the PMF. Small probabilities indicate evidence against the null hypothesis. We will demonstrate this process shortly in an example.

To estimate β_p, a conditional maximum likelihood estimate (CMLE) can be obtained by maximizing Equation 6.5. However, when the random variable $\sum_{i=1}^n Y_i x_{ip}$ takes on its minimum or maximum possible observed value, then a finite estimate does not exist. In these cases, a *median unbiased estimate* (MUE) can be found instead, and this estimate will always be finite. The estimate is calculated essentially by finding the median of the values $t_1, \ldots, t_u$ using the conditional distribution in Equation 6.6. Specific calculation details are available in Mehta and Patel (1995).

Exact logistic regression is performed by the `elrm()` function of the `elrm` package (Zamar et al., 2007). This function provides approximations to exact logistic regression using

Markov Chain Monte Carlo (MCMC) simulation methods.[10] We will not discuss the details on how the MCMC methods work, but rather refer the reader to the corresponding paper on the package and also to Section 6.6 for an introduction to MCMC methods. It is important to note that the `elrm()` function requires the data to be in explanatory variable pattern (EVP; see Section 5.2) form. We discuss how to use this function and other aspects of the `elrm` package in our next example.

Example: Cancer free (MP5.1.R)

Mehta and Patel (1995) provide a series of examples demonstrating exact logistic regression. In particular, Section 5.1 of their paper examines the proportion of individuals (`w/n`) who were in remission from osteosarcoma over a three-year period. The explanatory variables included within a logistic regression model were lymphocytic infiltration (`LI`, 1 = present and 0 = not present), gender (`gender`, 1 = male and 0 = female; no other genders reported), and any osteoid pathology (`AOP`, 1 = yes and 0 = no). Below are the data.

```
> # Number disease free (w) out of total (n) per EVP
> set1 <- data.frame(LI = c(0,0,0,0,1,1,1,1), gender =
    c(0,0,1,1,0,0,1,1), AOP = c(0,1,0,1,0,1,0,1), w =
    c(3,2,4,1,5,3,5,6), n = c(3,2,4,1,5,5,9,17))
> set1
  LI gender AOP w  n
1  0      0   0 3  3
2  0      0   1 2  2
3  0      1   0 4  4
4  0      1   1 1  1
5  1      0   0 5  5
6  1      0   1 3  5
7  1      1   0 5  9
8  1      1   1 6 17
```

The data show that everyone who does not have lymphocytic infiltration present is disease free for three years. Thus, there is complete separation, and this causes problems when fitting a logistic regression model through maximum likelihood estimation.

```
> mod.fit <- glm(formula = w/n ~ LI + gender + AOP, data = set1,
    family = binomial(link = logit), weights = n, trace = TRUE,
    epsilon = 1e-8)  # Default epsilon value specified
Deviance = 2.943682 Iterations - 1
Deviance = 2.039803 Iterations - 2

<OUTPUT EDITED>

Deviance = 1.627772 Iterations - 19
Deviance = 1.627772 Iterations - 20
```

[10]Monte Carlo simulation methods are used in other software packages as well because of the complicated computations. For example, the LogXact software implements these methods as given by Mehta et al. (2000).

```
> round(summary(mod.fit)$coefficients, 4)
             Estimate Std. Error z value Pr(>|z|)
(Intercept)   23.4920 11084.3781  0.0021   0.9983
LI           -21.3842 11084.3781 -0.0019   0.9985
gender        -1.6362     0.9123 -1.7935   0.0729
AOP           -1.2204     0.7712 -1.5825   0.1135
```

Interestingly, `glm()` concludes its iterative numerical procedure after 20 iterations without any warning messages, because the relative change in the residual deviance is below the default value for `epsilon`. This can lead one to believe that convergence of the regression estimates has occurred. However, one should have strong concerns about the regression estimates because of the extremely large values of $\widehat{Var}(\hat{\beta}_0)^{1/2}$ and $\widehat{Var}(\hat{\beta}_1)^{1/2}$. In fact, when we make the convergence criterion more strict by decreasing the value of `epsilon` from its default, the estimates of β_0 and β_1 and their respective variances continue to diverge from zero without bound. Therefore, the output given by `glm()` should not be used.

We next perform the MCMC approximation to exact logistic regression using the `elrm()` function. As with `glm()`, a `formula` argument specifies the model; however, a `weights` argument is not needed even though the data are in EVP form. A seed is set prior to invoking `elrm()` so that the same results can be reproduced. Below is the code and output.

```
> library(package = elrm)
> set.seed(8718)
> mod.fit.elrm.LI <- elrm(formula = w/n ~ LI + gender + AOP,
    interest = ~ LI, iter = 101000, dataset = set1, burnIn =
    1000, alpha = 0.05)
Generating the Markov chain ...
Progress: 100%
Generation of the Markov Chain required 6 secs
Conducting inference ...
Inference required 0 secs

> summary(mod.fit.elrm.LI)

Call:
elrm(formula = w/n ~ LI + gender + AOP, interest = ~ LI, iter =
    101000, dataset = set1, burnIn = 1000, alpha = 0.05)

Results:
   estimate p-value p-value_se mc_size
LI -1.85724 0.06285    0.00234   1e+05

95% Confidence Intervals for Parameters
   lower     upper
LI  -Inf 0.1732663
```

The `interest` argument specifies which specific regression parameters are of interest for estimation and inference. Only those explanatory variables identified in this argument have information returned on them. In this case, we only included the lymphocytic infiltration variable. The estimate for β_1 in the model $\text{logit}(\pi) = \beta_0 + \beta_1\texttt{LI} + \beta_2\texttt{gender} + \beta_3\texttt{AOP}$ is $\hat{\beta}_1 = -1.8572$.

More than one explanatory variable can be specified in the `interest` argument by using the usual syntax for a `formula` argument. For that case, a joint test involving the corresponding regression parameters is also performed. Estimation and inference for the intercept term β_0 is not available through `interest` or any other argument. This is likely because inference is usually focused on the parameters representing effects of the explanatory variables.

The MCMC methods implemented by `elrm()` iteratively produce a dependent random sequence of possible sufficient statistic values for the regression parameter of interest, where the values in the sample are approximately at the same relative frequency as the probabilities given by Equation 6.6. The `iter` argument within `elrm()` specifies the total number of iterations in the random sequence, and the `burnIn` argument specifies the size of the initial iterations to discard (this helps to ensure that the sample is representative of the corresponding distribution). From the `summary()` output, we see that this results in a sequence of length 100,000 for inference purposes. The `plot()` method function that can be used with `mod.fit.elrm.LI` displays the history of the sampling process and the frequency histogram for the distribution of the sufficient statistic values. While we do not show these plots here, we do display the relative frequency distribution below associated with the sufficient statistic of the lymphocytic infiltration variable.

```
> names(mod.fit.elrm.LI)
[1]  "coeffs"        "coeffs.ci"       "p.values"
[4]  "p.values.se"   "mc"              "mc.size"
[7]  "obs.suff.stat" "distribution"    "call.history"
[10] "dataset"       "last"            "r"
[13] "ci.level"

> mod.fit.elrm.LI$obs.suff.stat
LI
19

> mod.fit.elrm.LI$distribution  # Sorted by second column
$LI
     LI    freq
[1,] 25 0.00162
[2,] 24 0.02298
[3,] 19 0.03825
[4,] 23 0.11684
[5,] 20 0.18500
[6,] 22 0.30061
[7,] 21 0.33470

> sum(mod.fit.elrm.LI$distribution$LI[1:3,2])  # Two tail test
[1] 0.06285
```

The observed value for the sufficient statistic is $t = 19$. Because it is the smallest value of t, this means that $\hat{\beta}_1 = -1.8572$ is an MUE. The probability of $t = 19$ is estimated to be 0.03825. Using the relative frequency distribution, we obtain an estimate of the exact p-value for a test of $H_0 : \beta_1 = 0$ vs. $H_a : \beta_1 \neq 0$ as $\hat{P}(T = 19|\, I) + \hat{P}(T \geq 24|\, I) = 0.06285$, which is also given in the output produced by the `summary()` function. Therefore, there is marginal evidence that lymphocytic infiltration is associated with remission.

The same code can be used for gender and osteoid pathology by changing the `interest` argument. Code for those computations are included in our program. In summary, we find CMLEs of $\hat{\beta}_2 = -1.5480$ and $\hat{\beta}_3 = -1.0949$. The p-value for $H_0 : \beta_2 = 0$ vs. $H_a : \beta_2 \neq 0$ is 0.1110, and the p-value for $H_0 : \beta_3 = 0$ vs. $H_a : \beta_3 \neq 0$ is 0.1658.

The `logistiX()` function of the `logistiX` package also performs exact logistic regression. Computations are performed without the need of approximation. Unfortunately, this package is no longer maintained by its authors and is now archived on R's website. This prevents the package from being downloaded and installed the usual way as described in Appendix A. Also, the package is no longer updated to reflect changes in R, which may lead to errors or unexpected results from running its functions. For those readers interested in installing the package and using its functions, we provide information and code in our program.[11]

Within the program for this example, we also provide code to implement the modified likelihood methods proposed by Firth (1993) using the `logistf()` function discussed in Section 2.2.7. The estimated logistic regression model is $\text{logit}(\hat{\pi}) = 4.2905 - 2.4611\text{LI} - 1.4153\texttt{gender} - 1.1039\text{AOP}$, which leads to similar conclusions as with the exact methods. Also, the hypothesis test involving $H_0 : \beta_1 = 0$ vs. $H_a : \beta_1 \neq 0$ results in a p-value of 0.0309, which again leads to similar conclusions as with exact logistic regression. For this example, the p-values obtained by testing the significance of each explanatory variable are a little smaller for the modified likelihood approach than for exact logistic regression.

Both exact logistic regression and the modified likelihood approach of Firth (1993) can replace standard maximum likelihood estimation when complete separation occurs. Heinze (2006) discusses the relative merits of the two procedures. The paper notes that the modified likelihood approach can be applied in situations when explanatory variables are truly continuous or when there are few observations per EVP. Exact methods generally cannot be used in these situations because of problems with evaluating the conditional distribution of a sufficient statistic (there may be only one permutation of $y_1, \ldots, y_n$ that satisfies I). In a very limited simulation study, Heinze (2006) also shows that both the modified likelihood and exact methods lead to conservative tests. However, the modified likelihood approach can produce more powerful tests. Heinze goes on to show that calculating p-values for exact tests using the mid p-value method (often used to limit the conservativeness of a test involving a discrete distribution, see Exercise 11) allows the two tests to have comparable power. However, tests based on a mid-p-value are no longer guaranteed to reject the null hypotheses at a rate less than or equal to the type I error level. This mid p-value is available in the `ciout` component of a model-fit object produced by `logistiX()`.

It is also important to remember that inference using the modified likelihood approach requires large samples in order for distributional approximations to be accurate. The exact inference methods do not have this requirement. While Heinze (2006) shows that the inferences based on the modified likelihood approach generally work as intended, these conclusions were based on a limited, although realistic, set of simulations where moderately sized samples were used.

[11]The `logistiX` package is an important illustration of a downside to using R or any other software package that relies on user contributions. If the package is no longer maintained, one may no longer be able to use the package. This is why we focus on packages within the default installation of R. We only use packages contributed by users when no other package exists for a particular statistical method or there are significant advantages from using them.

6.2.4 Additional exact inference procedures

Exact inference methods are used in many other places involving categorical data, and a large number of R packages include functions to perform these methods. A general survey of current R packages can be found simply by doing a search on "exact" at the CRAN web page that lists all R packages: `http://cran.r-project.org/web/packages/available_packages_by_name.html`. For example, the `exact2x2` package provides a function named `exact2x2()` that not only performs Fisher's exact test but also provides a Blaker-based alternative that is less conservative (the Blaker interval for a single binomial probability was discussed in Section 1.1.2), and the `mcnemar.exact()` function that provides an exact version of McNemar's test discussed in Section 1.2.6. Also, the `exactLogLinTest` package can perform exact inference for Poisson regression models.

6.3 Categorical data analysis in complex survey designs

All of the statistical methods that we have examined so far assume that we have a simple random sample of units (animals, free throws, etc.) from an infinite population. This assumption allows us to form our usual models: the binomial, the Poisson, and the multinomial. These models are convenient for the mathematical development of analysis methods and sampling distributions, but they may not reflect how data are actually gathered. More complex methods of sampling are often used in surveys, particularly when they are done on human populations. For example, large national surveys, like the Centers for Disease Control and Prevention's National Health Interview Survey or Statistics Canada's General Social Survey, feature complexities such as stratification, clustering, and unequal-probability sampling.

When complex sampling designs are used, the resulting observations are no longer independent, and they may have other features that further invalidate our usual models. The purpose of this section is to show how to perform many of the same types of statistical analyses from past chapters when sampled units are selected using a complex sampling design. We present methods for computing confidence intervals for a proportion, performing tests for independence, and fitting regression models.

Analysis of complex survey data is a broad topic. We provide only basic introductions to the important aspects of these analyses. Other resources provide more in-depth discussions. In fact, there are whole textbooks on this subject. We recommend that readers who are interested in that level of detail examine resources such as Korn and Graubard (1999), Thompson (2002), Lumley (2011), Skinner (2019), and Lohr (2022).

6.3.1 The survey sampling paradigm

Sampling designs are often constructed to increase the convenience of the sampling process and to reduce variability of particular statistics of interest. We briefly discuss below the specific features of a sampling design that enable us to accomplish these goals.

Stratification and Clustering

Two key features in complex survey designs are *stratification* and *clustering*. Stratification is the labeling of population units into groups, called strata, according to some similar, quantifiable feature(s) prior to sample selection. For example, people in a population could be stratified by gender, income, marital status, neighborhood, or any number of possible features. Sampling is then done separately within each stratum according to some probability-based design. The ability to control sample sizes within each stratum can ensure that estimated quantities from each stratum have reasonable precision.

Clustering also represents a grouping of units, but for a completely different purpose. Clusters are generally formed for convenience: units are grouped in such a way that those within a cluster are easily sampled together. For example, in door-to-door surveys it is much easier to sample 10 households in a particular neighborhood than to select one household from each of 10 different neighborhoods. The drawback with clustering is that the units within a cluster tend to be similar in some ways and, hence, their responses tend to be more similar to one another than to responses from other clusters. This means that a sample consisting of only one cluster is not an adequate representation of a whole population, while a sample consisting of multiple clusters contains groups of data that are not independent.

Thus, stratification makes sampling a bit less convenient but improves precision, while clustering makes sampling more convenient but diminishes precision. In practice, these tools are often used together, and often in ways that make it very difficult to determine a good distributional model to represent the data. The usual binomial, multinomial, and Poisson models that assume independent observations are no longer valid and may, in fact, very poorly represent the actual structure of the sample. In particular, parameter estimates may be biased, and standard errors may be poorly estimated, sometimes severely so.

To summarize the impact that a sampling design has on the variance of an estimated parameter, *design effects* are often calculated for a survey. The design effect, or *deff*, is the ratio of the variance for the estimate under the survey design to the variance that the same sample size would provide under simple random sampling. For example, under simple random sampling from a Bernoulli distribution, we showed in Section 1.1.2 that the variance for the sample proportion is $\widehat{Var}_{SRS,n}(\hat{\pi}) = \widehat{Var}(\hat{\pi}) = \hat{\pi}(1-\hat{\pi})/n$. Under a complex survey design, the variance for the sample proportion, $\widehat{Var}_{DES,n}(\hat{\pi})$, could be something quite different. The deff measures $\widehat{Var}_{DES,n}/\widehat{Var}_{SRS,n}$. In most surveys, deffs are greater than 1.

Survey weights

Analysis of complex survey data is complicated by the presence of *survey weights*. Surveys—especially those with stratification—are often conducted with unequal probability sampling, meaning that not all units have the same probability of being included in the sample. For example, suppose that a regional government wants to build a new prison in a small town and wants to gauge support for the project by surveying the population. Furthermore, suppose that 1,000 people live in the town and that there are 1,000,000 people in the government's region outside the town. Naturally, they want to be certain to include both town residents and other regional residents in the sample. If a sample of total size 50 were to be drawn by simple random sampling, there is approximately a 95% chance that *no* town people would be selected. Instead, it might make sense to sample, say, 10 people from the town and 40 from the rest of the region, to ensure an adequate sample size to estimate parameters for both groups. Then each town person in the sample represents 100 members of the town population, while each outsider in the sample represents 25,000 members of the rest of the population. For constructing estimates of population totals, multiplying town responses by 100 and outsider responses by 25,000 will tend to give unbiased estimates of population totals for the entire region. This is the basis for survey weights: for each sampled unit, its weight estimates how many members of the population that unit represents.[12]

Weights can be constant if all members of the overall population have equal probability of being included in the sample, or weights can vary considerably according to stratum sizes and other features of the survey design. Weights are often scaled so that their sum across the entire sample is approximately equal to the population size. Sometimes they are instead scaled so that their sum is equal to the sample size, in which case they vary around 1. The scales are important only for the interpretation of count estimates that they may produce, i.e., whether they reflect counts for the population or for the sample. One can convert weights or counts from one scale to the other by multiplying the weights by a constant that depends on the population and sample sizes. We generally assume in this section that weights are scaled to produce population counts.

6.3.2 Overview of analysis approaches

Two main approaches have emerged for analyzing data from complex surveys. They are called *design-based* and *model-based* analysis. They have different uses, different assumptions, and can give substantially different results, so it is important to assess first which method is most appropriate for a given problem. Binder and Roberts (2003, 2009) give excellent discussions of the differences between the two approaches. Design-based inference assumes that there exists a fixed, finite population from which the sample was drawn and about which inferences are to be made. Model-based inference assumes that the population is a temporary thing that is constantly changing and itself is sampled from a "super-population" according to some model.

As an example to highlight the differences between these two approaches, suppose one does a survey on a sample selected from first-year students at a particular university. If the interest is specifically in *that* year's first-year students at *that* university, then a design-based approach would be appropriate. If, instead, the interest is to learn about first-year students more generally, and the results will be applied to other years or other universities, then the first-year students from whom the sample is selected are themselves a sample from this larger super-population. A model-based approach would likely be more appropriate in this case. Using the wrong approach can lead to estimates that are biased and inconsistent, meaning that they estimate the wrong population parameter. Also, standard errors calculated under the wrong approach can differ drastically from the correct ones, giving inappropriate confidence intervals or test results. Selecting the appropriate approach is therefore a crucial first step.

Design-based analysis

If units are selected from a fixed population with known probabilities, then these *inclusion probabilities* can be used to construct the survey weights. The methods for constructing weights are covered in standard texts on survey sampling and are beyond our present scope. Given the weights, the general approach to design-based inference with categorical data proceeds as follows.

First, the weights are used to form weighted counts of responses in each response category by simply summing the weights for all units in that category. These weighted counts estimate the corresponding population total counts. The covariance matrix of the weighted counts is found using one of many possible methods that are standard in survey analysis (see, e.g., Korn and Graubard, 1999 or Lohr, 2022). Next, any statistics or formulas that are normally written as functions of sample counts—including most statistics covered in

[12]Survey weights may also include adjustments for missing data and/or nonresponse.

Chapters 1–4—are instead written in terms of the weighted counts. The delta method is used as described in Appendix B to "linearize" these functions with respect to the weighted counts. Such linear approximations make finding the variance of any statistic relatively simple by using Equation B.7. Approximate normality of many statistics is established using forms of the central limit theorem that are appropriate for finite population problems (see, e.g., Binder, 1983). Wald-type tests and confidence intervals are usually constructed from the resulting estimates. Additional analyses based on Pearson and likelihood-ratio chi-square test statistics can be replaced by similar calculations on the weighted estimates. Approximate sampling distributions for the statistics are derived (Imrey et al., 1982; Rao and Scott, 1981, 1984) that account for both the variability of weights and the correlations among observations that are created by the survey design.

Alternatively, *resampling methods*—in particular, the jackknife, the bootstrap , and balanced repeated replication—are often used to generate replicate sets of sampling weights, especially in large government-based surveys. The statistic of interest is calculated on each replicate set, so that the variability among these replicate statistics can be used to estimate the standard error of the statistic. This avoids the need to derive approximations using the delta method.

It should be noted that hypothesis tests are not commonly performed in strict design-based analysis. This is because it is unreasonable to expect null hypotheses to be exactly true for the finite population to which they are to apply. For example, a null hypothesis that a population proportion is 0.5 cannot possibly be true if the total size of the population is odd! When we conduct hypothesis tests in a design-based analysis, then, we are generally implying that there is a super-population that created the current finite population (e.g., there is a constant flow of people into and out of various age groups), and the hypotheses are meant to apply to this super-population. The decision to use design-based inference in this case can still be justified when we do not believe that a model can adequately account for all features of the design. See Binder and Roberts (2009) for an excellent discussion.

Model-based analysis

Using methods from Section 6.5, it is possible to explicitly account for clustering by including a corresponding random effect factor in a model. Furthermore, strata can be included as fixed effects in a model to allow separate estimates to be calculated for each stratum. Thus, it may be possible to account for many features of a complex sampling design by augmenting a model appropriately. If the augmented model is correctly specified, then the usual analysis procedures (e.g., generalized linear (mixed) models) meant for simple random samples can be used, and the fact that the data came from a different design is ignored. Estimates and standard errors obtained from the model are then appropriate estimates of their corresponding population quantities. Binder and Roberts (2009) suggest comparing the sizes of design-based and model-based standard errors. If they are similar, then the model seems to account for features of the design successfully and resulting inferences are more likely to be reliable than if the standard errors differ substantially.

Comparison

Design-based estimates of population counts are unbiased, and subsequent estimates based on these counts can be expected to be unbiased, or nearly so, depending on the structure of the estimate. Model-based estimates that ignore weights can be biased—sometimes very badly—as estimates of parameters for the finite population from which a sample is drawn. However, design-based estimates tend to have higher variability than do model-based estimates, and the difference becomes worse the more variability there is in the weights. There is no universal preference for one approach over the other, as the bias vs. variance trade-off is very much case-dependent. Nonetheless, Korn and Graubard (1999, p. 143) note: "In summary, weighted estimation used with variance estimation based on the survey design will yield appropriate analysis in most situations."

Because model-based analyses use existing techniques that we cover elsewhere, we discuss only design-based analysis methods here. We begin our discussion by introducing an example and the R package that is used for design-based analyses in this section.

Example: NHANES 1999–2000 (SurveySmoke.R, SmokeRespAge.csv)

The National Health and Nutrition Examination Survey (NHANES) from the U.S. Centers for Disease Control and Prevention is a program that has run a series of surveys on various health issues since the 1960s. Among the many questions asked in 1999–2000 are five questions about lifetime tobacco use and four about respiratory symptoms. Specifically, subjects are asked whether they have used at least 100 cigarettes, smoked a pipe at least 20 times, smoked a cigar at least 20 times, used snuff at least 20 times, or used chewing tobacco at least 20 times. The responses are recorded as a binary variable for each of the five tobacco items. Subjects are also asked about respiratory symptoms: a persistent cough, bringing up phlegm, wheezing/whistling in chest, and dry cough at night. Presence of each of the four symptoms is also recorded as a separate binary response for each symptom. Finally, there are weights associated with each subject, and an additional set of 52 jackknife replicate weights that are to be used for variance estimation (Section 6.3.4 gives an example of how these are used). The population about which inferences can be drawn is all adults aged 20 or older in the United States at the time the survey was conducted. See `http://www.cdc.gov/nchs/nhanes.htm` for a full description of the NHANES program, questionnaires, and sampling methods used. We will use these data to demonstrate various forms of analyses by examining relationships among tobacco-use and respiratory-symptom variables that might naturally be of interest to a researcher in public health.

The code and output below show a small portion of the data. Age is listed first. Next come the five smoking indicators (prefixed by `sm_`), followed by the four respiratory symptom indicators (prefixed by `re_`). The survey weights supplied with the data are next (`wtint2yr`), and last are the 52 sets of jackknife replicate weights, `jrep01`–`jrep52` (only two are shown below). Notice that these replicate weights are already on the same scale as the original survey weights; in some surveys, these replicate weights are merely "adjustment factors" (numbers varying around 1) that need to be multiplied by the original weights to create new combined weights. It is apparent from the weights that some subjects represent substantially more members of the population than others. For example, observation #5 represents 92,603 individuals in the population, and observation #6 represents just 1,647.

```
> smoke.resp <- read.csv(file = "C:\\data\\SmokeRespAge.csv")
> smoke.resp[1:6, 1:13]
  age sm_cigs sm_pipe sm_cigar sm_snuff sm_chew re_cough
1  77       0       1        1        0       0        0
2  49       1       1        1        0       1        0
3  59       1       0        0        0       0        0
4  43       1       0        0        0       0        0
5  37       0       1        1        0       0        0
6  70       1       0        0        0       0        0
```

```
  re_phlegm re_wheez re_night   wtint2yr    jrep01     jrep02
1         0        0        0 26678.636 26923.62 26824.718
2         0        0        0 91050.847 92772.79 91672.817
3         0        0        0 22352.089 22495.11 24238.166
4         0        0        0 21071.164 21209.55 22343.413
5         0        1        0 92603.187 94354.49 93235.761
6         0        0        0  1647.446  1681.59  1648.714
> nrow(smoke.resp)  # Sample size
[1] 4852
```

The survey package in R performs a wide range of statistical analyses for data from a complex sampling design. The package contains functions that perform several common analyses for categorical data. In particular, there are capabilities for the analysis of single proportions, contingency tables, and logistic, Poisson, and certain multinomial regression models for ordinal responses. Multinomial regression models for nominal responses and automated model selection techniques are not available, and there is very limited capacity for model diagnostics. Lumley (2011) describes the package in much more detail than we can cover here. Also, the package is frequently updated, so one should check the documentation for this package before using the programs below in case functions have been changed or added. Version 4.2-1 was used for the calculations in this section.

After loading the survey package, we need to convert our data into an object that contains information about the design so that the analysis functions can properly calculate statistics and their variances. We use svrepdesign() here because our design has replicate weights; other functions allow the user to instead specify stratification and clustering identifiers that can be used in variance estimation. Arguments identify the columns of the data frame that contain the analysis variables (data), the original survey weights (weights), and the replicate weights (repweights). The description of the survey design (http://www.cdc.gov/nchs/data/nhanes/guidelines1.pdf) recommends using the leave-one-out version of jackknife, which is selected here using the argument value type = "JK1". As noted above, the replicate weights are on the same scale as the survey weights, so combined.weights = TRUE. The scale = 51/52 relates the presence of 52 replicate weights to the formula for leave-one-out jackknife estimates of variance (Lohr, 2022, p. 378). The code below creates an object of the class svyrep.design, which can be used for all analyses within survey.

```
> library(package = survey)
> jdesign <- svrepdesign(data = smoke.resp[, 1:10], weights =
   smoke.resp[,11], repweights = smoke.resp[,12:63], type =
   "JK1", combined.weights = TRUE, scale = 51/52)
> class(jdesign)
[1] "svyrep.design"
```

6.3.3 Weighted cell counts

The first and most fundamental step in an analysis of categorical data from a complex survey is estimating the population count of units in each category. Not only might estimating these counts be one of the main goals of the analysis, but also any further calculations, such as those described in Sections 6.3.4–6.3.6, are based upon these counts.

Suppose that the population consists of N units and that the response Y has categories $1, \ldots, I$, containing $N_1, N_2, \ldots, N_I$ members of the population, respectively, where $\sum_{i=1}^{I} N_i = N$. Let the sample consist of n units selected from the population according to some design. For each sampled unit, define its survey weight to be w_s, $s = 1, \ldots, n$ and its observed response to be y_s. For example, on p. 463 we show information on sampled units $s = 1, \ldots, 6$ from a sample of $n = 4852$ for the NHANES data set. The `wtint2yr` variable in that example corresponds to w_s.

Because the survey weight w_s is the number of population units represented by sampled unit s, the estimate of N_i is simply the sum of weights for all sampled units in category i:

$$\hat{N}_i = \sum_{s=1}^{n} w_s I(y_s = i),\ i = 1, \ldots, I \tag{6.7}$$

where $I(y_s = i)$ indicates whether or not unit s is in category i (1 for yes, 0 for no). We can similarly estimate the population total as $\hat{N} = \sum_{s=1}^{n} w_s = \sum_{i=1}^{I} \hat{N}_i$.

For convenience, collect the weighted cell counts into a vector, $\hat{\mathbf{N}} = (\hat{N}_1, \hat{N}_2, \ldots, \hat{N}_I)'$. The covariance matrix of $\hat{\mathbf{N}}$, $\widehat{Var}(\hat{\mathbf{N}})$, not only measures the variability of $\hat{\mathbf{N}}$, but also serves as the foundation for variances, confidence intervals, and test-statistic calculations for functions of $\hat{\mathbf{N}}$. This covariance matrix can be found using linearization, jackknife, and bootstrap methods as described earlier and discussed in detail in standard texts on analysis of surveys (e.g., Lohr, 2022). We generally do not provide the details for these methods here, but instead rely on functions from the `survey` package to complete the calculations when possible.

Degrees of freedom for $\widehat{Var}(\hat{\mathbf{N}})$, say κ, are not always straightforward to find. The default is to use (Number of clusters)−(Number of strata), although Rust and Rao (1996) indicate that this is generally an overestimate. In the NHANES example, the 52 replicate weights were formed from 52 random subgroups ignoring the design's clustering and stratification. This leads to $\kappa = 52 - 1 = 51$ degrees of freedom for $\widehat{Var}(\hat{\mathbf{N}})$.

Example: NHANES 1999–2000 (SurveySmoke.R, SmokeRespAge.csv)

For each of the nine binary responses for tobacco use and respiratory symptoms, we can estimate the population total with the weighted count of positive responses. The code below demonstrates the calculations for the first tobacco item, cigarette use. The result is the estimated number of members of the population who have smoked at least 100 cigarettes in their lifetimes.

```
> wt_cig <- smokerespj$sm_cigs * smokerespj$wtint2yr
> totcigwt <- sum(wt_cig)
> totcigwt
[1] 94480990
```

The variance of this estimate can be found by repeating the calculation on each of the 52 sets of jackknife replicate weights. Let $\hat{N}_i^{(r)}$ be the estimated population count for category i from replicate weight set r. Then the variance is found from the formula for leave-one-out jackknife estimates of variance (Lohr, 2022, p. 378),

$$\widehat{Var}(\hat{N}_i) = [(R-1)/R] \sum_{r=1}^{R} (\hat{N}_i^{(r)} - \hat{N}_i)^2. \tag{6.8}$$

More generally, the variance of a full vector of counts is $\widehat{Var}(\hat{\mathbf{N}}) = [(R-1)/R]\sum_{r=1}^{R}(\hat{\mathbf{N}}^{(r)} - \hat{\mathbf{N}})(\hat{\mathbf{N}}^{(r)} - \hat{\mathbf{N}})'$. Equation 6.8 can be calculated in R as shown below.

```
> Nhat.reps <- numeric(length = 52)
> for(r in 1:52){
    Nhat.reps[r] <- sum(smoke.resp$sm_cigs * smoke.resp[,j+11])
  }
> sum.sq <- var(Nhat.reps) * 51
> var.tot <- sum.sq * (51/52)
> sqrt(var.tot)  # Estimated standard deviation
[1] 2483739
```

The function `svytotal()` performs these calculations automatically, where we use the `design` argument to specify the data and its associated design elements.

```
> svytotal(x = ~ sm_cigs, design = jdesign)
             total      SE
sm_cigs 94480990 2483739
```

Documentation for NHANES 1999–2000 indicates that the jackknife procedure used to estimate variances may produce an underestimate of the true sampling variance for some statistics. It cautions to be careful when interpreting "marginally significant" results when these variances are used in hypothesis tests or confidence intervals.

6.3.4 Inference on population proportions

Analyses of categorical survey data often focus on population proportions rather than total counts (e.g., see Skinner (2019) for a discussion). Define the population proportions for each category (also called *cell probabilities*) as $\pi_i = N_i/N$, $i = 1, \ldots, I$. The population proportions are estimated as

$$\hat{\pi}_i = \hat{N}_i/\hat{N},\ i = 1, \ldots, I.$$

We use $\hat{N}$ in the denominator even when N is known, so that the estimated proportions sum to 1 across all categories.

Note that $\hat{\pi}_i$ is a ratio of two weighted estimates based on correlated observations. Its variance estimate is not simply the usual formula for the variance of a binomial proportion, $\hat{\pi}(1-\hat{\pi})/n$, as given in Equation 1.3. Instead, using the delta method from Appendix B.4.2, one can show that

$$\widehat{Var}(\hat{\pi}_i) = \frac{\widehat{Var}(\hat{N}_i) + \hat{\pi}_i^2\widehat{Var}(\hat{N}) - 2\hat{\pi}_i\widehat{Cov}(\hat{N}_i, \hat{N})}{\hat{N}^2}, \quad (6.9)$$

where the needed design-based estimates of variances and covariances are all found from $\widehat{Var}(\hat{\mathbf{N}})$. Alternatively, this variance can be calculated directly using a resampling method. For example, the leave-one-out jackknife with R replicate weights results in the formula

$$\widehat{Var}(\hat{\pi}_i) = [(R-1)/R]\sum_{r=1}^{R}(\hat{\pi}_i^{(r)} - \hat{\pi}_i)^2, \quad (6.10)$$

where $\hat{\pi}_i^{(r)}$ is estimated from the rth set of replicate weights.

Methods for forming confidence intervals for the true proportions are analogous to those from Section 1.1.2. The Wald interval provides a simple calculation,

$$\hat{\pi}_i \pm Z_{1-\alpha/2}\sqrt{\widehat{Var}(\hat{\pi}_i)}$$

but with deficiencies noted in Section 1.1.2. Because $\widehat{Var}(\hat{\mathbf{N}})$ is based on a sum-of-squares calculation, a $t_{\kappa,1-\alpha/2}$ critical value can be used in place of $Z_{1-\alpha/2}$. However, this makes little practical difference unless $\kappa < 30$.

Kott and Carr (1997) propose a more accurate confidence interval for $\hat{\pi}_i$ by using a modification of the Wilson score interval given by Equation 1.4. Making an analogy to the binomial variance, $\hat{\pi}(1-\hat{\pi})/n$, they define the *effective sample size* for estimating proportions π_i, $i = 1, \ldots, I$, to be the value $n_i^* = \hat{\pi}_i(1-\hat{\pi}_i)/\widehat{Var}(\hat{\pi}_i)$. Then a $100(1-\alpha)\%$ confidence interval for π_i is

$$\frac{2n_i^*\hat{\pi}_i + t^2_{\kappa,1-\alpha/2} \pm t_{\kappa,1-\alpha/2}\sqrt{t^2_{\kappa,1-\alpha/2} + 4n_i^*\hat{\pi}_i(1-\hat{\pi}_i)}}{2(n_i^* + t^2_{\kappa,1-\alpha/2})} \tag{6.11}$$

Surveys are often more concerned with estimating population quantities rather than with testing hypotheses, but should a test of $H_0 : \pi_i = \pi_{i0}$ be needed, the usual Wald tests are used (see Section 1.1.2). The test statistic is

$$Z_W = \frac{\hat{\pi}_i - \pi_{i0}}{\sqrt{\widehat{Var}(\hat{\pi}_i)}}$$

where the quantities $\hat{\pi}_i$ and $\widehat{Var}(\hat{\pi}_i)$ are estimated from the survey using methods described above. The decision rule is found by comparing Z_W to the appropriate critical value from the standard normal (or t_κ) distribution according to the alternative hypothesis being tested.

Example: NHANES 1999–2000 (SurveySmoke.R, SmokeRespAge.csv)

We estimate the proportion of the population who have smoked at least 100 cigarettes in their lifetimes using a 95% confidence interval. The `svymean()` function estimates means and standard errors while accounting for the survey design. Since proportions are means of binary variables, we can use this function to estimate π_i.

```
> cigprop <- svymean(x = ~ sm_cigs, design = jdesign)
> cigprop
           mean      SE
sm_cigs 0.49434 0.0131
```

We find $\hat{\pi}_i = 0.49$, so nearly half of the population had used at least 100 cigarettes. The estimated standard error is $\widehat{Var}(\hat{\pi}_i)^{1/2} = 0.013$. The `svymean()` function recognizes the contents of the design `jdesign` and uses Equation 6.10 rather than Equation 6.9 for this calculation. A t-based confidence interval is calculated as shown below.

```
> confint(object = cigprob, level = 0.95, df = 51)
            2.5 %    97.5 %
sm_cigs 0.4679677 0.5207125
```

A Wald interval could be found from the code above by omitting the df argument. A Wilson score interval requires a few more calculations. We must first save the estimated proportion and standard error in an object and then calculate the effective sample size. Then, the interval is calculated according to Equation 6.11.

```
> # Wilson score interval preliminary calculations
> pihat <- coef(cigprop)
> # Effective sample size for Wilson score interval
> # Need as.numeric() to convert vcov() result from matrix
> eff_sample <- pihat*(1-pihat)/as.numeric(vcov(cigprop))
> round(eff_sample, 3)
 sm_cigs
1448.545
>
> tcrit <- qt(p = c(0.025, 0.975), df = 51)
> # Wilson score interval
> j.Wilsonci_wt <- (2*eff_sample*pihat + tcrit[2]^2 +
    tcrit*sqrt(tcrit[2]^2 + 4*eff_sample*pihat*(1-pihat))) /
    (2*(eff_sample + tcrit[2]^2))
> round(j.Wilsonci_wt, digits = 3)
[1] 0.468 0.521
```

The confidence intervals are essentially the same, roughly 0.468 to 0.521, because of the large sample size for this survey.

6.3.5 Contingency tables and loglinear models

Methods for analyzing contingency tables are introduced in Sections 1.2, 3.2, and 4.2.4. Simpler analyses for two-dimensional tables focus on tests for independence between the row and column variables. Loglinear models for contingency tables of any size allow for more detailed analyses, including tests for various forms of association and estimates of proportions or odds ratios.

The same analyses are often done when the data arise from a complex survey. The main difference is in how the calculations are performed. First, the table is constructed from survey-weighted cell counts $\hat{\mathbf{N}}$ or corresponding proportions rather than the sample counts or proportions. For example, a loglinear model for two factors under independence is

$$\log(N_{ij}) = \beta_0 + \beta_i^X + \beta_j^Z,\ i = 1, \ldots, I,\ j = 1, \ldots, J,$$

which is simply Equation 4.5, except modeling the log of the unknown population count, $\log(N_{ij})$, instead of $\log(\mu_{ij})$. Second, $\widehat{Var}(\hat{\mathbf{N}})$ is calculated based on the survey design rather than a Poisson model. Finally, the error degrees of freedom for any test or confidence interval from a model must be adjusted to $\kappa - p - 1$, where p is the number of parameters estimated in the model. Note that this last condition limits the size of models that can be considered without using special estimation techniques (Korn and Graubard, 1999, Section 5.2).

We start by considering a two-way contingency table formed from variables X with I categories and Y with J categories. Using the same notation as in Section 3.2.1, let π_{ij} be the proportion of the population satisfying $X = i, Y = j$. Testing independence of X and Y implies $H_0 : \pi_{ij} = \pi_{i+}\pi_{+j}$, where π_{i+} and π_{+j} are the population marginal proportions for $X = i$ and for $Y = j$, respectively. Ordinarily, a Pearson or LR test for independence would be used to compare the observed counts to those expected under the null hypothesis, as discussed in Sections 1.2.3, 3.2.3, and 4.2.4. In the context of a loglinear model for survey data, these two sets of counts are obtained from separate model fits to weighted data. The "observed" counts come from a saturated loglinear model, while the estimated expected counts are from a model fit that satisfies the null hypothesis. While it is straightforward to compute these statistics with weighted survey data, the survey design makes finding the probability distribution of the resulting statistics more difficult.

Tests of independence: Rao-Scott methods

The use of a χ^2 distribution for Pearson and LRT statistics is typically justified by the appropriateness of the underlying Poisson or multinomial sampling model for the data. However, these sampling models may be poor representations of the data from complex survey designs, so the resulting χ^2 distribution should not be expected to accurately approximate the true probability distribution of any test statistic. Indeed, there is much evidence to indicate that naive use of χ^2 distributions with Pearson statistics calculated on survey data can result in very poor analyses (e.g., Rao and Scott, 1981; Scott and Rao, 1981; Thomas and Rao, 1987; Thomas et al., 1996). In particular, in many designs the test tends to reject the null hypothesis much more often than the specified α level when the null hypothesis is true.

Because Pearson statistics are used so broadly in categorical data analysis, Rao and Scott (1981; 1984) developed corrections to improve their performance in complex survey designs. The corrections are similar to the well-known Satterthwaite (1946) corrections that are common in linear models analysis. Let X^2 be the Pearson test statistic for a particular test, and suppose that X^2 would have ν degrees of freedom under simple random sampling (e.g., in a two-way contingency table, $\nu = (I-1)(J-1)$). Rao-Scott methods compare the mean and variance of X^2 under the actual sampling design to the means and variances from members of the χ^2-distribution family. Using the fact that the mean and variance of the χ^2_ν distribution are ν and 2ν, respectively, relatively simple adjustments are then made to the test statistic and to the degrees of freedom that improve how well a χ^2 distribution matches the actual distribution of X^2.

First-order correction The simplest form of a Rao-Scott correction ensures that the mean of the distribution for the test statistic and the mean of the chosen sampling distribution are the same. This is called a *first-order correction*, because it is based on matching the means (first moments) of the two distributions. For ease of notation and to allow extension to larger loglinear models, let K be the number of cells in the table and relabel the probabilities π_{ij}, $i = 1, \ldots, I$; $j = 1, \ldots, J$ as π_k, $k = 1, \ldots, K$. For each π_k let its value under the null hypothesis be denoted by π_{0k}. Recalling the definition of a deff from Section 6.3.1, let d_k^0 be the deff corresponding to $\hat{\pi}_k$ under the null hypothesis:

$$d_k^0 = nVar_0(\hat{\pi}_k)/[\pi_{0k}(1-\pi_{0k})]$$

where $Var_0(\hat{\pi}_k)$ is the variance of $\hat{\pi}_k$ under the null hypothesis. Suppose that each probability estimate has the same deff, $d_k^0 = d^0$, and, further, that the survey-estimated covariances follow $\widehat{Cov}(\hat{\pi}_{k_1}, \hat{\pi}_{k_2}) = d^0(-\pi_{0k_1}\pi_{0k_2})$ for all pairs (k_1, k_2). Under these conditions, X^2 has a distribution that is asymptotically equivalent to that of d^0W, where $W \sim \chi^2_\nu$. Hence, the distribution of X^2/d^0 is approximately χ^2_ν. Note that under simple random sampling, $d^0 = 1$, so the test statistic remains X^2.

In practice, $Var_0(\hat{\pi}_k)$ is rarely known in advance, and not all cells k have exactly the same deff due to chance, so this simple adjustment is not generally accurate. Instead, one can estimate an average deff across all of the cells using matrix techniques outlined in Rao and Scott (1981) and Scott (2007). Call the resulting estimate $\bar{d}$. Then the first-order Rao-Scott test approximates p-values and critical values for tests involving X^2 by comparing

$$X^2_{RS1} = X^2/\bar{d}$$

to a χ^2_ν distribution. This procedure works very well as long as there is not much variability among the d^0_ks. This test is generally available as the "Chisq" argument value in testing procedures within the survey package.

Second-order correction More generally, when the deffs are not all approximately equal, X^2 has a distribution that is asymptotically equivalent to $\sum_{l=1}^{\nu} \delta_l W_l$, where W_l, $l = 1, \ldots, \nu$, are independent χ^2_1 random variables and δ_l, $l = 1, \ldots, \nu$, are quantities called *generalized deffs*. The δ_l's are found using matrix techniques described in Rao and Scott (1981) and Scott (2007) and have the same mean as the deffs, $\bar{d}$. Let c be the coefficient of variation among the δ_l's: $c^2 = \sum_{l=1}^{\nu}(\delta_l - \bar{d})^2/(\nu\bar{d}^2)$. Then, the second-order-corrected Rao-Scott test compares

$$X^2_{RS2} = X^2/[\bar{d}(1+c^2)]$$

to a $\chi^2_{\nu/(1+c^2)}$ distribution.

Additional approaches Thomas and Rao (1987) propose modified versions of the Rao-Scott tests that use F distributions rather than χ^2. In particular, they compare the statistic

$$F_{TR} = X^2_{RS1}/\nu = X^2/[\nu\bar{\delta}] \tag{6.12}$$

to an $F_{\nu/(1+c^2),\kappa\nu/(1+c^2)}$ distribution, where κ is again the degrees of freedom associated with the variance estimate $\widehat{Var}(\hat{\mathbf{N}})$. Simulations in Thomas and Rao (1987), Thomas et al. (1996), and Rao and Thomas (2003) show that this F-test works well enough under most conditions to be used as a primary procedure for Pearson tests with complex survey data. In fact, this is the default procedure for functions that perform Pearson-type tests in survey, and it can also be specified as the "F" argument value for testing arguments in these functions.

Other methods exist for computing p-values and critical values from X^2. In particular, survey can compute a "saddlepoint approximation" to the linear combination of chi-squares with estimated deffs, $\sum_{l=1}^{\nu} \hat{\delta}_l W_l$ (the saddlepoint argument value in tests, Kuonen, 1999). We have not investigated the use of this method for approximating the distribution of Pearson statistics in this context, but we expect that it would be effective when the sample size is large, so that the generalized deff estimates are fairly accurate.

The difference between model deviances—which is an LRT statistic when standard sampling models are used—can be corrected in the same manner as the Pearson statistic to provide a viable model-comparison test for comparing any two nested models. Thus, the mainstays of categorical modeling and contingency table analyses have useful analogs that can be applied in the complex-survey setting. These procedures are demonstrated in the examples that follow.

Lipsitz et al. (2015) also develop tests for independence in contingency tables. Their tests are particularly useful in cases where the table contains a zero count.

Inference on odds ratios

Odds ratios are introduced in Section 1.2.5 and discussed further in the context of a loglinear model in Section 4.2.4. For ease of notation, suppose X and Y are binary, although they could alternatively represent any two rows or columns from a larger table. Create the resulting 2×2 table of estimated population totals using the same approach as given in Equation 6.7. The population odds ratio is $OR = [\pi_{11}/(1-\pi_{11})]/[\pi_{21}/(1-\pi_{21})] = (N_{11}N_{22})/(N_{21}N_{12})$. Define $\hat{N}_{ij}$ to be the weighted count for $X = i, Y = j$, $i = 1, 2$; $j = 1, 2$. The population odds ratio is then estimated by

$$\widehat{OR} = \frac{\hat{N}_{11}\hat{N}_{22}}{\hat{N}_{21}\hat{N}_{12}}.$$

As was the case in Section 1.2.5, it is generally more accurate to base inference on $\log(\widehat{OR})$ than on $\widehat{OR}$ itself. Using the delta method to estimate the variance of $\log(\widehat{OR})$ leads to the formula

$$\widehat{Var}(\log(\widehat{OR})) = \mathbf{a}'\widehat{Var}(\hat{\mathbf{N}})\mathbf{a},$$

where $\hat{\mathbf{N}} = (\hat{N}_{11}, \hat{N}_{12}, \hat{N}_{21}, \hat{N}_{22})'$, $\widehat{Var}(\hat{\mathbf{N}})$ is the 4×4 covariance matrix computed using some method that accounts for the survey design, and $\mathbf{a} = (1/\hat{N}_{11}, -1/\hat{N}_{12}, -1/\hat{N}_{21}, 1/\hat{N}_{22})'$ is the vector of derivatives of $\log(OR)$ with respect to $N_{11}, N_{12}, N_{21}, N_{22}$ evaluated at $\hat{\mathbf{N}}$. Inference proceeds by computing Wald intervals for $\log(OR)$ as in Section 1.2.5, leading to the confidence interval

$$\exp\left[\log\left(\widehat{OR}\right) \pm Z_{1-\alpha/2}\sqrt{\widehat{Var}(\log(\widehat{OR}))}\right].$$

As was the case for a confidence interval for π_i, a $t_{\kappa,1-\alpha/2}$ critical value can be used in place of $Z_{1-\alpha/2}$. Also, $\widehat{Var}(\log(\widehat{OR}))$ can be estimated directly using a resampling method where available. Hypothesis tests can be conducted using a Wald test statistic and a rejection region or p-value based on the standard normal (or t_κ) distribution. This is equivalent to determining whether the confidence interval contains the hypothesized value.

The calculations for odds ratios and their confidence intervals or hypothesis tests are most easily carried out within the context of a loglinear model. As noted earlier in this section, the models are parameterized exactly the same as in Section 4.2.4, although they are applied to the population totals N_{ij} in an $I \times J$ table instead of the population means μ_{ij}. A model with only main effects represents independence between X and Y. A saturated model allows odds ratios between each pair of rows and columns to be estimated by the model through the XY interaction parameters.

Example: NHANES 1999–2000 (SurveySmokeLoglinear.R, SmokeRespAge.csv)

We examine the association between Y = *any* respiratory symptoms and X = *any* tobacco use. We define "any respiratory symptoms" (`anyresp`) as `TRUE` (1) if any of the binaries for persistent cough, bringing up phlegm, wheezing/whistling in chest, or dry cough at night are 1, and `FALSE` (0) if all of the binaries are zero. Similarly, "any tobacco use" (`anysmoke`) is `TRUE` (1) if any of the binaries for cigarettes, pipe, cigar, snuff, or chewing tobacco are 1, and `FALSE` (0) if all are 0. We first create the new binaries by applying `update()` to the existing object created by `svyrepdesign()`. We then compute the table totals using `svytable()`. The summary of this object gives the estimated totals and a test for independence using the default Thomas-Rao F-test. Tests for independence can also be conducted using `svychisq()`, where the `statistic` argument controls the type of test used.

```
> jdesign <- update(object = jdesign, anytob = sm_cigs + sm_pipe
    + sm_cigar + sm_snuff + sm_chew > 0, anyresp = re_cough +
    re_phlegm + re_wheez + re_night > 0)
> head(jdesign$variables, n = 2)
  age sm_cigs sm_pipe sm_cigar sm_snuff sm_chew re_cough
1  77       0       1        1        0       0        0
2  49       1       1        1        0       1        0
  re_phlegm re_wheez re_night anytob anyresp
1         0        0        0   TRUE   FALSE
2         0        0        0   TRUE   FALSE

> # Table of weighted counts
> wt.table <- svytable(formula = ~ anytob + anyresp, design =
    jdesign)
> summary(wt.table)
       anyresp
anytob     FALSE     TRUE
  FALSE 73873587 13006040
  TRUE  75870428 28375419

        Pearson's X^2: Rao & Scott adjustment

data:  svychisq(~anytob + anyresp, design = jdesign, statistic =
    "F")
F = 40.1833, ndf = 1, ddf = 51, p-value = 6.016e-08

> svychisq(formula = ~ anytob + anyresp, design = jdesign,
    statistic = "F")
        Pearson's X^2: Rao & Scott adjustment
data:  svychisq(~anytob + anyresp, design = jdesign, statistic =
    "F")
F = 40.1833, ndf = 1, ddf = 51, p-value = 6.016e-08
```

The totals show that there are similar numbers of people who experience no respiratory symptoms (anyresp = FALSE) regardless of tobacco use. However, there is a somewhat greater number of people who have used tobacco and experienced respiratory symptoms than who have *not* used tobacco and have experienced respiratory symptoms. This suggests that there may be an association between tobacco use and respiratory symptoms. The Rao-Scott tests confirms this with $F_{TR} = 40.1$ and a p-value of 6.0×10^{-8}.

In the program for this example, we show that the cell totals can be converted to proportions by dividing by the estimated total population size. Here, we present an alternative method for computing these proportions and the odds ratio directly using svymean(). This function also has the added feature to save out the proportion calculations on each jackknife replicate, allowing us to estimate the variance of the log-odds ratio values so that we can form a confidence interval. We use the replicate proportions to compute the odds ratio estimate for each replicate, and then amend Equation 6.10 to estimate the variance of $\log(\widehat{OR})$ instead of $\hat{\pi}_i$. Finally, a confidence interval is created using the t_{51}-distribution approximation to the sampling distribution of $\log(\widehat{OR})$. Notice that the input to svymean() uses the interaction() function. This function creates a new factor from all combinations of factors that are supplied as its argument values, and must be used here instead of other interaction-forming operators, ":" or "*".

```
> props22 <- svymean(x = ~ interaction(anytob, anyresp), design
   = jdesign, return.replicates = TRUE)
> props22
                                               mean      SE
interaction(anytob, anyresp)FALSE.FALSE 0.38652 0.0113
interaction(anytob, anyresp)TRUE.FALSE  0.39697 0.0114
interaction(anytob, anyresp)FALSE.TRUE  0.06805 0.0055
interaction(anytob, anyresp)TRUE.TRUE   0.14847 0.0089

> names(props22)
[1] "mean"        "replicates"

> # Odds ratio; order of terms is available from output
> ORhat <- props22$mean[1] * props22$mean[4]
       / (props22$mean[2] * props22$mean[3])
> ORhat
interaction(anytob, anyresp)FALSE.FALSE
                                2.12429

> OR.reps <- props22$replicates[,1]*props22$replicates[,4]
         / (props22$replicates[,2]*props22$replicates[,3])
> reps <- length(OR.reps)
# Calculating sum of squares using (n-1)*var()
> var.logOR <- (reps - 1)*var(log(OR.reps))*((reps - 1)/reps)

> # t-based confidence interval
> exp(log(ORhat) + qt(p = c(0.025, 0.975), df = reps-1) *
   sqrt(var.logOR))
[1] 1.667768 2.705778
```

The proportions reflect the same pattern noted above. We calculate $\widehat{OR} = 2.1$ with a t-based confidence interval from 1.7 to 2.7. This is strong evidence of a positive association between use of any tobacco product and presence of any respiratory symptoms.

This analysis can instead be conducted by fitting a saturated 2×2 loglinear model, finding a confidence interval for the association parameter—which is equivalent to $\log(\widehat{OR})/4$—and rescaling and exponentiating the results. We show how to perform this analysis in the program for this example and provide a similar type of analysis in the next example.

Higher-dimensional tables

The loglinear-modeling approach to the analysis of tables constructed from more than two variables was described in Section 4.2.5. The same analyses can be applied to data from a complex survey with modifications as described above using the `svyloglin()` function from the `survey` package. Information criteria are *not* generally available for use in model selection, because design-based model fitting does not use true likelihoods. Instead, combinations of forward and backward selection procedures are used that rely on hypothesis tests for their decisions to include or exclude variables. The example below demonstrates one possible approach to model selection, first choosing the complexity of possible interactions in the model and then determining which interactions of that level are needed.

Odds ratios can be computed for any comparison that can be expressed as a combination of model parameters. As always, the key step is to identify which model parameters are related to the odds ratios in question. Odds ratios are estimated primarily from parameters representing two-way interactions, although they can be impacted by higher-order interactions as well. Reiterating Equation 4.7, the odds ratio between levels i and i' of variable X and levels j and j' of variable Y is

$$OR_{ii',jj'} = \exp(\beta_{ij}^{XY} + \beta_{i'j'}^{XY} - \beta_{i'j}^{XY} - \beta_{ij'}^{XY}),$$

where each β_{ij}^{XY} is a parameter from the `X:Y` interaction in a model from `svyloglin()`.[13] Some care is needed when there are also higher-order interactions in a model. Recall the basic rules that govern the identification of terms involved in odds ratios:

1. Odds ratios are always 1 for any pair of variables that are not involved together in two-factor or higher interactions.

2. If a pair of variables is involved in a two-way interaction but does not appear together in higher-order interactions, then odds ratios between them are estimated from the two-way interaction parameters and their values are constant across all levels of other variables.

3. If a pair of variables appears together in both a two-way interaction *and* three-factor or higher-order interactions, then the odds ratios between them vary depending on the levels of the other factors with which they appear in model terms. One must then estimate odds ratios separately for each level of these other factors (but not for variables with which they are not involved in three-way or higher-order interactions).

See Table 4.5 for details.

Example: NHANES 1999–2000 (SurveySmokeLoglinear.R, NewSvyloglin-Pvalue.R, SmokeRespAge.csv)

To what extent does use of one tobacco product relate to use of others? Do people who use one product do so essentially independently of other products, or are there certain products that tend to be used together or used separately?

To address these questions, a loglinear model is sought that describes the associations among the five different tobacco-use binary variables. The counts for certain cells of the full 2^5 table are given below.

```
> alltob <- svytable(formula = ~ interaction(sm_cigs, sm_pipe,
    sm_cigar, sm_snuff, sm_chew), design = jdesign)
> cbind(alltob)
               alltob
0.0.0.0.0 86879626.67
1.0.0.0.0 64660110.33
0.1.0.0.0  1248006.32
```

[13]Note that `svyloglin()` uses a different arrangement of the factors prior to estimation than most other model-fitting functions use. Specifically, it uses a "sum-to-zero" set of contrasts to represent factors rather than the usual "set-first-to-zero" set as described in Section 2.2.6. See the program SurveySmokeLoglinear.R for an important note on the computational implications of this parameterization.

```
<OUTPUT EDITED>

0.1.0.0.1          0.00

<OUTPUT EDITED>

0.1.0.1.1          0.00

<OUTPUT EDITED>
```

For example, we estimate that more than 86 million people have used none of the tobacco products in the required amounts (combination 0.0.0.0.0), while more than 64 million have used cigarettes only (combination 1.0.0.0.0). Notice that there are two zero counts, both involving positive pipe and chew response combined with negative cigarette and cigar response. We will see shortly that this causes problems with comparisons involving models containing the four-way interaction of these four variables.

Model selection is based initially on comparing models of increasing orders, as these are conveniently constructed in svyloglin(). We start with main effects only, then add interactions by updating the model object's formula argument to include all two-way interactions, all three-way interactions, all four-way interactions, and finally, the five-way interaction (the saturated model). We use the update() function to build the new models based on the original model, as recommended by the survey package author (Lumley, 2011, p. 124). It makes constructing complicated models easier and is computationally more efficient than rewriting entire models in multiple svyloglin() fits. Comparisons between models are done using deviance or Pearson (called "Score" in the output) statistics calculated in the anova() method function. The p-values use the Thomas-Rao F-corrections described in Equation 6.12 by default.

Unfortunately, as of Version 4.2-1 of the survey package, these p-values, as well as those from the Rao-Scott second-order test are not correct. We have written a function called newpval() (in NewSvyloglinPvalue.R) that produces the correct p-values for these two tests. We first show tests with the incorrect p-values in the output that follows. Subsequently, we run all tests with newpval() to obtain the correct p-values. We use print() to print the results from anova() or newpval() because this allows us to change the type of test using the pval argument, if desired. Also, the default output from these functions shows all of the terms of the two models being compared. We have edited the results for brevity.

```
> ll.tob1 <- svyloglin(formula = ~ sm_cigs + sm_pipe + sm_cigar
    + sm_snuff + sm_chew, design = jdesign)
> ll.tob2 <- update(object = ll.tob1, formula = ~ .^2)
> ll.tob3 <- update(object = ll.tob1, formula = ~ .^3)
> ll.tob4 <- update(object = ll.tob1, formula = ~ .^4)
> ll.tob5 <- update(object = ll.tob1, formula = ~ .^5)

> test45 <- anova(ll.tob4, ll.tob5)
> print(x = test45)
Analysis of Deviance Table

<OUTPUT EDITED>
```

```
Deviance= 4.265026e-12 p= 0.9339042
Score= 5.584675e-12 p= 0.9324719

> test35 <- anova(ll.tob3, ll.tob5)
Error in solve.default(t(wX2) %*% Psat %*% wX2, t(wX2) %*% V %*%
    wX2) :
  system is computationally singular: reciprocal condition
     number = 8.92332e-18

> test25 <- anova(ll.tob2, ll.tob5)
Error in solve.default(t(wX2) %*% Psat %*% wX2, t(wX2) %*% V %*%
    wX2) :
   system is computationally singular: reciprocal condition
      number = 3.03614e-18

> test15 <- anova(ll.tob1, ll.tob5)
Error in solve.default(t(wX2) %*% Psat %*% wX2, t(wX2) %*% V %*%
    wX2) :
   system is computationally singular: reciprocal condition
      number = 4.32276e-18

> test34 <- anova(ll.tob3, ll.tob4)
> print(x = test34)
Analysis of Deviance Table

<OUTPUT EDITED>

Deviance= 8.033536 p= NaN
Score= 7.317034 p= NaN

Warning messages:
1: In pf(dev/tr(Delta), tr(Delta)^2/tr2(Delta), dfnull *
    tr(Delta)^2/tr2(Delta),  :
  NaNs produced
2: In pchisq(pearson * ncol(wX2)/tr(Delta), df =
    tr(Delta)^2/tr2(Delta),  :
  NaNs produced
3: In pf(pearson/tr(Delta), tr(Delta)^2/tr2(Delta), dfnull *
    tr(Delta)^2/tr2(Delta),  :
  NaNs produced
```

Several tests comparing lower-order models to the saturated model report computational errors. The test comparing the model with four-way interactions gives essentially the same fit as the saturated model ($X^2 = 5.6 \times 10^{-12}$, p-value = 0.93, although this is using the function that produces incorrect p-values). Next, trying to compare the four-way and three-way interaction models results in warnings. The reason for this becomes evident upon looking at the parameter estimates and confidence intervals for `ll.tob4` (not shown, see program for code). All parameter estimates for terms that can be formed from combinations of `sm_cigs`, `sm_pipe`, `sm_cigar`, and `sm_chew` have enormous standard errors that would produce confidence limits beyond ± 100. These values are nonsensical, considering that they would be exponentiated to form cell counts and odds ratios. They are another reflection of the adverse impact of zero counts on the loglinear model. We refit the model excluding the four-way interaction term for these variables below and perform model comparisons using corrected p-values below.

```
> ll.tob41 <- update(object = ll.tob4, formula = ~. -
    sm_cigs:sm_pipe:sm_cigar:sm_chew)
> source('NewSvyloglinPvalue.R')

> test45a <- newpval(ll.tob41, ll.tob5)
> print(x = test45a)
Analysis of Deviance Table

<OUTPUT EDITED>

Deviance= 6.0819 p= 1
Score= 4.292426 p= 1

> test34a <- newpval(ll.tob3, ll.tob41)
> print(x = test34a)
Analysis of Deviance Table

<OUTPUT EDITED>

Deviance= 1.951636 p= 0.8767901
Score= 2.110058 p= 0.8639808

> test23a <- newpval(ll.tob2, ll.tob3)
> print(x = test23a)
Analysis of Deviance Table

<OUTPUT EDITED>

Deviance= 45.02365 p= 0.03494121
Score= 55.94341 p= 0.009532241

> test12a <- newpval(ll.tob1, ll.tob2)
> print(x = test12a)
Analysis of Deviance Table

<OUTPUT EDITED>

Deviance= 2227.541 p= 2.74247e-64
Score= 29924.8 p= 1.60375e-197
```

The new model with four-way interactions is now seen to provide very little improvement over the third-order model ($X^2 = 2.11$, p-value $= 0.86$). However, the model with three-way interactions is a significantly better fit to the data than the one with just two-way interactions ($X^2 = 55.94$, p-value $= 0.009$), so we continue working with the third-order model.

Several possibilities exist for developing a final model. We could simply use the full third-order model as our final model, particularly if we were not interested in a parsimonious explanation for the associations in the data. Or, we could apply backward elimination or forward selection to the third-order terms from the model, keeping in mind that the choice of significance level for adding or deleting a term has nothing to do with type I error rates in any tests (see discussion on p. 345). We show the results of backward elimination in the program for this example. This shows that there is clearly a `sm_pipe:sm_snuff:sm_chew` interaction, and that there could be two additional three-way interactions if a threshold level of 0.10 or higher is used. For the sake of simplicity in our demonstration, we choose the model with just the one added three-way interaction. This model is named `ll.tob39` in the program.

Model-estimated odds ratios can be computed between each pair of factors either by manual manipulation of selected model-parameter estimates, or by the more automated calculations available in `svycontrast()`. The ordering of the parameter estimates is determined by using either `summary(ll.tob39)` or `coef(ll.tob39)`. This reveals that the parameter estimates are in the order 1, 2, 3, 4, 5, (12), (13), (14), (15), (23), (24), (25), (34), (35), (45), (245), where 1 = `sm_cigs`, 2 = `sm_pipe`, 3 = `sm_cigar`, 4 = `sm_snuff`, 5 = `sm_chew`, and two or more numbers in parentheses means an interaction among the corresponding variables. Coefficients on the parameter estimates must be assigned to match this ordering. Because `svyloglin()` uses sum-to-zero constraints and all of our variables have two levels each, the parameter estimates for the second level of each main effect are simply −1 times the estimate for the first level (e.g., $\hat{\beta}_2^X = -\hat{\beta}_1^X$). However, the function suppresses recording of the redundant second parameter estimate, so only one parameter is shown. Similarly, parameter estimates for levels of two-factor interaction parameters are multiplied by −1 for each factor level that changes from the (1,1) level (e.g., $\hat{\beta}_{12}^{XY} = -\hat{\beta}_{11}^{XY}, \hat{\beta}_{21}^{XY} = -\hat{\beta}_{11}^{XY}, \hat{\beta}_{22}^{XY} = \hat{\beta}_{11}^{XY}$), and only the result for the (1,1) level is retained. Thus, to compute an odds ratio, we need to multiply the given parameter estimates by 4 before exponentiating:

$$OR_{12,12} = \exp(\beta_{11}^{XY} + \beta_{22}^{XY} - \beta_{21}^{XY} - \beta_{12'}^{XY}) = \exp(4\beta_{11}^{XY}).$$

For example, the cigarette-pipe (12) odds ratio is determined by the coefficient vector $(0, 0, 0, 0, 0, 4, 0, 0, 0, 0, 0, 0, 0, 0, 0, 0)$. Note that when there are three or more levels of a factor, then `svyloglin()` does provide parameter estimates for each level, and so interaction parameters need to combined using ±1 coefficients to estimate an odds ratio.

The coefficients for estimating all odds ratios involving cigarettes—which are not involved in a three-way interaction—are entered in `svycontrast()` as shown below. We also probe the three-way interaction by estimating the odds ratio between snuff and chew separately for pipe non-users and users. The abbreviations used are `ct` = cigarettes, `pi` = pipes, `cr` = cigars, `sn` = snuff, and `ch` = chew. Confidence intervals can be found using either manually as shown below or using `confint()` as shown in the program.

```
> logORs <- svycontrast(stat = ll.tob39, contrasts = list(
    ct.pi  = c(0,0,0,0,0,4,0,0,0,0,0,0,0,0,0,0),
    ct.cr  = c(0,0,0,0,0,0,4,0,0,0,0,0,0,0,0,0),
    ct.sn  = c(0,0,0,0,0,0,0,4,0,0,0,0,0,0,0,0),
    ct.ch  = c(0,0,0,0,0,0,0,0,4,0,0,0,0,0,0,0),
    sn.ch.pi0  = c(0,0,0,0,0,0,0,0,0,0,0,0,0,0,4, 4),
    sn.ch.pi1  = c(0,0,0,0,0,0,0,0,0,0,0,0,0,0,4,-4)
  ))

> ORs <- as.data.frame(logORs)
> ORs$OR <- exp(ORs$contrast)
> # Calculate degrees of freedom for t distribution
```

```
> df <- ll.tob39$df.null - length(coef(ll.tob39)) - 1
> # Confidence intervals
> ORs$lower.CI <- exp((ORs$contrast + qt(0.025, df)*ORs$SE))
> ORs$upper.CI <- exp((ORs$contrast + qt(0.975, df)*ORs$SE))
> round(ORs[,3:5], 2)
              OR lower.CI upper.CI
ct.pi       3.37     2.39     4.74
ct.cr       2.70     2.01     3.62
ct.sn       2.01     1.20     3.37
ct.ch       1.32     0.85     2.05
sn.ch.pi0  52.61    22.80   121.41
sn.ch.pi1   7.43     3.63    15.24
```

All of these estimated associations are positive (although the confidence interval for the cigarette-chew odds ratio does include 1). People who have used cigarettes are generally more likely to have used pipes, cigars, or snuff than are people who have not used cigarettes. For the association between snuff and chew, the two forms of tobacco that are not smoked, the odds ratio for non-pipe users (`sn.ch.pi0`) is much larger than for pipe users (`sn.ch.pi1`). The estimated odds of snuff use are estimated to be 52 times as large among chew users as among non-chew users (95% confidence interval of 23 to 121) when the person is not also a pipe user, but the estimated odds of snuff use are only 7.4 times as large among chew users as among non-chew users (95% confidence interval of 3.6 to 15.4) when the person *is* also a pipe user. Keep in mind that these confidence levels are approximate because the confidence intervals are computed on the same data as was used to select the terms into the model. The actual confidence level may be lower than the one stated for the interval (see Section 5.5).

Alternative model estimation and inference methods

An alternative method has been proposed by Clogg and Eliason (1987) to account for survey weights in a loglinear model analysis. The idea is to pool together the survey weights corresponding to subjects in each cell of the table being analyzed. The mean weight is used as an offset for a Poisson rate regression model (see Section 4.3), and the analysis is carried out as if the data had arisen from a simple random sample. This suggestion was recommended by Agresti (2002, p. 391) and used in some research (e.g., Schwartz and Mare, 2005, Vermunt and Magidson, 2007, and Beller, 2009).

This approach has been scrutinized in Skinner and Vallet (2010) and Loughin and Bilder (2010), who find that it does not work as intended. The approach has two main flaws:

1. Although the survey weights are incorporated into the analysis, other features of the survey design (clustering and stratification) are not. This means that standard errors cannot be estimated properly and, hence, inferences cannot be trusted.

2. Even in rare cases where the survey design has little effect on the standard errors, the method is based on an approximation that assumes that survey weights are constant among all members of a given table cell (Loughin and Bilder, 2010). The procedure suffers badly when there is variability among the weights within a cell. Standard errors and model tests are adversely affected by this variability, resulting in inferences that are too liberal, by potentially large amounts.

Considering these problems, we strongly caution against using this procedure as a quick substitute for a full design-based analysis.

6.3.6 Logistic regression

Logistic regression is covered in detail in Chapter 2. The same principles given there apply to an analysis of weighted survey data. We describe here how to implement a logistic regression analysis using the `survey` package. Further information is available in Lumley (2010, Chapter 6). A detailed discussion of the theory is given in Heeringa et al. (2010).

To begin, plots of binary responses or estimated proportions can be made against chosen explanatory variables. The important thing to remember is that each sampled unit represents different numbers of members of the population according to the survey weights. Bubble plots like Figure 2.5 in Section 2.2.4 can be formed using total weights rather than sample sizes to define the bubble widths. The `survey` package has a plotting function `svyplot()` that can do this automatically.

A logistic regression model is fit using the `svyglm()` function. This function mimics the `glm()` function described in Section 2.2.1, except that model fitting does not use maximum likelihood estimation, because the binomial likelihood is inappropriate for weighted, correlated survey data. Instead, a *pseudo-likelihood* function is used that resembles the binomial likelihood applied to weighted counts rather than to sample counts (see Section 6.4.3 for another use of pseudo-likelihood). Tests for the regression parameters based on Wald statistics are available in the `summary()` of the model, while confidence intervals are found using `confint()`. Standard errors are computed internally either by linearization or by resampling methods. Tests are versions of Rao-Scott second-order-corrected t-tests: the t statistic is found in the usual manner as the ratio of the estimate to its standard error. The p-value is calculated by comparing the square of this statistic to an F distribution. The default confidence interval is a Wald-type interval; t-intervals can also be created.

There is an `anova()` method that can perform LR-like or Wald-based model-comparison tests on objects resulting from `svyglm()`. This requires that both the full and reduced (null-hypothesis) model are fit to the data. Alternatively, these tests can be performed directly on the full model using the `regTermTest()` function. Although it is cumbersome to use this function for testing individual parameters in a large model, it is especially useful for testing the significance of groups of regression parameters representing a single categorical explanatory variable, or for testing whether all regression parameters for terms of a certain order or associated with a particular variable are different from 0.

Information criteria for selecting models as described in Section 5.1.2 have been adapted for design-based analysis of logistic regression and other GLMs (Lumley and Scott, 2015). While the BIC version has a somewhat different form from the usual BIC, the AIC version uses a simple adjustment that multiplies the penalty term by the design effect for the model. The functions `AIC()` and `BIC()` in the `survey` package compute these quantities for models fit using `svyglm()`.

As with loglinear models, odds-ratio and probability-of-success computation requires additional work. The parameterization of `svyglm()` is exactly the same as in `glm()`, so the process of estimating an odds ratio or a probability of success from model parameters is exactly the same as was covered in Section 2.2.3–2.2.5. Log odds ratios are estimated using the `svycontrast()` function to compute the proper linear combinations of the coefficients and corresponding standard errors. From these quantities, the odds ratios and corresponding confidence intervals are found, again using a t distribution with $\kappa - p + 1$ degrees of freedom, where p is the number of parameters in the model (Korn and Graubard, 1999). Predicted probabilities and confidence intervals are found in a similar manner by using `svycontrast()` to compute the logits for the desired combinations of explanatory variables and their corresponding standard errors. The inverse logit link, $\exp(\cdot)/(1+\exp(\cdot))$, is applied to these logits and also to confidence interval endpoints computed in the logit scale. Alternatively, the `emmeans` package has the capacity to produce Wald-based inferences from models fit with `svyglm()`. Finally, if interest is restricted to viewing fitted probabilities for combinations of explanatory variables that are present in the data, these can be obtained using the `predict()` method.

Example: NHANES 1999–2000 (SurveySmokeLogistic.R, SmokeRespAge.csv)

We model presence of any respiratory symptoms (`anyresp`) against `age`, which is a continuous covariate, and the binary factors for each tobacco product. We begin by fitting a series of models, starting with `age` by itself (model `M1`), then adding all five tobacco-use factors as main effects (`M2`), and finally including interactions between `age` and each of the tobacco binaries (`M3`).[14] Note the use of a quasi-binomial family in `svyglm()`. This is needed because we are modeling survey-weighted counts of successes that may not be integers.[15]

Sequential comparisons of these three models gives us a starting point for further backward elimination until a satisfactory reduced model is obtained. We then use AIC in our backward elimination, stopping when removing a term causes AIC to increase compared to the previous model. Below are the initial model comparisons of `M1` versus `M2` and `M2` versus `M3`. We show only the `anova()` function, but additional calculations from `regTermTest()` are available in the program for this example.

```
> m1 <- svyglm(formula = anyresp ~ age, design = jdesign, family
   = quasibinomial(link = "logit"))
> m2 <- svyglm(formula = anyresp ~ age + sm_cigs + sm_pipe +
   sm_cigar + sm_snuff + sm_chew, design = jdesign, family =
   quasibinomial(link = "logit"))
> anova(m1, m2)
Working (Rao-Scott+F) LRT for sm_cigs sm_pipe sm_cigar sm_snuff
   sm_chew
 in svyglm(formula = anyresp ~ age + sm_cigs + sm_pipe +
    sm_cigar + sm_snuff + sm_chew, design = jdesign, family =
    quasibinomial(link = "logit"))
Working 2logLR =  60.89088 p= 2.9541e-06
(scale factors:  1.8 1.1 1 0.73 0.33 );  denominator df= 45

> m3 <- svyglm(anyresp ~ age * (sm_cigs + sm_pipe + sm_cigar +
   sm_snuff + sm_chew), design = jdesign, family =
   quasibinomial())
> anova(m2, m3)
Working (Rao-Scott+F) LRT for age:sm_cigs age:sm_pipe
   age:sm_cigar age:sm_snuff age:sm_chew
 in svyglm(formula = anyresp ~ age * (sm_cigs + sm_pipe +
    sm_cigar + sm_snuff + sm_chew), design = jdesign, family =
    quasibinomial())
```

[14] The `formula` argument `A*(B+C)` produces the model terms `A + B + C + A:B + A:C`. The same model would be produced by `A*B + A*C`.

[15] The help for `svyglm()` explains: "For binomial and Poisson families use `family = quasibinomial()` and `family = quasipoisson()` to avoid a warning about non-integer numbers of successes. The 'quasi' versions of the family objects give the same point estimates and standard errors and do not give the warning."

```
Working 2logLR =  10.75146 p= 0.088422
(scale factors:  2.1 1.1 0.87 0.56 0.41 );  denominator df= 40

> AIC(m1)
      eff.p          AIC    deltabar
   1.347005 5065.777010    1.347005
> AIC(m2)
      eff.p          AIC    deltabar
  12.632526 4951.275326    2.105421
> AIC(m3)
      eff.p          AIC    deltabar
  23.859266 4950.422533    2.169024
```

The evidence from the comparison of models m1 and m2 strongly indicates that at least one of the tobacco use factors contributes significantly to the model already containing age (LRT statistic $= 60.9$, p-value ≈ 0). The t-tests associated with parameter estimates (not shown) point to cigarettes as the strongest contributor by far ($t = 6.4$, p-value ≈ 0, next largest has $t = 1.47$, p-value $= 0.15$). The inclusion of age-by-tobacco interactions in m3 results in a more modest improvement over the model with main effects only (LRT statistic $= 10.8$, p-value $= 0.09$). These conclusions are supported by AIC, which shows a considerably smaller value for m2 than m1, 4,951 vs. 5,065, and a further, much smaller reduction for m3 to 4,950.

Therefore, we consider m3 for backward elimination of individual terms. We maintain hierarchy in the process, allowing the main effect of a tobacco-use factor to be considered for elimination only after its interaction with age has already been eliminated. The AIC comparisons support the model shown below, for which the AIC is 4931. We add `return.replicates = TRUE` to `svyglm()` to ensure that subsequent follow-up calculations from other functions properly use the jackknife weights for standard errors.

```
> m.final <- svyglm(formula = anyresp ~ age + sm_cigs + age *
    (sm_pipe + sm_chew), design = jdesign, family =
    quasibinomial(link = "logit"), return.replicates = TRUE)
> summary(m.final)

Call:
svyglm(formula = anyresp ~ age + sm_cigs + age * (sm_pipe +
    sm_chew), design = jdesign, family = quasibinomial(link =
    "logit"), return.replicates = TRUE)

Survey design:
update(object = jdesign, anyresp = (re_cough + re_phlegm +
    re_wheez + re_night > 0))

Coefficients:
             Estimate Std. Error t value Pr(>|t|)
(Intercept) -2.039511   0.148635 -13.722  < 2e-16 ***
age          0.006693   0.002360   2.836  0.00682 **
sm_cigs      0.705353   0.110126   6.405 7.81e-08 ***
sm_pipe      1.654550   0.611060   2.708  0.00954 **
sm_chew      1.461494   0.829967   1.761  0.08505 .
age:sm_pipe -0.025002   0.009924  -2.519  0.01538 *
```

```
age:sm_chew -0.022896   0.014350  -1.596  0.11760
---
Signif. codes:
0 '***' 0.001 '**' 0.01 '*' 0.05 '.' 0.1 ' ' 1

(Dispersion parameter for quasibinomial family taken to be
   4852.419)

Number of Fisher Scoring iterations: 4

> AIC(m.final)
      eff.p          AIC     deltabar
  13.420250 4930.830503     2.236708
```

This model indicates that the relationship between presence of any respiratory symptoms and age is influenced by use of pipe and chew, but not other products. Cigarette use has a very significant relationship with presence of respiratory symptoms, independent of age. Notice that the signs of the two interaction coefficients are negative and have larger magnitudes than the age main effect. This implies that the model is estimating that there is actually a *decreasing* relationship between presence of respiratory symptoms and age for users of pipes and chew, which is counterintuitive.

Exploring these features further, we calculate odds ratios relating the change in presence of respiratory symptoms to a 10-year increase in age separately for (a) people whose tobacco use does not include pipes or chew and people who have used (b) pipe tobacco only and (c) chew only. The `emmeans` package can estimate linear predictors and perform various comparisons among these linear predictors, as shown in previous chapters. We add here the `by` argument to `emmeans()` to request that linear predictors be computed for the ages separately at levels of the variables listed in the `by` argument, `sm_pipe` and `sm_chew`. This will ease the process of comparing log-odds at different `age` levels separately for levels of these other variables. Note that we must specify `df` in the `confint()` function so that `emmeans` uses the proper degrees of freedom for t-based critical values. Alternatively, calculations can be performed more manually analogous to `mcprofile` using the `svycontrast()` function. Numbering parameters based on the ordering of the parameter estimates in the previous output, with β_0 representing the intercept, the estimated odds ratios for these three cases are (a) $\exp(10\hat{\beta}_1)$, (b) $\exp(10\hat{\beta}_1 + 10\hat{\beta}_5)$, and (c) $\exp(10\hat{\beta}_1 + 10\hat{\beta}_6)$. We show the simpler approach using `emmeans()` below, and include the `svycontrast()` calculations in the program for this example.

```
> emm.age <- emmeans(object = m.final, specs= ~ age, at =
   list(age = c(30, 40)), by = c("sm_pipe", "sm_chew"))
> logOR.age <- contrast(object = emm.age, method = "consec")
> confint(object = logOR.age, type = "response", df =
   m.final$df.residual)
sm_pipe = 0, sm_chew = 0:
 contrast      odds.ratio     SE df lower.CL upper.CL
 age40 / age30      1.069 0.0252 45    1.020    1.121

sm_pipe = 1, sm_chew = 0:
 contrast      odds.ratio     SE df lower.CL upper.CL
 age40 / age30      0.833 0.0848 45    0.678    1.022
```

```
sm_pipe = 0, sm_chew = 1:
 contrast        odds.ratio      SE df lower.CL upper.CL
 age40 / age30        0.850 0.1203 45    0.640    1.131

sm_pipe = 1, sm_chew = 1:
 contrast        odds.ratio      SE df lower.CL upper.CL
 age40 / age30        0.662 0.1002 45    0.488    0.898

Results are averaged over the levels of: sm_cigs
Degrees-of-freedom method: user-specified
Confidence level used: 0.95
Intervals are back-transformed from the log odds ratio scale
```

Each odds ratio is interpreted as the multiplicative change in odds of respiratory symptoms for a 10-year increase in age for a person who has used the specified tobacco product. The estimated odds ratios are above 1 for non-pipe or chew users, although the confidence interval for this odds ratio is only 1.02 to 1.12, suggesting that there is a statistically significant, but very small increase in probability of symptoms with age. The overall effect of age for pipe and/or chew users is not clear. The associated estimated odds ratios are all below 1, indicating lower probability of respiratory symptoms in older people, but two of the three confidence intervals contain 1. A subject-matter expert may want to explore some of these potentially counter-intuitive findings further. Additional odds ratios focusing on tobacco use at specific ages are shown in the program for this example.

We also compute confidence intervals for probabilities of respiratory symptoms for each combination of cigarette, pipe, and chew use for 50-year-olds.

```
> emm.preds <- emmeans(object = m.final, specs = ~ sm_cigs +
    sm_pipe + sm_chew + age, at = list(age = 50))
> ci.preds <- confint(object = emm.preds, type = "response", df
    = m.final$df.residual)
> round(ci.preds, 2)
 sm_cigs sm_pipe sm_chew age prob   SE df lower.CL upper.CL
       0       0       0  50 0.15 0.01 45     0.13     0.18
       1       0       0  50 0.27 0.01 45     0.24     0.30
       0       1       0  50 0.21 0.03 45     0.16     0.28
       1       1       0  50 0.36 0.04 45     0.28     0.44
       0       0       1  50 0.20 0.03 45     0.14     0.28
       1       0       1  50 0.34 0.05 45     0.25     0.43
       0       1       1  50 0.27 0.04 45     0.20     0.37
       1       1       1  50 0.43 0.05 45     0.33     0.53

Degrees-of-freedom method: user-specified
Confidence level used: 0.95
Intervals are back-transformed from the logit scale
```

Clearly, tobacco use is associated with greater probabilities of respiratory symptoms, although no causal relationship can be inferred from this analysis.

Plots of the model-predicted probabilities and the estimated proportions are given in Figure 6.3 (code is available in our program). Each bubble represents the weighted sample estimate for the proportion of the population with respiratory symptoms at

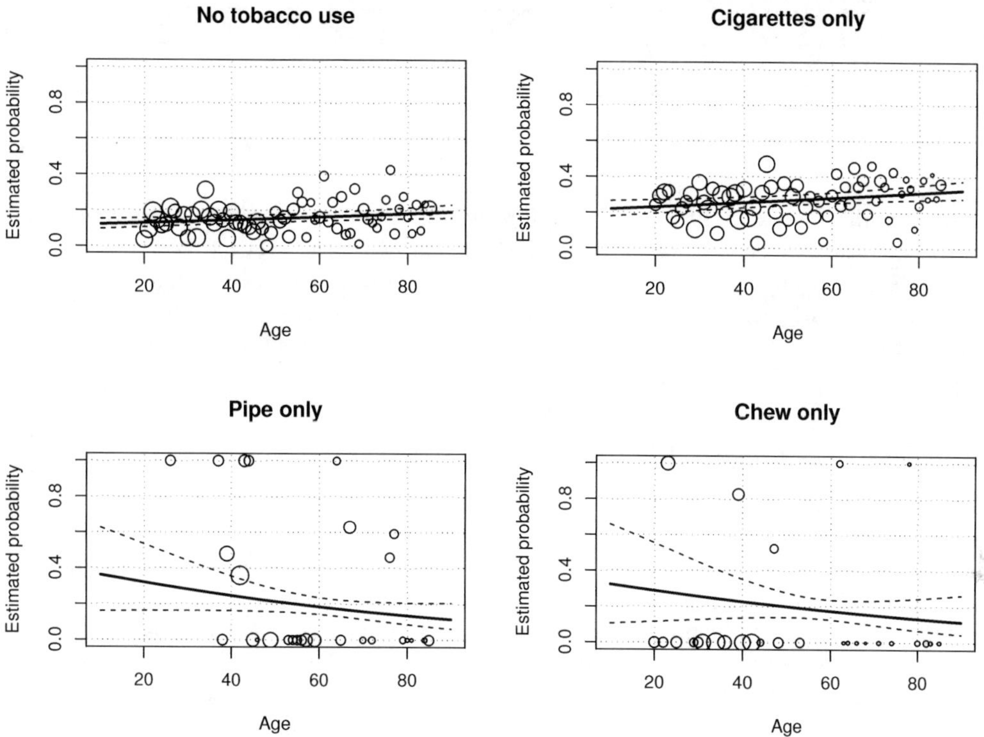

Figure 6.3: Final model fitted to data for four tobacco-use cases. The solid line is the fitted model and the dashed lines are pointwise 95% confidence intervals for the probability of respiratory symptoms at each age. Bubble sizes reflect relative survey weight for each year. Bubbles sizes in different panels are not comparable.

each age (one bubble per year). Clearly, there is not much data with which to estimate the trends for people who have used chew or pipes, so the confidence intervals here are much wider and appear not to preclude a line of zero slope. We reiterate the caution from Section 5.5 that all confidence intervals from this example may have unknown true confidence level, because they are computed using the same data that was used to select the final model (see Section 5.5).

6.4 "Choose-all-that-apply" data

Survey questions often include the phrase "choose all that apply" or "pick any" when requesting individuals to select from a pre-defined list of response options or *items*. Individuals may select any number of these items in their responses. For this reason, categorical variables summarizing these responses are referred to as *multiple response categorical variables*

Table 6.6: Joint positive responses for the swine management data.

	Lagoon	Pit	Natural drainage	Holding tank
Nitrogen	27	16	2	2
Phosphorus	22	12	1	1
Salt	19	6	1	0

(MRCVs; Bilder and Loughin, 2004).[16] Allowing individuals to provide multiple responses to a single question creates correlation among these responses. For this reason, analyzing MRCV data is more challenging than analyzing single-response categorical variables (SRCVs). The purpose of this section is to show how the Pearson chi-square test can be extended to these data situations and to examine how regression models can be used to explore relationships among MRCVs.

6.4.1 Item response table

Table 6.6 summarizes the responses of 279 Kansas swine farmers who were asked questions in a survey about their swine waste management practices (Richert et al., 1995; Bilder and Loughin, 2007). One question asked farmers to choose all swine waste storage methods that they use among four items: lagoon, pit, natural drainage, and holding tank. Another question asked farmers to choose all contaminants that they test for among three items: nitrogen, phosphorus, and salt. Although the summary of these counts in Table 6.6 looks like an ordinary contingency table, there are several important distinctions. First, because farmers could choose all responses that applied to them, an individual farmer may contribute to more than one cell count in the table. Second, negative answers are not adequately represented in Table 6.6. Farmers who used none of the listed storage methods are not counted, regardless of what contaminants they test for, and vice versa. Thus, there is important information missing in this representation.

A better way to represent these data is to consider both positive and negative responses for each item. Specifically, let W_i denote the binary response ($1 =$ positive, $0 =$ negative) to row item i for $i = 1, \ldots, I$, and let Y_j denote the binary response to column item j for $j = 1, \ldots, J$. For example, in the swine waste management data, $Y_1 = 1$ means that a farmer uses a lagoon for waste storage. Then, 2×2 contingency tables can be formed from the binary variables for each combination of a row item and a column item. Table 6.7 shows two of the $IJ = 12$ possible tables for the swine waste management example. It is important to note that each farmer appears in exactly one cell in each of the tables; thus, the counts in each table sum to 279, the number of farmers participating in the survey. Also, notice that the counts in Table 6.6 are merely the cell counts for the $W_i = 1$ and $Y_j = 1$ response combination from these 12 tables. We refer to the full set of IJ 2×2 tables as the *item response table* for the data. The item response table is the summary upon which inferences are constructed in the next two sections.

6.4.2 Testing for marginal independence

Procedures to test for independence between two MRCVs were developed in a series of papers by Loughin and Scherer (1998), Agresti and Liu (1999), Thomas and Decady (2004), and Bilder and Loughin (2004). These papers identified two main complications that need

[16] In the social sciences literature, the phrase "pick any/n" (said as "pick any out of n") is often used to describe the resulting data when there are n items in the list. Coombs (1964) is likely the first to coin this terminology.

Table 6.7: Two 2×2 contingency tables for the swine waste management data.

		Lagoon 0	Lagoon 1			Pit 0	Pit 1
Nitrogen	0	123	116	Nitrogen	0	175	64
	1	13	27		1	24	16

to be addressed when testing for independence. First, the usual notion of independence needs to be redefined in the context of an MRCV. Second, constructing and finding the probability distribution of any test statistic under a null hypothesis is complicated by the fact that many responses in the cells of either Table 6.6 or the item response table are provided by the same individuals. None of the usual sampling models are appropriate for these counts—they are not multinomial counts, and they are not independent Poisson counts—so the theory that leads to using Pearson statistics and chi-square distributional approximations no longer applies. We show below how these issues can be overcome.

Marginal independence hypotheses

The primary interest in many problems involving "choose-all-that-apply" questions is determining whether the probability of a positive response to each item changes depending on the responses to other questions. For example, does the probability of testing for a particular contaminant depend on what waste-storage methods are used? In other words, are each of the 12 2×2 tables in the item-response table consistent with an independence hypothesis, or does at least one indicate an association between testing a contaminant and using a waste storage method? Each of these 2×2 tables is a *marginal* table because it summarizes data without regard to the responses to all other items. The hypothesis that independence holds in each of these tables is therefore termed *simultaneous pairwise marginal independence* (SPMI), as first named by Agresti and Liu (1999).

Define $\gamma_{ij} = P(W_i = 1, Y_j = 1)$ as a pairwise joint probability for $i = 1, \ldots, I$ and $j = 1, \ldots, J$. Also, define $\gamma_{i+} = P(W_i = 1)$ and $\gamma_{+j} = P(Y_j = 1)$ as marginal probabilities of a positive response to the corresponding items. Then the hypotheses for SPMI are

$$H_0 : \gamma_{ij} = \gamma_{i+}\gamma_{+j} \text{ for } i = 1, \ldots, I \text{ and } j = 1, \ldots, J$$
$$H_a : \gamma_{ij} \neq \gamma_{i+}\gamma_{+j} \text{ for a least one } (i, j) \text{ pair.}$$

One can show that $\gamma_{ij} = \gamma_{i+}\gamma_{+j}$ implies that $P(W_i = w_i, Y_j = y_j) = P(W_i = w_i)P(Y_j = y_j)$ for all combinations of $w_i = 0$ or 1 and $y_j = 0$ or 1, so this is why the null hypothesis can be written using just one cell of each 2×2 table.

Instead of writing hypotheses about the pairwise joint probabilities γ_{ij}, one could instead examine the full joint probabilities $P(W_1 = w_1, \ldots, W_I = w_I, Y_1 = y_1, \ldots, Y_J = y_J)$ for each of the 2^{I+J} binary response combinations. A different form of independence involving the joint probabilities, known as *joint independence*, occurs when the combination of responses to one question is independent of the combination of responses to the other: $P(W_1 = w_1, \ldots, W_I = w_I, Y_1 = y_1, \ldots, Y_J = y_J) = P(W_1 = w_1, \ldots, W_I = w_I)P(Y_1 = y_1, \ldots, Y_J = y_J)$. One can show that if joint independence holds, then SPMI does as well, but it is possible for SPMI to be true when joint independence is not. Berry and Mielke (2003) provide details on testing for joint independence. However, we have found testing for joint independence to be of limited use, unless there is interest in the specific response combinations of $w_1, \ldots, w_I$ and $y_1, \ldots, y_J$.

Test Statistics

We next develop tests for SPMI. Treating each 2×2 table separately, MLEs of the pairwise joint probabilities are $\hat{\gamma}_{ij} = m_{ij}/n$, $\hat{\gamma}_{i+} = m_{i+}/n$, and $\hat{\gamma}_{+j} = m_{+j}/n$, where m_{ij} represents the number of positive responses for $W_i = 1$ and $Y_j = 1$, and n is the sample size. Let $X^2_{S,ij}$ be the usual Pearson statistic for testing independence within the $(i,j)^{th}$ 2×2 contingency table. Then the sum of these Pearson statistics is a natural statistic to use for testing simultaneous independence in all tables. We can write this as

$$\begin{aligned}
X^2_S &= \sum_{i=1}^{I}\sum_{j=1}^{J} X^2_{S,ij} \\
&= \sum_{i=1}^{I}\sum_{j=1}^{J}\left[\frac{(m_{ij} - m_{i+}m_{+j}/n)^2}{m_{i+}m_{+j}/n} + \frac{(m_{i+} - m_{ij} - m_{i+}(n - m_{+j})/n)^2}{m_{i+}(n - m_{+j})/n}\right. \\
&\quad + \frac{(m_{+j} - m_{ij} - m_{+j}(n - m_{i+})/n)^2}{m_{+j}(n - m_{i+})/n} \\
&\quad \left. + \frac{(n - m_{i+} - m_{+j} + m_{ij} - (n - m_{i+})(n - m_{+j})/n)^2}{(n - m_{i+})(n - m_{+j})/n}\right] \\
&= n\sum_{i=1}^{I}\sum_{j=1}^{J}\frac{(\hat{\gamma}_{ij} - \hat{\gamma}_{i+}\hat{\gamma}_{+j})^2}{\hat{\gamma}_{i+}\hat{\gamma}_{+j}(1 - \hat{\gamma}_{i+})(1 - \hat{\gamma}_{+j})}. \qquad (6.13)
\end{aligned}$$

If the collection of $X^2_{S,ij}$ statistics were independent, then X^2_S would have a large-sample χ^2_{IJ} distribution and SPMI would be rejected if $X^2_S > \chi^2_{IJ,1-\alpha}$. Unfortunately, the individual test statistics are likely to be correlated, because the tables upon which they are based are merely different summaries of the same set of responses. For example, each farmer is represented in each of the IJ 2×2 contingency tables in the swine management data example. Bilder and Loughin (2004) show that ignoring the correlation and performing the test using a χ^2_{IJ} distribution leads to a liberal test—one that rejects H_0 too often when SPMI is true—when there is moderate-to-strong association among an individual's responses within one MRCV (e.g., if farmers who test for one contaminant also tend to test for another). Alternative testing methods proposed by Bilder and Loughin (2004) are described next.

Bonferroni adjustment

Rather than combine the individual 2×2 table statistics into a sum, one can compute a p-value to test for independence in each table and combine these results into a single test. The simplest way to do this is to use a Bonferroni adjustment on the individual tests. For each $X^2_{S,ij}$, calculate a p-value, say p_{ij}, using the usual χ^2_1 approximation or an exact method from Section 6.2, and reject SPMI if any p_{ij} is less than α/IJ. Equivalently, an adjusted p-value (Westfall and Young, 1993) for the overall SPMI test can be calculated as $\tilde{p} = IJ\times \min_{ij}(p_{ij})$, where $\tilde{p}$ is truncated to 1 if this product is greater than 1. Then SPMI is rejected if $\tilde{p} < \alpha$. As is often the case with a Bonferroni adjustment, the overall SPMI test can be conservative—it does not reject often enough when SPMI is true—if the number of 2×2 tables is more than a few.

Bootstrap approximation

A second approach proposed by Bilder and Loughin (2004) involves the bootstrap. The bootstrap is a commonly used technique to estimate a statistic's probability distribution when its distribution may be difficult to obtain mathematically or when large-sample approximations may be poor. We focus here on the application of the bootstrap rather than on details for why it works well in this situation. Interested readers can examine Davison and Hinkley (1997) for more details on the bootstrap.

To implement the bootstrap, data are "resampled"[17] by randomly selecting an observed row-response combination $(w_1, \ldots, w_I)$ and matching it with an independently chosen column-response combination $(y_1, \ldots, y_J)$. For example, one farmer's waste-storage responses may be randomly matched to another farmer's contaminant-testing responses. Repeating this process n times forms a "resample" with very similar characteristics to the original sample—e.g., sample size, proportions of positive responses to each item, and associations among the items within each MRCV—while ensuring that the items from the two different MRCVs are independent of each other.[18] A large number of resamples, say B, are created in this way. For each resample, $b = 1, \ldots, B$, we calculate the test statistic, denoted now by $X^{2*}_{S,b}$, where we use the asterisk in the superscript to differentiate it from the original statistic's observed value. The p-value for the test is the proportion of the $X^{2*}_{S,b}$ values greater than or equal to the observed X^2_S, $B^{-1}\sum_{b=1}^{B} I(X^{2*}_{S,b} \geq X^2_S)$, where $I(\cdot)$ is an indicator function with a value of 1 when $X^{2*}_{S,b} \geq X^2_S$ and a value of 0 otherwise. This p-value measures how extreme the original test statistic is relative to the distribution of the same statistic under the null hypothesis. Small p-values indicate evidence against SPMI. Bilder and Loughin (2004) show that this test holds the correct size; i.e., the test rejects at the stated α-level when SPMI is true.

Rao-Scott corrections

In large samples, the test statistic X^2_S has a distribution that is approximated by a linear combination of independent χ^2_1 random variables (see Appendix A of Bilder and Loughin (2004)). However, the coefficients of the linear combination are unknown, so the distribution of X^2_S must be further approximated somehow. One approach is to use the Rao-Scott methods described in Section 6.3.5. A first-order correction adjusts X^2_S so that it has the same mean as a χ^2_{IJ} random variable; in other words, determine a value of c such that the expected value of the statistic $X^2_{RS1} = cX^2_S$ is IJ. Interestingly, Thomas and Decady (2004) and Bilder and Loughin (2004) show that this adjustment factor is simply $c = 1$. Thus, X^2_{RS1} can be compared to a χ^2_{IJ} distribution for the first-order correction, which results in the same testing method as from naively treating each $X^2_{S,ij}$ as independent. However, as mentioned earlier, performing the test in this manner leads to a liberal test.

A second-order correction further adjusts X^2_S so that it also has the same variance as a χ^2 random variable with adjusted degrees of freedom. For $p = 1, \ldots, IJ$, define λ_p to be the coefficients of the linear combination in the large-sample distributional approximation for X^2_S described in the previous paragraph. Then the distribution of the statistic $X^2_{RS2} = IJ\, X^2_S / \sum_{p=1}^{IJ} \lambda_p^2$ can be approximated by a χ^2 random variable with $I^2J^2/\sum_{p=1}^{IJ} \lambda_p^2$ degrees of freedom (this may not be an integer). Values of X^2_{RS2} greater than the $1 - \alpha$ quantile of

[17]Resampling is similar to permuting data, which was discussed in Section 6.2. Permutations of observations result from sampling "without replacement" from an original set of data. Thus, the same observations are retained, but their order is randomized. Resamples of observations result from sampling "with replacement" from an original set of data. Thus, the observations in a data set constructed by resampling take the same values as those in the original data, but these values occur with different frequencies. Some observations may occur multiple times, while others may not appear at all within a resample of n observations.

[18]This form of independence is actually joint independence. Because joint independence implies SPMI, the SPMI null hypothesis is guaranteed to hold under this resampling approach.

this χ^2 distribution indicate evidence against SPMI. Bilder and Loughin (2004) show that this testing procedure is generally pretty good but can be a little conservative.

The key to implementing the second-order correction is to estimate λ_p for $p = 1, \ldots, IJ$. These coefficients are functions of the large-sample covariance matrix for the quantities $\hat{\gamma}_{ij} - \hat{\gamma}_{i+}\hat{\gamma}_{+j}$. Details of their calculation are given in Thomas and Decady (2004) and Bilder and Loughin (2004). We implement the calculations here using the function `MI.test()` from the `MRCV` package (Koziol and Bilder 2014). The computational algorithm used in this function can cause R to run out of memory when $I + J$ is large. Surendra (2016) and Chankov (2017) propose alternative computational approaches that lessen these memory requirements. Neither of these approaches are available in an R package. However, we illustrate in Exercise 7 the use of an R function created by Chankov (2017) that implements his alternative.

Example: Swine waste management data (Swine.R)

We now formally examine the swine waste management data first described in Section 6.4.1. Both the data and the analysis functions we use are contained in the `MRCV` package. The `farmer2` data frame within the package contains the response combinations $(w_1, w_2, w_3, y_1, y_2, y_3, y_4)$ for each of the $n = 279$ farmers, where the indices on the items correspond to the ordering shown in Table 6.6. The first few observations are shown below.

```
> library(package = MRCV)
> head(farmer2)
  w1 w2 w3 y1 y2 y3 y4
1  0  0  0  0  0  0  0
2  0  0  0  0  0  0  1
3  0  0  0  0  0  0  1
4  0  0  0  0  0  0  1
5  0  0  0  0  0  0  1
6  0  0  0  0  0  0  1
```

For example, the second farmer did not test for any of the three contaminants and uses only a holding tank. Summary tables as given in Tables 6.6 and 6.7 are obtained using the `marginal.table()` and `item.response.table()` functions, respectively.

```
> marginal.table(data = farmer2, I = 3, J = 4)
   y1         y2         y3         y4
   count %    count %    count %    count %
w1 27     9.68 16     5.73 2      0.72 2      0.72
w2 22     7.89 12     4.30 1      0.36 1      0.36
w3 19     6.81  6     2.15 1      0.36 0      0.00

> item.response.table(data = farmer2, I = 3, J = 4)
      y1       y2      y3      y4
      0   1    0   1   0   1   0   1
w1 0 123 116 175 64 156 83 228 11
   1  13  27  24 16  38  2  38  2
w2 0 128 121 181 68 165 84 237 12
   1   8  22  18 12  29  1  29  1
w3 0 134 124 184 74 174 84 245 13
```

```
1   2  19  15   6  20   1  21   0
```

When using these functions, it is important to properly order the columns within the data frame given by `data`. The value for the `I` argument indicates that the first three columns in the data frame are for the first MRCV. The value for the `J` argument indicates that the next four columns in the data frame are for the second MRCV.

We test the null hypothesis of SPMI, meaning that each item among the storage methods is independent of each item among the tests for contaminants. To perform this test using the Bonferroni adjustment, we use the `MI.test()` function with the `type = "bon"` argument value. The `add.constant = FALSE` argument value in `MI.test()` specifies that no constants should be added to cells in the item-response table that have zero counts (adding small constants may be helpful when many 0 cell counts are present).

```
> MI.test(data = farmer2, I = 3, J = 4, type = "bon",
    add.constant = FALSE)
Test for Simultaneous Pairwise Marginal Independence (SPMI)

Unadjusted Pearson Chi-Square Tests for Independence:
X^2_S = 64.83
X^2_S.ij =
      y1   y2    y3   y4
w1  4.93 2.93 14.29 0.01
w2  6.56 2.11 11.68 0.13
w3 13.98 0.00  7.08 1.11

Bonferroni Adjusted Results:
p.adj = 0.0019
p.ij.adj =
   y1     y2     y3     y4
w1 0.3163 1.0000 0.0019 1.0000
w2 0.1253 1.0000 0.0076 1.0000
w3 0.0022 1.0000 0.0934 1.0000
```

The output gives $X_S^2 = 64.83$ and the individual $X_{S,ij}^2$ values. For example, $X_{S,13}^2 = 14.29$ for nitrogen testing and natural drainage, which is the largest value among the 2×2 tables. With a χ_1^2 approximation for the largest X_{ij}^2 value, we obtain a Bonferroni-adjusted p-value of $\tilde{p} = 0.0019$, indicating strong evidence against SPMI; hence, not every contaminant tested is independent of every waste storage method. The output also provides each of the p-values for the 2×2 tables, so that we can better understand the nature of any associations between items from different variables. These p-values are Bonferroni-adjusted using $\min(IJ \times p_{ij}, 1)$. At the $\alpha = 0.05$ level, the combinations with significant association are salt and lagoon, nitrogen and natural drainage, and phosphorus and natural drainage.

For the bootstrap approximation, resamples are taken from the observed data vectors (w_1, w_2, w_3) and (y_1, y_2, y_3, y_4) independently with replacement from the original data and then $X_{S,b}^{2*}$ is calculated for each resample. This process is demonstrated below for one resample of size $n = 279$.

```
> I <- 3
> J <- 4
> n <- nrow(farmer2)

> set.seed(7812)
> iW <- sample(x = 1:n, size = n, replace = TRUE)
> iY <- sample(x = 1:n, size = n, replace = TRUE)

> # Use resampled index values to form resample
> farmer2.star <- cbind(farmer2[iW, 1:I], farmer2[iY,
   (I+1):(I+J)])
> head(farmer2.star)  # The row numbers here are from iW
    w1 w2 w3 y1 y2 y3 y4
75   0  0  0  1  0  0  0
9    0  0  0  0  0  1  0
256  1  1  0  1  0  0  0
157  0  0  0  0  0  0  0
242  1  0  0  0  1  0  0
147  0  0  0  1  1  0  0

> MI.stat(data = farmer2.star, I = I, J = J, add.constant =
   FALSE)
$X.sq.S
[1] 17.44788

$X.sq.S.ij
          y1       y2        y3          y4
w1 0.9412803 2.023367 0.4167563 0.007350586
w2 0.5177964 2.500335 0.6728595 0.154705323
w3 0.7953081 8.950444 0.4136119 0.054060485

$valid.margins
[1] 12
```

The MI.stat() function is used to calculate $X^{2*}_{S,b} = 17.45$. We repeat this same resampling process by using the function MI.test(), where type = "boot" specifies the bootstrap test and B specifies the number of resamples.

```
> set.seed(7812)
> MI.test(data = farmer2, I = 3, J = 4, B = 5000, type = "boot",
   add.constant = FALSE, plot.hist = TRUE)
Test for Simultaneous Pairwise Marginal Independence (SPMI)

Unadjusted Pearson Chi-Square Tests for Independence:
X^2_S = 64.83
X^2_S.ij =
      y1   y2    y3   y4
w1  4.93 2.93 14.29 0.01
w2  6.56 2.11 11.68 0.13
w3 13.98 0.00  7.08 1.11

Bootstrap Results:
```

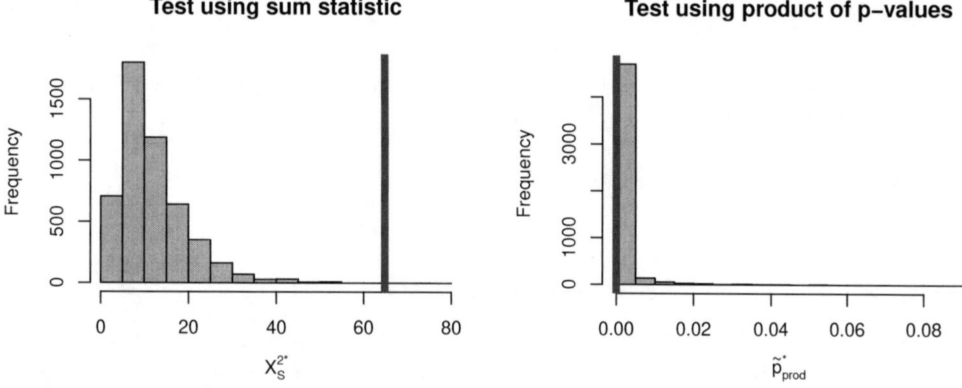

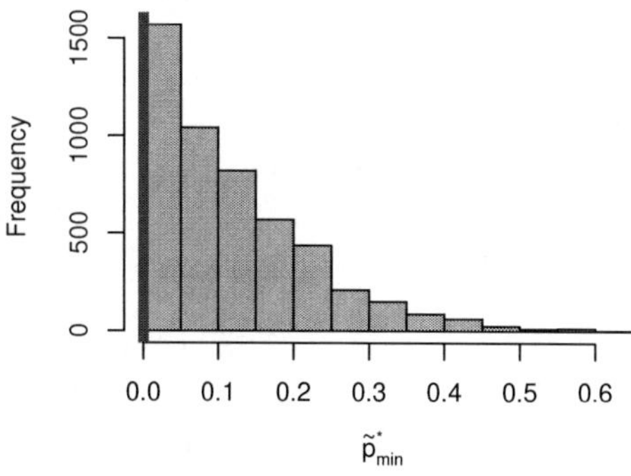

Figure 6.4: Histograms of the estimated probability distributions under SPMI. The vertical lines denote the observed statistic value for the original data.

```
Final results based on 5000 resamples
p.boot = 0.0002
p.combo.prod = 0.0002
p.combo.min = 0.0038
```

Similar to Section 6.2, we specify a seed number before running `MI.test()` to reproduce the same results at a later time if needed. Out of $B = 5000$ resamples, only one resulted in a $X^{2*}_{S,b}$ that was larger than the observed X^2_S, so the p-value (`p.boot`) is 0.0002. Figure 6.4 (upperleft) provides the bootstrap-estimated probability distribution for X^2_S. The location of the test statistic's observed value is indicated by the vertical line. Obviously, X^2_S is an extreme value relative to this distribution, resulting in rejection of the SPMI hypothesis.

In addition to performing the test with X^2_S, the function also performs a bootstrap test for SPMI using two p-value combination methods (Loughin, 2004). For these tests, the p-values from each $X^2_{S,ij}$ are found using a χ^2_1 approximation. The IJ p-values are "combined" by multiplying them together or by finding their minimum value. For both p-value combination methods, small observed values relative to the estimated distribution indicate evidence against the null hypothesis. In both cases, there is substantial evidence against SPMI.

The `type` argument in `MI.test()` can also have a value of `"rs2"` for the second-order Rao-Scott correction and `"all"` for all testing procedures. For the second-order Rao-Scott correction, $X^2_{RS2} = 36.62$ and the degrees of freedom for the χ^2 distribution is 6.78 (results not shown). This leads to a p-value of 4.5×10^{-6}, which again indicates strong evidence against SPMI.

Other tests for independence involving MRCVs

The testing procedures discussed in this section can be extended to other situations. For example, in the case of one MRCV and one ordinary SRCV, a test for *multiple marginal independence* (MMI) can be performed (see Exercise 5). Suppose that the SRCV is the row variable with I levels and the MRCV is the column variable with J items. Then, the item-response table consists of J different $I \times 2$ contingency tables, each of which summarizes counts for one item across the levels of the SRCV. Analogous to X^2_S, Pearson chi-square statistics can be calculated for each of the J contingency tables, and the sum of these individual statistics provides an overall test statistic for MMI (Agresti and Liu, 1999). The Bonferroni, bootstrap, and Rao-Scott approaches can all be applied, and each has properties similar to when they are used to test for SPMI (Bilder et al., 2000). Additionally, when there is a third SRCV, a test can also be performed for *conditional multiple marginal independence* given this third variable (Bilder and Loughin, 2002).

6.4.3 Regression modeling

As was shown in Sections 4.2.3 and 4.2.4, loglinear Poisson regression models are useful for describing relationships among categorical variables and for estimating the strength of associations using odds ratios. In the context of MRCVs, we can adapt these models both to test for SPMI and to describe associations between MRCVs when SPMI does not hold. This flexibility gives models a distinct advantage over the methods in the previous section.

Consider first the regression model for a single 2×2 table that assumes independence between items W_i and Y_j,

$$\log(\mu_{ab}) = \beta_0 + \beta_a^W + \beta_b^Y,\ a = 1, 2,\ b = 1, 2, \qquad (6.14)$$

where a and b represent levels of items W_i and Y_j, respectively. A model that simultaneously assumes independence between all pairs of W-items with Y-items (i.e., SPMI) can be written as

$$\log(\mu_{ab(ij)}) = \beta_{0(ij)} + \beta_{a(ij)}^W + \beta_{b(ij)}^Y,\ a = 1, 2,\ b = 1, 2,\ i = 1, \ldots, I,\ j = 1, \ldots, J, \qquad (6.15)$$

where $\mu_{ab(ij)}$ is the expected count for row a and column b of the $(i, j)^{th}$ 2×2 contingency table summarizing the pairwise joint responses. This is just Equation 6.14 with extra subscripts (i, j) to identify the two items being modeled. Odds ratios for each 2×2 table are equal to 1 for Equation 6.15.

Alternatively, we can consider models that allow odds ratios to not be equal to 1. Parameters can be added to Equation 6.15 that relate the odds ratios to the row and column items in much the same way that two-way ANOVA models relate means to levels of factors. These regression models include

1. Homogeneous association: $\log(\mu_{ab(ij)}) = \beta_{0(ij)} + \beta_{a(ij)}^W + \beta_{b(ij)}^Y + \lambda_{ab}$

2. W-main effects: $\log(\mu_{ab(ij)}) = \beta_{0(ij)} + \beta^W_{a(ij)} + \beta^Y_{b(ij)} + \lambda_{ab} + \lambda^W_{ab(i)}$

3. Y-main effects: $\log(\mu_{ab(ij)}) = \beta_{0(ij)} + \beta^W_{a(ij)} + \beta^Y_{b(ij)} + \lambda_{ab} + \lambda^Y_{ab(j)}$

4. W- and Y-main effects: $\log(\mu_{ab(ij)}) = \beta_{0(ij)} + \beta^W_{a(ij)} + \beta^Y_{b(ij)} + \lambda_{ab} + \lambda^W_{ab(i)} + \lambda^Y_{ab(j)}$

5. Saturated: $\log(\mu_{ab(ij)}) = \beta_{0(ij)} + \beta^W_{a(ij)} + \beta^Y_{b(ij)} + \lambda^{WY}_{ab(ij)}$

Each of these models allows for the odd ratios, OR_{ij}, say, in each 2×2 contingency table to vary in a particular way. The homogeneous association model uses the association parameter λ_{ab}, which is the same for all 2×2 tables. This leads to the equality of each odds ratio, i.e., $OR_{11} = \cdots = OR_{IJ}$, but these odds ratios are not necessarily equal to 1. The Y-main effects model allows the odds ratios to vary across levels of the column MRCV, Y, but they are equal across levels of the row MRCV, W; i.e., $OR_{1j} = \cdots = OR_{Ij}$ for $j = 1, \ldots, J$. The W-main effects model allows odds ratios to vary analogously across levels of W. The W- and Y-main effects allows these odds ratios to vary across levels of both variables, but requires that differences in logs of the odds ratios between two levels of one of the variables are constant across all levels of the other variable. Finally, the saturated model does not assume any type of structure among the odds ratios between W items and Y items and allows them to vary without restriction. This model is essentially Equation 4.6 with extra subscripts (i, j) to identify the two items being modeled. The previous four models specify an association that lies between independence and complete dependence by allowing odds ratios to vary in a structured way.

A separate likelihood function based on the Poisson distribution can be written for each of the 2×2 tables. However, the parameters for the odds ratios in most of these models apply to several tables at once, so we need a likelihood function that can simultaneously model these tables. A full likelihood function across all 2×2 tables is not as easy to write out, because this would entail specifying a model for the associations among all items within each MRCV. Instead, Bilder and Loughin (2007) propose to use a *pseudo-likelihood* function, which is a construct developed by Rao and Scott (1984) for fitting a loglinear model to contingency table data arising through complex survey sampling (see Section 6.3.5).[19] For the MRCV setting, Bilder and Loughin (2007) form a pseudo-likelihood function simply by multiplying each of the separate IJ Poisson likelihood functions together. The parameter estimators resulting from maximizing the pseudo-likelihood function are consistent and approximately normally distributed in large samples, just like MLEs (see Appendix B.3.3), as long as the models for the individual tables are correct. The estimates can be found using the `glm()` function with `family = poisson(link = "log")`.

There are some drawbacks to this model-fitting approach. First, the standard errors given by `glm()` are likely to be wrong because the likelihood used in the model does not account for the correlations among the counts in the IJ tables. Fortunately, these standard errors can be corrected using "sandwich" methods similar to those used for generalized estimating equations (see Section 6.5.5). The `genloglin()` function of the `MRCV` package makes the necessary corrections and will be discussed in an example shortly. Second, the model parameters and corresponding odds ratios are likely to be estimated with reduced precision relative to what could be achieved with a full likelihood. This is unavoidable, unless a more complete model for the correlations among items within the same MRCV can be accurately developed and used.

Information criteria have not been developed to aid in model selection for choose-all-that-apply data. For this reason, we focus on hypothesis tests to compare two models at a time.

[19]This is the same principle used in the generalized estimating equation methods of Zeger and Liang (1986), where a working correlation matrix under independence is used. See also Section 6.5.5.

While this is not ideal for model selection (see discussion on p. 345), there currently is not a better alternative. To implement the approach, a Pearson statistic can be calculated to compare two models where one is nested within the other (i.e., the null hypothesis model is a special case of the alternative hypothesis model with constraints on possible parameter values such as hypothesizing that some are equal to 0). The statistic has the form

$$X^2_M = \sum_{a,b,i,j} \frac{\left(\hat{\mu}^{(a)}_{ab(ij)} - \hat{\mu}^{(0)}_{ab(ij)}\right)^2}{\hat{\mu}^{(0)}_{ab(ij)}},$$

where $\hat{\mu}^{(0)}_{ab(ij)}$ and $\hat{\mu}^{(a)}_{ab(ij)}$ are the model-predicted counts for the models under the null and alternative hypotheses, respectively. Once again, the large-sample distribution of X^2_M is a linear combination of independent χ^2_1 random variables. First- and second-order Rao-Scott corrections can be calculated to evaluate the extremeness of X^2_M under the null hypothesis. Similar to Section 6.4.2, the first-order correction can lead to a liberal test and the second-order correction can be conservative at times (Bilder and Loughin, 2007). Analogous calculations can be applied to statistics created using a likelihood-ratio formulation.

As an alternative to a Rao-Scott correction, the bootstrap can be used to estimate the distribution of X^2_M. Bilder and Loughin (2007) give details of a semi-parametric resampling approach to estimate this distribution that uses the Gange (1995) algorithm for generating correlated binary data. This form of resampling differs from what was presented in Section 6.4.2, because the test statistics are computed from a null-hypothesis model that does not necessarily assume SPMI. Both the null- and alternative-hypothesis models are fit to the resampled data sets, and $X^{2\star}_{M,b}$ is calculated for each resample. The p-value for the test is the proportion of the $X^{2*}_{M,b}$ values greater than or equal to the observed X^2_M; i.e., calculate $B^{-1}\sum_{b=1}^{B} I(X^{2*}_{M,b} \geq X^2_M)$. Small p-values indicate evidence that the null-hypothesis model is too simple.

Once an adequate model is found, model-estimated odds ratios and their corresponding confidence intervals are calculated to examine associations between MRCVs. However, it is important to note that inferences performed after model selection may not achieve the desired confidence level (see Section 5.5). Standardized residuals are also calculated to determine if deviations from the specified model exist. We illustrate calculations of odds ratios and standardized residuals in the next example.

Example: Swine waste management data (Swine.R)

To demonstrate how `genloglin()` works, we summarize the original MRCV binary responses into 2×2 table counts using `item.response.table()`. This time, however, the `create.dataframe = TRUE` argument value is added to `item.response.table()` so that the 2×2 table counts are now given in one column. The columns `W` and `Y` contain the names of the items involved in a particular table, corresponding to the (i, j) model indices, and `wi` and `yj` identify the four 0-1 combinations of the table, the a and b indices. Below is the code and output.

```
> mod.data.format <- item.response.table(data = farmer2, I = 3,
   J = 4, create.dataframe = TRUE)
> head(mod.data.format)
   W  Y wi yj count
1 w1 y1  0  0   123
2 w1 y1  0  1   116
```

```
3 w1 y1  1  0    13
4 w1 y1  1  1    27
5 w1 y2  0  0   175
6 w1 y2  0  1    64
```

We apply the genloglin() function to fit models with data in this format. The function automatically reformats the data so that we can simply use the farmer2 data frame in its data argument. Due to the naming conventions of genloglin(), we recommend in general applications to always use w and y as the leading letters representing items for the MRCVs in the data frame given by data. The genloglin() function uses glm() on the summarized counts to fit the pseudo-likelihood model described earlier, and it corrects the standard errors using a sandwich estimator of the covariance matrix.

The genloglin() function offers two ways to specify a model. First, the model argument accepts the names "spmi", "homogeneous", "w.main", "y.main", "wy.main", or "saturated", where the "w" and "y" designations correspond to the names assigned to the variables by item.response.table(). Alternatively, the model argument can take on a formulation similar to the formula argument of glm(). Below we use the first way to specify the Y-main-effects model.

```
> set.seed(8922)
> mod.fit1 <- genloglin(data = farmer2, I = 3, J = 4, model =
    "y.main", boot = TRUE, B = 2000)
> summary(mod.fit1)
Call: glm(formula = count ~ -1 + W:Y + wi %in% W:Y + yj %in% W:Y
    + wi:yj + wi:yj %in% Y, family = poisson(link = log), data =
    model.data)

Deviance Residuals:
       Min         1Q     Median          3Q         Max
  -1.58007   -0.13272    0.00043    0.10282    0.79587

Coefficients:
           Estimate    RS SE  z value  Pr(>|z|)
Ww1:Yy1     4.83360  0.06535   73.969   < 2e-16 ***
Ww2:Yy1     4.85571  0.06387   76.023   < 2e-16 ***
Ww3:Yy1     4.87418  0.06314   77.199   < 2e-16 ***
Ww1:Yy2     5.15802  0.04696  109.838   < 2e-16 ***
Ww2:Yy2     5.19427  0.04411  117.750   < 2e-16 ***

<OUTPUT EDITED>

Yy2:wi:yj  -0.70644  0.63025   -1.121   0.26233
Yy3:wi:yj  -3.56978  0.88623   -4.028  5.62e-05 ***
Yy4:wi:yj  -1.39762  0.85852   -1.628   0.10354
---
Signif. codes:  0 '***' 0.001 '**' 0.01 '*' 0.05 '.' 0.1 ' ' 1

(Dispersion parameter for poisson family taken to be 1)

Null deviance: 25401.0663    Residual deviance:      5.8825
Number of Fisher Scoring iterations: 4
```

The output begins with the actual call to `glm()`. The model syntax given in the `formula` argument shows the alternative way that the Y-main-effects model could have been specified, where the variable names are derived from the data frame created by `item.response.table()`. This syntax uses *nested effects* to specify first a loglinear model under independence for each 2×2 table using the `W:Y + wi%in%W:Y + yj%in%W:Y` portion of the code. This ensures that the marginal counts for each 2×2 table match between the data and the model, as is typically required for loglinear modeling of associations. The `wi:yj + wi:yj%in%Y` syntax then allows the association within each 2×2 table to vary in the specified manner across the entire item response table. As an additional example of this syntax, the saturated model can be specified using `model = count ~ -1 + W:Y + wi%in%W:Y + yj%in%W:Y + wi:yj%in%W:Y` in `genloglin()`. This formulation expands the use of nested effects further by essentially estimating Equation 4.6 to each 2×2 table.

Continuing in the output, we can relate the estimated parameters to the model specification. For example, `W:Y` in the `formula` argument of `glm()` corresponds to the intercept terms $\beta_{0(ij)}$, which are listed as `Ww1:Yy1` to `Ww3:Yy4` in the output. Also, `wi:yj%in%Y` corresponds to the Y-main effect terms $\lambda^{Y}_{ab(j)}$ which are listed as `Yy2:wi:yj` to `Yy4:wi:yj` in the output (`Yy1:wi:yj` is not estimated because $\lambda^{Y}_{ab(1)}$ is set to 0).

We next check the fit of the Y-main-effects model by comparing it to the saturated model to determine if the simpler association structure is adequate. The `boot` argument of `genloglin()` was set to `TRUE` (the default value) previously, so we are able to estimate the distribution of X^2_M using the bootstrap with 2,000 resamples (`B` argument). By specifying `type = "all"` in the `anova()` method function below, we are able to view the results from both the bootstrap and the second-order Rao-Scott approaches.

```
> comp1 <- anova(object = mod.fit1, model.HA = "saturated", type
    = "all")
> comp1
Model comparison statistics for
H0 = y.main
HA = saturated

Pearson chi-square = 5.34
LRT = 5.88

Second-Order Rao-Scott Adjusted Results:
Rao-Scott Pearson chi-square statistic = 10.85, df = 5.23, p =
    0.0624
Rao-Scott LRT statistic = 11.96, df = 5.23, p = 0.0409

Bootstrap Results:
Final results based on 2000 resamples
Pearson chi-square p-value = 0.031
LRT p-value = 0.017
```

From the output, $X^2_M = 5.34$ and the corresponding p-value is 0.031 using the bootstrap approach. The p-value using the second-order Rao-Scott approach is 0.0624.

Both p-values indicate that there is marginal evidence of fit problems with the simpler Y-main-effects model.

Observed odds ratios and model-predicted odds ratios for each (W_i, Y_j) combination are found using the `predict()` function.

```
> OR.mod <- predict(object = mod.fit1, alpha = 0.05)
> OR.mod
Observed odds ratios with 95% asymptotic confidence intervals
   y1                  y2                 y3
w1 2.2   (1.08, 4.47)  1.82 (0.91, 3.65) 0.1 (0.02, 0.42)
w2 2.91  (1.25, 6.78)  1.77 (0.81, 3.88) 0.07 (0.01, 0.51)
w3 10.27 (2.34, 44.98) 0.99 (0.37, 2.66) 0.1 (0.01, 0.78)
   y4
w1 1.09 (0.23, 5.12)
w2 0.68 (0.09, 5.43)
w3 0.45 (0.03, 7.83)

Model-predicted odds ratios with 95% asymptotic confidence
    intervals
   y1                y2                y3
w1 3.18 (1.54, 6.57) 1.57 (0.78, 3.18) 0.09 (0.02, 0.44)
w2 3.18 (1.54, 6.57) 1.57 (0.78, 3.18) 0.09 (0.02, 0.44)
w3 3.18 (1.54, 6.57) 1.57 (0.78, 3.18) 0.09 (0.02, 0.44)
   y4
w1 0.79 (0.21, 2.98)
w2 0.79 (0.21, 2.98)
w3 0.79 (0.21, 2.98)

Bootstrap Results:
Final results based on 2000 resamples
Model-predicted odds ratios with 95% bootstrap BCa confidence
    intervals
   y1               y2                y3
w1 3.18 (1.53, 7.3) 1.57 (0.72, 3.16) 0.09 (0.03, 0.33)
w2 3.18 (1.53, 7.3) 1.57 (0.72, 3.16) 0.09 (0.03, 0.33)
w3 3.18 (1.53, 7.3) 1.57 (0.72, 3.16) 0.09 (0.03, 0.33)
   y4
w1 0.79 (0.27, 3.5)
w2 0.79 (0.27, 3.5)
w3 0.79 (0.27, 3.5)
```

The first table gives the observed odds ratios and associated confidence intervals for each 2×2 table using Equation 1.10. The second and third tables give the model-predicted odds ratios and corresponding confidence intervals. These values are equal within each column because no W effects are estimated in the Y-main-effects model. In the second table, the confidence interval is a Wald interval using model-based estimates of the odds ratio and variance. The last table uses bootstrap-based intervals calculated by the BC_a method; see Davison and Hinkley (1997) for details on the calculation of this type of interval.[20]

[20]The bootstrap BC_a confidence interval method is one of two bootstrap-based methods recommended for general practice. The lower and upper bounds of this interval are specific lower and upper quantiles from

The fit of the model within each 2×2 table is assessed through standardized Pearson residuals, which are found using the `residuals()` function. Variances for the residuals are calculated using either model-based, large-sample approximations or bootstrap estimates. The bootstrap version is shown below.

```
> resid.mod <- residuals(object = mod.fit1)
> resid.mod$std.pearson.res.boot.var
      y1            y2            y3            y4
W     1     0       1     0       1     0       1     0
w1 1 -2.39  2.39   1.03 -1.03   0.31 -0.31   1.13 -1.13
   0  2.39 -2.39  -1.03  1.03  -0.31  0.31  -1.13  1.13
w2 1 -0.59  0.59   0.91 -0.91  -0.67  0.67  -0.44  0.44
   0  0.59 -0.59  -0.91  0.91   0.67 -0.67   0.44 -0.44
w3 1  2.83 -2.83  -1.73  1.73   0.24 -0.24  -0.87  0.87
   0 -2.83  2.83   1.73 -1.73  -0.24  0.24   0.87 -0.87
```

Using standard guidelines (see Section 5.2.1), all of the standardized residuals are reasonably small (below 2 in absolute value) except those for (W_1, Y_1) and (W_3, Y_1), which leads us to a similar overall fit conclusion as when using X^2_M. These residuals specifically show that the associations across contaminants for the lagoon waste storage method do not appear to be as homogeneous as those within the other waste storage methods. This suggests that a new model which takes this heterogeneity into account could potentially fit better. We construct this model by adding an additional term that forces a perfect fit to the counts in the (W_3, Y_1) combination table.

```
> set.seed(9912)
> mod.fit.final <- genloglin(data = farmer2, I = 3, J = 4, model
    = count ~ -1 + W:Y + wi%in%W:Y + yj%in%W:Y + wi:yj +
    wi:yj%in%Y + wi:yj%in%W3:Y1, boot = TRUE, B = 2000)
> comp.final1 <- anova(object = mod.fit.final, model.HA =
    "saturated", type = "all")
> comp.final1

Model comparison statistics for
H0 = count ~ -1 + W:Y + wi %in% W:Y + yj %in% W:Y + wi:yj +
    wi:yj %in% Y + wi:yj %in% W3:Y1
HA = saturated

Pearson chi-square statistic = 1.81
LRT statistic = 1.86

Second-Order Rao-Scott Adjusted Results:
Rao-Scott Pearson chi-square statistic = 4.29, df = 5.07, p =
    0.5178
Rao-Scott LRT statistic = 4.4, df = 5.07, p = 0.503

Bootstrap Results:
```

the B statistics calculated on the resampled data sets. Rather than using the $\alpha/2$ and $1-\alpha/2$ quantiles, adjusted quantiles are found that take into account possible bias and unstable variance in the statistic being studied.

```
Final results based on 2000 resamples
Pearson chi-square p-value = 0.404
LRT p-value = 0.388
```

This new model fits considerably better than the Y-main-effects model. Comparing the new model to the saturated model, we obtain $X^2_M = 1.81$ with a bootstrap p-value of 0.404. Also, the standardized residuals are all relatively small in absolute value (not shown). Therefore, this new model appears to reasonably explain the associations between the contaminants and the waste storage methods. The model-estimated odds ratios are 2.48 for (W_1, Y_1) and (W_2, Y_1) with a 95% bootstrap BC_a interval of (1.17, 5.42). Of course, the model estimated odds ratio for (W_3, Y_1) and corresponding confidence interval is essentially the same as for the observed values as calculated earlier. All of the other model-estimated odds ratios and standardized residuals are essentially the same as in the Y-main-effects model. In summary, there is moderate to large positive association between lagoon waste storage and testing for each contaminant. Also, there tends to not be contaminant testing when natural drainage is used. For pit and holding tank, there is no clear association with contaminant testing.

We conclude this example with a reminder of the concerns expressed in Section 5.5 regarding validity of inferences following model selection. When the same data are used for both model selection and for post-selection inference, statistical research has grown to acknowledge that there is more uncertainty present for inferences than stated in a confidence level or a type I error rate. For example, the 95% bootstrap BC_a intervals stated here may actually have a lower level of confidence. Unfortunately, it would be difficult to select new farmers for this survey so that the model can be applied to new data. Furthermore, splitting the data into two separate parts—one for model selection and one for post-selection inference—dilutes our ability to construct a good model given the sample size. The current standard of practice is what we present here. We recommend anyone using this approach for research to include caveats in their work such as those stated in this paragraph and in Section 5.5.

Models for three or more MRCVs are possible as well. For example, consider the case of three MRCVs where the third variable is represented by binary responses Z_k for $k = 1, \ldots, K$. Rather than the IJ 2×2 tables as we saw previously, we now have IJK $2 \times 2 \times 2$ tables. The model under complete independence is

$$\log(\mu_{abc(ijk)}) = \beta_{0(ijk)} + \beta^W_{a(ijk)} + \beta^Y_{b(ijk)} + \beta^Z_{c(ijk)}, \tag{6.16}$$

for $a = 1, 2$, $b = 1, 2$, $c = 1, 2$, $i = 1, \ldots, I$, $j = 1, \ldots, J$, and $k = 1, \ldots, K$. This is the usual Poisson regression model for complete independence between three binary categorical variables, with additional (i, j, k) subscripts added to denote the specific W, Y, Z combination $2 \times 2 \times 2$ table. More complicated models can be formed in which main effects and interactions are added across levels of W, Y, and Z. For example, consider a model that contains the terms included in Equation 6.16 and

$$\lambda_{ab} + \lambda^W_{ab(i)} + \lambda^Y_{ab(j)} + \lambda^{WY}_{ab(ij)} + \delta_{bc} + \delta^Y_{bc(j)} + \omega_{ac}.$$

The λ-parameters allow odds ratios between W and Y to be different for each W, Y combination but remain constant across levels of Z; the δ-parameters allow odds ratios between Y and Z to be different for each Y level, but the same for all W and Z levels; and the ω-parameter allows odds ratios between W and Z to be different from 1, but the same for all levels of W, Y, and Z. Model fitting, hypothesis testing, and odds ratio estimation follow in the same manner as described for the two MRCV case. However, as the number of MRCVs increases, finding a suitable model and interpreting its parameters can become more difficult.

6.5 Mixed models and estimating equations for correlated data

All of the models that were introduced in Chapters 2–4 make the critical assumption that the data to which they are fit consist of independent observations. In real applications, data are often collected in clusters, or multiple measurements may be made on the same subject. With such data, it is typical that measurements from the same cluster or unit are more similar to one another than to observations made on different clusters or units. In other words, the data within these groupings are *correlated*.

Performing a statistical analysis on correlated data as if they were independent is not recommended. In particular, when data are positively correlated—i.e., observations from the same group are more similar than those from different groups—a statistical analysis that assumes independence tends to be too liberal. To see why this is true, consider an extreme situation where we have a sample of g groups, each with m observations that all provide identical results, while observations in different groups may vary. Then, the multiple measurements in each group are superfluous, and a single measurement from each group is all that is needed. This means that instead of having a sample of $n = gm$ observations, we effectively have a sample of only $g = n/m$ meaningful observations; the other data tell us nothing new. A statistical analysis should be conducted only on these g numbers, meaning that the sample is effectively much smaller than it appears. If an analysis is done instead on all n observations assuming independence, estimated variances for statistics will be much smaller than they should be. The resulting confidence intervals will be narrower than they should be to provide the stated level of confidence, and p-values smaller than they should be, resulting in excessive type I error rates.

This is a serious issue that should not be ignored in an analysis. There are two basic approaches that account for the effects of analyzing correlated data. The first is to change the statistical model so that it correctly reflects the grouping structure of the data. There are numerous ways to achieve this, and in this section we focus on one of the more commonly used models for correlated counts and proportions, the *generalized linear mixed model*. In Section 6.5.5, we briefly discuss the second approach, *generalized estimating equations*, in which models are fit to the data assuming independence, but the inferences resulting from the model are adjusted to account for correlation.

6.5.1 Random effects

Populations of measurements often fall naturally into groups. For example, several different hospitals may participate in a medical study, each with its own set of doctors and patients. In a quality-assurance study, we may sample products from several different production runs (or "batches"). In the placekicking example introduced in Chapter 2, placekicks are attempted by different kickers and in different games.

Data that are sampled in groups of this type are called *clustered*. In each of these examples, the grouping variable—hospital, batch, kicker, or game—may be an unavoidable part of the study. Sampling in groups or batches may be needed to ensure that a large enough sample can be collected within a reasonable time frame or budget. Sampling from *multiple* groups helps to ensure that the results of the study apply to units in a larger population of such groups, rather than to just those in one particular group.

In any case, these grouping variables are not of primary interest in the study. They may nonetheless have an impact on the mean or probability of a measured response because different levels of the variables may have naturally higher or lower response potential (e.g., different kickers may have higher or lower success probabilities than other kickers across all placekicks). This variability imparts a correlation onto the measurements, because all measurements from the same cluster are exposed to that cluster's unique response potential, and therefore tend to be more similar to one another than they are to measurements from different clusters that have different response potential.

Another way in which measurements can be grouped is when units are independent but multiple observations are made on each unit at different times or different locations on the unit. These observations are called *longitudinal data* when measured temporally, or more generally *repeated measures*, and the units upon which the repeated measurements are taken are generically referred to as "subjects." Repeated measures are quite common in medical studies, where measurements of things like vital signs, disease status, and side effects may be recorded on subjects at specific times. Repeated measures are also used in areas such as studying growth of plants or animals over time, or in measuring soil nutrients and contamination at different prescribed depths. The goal is then to understand how the measurements within subjects change across time or location, possibly interacting with some properties of the subjects. Notice that repeated measures produce correlated data: if a subject's measurement is above average at a particular time or depth interval, it may be likely to remain above average at neighboring times or depths. As with clustered data, correlation among repeated measures taken on the same subject can be alternatively expressed as a different overall response potential for each subject, causing higher- or lower-than-average responses for all measurements on the same subject.

Thus, clustered data and repeated measures are two different versions of the same phenomenon of grouped measurements. However, there are characteristics that distinguish the two structures. First, repeated measures are usually measured in a specific order (in time or space) that may be common to all subjects. For example, measurements of growth might be taken at specific points in time for all subjects. Soil properties might be measured on a core sample at specific depths below the surface. This creates an ordering to the measurements. A result of this ordering is that pairs of measurements that are temporally or spatially close to each other tend to be more similar—more highly correlated—than pairs that are farther apart. On the other hand, with clustered data the measurements within a cluster are often *not* arranged in any particular order that affects their correlations. They are usually *exchangeable,* meaning that one could rearrange the measurements' identifying labels randomly and not lose information.

Repeated measures and clustered data can also differ in how explanatory variables are measured. With clustered data, both the clusters and the individual units within a cluster may vary in ways we can measure. Thus, we can have explanatory variables that are the same for all units in a cluster and also those that are different for each unit in a cluster. On the other hand, in repeated measures we have the same measurement taken on the same subject multiple times. Thus, the time or location is usually the only thing that changes among measurements within a subject. Explanatory variables measured on the subject are constant for all measurements within the subject. When explanatory variables measured on subjects *can* change across time or location, they are called time-varying. For example, the extended version of the alcohol data from Chapter 4 contained daily measurements of alcohol consumption as a response. Explanatory variables included both time-invariant measures of the individual subjects recorded at the start of the study, and time-varying measures recorded daily along with alcohol consumption. Depending on the modeling goals, time-varying explanatory variables may require special care. More details on explanatory variables for repeated measures and clustered data are available in Singer and Willett (2003) and Fitzmaurice et al. (2009).

Consider next the population of all possible levels of a particular grouping variable (e.g., all possible hospitals in a medical study or all possible trees in a growth study). Some levels naturally have higher mean response values and others have lower means, where the notion of "mean" here includes probabilities for simplicity. We can imagine that the population has an overall mean and that each level has a constant value associated with it that gets added to the overall mean or probability, raising or lowering it by a unique amount. This constant value is called the *random effect* of the level. Because variables identifying levels, like hospitals or tree labels, are always categorical, they are often called *random-effects factors.* The levels of a random-effects factor that are used in a particular study are assumed to have been sampled from a population of possible levels.

All of the categorical explanatory variables we have studied prior to this section are *fixed-effects factors.* Their levels are either chosen deliberately or observed among a sample, and comparisons of means or probabilities among those specific levels are of interest. In particular, there is not assumed to be some larger, potentially infinite, population of categories that we have not yet observed.

For the purposes of modeling, we typically assume that the random effects relating to the population of factor levels follow some known distribution with unknown parameters. Most commonly, we assume they are normally distributed with mean 0 and unknown variance. The zero mean ensures that, on average, the random effects don't change the overall mean of the population, while the unknown variance creates a new parameter in the model, representing the amount by which the measurements from different hospitals or trees may differ from one another. The variance associated with a particular random-effects factor is called its *variance component.*

Understanding the variance component of a random-effects factor can sometimes be the focus of a study. For example, there are many labs that can analyze blood samples, all supposedly using techniques that follow the same protocols and staffed by technicians that have the same training. A regulatory agency might be interested in assessing whether the protocols and training are adequate. If they are not, then we might expect that there is noticeable variability among labs in the results that they produce on the same samples, and that some labs use methods or technicians that produce consistently higher or lower responses than others. In this case, we are interested in assessing whether there is variability among all labs in the population rather than in comparing a specific lab to another. In other words, we want to know whether the variance component for the lab random-effect factor is zero.

A given study can have more than one random-effects factor, and these factors can be nested or crossed. For example, there may be many labs that can process and analyze blood samples, and many technicians working at each lab. We could gather multiple samples from each of a large number of donors, and send three samples from each donor to a set of randomly selected labs to be analyzed by three different technicians within those labs. Then, we have random-effects factors representing labs, technicians nested within the labs, and donors crossed with both labs and technicians.

6.5.2 Mixed-effects models

Models may contain only fixed effects, only random effects, or a combination of fixed and random effects. The latter are called *mixed-effects models*, or "mixed models" for short. Recall from Sections 2.3 and 4.2.1 that logistic, Poisson, and many other regression models for counts or proportions are different types of *generalized linear models* (GLMs). When a mixed model is applied to data whose fixed effects would normally be modeled using a generalized linear model, then we create a *generalized linear mixed model* (GLMM).

Consider first a simple problem in which the only explanatory variable is a random-effects factor with a sampled levels. For example, perhaps we have measured the number of pine beetles on the t closest pine trees to each of a different random locations in a region, and we want to understand how variable beetle counts are throughout the region. It is possible that trees in different areas have higher or lower mean counts for a variety of reasons that we have not measured. If we use a Poisson GLM for this problem, and assume that all locations have the same mean, the variability in means among locations would likely cause overdispersion (see Section 5.3). If we instead add location to the model in the usual way, as a fixed-effect categorical variable, then we can estimate the means from these a locations, but we cannot guess with any known probability the range within which means at other locations might lie.

As an alternative, we can fit a Poisson model that allows the mean to vary randomly among different locations in the population, and we can estimate the variability among those means. Using a log link, $g(\mu_{ik}) = \log(\mu_{ik})$, where μ_{ik} is the mean beetle count for tree k at location i. We can write the linear predictor as

$$g(\mu_{ik}) = \beta_0 + b_{0i}, \tag{6.17}$$

where β_0 is the value of the linear predictor (the log-mean) at the average location, and b_{0i}, $i = 1, \ldots, a$ is the random effect of location i. The random effects $b_{01}, \ldots, b_{0a}$ are assumed to be a random sample from $N(0, \sigma^2_{b0})$. For example, the variability of beetle counts at different locations can be found by estimating σ^2_{b0}.

Note that Equation 6.17 is a random-effects model, because the only variable in the model is location, which is random. Notice also that we could write the model a little differently by combining the average log mean with the random effect for each location into a single symbol, say $\tau_i = \beta_0 + b_{0i}$, so that it is apparent that the model represents a *random intercept* for each location. In general, a random-effects model combines random effects with their respective fixed "average" parameter to create random parameter values for each subject or cluster.

We can extend this to a generalized linear *mixed* model if we have additional measurements. For example, trees in some locations may be larger than those in other locations because of recent logging, fires, or other impacts, and larger trees might harbor more beetles simply because of the larger volume in which beetles could reside. Suppose that we measure the circumference of each tree in the study, denoted by $x_{ik}, i = 1, \ldots, a$, $k = 1, \ldots, t$, and believe that the log-mean beetle count might change linearly with the circumference. A fixed-effects generalized linear model ignoring locations would use $g(\mu_{ik}) = \beta_0 + \beta_1 x_{ik}$, where β_0 is the intercept (technically, the log-mean beetle count on a tree with 0 circumference) and β_1 is the slope of the log-mean (the change in log-mean beetle count per 1-unit increase in circumference).

If we want to allow for the effects of locations in this model, there are several ways to do so. Picture fitting a separate Poisson GLM to the beetle-count/tree-size relationship in each location. We would estimate a different intercept and slope in each case, but the estimates may or may not all be estimating the same underlying population quantity. If we believe that the underlying slopes should all be the same—i.e., the log-mean count changes at the same rate with increasing circumference at all locations—then the random location effects act only on the intercepts, and the linear predictors are parallel. The corresponding GLMM uses the linear predictor,

$$g(\mu_{ik}) = \beta_0 + \beta_1 x_{ik} + b_{0i}, \tag{6.18}$$

where $b_{01} \ldots, b_{0a}$ are assumed to be a random sample from $N(0, \sigma^2_{b0})$. A consequence of this model is that the ratio between mean counts for trees of a given size at two different locations is the same for every size of tree, or equivalently that the ratio of means between two particular sizes of tree is the same at each location. In a logistic regression model, the equivalent assumption would be that the odds ratios between any two levels of the random effect are the same for all values of x or vice versa. Notice that we could rearrange this model as $g(\mu_{ik}) = (\beta_0 + b_{0i}) + \beta_1 x_{ik}$, making it a little more clear that this model represents random intercepts for each location with a common slope for all locations.

Alternatively, if we expect that the true mean ratios (or odds ratios) between levels of x change depending on the level of the random effect, then the random effect is changing the slope as well. For the Poisson regression model, this results in the linear predictor

$$g(\mu_{ik}) = \beta_0 + \beta_1 x_{ik} + b_{0i} + b_{1i} x_{ik}, \tag{6.19}$$

where the additional random effects $b_{11}, \ldots, b_{1a}$ are a random sample from $N(0, \sigma^2_{b1})$. The random effect b_{1i} measures the amount by which the slope of x in the linear predictor for location i differs from the average slope, β_1. Notice that there are now two variance components, σ^2_{b0} and σ^2_{b1}, that are estimated when the model is fit to the data. Rearranging terms in this model yields $g(\mu_{ik}) = (\beta_0 + b_{0i}) + (\beta_1 + b_{1i}) x_{ik}$. Thus, it is apparent that this model consists of both random intercepts and random slopes for each location.

One further model feature that can be considered is the relationship between b_{0i} and b_{1i}, $i = 1, \ldots, a$. In the model above, we have specified a separate normal distribution for the random effects on slope and intercept, which implies that we treat them as independent random variables. However, it might be that locations with very low intercepts have very low slopes as well, indicating that the mean beetle count is low for small trees and does not increase by much for larger ones. In other regions where small trees have high counts, it might be that the beetle populations increase even more rapidly in larger trees. Thus, we would have a positive correlation between slope and intercept random effects. This can be specified by adding to Equation 6.19 the assumption that (b_{0i}, b_{1i}), $i = 1, \ldots, a$, have correlation ρ_{01}.

Much more general models can be developed for problems with multiple fixed and random effects. The key in each case is to carefully specify the ways in which each random effect might affect the responses. There are several books on both linear mixed models and generalized linear mixed models that cover this process in detail. See, e.g., Raudenbush and Bryk (2002), Molenberghs and Verbeke (2005), Littell et al. (2006), and Stroup et al. (2018).

Example: Falls with head impact (FallsGLMM.R, FallHead.csv)

Falls are a serious problem among the elderly, resulting in injuries, medical expenses, and sometimes death. In particular, head impacts during a fall can have severe consequences. Schonnop et al. (2013) studied video footage of 227 falls among 133 residents at two long-term care facilities in British Columbia, Canada. They recorded numerous attributes of each fall, including the direction of the fall, whether there was impact with any of several body parts during the fall, and the age and gender of the resident. Footage of some falls had obstructed views, and so not all falls have complete data records. We consider a reduced version of this data set[21] consisting of the 215 falls with recorded values for all of the following variables:

[21] Data kindly provided by Dr. Steve Robinovich, Department of Biomedical Physiology and Kinesiology, and School of Engineering Science, Simon Fraser University.

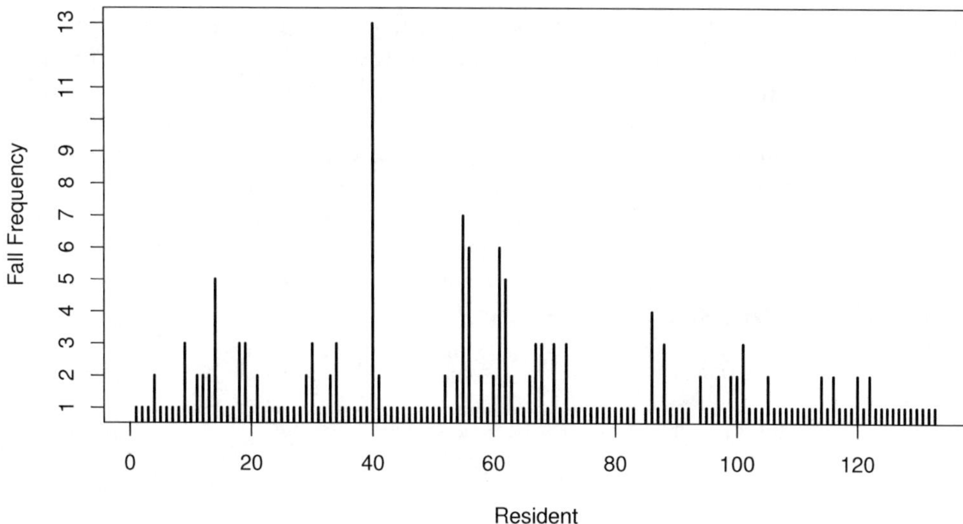

Figure 6.5: Number of falls per resident in the data on falls with head impact. Note that falls for two residents had incomplete data and were removed.

- `resident`: A numerical identification code for the resident whose fall was recorded.
- `initial`: A four-level categorical variable with levels `Backward`, `Down`, `Forward`, and `Sideways` describing the initial direction of the fall.
- `head`: A binary variable indicating whether the fall resulted in the resident's head impacting the floor (1 = yes, 0 = no)

The first few lines of data are shown below, along with the code that plots the number of falls for each subject (Figure 6.5).

```
> fall.head <- read.csv("C:\\Data\\FallHead.csv")
> head(fall.head)
  resident  initial head
1       56 Sideways    0
2        9 Backward    0
3       30  Forward    0
4        9     Down    0
5       70 Sideways    0
6       21 Sideways    1

> fallFreq <- table(fall.head$resident)
> plot(x = as.numeric(names(fallFreq)), y = fallFreq, xlab =
   "Resident", ylab = "Fall Frequency", type = "h", lwd = 2)
```

In this example, we model the probability of head impact as a function of the initial direction of the fall. The binary variable `head` is our response variable. The direction of the fall, `initial`, is a fixed effect because (a) the four levels—forward, backward, sideways, and straight down—are the only four levels possible, and (b) we are interested in comparing these four levels against one another for the possibility of different effects on the probability of head impact.

From Figure 6.5, it is clear that most residents were observed falling only once, but some fell numerous times over the course of the study. It is possible that there is a common underlying mechanism behind the multiple falls experienced by a given resident—e.g., a tendency for dizziness or weakness in a particular leg—that causes the resident to impact the floor in similar ways each time. For this reason, we consider `resident` to be a grouping factor representing clusters of falls in any analysis we undertake. It is a random-effects factor because the goals of the study are not merely to understand falling behavior among *these specific residents.* Rather, the observed residents are intended to represent a wider population of residents at similar long-term care facilities. The resulting model has the form,

$$\text{logit}(\pi_{ik}) = \beta_0 + \beta_2 x_{2ik} + \beta_3 x_{3ik} + \beta_4 x_{4ik} + b_{0i}, \tag{6.20}$$

where π_{ik} is the probability that fall k for resident i has a head impact ($k = 1, \ldots, t_i; i = 1, \ldots, a$); β_0 is the log-odds of head impact fall for the average person with level 1 of `initial` (`"Backward"`); $x_{2ik}, x_{3ik}, x_{4ik}$ are indicator variables for levels `"Down"`, `"Forward"`, and `"Sideways"` of `initial`, respectively; β_j, $j = 2, 3, 4$ is the difference in log-odds of head impact between level j and level 1 of `initial`; and b_{0i} is the random effect of resident i upon the log-odds of head impact. We assume that the b_{0i}'s are independent and normally distributed with mean 0 and variance σ^2_{b0}.

Although residents were not randomly sampled in any way from the supposed population of residents of long-term care facilities, we make the assumption that their falling experiences are representative of such individuals. This assumption should not be taken lightly. If there is something different about the residents at the sampled facilities—for example, if they have different diets, exercise habits, or ethnicity that somehow relates to their falling experience—then the legitimacy of our assumption is questionable, and the results of the study may not apply to the larger population in the way we hope. We could partially check the representativeness of the sample by comparing relevant demographic characteristics of the sample to those of the population (to the extent that these are known). Any obvious differences would serve as a warning that inferences from the sample may not apply to the desired population. See McLean et al. (1991) for a discussion of the different populations that can be implied by the use of different random-effects structures.

6.5.3 Model fitting

A GLMM is fit to data using maximum likelihood estimation. Writing out a likelihood function turns out to be difficult for a GLMM, because the model contains unobserved random variables—the exact values of the random effects—in addition to the observed response variable, Y. The mathematical details of the process are provided in numerous books, including Raudenbush and Bryk (2002), Molenberghs and Verbeke (2005), and Stroup et al. (2018). Here, we give an overview of the process.

Conditional on the random effects (i.e., for a specific cluster or subject), the distribution function of Y has the familiar form of a generalized linear model, like binomial or Poisson. The mean of the conditional distribution for Y changes for each observation depending on the explanatory variables and on the unknown random effects associated with that observation. Estimating the regression parameters from the conditional distribution for each cluster or subject results in estimates that apply only to the clusters or subjects that were present in the data. However, we want our parameter estimates to apply to the entire populations of random-effects levels, not just to those observed. We need to remove this dependence on the random effects somehow.

There are several ways to do this, but the most common one is to base the maximum likelihood estimates on the marginal distribution of Y; that is, on the average distribution over all possible values of the random effects. To do this, we first form the joint distribution between the response and the random effects. We then obtain the marginal distribution of Y by integrating the joint distribution with respect to the random effects, and form the likelihood from this marginal distribution. The parameters in this likelihood are the regression coefficients and the variance components for the random effects, as well as any modeled correlations between random effects.

For example, the model described in Equation 6.20 specifies that the conditional distribution of the binary response of head impact, given the resident's random effect, follows a binomial distribution with one trial and with probability of head impact based on the logit form given in the equation. Denote this distribution by $f(y_{ik}|b_{0i};\boldsymbol{\beta})$ for resident i and impact k, where $\boldsymbol{\beta}$ is a vector that contains all of the fixed-effect regression parameters. This distribution applies only to resident i's unique random effect value. The joint distribution of both the random effect and the responses for resident i is $f(\mathbf{y}_i, b_{0i};\boldsymbol{\beta},\sigma^2_{b0}) = \prod_{k=1}^{t_i} f(y_{ik}|b_{0i};\boldsymbol{\beta})g(b_{0i};\sigma^2_{b0})$, where $\mathbf{y}_i = (y_{i1},\ldots,y_{it_i})$ is a vector of responses for resident i and $g(b_{0i};\sigma^2_{b0})$ represents the normal density function with mean 0 and variance σ^2_{b0}, evaluated at resident i's random effect value b_{0i}. Because b_{0i} is a random variable, we obtain the marginal distribution of the response, $h(\mathbf{y}_i;\boldsymbol{\beta},\sigma^2_{b0})$, by averaging across all possible values of b_{0i} using an integral:

$$\begin{aligned} h(\mathbf{y}_i;\boldsymbol{\beta},\sigma^2_{b0}) &= \int_{-\infty}^{\infty} \prod_{k=1}^{t_i} f(y_{ik}|b_{0i};\boldsymbol{\beta})g(b_{0i};\sigma^2_{b0})\,\mathrm{d}b_{0i} \\ &= \int_{-\infty}^{\infty} \prod_{k=1}^{t_i} \pi_{ik}^{y_{ik}}(1-\pi_{ik})^{1-y_{ik}} \frac{1}{\sqrt{2\pi\sigma^2_{b0}}} \exp\left(-\frac{b_{0i}}{2\sigma^2_{b0}}\right) \mathrm{d}b_{0i}, \end{aligned}$$

where π_{ik} is found from Equation 6.20 and is a function of $\boldsymbol{\beta}$ and b_{0i}. Finally, the likelihood function for the parameters given all of the data $\mathbf{y} = (\mathbf{y}_1,\ldots,\mathbf{y}_a)$ is the product $L(\boldsymbol{\beta},\sigma^2_{b0}|\mathbf{y}) = \prod_{i=1}^{a} h(\mathbf{y}_i;\boldsymbol{\beta},\sigma^2_{b0})$.

Unfortunately, the integral in this likelihood cannot generally be evaluated mathematically. We must instead use some alternative method to evaluate the integral, and then maximize this evaluation to find the parameter estimates. There are three general approaches to performing the evaluation:

1. *Penalized quasi-likelihood* or *pseudo-likelihood* ("approximate the model"): Using a Taylor-series approximation, find convenient approximations to the inverse link function that allow the model to be written as "mean + error," the way a normal linear regression model is typically expressed. The mean and error portions are both evaluated from quantities that are estimated, resulting in "pseudo-data" that are approximately normally distributed. Using the pseudo-data in the approximate model results in an integral that can be evaluated relatively easily using iterative numerical techniques. The pseudo-data are updated with each iteration.

2. *Laplace approximation* ("approximate the integrand"): Because we assume normal distributions for any random effects, the integrand has a form that can be approximated by a simpler function that is easier to integrate mathematically. The resulting function is maximized using iterative numerical techniques.

3. *(Adaptive) Gaussian quadrature* ("approximate the integral"): An integral in one dimension is an area under a curve. This area can be approximated by a sequence of

rectangles, very much like approximating a density function with a histogram. Gaussian quadrature is a method for performing this type of computation in any number of dimensions. The locations of the rectangle edges are referred to as the "points of quadrature." Using a large number of quadrature points represents the shape of the integrand better than using fewer wide rectangles, but is also more computationally intensive. Also, using the shape of the function to help select the locations of the quadrature points, a procedure called *adaptive* quadrature, can result in a better approximation than using a preselected grid of points that may be far from any peaks in the function.

The penalized quasi-likelihood (PQL) approach is applicable to fairly general problems, including repeated measures. However, because it is based on an altered version of the data, there are limits to the inferences that can be derived from the model and its estimates. For example, likelihood ratio inference and information criteria are not valid for comparing models with different sets of fixed effects, because these models produce different pseudo-data. Also, PQL produces estimates that are biased, and this bias may be extreme in models for binomial responses that have small numbers of trials (Breslow and Lin, 1995). In general, it works best when the models are reasonably approximated by normal distributions, which argues against its use for binomial data involving small numbers of trials or many extreme probabilities and for Poisson data with very small means.

Adaptive Gaussian quadrature (AGQ) with many quadrature points provides the closest approximation to the likelihood, but is also computationally the most difficult. For problems with more than one or two random-effect factors, the computing time of AGQ can become prohibitive with multiple quadrature points. It turns out that AGQ with only one quadrature point is mathematically equivalent to using the Laplace approximation. Therefore, Laplace can be used on problems with more complex random-effects structures than AGQ can handle. However, Laplace does not always provide very accurate estimates of the MLEs. Overall, we recommend using AGQ where possible, with as many quadrature points as feasible, and Laplace otherwise. If Laplace cannot be performed, then PQL can be used, but inferences should be viewed as very approximate.

Regardless of the fitting method used, the quantities that are returned are essentially the same. Primarily, there are estimates for the regression parameters and the variance components along with their standard errors, and the residual deviance corresponding to the log-likelihood function evaluated at the parameter estimates. Secondarily, the estimated values of the random effects for each cluster or subject may be produced. These are sometimes called *conditional modes.*

Fitting GLMMs in R

There are numerous R functions that use various implementations of one of these three methods to fit GLMMs. The `glmer()` function of the `lme4` package can perform AGQ on binomial and Poisson models with a single random effect. It can also use Laplace approximations on the same distributions with more complicated random-effect structures, including both nested and crossed effects. The `glmmPQL()` function in the `MASS` package applies PQL to any model family available in `glm()`, and can furthermore handle a variety of repeated-measures correlation structures, which `glmer()` cannot do. In this introductory section on GLMMs, we focus on simpler models and, hence, use `glmer()`. We recommend consulting more comprehensive resources before attempting to fit models. Research has shown that failing to properly account for the random-effects structure of a complex model can result in very poor inference, such as type I error rates that are arbitrarily high; see Loughin et al. (2007) and Fang and Loughin (2012) for examples.

Example: Falls with head impact (FallsGLMM.R, FallHead.csv)

The model for our analysis shown in Equation 6.20 contains just one random effect, `resident`. Therefore, we use the AGQ method available in `glmer()`. The fixed-effects portion of the `formula` argument uses the same syntax as `glm()` and other regression modeling functions (e.g., `Y ~ x`). Random effects are incorporated into the `formula` argument value by adding terms of the form `(a|b)`, where `b` is replaced by the name of the random-effects factor and `a` is replaced by one or more terms in `formula` form that indicate the variables whose coefficients are to be taken as random. For example,

- `(1|b)`: Random effects are added to the intercept for each level of `b` (e.g., Equation 6.18).
- `(x|b)`: Random effects are added to the intercept and to the regression coefficient for numeric `x`, and these random effects are correlated.
- `(1|b)+(0+x|b)`: Both the intercept and the regression coefficient for numeric `x` have *independent* random effects (e.g., Equation 6.19).
- `(x1+x2|b)`: The intercept and the regression coefficients for numeric `x1` and `x2` have *correlated* random effects.

In our falls example, Equation 6.20 is represented by the syntax, `head ~ initial + (1|resident)`.

To show the effects of the number of quadrature points on the parameter estimates, we fit the model for 1, 2, 3, 5, and 10 points using the `nAGQ` argument. The `glmer()` function produces S4 objects (first discussed in Section 2.2.4) of class `glmerMod`. Numerous familiar S3 method functions (e.g., `summary()`, `confint()`, `vcov()`, and `predict()`) can be applied to access information that is typically needed in an analysis. We print the variance component estimates ($\hat{\sigma}^2_{b0}$) for each model by accessing the `varcor` component of the list produced by `summary()`. This produces a list of matrices containing the variances and covariances associated with each random effect in the model. The default printout consists of the corresponding standard deviations (square-root of the variance component) and correlations. Here, we have only one random effect, so just one number, the square root of the variance component, $\hat{\sigma}_{b0}$, is printed.[22] The results are below.

```
> library(package = lme4)

> mod.glmm.1 <- glmer(formula = head ~ initial + (1|resident),
    nAGQ = 1, data = fall.head, family = binomial(link = "logit"))
> summary(mod.glmm.1)$varcor
Groups   Name        Std.Dev.
resident (Intercept) 0.25192

> mod.glmm.2 <- glmer(formula = head ~ initial + (1|resident),
    nAGQ = 2, data = fall.head, family = binomial(link = "logit"))
> summary(mod.glmm.2)$varcor
Groups   Name        Std.Dev.
```

[22] If further access to these quantities is needed, they can be accessed using `summary(obj)$varcor[[q]][i,j]` to produce element (i, j) of the covariance matrix for the qth random effect term given in the `formula` argument.

```
resident (Intercept) 0.26586

> mod.glmm.3 <- glmer(formula = head ~ initial + (1|resident),
    nAGQ = 3, data = fall.head, family = binomial(link="logit"))
> summary(mod.glmm.3)$varcor
Groups    Name        Std.Dev.
resident (Intercept) 0.3034

> mod.glmm.5 <- glmer(formula = head ~ initial + (1|resident),
    nAGQ = 5, data = fall.head, family = binomial(link = "logit"))
> summary(mod.glmm.5)$varcor
Groups    Name        Std.Dev.
resident (Intercept) 0.30342

> mod.glmm.10 <- glmer(formula = head ~ initial + (1|resident),
    nAGQ = 10, data = fall.head, family = binomial(link =
    "logit"))
> summary(mod.glmm.10)$varcor
Groups    Name        Std.Dev.
resident (Intercept) 0.30342
```

Notice that using nAGQ = 1 (equivalent to the Laplace approximation) leads to considerably different variance component estimates than does AGQ with three or more points. The last three methods yield very similar estimates, and the computational time in this problem is negligible, so we choose the model with five quadrature points to continue our analysis.

We next produce a summary of the results and save the summary as an object for later use before printing it. The relevant portion of the results is below.

```
> summ <- summary(mod.glmm.5)
> summ
Generalized linear mixed model fit by maximum likelihood
  (Adaptive Gauss-Hermite Quadrature, nAGQ = 5) [glmerMod]
 Family: binomial  ( logit )
Formula: head ~ initial + (1 | resident)
   Data: fall.head

     AIC      BIC   logLik deviance df.resid
   279.5    296.4   -134.8    269.5      210

Scaled residuals:
    Min      1Q  Median      3Q     Max
-1.2569 -0.7133 -0.6623  0.8608  2.4598

Random effects:
 Groups   Name        Variance Std.Dev.
 resident (Intercept) 0.09206  0.3034
Number of obs: 215, groups:  resident, 131

Fixed effects:
            Estimate Std. Error z value Pr(>|z|)
(Intercept)  -0.6447     0.2469  -2.612  0.00901 **
initialDown  -1.1705     0.6783  -1.726  0.08440 .
```

```
initialForward      0.9581        0.3689    2.597  0.00940 **
initialSideways    -0.1208        0.3768   -0.321  0.74855
```

The output lists the information criteria and model fit summaries, followed by the variance component estimates, the coefficients for the fixed-effects portion of the regression, and the correlation among the fixed-effect parameter estimates (not shown). The estimated log-odds of head impact for fall k from resident i is

$$\text{logit}(\hat{\pi}_{ik}) = -0.65 - 1.17x_{2ik} + 0.96x_{3ik} - 0.12x_{4ik} + b_{0i},$$

where $x_{2ik}, x_{3ik}, x_{4ik}$ are indicator variables for levels 2, 3, and 4 of `initial` (straight-down, forward, and sideways falls, respectively), and b_{0i} is a random variable drawn from a normal distribution with mean 0 and variance 0.092.

Note that the average value of b_{0i} is zero. Thus, for an average resident with a backward initial direction, $\text{logit}(\hat{\pi}) = -0.6447$, which results in $\hat{\pi} = 0.34$. Similarly for an average resident with a downward initial direction of fall, $\text{logit}(\hat{\pi}) = -0.6447 - 1.1705 = -1.8152$, which translates to $\hat{\pi} = 0.14$. The intercept for different residents varies according to a normal distribution with estimated standard deviation $\hat{\sigma}_{b0} = 0.303$. Thus, we can estimate that approximately 95% of residents who fall backward initially have log-odds of head impact within $\hat{\beta}_0 \pm 2\hat{\sigma}_{b0} = -0.6447 \pm 0.606 = -1.25$ to -0.04. This translates to probabilities between 0.22 and 0.49. Similarly, approximately 95% of residents with downward falls have log-odds of head impact within -1.8152 ± 0.606, which corresponds to probabilities between 0.08 and 0.23. Notice that these intervals are not confidence intervals on a statistic, but rather represent a range of estimated probabilities of head impact for different residents in the population.

There are many method functions that can provide useful summary information from the saved object resulting from `glmer()`. Below are a few examples. One can explore them further by using `methods(class = "merMod")` to identify them and `names()` on objects created by these functions.

```
> # Estimates of fixed-effect parameters
> fixef(mod.glmm.5)
     (Intercept)      initialDown   initialForward
      -0.6446629       -1.1704700        0.9581401
 initialSideways
      -0.1207953

> # Conditional Modes (b_0i) for random effects listed **BY
   resident ID**, not in data order.
> head(ranef(mod.glmm.5)$resident)
  (Intercept)
1  0.05913603
2 -0.03104570
3 -0.03104570
4 -0.01369701
5  0.05913603
6 -0.02865810

> # Show that there is one element per resident
> nrow(ranef(mod.glmm.5)$resident)
[1] 131
```

```
> # coef = Fixed + Random effects. Listed **BY resident ID**,
   not in data order.
> head(coef(mod.glmm.5)$resident)
  (Intercept) initialDown initialForward initialSideways
1  -0.5855269    -1.17047      0.9581401      -0.1207953
2  -0.6757086    -1.17047      0.9581401      -0.1207953
3  -0.6757086    -1.17047      0.9581401      -0.1207953
4  -0.6583599    -1.17047      0.9581401      -0.1207953
5  -0.5855269    -1.17047      0.9581401      -0.1207953
6  -0.6733210    -1.17047      0.9581401      -0.1207953
```

The fixed-effect ML parameter estimates are obtained using `fixef()`. Predicted values for the random effects b_{0i}, $i = 1, \ldots, a$, are obtained with `ranef()` using a series of calculations described in Bates (2010). Combined fixed and random predictions for each resident are found from `coef()`. Note that the intercept, which is $\hat{\beta}_0 + b_{0i}$, changes for each resident, but the other three parameter estimates are not associated with random effects and remain constant for all residents.

Next we use various calls to `predict()` to show the predicted logit, $g(\hat{\pi}_{ik})$, for each resident, the estimated average logit assuming $b_{0i} = 0$, and the respective predicted probabilities of head impact ($\hat{\pi}_{ik}$) and estimates for an average resident. The argument `re.form` determines which random effects to include in the predictions, where `NULL` includes all by default, computing predictions for each subject using its predicted random effect values, and `NA` includes none.

```
> # Predicted logit for each observation
> logit.i <- round(predict(object = mod.glmm.5, newdata =
   fall.head, re.form = NULL, type = "link"), digits = 3)
> # Estimated mean logit for each fall direction, by observation
> logit.avg <- round(predict(object = mod.glmm.5, newdata =
   fall.head, re.form = NA, type = "link"), digits = 3)

> # Predicted probability for each observation
> pi.hat.i <- round(predict(object = mod.glmm.5, newdata =
   fall.head, re.form = NULL, type = "response"), digits = 3)
> # Estimated average probability for each fall direction, by
   observation
> pi.hat.avg <- round(predict(object = mod.glmm.5, newdata =
   fall.head, re.form = NA, type = "response"), digits = 3)

> # Conditional modes listed in order of the original data (they
   are currently ordered by resident)
> ranefs <- round(ranef(mod.glmm.5)$resident[fall.head$resident,
   ], digits = 3)

> # Print of all predictions and mean estimates together
> head(cbind(fall.head, ranefs, logit.i, logit.avg, pi.hat.i,
   pi.hat.avg))
  resident  initial head ranefs logit.i logit.avg pi.hat.i
1       56 Sideways    0 -0.083  -0.849    -0.765    0.300
2        9 Backward    0 -0.073  -0.717    -0.645    0.328
3       30  Forward    0 -0.018   0.295     0.313    0.573
4        9     Down    0 -0.073  -1.888    -1.815    0.132
```

```
5          70 Sideways    0 -0.001    -0.766        -0.765      0.317
6          21 Sideways    1  0.098    -0.668        -0.765      0.339
  pi.hat.avg
1      0.317
2      0.344
3      0.578
4      0.140
5      0.317
6      0.317
```

Note that `logit.i` differs from `logit.avg` by exactly `ranefs`, showing the effect of the random effect values (conditional modes) on the logits, and hence on the probabilities in `pi.hat.i` relative to `pi.hat.avg`. Thus, the "average" quantities are constant for each fall with the same initial fall direction, whereas the predictions vary a bit for each resident. We can determine from this that, on average, the estimated probability of head impact is highest for forward initial direction (0.58) and smallest for falls that are straight down (0.14).

6.5.4 Inference

The fitted model contains estimates for the fixed effects and variance components, along with the conditional modes. All of these may be desired targets for tests and/or confidence intervals. However, inference in GLMMs must be done carefully. Because we are using ML estimation, parameter estimates should all have approximate normal distributions in large samples, with variances that are not difficult to compute. However, in practice, what constitutes a "large sample" may depend in part on the number of clusters or subjects, the number of observations within clusters or subjects, and the shape of the distribution for the response variables being modeled.

Tests and confidence intervals for fixed-effect parameters

The fixed-effect portion of a GLMM consists of a model like those presented in Chapters 2 and 4. Therefore, the types of analyses that are likely to be needed are the same as what was presented in those chapters. Hypothesis tests for individual parameters or for comparisons of parameters are common, and confidence intervals for these same quantities and for certain means or probabilities are often desired. In those fixed-effects models, LR methods were generally recommended for large samples, where the requirements for "large" were not too severe. Alternatively, Wald methods based on the large-sample normality of the MLE could be used, but the results were often not as accurate as those from LR methods at comparable sample sizes.

In a GLMM, inferences are less accurate than in fixed-effect models, because the process of estimating the variance components changes the sampling distribution of LR and Wald test statistics, similar to what was described in Section 5.3.3. Unfortunately, an explicit form for the adjustment that is common in linear mixed models is not theoretically supported by GLMMs, and not implemented in `glmer()`. Thus, inferences must be taken to be very approximate.

With this in mind, LR and Wald methods are both available to varying degrees. As usual, profile LR confidence intervals require iterative re-evaluation of the model to find interval endpoints. This takes more computing time than in fixed-effect models, because the likelihood can be much more difficult to evaluate, particularly with complicated models or large data sets. The mcprofile package has not been extended to work with these models. Wald test statistics and confidence intervals can be created using emmeans in similar ways as in previous chapters. Wald-based inferences can also be performed with glmerMod-class objects using the multcomp package (introduced in Exercise 20 of Chapter 2).

Our preferred approach is an alternative to these methods that is valid even when samples are not large, the parametric bootstrap. Bootstrap methods are discussed in Sections 3.2.3 and 6.4.2 and are covered more completely in Davison and Hinkley (1997). For a hypothesis test using a parametric bootstrap, data sets (called "resamples") are simulated in a manner similar to what was done in Section 2.2.8 using a version of the GLMM whose parameters are estimated assuming that the null hypothesis is true. For example, we might remove a term from the model and simulate data from the new fitted model to test the significance of the deleted term. The full GLMM is refit to each resample, and a test statistic—commonly a Wald statistic or some other easy-to-compute quantity—is computed on each refit. A p-value is computed as the proportion of these simulated test statistic values that are at least as extreme (i.e., that favor the alternative hypothesis at least as much) as the one computed on the original data.

To find a parametric bootstrap confidence interval for a parameter, the simulation model is the original fitted model with all effects intact. The parameter is estimated for each resample, and the set of simulated parameter estimates is manipulated into interval endpoints using some appropriate technique, several of which are described in Davison and Hinkley (1997). Note that the parameter in question may be a regression-model parameter or some other quantity computed from them, such as an odds ratio or a specific mean or probability. For inferences on the regression parameters, the confint() function can be applied directly to the model-fit object from glmer() to obtain profile LR, Wald, or bootstrap confidence intervals. Examples are given in the program accompanying the example for this section. For inferences on other parameters, such as odds ratios and mean ratios, the simulation must be programmed manually.

Example: Falls with head impact (FallsGLMM.R, FallHead.csv)

We first show the overall test for equality of probabilities across the four initial fall directions, $H_0 : \beta_2 = \beta_3 = \beta_4 = 0$ in Equation 6.20. The alternative hypothesis is simply that not all initial directions have the same probability, H_a : not all of $\beta_2, \beta_3, \beta_4$ are zero. We conduct this test two ways: using the LRT and using a parametric bootstrap. The LRT for each fixed-effects term in the model is conducted using the drop1() function.

```
> lrt <- drop1(mod.glmm.5, test = "Chisq")
> lrt
Single term deletions

Model:
head ~ initial + (1 | resident)
        Df    AIC    LRT  Pr(Chi)
<none>      279.52
initial  3 289.16 15.638 0.001345 **
```

The LRT gives a test statistic of 15.6, and a p-value of 0.001 based on the large-sample χ^2_3 approximation. We next compare these results to those obtained using the bootstrap by simulating 1,000 data sets from the null-hypothesis model using the simulate() function in the lme4 package. On each simulated data set, we fit each model and use drop1() to compute the LRT statistic.[23] The p-value is estimated by computing the proportion of times the simulated test statistic is at least as large as the observed value.

```
> names(lrt)
[1] "Df"       "AIC"      "LRT"       "Pr(Chi)"
> orig.LRT <- lrt$LRT[2]  # Saves LR Test statistic

> # Fit null model
> mod.glmm0 <- glmer(formula = head ~ (1|resident), nAGQ = 5,
    data = fall.head, family = "binomial")
> sims <- 1000
> simfix.h0 <- simulate(mod.glmm0, nsim = sims, seed = 9245982)
> # Fit model and compute test statistic
> LRT0 <- numeric(length = sims)
> for (i in 1:sims){
    m1 <- glmer(formula = simfix.h0[,i] ~ initial +
        (1|resident), nAGQ = 5, data = fall.head, family =
        "binomial")
    LRT0[i] <- drop1(m1, test="Chisq")$LRT[2]
  }
boundary (singular) fit: see help('isSingular')
boundary (singular) fit: see help('isSingular')

<OUTPUT EDITED>

boundary (singular) fit: see help('isSingular')
boundary (singular) fit: see help('isSingular')
> summary(LRT0)
    Min.  1st Qu.   Median     Mean  3rd Qu.     Max.
 0.04826  1.26985  2.32598  3.04838  4.28339 17.16812  > pval <-
     mean(LRT0 >= orig.LRT)
> pval
[1] 0.003
```

The parametric bootstrap code runs for several minutes and produces numerous warnings about a "singular fit." These warnings happen because some of the simulated data sets do not demonstrate any excess variability in the log-odds of head impact across different residents. Thus, the variance for the normal distribution of the random effects is estimated to be zero. These warnings do not affect the calculation of the LRT statistic. The p-value for the bootstrap test is 0.003—3 out of 1,000 simulated data sets produced an LRT statistic that was at least as large as the 15.6 that we originally observed. Both the simulated and the large-sample p-values lead to the same decision to reject the null hypothesis and conclude that the probabilities of head impact are not the same for all four initial falling directions.

[23]The version of lme4 that we used, 1.1-30, contains a function called bootMer() that can perform the parametric bootstrap. The function does the simulation step internally and requires a user-specified function to perform an analysis on each resample. It produces an object that can be processed into various summary statistics, like p-values, critical values, and confidence intervals. Ultimately, it does not reduce coding effort appreciably compared to our code.

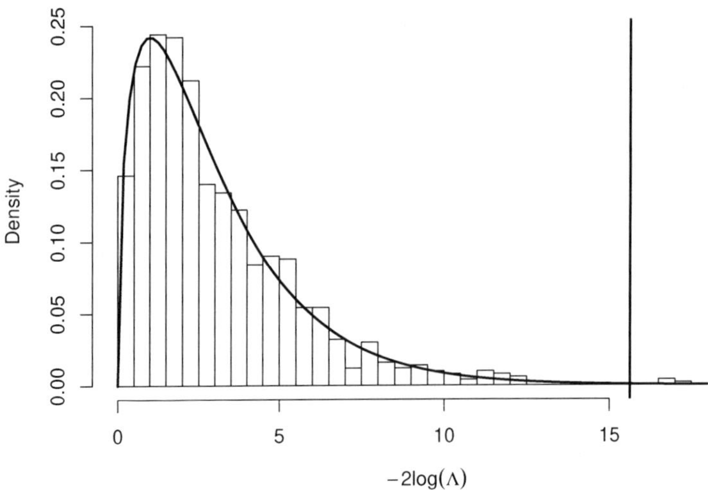

Figure 6.6: Comparison of parametric bootstrap estimate of the distribution of $-2\log(\Lambda)$ with the large-sample χ^2_3 distribution. The vertical line marks the value of the $-2\log(\Lambda)$ in the original data.

Figure 6.6 shows the histogram of the 1,000 simulated values of $-2\log(\Lambda)$ from the parametric bootstrap and the χ^2_3 distribution that is used to approximate the distribution of the test statistic for the LRT. The two distributions are in reasonable agreement, although this is not always the case for other data sets. The reasonably large sample size and the relatively moderate binomial probabilities likely allow the χ^2_3 approximation to serve well in this problem.

To determine which initial fall directions have higher or lower probabilities of head impact, we construct confidence intervals for all pairwise comparisons using the `emmeans` package. Our program also shows how to use the `multcomp` package. The code below produced confidence intervals for the probability of a head impact, as well as odds ratio comparisons between the initial fall directions. The overall familywise confidence level is controlled at 95% using a Tukey adjustment for multiple inferences (similar to the single-step method used in Section 2.2.6).

```
> library(package = emmeans)
> emm1 <- emmeans(object = mod.glmm.5, specs = ~ initial)
> confint(object = emm1, type = "response")
 initial    prob     SE  df asymp.LCL asymp.UCL
 Backward 0.344 0.0557 Inf    0.2444     0.460
 Down     0.140 0.0768 Inf    0.0445     0.362
 Forward  0.578 0.0676 Inf    0.4428     0.702
 Sideways 0.317 0.0641 Inf    0.2066     0.454

Confidence level used: 0.95
Intervals are back-transformed from the logit scale

> confint(object = contrast(object = emm1, method = "pairwise"),
   type = "response", adjust = "tukey")
 contrast            odds.ratio     SE  df asymp.LCL asymp.UCL
 Backward / Down          3.224 2.1864 Inf    0.5644    18.411
```

```
 Backward / Forward            0.384 0.1415 Inf      0.1487       0.990
 Backward / Sideways           1.128 0.4252 Inf      0.4286       2.971
 Down / Forward                0.119 0.0825 Inf      0.0200       0.707
 Down / Sideways               0.350 0.2423 Inf      0.0591       2.072
 Forward / Sideways            2.942 1.1783 Inf      1.0512       8.232

Confidence level used: 0.95
Conf-level adjustment: tukey method for comparing a family of 4
   estimates
Intervals are back-transformed from the log odds ratio scale

> confint(object = contrast(object = emm1, method =
   "revpairwise"), type = "response", adjust = "tukey")
 contrast                odds.ratio    SE  df asymp.LCL asymp.UCL
 Down / Backward              0.310 0.210 Inf    0.0543     1.772
 Forward / Backward           2.607 0.962 Inf    1.0105     6.725
 Forward / Down               8.403 5.827 Inf    1.4151    49.901
 Sideways / Backward          0.886 0.334 Inf    0.3366     2.333
 Sideways / Down              2.857 1.977 Inf    0.4826    16.910
 Sideways / Forward           0.340 0.136 Inf    0.1215     0.951

Confidence level used: 0.95
Conf-level adjustment: tukey method for comparing a family of 4
   estimates
Intervals are back-transformed from the log odds ratio scale
```

The estimates for the probability of head impact for each initial fall direction match those obtained previously using the `predict()` function, except that we now have 95% confidence intervals for each probability. The confidence intervals for the probability of head impact for `Down` and `Forward` are $(0.04, 0.36)$ and $(0.44, 0.70)$, respectively. Note that these do not overlap, lending strong support to the previous result that not all initial fall directions have the same probability of head impact.

Further evidence is supplied more directly by the confidence intervals for the odds ratios. The confidence intervals for the three odds ratios involving forward falls all exclude 1. Specifically, re-expressing the estimates and confidence intervals so that forward falls are in the numerator, we find that the odds of head impact from a forward fall are estimated to be 2.6 times as large as from a backward fall, with confidence interval $(1.0, 6.7)$; 8.4 times as large as from a downward fall $(1.4, 49.4)$; and 2.9 times as large as from a sideways fall $(1.1, 8.2)$. These confidence intervals are all fairly wide, indicating that there remains considerable uncertainty in the estimation of these odds ratios. This is partly because we used the Tukey adjustment for simultaneous inferences in `emmeans`, which ensures that we are 95% confident that *all six* confidence intervals will contain their respective true odds ratios. Of course, this probability is approximate because Wald inferences in mixed models require large samples for accuracy. With 215 falls spread among 131 residents, this is perhaps a "moderate" sample size—in particular, a relatively small number of residents had multiple falls, which may reduce accuracy of the variance component estimate. Therefore, we view these results cautiously.

Alternatively, we can use the parametric bootstrap to conduct tests and compute confidence intervals. The code for these calculations is included in the program for this example. Westfall and Young (1993) provide additional algorithms that can be used to control familywise confidence levels for resampling-based confidence intervals.

Tests and confidence intervals for variance components

Inference for variance components of random effects is complicated by the fact that we cannot have negative variance, so the variance component parameter is bounded below by zero. In practice, estimates of variance components are often exactly equal to zero, as we have seen in our example. This is a clear indication that the corresponding random effect variance component plays a negligible role in the model and can be eliminated. Even when a non-zero estimate is obtained for one sample, the probability distribution of estimates from all possible samples from the population may still place sizable probability mass on zero. Furthermore, the sampling distribution for variance components tends to be highly skewed to the right. That is, most samples from a population yield variance component estimates that are slightly below their true values, and may often be zero, but occasionally they are considerably higher. This means that Wald-based inference, which assumes a normal distribution for the variance component estimate, is totally inadequate for variance parameters in GLMMs.[24] Except with enormous sample sizes, the lower endpoints of Wald confidence intervals are often negative, and Wald tests are horribly conservative.

At the same time, the lower bound at zero complicates LR inference. In particular, testing the significance of the variance component using $H_0 : \sigma_b^2 = 0$ vs. $H_a : \sigma_b^2 > 0$ involves inference for a boundary value of a parameter, which invalidates the theory that leads to a chi-square distribution for LRT statistics. This does not affect profile LR confidence intervals the same way it does hypothesis tests, but as previously noted, profiling the likelihood to form confidence intervals can be a difficult computational problem.

This leaves the parametric bootstrap as the best alternative for inference. Tests of the hypothesis $H_0 : \sigma_b^2 = 0$ are relatively straightforward to conduct, provided the run time for repeatedly estimating the model on a large number of resamples is feasible. The process uses the same approach that was used for fixed effects: simulate a large number of new sets of responses from the fitted model that assumes H_0 to be true, fit the full model to each resample, and calculate some test statistic that tests H_0 using the fit of this model. The variance component estimate itself can be used as a test statistic. We then compute the p-value as the proportion of resamples that lead to a variance component estimate at least as large as the one from the original sample, because large estimates of the variance component favor H_a. Alternatively, the LRT statistic can be computed on each resample and compared to the value from the original sample. However, this requires fitting two models for each resample and, hence, takes longer to conduct. To reduce computational time in either case, it may be possible that a relatively small number of simulations may determine conclusively whether the null hypothesis should be rejected. For example, if 50 simulations are done and 40 of them lead to a more extreme LRT statistic than in the original sample (i.e., a p-value of 0.80), then it is already clear that the null hypothesis should not be rejected and there is little reason to performing more simulations.

If a problem is too large to allow a parametric bootstrap to be done feasibly, then the usual LRT statistic can be calculated, but the p-value that would normally be found from a chi-square distribution must be evaluated differently. In particular, the actual distribution of the LRT statistic places more probability mass at 0 than the chi-square approximation allows, so that the standard p-value is at least twice as large as it should be (Littell et al., 2006). Therefore, an approximate p-value can be found by dividing the original p-value by 2 if the test statistic is greater than 0; otherwise, the p-value is 1 because this is the probability that the LRT statistic is 0 or larger.

[24]Stroup (2013) p. 182 claims, "No literate data analyst should use the Wald covariance component test, with the possible exception of extremely large studies with tens of thousands or more observations."

Confidence intervals for variance components represent a much more difficult problem. As long as the profiling can be done reasonably quickly, profile LR intervals can be formed. However, their properties are not well known when the sampling distribution of the variance component may contain probability mass at zero.

Parametric bootstrap confidence intervals for variance components in GLMMs have also not been well researched. Creating resamples from the estimated model and computing variance-component estimates on each resample is reasonably straightforward, assuming that the problem is small enough that a resampling simulation can be applied. However, it is not clear how the resulting estimated probability distribution of the variance components should be used. Davison and Hinkley (1997) describe numerous types of confidence intervals that can be calculated from a bootstrap estimate of a sampling distribution in general—e.g., percentile, bootstrap-t, BC_a, and pivot-based intervals—but all of these require assumptions about the statistic's true probability distribution that are not met when the distribution places considerable probability mass on a boundary value. Limited simulations in Austin and Leckie (2020) suggest that basic percentile intervals created from a parametric bootstrap can often provide coverage close to the stated confidence level. However, performance can become erratic when the number of clusters is small ("small" in their study was 25 clusters) or when the true variance component is closer to zero.

Therefore, we cannot provide uniform guidance for computing parametric bootstrap confidence intervals for variance components when the estimated sampling distribution contains probability mass at zero. However, based on general knowledge of how the bootstrap works, we can offer some suggestions. We expect that a percentile or BC_a interval will still provide better true confidence levels than a Wald interval. The percentile interval is particularly easy to compute—simply set the lower confidence limit at the $\alpha/2$ quantile of the simulated variance-component estimates and the upper confidence limit at the $1-\alpha/2$ quantile of these estimates—although it often provides true confidence levels somewhat below the stated level. The BC_a interval requires some small additional calculation (see Davison and Hinkley, 1997 for details), but in many problems provides confidence intervals whose true confidence levels are close to the stated level. We recommend the BC_a interval when it can be calculated, and the profile LR interval otherwise.

These intervals can also be used when the variance component estimate is far enough from zero that the boundary problem is not an issue. Indeed, we would anticipate that their ability to provide an appropriate interval would be better in this case, and the simulations in Austin and Leckie (2020) suggest that this may be true in limited circumstances. Demonstrating this more generally remains an open research topic.

For model selection, we have previously used information criteria to identify suitable models from a collection of candidate models (see Section 5.1.2). Using ICs with GLMMs is somewhat more complicated than with fixed-effect models. Information criteria can be computed for GLMMs in the usual way based on ML estimates of the model parameters, as long as the estimation method is not penalized quasi-likelihood. However, Greven and Kneib (2010) warn that the boundary problem affects AIC in linear mixed models, causing it to favor a model without a random effect over one with the random effect more often than it should. Whether this problem carries over into GLMMs is not yet known. More generally, with mixed models there are different criteria depending on whether one wishes to use the models for future data conditional on the random effects (e.g., observing more falls on the same residents) or with new random effect values (e.g., observing falls on new residents). See Saefken et al. (2014) for further discussion.

Example: Falls with head impact (FallsGLMM.R, FallHead.csv)

The estimated variance component for the `resident` effect is $\hat{\sigma}^2_{b0} = 0.092$. Although understanding the variability among residents was not the primary focus of this study, learning about between-resident variability may provide useful information. For example, significant variability among residents may suggest that some residents are able to reduce their probability of head impact by taking preventative measures as they are falling.

To perform the parametric bootstrap test of $H_0 : \sigma^2_{b0} = 0$ vs. $H_a : \sigma^2_{b0} > 0$, we demonstrate two approaches, one based on estimating the sampling distribution of the estimated variance component, $\hat{\sigma}_{b0}$, and the other based on estimating the sampling distribution of the LRT statistic. For the former approach, we first extract the variance component estimate from the original data and save it into an object named `orig.vc`. Next, we simulate 1,000 data sets from the GLM with no random effect (i.e., the model under H_0) and refit the model for each one so that we can obtain the variance component estimate. These estimates are saved into a one-column matrix named `varcomps0`. The p-value is computed and a histogram of the variance component estimates is plotted. The LR version of the test is performed similarly, by computing the statistic—the difference in residual deviances between the models with and without the variance component—on the original data (`orig.LR`) and on each bootstrap resample (`LR0`). The bootstrap estimate of the p-value is compared to the one using the large-sample chi-square approximation.

```
> mod.glm <- glm(formula = head ~ initial, data = fall.head,
    family = binomial(link = "logit"))
> sims <- 1000
> simmod.h0 <- simulate(mod.glm, nsim = sims, seed = 28662819)
> orig.vc <- summary(mod.glmm.5)$varcor[[1]][1,1]
> varcomps0 <- matrix(data = NA, nrow = sims, ncol = 1)
> LR0 <- matrix(data = NA, nrow = sims, ncol = 1)
> for (i in c(1:sims)){
    mm <- glmer(formula = simmod.h0[,i] ~ initial +
         (1|resident), nAGQ = 5, data = fall.head, family =
         binomial(link = "logit"))
    varcomps0[i,] <- summary(mm)$varcor[[1]][1,1]
    m0 <- glm(formula = simmod.h0[,i] ~ initial, data =
         fall.head, family = binomial(link = "logit"))
    LR0[i,] <- -2*logLik(m0) + 2*logLik(mm)
}

> # Bootstrap p-value using variance component estimate
> varcomps0 <- na.omit(varcomps0)
> nrow(varcomps0)
[1] 1000
> pval <- sum(varcomps0 >= orig.vc)/nrow(varcomps0)
> pval
[1] 0.328

> # Bootstrap p-value using LRT statistic
> # Original data LR test stat
> orig.LR <- -2*logLik(mod.glm) + 2*logLik(mod.glmm.5)
> pval.lr <- sum(LR0 >= orig.LR)/nrow(LR0)
```

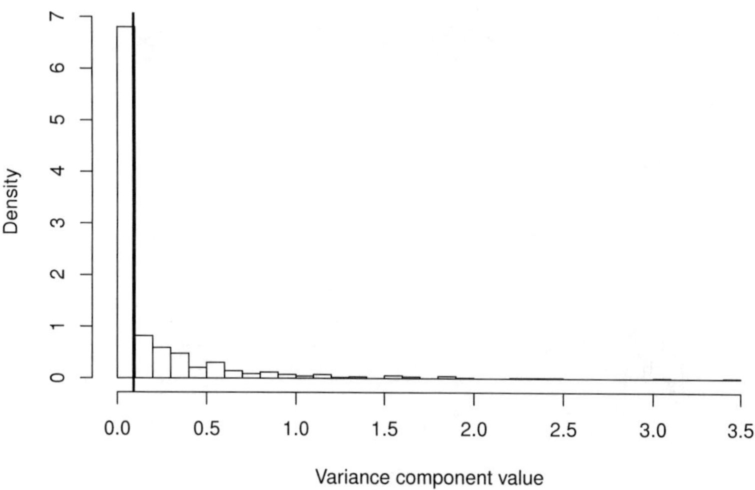

Figure 6.7: Histogram of variance component estimates using parametric resampling under $H_0 : \sigma_{b0}^2 = 0$ for the falls example. The thick vertical line indicates the location of the estimate from the original data.

```
> pval.lr
[1] 0.318

> # Large sample p-value
> as.numeric(1 - pchisq(orig.LR, df = 1))/2
[1] 0.3933802
```

The p-value for the parametric bootstrap test using the variance component estimate as the test statistic is 0.328. There is insufficient evidence to reject the null hypothesis that the variance component is zero. In other words, there is not substantial variability among residents in their log-odds of head impact, given their initial fall direction. The histogram given in Figure 6.7 shows the highly skewed distribution of the variance component estimates under the null hypothesis. Clearly, no normal approximation should be used with this distribution, so a Wald test would be highly inappropriate. The bootstrap estimate of the p-value for the LRT is 0.318, while using a χ_1^2 distribution produces a p-value of 0.393. Although this latter p-value is in reasonable agreement with the bootstrap result, the bootstrap estimate of the sampling distribution of the LRT statistic is very different from the χ_1^2. The histogram given in Figure 6.8 compares the two approximations. Ignoring the 50% of resamples that are expected to produce values of 0 under the null hypotheses, the remaining resamples suggest that the LRT statistic has much greater skewness than the χ_1^2.

The code for computing confidence intervals for σ_{b0}^2 is provided in the program for this example.

6.5.5 Marginal modeling using generalized estimating equations

As described in Section 6.5.3, estimating parameters in GLMMs can be a complicated and difficult process computationally. The results of a GLMM include predictions for each observed cluster or subject (for ease of discussion, in the remainder of this section we use

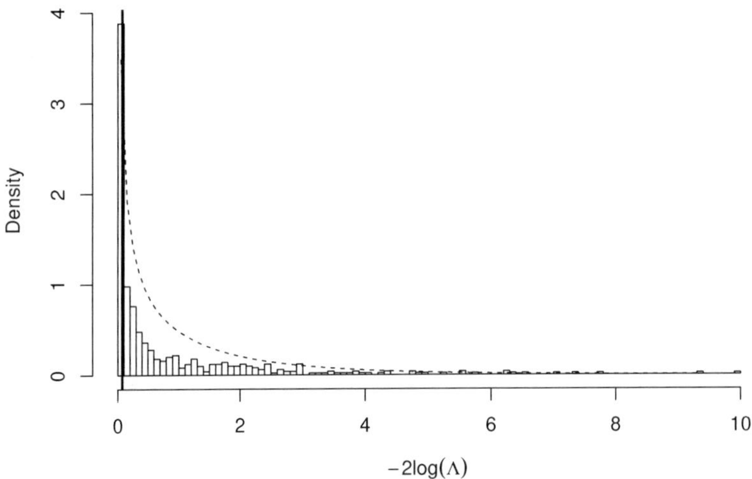

Figure 6.8: Histogram of LRT statistics using parametric resampling under $H_0 : \sigma^2_{b0} = 0$ for the falls example. The histogram excludes 500 resamples with LRT statistic values of zero that are expected to occur by chance under H_0. The thick vertical line indicates the location of the statistic from the original data. The dashed curve is the χ^2_1 approximation.

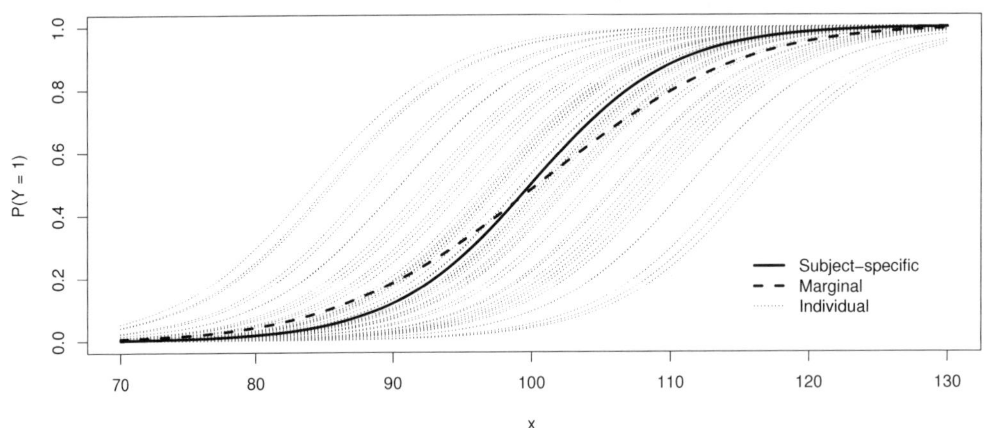

Figure 6.9: Subject-specific model and marginal model for simulated sample of 50 subjects from a population with probability of success following a logistic curve with common slope but different intercepts. The code to produce the plot is in LogisticSim.R.

"subject" to refer to either a subject or a cluster). They also include estimates of various regression parameters for an "average subject" (one with all random effects equal to zero). A GLMM is referred to as a *subject-specific model*, because the addition of random effects allows each subject to have its own different parameter values as described in Section 6.5.2. An alternative approach is to directly model how an explanatory variable relates to the population average response, rather than how it relates to individuals in the population. A direct model for the population average is referred to as a *marginal model*, because the population average is derived from the marginal distribution of the outcome.

The marginal model is fundamentally different from the GLMM as is seen in Figure 6.9. The plot shows the simulated probability of success, $P(Y = 1)$, for a sample of 50 subjects whose probability follows $\text{logit}(\pi_i) = -20 + b_{0i} + 0.2x_i$, $i = 1, \ldots, 50$, where the b_{0i}'s are independent $N(0, 1.5^2)$ random variables. The thick solid curve is the probability curve for the "average subject"; i.e., for $b_{0i} = 0$. This is the quantity that is estimated by the fixed-effects portion of a GLMM. The dashed curve is the average probability at each value of x, which is the quantity that is estimated by a marginal model. The difference between them occurs because the subject-specific model averages horizontally and the marginal model averages vertically. In linear regression models, this distinction does not matter, but it does matter when the response curve is nonlinear.

In some applications, a marginal model is more relevant than a subject-specific model. For example, in government studies on the association between vaccination frequency and prevalence of a disease, understanding the effects on the population average is more useful for informing national medical policy than understanding the effects on individuals. Further summaries of the details and interpretations of marginal models are given in Agresti (2013) and Molenberghs and Verbeke (2005).

Creating and estimating marginal models

Suppose that we have a subjects, and for convenience, suppose that there are t response measurements per subject (in some settings, the number of responses per subject can vary among subjects). Our typical interest is in modeling how the mean (or probability) may relate to variables measured on subjects. Modeling a marginal distribution ordinarily involves first specifying a full joint distribution among all t correlated responses within a subject. This requires formulating models not only for the mean, but also for two-way associations among pairs of responses, for three-way associations, and so forth up to the t-way associations. We may have very little idea how these higher-order associations should be modeled. Thus, we consider these associations to be a "nuisance" to our modeling effort.

As an approximate model, suppose we assume that all associations among responses are zero; that is, the responses within a subject are independent. This means that $n = at$ responses are treated as independent rather than as having come from clusters of responses. It is quite likely that this independence assumption is wrong in many applications, but it allows us to create a "working model" that is easily specified and fit to the data, similar to the pseudo-likelihood approach used in Sections 6.3.6 and 6.4.3. Liang and Zeger (1986) fit this model using a technique closely resembling ML estimation, but does not use a likelihood based on a fully specified joint distribution of responses. They instead propose solving a set of *generalized estimating equations* (GEEs) to obtain the parameter estimates.[25] Estimators found by solving GEEs have many of the properties that MLEs enjoy. In particular, they are approximately normally distributed in large samples and they are consistent (see Appendix B.3.3) as long as the model for the means is correct, even when the working model for the associations is wrong. However, the variances of the parameter estimates depend on the association structure that we are ignoring in the model. Specifically, variance estimates produced by fitting a standard GLM tend to be too small when responses within subjects are positively correlated. Liang and Zeger (1986) develop a method to correct the variances. The resulting variance estimators are known as "sandwich" estimators because of their mathematical form as a product of three matrices, where the first and third matrices are the same. They are also called "robust" or "empirically corrected" variances. Further details on their calculation are given in Liang and Zeger (1986), Molenberghs and Verbeke (2005), and Agresti (2013) .

[25]"Estimating equations" are functions of parameters that are set to zero to find parameter estimates, like the score function described in Appendix B.3.1 for finding MLEs.

The GEE approach can be applied using other assumed association structures for the responses instead of independence. In fact, the efficiency (precision) of the GEE parameter estimates improves when a correlation structure is used that more closely resembles that of the true joint distribution of responses. However, efficiency can also suffer if a correlation model is chosen that requires estimating more parameters than are necessary. Therefore, preference is generally given to simpler models for the association that require fewer parameters.

A popular choice is to assume that every pair of responses has the same non-zero correlation. This is called an *exchangeable* correlation structure and is often appropriate for clustered data. Or, if measurements are made across time, as in repeated measures, we could assume that the correlation between a pair of responses depends on how far apart the measurements were taken. In particular, if the correlation decreases exponentially with distance, then this is an *auto-regressive* (AR) correlation structure. These two structures are popular because they can both be estimated with a single parameter that is separate from the regression parameters. Alternatively, a completely unspecified correlation structure can be chosen so that there is one correlation parameter estimated for each different pair of responses within a subject. This structure is often referred to as *unstructured*. Note that this results in $t(t-1)/2$ correlation parameters in total, so it is best used only when there is substantial doubt that a simpler structure is appropriate, and there is a sufficiently large number of subjects to provide precise estimates of each correlation parameter. Additional correlation structures can be considered too, but it is typically difficult to discern very finely among them, particularly if the number of subjects is not very large.

Inferences from GEEs are performed using Wald methods, so in all cases a "large" sample size is needed. How large is "large" depends on the number of subjects, measurements per subject, explanatory variables, and the sizes of the means or probabilities being estimated. Unlike in other problems, where we could check the quality of the inferences using a simulation based on the estimated model, GEEs are based on an incomplete model for the responses, so it is not entirely clear what model should be used for simulations. When there is concern regarding the adequacy of the Wald inferences and/or variance estimates, a nonparametric bootstrap approach (Davison and Hinkley, 1997) can be used instead. Resampling should be applied to the a subjects (with all t responses included whenever a particular subject is resampled), rather than individual responses, so that the within-subject correlation structure is retained when the GEE method is applied to the resamples.

The GEE approach is popular because of its relative simplicity. The computations are all straightforward and the variance correction works fairly well in large samples, although it has a tendency to underestimate the true variance of parameter estimates, resulting in liberal inferences. However, the approach is limited primarily to problems with repeated measures or simple clustering. It is not easily applicable to problems with nested or crossed random effects.

Example: Falls with head impact (FallsGEE.R, FallHead.csv)

We reanalyze the falls data with a GEE approach, using a model

$$\text{logit}(\pi_{ik}) = \beta_0 + \beta_2 x_{2ik} + \beta_3 x_{3ik} + \beta_4 x_{4ik}$$

for fall k from resident i. Note that there are no special parameters or effects in this model that distinguish the falls from resident i from those of another resident. That is, falls for any residents that have the same values of x_2, x_3, and x_4 have the same probability of head impact.

The model is fit using the `geeglm()` function from the `geepack` package (Halekoh et al., 2006). This function fits GEEs to model families available in `glm()`. The `emmeans` package works well with models produced by `geeglm()` to perform Wald inferences using the corrected variances. Also, there is an `anova()` method function that can be used with the resulting model-fit object. The function computes Wald tests for regression parameters involving numerical explanatory variables, and multi-parameter Wald tests for those from categorical explanatory variables (see Appendix B.5.1).[26]

The data for `geeglm()` must first be arranged so that all data from each cluster are consecutive observations. The function assumes that a new cluster begins with each change in value of the variable listed in its `id` argument. It automatically estimates a scale parameter as in quasi-likelihood methods (see Section 5.3.3), unless `scale.fix = TRUE` is specified. Finally, correlation structures are given in the `corstr` argument, with options including `"independence"`, `"exchangeable"`, and `"ar1"`.

We fit the model for the probability of head impact with initial direction of fall as the explanatory variable. Following the summary of the model fit, we explore the contents of the `summary()` results and use the `anova()` function to test whether all levels of `initial` have equal probability of head impact.

```
> library(package = geepack)
> # Need to sort data in order of the clusters first.
> fall.head.o <- fall.head[order(fall.head$subject),]
> # Need to specify scale.fix = TRUE, or else quasi-binomial
   will be fit.
> mod.gee.i <- geeglm(formula = head ~ initial, id = subject,
   data = fall.head.o, scale.fix = TRUE, family = binomial(link
   = "logit"), corstr = "independence")
> summ <- summary(mod.gee.i)
> summ
Call: geeglm(formula = head ~ initial, family = binomial(link =
   "logit"), data = fall.head.o, id = subject, corstr =
   "independence", scale.fix = TRUE)
 Coefficients:
                Estimate Std.err  Wald Pr(>|W|)
(Intercept)      -0.6360  0.2475 6.601  0.01019 *
initialDown      -1.1558  0.6510 3.152  0.07583 .
initialForward    0.9544  0.3401 7.875  0.00501 **
initialSideways  -0.1085  0.3435 0.100  0.75222
---
Signif. codes:  0 '***' 0.001 '**' 0.01 '*' 0.05 '.' 0.1 ' ' 1
Scale is fixed.
Correlation: Structure = independence
Number of clusters:   131   Maximum cluster size: 13

> names(summ)
[1] "call"             "terms"            "family"
```

[26]Unfortunately, the `anova()` function computes only "sequential" tests. Each term is tested assuming that only the terms listed before it in the `formula` are also in the model. This means that the test results may change depending on the order of the terms in the model, which is usually not a good feature. The Wald tests produced by `summary()` are the more common "partial" tests, which assume that *all* other terms are present before adding the tested term. To get partial tests for a categorical variable in a multivariable model, refit the model without the categorical variable and use `anova(mod1, mod2)`, where `mod1` is the full model and `mod2` is the smaller model. Exercise 4 provides an example of this process.

```
[4] "contrasts"       "deviance.resid" "coefficients"
[7] "aliased"         "dispersion"      "df"
[10] "cov.unscaled"    "cov.scaled"      "corr"
[13] "corstr"          "scale.fix"       "cor.link"
[16] "clusz"           "error"           "geese"
> summ$cov.scaled
         [,1]     [,2]     [,3]     [,4]
[1,]  0.06128 -0.06660 -0.05763 -0.05683
[2,] -0.06660  0.42378  0.09347  0.04674
[3,] -0.05763  0.09347  0.11568  0.05336
[4,] -0.05683  0.04674  0.05336  0.11800

> # anova() method performs Wald chi-square test
> anova(mod.gee.i)
Analysis of 'Wald statistic'
Table Model: binomial, link: logit
Response: head
Terms added sequentially (first to last)
        Df   X2 P(>|Chi|)
initial  3 21.5    8.2e-05 ***
```

The model-fit summary looks much like other regression summaries. Standard errors are computed from the robust sandwich estimates of each parameter estimate's variance. These are also available from the square roots of the diagonal elements of `cov.scaled` from the resulting object for `summary()`, which contains the estimated variances and covariances of the regression parameter estimates. The Wald test for equal probability of head impact for each direction is clearly significant, with a test statistic 21.5 and a very small p-value.

The program for this example contains the `emmeans` code for pairwise comparisons among the levels of `initial`. These are carried out in the same manner as they were for the GLMM analysis and yield the same interpretations with slightly different values.

6.6 Bayesian methods for categorical data

Analyses in Chapters 1–5 use *frequentist methods* to estimate a fixed, unknown parameter, such as a population prevalence π, an odds ratio OR, or a regression parameter β_r. This involves taking a random sample from a population and assuming a joint probability distribution for the corresponding random variables. Using this information, an MLE is found for the parameter of interest. Confidence intervals are constructed using various properties of the likelihood function and MLEs discussed in Appendix B. The stated confidence level with these intervals is derived from the hypothetical process of repeating the same sampling and calculations over and over again (from which the name "frequentist" is derived) so that $(1-\alpha)100\%$ of the intervals contain the parameter. Notice that the interpretation of a confidence level does not mean that one specific interval has a $1-\alpha$ probability of containing the parameter. Once an interval is constructed, it either contains the parameter or it does not, so the probability, which is unknown to us, is either 1 or 0. Currently, frequentist methods are the most predominant approach to statistical inference.

Bayesian methods present an alternative paradigm for statistical inference. The popularity of these methods has grown immensely over the last several decades because of the availability of new computational algorithms and faster computers. An advantage to using Bayesian methods is that prior knowledge about an underlying problem can be incorporated into an analysis. This is done by treating a parameter as a random quantity with a probability distribution that we initially specify completely. This is called a *prior distribution* (or just the "prior") because it represents what we believe about the parameter prior to considering the data. For example, if a goal is to estimate the prevalence of HIV in a population, one would expect that this prevalence is low even before collecting the data, so a chosen distribution for the probability of HIV in a randomly selected person would concentrate mainly on values near zero. After data are collected, this prior is updated to form a new distribution, known as a *posterior distribution* (or just the "posterior"), to reflect information gained by the data. We obtain an estimate of the parameter by using summaries of its posterior, such as the mean or mode, which represent likely values. We can also use the posterior to construct an interval of likely values for the parameter. This interval has a $1-\alpha$ probability of containing the parameter, in contrast to what a confidence interval represents.

The application of *Bayes' rule* plays a fundamental role for Bayesian methods. The rule states that for two events A and B, the conditional probability of B given A is

$$P(B|A) = \frac{P(A \cap B)}{P(A)} = \frac{P(A|B)P(B)}{P(A \cap B) + P(A \cap \bar{B})}, \tag{6.21}$$

where $\bar{B}$ denotes the compliment (or opposite) of B. In our context, one can roughly think of B as representing the parameter and A as representing the data. Then, $P(B)$ represents the prior knowledge that one may have about the parameter, and we update this knowledge with information obtained about the data to calculate the posterior probability $P(B|A)$.

A more general version of Bayes rule for inference on parameters can be stated as

$$p(\theta|y) = \frac{f(y|\theta)p(\theta)}{f(y)}, \tag{6.22}$$

where θ is the parameter of interest, y represents the observable data, and $p(\cdot)$ and $f(\cdot)$ represent particular probability distributions for θ and Y, respectively. Equation 6.22 is the posterior distribution for θ given information observed about Y, with $p(\theta)$ denoting the prior for θ. For clarity here, note that not only is Y a random variable, but θ is a random variable as well; we forgo the formality of using a capital version of θ for ease of exposition.

The purpose of this section is to discuss the application of Bayesian methods for a number of categorical data settings. We begin with an introduction to the Bayesian paradigm by estimating one probability of success parameter in the same context as in Section 1.1.1. Next, we extend these ideas to more computationally complex situations involving the regression models discussed in Chapters 2–4. Finally, we provide additional resources for the application of Bayesian methods with R.

6.6.1 Estimating a probability of success

We next discuss in detail how Equation 6.22 is applied in the context of estimating a probability-of-success parameter π in a binomial model. The data is represented by the number of successes W out of a fixed number of trials n. Our goal is to obtain the posterior for π given information observed about W.

Throughout Chapters 1 and 2, we used a binomial distribution that was first given in Equation 1.1. We now re-express this distribution as

$$f(w|\pi) = \binom{n}{w} \pi^w (1-\pi)^{n-w},$$

where we use $f(\cdot)$ to represent the PMF for W and add "$|\pi$" to emphasize that probabilities are conditioned on a value for π. This conditional PMF is used in the numerator of Equation 6.22.

Knowledge about π before a sample is taken can be accounted for by the prior distribution, $p(\pi)$, representing likely values for the parameter. There are many possible priors that could be taken, but the one most often used in this situation is a beta distribution (see Equation 1.5). This distribution allows for $0 < \pi < 1$, is quite flexible in its shape, and leads to a mathematically nice result (to be shown shortly). For example, if we believe that π should be small as in the HIV prevalence example, a beta distribution with $a = 1$ and $b = 10$ may be reasonable because the distribution is right-skewed with much of its probability close to 0.[27] The prior distribution for π is also used in the numerator of Equation 6.22.

Finding the denominator in Equation 6.22 is typically the hardest part of a Bayesian analysis. To compute $f(w)$, we first need to find the joint probability density function of W and π. Assuming a beta prior with parameters $a > 0$ and $b > 0$, we obtain

$$\begin{aligned} f(w,\pi) &= f(w|\pi)p(\pi) \\ &= \binom{n}{w} \pi^w (1-\pi)^{n-w} \frac{\Gamma(a+b)}{\Gamma(a)\Gamma(b)} \pi^{a-1}(1-\pi)^{b-1} \\ &= \frac{\Gamma(n+1)}{\Gamma(w+1)\Gamma(n-w+1)} \frac{\Gamma(a+b)}{\Gamma(a)\Gamma(b)} \pi^{w+a-1}(1-\pi)^{n+b-w-1} \end{aligned}$$

for $w = 0, \ldots, n$ and $0 < \pi < 1$. Note that we use the relation $\Gamma(c+1) = c!$ for an integer c. To obtain $f(w)$, we integrate over all possible values of π:

$$\begin{aligned} f(w) &= \int_0^1 \frac{\Gamma(n+1)}{\Gamma(w+1)\Gamma(n-w+1)} \frac{\Gamma(a+b)}{\Gamma(a)\Gamma(b)} \pi^{w+a-1}(1-\pi)^{n+b-w-1} d\pi \\ &= \frac{\Gamma(n+1)}{\Gamma(w+1)\Gamma(n-w+1)} \frac{\Gamma(a+b)}{\Gamma(a)\Gamma(b)} \frac{\Gamma(w+a)\Gamma(n+b-w)}{\Gamma(n+a+b)}. \end{aligned}$$

Putting all the pieces together, we obtain the posterior distribution of π given information observed about W:

$$\begin{aligned} p(\pi|w) &= \frac{f(w|\pi)p(\pi)}{f(w)} \\ &= \frac{\Gamma(n+a+b)}{\Gamma(w+a)\Gamma(n-w+b)} \pi^{w+a-1}(1-\pi)^{n+b-w-1} \end{aligned} \quad (6.23)$$

for $0 < \pi < 1$. Interestingly, Equation 6.23 is a beta distribution just like the prior, but now with parameter values $w+a$ and $n+b-w$ controlling the shape of the distribution. When a posterior is in the same family of distributions as the prior, these distributions are said to belong to a *conjugate family*. Thus, the beta distribution is conjugate to the binomial distribution.

---

[27] Use `curve(expr = dbeta(x = x, shape1 = 1, shape2 = 10), xlim = c(0,1))` to plot the density function in R.

Bayes estimate

The posterior distribution is used to obtain an estimate of π. One reasonable estimate is the mean $E(\pi|w)$, which can be shown in this case to be $(w+a)/(n+a+b)$ by using properties of beta distributions (see p. 107 of Casella and Berger, 2002). This mean is known as a *Bayes estimate*, and we denote it as $\hat{\pi}_B$. As an alternative, we can use the median or the mode of the posterior as an estimate. Note that finding the mode of the posterior is analogous to finding the MLE from the likelihood function, and it is guaranteed to equal the mean only when the posterior is symmetric and unimodal.

Before observing the data, a sensible estimate for π would have been $E(\pi) = a/(a+b)$ obtained from the prior distribution. Interestingly, the Bayes estimate in this problem can be factored into two components—one that uses $E(\pi)$ and one that uses the MLE $\hat{\pi} = w/n$:

$$\hat{\pi}_B = \left(\frac{n}{n+a+b}\right)\hat{\pi} + \left(\frac{a+b}{n+a+b}\right)E(\pi). \tag{6.24}$$

Thus, the Bayes estimate is a weighted average of the MLE and the mean from the prior distribution. A larger sample size results in more relative weight on the MLE, while a smaller sample size results in more relative weight on the prior mean. In Bayesian models for other parameters, the same principle holds: as the sample size grows, the Bayes estimate relies progressively less on the prior and more on the data. However, showing the relationship in a mathematical closed form is not often as easy as it is in this setting.

Choosing a prior distribution

The choice of a prior distribution obviously plays an important role in any Bayesian analysis. Past experience may suggest a range of possible values of a parameter, which then leads to the use of a prior that concentrates on this range. The prior has parameters of its own, called *hyperparameters*. For example, our beliefs about π suggest that we can use a beta distribution with particular values of a and b. Selecting a prior in this way is both one of the advantages and one of the criticisms of Bayesian analysis, because different analysts may choose different priors based on different beliefs. These different *subjective* choices could conceivably affect the outcome of the analysis in important ways.

As an alternative to choosing a prior subjectively, a *noninformative prior distribution* is frequently used in practice. This is a distribution that places equal probability or density on all possible values of the parameter. For our binomial model, a beta distribution with $a = b = 1$ (equivalent to a uniform distribution on the interval $(0, 1)$) is a noninformative prior for estimating π. Of course, this is still specifying a belief about the parameter—specifically that it has an equal chance of being any allowed value—so it is not the same as a frequentist method that specifies no such belief.

While a uniform distribution is noninformative for estimating π, it is no longer noninformative if the goal is to estimate a monotone transformation of π, like an odds. Thus, an analyst wishing to use the prior may need to choose a different one for each different way that the parameter might be expressed. One way to avoid this problem is to use a Jeffreys' prior, which is a distribution that is invariant to a monotone transformation of the parameter (e.g., see Robert, 2007 and Tanner, 1996 for further details). In general for a parameter θ and data y, the prior is chosen to be proportional to

$$\left[E\left(\frac{\partial^2}{\partial\theta^2}\log[f(y|\theta)]\right)\right]^{1/2}.$$

The quantity within the square root is often referred to as *Fisher information*, and it is very closely related to the variance of a maximum likelihood estimator (see Equation B.5).

Fisher's information is conceptualized as the amount of information about the parameter that is available from the data. Thus, basing the prior on Fisher's information allows for more weight to be assigned to values of θ that have larger Fisher's information. This leads to a different way to think about the notion of "noninformative," in that Jeffreys' prior agrees more with the data than other prior distributions might (Robert, 2007, p. 130).

There are other ways to choose a prior distribution. In particular, probability distributions known as *hyperpriors* can be specified for each hyperparameter of a prior distribution. The resulting analysis is called *hierarchical Bayesian* because of the nested prior specifications. Another way to choose a prior is to estimate it using the observed data through *empirical Bayesian methods.* A problem with this approach is that the prior cannot be fully specified before the data collection, which is somewhat contrary to the Bayesian philosophy of fully stating "prior" beliefs. Nonetheless, some analysts like it for its objectivity. For more information on these and other ways to specify prior distributions, please see full textbooks on Bayesian methods, such as Gelman et al. (2013) and Carlin and Louis (2008).

Credible intervals

In a Bayesian analysis, interval estimates are based on quantiles of the posterior for a parameter. These intervals are called *credible intervals*. The interpretation of a credible interval is that it has $1-\alpha$ probability of containing the parameter because it is based on the posterior distribution of the parameter. The most popular type of credible interval is an *equal-tail interval*, which finds limits such that the probability to the left of the lower limit is $\alpha/2$ and the probability to the right of the upper limit is $\alpha/2$. For example, based on the posterior for π that we derived earlier, we obtain a $(1-\alpha)100\%$ equal-tail credible interval of

$$\text{beta}\,(\alpha/2;\ w+a,\ n+b-w) < \pi < \text{beta}\,(1-\alpha/2;\ w+a,\ n+b-w),$$

where $\text{beta}\,(\gamma;\ w+a,\ n+b-w)$ is the γth quantile from the beta distribution with parameter values $w+a$ and $n+b-w$. Equal-tail intervals are relatively easier to calculate than other intervals, and they perform best when the posterior is fairly symmetric.

When the posterior is not symmetric, a better choice is to use a highest posterior density (HPD) credible interval. This interval is calculated by finding the lower and upper limits for an interval that correspond to the region with the highest posterior density, i.e., the most "probable" region. The resulting interval is generally the narrowest possible interval that contains $(1-\alpha)100\%$ of the posterior density. There usually are not closed-form expressions for the calculation of HPD intervals, so we show how to find this type of interval in the next example. It is important to note that an HPD interval is not invariant to transformations. Thus, simply transforming its limits for π does not necessarily lead to an HPD interval for the same transformation of π.

Example: Bayes estimate and credible intervals (CIpiBayes.R)

Suppose $w = 4$ successes are observed out of $n = 10$ trials. If a noninformative beta distribution with $a = b = 1$ is used as the prior, we obtain the Bayes estimate $\hat{\pi}_B = (4+1)/(10+1+1) = 0.4167$ and a 95% equal-tail credible interval of $0.1675 < \pi < 0.6921$. The computations in R are straightforward, where one can simply use `qbeta(p = c(0.05/2, 1-0.05/2), shape1 = 4 + 1, shape2 = 10 + 1 - 4)` to compute the interval. If Jeffreys' prior is used instead, one can first show that this prior is proportional to $\pi^{-1/2}(1-\pi)^{-1/2}$, which leads to the use of a beta distribution with $a = b = 1/2$. The Bayes estimate is 0.4091, and the equal-tail credible interval is $0.1531 < \pi < 0.6963$, providing a lower bound that is a little different than

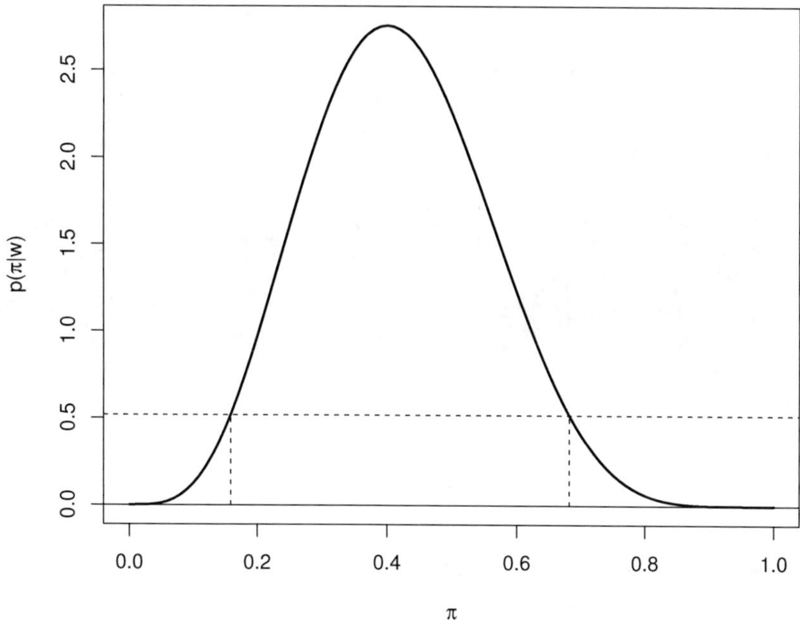

Figure 6.10: Posterior density plot with a horizontal line drawn at $p(\pi|w) = 0.5184$ and vertical lines indicating the limits for the 95% HPD interval.

when using $a = b = 1$. The `binom.confint()` function of the `binom` package calculates these estimates and intervals as well when `methods = "bayes"`, `type = "central"`, `prior.shape1 = a`, and `prior.shape2 = b` are specified.

These estimates and intervals are quite similar to the MLE $\hat{\pi} = 0.4$ and Wilson interval $0.1682 < \pi < 0.6873$ obtained in an example from Section 1.1.1. Part of the reason can be seen by examining the expression for $\tilde{\pi} = (w + Z^2_{1-\alpha/2}/2)/(n + Z^2_{1-\alpha/2})$, the estimator for π used in constructing the Wilson and Agresti-Coull confidence intervals. If $a = b = 1.96/2 = 0.98$, $\hat{\pi}_B$ and $\tilde{\pi}$ are the same.

To find the HPD interval, suppose a horizontal line is drawn across a plot of the posterior distribution $p(\pi|w)$, and suppose the corresponding values of π where the line intersects $p(\pi|w)$ are denoted as π_{lower} and π_{upper}. Iterative numerical procedures are used then to find the one horizontal line such that the area between π_{lower} and π_{upper} is $1 - \alpha$. Figure 6.10 shows the posterior with $a = b = 1$ and a horizontal line at $p(\pi|w) = 0.5184$. The corresponding HPD interval is $0.1586 < \pi < 0.6818$ with a probability of 0.95 between these two limits. This interval is fairly similar to the equal-tail credible interval because there is only slight right-skeweness in the posterior.

The computational process for the HPD interval is performed by `binom.confint()` with `methods = "bayes"`, `type = "highest"`, `prior.shape1 = a`, and `prior.shape2 = b`.

```
> w <- 4
> n <- 10
```

```
> alpha <- 0.05
> a <- 1
> b <- 1

> library(package = binom)
> save.hpd <- binom.confint(x = w, n = n, conf.level = 1-alpha,
    methods = "bayes", type = "highest", prior.shape1 = a,
    prior.shape2 = b)
> save.hpd

  method x  n      mean     lower     upper
1  bayes 4 10 0.4166667 0.1585724 0.6817678

> # Verify lower and upper limits are at the same p(pi|w) values
> dbeta(x = c(save.hpd$lower, save.hpd$upper), shape1 = w + a,
    shape2 = n + b - w)
[1] 0.5183494 0.5183494

> # Verify area between limits is 1-alpha
> pbeta(q = save.hpd$upper, shape1 = w + a, shape2 = n + b - w)
    - pbeta(q = save.hpd$lower, shape1 = w + a, shape2 = n + b -
    w)
[1] 0.95
```

We also provide our own HPD interval code in the program to give additional insight into the computational process.

Of course, Bayes estimates and credible intervals are affected by the choice of a and b in the prior distribution. For example, when we choose $a = b = 1$, we can rewrite the Bayes estimate as

$$\begin{aligned}\hat{\pi}_B &= \left(\frac{n}{n+a+b}\right)\left(\frac{w}{n}\right) + \left(\frac{a+b}{n+a+b}\right)\left(\frac{a}{a+b}\right)\\ &= \left(\frac{10}{12}\right)0.4 + \left(\frac{2}{12}\right)0.5\end{aligned}$$

using Equation 6.24. This helps to show that this prior leads to a slightly larger estimate of π than does the MLE. Exercise 2 investigates further how Bayes estimates and credible intervals change for different values of a and b.

6.6.2 Regression models

To apply Bayesian methods to regression modeling, one needs to begin with the specification of a prior distribution for the regression parameters $\beta_0, \ldots, \beta_p$. For convenience, this typically involves first assuming that the parameters are independent so that a more complex joint probability distribution is not needed. Also, while a subjective prior could be used for each parameter, more often a noninformative prior or a *weakly informative prior* is chosen that distributes probability rather evenly over a large range. For example, a weakly informative prior could be a normal distribution with a very large variance.

The distribution for the data given the parameters is typically chosen to be some common model, such as those presented in Chapters 2–4. The posterior distribution for $\beta_0, \ldots, \beta_p$

given the data $\mathbf{y} = (y_1, \ldots, y_n)$ has the form

$$p(\beta_0, \ldots, \beta_p|\mathbf{y}) = \frac{f(\mathbf{y}|\beta_0, \ldots, \beta_p)p(\beta_0, \ldots, \beta_p)}{f(\mathbf{y})}$$
$$= \frac{\left[\prod_{i=1}^{n} f(y_i|\beta_0, \ldots, \beta_p)\right]\left[\prod_{r=0}^{p} p(\beta_r)\right]}{\int \cdots \int \left[\prod_{i=1}^{n} f(y_i|\beta_0, \ldots, \beta_p)\right]\left[\prod_{r=0}^{p} p(\beta_r)\right] d\beta_0 \cdots d\beta_p} \quad (6.25)$$

when we take $\beta_0, \ldots, \beta_p$ to be independent. Very often, this expression for the posterior cannot be evaluated in a simple closed form, unlike those given in the previous section.

A variety of general-purpose, Monte Carlo simulation methods can be used to evaluate Equation 6.25 without giving its complete closed-form expression. Unlike previous simulations throughout this book, it is mathematically and computationally convenient to generate a sampled value from the posterior by basing it on the previously sampled value rather than by generating a new, independent value. This results in a sequence of simulated values that are serially correlated—each depending on the previous value—producing what is known as a Markov chain. After generating a large sample of these values, the posterior and associated quantities can be estimated despite this dependence. These corresponding simulation methods are known as *Markov chain Monte Carlo* (MCMC). Our focus is on one MCMC method known as the *Hamiltonian algorithm*. We use a popular implementation of this algorithm that is coded in the Stan programming language (Carpenter et al., 2017). For a detailed discussion on the *Hamiltonian* algorithm and its implementation, please see the Stan website (`http://mc-stan.org/rstan`), McElreath (2020), and Thomas and Tu (2021).

Stan can be accessed within R through several packages, including `rstan`, `rstanarm`, and `brms`. Our focus is on `rstanarm` because it provides a `glm()`-function like interface for specifying models so that users do not need to learn the programming language. The `rstanarm` package is a user-contributed package that needs to be downloaded and installed prior to use, like many other packages used in the textbook. However, unlike these other packages, Windows-based computers need to install the Rtools executable program outside of R before installing `rstanarm`. This program is available from the same CRAN web page as R itself and provides a number of additional tools required for some packages. In particular, `rstanarm` uses the C++ compiler in Rtools to create executable programs in that language. Users of `rstanarm` do not need any C++ experience. Functions in this package create the C++ code and executable programs on their own. The authors of the package chose to work with C++ rather than R code alone to increase speed of execution.

Example: Placekicking (PlacekickBayes.R, Placekick.csv)

We consider again the placekicking example from Chapter 2, where our purpose is to examine the model $\text{logit}(\pi) = \beta_0 + \beta_1$`distance` using Bayesian methods and to compare our results to those obtained in Chapter 2. The `stan_glm()` function in `rstanarm` estimates the regression parameters in the logistic regression model. Familiar arguments within `stan_glm()` are `formula`, `data`, and `family`, which serve the same purpose as in `glm()`. Additional arguments include `seed` for the seed number to use with the simulation, `chains` for the number of separate sampling chains (sets of samples) to include, and `iter` for the sample size of each chain. It is important to note that the seed number set may only lead to the exact same samples for the same version of Stan being used. If the samples are different across versions of Stan, the overall conclusions reached by the analysis will be the same as long as convergence to the posterior occurs (see p. 541 for a discussion about convergence).

By default, the `stan_glm()` function will use an initial sample, known as a *warm-up sample*, for each chain that is half as large as `iter`. These warm-up samples allow for an initial exploration of the posterior to optimize the process for the actual samples that will be used for the analysis. Also by default, the `stan_glm()` function uses four separate chains, where the four chains are put together into one at the end. This allows for each chain to be run in parallel on separate computer processor cores, which speeds up the sampling process. While the default used by the function is one core, multiple cores can be used by specifying the desired number in the `cores` argument.

The priors for regression parameters are specified with `prior_intercept` for the intercept and `prior` for the coefficients on the explanatory variables. Weakly informative normal priors are commonly used for these parameters. They are specified by using the `normal()` function with hyperparameters `scale` set to some large number for the standard deviation and `location` usually set to 0 for the mean. Other priors can be used instead—see the list in the help for `stan_glm()`.

Below is our code and output from `stan_glm()`. While `iter` is given as 25000, the warm-up uses half of these iterations leading each chain to contribute 12,500 samples to the posterior. Thus, the total sample size from the four chains is 50,000. We chose this large number of samples to guarantee a good representation of the posterior distribution. The entire process ran in less than two minutes on our computers, and could have run faster had we chosen to use two or more cores.

```
> library(package = rstanarm)

> norm_prior <- normal(location = 0, scale = 100)
> mod.fit.stan <- stan_glm(formula = good ~ distance, data =
   placekick, family = binomial(link = "logit"), prior =
   norm_prior, prior_intercept = norm_prior, seed = 7814, iter =
   25000)

SAMPLING FOR MODEL 'bernoulli' NOW (CHAIN 1).
Chain 1:
Chain 1: Gradient evaluation took 0 seconds
Chain 1: 1000 transitions using 10 leapfrog steps per transition
   would take 0 seconds.
Chain 1: Adjust your expectations accordingly!
Chain 1:
Chain 1:
Chain 1: Iteration:     1 / 25000 [  0%]  (Warmup)
Chain 1: Iteration:  2500 / 25000 [ 10%]  (Warmup)
Chain 1: Iteration:  5000 / 25000 [ 20%]  (Warmup)
Chain 1: Iteration:  7500 / 25000 [ 30%]  (Warmup)
Chain 1: Iteration: 10000 / 25000 [ 40%]  (Warmup)
Chain 1: Iteration: 12500 / 25000 [ 50%]  (Warmup)
Chain 1: Iteration: 12501 / 25000 [ 50%]  (Sampling)
Chain 1: Iteration: 15000 / 25000 [ 60%]  (Sampling)
Chain 1: Iteration: 17500 / 25000 [ 70%]  (Sampling)
Chain 1: Iteration: 20000 / 25000 [ 80%]  (Sampling)
Chain 1: Iteration: 22500 / 25000 [ 90%]  (Sampling)
Chain 1: Iteration: 25000 / 25000 [100%]  (Sampling)
Chain 1:
Chain 1:  Elapsed Time: 29.949 seconds (Warm-up)
Chain 1:                12.117 seconds (Sampling)
Chain 1:                42.066 seconds (Total)
Chain 1:
```

```
SAMPLING FOR MODEL 'bernoulli' NOW (CHAIN 2).
Chain 2:

<OUTPUT EDITED>

Chain 4: Iteration: 20000 / 25000 [ 80%]  (Sampling)
Chain 4: Iteration: 22500 / 25000 [ 90%]  (Sampling)
Chain 4: Iteration: 25000 / 25000 [100%]  (Sampling)
Chain 4:
Chain 4:  Elapsed Time: 25.892 seconds (Warm-up)
Chain 4:                13.802 seconds (Sampling)
Chain 4:                39.694 seconds (Total)
Chain 4:

> summary(mod.fit.stan)
Model Info:
 function:     stan_glm
 family:       binomial [logit]
 formula:      good ~ distance
 algorithm:    sampling
 sample:       50000 (posterior sample size)
 priors:       see help('prior_summary')
 observations: 1425
 predictors:   2

Estimates:
              mean   sd   10%   50%   90%
(Intercept)  5.8    0.3  5.4   5.8   6.3
distance    -0.1    0.0 -0.1  -0.1  -0.1

Fit Diagnostics:
           mean   sd   10%   50%   90%
mean_PPD  0.9    0.0  0.9   0.9   0.9

The mean_ppd is the sample average posterior predictive
   distribution of the outcome variable (for details see
   help('summary.stanreg')).

MCMC diagnostics
              mcse Rhat n_eff
(Intercept)   0.0  1.0   9482
distance      0.0  1.0  13468
mean_PPD      0.0  1.0  18969
log-posterior 0.0  1.0   9354

For each parameter, mcse is Monte Carlo standard error, n_eff is
   a crude measure of effective sample size, and Rhat is the
   potential scale reduction factor on split chains (at
   convergence Rhat=1).
```

The output is quite verbose! Much of the verboseness is to provide the user indications of progress during the sampling process. The `summary()` function gives a general summary of the model fit. The `mean` column of the `Estimates` table shows the Bayes estimates. Because the default number of significant digits is small, we use the `stan_summary` component of the returned object to display these estimates with more precision.

```
> mod.fit.stan$stan_summary
                      mean       se_mean         sd
(Intercept)      5.8335013 3.381344e-03 0.32926781
distance        -0.1154814 7.246227e-05 0.00840949
mean_PPD         0.8855856 7.694691e-05 0.01059772
log-posterior -395.3258214 1.043994e-02 1.00970352
                      2.5%          10%          25%
(Intercept)      5.1965178    5.4142368    5.6101288
distance        -0.1322880   -0.1262604   -0.1210882
mean_PPD         0.8645614    0.8722807    0.8785965
log-posterior -398.0489628 -396.6508381 -395.7134839
                       50%          75%          90%
(Intercept)      5.8306246    6.0520606    6.2562219
distance        -0.1154356   -0.1098165   -0.1047706
mean_PPD         0.8856140    0.8926316    0.8989474
log-posterior -395.0161638 -394.6057553 -394.4221240
                     97.5%     n_eff     Rhat
(Intercept)     6.49126371  9482.433 1.000346
distance       -0.09912972 13468.382 1.000260
mean_PPD        0.90596491 18968.908 1.000048
log-posterior -394.34071241  9353.871 1.000066
```

The Bayes estimates (`mean` column) and the standard deviations (`sd` column) are very similar to their respective MLEs and standard errors obtained in Section 2.2.1. Overall, these similarities are not surprising because of the large sample size for the placekicking data and the weakly informative priors. In general, the estimates and variances need not be this similar. The 2.5% to 97.5% columns provide information about the posterior distribution. From these columns, we obtain the 95% equal-tail credible interval for the distance regression parameter as $-0.1323 < \beta_1 < -0.0991$ (`2.5%` and `97.5%` columns), which is again quite close to the 95% confidence interval for β_1 given in Section 2.2.3.

Estimates for functions of the regression parameters, such as odds ratios and probabilities of success, can be found using the emmeans package.

```
> library(package = emmeans)

> # Estimate pi
> calc.est <- emmeans(object = mod.fit.stan, specs = ~ distance,
   at = list(distance = c(20,30)))
> summary(calc.est, type = "response")
 distance  prob lower.HPD upper.HPD
       20 0.971     0.961     0.980
       30 0.914     0.896     0.931

Point estimate displayed: median
Results are back-transformed from the logit scale
HPD interval probability: 0.95
```

```
> # Estimate odds ratio
> confint(object = contrast(object = calc.est, method =
    "pairwise", reverse = TRUE), level = 0.95, type = "response")
 contrast                  odds.ratio lower.HPD upper.HPD
 distance20 / distance30         3.17      2.66      3.71

Point estimate displayed: median
Results are back-transformed from the log odds ratio scale
HPD interval probability: 0.95
```

The package provides median estimates, rather than means, and HPD credible intervals, rather than equal tail. For example, the estimated odds ratio for a 10-yard decrease in the distance is 3.17 with a 95% HPD credible interval of (2.66, 3.71). Overall, these estimates and intervals are very similar to those found in Sections 2.2.3 and 2.2.4.

The mod.fit.stan object returned from stan_glm() contains the samples. To compute further summaries that are not offered in the package, we can extract the MCMC samples from the model object using as.data.frame().[28]

```
> my.sample <- as.data.frame(mod.fit.stan)
> head(my.sample)
  (Intercept)    distance
1     5.531287 -0.1081238
2     5.424652 -0.1042272
3     5.490935 -0.1074596
4     5.650419 -0.1130099
5     5.501880 -0.1075973
6     5.535950 -0.1086675

> tail(my.sample)
      (Intercept)    distance
49995    5.543646 -0.1131418
49996    5.207068 -0.1012164
49997    5.999888 -0.1166932
49998    6.099978 -0.1225442
49999    6.077347 -0.1217020
50000    5.967354 -0.1175842
```

The samples from the four separate chains are concatenated for the my.sample object. Next, we confirm some of the calculations performed previously as well as other calculations not readily available.

```
> # Mean of sample - see summary(mod.fit.stan)
> colMeans(my.sample)
(Intercept)    distance
```

[28]An alternative way to extract this same information is to examine the contents of the list returned by stan_glm(), similar to what was first shown for glm() in Chapter 2. Unfortunately, the package's authors do not make this easy because of their use of a multiple-level list that combines S3 and S4 class structures. For example, one needs to use mod.fit.stan$stanfit@sim$samples[[1]]$beta for this example to examine the sampled values of β_1.

```
  5.8335013   -0.1154814
> # Standard deviation of sample, see summary(mod.fit.stan)
> apply(X = my.sample, MARGIN = 2, FUN = sd)
(Intercept)    distance
 0.32926781  0.00840949

> # Is beta1 >= 0?
> beta1 <- my.sample[,2]
> min(beta1)
[1] -0.1529212
> max(beta1)
[1] -0.0794323
> mean(beta1 >= 0)
[1] 0

> # Equal-tail credible interval for beta1
> quantile(x = beta1, probs = c(0.025, 0.975))
       2.5%        97.5%
-0.13228796 -0.09912972

> # Estimate pi at distance = 20, see summary(calc.est, type =
   "response")
> # Median estimate and equal-tail credible interval
> quantile(x = plogis(q = beta0 + beta1*20), probs = c(0.025,
   0.5, 0.975))
     2.5%       50%     97.5%
0.9603139 0.9713155 0.9796349
> # HPD credible interval
> library(package = HDInterval)
> hdi(obj = plogis(q = beta0 + beta1*20), credMass = 0.95)
    lower     upper
0.9609884 0.9800676
attr(,"credMass")
[1] 0.95

> # Estimate OR for a 10-yard decrease in distance, see confint()
> OR10.stan <- exp(-10*my.sample[,2])
> # Bayes estimate
> mean(OR10.stan)
[1] 3.184687
> # Median estimate and equal-tail credible interval
> quantile(x = OR10.stan, probs = c(0.025, 0.5, 0.975))
    2.5%      50%    97.5%
2.694728 3.171979 3.754217
> # HPD credible interval
> hdi(obj = OR10.stan, credMass = 0.95)
   lower    upper
2.661447 3.712580
attr(,"credMass")
[1] 0.95
```

For example, the estimate of $P(\beta_1 \geq 0)$ is less than $< 1/50000$, indicating that it is very likely that β_1 is negative. We also use in the code `hdi()` of the `HDInterval` package as a general purpose HPD credible interval calculation tool.

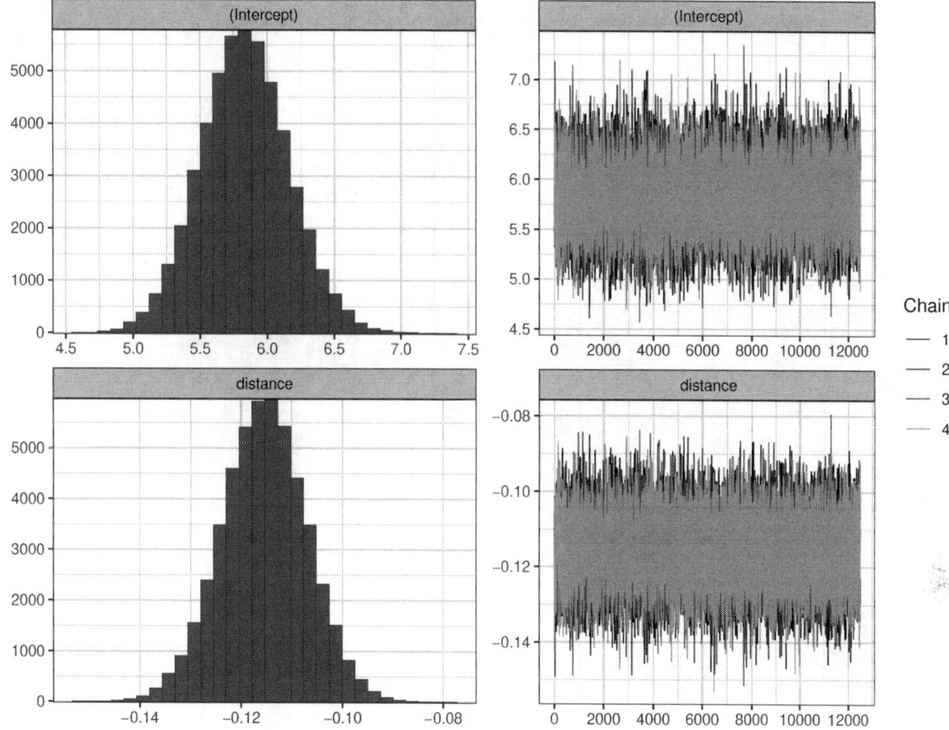

Figure 6.11: Histograms and trace plots for the β_0 and β_1 sampled values.

Convergence

An MCMC method is said to have converged to the posterior distribution when the empirical distribution becomes "stable" in the sense that different sequences of the chain have empirical distributions that closely resemble one another. One would not want to use the sample without convergence because it may provide a poor representation of the posterior. Therefore, an important aspect to using any MCMC method is knowing whether convergence has been achieved. Plummer et al. (2006), Robert and Casella (2010), and McElreath (2020) provide nice introductory discussions on diagnostic measures for this purpose. More in-depth discussions on diagnostic measures are available in Cowles and Carlin (1996) and Robert and Casella (2004). This subsection provides a summary of these references along with our own insights.

A trace plot is often the first tool used to assess convergence. This plot simply shows the simulated parameter values in the chain plotted against an index representing the order in which these values were obtained. These simulated parameter values are joined by a line. If the algorithm has converged, both the mean and the variability of the simulated values should remain fairly constant throughout the trace plot. Also, the chain should "mix" well, meaning that the sampled values visit all regions of the posterior very frequently, rather than sticking within regions for long sequences of iterations. Convergence is often characterized in a trace plot by (1) a thick band of values that is undulating gently within a fairly constant range and (2) spikes of varying lengths coming out of the band. The trace plots in Figure 6.11 (second column) are examples where there appears to be convergence because the central band (gray region) and random spikes appear relatively consistent across the ordering of the sampled values.

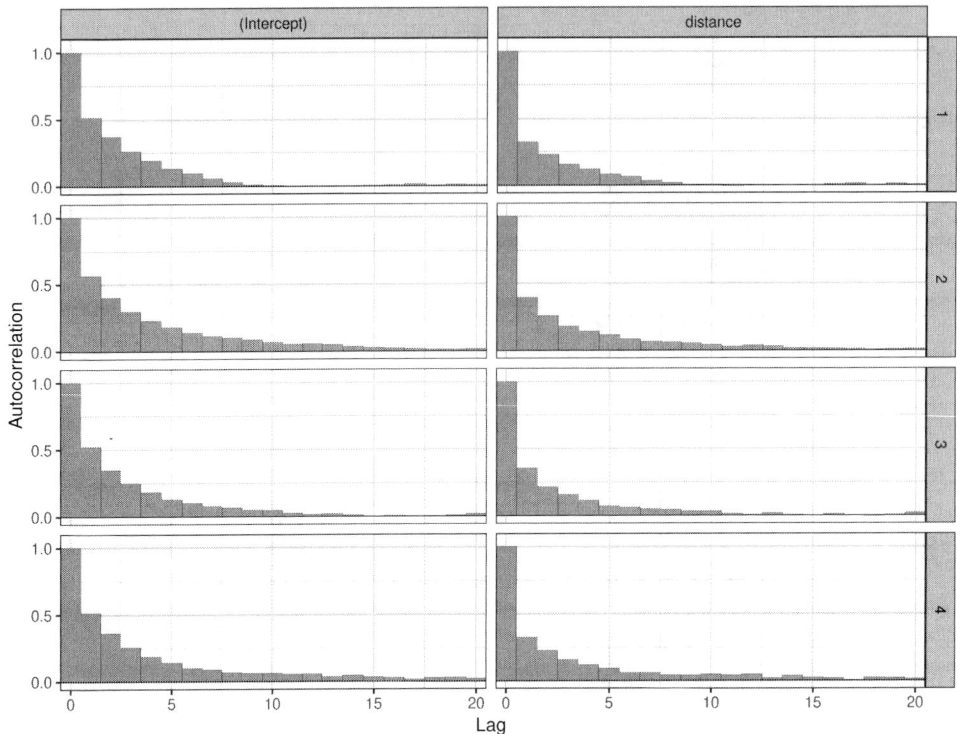

Figure 6.12: Bar plots of the autocorrelations for each chain.

A potential problem can occur with MCMC when a chain tends to stay in one region for a long sequence of sampled values (i.e., it does not mix well). On a trace plot, poor mixing is characterized by bands with central tendency that clearly moves and spikes that are much narrower than the full range of parameter values. This poor mixing is also characterized by large positive *autocorrelations* between successive simulated values. An autocorrelation, in general, is very similar to the Pearson correlation coefficient between two variables, but it represents how similar sampled values close together in a time sequence tend to be, relative to the overall mean. In the context of an MCMC sample, if $\beta_r^{(t)}$ represents the rth parameter indexed by the iteration t it was obtained in the sample, the autocorrelation at *lag* 1 essentially involves calculating the Pearson correlation coefficient using the pairs $(\beta_r^{(2)}, \beta_r^{(1)}), (\beta_r^{(3)}, \beta_r^{(2)}), \ldots, (\beta_r^{(n)}, \beta_r^{(n-1)})$ for an MCMC sample of size n. The lag represents the time separation within the pairs, and autocorrelations can be calculated for other lags as well. If the autocorrelations quickly decrease toward 0 as a function of increasing lag values, this suggests the chain is mixing well.[29] The bar plots of autocorrelations in Figure 6.12 are examples of autocorrelations decreasing to 0 fairly rapidly as the lag increases.

Because of the positive autocorrelation among elements within the chain, statistics computed from the chain have more variability than they would have if the elements were independent. The *effective sample size* is roughly defined as the size of a random sample

[29]Autocorrelations are widely used in the context of time series analysis to understand the dependence of a random variable over time (e.g., see Shumway and Stoffer, 2017). If autocorrelations do not quickly decrease toward 0 as a function of the lag, this suggests non-stationarity for the variable (i.e., mean, variance, and other aspects are not consistent across time). The examination of autocorrelations relative to MCMC methods is done in a very similar context to determine non-convergence.

that would provide an estimate of the mean that has the same precision as the chain's mean estimate. The effective sample size is generally somewhat smaller than the chain's sample size. The stronger the correlation between consecutive elements in the chain, the smaller the effective sample size is. If the effective sample size is a small fraction of the total number of iterations, this may again suggest non-convergence.

At each iteration of the Hamiltonian algorithm, a value is proposed to include into the sample. This proposal is accepted for the sample with a particular probability that is dependent on components of the posterior distribution and the most recent previous sample value. If a proposed value is not accepted, this most recent previous sample value becomes the "new" sampled value for that iteration. Thus, the series of accepted values can have repeats. The rate at which an algorithm accepts new proposals into the sample is a frequently used diagnostic tool for convergence. Acceptance rates that are too low can lead to a poor representation of the posterior. One advantage of the Hamiltonian algorithm over most algorithms is its high rate of acceptance. Other algorithms, like the Metropolis-Hastings algorithm, can have acceptance rates in the range of 0.2 to 0.5, while the Hamiltonian algorithm will likely be much higher for the same model and data. For example, the target acceptance rate for `stan_glm()` is 95% (`adapt_delta` argument sets the value) for the priors examined in our textbook. This efficiency means that an accurate estimate of the posterior can be obtained using a smaller MCMC sample size than is required with other algorithms.

As described earlier, the `stan_glm()` function uses four separate chains by default. An additional advantage of running multiple chains is that they provide several independent estimates of the posterior. If they all appear to very similar, this is further evidence that each one has converged to the same distribution. Additional diagnostic tools are available with the running of multiple chains. For example, one can compare the variability of the means from different chains against the variability of samples within each chain, using an analysis-of-variance test statistic. This statistic is commonly known in this context both as the Gelman-Rubin convergence statistic and as the potential scale reduction statistic (Gelman and Rubin, 1992), while symbolically often denoted as $\hat{R}$. Ideally, we would like this ratio to be as close to 1 as possible. Gelman and Shirley (2011) suggest samples resulting in ratios larger than 1.1 should not be used.

If problems with convergence are found through the aforementioned diagnostic measures, there are potential remedies. A larger MCMC sample size may be necessary to obtain an accurate representation of the posterior distribution. Also, different priors can be used. In particular, priors that are more informative may be needed at the consequence of greater subjectivity in the analysis. Another potential solution is to *thin* the chain. Thus, rather than using every value that was accepted into the chain, one could use every kth value instead. This generally reduces the autocorrelation between the sampled values included in the analysis, but a larger MCMC sample size must be taken to achieve a comparable level of precision in the estimates. This is because $k - 1$ out of a every k sampled values are excluded. And, while thinning is commonly used as a remedy for convergence problems, not all statistics researchers recommend using it (Christensen et al., 2010). One can instead take a very large MCMC sample without thinning in the hopes that the size itself will allow the chain to visit all regions of the posterior even with large autocorrelations present.

An additional potential problem that is unique to the Hamiltonian algorithm is *divergent transitions*. This can occur when the sampling process has difficulty exploring the posterior distribution. The `stan_glm()` function will print a warning message if divergent transitions occur. One should check the previously mentioned diagnostics to make sure everything is fine. Additionally, one can increase the `adapt_delta` argument of `stan_glm()`. Increasing this value, say to 99%, often will solve the problem (see Exercise 9 for an example). This increased target acceptance rate instructs the algorithm to explore the posterior distribution more carefully to ensure a higher acceptance rate. Additional solutions are available in the more general rstan package. See McElreath (2020) and the Stan website for more information.

Example: Placekicking (PlacekickBayes.R, Placekick.csv)

The rstanarm package provides tools to use functions within the bayesplot package that produce graphical and numerical assessments of convergence. These graphical assessments are accessible through the generic plot() function. Below are two examples with the corresponding plots displayed in Figures 6.11 and 6.12.

```
> plot(mod.fit.stan, plotfun = "combo", combo = c("hist",
   "trace"))
> plot(mod.fit.stan, plotfun = "acf_bar")
```

The plotfun argument of plot() specifies functions of interest in bayesplot, where multiple plots can be specified using the combo argument. For example, the first call to plot() produces both histograms and trace plots for the β_0 and β_1 sampled values. It is important to note that bayesplot makes use of the ggplot2 package (see Appendix A.7.2) rather than the main graphics package to produce its plots. Therefore, this package must be installed prior to using bayesplot.

The trace plots in Figure 6.11 (second column) overlay the samples for the four chains. Each has a thick, central band in the middle with a fairly constant range of parameter values. Also, the values in the trace plot are rapidly changing over short spans of the samples. Therefore, these plots suggest that convergence has been achieved. In general, because of the large number of sampled values displayed in trace plots, it can be useful to look at a smaller range of these values in the order they were taken. For example, including window = c(1,1000) in plot() will show only the first 1,000 values for each chain. It also can be useful to examine one chain at a time. Unfortunately, a direct argument is not available in plot() for this purpose. Instead, the sample can be extracted from the object saved from stan_glm() and plotted. We illustrate how this is done in the program.

The samples for all four chains are combined for the histograms in Figure 6.11 (first column). The histogram for β_0 (Intercept) is above the histogram for β_1 (distance). These histograms show fairly symmetric distributions, which is why equal-tail and HPD credible intervals gave similar lower and upper bounds for each regression parameter. The bar plots in Figure 6.12 provide information about the estimated autocorrelations for each chain. The bar heights (y-axis) represent the numerical value of the autocorrelations for each lag (x-axis). The lag 0 autocorrelation of 1 should be ignored because it is 1 by definition: observations are perfectly correlated with themselves. Overall, the autocorrelations quickly decrease toward 0 for each regression parameter, which supports our previous conclusions regarding convergence. If autocorrelations remained large, thinning could be accomplished through a thin argument in stan_glm(). For example, thin = 5 means every 5th simulated value is included in the sample.

Two measures of interest from the previous example are given in the MCMC diagnostics table of the summary(mod.fit.stan) output and the displayed information for mod.fit.stan$stan_summary. First, the n_eff column gives the effective sample size for each parameter. For β_0 and β_1, these values are 9,482 and 13,468, respectively. While these sample sizes are much less than the 50,000 in total obtained from the four separate chains, this is to be expected due to the autocorrelation from the MCMC sampling process. Also, these effective sample sizes are more than enough to accurately approximate all but the most extreme quantiles from the posterior.[30] Second, the `Rhat` column provides the Gelman-Rubin convergence statistic. The values are all close to 1 for β_0 and β_1, which provides strong evidence for convergence.

An intuitive way to determine the actual acceptance rate is by examining if successive values in the chain are equal. We illustrate this process in our program and find an acceptance rate of 99.3%. This is a very satisfactory value.

An alternative way to summarize the samples and to examine diagnostic measures is with the `launch_shinystan()` function in `rstanarm`. This function uses the `shiny` package to produce a web browser-based app for this purpose. Once the app is open, the user can select menu items like Estimate and Diagnose to summarize the samples and diagnose potential problems with the sampling process, respectively. While we applaud its authors for creating this tool, we currently recommend against using it because of its slow opening time and the fact that the R session cannot be used while the app is open.

In addition to convergence of the MCMC methods, another concern when applying Bayesian methods is whether inferences might change if some other justifiable priors were chosen instead. This is referred to as *sensitivity analysis*. We investigated this issue by examining different prior specifications as shown in our program. The results were very similar to those with the original priors, so this alleviated any of our concerns. This is not surprising because the large sample size for the data reduces the impact of the priors.

6.6.3 Alternative computational tools

Until the fairly recent introduction of Stan and its implementation of the Hamiltonian algorithm, the Metropolis-Hastings and Gibbs sampling algorithms were the primary MCMC methods used in practice. Implementations of these algorithms are available in the `MCMCpack` of R. This package provides functions, such as `MCMClogit()` to estimate the regression parameters of a logistic regression model using an interface similar to `glm()` for specifying models. Bayesian versions of most models examined in our book can be implemented with this package. Examples are available in the first edition of our textbook.

Two general software implementations of Gibbs sampling are Bayesian Inference Using Gibbs Sampling (BUGS) and Just Another Gibbs Sampler (JAGS; `https://mcmc-jags.sourceforge.io`). Similar to Stan, R has packages—including `R2OpenBUGS` for BUGS and `R2jags` for JAGS—that allow R to use these software packages. We chose to use Stan and the Hamiltonian algorithm in this edition of our book because of their faster speed, better convergence properties, and ease of use.

The Integrated Nested Laplace Approximations (INLA; `http://www.r-inla.org`) approach of Rue et al. (2009) provides an alternative to MCMC. This method uses mathematical approximations to integrals rather than simulation. A package to use the software in R is not available from CRAN, but can be downloaded from its website.

[30]It is important to make sure the quantile is not too close to the smallest or largest values in the sample. For example, a 0.99 quantile from a sample of size 100 may be a poor estimate of the true quantile because it is approximately the 99th largest value out of 100. If additional samples of the same size were taken, there would likely be a significant amount of variability in the estimates obtained. Alternatively, a sample size of 10,000 would provide a much better estimate because the 0.99 quantile would be the 9,900th largest value. There would be more information available (more observations) in the extreme portion of the distribution, which would lead to a good estimate.

There are additional R packages built upon Stan and the other computational tools mentioned here. More are being developed because Bayesian computation is an active area of statistical research. The Bayesian inference task view on CRAN at `http://cran.r-project.org/web/views/Bayesian.html` is a good starting place for readers who would like more information on these alternatives and new developments.

6.7 Group testing

A very important tool for combating infectious disease outbreaks is widespread testing to identify which individuals have the disease. This allows infected individuals to receive treatment and to isolate themselves from others. The typical testing process involves obtaining specimens, such as blood or saliva, from people and testing each specimen separately for the presence of the pathogen causing the infection. However, when laboratory testing capacity is insufficient to test all specimens in this way, or when the cost of testing is large, laboratories often use *group testing*[31]—combining portions of specimens from several individuals into one for a single test evaluation—to complete the testing process efficiently.

During the COVID-19 pandemic, group testing was first used for community-wide testing in the United States by the Nebraska Public Health Laboratory (Abdalhamid et al., 2020). At this laboratory, portions of five specimens at a time were combined into a single specimen and tested. If a "grouped" specimen tested negative, all five corresponding individuals were declared negative. Thus, only one test was needed to determine the disease outcome for five individuals. If a grouped specimen tested positive, each of the five specimens was retested individually to determine its disease outcome. Because disease prevalence in the community was relatively low, the majority of group tests gave negative results. Thus, significantly fewer tests were required overall than if each specimen had initially been tested separately. Because fewer tests were needed, the laboratory could handle more specimens overall, thereby increasing its testing capacity. This same type of testing process was adopted by laboratories across the world during the pandemic.

In addition to COVID-19 testing, infectious disease applications of group testing include screening blood donations for infections (Dodd, 2019; American Red Cross, 2023) and testing for sexually transmitted diseases (Lewis et al., 2012; Ando et al., 2021). Group testing is used in many other areas, such as testing aquatic animals for infectious diseases (Laurin et al., 2019), monitoring West Nile virus in mosquitoes (White et al., 2001), detecting bacteria in food (Mester et al., 2017), discovering chemical compounds for new pharmaceuticals (Kainkaryam and Woolf, 2009), and checking sensors in electronic networks (Lo et al., 2013). In the most general sense, each application involves grouping items together and testing the group for a binary trait. For simplicity, in this section we use the terminology associated with infectious disease testing: grouping "specimens" and labeling the binary outcome as "positive" or "negative." We deviate from this convention as needed for specific applications.

Group testing has two separate goals—identification and estimation—where one or both may be relevant for a particular application. Identification involves determining which specimens are positive. Section 6.7.1 discusses how to choose the best group sizes for this goal. On the other hand, estimation involves determining the probability of disease for an individual. This probability may represent an overall disease prevalence for a population or

[31]Group testing is known by many names, including pooled testing, specimen pooling, batch testing, and bulk testing.

an individual-specific probability as estimated by a regression model. In some applications, estimation may the only goal. For example, knowing which particular mosquito is infected with West Nile virus is not important. Rather, the objective is to know whether the virus is present in an area and, if so, estimate its overall prevalence. In fact, for this application and others, a group test may destroy specimens so that further testing is impossible.[32] Section 6.7.2 discusses estimation for a number of scenarios.

6.7.1 Identification

The testing process implemented by the Nebraska Public Health Laboratory is often referred to as "Dorfman's algorithm" in honor of Robert Dorfman and his seminal paper (Dorfman, 1943) that introduced the algorithm. The word *algorithm* is used because testing in groups and subsequently retesting specimens corresponding to positive groups involves a sequence of steps. Other algorithms exist, but we will focus for now on Dorfman's algorithm because it is implemented the most in practice.

While the Nebraska Public Health Laboratory used groups of size five for their testing, other implementations of Dorfman's algorithm can use different group sizes. For example, the American Red Cross uses groups of size 16 when screening blood donations. Laboratories, in general, would like to use the group size that leads to the smallest number of tests. Choosing group sizes too large or too small leads to using more tests than necessary to get the same results.

Optimal testing configuration

To determine the best group size, suppose a laboratory has essentially a continuous series of independent specimens that need to be tested. They wish to select the group size I with which to perform the tests ($I \geq 2$), and can create K groups of this size using the specimens currently available. A particular group, say group k, requires T_k tests to determine the positive/negative outcomes within the group, depending on the initial outcome. When the group test is negative, all specimens within it are declared negative so that only $T_k = 1$ test is needed. When the group test is positive, each specimen needs to be retested individually to differentiate between the positives and negatives so that $T_k = I + 1$ tests are required. Our goal is to select the I that makes the total number of tests performed as small as possible. Therefore, we would like to know or estimate the expected number of tests $E(T_k)$ for a given I.

A number of quantities need to be defined to develop the expression for $E(T_k)$. For each $k = 1, \ldots, K$, define Z_k as a binary random variable denoting the outcome for the test on the kth group, where we always use a 1 to denote a positive and a 0 to denote a negative. Because tests may not be 100% accurate, we also need to define $\tilde{Z}_k$ as the true binary status for the kth group. The accuracies of these tests are given by the sensitivity $S_e = P(Z_k = 1|\tilde{Z}_k = 1)$ and the specificity $S_p = P(Z_k = 0|\tilde{Z}_k = 0)$.[33] Lastly, define $Y_{i(k)}$ for $i = 1, \ldots, I$ to denote whether the algorithm identifies the ith specimen within the kth group as positive or negative, and define $\tilde{Y}_{i(k)}$ as the true binary status for this specimen. The probability that specimen i in group k is *truly* positive is $P(\tilde{Y}_{i(k)} = 1) = \tilde{\pi}$.

---

[32]When testing mosquitoes for West Nile virus, traps are set across a large geographical area. Mosquitoes caught within a trap are pulverized together at the laboratory and subsequently tested as a group. No retesting is possible.

[33]Section 6.1 provides more information about the sensitivity and specificity of diagnostic tests.

There is only one test performed when $Z_k = 0$, and there are $I+1$ tests performed when $Z_k = 1$. Thus, the expected number of tests for group k is

$$\begin{aligned} E(T_k) &= 1 \times P(Z_k = 0) + (I+1) \times P(Z_k = 1) \\ &= 1 + I \times P(Z_k = 1). \end{aligned}$$

Using the definitions of marginal and conditional probabilities, we obtain

$$\begin{aligned} P(Z_k = 1) &= P(Z_k = 1 \text{ and } \tilde{Z}_k = 0) + P(Z_k = 1 \text{ and } \tilde{Z}_k = 1) \\ &= P(Z_k = 1|\tilde{Z}_k = 0)P(\tilde{Z}_k = 0) + P(Z_k = 1|\tilde{Z}_k = 1)P(\tilde{Z}_k = 1) \\ &= (1 - S_p)(1 - \tilde{\pi})^I + S_e\left[1 - (1 - \tilde{\pi})^I\right] \\ &= S_e + (1 - S_e - S_p)(1 - \tilde{\pi})^I. \end{aligned}$$

Note that $P(\tilde{Z}_k = 0) = (1 - \tilde{\pi})^I$ because each specimen is independent and each individual has a probability of $\tilde{\pi}$ for being positive. The expected number of tests for group k becomes

$$E(T_k) = 1 + I\left[S_e + (1 - S_e - S_p)(1 - \tilde{\pi})^I\right]. \tag{6.26}$$

We want to find the I that minimizes the expected number of tests needed across all groups. Because each T_k has the same probability distribution (each group has the same size and the probability a specimen is positive is the same across all specimens), we can focus only on $E(T_k)$. However, we need to find the I that minimizes the expected number of tests per individual tested, i.e., minimize $E(T_k)/I$, rather than $E(T_k)$ alone. This is because $E(T_k)$ increases as a function of I. Putting the objective function on a per-individual basis allows for fair comparisons across group sizes. The value of I that leads to the minimization of $E(T_k)/I$ is referred to as the *optimal testing configuration* (OTC). A closed form solution for the exact OTC is not available. Instead, $E(T_k)/I$ can be evaluated over a range of potential values for I to find the corresponding value that results in the minimization.

Accuracy

In addition to the expected number of tests, the accuracy of the group testing algorithm is of interest. There are multiple ways to measure accuracy. Each measure uses the positive and negative outcomes produced by the algorithm in different ways, and each is affected by testing error that can occur at different stages of the algorithm. For this reason, it is important to clearly define accuracy measures and evaluate them relative to the group testing algorithm.

Suppose group k tests positive so that retests are needed for each individual. Define the retest binary outcome of the ith specimen within the kth group, $R_{i(k)}$, to be 1 when individual i is positive and 0 otherwise. The sensitivity and specificity for the retest are denoted by $S_{eR} = P(R_{i(k)} = 1|\tilde{Y}_{i(k)} = 1)$ and $S_{pR} = P(R_{i(k)} = 0|\tilde{Y}_{i(k)} = 0)$. These values may be different from the previously defined S_e and S_p, respectively, for a number of reasons, including the use of a different type of test for a retest than for a group test. Note that only members of a positive group ($Z_k = 1$) have a value for $R_{i(k)}$. For these individuals, $R_{i(k)} = Y_{i(k)}$. When $Z_k = 0$, all individuals within the kth group are declared negative, so that $Y_{i(k)} = 0$ for $i = 1, \ldots, I$.

One measure of the accuracy for an algorithm is the pooling sensitivity: $PS_e = P(Y_{i(k)} = 1|\tilde{Y}_{i(k)} = 1) = P(Z_k = 1 \text{ and } R_{i(k)} = 1|\tilde{Y}_{i(k)} = 1)$. To simplify this expression, an assumption of conditional independence between the tests across the two stages of testing (stage 1 represents the group test and stage 2 represents the retests) is most often made. This leads to

$$PS_e = P(Z_k = 1|\tilde{Y}_{i(k)} = 1)P(R_{i(k)} = 1|\tilde{Y}_{i(k)} = 1)$$
$$= S_e S_{eR}.$$

Empirical findings have found this conditional assumption to be reasonable (Litvak et al., 1994). A different measure of accuracy is the pooling specificity:

$$PS_p = P(Y_{i(k)} = 0|\tilde{Y}_{i(k)} = 0)$$
$$= 1 - \frac{P(Y_{i(k)} = 1) - PS_e\tilde{\pi}}{1 - \tilde{\pi}}, \quad (6.27)$$

where

$$P(Y_{i(k)} = 1) = (1 - S_p)(1 - S_{pR})(1 - \tilde{\pi})^I + S_e S_{eR}\tilde{\pi}$$
$$+ S_e(1 - S_{pR})(1 - \tilde{\pi})\left[1 - (1 - \tilde{\pi})^{I-1}\right].$$

Exercise 1 examines the derivation. We briefly examine other measures of accuracy in the next example.

Example: HIV screening (HIVGroupSize.R)

Sherlock et al. (2007) describe the HIV testing practices for a number of cities across the United States. In particular, Washington, DC, used Dorfman's algorithm with groups of size 20. The overall prevalence for those tested during the examined time period was 0.0041. The manufacturer of the diagnostic test reported a specificity of 0.976 from its clinical trials. The reported sensitivity was split into three different cases based on an individual's CD4 protein count level: 0.967 for individuals with a very low count, 0.884 for individuals with a low count, and 0.653 for individuals with a normal count.[34] We use the overall prevalence and accuracy values as the values for $\tilde{\pi}$, S_e, S_{eR}, S_p, and S_{pR} that are needed to calculate $E(T_k)$, PS_e, and PS_p. In actual practice, the resulting calculations could provide information for future testing so that group testing can be applied efficiently.

The `binGroup2` package (Hitt et al., 2023) provides functions to perform the needed calculations. To find the expected number of tests, we need to first construct a *group membership matrix* that details the group testing process. Each row in this matrix represents a testing stage, and each column of the matrix represents a specimen. Each element within the matrix specifies the group number within a stage to which each specimen belongs. Below we construct this matrix for one group of 20 specimens using `matrix()`.

```
> library(package = binGroup2)

> I <- 20
> group.member <- matrix(data = c(rep(1, times = I), 1:I), nrow
   = 2, ncol = 20, byrow = TRUE)
> group.member
     [,1] [,2] [,3] [,4] [,5] [,6] [,7] [,8] [,9] [,10] [,11]
[1,]    1    1    1    1    1    1    1    1    1     1     1
```

[34] We will focus on the VERSANT HIV-1 RNA 3.0 Assay (bDNA), which was one of two nucleic acid amplification assays used. Accuracy information is available at `https://www.fda.gov/vaccines-blood-biologics/approved-blood-products/versant-hiv-1-rna-30-assay-bdna`.

```
[2,]    1    2    3    4    5    6    7    8    9   10   11

     [,12] [,13] [,14] [,15] [,16] [,17] [,18] [,19] [,20]
[1,]     1     1     1     1     1     1     1     1     1
[2,]    12    13    14    15    16    17    18    19    20
```

The group membership matrix here indicates that 20 specimens are combined into one group for the first stage. In the second stage, each specimen is tested separately; thus, they are each in their own group of size 1. For Dorfman's algorithm, the first row is always all 1's, and the second stage is always $1, \ldots, I$.

The `opChar1()` function of the `binGroup2` package calculates $E(T_k)$, PS_e, and PS_p. Setting its `algorithm` value to `"D2"` indicates that Dorfman's two-stage algorithm is to be used. Other arguments include: `p` for the overall prevalence ($\tilde{\pi}$), `Se` for the sensitivity, `Sp` for the specificity, and `hier.config` for the group membership matrix. For the sensitivity and specificity, we will assume these values are equal across the two stages and just specify one value for the arguments. If they were not equal, a vector of length two can be used to indicate the first and second-stage accuracy measures (e.g., numerical values for S_e and S_{eR} would be put into a vector for `Se`). We obtain the following from R.

```
> pitilde <- 0.0041
> test.vlow <- opChar1(algorithm = "D2", p = pitilde, Se =
   0.967, Sp = 0.976, hier.config = group.member)

 Number of minutes running:  0

> summary(test.vlow)

Algorithm: Non-informative two-stage hierarchical testing

Testing configuration:
Stage 1: 20

Expected number of tests: 2.97
Expected number of tests per individual: 0.1484
Accuracy for individuals:
     PSe    PSp   PPPV   PNPV Individuals
1 0.9351 0.9977 0.6285 0.9997         All

Overall accuracy of the algorithm:
     PSe    PSp   PPPV   PNPV
1 0.9351 0.9977 0.6285 0.9997

PSe denotes the pooling sensitivity.
PSp denotes the pooling specificity.
PPPV denotes the pooling positive predictive value.
PNPV denotes the pooling negative predictive value.

> test.low <- opChar1(algorithm = "D2", p = pitilde, Se = 0.884,
   Sp = 0.976, hier.config = group.member)
> test.norm <- opChar1(algorithm = "D2", p = pitilde, Se =
   0.653, Sp = 0.976, hier.config = group.member)
```

```
> # Expected number of tests per individual
> data.frame(CD4.vlow = test.vlow$ET/I, CD4.low = test.low$ET/I,
   CD4.norm = test.norm$ET/I)
  CD4.vlow   CD4.low  CD4.norm
1 0.148387 0.1418397 0.1236176
```

The expected number of tests per individual are between 0.12 and 0.15. Thus, the expected number of tests are significantly lower from using group testing than what would occur from individual testing (i.e., only testing each specimen separately, resulting in one test per person in need of testing). Furthermore, the laboratory could use the largest expected number of tests here or a weighted average of the expected number of tests (weighted by proportion of people expected with very low, low, or normal CD4 counts) to plan for the costs and time involved with testing.

Pooling sensitivities and specificities for the very low case are provided in the `Overall accuracy of the algorithm` table of the output. In addition to these accuracy measures, *predictive values* are given to provide a different perspective on accuracy. The pooling positive predictive value (`PPPV`) is $P(\tilde{Y}_{i(k)} = 1|Y_{i(k)} = 1)$ and the pooling negative predictive value (`PNPV`) is $P(\tilde{Y}_{i(k)} = 0|Y_{i(k)} = 0)$ (see Kim et al. (2007) and Black et al. (2015) for expressions). For example, the positive predictive value is 0.6285. Thus, an individual receiving a positive test result still has a substantial probability (0.3715) that they are not actually positive. This individual should be retested and/or continuously monitored to determine if they are truly positive.

The OTC is found by the `OTC1()` function of the `binGroup2` package. Important arguments for the function are (1) `algorithm = "D2"` for Dorfman's algorithm, (2) `group.sz` representing a vector of group sizes to search over, and (3) `obj.fn = "ET"` for the minimization of $E(T_k)/I$. Below is how the `OTC1()` function is used for very low CD4 counts.

```
> save.config.vlow <- OTC1(algorithm = "D2", p = pitilde, Se =
   0.967, Sp = 0.976, group.sz = 3:30, obj.fn = "ET")
Initial Group Size = 3
Initial Group Size = 4

<OUTPUT EDITED>

Initial Group Size = 30

 Number of minutes running:  0

> summary(save.config.vlow)

Algorithm: Non-informative two-stage hierarchical testing

Optimal testing configuration:
   Stage 1
ET      17

Expected number of tests:
   E(T)  Value
ET 2.49 0.1464
```

```
E(T) denotes the expected number of tests.
Value denotes the objective function value per individual.

Overall accuracy of the algorithm:
      PSe    PSp    PPPV   PNPV
ET 0.9351 0.9980 0.6563 0.9997

PSe denotes the pooling sensitivity.
PSp denotes the pooling specificity.
PPPV denotes the pooling positive predictive value.
PNPV denotes the pooling negative predictive value.

> # Proportion reduction in tests by using OTC
> (test.vlow$ET/I - save.config.vlow$opt.ET$value)/
   (test.vlow$ET/I)
[1] 0.01313203
```

The OTC uses a group size of 17 with an expected number of tests per individual of $E(T_k)/I = 0.1464$. Therefore, future testing should consider using this smaller group size. However, there is only a 1.3% decrease in the expected number of tests per individual when compared to a group size of 20, so the benefits from changing the testing algorithm may not be enough to outweigh any costs in making this change.

Each individual potentially could have a different probability of being positive for a disease. For many situations, it may not be possible to acknowledge this heterogeneity, because there may not be a way to find these different probabilities. However, when it is possible, we can generalize the probability of disease for an individual to be $P(\tilde{Y}_{i(k)} = 1) = \tilde{\pi}_{i(k)}$. With this generalization, the expected number of tests becomes

$$E(T_k) = 1 + I\left\{S_e + (1 - S_e - S_p)\left[\prod_{i=1}^{I}(1 - \tilde{\pi}_{i(k)})\right]\right\}. \tag{6.28}$$

The pooling specificity is generalized to

$$\begin{aligned} PS_{p,i(k)} &= P(Y_{i(k)} = 0|\tilde{Y}_{i(k)} = 0) \\ &= 1 - \frac{P(Y_{i(k)} = 1) - PS_e\tilde{\pi}_{i(k)}}{1 - \tilde{\pi}_{i(k)}}, \end{aligned} \tag{6.29}$$

where

$$\begin{aligned} P(Y_{i(k)} = 1) = (1 - S_p)(1 - S_{pR})\prod_{i=1}^{I}(1 - \tilde{\pi}_{i(k)}) + S_eS_{eR}\tilde{\pi}_{i(k)} \\ + S_e(1 - S_{pR})(1 - \tilde{\pi}_{i(k)})\left[1 - \prod_{i'=1, i' \neq i}^{I}(1 - \tilde{\pi}_{i'(k)})\right], \end{aligned}$$

and the pooling sensitivity remains the same as before. The pooling specificity can now be different for each individual because of $\tilde{\pi}_{i(k)}$ in Equation 6.29. For this reason, functions in `binGroup2` also report an overall pooling specificity measure based on a weighted average of the $\tilde{\pi}_{i(k)}$ and $PS_{p,i(k)}$ values for the individuals tested (Hitt et al., 2019).

The benefits from using group testing can be increased further by taking advantage of the heterogeneity among the probabilities of disease. The process is known as *informative group testing* (Bilder et al., 2010), and a number of different algorithms have been proposed to implement it (Bilder and Tebbs, 2012). We focus on the pool-specific optimal Dorfman (PSOD) algorithm of McMahan et al. (2012) here because it most closely coincides with Dorfman's algorithm. This algorithm considers a set or *block* of N specimens to split into K groups of size I_k for $k = 1, \ldots, K$. The specimens are ordered by their probabilities of being positive from smallest to largest, and sequentially put into groups of equal or smaller sizes until all specimens are in a group. Thus, specimens in the first group have the I_1 smallest probabilities, and specimens in the last group have the I_K largest probabilities, where $I_1 \geq ... \geq I_K$. The expected number of tests for the block follows from using Equation 6.28 for $E(T_k)$, but now using the ordered probabilities and summing these expectations over all of K groups. The PSOD algorithm performs a numerical iterative search to find the OTC, now defined as the group sizes and number of groups that minimize $\sum_{k=1}^{K} E(T_k)$.

Generally, as the variability in the disease probabilities increase, the expected number of tests for the block will decrease. This is because informative group testing optimally assigns specimens to groups. For example, consider the situation without testing error and every individual has a probability of 0 or 1 for having a disease. This situation represents the maximum possible amount of variability in the disease probabilities. The PSOD algorithm would place all corresponding specimens with a 0 probability in one group and all corresponding specimens with a 1 probability in separate groups of size 1. No retests would be needed. If we reduce the variability by allowing disease probabilities to slightly deviate from 0 and 1, specimens with "low" probabilities should be assigned to large groups and specimens with "high" probabilities should be assigned to small groups. Some retests will likely be needed, because of uncertainty in who is positive or negative, and the overall expected number of tests will likely be larger than in the maximum possible variability situation.

Example: HIV screening (HIVGroupSize.R)

When people visit a medical clinic to be tested for HIV or other infectious diseases, they often fill out a questionnaire regarding their individual risk factors. Clinical observations, such as symptoms or particular medical conditions, may be also observed by medical practitioners. When these types of information are available, individual risk probabilities of being positive can be estimated with the help of a binary regression model. These estimated probabilities can then be incorporated into an informative group testing algorithm as an approximation of the true probabilities of being positive.

Because individual-specific information was not available for the HIV example here, we will suppose the individual risk probabilities have a mean equal to $\tilde{\pi} = 0.0041$ and follow a beta distribution with parameters a and $b = a(1 - \tilde{\pi})/\tilde{\pi}$. The expected values of the order statistics[35] represent the average values that a sample of size N from this beta distribution would take on. The `expectOrderBeta()` function of `binGroup2` calculates these values. With a block size (`size`) of N= 35 and a beta distribution parameter (`alpha`) of $a = 0.5$, we obtain the following values.

```
> p.beta1 <- expectOrderBeta(p = pitilde, alpha = 0.5, size = 35)
> head(p.beta1, n = 4)
```

[35] Order statistics are the ordered random variables for a sample. For example, the minimum value for a sample is the first order statistic, and the largest value for a sample is the last order statistic. See Section 5.4 of Casella and Berger (2002).

```
[1] 8.188579e-10 1.796176e-08 1.912028e-07 1.315783e-06
> tail(p.beta1, n = 4)
[1] 0.007379782 0.012217066 0.016845808 0.023854869
```

Therefore, the largest probability on average is 0.0239. These values in p.beta1 provide us with potential disease probabilities that we can use to understand how well PSOD would work in practice.

We use the OTC1() function to find the OTC for PSOD. Disease probabilities can be included as a vector in the function call using the function's p argument. If the algorithm is used to develop a plan prior to knowing these probabilities, then the argument value can be an object containing the N expected order statistics, as calculated in p.beta1 above. Alternatively, if p is given the single value $\tilde{\pi}$, then expectOrderBeta() is used internally by the function to find the hypothetical probabilities with the same overall prevalence (this is what we do in our subsequent code). The argument algorithm = "ID2" in OTC1() specifies that PSOD is used (ID2 stands for "Informative Dorfman over two stages") and group.sz gives the value for N. Next, we show how to use OTC1() when $N = 35$ and $a = 0.5$ for very low CD4 counts.

```
> PSOD.0.5 <- OTC1(algorithm = "ID2", p = pitilde, Se = 0.967,
   Sp = 0.976, group.sz = 35, obj.fn = "ET", alpha = 0.5)
Block Size = 35

 Number of minutes running:  0.17

> summary(PSOD.0.5)

Algorithm: Informative two-stage hierarchical testing

Optimal testing configuration:
   Block size Group sizes
ET         35       25,10

Expected number of tests:
   E(T)  Value
ET 4.52 0.1291

E(T) denotes the expected number of tests.
Value denotes the objective function value per individual.

Overall accuracy of the algorithm:
      PSe    PSp   PPPV   PNPV
ET 0.9351 0.9984 0.6991 0.9997

PSe denotes the pooling sensitivity.
PSp denotes the pooling specificity.
PPPV denotes the pooling positive predictive value.
PNPV denotes the pooling negative predictive value.

> # Reduction in expected number of tests per individual
```

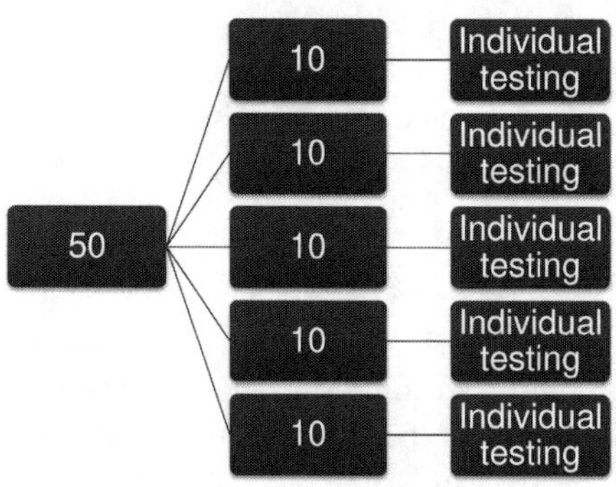

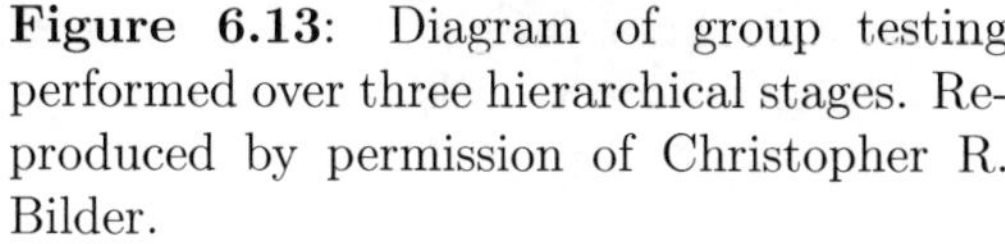

Figure 6.13: Diagram of group testing performed over three hierarchical stages. Reproduced by permission of Christopher R. Bilder.

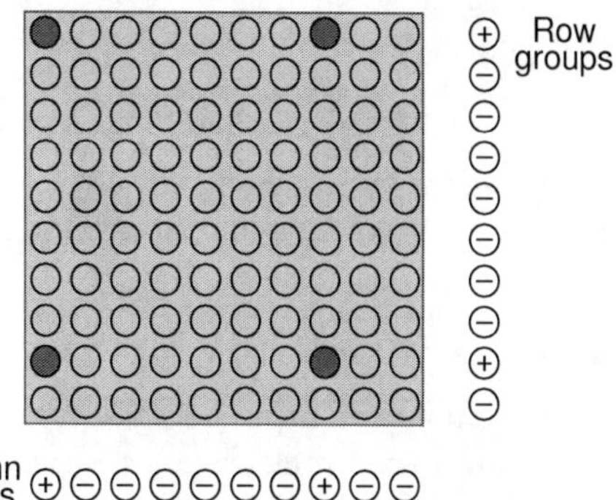

Figure 6.14: Diagram of group testing performed in a two dimensional array-like structure. Filled-in circles indicate which specimens need to be retested because of positive row and column tests. Reproduced by permission of Christopher R. Bilder.

```
> (save.config.vlow$opt.ET$value - PSOD.0.5$opt.ET$value) /
    save.config.vlow$opt.ET$value
[1] 0.1187173
```

The OTC has two groups: the first group contains individuals with the 25 lowest probabilities and the second group contains the remaining 10 individuals. The expected number of tests per individual is 0.1291. Compared to the 0.1464 obtained earlier under *non-informative group testing* (every individual had the same probability), we see that there is a 11.9% reduction in the expected number of tests per individual by using informative group testing.

While Dorfman's algorithm and its variants are most often used, other algorithms frequently lead to a smaller number of tests. Summaries of these algorithms are available in resources including Hughes-Oliver (2006) and Bilder (2019b). All of these algorithms can be categorized as *hierarchical* or *non-hierarchical.* Dorfman's algorithm is a hierarchical algorithm because specimens can be tested only once per stage. More stages are possible with hierarchical group testing algorithms. For example, Figure 6.13 demonstrates a three-stage hierarchical testing algorithm used for HIV testing in San Francisco (Sherlock et al., 2007). In the first stage, a group from 50 specimens is formed. If this group tests positive, 5 subgroups of size 10 are formed in stage 2 from these 50 individuals. If any of these subgroups test positive, individuals are retested separately. Whenever a group tests negative at any stage, its corresponding individuals are declared to be negative. Other variations are possible. For example, some algorithms require positive groups to be halved (Litvak et al., 1994; Black et al., 2012), while others allow positive testing groups to be unequal in size (Black et al., 2015).

Non-hierarchical algorithms allow for specimens to be tested multiple times within a stage in an effort to reduce the number of retests needed. *Array testing* (also known as *matrix pooling*) is a commonly used form of this type of algorithm. Figure 6.14 provides a diagram demonstrating how array testing works. Specimens (circles) within the square are arranged into 10 rows and 10 columns. For example, this diagram could represent a microplate used in a laboratory with specimens within its wells. Groups of specimens are formed for each row and for each column. Specimens in rows *or* columns that test negative are declared negative. Specimens in rows *and* columns that test positive need to be retested separately to decode the positives from the negatives. When testing error is possible, columns (rows) may test positive without any rows (columns) testing positive. In this situation, all specimens in a column (row) should be retested (Kim et al., 2007).

6.7.2 Estimation

We change our focus now to estimating the probability an individual has a disease using data obtained through group testing. When individuals in a population are homogeneous, this probability is simply $\tilde{\pi}$ and also represents the overall disease prevalence. In other situations, we may expect differences among individuals. We can still estimate the overall prevalence in this situation and continue to denote it as $\tilde{\pi}$. Later in this section, we will examine how to take into account observable differences among individuals and use regression models to estimate the probability an individual has a disease ($\tilde{\pi}_{i(k)}$).

Overall prevalence

The simplest estimation setting involves estimating the overall prevalence when only the group responses are observed for Dorfman's algorithm. Thus, no retests are performed for positive groups, and no individuals are declared positive. This approach may be of interest in public health studies when tests are expensive or have limited availability. We will focus instead on a closely related and widely used application known as *multiple vector transfer design experiments*, where the goal is to estimate the probability that insect vectors (disease-carrying insects) will infect a plant with a pathogen (Swallow, 1985). Thus, $\tilde{\pi}$ represents the transmission rate from insects to plants.

The typical experiment set-up is shown in Figure 6.15. A greenhouse (rectangle) contains individually enclosed plants (circles) that are initially known to be uninfected. Multiple insect vectors are transferred to each of these plants. Whether a plant becomes infected from at least one insect is the group response. There is no interest in knowing which particular insects infect the plant. Rather, the purpose is to estimate the overall transmission rate from insect to plant. The plants are observed over a sufficiently long period of time to ensure that no testing error is made.

In the context of our previous notation, $\tilde{Y}_{i(k)}$ represents whether the ith insect (specimen) on the kth plant (group) transfers the virus to the plant (1) or does not transfer the virus (0). Each plant contains I insects, and all insects are assumed to be independent within and across groups. We use $\tilde{Z}_k$ to indicate whether or not the kth plant is infected. The probability that a randomly chosen plant is infected is denoted by $\tilde{\theta}$; i.e., $\tilde{\theta} = P(\tilde{Z}_k = 1)$. Our goal is to estimate $P(\tilde{Y}_{i(k)} = 1) = \tilde{\pi}$, the overall transmission rate, by observing only the K plant responses.

Estimation of $\tilde{\pi}$ is possible because $\tilde{\theta}$ is a simple function of the transmission rate:

$$\begin{aligned}\tilde{\theta} &= 1 - P(\tilde{Z}_k = 0)\\ &= 1 - P(\tilde{Y}_{1k} = 0, \ldots, \tilde{Y}_{Ik} = 0)\\ &= 1 - \prod_{i=1}^{I} P(\tilde{Y}_{ik} = 0)\\ &= 1 - (1 - \tilde{\pi})^I. \end{aligned} \qquad (6.30)$$

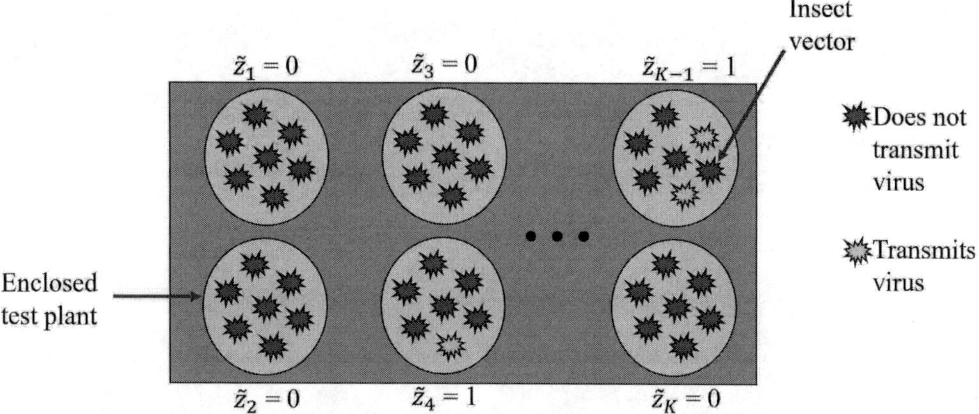

Figure 6.15: Greenhouse diagram for a multiple vector transfer design experiment. Reproduced by permission of Christopher R. Bilder.

We can rewrite Equation (6.30) as $\tilde{\pi} = 1 - (1 - \tilde{\theta})^{1/I}$, so that an estimate of $\tilde{\pi}$ becomes

$$\hat{\tilde{\pi}} = 1 - (1 - \hat{\tilde{\theta}})^{1/I}, \tag{6.31}$$

where $\hat{\tilde{\theta}} = \sum_{k=1}^{K} \tilde{Z}_k / K$ is the observed proportion of positive groups. Because $\hat{\tilde{\theta}}$ is the MLE of $\tilde{\theta}$, the invariance property of MLEs (see Appendix B.4) ensures that $\hat{\tilde{\pi}}$ in Equation (6.31) is the MLE of $\tilde{\pi}$. The $(1-\alpha)100\%$ Wald interval for $\tilde{\pi}$ is

$$\hat{\tilde{\pi}} \pm Z_{1-\alpha/2} \sqrt{\frac{\left[1 - (1 - \hat{\tilde{\pi}})^I\right](1 - \hat{\tilde{\pi}})^{2-I}}{I^2 K}},$$

where the variance used in the interval is derived via the delta method (see Exercise 11).

As we have seen elsewhere, Wald-based confidence intervals often fail to provide intervals that achieve the stated confidence level. Unfortunately, the same is true here. Tebbs and Bilder (2004) recommend using a Wilson interval variant instead. This interval is based on using the standard Wilson interval for $\tilde{\theta}$ (Equation 1.4) and then using the relationship between $\tilde{\pi}$ and $\tilde{\theta}$ to obtain an interval for $\tilde{\pi}$. Thus, the $(1-\alpha)100\%$ Wilson interval for $\tilde{\pi}$ is

$$1 - (1 - \tilde{\theta}_L)^{1/I} < \pi < 1 - (1 - \tilde{\theta}_U)^{1/I},$$

where $\tilde{\theta}_L$ and $\tilde{\theta}_U$ are the lower and upper limits, respectively, for a $(1-\alpha)100\%$ Wilson interval for $\tilde{\theta}$. Other confidence intervals can be constructed using this same process. For example, a Clopper-Pearson interval can be formed in this manner and still retain the quality that its true confidence level is equal to or greater than the stated level.

Example: Transmission of the Mal Rio Cuarto virus (MalRioCuarto.R)

Ornaghi et al. (1999) perform a multiple-vector transfer design experiment designed to estimate the probability that infected planthoppers transmit the Mal Rio Cuarto virus to maize plants. Each of 24 separately enclosed plants had seven female planthoppers placed upon it. Of these plants, three became infected.

The `propCI()` function of the `binGroup2` package calculates confidence intervals for $\tilde{\pi}$. Its `x` argument denotes the number of positive groups, the `m` argument denotes the group size, and the `n` argument denotes the number of groups. Below is our use of the function along with additional calculations.

```
> sum.z <- 3  # Infected plants
> I <- 7  # Group size
> K <- 24  # Number of plants
> alpha <- 0.05

> # MLE of theta
> theta.tilde.hat <- sum.z/K
> theta.tilde.hat
[1] 0.125

> # MLE of pi
> pi.tilde.hat <- 1 - (1 - theta.tilde.hat)^(1/I)
> pi.tilde.hat
[1] 0.01889512

> library(package = binGroup2)

> # Wilson interval
> propCI(x = 3, m = 7, n = 24, ci.method = "score", conf.level =
   0.95)

95 percent Wilson score confidence interval:
 [ 0.006325, 0.05164 ]
Point estimate (Maximum Likelihood): 0.0189

> # Clopper-Pearson interval
> propCI(x = 3, m = 7, n = 24, ci.method = "CP", conf.level =
   0.95)

95 percent Clopper-Pearson confidence interval:
 [ 0.003838, 0.05432 ]
Point estimate (Maximum Likelihood): 0.0189

> # Wald interval; function replaces negative lower bound with 0
> propCI(x = 3, m = 7, n = 24, ci.method = "Wald", conf.level =
   0.95)

95 percent Wald confidence interval:
 [ 0, 0.04009 ]
Point estimate (Maximum Likelihood): 0.0189

> var.pi.tilde.hat <- (1 - (1 - pi.tilde.hat)^I ) * (1 -
   pi.tilde.hat)^(2 - I) / (I^2 * K)
> pi.tilde.hat + qnorm(p = c(alpha/2, 1-alpha/2)) *
   sqrt(var.pi.tilde.hat)
[1] -0.002298779  0.040089018
```

The 95% Wilson interval (`method = "score"`) for $\tilde{\pi}$ is $(0.0063, 0.0516)$. Similar to what was seen in Section 1.1.2, the Clopper-Pearson interval (`method = "CP"`) is wider than the Wilson interval. The Wald interval has difficulties with small proportions and results in a negative lower bound here.

This example helps to illustrate the advantages of group testing. If we had instead designed the experiment with only one planthopper per plant, we would have needed a much larger number of plants to perform the experiment adequately because of the low transmission rate. For example, if we momentarily take $\tilde{\pi}$ to be equal to its estimate $\hat{\tilde{\pi}}$, the expected number of positive plants would be only $24 \times 0.0189 = 0.45$ for individual testing. Thus, there would be a large probability of observing few if any plants as positive. Instead, if we used the same number of planthoppers with only one per plant, a total of $24 \times 7 = 168$ plants and a much larger greenhouse space would be needed.

The MLE has a small amount of bias for fixed sample sizes. Hepworth and Biggerstaff (2017) propose alternative estimators, and their methods are available within `propCI()` through its `pt.method` argument. Estimation and inference for the prevalence in general group testing applications can be done in more complicated situations. For example, unequal group sizes could be present (Hepworth, 2005; Dres et al., 2015) and/or differences of two probabilities may be of interest (McCann and Tebbs, 2007; Biggerstaff, 2008). Situations like these are discussed in Bilder (2019a). The next subsection looks at these situations as well, but in a regression context.

Regression

Estimation of the probability that a specimen is positive can also be done as a function of explanatory variables. For example, suppose group testing is used to screen individuals for a sexually transmitted disease. Understanding which risk factors lead to a higher probability of disease could be the goal, in addition to determining who is positive or negative for the disease.

In our development here, we return to the situation where testing error is possible. We define $\tilde{Y}_{i(k)}$ as before to represent the true positive/negative status of a specimen. Our goal is to estimate $P(\tilde{Y}_{i(k)} = 1) = \tilde{\pi}_{i(k)}$ using a logistic regression model:

$$\text{logit}(\tilde{\pi}_{i(k)}) = \beta_0 + \beta_1 x_{i(k)1} + \cdots + \beta_p x_{i(k)p}. \tag{6.32}$$

The log-likelihood function is written as

$$\log(L(\beta_0, \ldots, \beta_p | \tilde{y}_{1(1)}, \ldots, \tilde{y}_{I(K)}) = \sum_{k=1}^{K} \sum_{i=1}^{I} \tilde{y}_{i(k)} \log(\tilde{\pi}_{i(k)}) + (1 - \tilde{y}_{i(k)}) \log(1 - \tilde{\pi}_{i(k)}). \tag{6.33}$$

Unfortunately, $\tilde{y}_{i(k)}$ cannot be observed perfectly because of the possibility of testing error, so methods of Section 2.2.1 cannot be used to obtain the MLEs. Also, these responses come about through observing group responses, so the methods of Section 6.1.2 cannot be used either.

Instead, we can use the expectation-maximization (EM) algorithm proposed by Xie (2001) to obtain the MLEs with group testing data. This algorithm alternates successively between two steps:

1. Expectation: Replace the unobservable outcomes in Equation (6.33) with expectations of their random variables conditional on the available data.

2. Maximization: Maximize the resulting function.

Convergence of these estimates occurs when there is very little change in the parameter estimates from successive iterations of both steps. For the first step, Equation (6.33) becomes

$$E\left[\log(L(\beta_0, \ldots, \beta_p | \tilde{y}_{1(1)}, \ldots, \tilde{y}_{I(K)})) | \mathcal{I}\right] = \sum_{k=1}^{K} \sum_{i=1}^{I} \omega_{i(k)} \log(\tilde{\pi}_{i(k)}) + (1 - \omega_{i(k)}) \log(1 - \pi_{i(k)}),$$

where $\mathcal{I}$ represents the data obtained from the tests performed by the group testing algorithm and $\omega_{i(k)} = E(\tilde{Y}_{i(k)}|\mathcal{I})$. Expressions for $\omega_{i(k)}$ are dependent on the group testing algorithm and are available in Zhang et al. (2013). It is important to note that all components of $\omega_{i(k)}$ are observable or functions of parameters to be estimated. Therefore, standard iterative numerical procedures can be used to maximize the resulting function in the second step.

The `gtReg()` function in the `binGroup2` package uses the EM algorithm to estimate the regression parameters in Equation 6.32. This function also calculates the estimated covariance matrix for the estimators. Because the EM algorithm is used to find the maximum likelihood estimates, standard methods discussed in Section 2.2.1 cannot be used directly to find this matrix. Instead, the matrix is estimated through methods developed by Louis (1982) and detailed by Xie (2001) for group testing.

Because fewer tests (i.e., smaller number of observations) are needed for group testing than for individual testing in appropriate settings, one may expect that some information is lost. In a statistical context, information equates to precision for estimators. Indeed, standard errors are larger on average when using data only from the initial group tests than when each specimen is tested separately. However, this is not the case when retesting data is included with the initial group tests. Thus, although fewer tests are performed with group testing, standard errors are similar and at times smaller on average compared to individual testing! This counterintuitive finding was first shown by Liu et al. (2012) and Zhang et al. (2013) and has been shown by others since. The reason for this result is that individual testing requires more tests, which leads to more chances for a false positive or a false negative test outcome. Thus, there is more uncertainty as a whole, meaning potentially larger standard errors for individual testing on average. It is important to note that estimators for both individual and group testing (including retests) have increasingly similar precision as the sensitivity and specificity increase. Estimators will have the same precision when tests are perfect.

Example: Prenatal infectious disease screening (HIVKenya.R, HIVKenya.csv)

An example from Section 6.1.2 focuses on data obtained from a prenatal infectious disease screening study in Kenya. The main purpose of this study was to evaluate how well group testing algorithms would work for disease surveillance in comparison to individual testing. While the group membership for each individual is not available, we have added simulated group memberships as given by the `gnum` variable to mimic a group-testing situation using Dorfman's algorithm.[36] Simulated group responses and retests are given by the `groupres` and `retests` variables, respectively.

```
> set1 <- read.csv(file = "c:\\data\\HIVKenya.csv")
> # 1st group tests negative so there are no retests
> # 2nd group tests positive and only individual 9 is positive
> head(set1[,c("age", "gnum", "groupres", "retest")], n = 10)
  age gnum groupres retest
1  16    1        0     NA
2  17    1        0     NA
3  26    1        0     NA
```

[36] Details on how to simulate group testing data from individual testing data are available in references including Zhang et al. (2013) and Bilder (2019a).

```
4   20   1   0   NA
5   18   1   0   NA
6   35   2   1    0
7   17   2   1    0
8   35   2   1    0
9   25   2   1    1
10  37   2   1    0
```

To illustrate how to use `gtReg()`, we show how to estimate $\text{logit}(\tilde{\pi}) = \beta_0 + \beta_1\texttt{age}$ with a sensitivity and specificity of 0.98 for each stage of the group testing algorithm. Initially, we only use the first-stage group responses in the estimation process by specifying `type = "sp"` (sometimes referred to as "simple pooling").

```
> library(package = binGroup2)
> mod.fit.gt1 <- gtReg(formula = groupres ~ age, data = set1,
   type = "sp", groupn = gnum, sens = 0.98, spec = 0.98, method
   = "Xie")

 Number of minutes running: 0

> summary(mod.fit.gt1)

Call: gtReg(type = "sp", formula = groupres ~ age, data = set1,
   groupn = gnum, sens = Se, spec = Sp, method = "Xie")

Deviance Residuals:
    Min        1Q    Median        3Q       Max
 -0.4177   -0.3314   -0.3132    0.6440    0.7167

Coefficients:
            Estimate Std. Error z value Pr(>|z|)
(Intercept) -3.54922    1.29435  -2.742  0.00611 **
age          0.04186    0.05070   0.826  0.40896
---
Signif. codes:  0 '***' 0.001 '**' 0.01 '*' 0.05 '.' 0.1 ' ' 1

    Null deviance: 347.9  on 133  degrees of freedom
Residual deviance: 168.7  on 132  degrees of freedom
AIC: 172.7

Number of iterations in EM: 29
```

The `formula` and `data` arguments work the same way as for `glm()`. Other arguments include `groupn` to specify the group numbers and `sens` and `spec` to indicate values for the sensitivity and specificity for the group tests, respectively. The estimated model using only the group responses is

$$\text{logit}(\hat{\tilde{\pi}}) = -3.5492 + 0.0419\texttt{age}.$$

There are some differences between this model and the one estimated using the individual testing data in Section 6.1.2. These differences may appear greater than they are in a practical sense because there is not sufficient evidence from either model that age is important. Interestingly, standard errors for the regression parameter estimates are only 1.7 times as large for the group-response model as for the individual-response model, despite using only about one-fifth the number of observations (groups are of size five, except for the last group which has a size of three).

Next, we include the second-stage retests in the estimation process. The `sens.ind` and `spec.ind` arguments of `gtReg()` specify the sensitivity and specificity of the individual tests in the second stage.

```
> mod.fit.gt2 <- gtReg(formula = groupres ~ age, data = set1,
   type = "sp", groupn  =  gnum, retest = retest, sens = 0.98,
   spec = 0.98, sens.ind = 0.98, spec.ind = 0.98, method = "Xie")

 Number of minutes running: 0

> summary(mod.fit.gt2)

Call:
gtReg(type = "sp", formula = groupres ~ age, data = set1, groupn
   = gnum, retest = retest, sens = 0.98, spec = 0.98, method =
   "Xie", sens.ind = 0.98, spec.ind = 0.98)

Deviance Residuals:
    Min         1Q     Median        3Q        Max
 -0.3655   -0.3303   -0.3150    0.6644    0.7081

Coefficients:
            Estimate Std. Error z value Pr(>|z|)
(Intercept) -1.87753    0.71509  -2.626  0.00865 **
age         -0.02750    0.02965  -0.927  0.35375
---
Signif. codes:  0 '***' 0.001 '**' 0.01 '*' 0.05 '.' 0.1 ' ' 1
AIC: 365.4

Number of iterations in EM: 6

> #Number of individual tests
> nrow(set1)
[1] 668

> # Number of tests from Dofrman's algorithm
> max(set1$gnum) + sum(set1$groupres)
[1] 354
```

The estimated model using the group responses and retests is

$$\text{logit}(\hat{\hat{\pi}}) = -1.8775 - 0.0275\text{age}.$$

This model is quite similar to the one estimated with data from individual testing. Remarkably, this model was based on approximately half as many observations as the model resulting from individual testing, yet its standard errors are quite similar (0.72 vs. 0.69 for the intercept and 0.030 vs. 0.029 for `age`).

Decisions regarding which testing algorithm to use, sizes of the groups, and which specimens to put into particular groups need to be made prior to implementing group testing. These choices are most often made with the goal of minimizing the expected number of tests needed to identify the positive/negative specimens, as shown in Section 6.7.1. However, when estimation is the only goal, one should focus on making the estimation as precise as possible while reducing the overall cost of testing. For estimating the overall prevalence, Swallow (1985) proposes to minimize the mean squared error[37] for a fixed number of groups and a prior estimate of the prevalence. The `designEst()` function of the `binGroup2` package implements this approach. For regression, Bilder and Tebbs (2009) show that putting individuals into groups as similar as possible (in terms of explanatory variable values) results in the smallest standard errors for regression parameter estimates when only initial group responses are observed. When retests are also available, Zhang et al. (2013) show that Dorfman's algorithm results in larger standard errors than three-stage halving or array testing.

6.7.3 Additional considerations and methods

Most group testing algorithms assume that the per-test sensitivity and specificity are fixed constants. In the context of infectious disease testing, this is because diagnostic testing companies perform large trials to determine the accuracy of their assay used for testing. However, one would still expect some variability in test properties from one trial to the next. McMahan et al. (2017) propose to acknowledge this variability by putting the regression model into a Bayesian context with prior distributions for the sensitivity and specificity. These priors can be noninformative or informative. In the latter case, priors can take advantage of past testing data, such as from the large trials performed by diagnostic testing companies.

The sensitivity and specificity in this section are also assumed to be constant across different group sizes. Depending on the group size and the test performed, a larger number of false negatives may occur because of what is known as the *dilution effect*. This is because positive specimens may be mixed with negative ones in a group, weakening the positive signal and causing a diagnostic test to potentially miss the positive result. Potential solutions to this problem include changing test thresholds for a positive/negative response for the group test (Bilder et al., 2021a) and employing a larger specimen volume for a grouped specimen (Chung et al., 2022). When the dilution effect cannot be overcome, statistical approaches tend to focus on using a continuous test response (e.g., viral load) that is transformed into the positive/negative binary response. Alternatively, Warasi et al. (2017) develop an innovative method for estimation that works directly with these positive/negative binary responses by using a model for the sensitivity that is embedded within the main regression model. This formulation assumes that the sensitivity is the same for individual testing and for a group that has all positive specimens, so that only one additional parameter needs to be estimated.

The focus in this section is on tests that determine the positive/negative outcome for one disease at a time. There are also assays available for the simultaneous testing of multiple diseases, such as the BD Max assay which tests for chlamydia, gonorrhea, and trichomoniasis. Applying group testing with these *multiplex assays* can dramatically reduce the number of tests needed when compared to testing for each disease separately. The first research article that explored the combination of multiplex assays and group testing was Tebbs et al. (2013) for Dorfman-like algorithms, and this was followed by Hou et al. (2017) for more general hierarchical algorithms. Bilder et al. (2019) and Bilder et al. (2021b) subsequently

[37]Mean squared error (MSE) is a quantity that is used to measure how far a statistic lies on average from its population value. See p. 330 of Casella and Berger (2002).

propose informative group testing algorithms for hierarchical and array testing algorithms using multiplex assays, respectively.

6.8 Splines and additive models

We have focused on linear predictors to relate explanatory variables to probabilities or means of responses. Linear predictors assume that the relationship between the link function of a generalized linear model and each explanatory variable is essentially a straight line, holding other variables constant. We can add some flexibility to these shapes by creating transformations of explanatory variables, such as polynomials, but the resulting curves remain fairly rigid and limited in shape. Sometimes, we are unsure how explanatory variables should relate to the link function. We may only be willing to assume that the trend is "smooth" in the sense that small changes in an explanatory variable should result in small changes of the function, rather than huge jumps, and in gradual changes of slope, rather than in sharp angles. A solution for these situations is to use *splines* in the link function. Splines provide a flexible representation of an explanatory variable in a model that can adapt smoothly and efficiently to any shape suggested by data.

6.8.1 Regression with splines

Suppose we would like to include some type of smooth, nonlinear function of x in a regression model. For example, we may want to write a logistic regression model as $\text{logit}(\pi) = f(x)$, where $f(x)$ contains our regression parameters and nonlinear transformations of x. A polynomial function like $f(x) = \beta_0 + \beta_1 x + \beta_2 x^2 + \beta_3 x^3$ may be a good approximation over limited ranges of x but not necessarily over the entire range. Instead, we could use piecewise polynomial approximations over multiple ranges. For example, we could have separate cubic functions for $f(x)$ over two distinct ranges:

$$f(x) = \begin{cases} \delta_0 + \delta_1 x + \delta_2 x^2 + \delta_3 x^3 & \text{for } x \leq k \\ \gamma_0 + \gamma_1 x + \gamma_2 x^2 + \gamma_3 x^3 & \text{for } x > k \end{cases} \tag{6.34}$$

for some regression parameters $\delta_0, \delta_1, \delta_2, \delta_3, \gamma_0, \gamma_1, \gamma_2, \gamma_3$, and k.

Fitting these separate functions to neighboring sets of data would likely result in a "jump discontinuity," where the two functions fail to meet at the boundary. For example, there is nothing in Equation 6.34 that forces the two cubic functions to be equal when $x = k$ unless the regression parameters are carefully chosen. This violates our belief that the function should be smooth. Furthermore, the two functions may approach the boundary from very different directions, violating our belief that slopes should change gradually. As an alternative, we can use Equation 6.34 with additional constraints on the parameters that force the two functions to join together smoothly at the boundary. Splines are piecewise polynomials that can be constructed to serve this purpose. They are popular in regression because they are easy to formulate and fit, and can be made as flexible as desired.

Cubic splines are especially popular and widely used in practice. Constraints added to impose smoothness on Equation 6.34 result in a function that can be represented as

$$f(x) = \beta_0 + \beta_1 x + \beta_2 x^2 + \beta_3 x^3 + \beta_4 (x - k_1)^3_+ + \cdots + \beta_{3+D}(x - k_D)^3_+, \tag{6.35}$$

where $\beta_0, \ldots, \beta_{3+D}$ are regression parameters; $k_1, \ldots, k_D$ are locations of the boundaries on the x-axis, known as knots; and $(x - k_d)_+^3$ is the "truncated power function" defined as

$$(x - k_d)_+^3 = \begin{cases} (x - k_d)^3 & \text{for } x > k_d \\ 0 & \text{for } x \leq k_d \end{cases}$$

for $d = 1, \ldots, D$. This cubic spline can be referred to as a *truncated power spline*. Notice that this formulation forces the spline to be continuous and smooth at the knots. For example,

$$f(x) = \beta_0 + \beta_1 x + \beta_2 x^2 + \beta_3 x^3 \text{ for } x \leq k_1$$

and

$$f(x) = \beta_0 + \beta_1 x + \beta_2 x^2 + \beta_3 x^3 + \beta_4 (x - k_1)^3 \text{ for } k_1 < x \leq k_2,$$

which implies that the two sides of the function are equal at $x = k_1$ and have equal slope and curvature there as well. Thus, the function looks smooth when plotted. Exercise 1 examines the relationship between Equations 6.34 and 6.35.

The expression for $f(x)$ can be written more generally as

$$f(x) = \sum_{j=0}^{3+D} \beta_j h_j(x), \tag{6.36}$$

where $h_0(x) = 1, h_1(x) = x, \ldots,$ and $h_{3+D}(x) = (x - k_D)_+^3$ are called *basis functions*. Equation 6.35 can be equivalently written in terms of alternative basis functions that are more numerically stable to work with when very large values of x are cubed. These are referred to as *basis splines* or *B-splines*. The basis functions are found through a recursive process with the functions for a cubic spline dependent on those for a quadratic spline and so on. Readers interested in the mathematical development can find details in numerical analysis books, including Burden et al. (2015). Specific expressions for the recursion are available in these books and in papers such as Perperoglou et al. (2019). The `bs()` function from the `splines` package (within the default installation of R) automatically creates these basis functions for a given set of values for x .

Cubic splines can provide poor approximations to the true relationship for extreme values of x in its range. To improve approximations at these extreme values of x, it is sometimes beneficial to replace the cubic function with a linear function in the first and last regions. These types of splines are referred to as *natural splines*. A recursive process is again used to evaluate natural spline basis functions at a specific value of x. This process is carried out in R using the `ns()` function from the `splines` package.

Equation 6.36 is a linear combination of basis functions with $4+D$ parameters. Therefore, its parameters are estimated through standard maximum likelihood estimation methods as described in previous chapters. Inferences for specific terms of the spline are not of interest. Rather, one may use standard inference methods on the entire spline to evaluate the importance of an explanatory variable. How well the model fits the data depends on the number and location of the knots. As the number of knots increases, the fitted model becomes more flexible and more closely approximates the data. While this may sound good, it can result in a model that works well for the sample itself but captures spurious relationships that are not present for the population. When this occurs, it is referred to as *overfitting* the data. Overfitting can be seen in a plot by a model that has a large amount of variability, also known as "wiggles," in it. Thus, the number of knots needs to balance between modeling potentially complex trends and wiggling unnecessarily. As for the location of the knots, these can be chosen a variety of ways, including particular sample quantiles of x, equally spaced points in the range of x, or values using the knowledge of the underlying data.

Example: Placekicking (PlacekickSplines.R, Placekick.csv)

We re-examine logistic regression models for the placekicking data that was introduced in Chapter 2. We previously found there was marginal evidence that a quadratic term was appropriate for the model using distance of the placekick as the only explanatory variable. We explore this potential nonlinearity some more by fitting linear, quadratic, and cubic logistic regression models for the probability of a successful placekick.

```
> mod.fit1 <- glm(formula = good ~ distance, family =
   binomial(link = logit), data = placekick)
> mod.fit2 <- glm(formula = good ~ distance + I(distance^2),
   family = binomial(link = logit), data = placekick)
> mod.fit3 <- glm(formula = good ~ distance + I(distance^2) +
   I(distance^3), family = binomial(link = logit), data =
   placekick)

> library(package = car)
> Anova(mod.fit3)
Analysis of Deviance Table (Type II tests)

Response: good
              LR Chisq Df Pr(>Chisq)
distance        6.9928  1   0.008184 **
I(distance^2)   4.2994  1   0.038126 *
I(distance^3)   3.4909  1   0.061707 .
---
Signif. codes:  0 '***' 0.001 '**' 0.01 '*' 0.05 '.' 0.1 ' ' 1

> data.frame(model1 = AIC(mod.fit1), model2 = AIC(mod.fit2),
   model3 = AIC(mod.fit3))
   model1   model2   model3
1 779.745 776.9546 775.4637
> data.frame(model1 = BIC(mod.fit1), model2 = BIC(mod.fit2),
   model3 = BIC(mod.fit3))
    model1   model2   model3
1 790.2689 792.7404 796.5114
```

The LRT shows there is marginal evidence that a cubic term should be included in the model given that the linear and quadratic terms are included. The AIC (see Section 5.1.2) suggests that the third-order model (linear, quadratic, and cubic terms) is best because it has the smallest AIC. However, the BIC (see Section 5.1.2) suggests that the first-order model (linear term only) is best because its value is the smallest.

To begin exploring a nonlinear effect for distance using splines, we begin with a truncated power spline using four evenly spaced knots at 25, 35, 45, and 55 yards, which seem like reasonable locations given the range of our data (18 to 66 yards). We later show simpler ways to achieve the same fit. Each truncated power term can be created using the `ifelse()` function. For example, `ifelse(test = placekick$distance > 25, yes = (placekick$distance - 25)^3, no = 0)` can be used to create the first term. Instead of coding three other `ifelse()` function calls like this, we wrote a function named `tp()` to create all truncated power terms. This function has two arguments: `x` is the explanatory variable and `knots` contains knot locations.

```
> tp <- function(x, knots) {
    numb.knots <- length(knots)
    numb.x <- length(x)
    tp.mat <- matrix(data = NA, nrow = numb.x, ncol = numb.knots)
    for(d in 1:numb.knots) {
      tp.mat[1:numb.x,d] <- ifelse(test = x > knots[d], yes = (x
          - knots[d])^3, no = 0)
    }
    colnames(tp.mat) <- paste("tp", knots, sep = "")
    as.data.frame(tp.mat)
  }

> knots4 <- seq(from = 25, to = 55, by = 10)
> placekick2 <- cbind(placekick, tp(x = placekick$distance,
    knots = knots4))
> head(placekick2[, c("distance", "tp25", "tp35", "tp45",
    "tp55")])
  distance tp25 tp35 tp45 tp55
1       21    0    0    0    0
2       21    0    0    0    0
3       20    0    0    0    0
4       28   27    0    0    0
5       20    0    0    0    0
6       25    0    0    0    0
> tail(placekick2[, c("distance", "tp25", "tp35", "tp45",
    "tp55")])
     distance  tp25 tp35 tp45 tp55
1420       44  6859  729    0    0
1421       20     0    0    0    0
1422       55 27000 8000 1000    0
1423       20     0    0    0    0
1424       35  1000    0    0    0
1425       50 15625 3375  125    0
```

The for() function inside of tp() evaluates the D truncated power terms on the data and stores them within a matrix. We use the paste() and colnames() functions to create a unique column labeling and use the as.data.frame() function to transform the resulting matrix into a data frame.

The logistic regression model using the cubic spline with the truncated power terms is estimated with glm().

```
> mod.fit.tp <- glm(formula = good ~ distance + I(distance^2) +
    I(distance^3) + tp25 + tp35 + tp45 + tp55, family =
    binomial(link = logit), data = placekick2)
> summary(object = mod.fit.tp)

Call:
glm(formula = good ~ distance + I(distance^2) + I(distance^3) +
    tp25 + tp35 + tp45 + tp55, family = binomial(link = logit),
    data = placekick2)

Deviance Residuals:
```

```
     Min       1Q   Median       3Q      Max  
 -2.7981   0.2008   0.2008   0.5172   1.2319  

Coefficients:
                 Estimate Std. Error z value Pr(>|z|)
(Intercept)    -1.388e+02  1.658e+02  -0.837    0.402
distance        1.862e+01  2.076e+01   0.897    0.370
I(distance^2)  -7.970e-01  8.590e-01  -0.928    0.353
I(distance^3)   1.115e-02  1.176e-02   0.948    0.343
tp25           -1.278e-02  1.274e-02  -1.003    0.316
tp35            2.100e-03  1.851e-03   1.135    0.256
tp45           -2.562e-04  2.186e-03  -0.117    0.907
tp55           -1.050e-02  1.592e-02  -0.659    0.510

(Dispersion parameter for binomial family taken to be 1)

    Null deviance: 1013.43  on 1424  degrees of freedom
Residual deviance:  764.38  on 1417  degrees of freedom
AIC: 780.38

Number of Fisher Scoring iterations: 7

> anova(mod.fit3, mod.fit.tp, test = "Chisq")
Analysis of Deviance Table

Model 1: good ~ distance + I(distance^2) + I(distance^3)
Model 2: good ~ distance + I(distance^2) + I(distance^3) + tp25
   + tp35 + tp45 + tp55
  Resid. Df Resid. Dev Df Deviance Pr(>Chi)
1      1421     767.46                     
2      1417     764.38  4   3.0881   0.5432

> AIC(mod.fit.tp)
[1] 780.3756
```

The estimated model is

$$\begin{aligned}\text{logit}(\hat{\pi}) = &-138.8 + 18.62x - 0.7970x^2 + 0.0112x^3 - 0.0128(x-25)^3_+ \\ &+ 0.00021(x-35)^3_+ - 0.0003(x-45)^3_+ - 0.0105(x-55)^3_+,\end{aligned}$$

where x represents the distance of the placekick. Both the LRT and AIC comparing the cubic spline model to the third-order model do not provide sufficient evidence that the truncated power terms should be included (LRT p-value is large and AIC is larger for the cubic spline model).

While the truncated power spline demonstrates the creation of basis terms, splines are more typically formulated as B-splines or natural splines in practice. Below we fit a logistic regression model using these cubic splines for the distance of the placekick.

```
> library(package = splines)
> mod.fit.bs <- glm(formula = good ~ bs(x = distance, knots =
   knots4), family = binomial(link = logit), data = placekick)
> summary(object = mod.fit.bs)
```

```
Call:
glm(formula = good ~ bs(distance, knots = knots4), family =
    binomial(link = logit), data = placekick)

Deviance Residuals:
    Min       1Q   Median       3Q      Max
-2.7981   0.2008   0.2008   0.5172   1.2319

                             Estimate Std. Error z value
(Intercept)                     3.068      1.787   1.717
bs(distance, knots = knots4)1   1.771      2.975   0.595
bs(distance, knots = knots4)2  -1.669      1.551  -1.076
bs(distance, knots = knots4)3  -1.177      2.018  -0.583
bs(distance, knots = knots4)4  -2.622      1.766  -1.485
bs(distance, knots = knots4)5  -3.257      2.167  -1.503
bs(distance, knots = knots4)6  -2.132      3.910  -0.545
bs(distance, knots = knots4)7 -14.803     14.482  -1.022
                             Pr(>|z|)
(Intercept)                     0.086 .
bs(distance, knots = knots4)1   0.552
bs(distance, knots = knots4)2   0.282
bs(distance, knots = knots4)3   0.560
bs(distance, knots = knots4)4   0.138
bs(distance, knots = knots4)5   0.133
bs(distance, knots = knots4)6   0.586
bs(distance, knots = knots4)7   0.307
---
Signif. codes:  0 '***' 0.001 '**' 0.01 '*' 0.05 '.' 0.1 ' ' 1

(Dispersion parameter for binomial family taken to be 1)

    Null deviance: 1013.43  on 1424  degrees of freedom
Residual deviance:  764.38  on 1417  degrees of freedom
AIC: 780.38

Number of Fisher Scoring iterations: 7

> mod.fit.ns <- glm(formula = good ~ ns(distance, knots =
    knots4), family = binomial(link = logit), data = placekick)
> summary(object = mod.fit.ns)

Call:
glm(formula = good ~ ns(x = distance, knots = knots4), family =
    binomial(link = logit), data = placekick)

Deviance Residuals:
     Min       1Q   Median       3Q      Max
  -3.0498   0.1984   0.1984   0.5046   1.6038

Coefficients:
                             Estimate Std. Error z value
(Intercept)                    4.6411     0.4077  11.384
ns(distance, knots = knots4)1 -2.7761     0.4987  -5.567
ns(distance, knots = knots4)2 -4.3256     0.6787  -6.373
```

```
ns(distance, knots = knots4)3   -3.4717       0.7210   -4.815
ns(distance, knots = knots4)4   -8.9331       1.8930   -4.719
ns(distance, knots = knots4)5   -6.0099       2.3966   -2.508
                               Pr(>|z|)
(Intercept)                     < 2e-16 ***
ns(distance, knots = knots4)1 2.59e-08 ***
ns(distance, knots = knots4)2 1.85e-10 ***
ns(distance, knots = knots4)3 1.47e-06 ***
ns(distance, knots = knots4)4 2.37e-06 ***
ns(distance, knots = knots4)5   0.0122 *
---
Signif. codes:  0 '***' 0.001 '**' 0.01 '*' 0.05 '.' 0.1 ' ' 1

(Dispersion parameter for binomial family taken to be 1)

    Null deviance: 1013.43  on 1424  degrees of freedom
Residual deviance:  766.01  on 1419  degrees of freedom
AIC: 778.01

Number of Fisher Scoring iterations: 6

> data.frame(model.bs = AIC(mod.fit.bs), model.ns =
    AIC(mod.fit.ns))
  model.bs model.ns
1 780.3756 778.0142
> data.frame(model.bs = BIC(mod.fit.bs), model.ns =
    BIC(mod.fit.ns))
  model.bs model.ns
1  822.471 809.5858
```

The AICs for the B-spline and the natural spline do not indicate a better model fit than the third-order or second-order models. The first-order model's AIC is slightly larger indicating a preference for the spline models. However, the BIC is larger for these spline models compared to the first-order model. Note also that the AIC for the B-spline matches the AIC for the truncated power spline, confirming that they are equivalent formulations of the same models.

The `predict()` function computes the estimated probability of success in the same manner as in Section 2.2.4. Using this function, Figure 6.16 provides comparisons among the different models we have explored. Overall, we see the models generally do not present estimated probabilities that are meaningfully different when placekicks are between 18 and 50 yards. The B-spline greatly deviates from the other models for distances greater than 55 yards, including unrealistic features like an increase in probability up to 0.5580 for placekicks at 57 yards and a dramatic decrease in probability to approximately 0 for placekicks greater than 62 yards. This represents the potential poor performance at extreme values as described earlier for these types of splines. On the other hand, the natural spline is much more stable for extreme distances and is relatively close to the estimates produced by the third-order model.

We used four knots in our previous work, but this decision was somewhat arbitrary. Also, the knot location did not take into account differences in the numbers of trials in different regions, which may partly explain the erratic performance of B-splines at large distances where data are sparse. A different knot placement could provide a better model fit. Finding the best number of knots and their locations for any given data set is typically a matter of trial and error. Additional code in our program

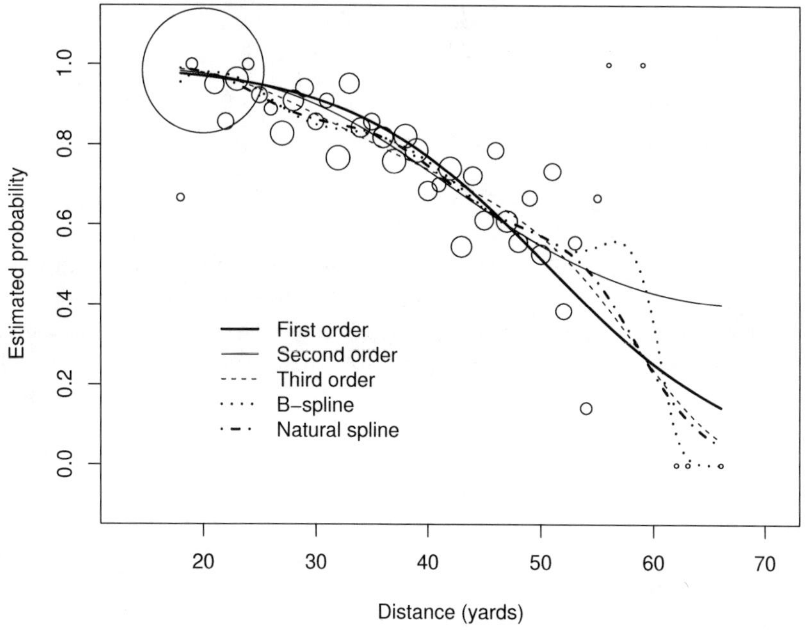

Figure 6.16: Spline models compared to first-, second-, and third-order models for the placekick data.

examines using the knot-selection process on natural splines (see Exercise 3 for B-splines). We continue using equally spaced knots, but also include knots chosen by quantiles. To use quantiles in the `bs()` and `ns()` functions, one specifies a desired degrees of freedom (number of regression parameters to estimate) for the spline in the `df` argument, which leads to a particular number of knots chosen depending on the type of spline used.[38] For example, using `ns(x = distance, df = 3)` in the `formula` argument for `glm()` leads to 3 parameters being estimated and knots placed at 20 (0.33 quantile of `distance`) and 29 yards (0.67 quantile of `distance`).[39] With this particular data set, we have an unusual complication in that there are a large number of placekicks attempted at 20 yards. This can lead to multiple quantiles being equal to this same distance and result in non-convergence of parameter estimates. For this reason, we focus on a relatively small number for the degrees of freedom.

Table 6.8 provides the AIC and BIC values for several different specifications for the knots. The AIC suggests the spline with 4 degrees of freedom is best, and the BIC suggests a spline with 3 degrees of freedom is best. Figure 6.17 includes the splines with specified knots in the top plot and the splines with specified degrees of freedom in the bottom plot, with first- and third-order models on both plots as references. In the top plot, we see the wiggling pattern that we described previously for the largest number of knots, indicating that there are probably more knots than needed. For the

[38]Excluding the intercept term, the basic and natural spline models estimate $D+3$ and $D+1$ parameters, respectively. These are the degrees of freedom for their respective models.

[39]Knot locations are provided in the `terms` component of the list returned by `glm()`.

Table 6.8: AIC and BIC values for models with a natural spline. For a third-order model (no knots), the AIC is 775.46 and the BIC is 796.51.

Specified knots	AIC	BIC	Specified degrees of freedom	AIC	BIC
25 to 55 by 3	790.80	859.20	5	771.98	803.55
25 to 55 by 6	781.68	823.77	4	771.95	798.26
25 to 55 by 10	778.01	809.59	3	775.16	796.21
25 to 55 by 15	777.47	803.78			
25 to 55 by 30	775.61	796.65			

smallest number of knots, we see the model is almost indistinguishable from a third-order model. In the bottom plot, the 5 degrees of freedom spline model leads to an almost perfect fit for the 18-yard placekicks, which we previously discussed was an unusual occurrence of only 2 successes out of 3 trials. While not shown, 4 degrees of freedom results in the same almost-perfect fit. It is not realistic to believe that probability of success decreases so drastically when the distance is decreased from 20 to 18 yards. The model is led to this shape because knots are placed at or around 20 yards, creating a region with very few observations at the end of the distance range. The apparent better fit as indicated by AIC may simply be because the model fits closely to an extreme point that other models miss, rather than a providing a better fit across all distances.[40] The spline model with 3 degrees of freedom (resulting in knots at 20 and 29 yards) has an AIC and BIC similar to that of the best model with specified knot locations, which are at 25 and 55 yards. We judge these two models as best among the spline-based models.

Overall, these types of cubic splines do not appear to improve our model in comparison to the third-order model. This statement should not dissuade readers from using cubic splines for other data! Rather, we chose this particular example to show how simpler models can work well and to provide further evidence that our past model choices are reasonable. For other data examples, splines can provide better fits to the data than a single polynomial.

Splines are frequently used as visual smoothing tools rather than as formal models for a phenomenon. Thus, numerical interpretation of the effect of an explanatory variable x on a response, by an odds ratio or a percentage change in the mean, may be less important than visualization by plotting the model as illustrated in our previous example. Also, one may have primary interest in one variable, like a treatment in a clinical trial, while accounting for the effects of one or more additional variables, which may be nonlinear. Thus, a logistic regression model with two explanatory variables, say x_1 and x_2, may be of interest, where x_1 is represented by a cubic spline and x_2 is represented as a linear term. One could use an odds ratio to interpret x_2 while holding x_1 constant.

When interpretation of x is of interest, the same types of calculations as shown in previous chapters can be used to estimate an odds ratio or percentage change in the mean. An important aspect here is that the odds ratio or percentage change in mean for a c-unit change in x may have a different value for each value of x. It can be helpful to go back to the basic principles from past chapters to better understand how calculations are performed.

[40]Such unexpected outcomes will not happen with all data sets, but this example shows that it can be useful to investigate knot locations when unexpected fits arise.

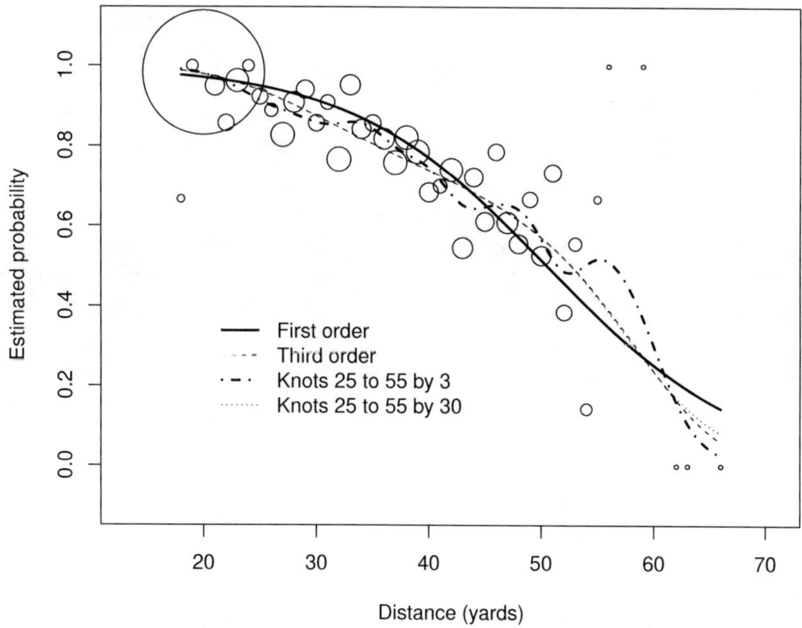

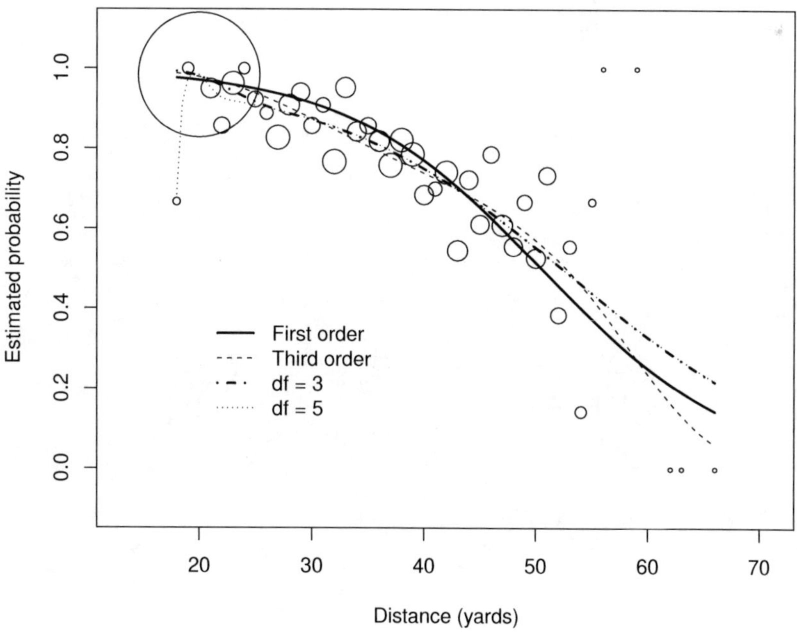

Figure 6.17: Natural spline models for a specified set of knots or a specified degrees of freedom when fit to the placekicking data. It is difficult to view the 5 degrees of freedom spline model for distances of 29 yards or larger because its fit is very close to the 3 degrees of freedom spline model.

In particular for a logistic regression model, the log-odds for the model are $\log(\pi/(1-\pi))$. The log of the odds ratio comparing two different values of x, say, a and b, with probabilities of success π_a and π_b, respectively, is

$$\log\left(\frac{\pi_a/(1-\pi_a)}{\pi_b/(1-\pi_b)}\right) = \log(\pi_a/(1-\pi_a)) - \log(\pi_b/(1-\pi_b))$$
$$= f(a) - f(b),$$

where $f(x)$ represents a linear combination of the basis functions and the regression parameters like in Equation 6.36. We can compute the difference term-by-term to write the expression as a new linear combination of regression parameters, and then use the processes described in Chapter 2 to compute profile LR confidence intervals using `mcprofile`. We can also obtain Wald confidence intervals using either manual calculation or the `emmeans` package. We illustrate the process in the next example.

Caution should be used when performing inferences on models where some variable selection has taken place (see Section 5.5). With splines, it is very common to try several different versions of the model, using different numbers of knots and/or knot placement. A preferred model may be one with the best value of AIC or BIC, or it may be chosen because it provides the most appealing smooth curve for the data. In either case, associated follow-up inferences may have larger type I error rates or lower confidence levels than the stated levels.

Example: Placekicking (PlacekickSplines.R, Placekick.csv)

We illustrate the odds ratio calculation process by focusing on a comparison of a 20-yard placekick to a 30-yard placekick using the natural spline model with knots at 25, 35, 45, and 55 yards. The estimated odds ratio can be found in multiple ways using the `predict()` function.

```
> prob <- predict(object = mod.fit.ns, newdata =
   data.frame(distance = c(20,30)), type = "response")
> prob
        1         2
0.9805018 0.8628003
> prob[1]/(1-prob[1]) / (prob[2]/(1-prob[2]))
      1
7.99643

> link <- predict(object = mod.fit.ns, newdata =
   data.frame(distance = c(20,30)), type = "link")
> link
       1        2
3.917741 1.838746
> exp(link[1] - link[2])
[1] 7.99643
```

The first set of code calculates the odds ratio directly with the estimated probabilities at distances of 20 and 30 yards. The second set of code exponentiates the difference of estimated link functions instead, using

$$\exp\left(\hat{f}(a) - \hat{f}(b)\right) = \exp\left(\sum_{j=0}^{5}\hat{\beta}_j h_j(a) - \sum_{j=0}^{5}\hat{\beta}_j h_j(b)\right)$$
$$= \exp\left(\sum_{j=0}^{5}\hat{\beta}_j (h_j(a) - h_j(b))\right), \quad (6.37)$$

where $a = 20$, $b = 30$, and $h_j(x)$ gives the value of the jth basis function at a distance x. Both sets of code calculate the estimated odds ratio to be 8.

The simplest way to calculate a confidence interval for OR is to use the `emmeans` package. The process is shown below. We also show how intervals for each π are easy to obtain from the `emmeans()` function.

```
> calc.est <- emmeans(object = mod.fit.ns, specs = ~ distance,
   params = "knots4", at = list(distance = c(20,30)), type =
   "response")
NOTE: Results may be misleading due to involvement in
   interactions

> summary(calc.est, level = 0.95)
 distance  prob      SE  df asymp.LCL asymp.UCL
       20 0.981 0.00476 Inf     0.969     0.988
       30 0.863 0.02643 Inf     0.802     0.907

Confidence level used: 0.95
Intervals are back-transformed from the logit scale

> confint(object = contrast(object = calc.est, method =
   "pairwise"), level = 0.95)
 contrast              odds.ratio   SE  df asymp.LCL asymp.UCL
 distance20 / distance30          8 2.81 Inf      4.02      15.9

Confidence level used: 0.95
Intervals are back-transformed from the log odds ratio scale
```

The `specs` argument uses `distance` because this is the explanatory variable that each basis function is computed on. A `params` argument must be included because `ns()` uses `knots4` to specify the locations of the knots. The note given can be ignored because there are no interactions. For splines fit with degrees of freedom rather than knot locations, the `params` argument is not needed and the note does not appear. We see that the estimated probabilities of success are the same as before. The 95% Wald confidence interval for 30-yard placekicks is (0.802, 0.907). The estimated odds of a successful placekick are 8 times as large for 20-yard placekicks as for 30-yard placekicks. The corresponding 95% Wald confidence interval is (4.02, 15.9).

The `mcprofile` package can be used to calculate profile LR confidence intervals. An important aspect of the process is to determine $h_j(x)$ for each corresponding regression parameter. These values are found using a method function for `predict()` that works with splines.

```
> distance.basis <- ns(x = placekick$distance, knots = knots4)
> func.val <- predict(object = distance.basis, newx = c(20,30))
> func.val[1:2,]
               1          2           3         4          5
[1,] 0.002489885 0.00000000 -0.06488768 0.1887642 -0.1238765
[2,] 0.427696078 0.02083333 -0.13810554 0.4017616 -0.2636560
```

Therefore, $h_1(20) = 0.0025$ and $h_5(30) = -0.2637$. The value of $h_0(x)$, the intercept term, is always 1 because `ns()` uses `intercept = FALSE` by default, and `glm()` includes the intercept when the model is fit. The `predict()` function also includes additional summary information in its output, so we exclude this here by specifying the two rows of interest.

We incorporate the basis function values into a matrix of regression parameter coefficients (contrast matrix, `K`). The remainder of the process to find the odds ratio estimate and corresponding confidence interval is identical to what was introduced in Chapter 2.

```
> # h(a) - h(b)
> K <- matrix(data = c(1, func.val[1,]) - c(1, func.val[2,]),
   nrow = 1, ncol = 6, byrow = TRUE)
     [,1]       [,2]        [,3]       [,4]       [,5]
[1,]    0 -0.4252062 -0.02083333 0.07321786 -0.2129974
          [,6]
[1,] 0.1397795
> # Estimated odds ratio
> exp(sum(mod.fit.ns$coefficients * K))
[1] 7.99643

> # mcprofile calculations
> linear.combo.OR <- mcprofile(object = mod.fit.ns, CM = K)
> ci.log.OR <- confint(object = linear.combo.OR, level = 0.95)
> exp(ci.log.OR)

   mcprofile - Confidence Intervals

level:           0.95
adjustment:      none

   Estimate lower upper
C1        8  4.05  16.2
```

The 95% confidence interval for the odds ratio is $4.05 < OR < 16.20$. The code can be modified for other odds ratios by changing the `newx` argument of `predict()` to the distances of interest. Because the spline is used in the model, odds ratios are not the same for 10-yard changes taken at other distances.

Smoothing splines

The previous splines require one to choose the number of knots and their positions. The optimal knots are rarely known in advance and difficult to ascertain even by looking at the data. *Smoothing splines* represent an alternative approach that does not require selecting knots. These splines seek a function of no predetermined shape that best fits the data, subject to a user-selected penalty on how much its slope is allowed to change across the range of the explanatory variable.

Specifically, a smoothing spline involving an explanatory variable x is a function $f(x)$ that maximizes

$$\log\left[L(f(x)|y_1,\ldots,y_n)\right] - \lambda J(f(x)), \tag{6.38}$$

where $\lambda J(f(x))$ represents the penalty. Firth's model-fitting approach from Section 2.2.7 and LASSO from Section 5.1.5 are also optimization problems of this type that involve penalties on a log-likelihood function. The function f may be any function satisfying certain minimal smoothness conditions. The $J(f(x))$ portion of Equation 6.38 is a measure of the total smoothness of the function, typically using $\int_a^b (\partial^2 f/\partial x^2)^2 dx$ for $a < x < b$. The λ coefficient on $J(f(x))$ is a nonnegative *tuning parameter* that needs to be determined. For $\lambda = 0$, the smoothness portion does not play a role, leading to a model fit that is highly variable (i.e., wiggly). For large λ, the model becomes more linear and potentially misses important nonlinear aspects of the relationship between x and the response variable. Methods to choose λ are examined in the next example.

Although the formulation of the smoothing spline is quite general, there is a mathematical simplification that results in a remarkable equivalence. A smoothing spline can be represented by a natural cubic spline that uses all unique values of the explanatory variable as knots (Perperoglou et al., 2019). This may seem impossible to fit, but the penalty coefficient λ shrinks the parameter estimates in this over-parameterized model to yield a curve that is as smooth or as wiggly as we need it to be. Most implementations of smoothing splines approximate this formulation by setting fewer than n knots in a grid across the range of x. The approximation gives very similar results unless the penalty is very small.

Because of the penalty term in the likelihood function, `glm()` and standard maximum likelihood estimation methods as described in previous chapters cannot be used. Penalized iteratively reweighted least squares, an extension of iteratively reweighted least squares introduced in Section 2.2.1, is used instead to find penalized MLEs. Wood (2017) provides details about the process, and computations can be completed with the `gam()` function of the `mgcv` package that is written by the same author. This package comes with the default installation of R, but users still need to invoke `library(package = mgcv)` prior to using its functions.

Example: Placekicking (PlacekickSplines.R, Placekick.csv)

We use a smoothing spline to continue examining the placekicking data. The syntax for `gam()` is based on the same syntax as for `glm()`. In the `formula` argument, we use `s(distance, bs = "cr")` to specify `distance` as the explanatory variable with a smoothing cubic spline. The `bs` argument is for basis, and the value `"cr"` is for cubic regression. This specification approximates the smoothing spline by fitting a natural spline with 9 knots and shrinking the parameters relative to their MLEs.

```
> library(package = mgcv)
> mod.fit.ps <- gam(formula = good ~ s(distance, bs = "cr"),
    family = binomial(link = "logit"), data = placekick)
> summary(mod.fit.ps)

Family: binomial
Link function: logit
```

```
Formula:
good ~ s(distance, bs = "cr")

Parametric coefficients:
            Estimate Std. Error z value Pr(>|z|)
(Intercept)    2.733      0.139   19.66   <2e-16 ***
---
Signif. codes:  0 '***' 0.001 '**' 0.01 '*' 0.05 '.' 0.1 ' ' 1

Approximate significance of smooth terms:
              edf Ref.df Chi.sq p-value
s(distance) 2.386  2.977  178.3  <2e-16 ***
---
Signif. codes:  0 '***' 0.001 '**' 0.01 '*' 0.05 '.' 0.1 ' ' 1

R-sq.(adj) =  0.196   Deviance explained = 24.1%
UBRE = -0.45541  Scale est. = 1         n = 1425
```

The output provides separate information regarding the spline and the usual intercept term included in a regression model. The spline portion (`smooth terms`) displays results for a test of the null hypothesis that the natural spline parameters are all zero. A Wald-like test statistic (details are provided in Wood (2013) and Wood (2017)) is approximated by a chi-square distribution with an *effective degrees of freedom* (EDF, `edf`) for the test.[41] Because the p-value is small, there is sufficient evidence for the importance of distance in this form. It is interesting to note the numerical value for the EDF is small. This provides a similar result as with our previous use of a natural spline; namely, a small value for the degrees of freedom (equivalently, the number of parameters needed for the spline) results in a good fit for these data.

Also included in the output is the unbiased risk estimator (UBRE, `UBRE`). This is a re-scaled AIC that includes the EDF for the usual degrees of freedom component of the statistic. The UBRE is used like an AIC to determine the most appropriate value for λ, although the estimate for λ is not given in the output.

The `predict()` function is used in the same manner as with `glm()` for finding estimates of the probability of success. With the help of this function, Figure 6.18 plots the model and provides comparisons to polynomial models. The code is found in our program. Overall, we see that there are no wiggles in the model. Once again, the model does not present estimated probabilities that are different in a practical context from non-spline models when placekicks are between 18 and 50 yards. Given that there are few observations past 55 yards, it is difficult to judge whether one model is preferred over another at these large distances.

The `emmeans` package provides a convenient way to form odds ratios with these models. The code and output below gives estimates for 10-yard increments of the distance for a placekick.

```
> library(package = emmeans)
```

[41]The EDF are calculated using a model representation based on a hat matrix and its relationship with degrees of freedom for a linear model (see Hastie and Tibshirani, 1990).

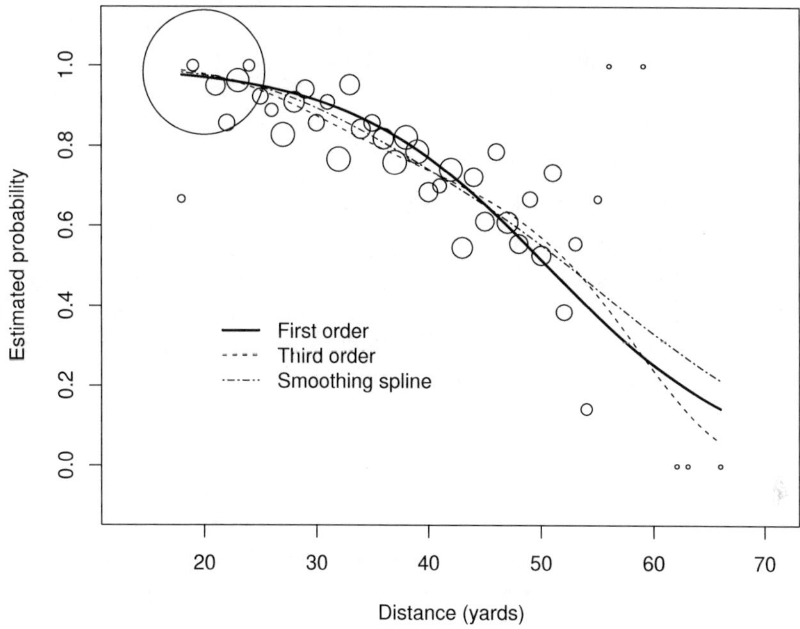

Figure 6.18: Smoothing spline model compared to first- and third-order models for the placekick data.

```
> calc.est <- emmeans(object = mod.fit.ps, specs = ~ distance,
   at = list(distance = c(20,30,40,50,60)))
> confint(object = contrast(object = calc.est, method =
   "consec", reverse = TRUE), level = 0.95, type = "response",
   adjust = "none")
 contrast              odds.ratio    SE   df lower.CL upper.CL
 distance20 / distance30     4.90 1.128 1422     3.12     7.70
 distance30 / distance40     2.96 0.502 1422     2.12     4.13
 distance40 / distance50     2.33 0.437 1422     1.61     3.36
 distance50 / distance60     2.51 0.994 1422     1.15     5.46

Confidence level used: 0.95
Intervals are back-transformed from the log odds ratio scale
```

These odds ratios change for each 10-yard increment because of the nonlinearity in the linear predictor of the spline. In Section 2.2.3 for a first-order logistic regression model, we found the estimated odds ratio to be 3.16 for every possible 10-yard increment. Therefore, from a practical prospective, it is questionable whether these differences in the estimated odds ratios are meaningful.

While splines are most commonly used for one explanatory variable at a time, multiple-variable splines can be created. For two-variable versions of these splines, this can be done by first creating basis functions for each variable. Then new terms are created by taking cross products between all of the basis functions for the variables, similar to how ordinary interaction terms are formed by multiplying explanatory variables. These basis-function products are referred to as *tensor products*. For example, if one variable uses 5 basis functions and another uses 7, there are 35 new tensor product terms in the interaction. This process can quickly result in very large models if more variables are available. As a result, multidimensional tensor-product splines are used much less often than one-variable splines. Exercise 5 examines the use of tensor products.

6.8.2 Generalized additive regression models

The previous subsection focused on models with one explanatory variable that was represented by a spline. For more than one explanatory variable, each can be represented as a linear term, a higher-order polynomial, a spline, or some other nonlinear format (e.g., a local regression model, see James et al., 2013). Therefore, we can consider an extension of the logistic regression model first given in Equation 2.3:

$$\log\left(\frac{\pi}{1-\pi}\right) = \beta_0 + f_1(x_1) + \cdots + f_p(x_p),$$

where $f_r(x_r)$ for $r = 1, \ldots, p$ is a transformation of x_r that may contain parameters to be estimated. Extensions of other models considered earlier in the textbook, such as a Poisson regression model, follow in a similar format. Models of this format are referred to as *generalized additive models* (GAMs). They have the same additive structure for the explanatory variables as a GLM but allow for more general representations of variables. These models were introduced by Hastie and Tibshirani (1986) and subsequently further explained in Hastie and Tibshirani (1990).

The `gam()` function of the `mgcv` package provides the tools to fit these models using smoothing spline representations for the explanatory variables. For example, `formula = Y ~ s(x1, bs = "cr") + s(x2, bs = "cr")` fits a separate smoothing spline for two separate explanatory variables named `x1` and `x2` with a response variable `Y`. A number of other spline representations are possible by specifying a different argument value for `bs`. Spline representations for interactions between explanatory variables are available with the `te()` and `ti()` functions to specify tensor product smoothers. Wood (2017) and the corresponding `mgcv` package documentation provide additional details.

Two alternative functions for GAMs are `gam()` from the `gam` package and `vgam()` from the `VGAM` package. The `gam` package is written by Trevor Hastie and corresponds to the original work in the area. This package provides smoothing spline and alternative nonlinear representations for explanatory variables in the model. Models are estimated using their backfitting algorithm, which iteratively cycles through estimating the parameters associated with each explanatory variable. This is required because an effort is made to find the optimal level of smoothing for each explanatory variable, which may depend on the shapes fit to the other variables. The `vgam()` function from the `VGAM` package provides the second main alternative with model-fitting options similar to `gam` and `mgcv`. We originally introduced this package in Chapter 3 to estimate models for multicategory responses. However, this package has a much greater functionality as described in its corresponding textbook Yee (2015). We provide examples of how to use these two packages in the program for the placekicking example of this section.

6.9 Exercises

Binary responses and testing error

1. Public health officials needed to know the SARS-CoV-2 infection prevalence during the COVID-19 pandemic. Many factors, including test availability, testing error, and asymptomatic infections, complicated the prevalence estimation process and led some governments to make decisions that were not fully informed. Ideally, public health officials would want to take a random sample of individuals, test each individual for the virus, and estimate the disease prevalence in a timely manner. Suppose public health officials were able to take this ideal approach. From a particular city, 1,000 individuals provided specimens and 50 tested positive for the virus. The assay used to test the specimens had $S_e = 0.95$ and $S_p = 0.995$, which was determined through a large clinical trial. Compute the prevalence estimate and the corresponding 95% Wald confidence interval.

2. Rogan and Gladen (1978) discuss a survey of 1 million individuals in the United States that was performed to estimate the prevalence of hypertension. For the individuals surveyed, 11.6% had a diastolic blood pressure above 95 mm Hg, which was considered to be hypertensive. Through an additional study, the sensitivity and specificity of the procedure for determining high blood pressure were estimated to be $S_e = 0.930$ and $S_p = 0.911$. Find the MLE for the overall true prevalence of hypertension (using > 95 mm Hg as the cutoff) and a corresponding 95% Wald confidence interval. Describe the effect that accounting for testing error has on understanding the prevalence of hypertension. What is a reason for the large or small effect?

3. Section 1.1.3 examined the true confidence levels for confidence intervals constructed for a probability of success π, where $n = 40$, $\alpha = 0.05$, and π ranged from 0.001 to 0.999 by 0.0005. Construct the same types of plots for a Wald confidence interval for $\tilde{\pi}$ ranging from 0.001 to 0.999 by 0.0005. Describe what happens to the true confidence levels as the amount of testing error changes and make comparisons to the case of no testing error. We recommend fixing one of S_e or S_p at 1 first while varying the other accuracy measure.

4. Derive $\hat{\tilde{\pi}}$ and its estimated variance using the process below.

   (a) Find the log of the likelihood function given in Equation 6.2.

   (b) Calculate the derivative of the log-likelihood function with respect to $\tilde{\pi}$.

   (c) Set the derivative found in (b) equal to 0 and solve for $\tilde{\pi}$. The resulting value is the MLE.

   (d) Calculate the second derivative of the log-likelihood function with respect to $\tilde{\pi}$.

   (e) Find the expected value of the result found in (d).

   (f) Invert the result from (e) and multiply by -1. This is $Var(\hat{\tilde{\pi}})$. Substitute $\hat{\pi}$ in for π to obtain $\widehat{Var}(\hat{\tilde{\pi}})$.

5. Consider again the example on prenatal infectious disease screening from Section 6.1.2.

   (a) Fit the logistic regression model that contains all of the explanatory variables available in the data set as linear terms. Make sure to include `marital.status` as a categorical variable.

(b) Perform appropriate LRTs on each of the explanatory variables, assuming the remaining variables are in the model.

(c) Determine the best fitting model for the data. Interpret the explanatory variables using odds ratios.

6. The methods described in Section 6.1.2 provide a convenient way to obtain a LR confidence interval for $\tilde{\pi}$ when there are no explanatory variables. Complete the following to obtain an interval for the hepatitis C prevalence among blood donors example from Section 6.1.1.

   (a) Construct a one-row data frame with variables named `positive` and `blood.donors` to represent the 42 out of 1,875 blood donors that tested positive for hepatitis C.

   (b) Fit the model $\text{logit}(\tilde{\pi}) = \beta_0$ using `glm()` with the `my.link()` function given in Section 6.1.2. Make sure to include `weights = blood.donors` within `glm()` because of the binomial form of the data.

   (c) Calculate the 95% LR confidence interval for β_0 using the `confint()` function.

   (d) Find the 95% LR confidence interval for $\tilde{\pi}$ using the relationship between $\tilde{\pi}$ and β_0. Compare this interval to the Wald interval calculated in Section 6.1.1.

   (e) Discuss why the calculation of a LR confidence interval for $\tilde{\pi}$ leads to limits between 0 and 1.

7. The purpose of this exercise is to examine what happens to $\widehat{Var}(\hat{\beta}_1)$ as the amount of testing error increases. For this exercise, perform all of your calculations using the prenatal infectious disease screening data from Section 6.1.2, where age is used as the explanatory variable to estimate the probability of HIV infection in a logistic regression model.

   (a) Fit the model with $S_e = S_p = 1$ using `optim()` or `glm()` with the `my.link()` function given in Section 6.1.2. Confirm that you obtain the same answer as when using `glm()` with `family = binomial(link = logit)`.

   (b) While holding S_p fixed at 1, examine what happens to $\widehat{Var}(\hat{\beta}_1)$ as S_e changes from 0.94 to 1 by 0.01.

   (c) While holding S_e fixed at 1, examine what happens to $\widehat{Var}(\hat{\beta}_1)$ as S_p changes from 0.94 to 1 by 0.01.

   (d) Compare your results from (b) and (c) and suggest reasons the variance did or did not substantially change. Discuss any convergence and/or calculation issues that arose.

8. Using the prenatal infectious disease screening data from Section 6.1.2 with age as the explanatory variable and HIV as the response variable, fit the logistic regression model using the MCSIMEX method as described below.

   (a) Create a new response variable in `set1` that transforms `hiv` into a variable with the class `factor`. Call this new variable `hiv.factor`. This change of class is needed to use the `mcsimex()` function of the `simex` package.

   (b) Fit a logistic regression model using `glm()` but do not account for testing error. Store the results from fitting the model in an object named `mod.fit.naive`. Include the argument value `x = TRUE` in `glm()` so that the **X** matrix as described in Section 2.2.1 is in the returned object (this will be needed for `mcsimex()` in part (d)).

(c) Construct a contingency table that contains S_p, $1 - S_p$, $1 - S_e$, and S_e.

```
test.err <- array(data = c(0.98, 1 - 0.98, 1 - 0.98, 0.98),
    dim = c(2,2), dimnames = list(obs.levels =
    levels(set1$hiv.factor), true.levels =
    levels(set1$hiv.factor)))
```

(d) Fit the logistic regression model using the MCSIMEX method with the `mcsimex()` function.

```
library(package = simex)
set.seed(1209)
mod.fit.mcsimex <- mcsimex(model = mod.fit.naive, mc.matrix
    = test.err, SIMEXvariable = "hiv.factor")
summary(mod.fit.mcsimex)
```

This implementation of the `mcsimex()` function uses Monte Carlo simulation to approximate the effects of testing error on the regression parameter estimators. Both large-sample (asymptotic) and jackknife-based[42] estimates of standard errors are produced. Compare these estimates and standard errors to those obtained in Section 6.1.2.

Exact inference

1. The hypergeometric distribution for a 2×2 table (see Table 6.2) can be found by starting from two independent binomial distributions and applying the following steps.

   (a) Write out the PMFs for W_1 and W_2 with binomial distributions involving n_1 and n_2 trials, respectively, and a common probability of success π. Note that n_1 and n_2 are known constants.

   (b) Find the joint PMF for W_1 and W_2 by multiplying the two PMFs from (a).

   (c) Discuss why the PMF for $W_+ = W_1 + W_2$ is binomial with $n_1 + n_2$ trials and probability of success π.

   (d) Calculate the conditional PMF for W_1 given W_+ as $P(W_1 = w_1|W_+ = w_+) = P(W_1 = w_1, W_2 = w_2)/P(W_+ = w_+)$ for $w_1 = 0, \ldots, w_+$ subject to $w_1 \leq n_1$ and $w_+ - w_1 \leq n_2$. Relate the PMF found to a hypergeometric distribution.

2. Perform a permutation test for independence with the fiber-enriched-crackers data set as in Section 6.2.2, but now use the LRT statistic $-2\log(\Lambda)$ rather than the Pearson statistic X^2. Compare your p-value to that obtained when using X^2. Is a χ^2_9 approximation appropriate for $-2\log(\Lambda)$?

3. Continue the analysis of the Larry Bird data set introduced in Section 1.2.

   (a) Perform Fisher's exact test.

[42]The jackknife is a procedure similar to the bootstrap in that resampling is used to take samples from the original sample. The difference between the two is that the jackknife uses a "leave-one-out" resampling approach so that n unique resamples are taken where each resample has the remaining $n - 1$ observations within it.

(b) Perform a permutation test for independence with $B = 10{,}000$ permutations using (i) `chisq.test()` and (ii) code that uses the `for()` function to find each permutation one-by-one. Set a seed number of 1938 at the start of each procedure using `set.seed()`. Compare your p-values to that obtained in part (a). Why are the p-values different?

(c) Is a χ^2_1 approximation for the Pearson statistic appropriate? Why or why not?

(d) Exercise 10 of Chapter 2 examines a logistic regression model fit to the data in the `bird` data frame. Fit this same model using exact logistic regression. Use the `elrm()` function with the additional argument of `r = 2`.[43] Perform a hypothesis test of $H_0 : \beta_1 = 0$ vs. $H_a : \beta_1 \neq 0$. Compare these results to those obtained for the model in Chapter 2 and with the inferences in parts (a) and (b).

4. Exercise 21 of Chapter 1 examined results from an HIV vaccine clinical trial. Using the data given in Table 1.9, apply the exact inference methods discussed in Sections 6.2.1 and 6.2.2 to determine if there is evidence that the vaccine prevents HIV infection.

5. Henao-Restrepo et al. (2017) made worldwide news with a new vaccine that was shown in a clinical trial to be effective in preventing Ebola virus infection. To obtain a sufficient sample size of people at risk of Ebola infection, the clinical trial sampled "clusters" of individuals who might have been exposed to it through a friend or relative. These clusters were put into two groups: individuals in group A clusters were immediately given the vaccine, while those in group B clusters were given the vaccine after 21 days. For the 70 group A clusters, no individuals within them became infected during the clinical trial. For the 116 group B clusters, 15 of them had at least one individual become infected.

(a) Construct a contingency table summarizing the data.

(b) Table 3 (column #1) in the paper gives a p-value of 0.0012 for a two-sided Fisher's exact test to test for independence between group and Ebola outcome. Reproduce this result.

(c) Perform a permutation test with X^2 for a two-sided test of independence. How well does the test statistic follow a χ^2_1 distribution?

(d) The hypothesis test performed in the paper was a two-sided test. Provide justification for why a one-sided test may be more appropriate. State what the hypotheses would be and perform the test using an exact inference method.

(e) A separate *New York Times* article discussing the results of the paper is available at `http://nyti.ms/2i8Ac4V`. The headline for this article is "New Ebola Vaccine Gives 100 Percent Protection." Discuss why the headline could potentially be misleading in the context of individuals worldwide.

6. Exercise 9 of Chapter 3 examined the relationship between classifications for dural arteriovenous fistulas (DAVFs) and symptoms. Complete the following parts using this data.

(a) When the computations for a Pearson chi-square test for independence are performed with `chisq.test()`, the output states `Chi-squared approximation may be incorrect`. Why is this message printed?

[43] The `r` argument controls how the MCMC sampling is performed. This argument needs to be less than or equal to the number of rows in the data frame, which is 2 here. The default value for `r` of 4 can be used with most other data frames that are larger in size.

(b) Perform a permutation test with X^2 for a two-sided test of independence.

(c) Should one be concerned about the χ^2_{36} distribution approximation that would be used with the Pearson chi-square test for independence? Use the distribution of the X^{2*} values needed for part (b) to answer this question.

7. Section 5.2 of Mehta and Patel (1995) examine a study involving HIV infection rates among infants when the mother was already infected during pregnancy. CD4 and CD8 blood counts were available for infants at 6 months of age along with their eventual HIV status. The data in EVP form is available in the file Ex5.2.csv (`w` is the number of HIV infections out of `n` individuals). Note that the CD4 and CD8 counts are coded as ordinal variables with levels of 0, 1, and 2. Mehta and Patel (1995) indicate that the researchers for this study considered these variables to be nominal with three levels each, so we do the same for this exercise.

   (a) Try to fit a logistic regression model using `glm()` with CD4 and CD8 counts as explanatory variables (without an interaction) to estimate the probability of HIV infection. Does R indicate the model has converged? If so, use a stricter convergence criteria and examine if the parameter estimates change.

   (b) Fit the same model as in (a) using exact logistic regression. Test the importance of CD4 and CD8 individually. The `factor()` function can be used with the `formula` argument in `elrm()` to create the needed indicator variables.

   (c) Provide the estimates of the exact distributions used in (b) and show how the p-values are calculated from these distributions.

8. Foxman et al. (1997) examine the relationship between type of contraceptive used and whether a woman had experienced a urinary tract infection (UTI). Data from the study were subsequently used in an advertisement for the LogXact software package and are given in the UTI.csv file. Explanatory variables representing contraception type are oral pill (`Oral`), condom (`Cond`), lubricated condom (`Lub`), spermicide (`Sperm`), and diaphragm (`Dia`), where a value of 1 indicates use and a value of 0 indicates non-use. An additional explanatory variable age is dichotomized into two groups based on less than 24 years of age (`Age24` = 1) or greater than or equal to 24 years of age (`Age24` = 0). The binomial response is the number of individuals infected (`Infected`) out of the number of individuals with the same EVP (`n`). Examine the data as described below without using interactions among the explanatory variables.

   (a) Attempt to fit a logistic regression model using `glm()` to estimate the probability of a UTI and show convergence is not obtained. Determine which explanatory variable leads to the non-convergence and describe why non-convergence occurs.

   (b) Fit the logistic regression model using `elrm()`. Perform hypothesis tests to determine the importance for each explanatory variable, given that the other variables are in the model.

   (c) Compare your results in (b) to fitting the same model using the modified likelihood methods of Firth (1993).

9. Using exact logistic regression methods, analyze the data on hepatitis C prevalence among healthcare workers from Exercise 19 of Chapter 2. Compare your results to those obtained from using MLEs.

10. Let $T_j = \sum_{i=1}^{n} Y_i x_{ij}$ denote the sufficient statistic for β_j, $j = 0, \ldots, p$. Using this additional notation, complete the following that examines two results from Section 6.2.3.

(a) Equation 6.5 expresses the conditional probability of $Y_1, \ldots, Y_n$ given I. Denote the p elements of I by $(I_0, I_1, \ldots, I_{p-1})$. Show how Equation 6.5 is found by first noting that

$$P(Y_1 = y_1, \ldots, Y_n = y_n, T_0 = I_0, \ldots, T_{p-1} = I_{p-1}) = \frac{\exp(\beta_0 I_0 + \cdots + \beta_{p-1} I_{p-1} + \beta_p \sum_{i=1}^n y_i x_{ip})}{\prod_{i=1}^n [1 + \exp(\beta_0 + \beta_1 x_{i1} + \cdots + \beta_p x_{ip})]}$$

and

$$P(T_0 = I_0, \ldots, T_{p-1} = I_{p-1}) = \sum_R \frac{\exp(\beta_0 I_0 + \cdots + \beta_{p-1} I_{p-1} + \beta_p \sum_{i=1}^n y_i^* x_{ip})}{\prod_{i=1}^n [1 + \exp(\beta_0 + \beta_1 x_{i1} + \cdots + \beta_p x_{ip})]}.$$

(b) Equation 6.6 expresses the conditional probability of $T_p = t$ given I. Discuss the role that $c(t)$ (and then $c(t_u)$, $u = 1, \ldots, U$) plays in the equation to justify the conditional probability.

11. A *mid p-value* is commonly used to lessen the conservativeness of a hypothesis-testing method involving a test statistic with a discrete probability distribution. A p-value is normally calculated as the sum of probabilities for all possible values of the test statistic that are *at least as extreme* as the observed value. The conservative nature of the test arises because in many cases there is no combination of test-statistic values whose probability sums to exactly α. A mid p-value is calculated the same way, except that only half of the probability for the observed value is added to the sum. Thus, a mid p-value is always smaller than an exact p-value. For example, the two-sided test mid p-value for the lymphocytic infiltration variable in the cancer-free example of Section 6.2.3, where the test statistic is $t = 19$, can be calculated as $0.5\hat{P}(T = 19|\, I) + \hat{P}(T \geq 24|\, I) = 0.0437$. Using the approximate exact distributions of $\sum_{i=1}^n Y_i x_{i2}$ and $\sum_{i=1}^n Y_i x_{i3}$ for the cancer-free example, calculate the mid p-values corresponding to tests of significance for gender and for any osteoid pathology. Compare these p-values to those obtained previously for these variables.

12. This exercise continues examining the data used for the complete separation example of Section 2.2.7.

    (a) Add a variable named `n` to `set1` and `set2` that contains a 1 for each observation. This variable represents the number of trials needed for `elrm()` when representing the EVP form of the data.

    (b) Use exact logistic regression to fit the model to the `set1` data frame. Compare the estimate of $\hat{\beta}_1$ and the corresponding p-value for $H_0 : \beta_1 = 0$ vs. $H_a : \beta_1 \neq 0$ to what was obtained from the modified likelihood methods of Firth (1993).

    (c) Complete (b) again but now use the `set2` data frame. Include a comparison to what was obtained using regular likelihood methods.

Categorical data analysis in complex survey designs

1. The `survey` package contains a data frame named `nhanes` that is part of the same large survey that provided the tobacco/respiratory-symptom data for the examples in Section 6.3. The version in the package does not have replicate weights. Instead, it contains variables for *primary sampling units* (PSUs; i.e., clusters), strata, and sampling weights. The `help(nhanes)` page describes the variables in more detail. It also provides the code for creating a `survey.design`-class object from these variables

(also available from `example(nhanes)`), so that the analysis functions in the package can be used.

(a) According to the documentation for NHANES (`http://www.cdc.gov/nchs/data/series/sr_02/sr02_161.pdf`), the PSUs are nested within strata. This means, for example, that PSU #1 in one stratum represents a different cluster from PSU #1 in a different stratum. Before starting the analysis, determine (i) how many strata there are and (ii) how many PSUs there are in the data. Based on these numbers, how many degrees of freedom should there be initially for tests and confidence intervals (i.e., what is κ)?

(b) Create a `survey.design` object using `svydesign()` as shown in the help for `nhanes`. Note that there are four categorical variables in the data set: a binary variable for high cholesterol (`HI_CHOL`, the response variable), and demographic variables for race (`race`), age (`agecat`), and gender (`RIAGENDR`). For each variable, use the survey design to obtain estimates of the proportions of the population in each class of the demographic variables, along with confidence intervals for these proportions.

(c) Fit a single logistic-regression model using `HI_CHOL` as the response variable and the three demographic variables as linear terms.

   i. Report the estimated model.

   ii. Test the effects of each explanatory variable using $\alpha = 0.05$. Draw conclusions from these tests.

   iii. Find a confidence interval for the odds ratio for each pair of age categories. Express all odds ratios so that the older age group is in the numerator. Interpret the results.

2. Refer to the example starting on p. 474. Perform a similar analysis of the associations among the four respiratory symptoms. Which symptoms appear to be related to one another? Use odds ratios to describe the associations.

3. As discussed in Section 6.3.5, one of the advantages of analyzing contingency tables with loglinear models is that it opens up a wider array of tests under the model-comparison framework. Tables with more variables, or with variables having ordinal or other structures, present no difficulty when analyzed as a loglinear model. The `anova()` function provides a Rao-Scott-type test for any pair of nested loglinear models. Thus, a wider class of hypothesis tests can be considered than merely tests for independence. For example, a linear-by-linear association model for two ordinal variables X and Y can be fit by using categorical versions of X and Y as main effects and `numeric` versions of the variables in an interaction.[44] The test of goodness-of-fit for the linear-by-linear association compares this model with the model containing the interaction in nominal form.

   Refer to the examples from that section and consider a new question: Does the number of respiratory symptoms experienced increase with the number of tobacco products used? This can be addressed using various loglinear models. Recall that

[44]For computational reasons in `svyloglin()`, this model is obtained by first fitting the independence model to the two factor variables (i.e., with no interaction) and then using `update()`, where the first argument is the previous model-fit object and the second argument is `formula = ~ . + as.numeric(X):as.numeric(Y)`, with the actual variable names used as `X` and `Y`.

there are four binary variables representing different respiratory symptoms and five representing tobacco products used. Summing each of these sets of variables leads to numerical variables `num.resp` $(0, ..., 4)$ and `num.tob` $(0, ..., 5)$. Three models are to be fit for this exercise: one with no interaction (i.e., independence of number of respiratory symptoms and number of tobacco products used), one with a numerical interaction (i.e., linear-by-linear association model; see Section 4.2.6), and one with a categorical interaction.

(a) Make a 6×5 table of the weighted counts and also proportions where the denominator in each row is that row's total weighted count. The following code achieves this.

```
wt.tab <- svytable(formula = ~ num.tob + num.resp, design =
    jdesign2)
sweep(x = wt.tab, MARGIN = 1, STAT = rowSums(wt.tab), FUN =
    "/")
```

Are any trends apparent? Use hypothesis tests and the previously described models to answer this question.

(b) Notice that the table has a zero count in it. This will cause problems for model fitting. Because the counts in the last two levels of `num.tob` are much smaller than those in other columns, create a new variable (`num.tob2`) that combines the 4 and 5 levels of `num.tob` (e.g., by using the minimum of `num_tob` and 4 for each observation). Print the new table.

(c) Fit three models to this table: independence, linear-by-linear association, and nominal association (saturated). Perform the tests of (i) no interaction versus linear-by-linear association and (ii) linear-by-linear association versus nominal association (the latter two should be constructed using the `update()` function based on the first model). Report the hypotheses, test statistics, p-values, and results of these tests. Interpret the two results.

(d) In the linear-by-linear association model, estimate the association with a point estimate and a 95% confidence interval. Interpret the results.

(e) In the nominal-association model:

   i. Estimate the odds ratio for each pair of consecutive rows and columns. That is, estimate $OR_{01,01}, OR_{01,12}, OR_{01,23}, OR_{01,34}, \ldots, OR_{34,23}, OR_{34,34}$, where in each symbol the first pair of numbers stands for the levels of `num.tob2` and the second pair stands for the levels of `num.resp`. Find confidence intervals in each case. Warning: finding the correct contrast coefficients for `svycontrast()` can be a challenge because of the parameterization used in `svyloglin()`. See p. 478 and the note in the SurveySmokeLoglinear.R program on the book's website, and use these to work out what the coefficients must be. Compare results to estimates obtained directly from `wt.tab`.

   ii. Do the confidence intervals suggest that the odds ratios are all similar, or are some clearly different? In particular, is there an overlap to all of the intervals? How does this finding relate to the tests performed in part (c)?

4. The `api` object in `survey` contains data from a census of 6,194 K-12 schools in California. The data were collected as part of an assessment of a now-defunct accountability measure known as the Academic Performance Index. The manual for the `survey`

package describes several subsets of this population that were drawn using different designs. This exercise uses a subset drawn using a 1-stage cluster sample, `apiclus1`, to explore the relationship between certain properties of the schools and whether the schools met certain achievement targets.

The response variable for this analysis is `comp.imp`, a binary measure of whether the school met a "comparable improvement" target, with values of `Yes` and `No`. The explanatory variables are

- `meals`: The percentage of children in the school who are eligible for subsidized meals.
- `ell`: The percentage of English language learners in the school.
- `full`: The percentage of teachers at the school who are fully qualified, as opposed to appointed on emergency status.

The manual for `survey` provides the following code that reads this data set into an object that `survey` functions can access.

```
library(package = survey)
data(api)
dclus1 <- svydesign(ids = ~dnum, weights = ~pw, data = apiclus1,
    fpc = ~fpc)
```

The `ids` argument lists the variable(s) used to select clusters; here it was the school district number, `dnum`. The `weights` argument refers to weights that were computed separately and supplied with the data. The `fpc` argument lists the "finite population correction", also computed separately and supplied with the data.

(a) Use the code above to create the required `svydesign`-class object, `dclus1`.

(b) Use `svymean()` to compute the mean values for each of the explanatory variables. Report the results. This will be useful later when we want to compute odds ratios representing meaningful differences in the levels of these variables.

(c) Fit a logistic regression model for the probability that a school meets the target against the linear effects of the three explanatory variables. Report the estimated model for the probability of success.

(d) Fit three reduced models, each dropping a different one of the explanatory variables. Use the `anova()` method function to test whether each of the parameters for these variables might be zero. Report the hypotheses, test statistics, p-values, and conclusions for the tests.

(e) Use `emmeans` with the model from part (c) to estimate the odds ratios and to compute the corresponding confidence intervals for an increase of 10% in each of the three variables. For the `at` argument in the `emmeans()` for each variable, round each variable's mean from part (a) down or up to the nearest 10 (e.g., for 17.5 use 10 and 20). Report the estimated ORs and their confidence intervals.

(f) In a short paragraph, draw conclusions about the effects of these variables on the probability that a school meets its target.

Table 6.9: Marginal table of swine waste storage and sources of veterinary information responses from 279 Kansas farmers.

	Consultant	Veterinarian	Extension service	Magazines	Feed companies
Lagoon	34	54	50	63	41
Pit	17	33	34	43	37
Natural drainage	6	23	30	49	34
Holding tank	1	4	4	6	2

"Choose-all-that-apply" data

1. For $I = 1$ and $J = 1$, show that Equation 6.13 is the Pearson statistic from Section 1.2.3.

2. Suppose the Pearson chi-square statistic,
$$\sum_{i=1}^{I}\sum_{j=1}^{J}\frac{(m_{ij} - m_{i+}m_{+j}/n)^2}{m_{i+}m_{+j}/n},$$
is applied to Table 6.6. Notice that this statistic is essentially one of four components in X_S^2. What would happen to this statistic's value if Table 6.6 instead summarized all of the negative responses rather than positive responses to both MRCVs? Why is this result undesirable?

3. Bilder and Loughin (2004) examine the Kansas farmer survey described in Section 6.4 as well, but they focus on the question about swine waste storage methods along with another question regarding the farmer's sources of veterinary information. Table 6.9 displays the positive responses to both questions. Note that source of veterinary information is an MRCV because farmers could choose up to five responses. The full data set is available in the data file BL2004.csv available on the textbook's website.

   (a) Construct a table like Table 6.9 for the data. There are examples in the research literature where a Pearson chi-square test for independence has been directly applied to a table like this. Discuss why this is a poor approach.

   (b) Construct all of the 20 possible 2×2 contingency tables for this example.

   (c) Compute $X_{S,ij}^2$ for $i = 1, \ldots, 4$ and $j = 1, \ldots, 5$ and sum them to form X_S^2. What would a naive χ_{IJ}^2 approximation to X_S^2 conclude in a test for SPMI?

   (d) Apply the Bonferroni adjustment, bootstrap, and second-order Rao-Scott correction to test for SPMI. Do all testing procedures reach the same conclusion?

   (e) In practice, only one of the procedures in (d) should be applied. Discuss the advantages and disadvantages for each of them in general practice.

4. Continuing Exercise 3, estimate the SPMI, homogeneous association, W-main effects, Y-main effects, W and Y-main effects, and saturated models. Which model is best? Interpret the fit of this model using odds ratios.

5. MRCVs are found in areas outside of surveys. For example, wildlife researchers are often interested in determining which types of foods are present in the diet of a particular species of animal. Scats are examined for their content, and the corresponding responses are summarized by MRCVs because each scat may contain more than one type of food.

Riemer et al. (2011) examine the scats of Steller sea lions along the Oregon and California coasts to better understand this species' diet. The `sealion` data frame in the `MRCV` package contains a portion of the data collected at the mouth of the Columbia River in August of 2004 and 2007. For each scat sample, the presence (1) or absence (0) of 11 different prey types are given; thus, prey type is the MRCV. With these data, we would like to determine if there are differences in diet across the two years. Because year is an SRCV, we are interested in testing for MMI as described at the end of Section 6.4.2.

(a) Construct 11 2×2 contingency tables to summarize the responses by year (`Date`) and prey presence or absence.

(b) For each 2×2 contingency table, compute a Pearson chi-square test statistic (say, X_j^2 for $j = 1, \ldots, 11$) and corresponding p-value to test for independence as given in Section 3.2.3.

(c) Perform an overall test for MMI by using a Bonferroni adjustment with the p-values in (b). State a conclusion about the comparison of diets.

(d) The `MI.test()` function can also perform the test for MMI. The same syntax as used for the SPMI test is also used for the MMI test; however, `I = 1` now needs to be set because `Date` is a SRCV. Use this function to perform the Bonferroni, bootstrap, and Rao-Scott approaches to test for MMI.

6. Wright (2010) examines two different ways to analyze data with one MRCV and one or more SRCVs. Read this paper and comment on the data analysis approaches presented.

7. Chankov (2017) re-examines data from a survey relating car attributes to car models that appeared in Meulders and De Bruecker (2017). The first example in Chankov (2017) focuses on 8 attributes (agile, economical, popular, practical, reliable, safe, sustainable, and versatile) for the Volkswagen Golf and the Ford Focus C-Max. The CarAttributes.csv file contains the responses from 78 individuals where respondents could choose all attributes that apply to each of the cars. Chankov (2017) uses these data to test for SPMI between the Golf attributes and the Focus C-Max attributes and illustrates the use of his own `multiple.chiRS2.test()` function for the test.

(a) Chankov (2017) displays a marginal table of the joint positive responses for the car models. Reproduce this table.

(b) Construct the item response table for the data.

(c) Perform a test for SPMI using the Bonferroni and bootstrap methods.

(d) Attempt to perform a test for SPMI using `MI.test()` and a second-order Rao-Scott correction. What error message is produced?

(e) Perform a test for SPMI using `multiple.chiRS2.test()` and a second-order Rao-Scott correction. This function is available in the mi.test_improvements.R program at `https://github.com/e-chankov/MItest_improvements` and requires the installation of the `data.table` and `checkmate` packages. Below is our code to run the function.

```
# set1 contains the data
# data.table() puts the data into the correct format for the
   function
```

```
multiple.chiRS2.test(data = data.table(set1), I = 8, J = 8,
    add.constant = 0)
```

(f) Compare the results among the SPMI testing methods.

8. Gregory et al. (2020) examine how researchers obtain and use secondary sources of data. Appendix A of this open access paper lists a number of choose-all-that-apply questions that were asked in a survey of researchers. The file SecondSource.csv contains a portion of the resulting data. The first 31 columns in the file represent an MRCV for researcher discipline that includes items like agriculture, chemistry, and math. The last five columns represent an MRCV for data type needed with items observational/empirical, experimental, derived/compiled, simulation, and other.

   (a) Figure 6 and Table B1 in the paper provide results from a test for SPMI between the two MRCVs using a Bonferroni adjustment. Describe what is presented in the figure and the table.

   (b) What is a downside to using the Bonferroni adjustment for the SPMI test here? How do the authors justify their use of it?

   (c) Perform the SPMI test between the two MRCVs using the Bonferroni adjustment and compare your results to the paper. Reproduce the p-values given in Table B1.

   (d) Use the bootstrap approximation and the second-order Rao-Scott correction to test for SPMI between the two MRCVs. If R has difficulty with the computations, describe the problems.

   (e) The full data set for the paper is available at the Data Archiving and Networked Services website at `https://doi.org/10.17026/dans-xsw-kkeq`. Use these data to reproduce other results for MRCVs given in Tables B2-B5 of the paper.

Mixed models and estimating equations for correlated data

1. In the placekicking example from Section 2.2.1, we ignored the possible effect of the kicker on the probability of success for a placekick. We now address whether kickers generally have different overall probabilities of success and/or different relationships between distance and the probability of success. The file PlacekickWithNames.csv contains the kicker's name for each observation in the placekicking data.

   (a) Determine the number of kickers and the number of attempts for each kicker. These numbers are all reasonably large, so we believe that LR inference should be acceptable here.

   (b) Fit a GLMM for estimating the probability of a successful placekick using the binary variable `good` as the response, `distance` as the numerical explanatory variable, and `kicker` as a random effect. Use the Laplace method for reasons that will be explained later.

      i. Write out the estimated logit model for this fit. Explain which of the regression parameters are held fixed and which are random according to this model.

      ii. Report the AIC, BIC, and residual deviance for this model.

(c) Fit an additional mixed model that allows the `distance` effect to vary randomly among kickers, independently of the intercept variation. This model has multiple random effects, and therefore cannot be fit in `glmer()` using more than one quadrature point.

i. Write out the estimated logit model for this fit.

ii. Report the AIC, BIC, and residual deviance for this model.

iii. Perform an LRT for the added variance component. What do the results imply about different kickers' responses to increasing distance of a placekick?

(d) Fit the fixed-effect model with only distance as the explanatory variable.

i. Report the AIC, BIC, and residual deviance for this model. Compare these to the other two models. What does this suggest regarding the effect of kicker on the probability of a successful placekick?

ii. Perform a likelihood ratio test for the intercept variance component relative to the model in part (b). What do the results imply about different kickers' responses to increasing distance of a placekick?

(e) Were we justified in ignoring the effects of different kickers in our analyses of this data set in Chapter 2 and elsewhere? Explain.

2. Exercise 22 in Chapter 2 examines the effect of picloram herbicide on tall larkspur. A grouping variable, `rep`, was excluded from that analysis. In this exercise, we reanalyze the data incorporating the `rep` as a random effect.

(a) Fit a mixed-effect logistic regression model for the number of weeds killed using the picloram amount as the fixed-effect variable and `rep` as the random effect. Use 1, 3, 5, and 10 points of quadrature. Report the estimated parameters and the model residual deviance for each fit.

(b) Compare the residual deviances to that from the fixed-effect model with picloram only. Are all of the mixed-model fits reasonable? Explain.

(c) Use the GLMM with five quadrature points here and for the remainder of this exercise.

i. Use the parametric bootstrap to test whether the `rep` effect variance component may be zero. Set the seed number to 2939002, and use 500 simulations. Interpret the results in terms of the experiment.

ii. Refer to the parametric bootstrap confidence interval code in the Falls-GLMM.R example program on the book website. Obtain parametric bootstrap confidence intervals for the fixed- and random-effect parameters in this model. Use the same simulation to perform both sets of computations. Set the seed number to 37919236, and use 2,000 simulations.

3. Continuing 2, refer to Exercise 21 in Chapter 2, where the concept of a *lethal dose level*, x_π, is introduced. In this exercise, we use the parametric bootstrap on the GLMM fit with 5 AGQ points to find a confidence interval for LD90, $x_{0.9} = [\text{logit}(0.9) - \beta_0]/\beta_1$.

(a) Use the delta method to show that $Var(\hat{x}_\pi) \approx [Var(\hat{\beta}_0) + \hat{x}_\pi^2 Var(\hat{\beta}_1) + 2\hat{x}_\pi Cov(\hat{\beta}_0, \hat{\beta}_1)]/\hat{\beta}_1^2$.

(b) Compute $\hat{x}_{0.9}$ from the GLMM fit.

(c) Compute $\widehat{Var}(\hat{x}_{0.9})$ from part (a). All of the needed variance and covariance estimates are available from the matrix computed by `vcov()` on the model fit.

(d) Refer to the FallsGLMM.R example program on the book website. Similar to the analysis of pairwise comparisons shown in that program, conduct a parametric bootstrap analysis of $x_{0.9}$ using the "bootstrap-t" approach described in the program. Base the Z-statistic on the estimate and variance of $\hat{x}_{0.9}$. Use 2,000 simulations and a seed number 37919236 (note that this is the same seed as in the previous problem, so the extra calculations can be added into that simulation loop if desired). Summarize the estimated sampling distribution of the Z-statistic. Does it appear close to normal?

(e) Find the 95% confidence interval limits using the critical values of the estimated distribution of Z. Interpret the results.

4. Refer to the alcohol-consumption example studied in Sections 4.2.2 and 5.4.2. The data presented in DeHart et al. (2008) covered a 30-day period for each subject. Our previous analyses considered only the first Saturday for each subject, so that there was only one count per person. The data file DeHartSimplified.csv contains data from the first full week for each subject. We now have multiple counts per subject, requiring a proper treatment of the grouped data.

   This exercise explores solutions to model-fitting problems that can occur when we fit GLMMs using `glmer()`.

   (a) Certain variables in the file have missing data, which the `geeglm()` function does not allow. Read in the data and find its number of rows. Then create a new data frame using `na.omit()` on the original data frame, which removes any observations that have missing data for one or more variables. How many rows are there in the new data frame? Use this data frame for the rest of this problem.

   (b) Write out a mathematical expression for a GLMM to model the number of drinks per day for each subject. Include in this model the explanatory variables represented by `nrel`, `prel`, `negevent`, `posevent`, `gender`, `rosn`, `age`, `desired`, and `state` in a linear form. Also include the `dayweek` variable in the model as a categorical explanatory variable with levels of 1 for Monday, ..., 7 for Sunday. And finally, take into account the dependence among responses for the same subject by including a random effect term for the intercept to represent the `id` variable that identifies each subject uniquely.

   (c) What does including only an intercept term for `id` in (a) imply about the different subjects' drinking habits on different days of the week?

   (d) Fit this model using 1 quadrature point. Note that R gives a warning stating that the model failed to converge. Report the AIC for the model (`extractAIC()` does this) and the estimated variance component for subjects.

      i. One of the package developers recommends changing the optimizer—the mathematical procedure for updating the iterative numerical estimation procedure—within `glmer()` when this occurs.[45] Rerun the model using an additional argument, `glmerControl(optimizer = "bobyqa")`. Does the error disappear? Report the AIC and the estimated variance component for subjects. Do they change very much from the previous fit using the default optimizer?

[45] http://stackoverflow.com/questions/21344555/convergence-error-for-development-version-of-lme4

ii. Increase the number of function evaluations in the optimizer by using `glmerControl(optimizer = "bobyqa", optCtrl = list(maxfun = 100000))`. Does the error disappear? Report the AIC and the estimated variance component for subjects. Do they change very much from the previous fit using the "bobyqa" optimizer?

iii. Compare the fixed-effect parameter estimates from the first and second fits. For each parameter, compute the percentage change in the estimate from the second fit relative to the estimate from the first fit. Which parameter experiences the largest magnitude of percentage change, and how much is the change?

iv. Repeat the percentage-change comparison of fixed effects for third fit relative to the second. Which parameter experiences the largest magnitude of percentage change, and how much is the change?

v. From this, conclude which step has the greatest effect on the estimation: changing the optimizer or increasing the number of function evaluations.

vi. Finally, which model fit has the smallest residual deviance, suggesting that it was able to achieve an overall better fit to the data?

5. Refer to the model used in the previous exercise.

   (a) Refit the model using various numbers of quadrature points and compare the residual deviances and variance component estimates across all fits. What seems to be the smallest number of quadrature points that achieves a "stable" fit, where the numbers change very little when more quadrature points are added?

   (b) Use `nAGQ = 5` for the remaining questions

      i. Test the significance of the variance component using an LRT. Report the results and indicate whether this test is likely to give reasonable results in this problem.

      ii. Test the significance of each fixed-effect variable using LRTs. Note that this requires refitting the model to correspond to each different null hypothesis tested. Report the results.

6. This problem applies a GEE approach to the analysis of the data in Exercise 4.

   (a) What is the marginal model that coincides with the subject-specific model of part (a) in Exercise 4? Use this model for the remainder of the problem.

   (b) Describe in terms of the data what the independence and exchangeable correlation structures mean. Why is exchangeable likely more appropriate?

   (c) Fit the model in (a) as a GEE using the exchangeable structure. Report the test results from `summary()` and `anova()`. What differences are there between the results?

   (d) Reverse the order of terms in the model and repeat the tests. Do the results change? Explain.

   (e) Carry out a test of significance for day of the week by fitting the model without the variable and using `anova()` to compare the resulting model to the full model. What hypotheses are being tested and what conclusions should be drawn from the test?

   (f) Using `emmeans()` with only day of the week in the `specs` argument, estimate the mean number of drinks consumed each day and include 95% confidence intervals. Which day(s) appear to involve the highest mean consumption of alcohol?

(g) Compute confidence intervals for the mean ratios involving each pair of days. Use the Tukey adjustment to control the familywise confidence level at 95%. Which days of the week have significantly different mean alcohol consumption?

7. Curtis et al. (2008) examine the role that shoe type and extra ankle protection had on whether an ankle sprain occurred for athletes playing on collegiate basketball teams. Athletic trainers from 22 schools recorded information on their athletes after every practice and game. This information included the shoe type (cushioned or non-cushioned), the ankle protection (braced, taped, braced and taped, or neither), and whether an ankle sprain occurred (yes or no). A total of 230 athletes contributed 44,560 observations. Unfortunately, the authors analyzed the data assuming each observation was independent. Propose how the data could be analyzed that takes into account the dependence among observations.

8. The purpose of this exercise is to examine a simple GLMM with Monte Carlo simulation. Consider a binomial response Y_i with n_i trials and probability of success parameter π_i for $i = 1, \ldots, a$. This probability of success parameter comes about through the model $\text{logit}(\pi_i) = \beta_0 + b_{0i}$ with the b_{0i} assumed to be independent random variables with a $N(0, \sigma^2_{b0})$ distribution. Assume $n_i = 10$ for all i, $\beta_0 = 1$, and $\sigma^2_{b0} = 4$ for this exercise.

   (a) Simulate $a = 1000$ observations for this scenario. Set a seed of 8182 right before simulating the data.

   (b) Compute the MLEs for β_0 and σ^2_{bo} from the data simulated in (a) using `glmer()` and AGQ with 5 points. Similar to using `glm()` with a binomial response, the `formula` argument of `glmer()` is written in as "success/trials" with a `weights` argument that specifies the variable for the trials.

   (c) Are the MLEs similar to the known values of the parameters? Suggest a reason for their similarity or dissimilarity.

   (d) Compute the 90% profile LR and bootstrap (parametric, percentile, 1,000 resamples) confidence intervals for σ^2_{b0}. Use the `confint()` function for these computations with `method = "profile"` for profile LR and `method = "boot", boot.type = "perc", nsim = 1000` for the bootstrap (the bootstrap intervals will take several minutes to compute).[46] Do the intervals contain the known value for σ^2_{b0}?

   (e) Calculate the 90% Wald confidence interval for σ^2_{b0}. The `confint()` function will not compute this interval. Instead, it can be computed using tools available from the `merDeriv` package. Below is the code where `mod.fit` contains results returned by `glmer()` from fitting the model.

```
library(package = merDeriv)
cov.mat <- vcov(mod.fit, full = TRUE)
sqrt(cov.mat[2,2])  # Standard error for sigma_{b0}^2
alpha <- 0.10
```

[46]There may be a small number of warnings about nonconvergence. These warnings often occur even when the iterative numerical procedures have achieved a reasonably good fit. We will ignore them here, because they will have little influence on the results across 1,000 simulations.

```
mod.fit@theta^2 + qnorm(p = c(alpha/2, 1-alpha/2)) *
   sqrt(cov.mat[2,2])
```

Compare the Wald interval to the intervals calculated in part (d). Please note that Wald intervals should not be used for σ^2_{b0}. The purpose here is to see the differences among the calculated intervals.

(f) Examine how well the profile LR and Wald confidence intervals capture σ^2_{b0}. Simulate 1,000 different data sets and calculate the estimated true confidence level for each interval. This may take as much as an hour to run, depending on the computer's processing capability. The bootstrap intervals could be included in this assessment as well, but the computation time would be much longer because of the resampling process.

9. Continue Exercise 8 by using the data simulated in its part (a).

(a) The `integrate()` function performs quadrature to integrate mathematical expressions written as an R function. To help understand how it works, execute the code below that evaluates a normal probability density function.

```
my.func.norm <- function(b, my.sigma) {
  dnorm(x = b, mean = 0, sd = my.sigma)
}

# P(-infinity < b < infinity)
integrate(f = my.func.norm, lower = -Inf, upper = Inf,
   my.sigma = 2)$value
pnorm(q = Inf, mean = 0, sd = 2)

# P(0 < b < infinity)
integrate(f = my.func.norm, lower = 0, upper = Inf, my.sigma
   = 2)$value
pnorm(q = Inf, mean = 0, sd = 2) - pnorm(q = 0, mean = 0,
   sd = 2)
```

The `integrate()` function performs integration over the first argument (`b`) given within the function specified by `f`. This integration is performed from the `lower` to the `upper` limits. Additional values (`my.sigma`) can be passed into the function needed for `f`.

(b) State the likelihood function expression. Use the symbols β_0 and σ^2_{b0} rather than their known values.

(c) Construct an R function that evaluates the likelihood function using the `integrate()` function. Use this function to verify the log-likelihood is -2272 at the MLEs obtained in part (b) of Exercise 8.

(d) With the function from (c), use the `optim()` function (see Section 2.2.1) to obtain the MLEs. Use the known values for β_0 and σ^2_{b0} as the starting values to simplify the process. Are the MLEs similar to those obtained by `glmer()`?

(e) Construct a contour plot (see Figure 2.3) of the log-likelihood function and show where the MLEs are located on it.

Bayesian methods for categorical data

1. For the hepatitis C example in Section 1.1.2, complete the following parts using a beta prior distribution with $a = b = 1$.

   (a) Plot the posterior distribution.

   (b) Compute the Bayes estimate for π.

   (c) Compute the 95% equal-tail and HPD intervals for π. Interpret the intervals.

   (d) Compare your results from (a) and (b) to the frequentist-based estimate and intervals. Provide reasons for their similarities or differences.

   (e) Hepatitis C prevalence would be expected to be small for blood donation settings like this. Propose a prior that would take this into account and complete (a)–(c) again. Do your results change? Discuss.

2. Investigate how the Bayes estimate and 95% credible intervals for π change for different values of a and b in a beta prior distribution. Consider the following observed data scenarios:

   (a) $w = 4, n = 10$

   (b) $w = 1, n = 10$

   (c) $w = 8, n = 20$

   (d) $w = 40, n = 100$

   Include comparisons across (a)–(d) in your responses.

3. Exercise 1 of Chapter 1 examined a number of simple experiments where the purpose was to estimate a probability of success. For each part, discuss how to determine appropriate priors and what they may be. Make sure to include priors that are informative in your discussion. Analyze the available data for each part using Bayesian methods.

4. With the placekicking data example in Section 6.6.2, try using smaller/larger number of MCMC samples, smaller/larger number of chains, and multiple cores of the processor to see the effect these changes have on model estimates, convergence measures, and computation time. One way within R to determine the number of cores on a computer is to use the `parallel` package and execute `detectCores(logical = FALSE)`.

5. Exercise 7 of Chapter 2 examined the probability of success for a placekick using data from two National Football League seasons. Complete the following parts using `Distance` as the only explanatory variable.

   (a) Discuss how the results from the placekicking example in Section 6.6, which used data from an earlier NFL season, could be used to help choose priors for the regression parameters.

   (b) Convert `Good` in the data set to a 0-1 variable because `stan_glm()` does not accept a response variable coded as `"Y"` or `"N"`. This can be done a number of ways, including with the `ifelse()` function.

   (c) Fit a logistic regression model using Bayesian methods. Comment on the effect that the distance has on the probability of success.

(d) Find the 95% equal-tail and HPD intervals for the probability of success when the distance is 20 yards. Interpret the intervals.

(e) Find the 95% equal-tail and HPD intervals for an odds ratio involving a 10-yard decrease in the distance. Interpret the intervals.

(f) Investigate convergence of the MCMC method.

6. Section 2.2.6 examined a data set involving the Tomato Spotted Wilt Virus. Analyze these data using a logistic regression model and the appropriate Bayesian methods. Include an infestation type and control method interaction within the model. Justify the choice of priors and provide evidence of convergence for the MCMC sampling. Compare your conclusions to those obtained using maximum likelihood methods.

7. The use of Bayesian methods can provide an alternative solution to the complete separation problem (see Section 2.2.7) that sometimes occurs in logistic regression. To illustrate this alternative approach, consider again the cancer-free example of Section 6.2.3.

   (a) Fit a logistic regression model using Bayesian methods. Include all three explanatory variables in the model as linear terms. Show that Bayes estimates are dependent on the priors by trying a few different normal distributions with the same mean but different amounts of variability. While Bayesian methods are touted as an alternative solution, this shows a downside of their application.

   (b) Despite the different Bayes estimates obtained in part (a), do the general conclusions change regarding the effects of the explanatory variables on the probability of being cancer-free?

   (c) Describe the shape of the histograms for the samples corresponding to each parameter. Relate these shapes to the trace plots.

   (d) Provide evidence that convergence to the posterior distribution occurs. Discuss how convergence here is different than the meaning of convergence when using a standard likelihood approach as presented in Section 2.2.1.

   (e) Compare the resulting model to what is found using the modified likelihood approach.

8. For the alcohol consumption example of Section 4.2.2, consider the model $\log(\mu) = \beta_0 + \beta_1\texttt{negevent} + \beta_2\texttt{posevent} + \beta_3\texttt{negevent} \times \texttt{posevent}$.

   (a) Estimate the regression parameters using Bayesian methods and weakly informative priors. The syntax here is the same for `stan_glm()` as for a logistic regression model but the `family` argument value is `poisson(link = "log")`.

   (b) Provide evidence that shows convergence to the posterior distribution has likely occurred.

   (c) Calculate the ratio of means for a 1-unit increase in negative events at each of the three positive-event quartiles. Interpret these values as percent changes, and find the corresponding 95% HPD credible intervals. Compare your answers, including the interpretations, to those given in Section 4.2.2.

9. For the wheat-data example of Section 3.4, consider the proportional odds regression model when `class`, `density`, `hardness`, `size`, `weight`, and `moisture` were used to estimate the log-odds of the cumulative probabilities for the kernel type.

(a) Estimate the same regression parameters but use `stan_polr()` for a Bayesian analysis. The function's syntax is identical to what was used with `polr()` but now with the addition of `seed` and `iter` as with `stan_glm()`. Use independent, noninformative priors for the regression parameters by specifying `prior = NULL`. It is important to note that there is no `prior_intercept` argument to specify priors separately for β_{j0} and the correct sign is used by `stan_polr()` for the regression parameters (β_r is estimated rather than η_r).

(b) Were there any divergent transitions indicated in the output? If there were, increase the `adapt_delta` argument value. We encountered this issue and including `adapt_delta = 0.99` in `stan_polr()` solved the potential problem.

(c) Evaluate convergence of the MCMC method.

(d) For the first observation, estimate the probability a kernel is healthy, scab, or sprout. State the corresponding 95% credible intervals.

(e) Estimate the odds ratios for kernel density with a decrease in value of 0.13 (1 standard deviation). State the corresponding 95% credible intervals.

(f) Compare your results for this problem to those obtained in Section 3.4 with maximum likelihood methods.

10. Continuing Exercise 9, consider now the multinomial regression model to estimate the log-odds for the kernel type. The `rstanarm` package does not have a function to estimate this model, so we will use instead the `brms` package. This alternative package also incorporates the Stan programming language to implement the Hamiltonian algorithm and has many similarities to `rstanarm`.

(a) After installing the `brms` package from CRAN, use the code below to estimate the regression parameters and save the results into an object named `mod.fit.brms`.

```
library(package = brms)
mod.fit.brms <- brm(formula = type ~ class + density +
    hardness + size + weight + moisture, data = wheat,
    family = categorical(link = "logit", refcat =
    "Healthy"), iter = 10000, seed = 5438)
summary(mod.fit.brms)
```

The `brm()` function specifies the model and components of the Hamiltonian algorithm much like `stan_glm()`. The `family` argument here specifies a multinomial regression model with the reference level of `"Healthy"` for the odds.

(b) Because no specification was given for the priors in (a), the default priors were used. Run `mod.fit.brms$prior` to see that these independent priors are noninformative for the regression parameters corresponding to the explanatory variables (`class = "b"`) and t distributions for the intercept regression parameters (`class = "Intercept"`). The `dpar` column in the output corresponds to whether the parameters are for the scab/healthy log-odds or the sprout/healthy log-odds.

(c) The process to specify priors for `brm()` is different than for `stan_glm()`. Rerun the code in (a) with the following prior specification inserted within `brm()`.

```
prior = c(set_prior(prior = "normal(0,100)", class = "b",
    dpar = c("muScab", "muSprout")), set_prior(prior =
```

```
"normal(0,100)", class = "Intercept", dpar = c("muScab",
"muSprout")))
```

This specifies weakly informative priors that have a normal distribution with mean 0 and standard deviation of 100 for all regression parameters. Verify the priors have changed by examining `mod.fit.brms$prior`.

(d) Compare the regression parameter estimates obtained in (a) and (c) with those obtained in Section 3.3 using maximum likelihood methods.

(e) Evaluate MCMC convergence for the model in (c). Below is code that produces helpful plots.

```
plot(mod.fit.brms, combo = c("hist", "trace"))
mcmc_plot(mod.fit.brms, type = "acf_bar")
```

Like `rstanarm`, `brms` makes use of the `bayesplot` package but through a different syntax. Also, the `summary(mod.fit.brms)` output states $\hat{R}$ and effective sample sizes. For the latter measure, two separate types are given with *bulk* corresponding to centrality measures for the posterior and *tail* corresponding to the 0.05 and 0.95 quantiles of the posterior.

(f) The `emmeans` package is not set up to work with multinomial models estimated by `brm()`. Instead, one must work directly with the samples to calculate summary measures. Extract these samples from the object returned by `brm()` in (c). This can be done with the `as.data.frame()` function in the same way as when working with `stan_glm()`. Using these samples, complete the following:

i. Estimate the probability a kernel is healthy, scab, or sprout using the observed explanatory variable values for the first observation. State the corresponding 95% credible intervals.

ii. Estimate the odds ratios for kernel density with a decrease in value of 0.13 (1 standard deviation). State the corresponding 95% credible intervals.

Compare the probabilities and odds ratios to those obtained in Section 3.3.

11. Continue the analysis of the Larry Bird data set introduced in Section 1.2 but now using Bayesian methods.

(a) Fit a logistic regression model using the second free throw outcome as the response and the first free throw outcome as the explanatory variable. Compare your model to the estimated model obtained in Exercise 10 of Chapter 2.

(b) Fit a Poisson regression model (see Exercise 8) that includes the first and second free throw outcomes as explanatory variables along with their interaction. Compare your model to the estimated model obtained in Section 4.2.4 for the same data example.

(c) After confirming convergence in (a) and (b), judge whether there is dependence between the first and second free throws.

Group testing

1. Derive the pooling specificity expressions given in Equations 6.27 and 6.29.

2. The Nebraska Public Health Laboratory used Dorfman's algorithm with groups of size 5 for COVID-19 testing. This group size was chosen based on a disease prevalence of 5% that was estimated by individually testing specimens prior to group testing implementation. Perfect accuracy was initially assumed.

    (a) Compute the expected number of tests per individual. What is the expected reduction in tests $(1 - E(T_k)/I)$ relative to testing each specimen separately?

    (b) Show that a group of size 5 is the OTC.

    (c) The lab chose the OTC early in the pandemic prior to having good estimates of the test sensitivity. For this reason, the laboratory also performed the same calculations using a sensitivity of 0.95. Find the OTC using this alternative sensitivity value. Does the OTC change?

    (d) Abdalhamid et al. (2020) cite the use of "A Shiny App for Pooled Testing" at `www.chrisbilder.com/shiny` as the tool used to perform the calculations for the laboratory. This is an online app that uses the `Shiny` package in R to create a web interface to the `binGroup2` package. Use this app to reproduce the calculations in parts (a) and (b). For part (a), select Hierarchical Testing > Calculation in the app. For part (b), select Hierarchical Testing > Optimal Configuration in the app.

3. In addition to the expected number of tests per individual, the expected increase in testing capacity relative to individual testing is an alternative metric used to measure group testing performance. This alternative metric is defined as $A = (I/E(T) - 1)100\%$. For example, a value of 150% represents that a laboratory will be able to test 150% more specimens on average (2.5 times the original testing capacity) using a group testing algorithm than testing each specimen separately.

    (a) Discuss why A represents the expected increase in testing capacity.

    (b) Compute A for part (a) of Exercise 2. Interpret the result.

    (c) Maximizing A rather than minimizing the expected number of tests per individual results in the same OTC. Why?

4. This exercise continues examining the example on HIV screening in Washington, DC for those people with very low CD4 counts.

    (a) Create an R function that calculates Equation 6.26. Using this function, calculate $E(T_k)$ when everyone has an equal probability of disease and $I = 20$. Verify that your answer matches what is obtained by `opChar1()`.

    (b) Are there group sizes other than $I = 17$ that are close to being optimal? Construct a plot of the expected number of tests per individual vs. group size to help answer this question. Assume all individuals have an equal probability of disease and examine possible group sizes of 3 to 30.

5. Sherlock et al. (2007) describe the HIV testing practices for a number of cities in addition to Washington, DC. One of these cities is San Francisco, where three-stage hierarchical testing is performed using the algorithm given in Figure 6.13. The overall prevalence was 0.01752 for the time period of the study. Assume this prevalence holds true for future testing. Also, assume the same assay accuracy as for the very low CD4 counts in the Washington, DC, example.

    (a) Equation 6.26 gives the expected number of tests when every individual has an equal probability of disease and each test has the same accuracy. For this same situation and the algorithm given in Figure 6.13, construct a basic equation for the expected number of tests. Leave the answer in terms of the probabilities that groups or subgroups test positive (i.e., P(Stage 1 group tests positive) and P(Stage 2 group tests positive)), rather than deriving these expressions in terms of the probability an individual has a disease.

(b) If every individual has an equal probability of being truly positive and the maximum possible group size is 50, complete the following parts.

i. What is the expected number of tests per individual for the group and subgroup sizes used by San Francisco? The `opChar1()` function can be used to help answer this question with the following code.

```
group.member <- matrix(data = c(rep(1, times = 50),
    c(rep(1:5, each = 10), 1:50)), nrow = 3, ncol = 50,
    byrow = TRUE)
test.vlow <- opChar1(algorithm = "D3", p = 0.01752, Se =
    0.967, Sp = 0.976, hier.config = group.member)
summary(test.vlow)
```

ii. Are there different group or subgroup sizes that may be more efficient than 50 and 10, respectively (i.e., less expected tests)? Use the `OTC1()` function to help answer this question with `algorithm = "D3"`.

iii. What is the OTC when using two-stage hierarchical testing? Compare the use of this algorithm to those in the previous two parts. Make sure to include overall accuracy in your comparison.

(c) If individuals could have different probabilities of being truly positive and the maximum possible group size is 50, complete the following parts.

i. Plot beta distributions for $a = 0.1, 0.5, 1$, and 2, and $b = a(1 - \tilde{\pi})/\tilde{\pi}$, where $\tilde{\pi}$ is the overall prevalence. Compare these distributions. How does a relate to the distribution of risks of having a disease?

ii. Find the OTC when using an initial group size of size 50, $a = 0.1$, and a three-stage hierarchical testing algorithm that takes advantage of the heterogeneity among disease probabilities. This can be done by using `OTC1()` with `algorithm = "ID3"` (the block for PSOD is now first tested as a group). Repeat finding the OTC, but now for $a = 2$. Compare these OTCs to what was found for three-stage hierarchical testing in part (a). Discuss the potential benefits of informative group testing.

iii. What are the OTCs when using two-stage hierarchical testing (PSOD) with $a = 0.1$ and with $a = 2$? Compare these results to those in the previous two parts. Make sure to include overall accuracy in your comparison.

Note that the `OTC1()` function may take a few minutes to complete its calculations.

(d) Were there benefits demonstrated for using group testing rather than individual testing in this exercise? Explain.

6. The science fiction genre frequently mixes real-world science with fictional elements to make stories believable to their audience. In *Battlestar Galactica*, an apocalyptic space-travel television series, this was done with the premise that artificial lifeforms

created by humans, known as Cylons, cause the almost complete destruction of humanity.[47] The 47,905 humans who survived joined together in a rag-tag fleet of spaceships to get away from these Cylons. Early during the series, the human leadership in the fleet discovers that some of the remaining humans may actually be Cylons. Therefore, determining who is human and who is Cylon becomes a top priority. In the episode "Tigh Me Up, Tigh Me Down," Dr. Gaius Baltar develops a blood test for this purpose. He soon is overwhelmed by the testing load because each test takes 11 hours to complete and each can only be performed one at a time. In the end, Baltar is so dismayed by the testing load that testing is not completed.

(a) Based on a 24-hour day and a 365-day year, how many years would it take to screen all individuals in the fleet?

(b) Bilder (2009) examines how group testing could have reduced the testing load. When this article was written, fans of the show knew there were seven Cylons in the fleet. Assuming perfect testing and no adjustments for the possibility of a remainder group,[48] complete the following with Dorfman's algorithm.

   i. Construct a plot of the expected number of tests per individual versus potential group sizes from 2 to 200.

   ii. What is the OTC among those group sizes used for part i?

   iii. Using this OTC, what is the expected amount of time it would take to screen all individuals in the fleet?

(c) The variability in the number of tests may be of interest as well in group testing applications. A large amount of variability may be undesirable because of the uncertainty in the actual overall number of tests needed. Complete the following to continue part (b).

   i. Show that the variance for the number of tests in a group is

$$Var(T_k) = I^2 P(Z_k = 1)P(Z_k = 0).$$

   ii. Compute Var(T_k) for the optimal group size. Assume perfect testing.

   iii. A rule of thumb in statistics is that most observed values for a random variable should fall within three standard deviations from its mean.[49] Using this rule of thumb, the OTC, and the previous variance calculation, find a range for the number of tests per individual and the amount of time it would take to screen *all* individuals in the fleet.

   iv. Because there are at most seven groups that are positive (each Cylon is in a different group) and at least one group will be positive, an alternative range for the total number of tests can be calculated using this information without taking into account the variance. Compute the minimum and maximum number of tests possible with the OTC. Use these values to compute the minimum and maximum screening time.

[47] We refer to the television series that aired from 2003 to 2009. An earlier version of the series with a somewhat different storyline aired from 1978 to 1980.

[48] When a finite number of specimens are screened, there may be one group that does not have a full size. For example, 47,905 individuals can be put into 4,790 groups of size 10 and one remainder group of size 5. As long as there are a large number of specimens being screened relative to the group size, accounting for a remainder group will not be practically important when computing the overall expected number of tests.

[49] This is based both on Chebychev's theorem and on the empirical rule. See pp. 122-3 of Casella and Berger (2002).

(d) Bilder (2009) examines other group testing algorithms in addition to Dorfman's algorithm. Use `OTC1()` to determine the OTC for one of these other algorithms, array testing. Search for this group size over the range of 3 to 50.[50] Determine the expected amount of time it would take to screen all individuals in the fleet and compare this value to the time found for Dorfman's algorithm.

7. Figure 1 of Bilder et al. (2020) provides a comparison of Dorfman's testing algorithm to three-stage hierarchical and array testing for prevalences between 0 and 0.10. The reduction in the expected number of tests and the expected increase in testing capacity (see Exercise 3) relative to individual testing were plotted for their corresponding OTC. Complete the following parts if you have access to the paper.

   (a) Which algorithm is the most efficient for testing?

   (b) Use the `OTC1()` function to reproduce some of the calculations at specific prevalences of interest. Three-stage hierarchical testing uses `algorithm = "D3"` and array testing uses `algorithm = "A2"` in this function.

8. This exercise continues the Mal Rio Cuarto virus example of Section 6.7.

   (a) Compute the 95% Wilson interval for $\tilde{\theta}$ and show how this is used to compute the 95% Wilson interval for $\tilde{\pi}$.

   (b) Ornaghi et al. (1999) also perform multiple vector transfer design experiments using male planthoppers. For 24 other plants, each with seven planthoppers, two of the plants were infected with the virus. Estimate the probability of transmission and an appropriate 95% confidence interval using these data. Interpret the interval.

9. Gildow et al. (2008) examine the transmission rate of the Cucumber Mosaic virus by aphids to snap bean plants. For each of 13 different species of aphids, the authors performed separate multiple vector transfer experiments that used 50 different plants and 10 aphid vectors per plant. After each transfer, aphids were allowed to feed on a plant for a 16-hour period. Each plant was then sprayed with an insecticide to kill the aphids and was observed for signs of infection over a four-week period. The number of plants that acquired the virus is given in Aphid.csv.

   (a) Calculate the MLE of the transmission rate for each species. Which species have larger or smaller transmission rates?

   (b) Calculate the 95% Wilson intervals for the transmission rate. Interpret the intervals.

   (c) Pairwise comparisons of the transmission rates can be made using the `propDiffCI()` function in `binGroup2`. Compare the *Aphis glycines* species to the *Myzus persicae* species using this function as given below.

      i. Find the difference in the MLEs by using the `pt.method = "mle"` argument.

      ii. Biggerstaff (2008) investigates a number of intervals, including a Wald and a score interval. Describe in general terms how expressions for these intervals could be formed.

[50] A larger group size than 50 is actually optimal. Because of the computational limits of `OTC1()` and the default memory size of R, we restrict the upper bound for the search.

iii. Calculate these intervals by specifying the correct argument value for the `ci.method` argument. Use a level of confidence equal to 95%. Compare these intervals.

iv. Is there sufficient evidence of a difference in transmission rates for these two species? Explain.

10. The MLE for $\tilde{\pi}$ has a small amount of bias. For this reason, Gildow et al. (2008) decide to calculate an estimator proposed by Bilder and Tebbs (2005) that was motivated by empirical Bayes methodology. Complete the following parts to find Bayesian-based estimates of the transmission rate for the Cucumber Mosaic virus.

 (a) Bilder and Tebbs (2005) work with one-parameter beta distributions for $\tilde{\pi}$ of the form $f(\tilde{\pi}|b) = b(1-\tilde{\pi})^{b-1}$. Plot a few of these distributions with $b \geq 1$ and discuss why this would be a good choice for a prior in situations when group testing is appropriate.

 (b) Define $T = \sum_{k=1}^{K} \tilde{Z}_k$ as a binomial random variable with probability of success parameter $\tilde{\theta}$. Using the prior from (a), show that the posterior distribution of $\tilde{\pi}$ given an observed value of T (denoted as t) and b is

$$f(\tilde{\pi}|t,b) = \frac{I\Gamma(K+b/I+1)}{\Gamma(K-t+b/I)\Gamma(t+1)}[1-(1-\tilde{\pi})^I]^t(1-\tilde{\pi})^{I(K-t)+b-1}$$

 for $0 < \tilde{\pi} < 1$. Note that the marginal distribution of T is

$$g(t|b) = \frac{b\Gamma(K+1)\Gamma(K-t+b/I)}{I\Gamma(K-t+1)\Gamma(K+b/I+1)}$$

 for $t = 0, \ldots, K$.

 (c) Show that the Bayes estimate for $\tilde{\pi}$ is

$$1 - \frac{\Gamma(K+b/I+1)\Gamma(K-t+b/I+1/I)}{\Gamma(K-t+b/I)\Gamma(K+b/I+1/I+1)}.$$

 (d) The Bayes estimate is a function of three known quantities: the group size I, the number of groups K, and the observed number of positives t. Unfortunately, b is not known; however, it can be estimated by finding the b that maximizes $g(t|b)$. Using this estimate with the Bayes estimate of $\tilde{\pi}$ results in what is known as an *empirical Bayes estimate.* Find the empirical Bayes estimate of the disease transmission rate for the *Aphis glycines* species. Below are two functions that will help with the computations.

```
# log(g(t|b))
gt.b <- function(b, t, I, K) {
  # Did not include lgamma(K + 1) - log(I) - lgamma(K - t +
     1) because these do not change as a function of b
     (maximization is not affected)
  log(b) + lgamma(K - t + b/I) - lgamma(K + b/I + 1)
}

find.est <- function(t, I, K, b.range = c(0.0001, 1000)) {

  # Find b that maximizes g(t|b)
  save.results <- optimize(f = gt.b, interval = b.range,
     maximum = TRUE, t = t, I = I, K = K)
```

```
  b.hat <- save.results$maximum

  # Empirical Bayes estimate (EB1 in Bilder and Tebbs) -
     used log gamma function because computational
     difficulties can occur with large gamma function values
  EB.max <- 1 - exp(lgamma(K - t + b.hat/I + 1/I) + lgamma(K
     + b.hat/I + 1) - lgamma(K + b.hat/I + 1 + 1/I) -
     lgamma(K - t + b.hat/I))

  # Gildow et al. used an empirical Bayes motivated
     estimator (EB2 in Bilder and Tebbs)
  EB.alt <- 1 - (1 - (t + 1) / (K + b.hat/I + 1))^(1/I)

  data.frame(b.hat = b.hat, EB.max = EB.max, EB.alt = EB.alt)
}
```

Note that Gildow et al. (2008) used the second estimator (`EB.alt`) calculated by `find.est()`. This estimator is derived using an empirical Bayes estimator for $\tilde{\theta}$, a modified prior distribution, and the relationship between $\tilde{\pi}$ and $\tilde{\theta}$. While this is not quite an empirical Bayes estimator, Bilder and Tebbs (2005) show that it has smaller bias than the MLE.[51]

(e) Find the 95% credible interval for the disease transmission rate of the *Aphis glycines* species. Substitute the estimate of b into the posterior distribution for your calculations. Include both the equal-tail and HPD intervals in your answer. Compare the intervals.

11. This problem examines $\widehat{Var}(\hat{\hat{\pi}})$ in more detail.

   (a) Derive $\widehat{Var}(\hat{\hat{\pi}})$ using the delta method and $\widehat{Var}(\hat{\hat{\theta}}) = \hat{\hat{\theta}}(1-\hat{\hat{\theta}})/K$.

   (b) What is $\widehat{Var}(\hat{\hat{\pi}})$ when all groups test positive? What happens to the Wald confidence interval for this situation?

12. Public health laboratories across the United States use group testing to test high volumes of clinical specimens for chlamydia. Joyner et al. (2020) discuss the testing performed by one of these laboratories, the State Hygienic Laboratory (SHL) at the University of Iowa. This laboratory tests female swab specimens in groups of size 4. If a group tests positive, individual retests are performed. Supplemental materials provided with the paper include a simulated data set based on the testing performed at the SHL (the actual data could not be released). We include part of their data in the file GT.SHL.csv. Each row of the data set provides the testing results and additional information on an individual. Below are descriptions of the variables:

   - `Age`: Age of individual
   - `Symptoms`: Symptoms of infection (1 if yes, 0 if no)
   - `New.partner`: New sexual partner in the last 90 days (1 if yes, 0 if no)
   - `Multiple.Partners`: Multiple sexual partners in the last 90 days (1 if yes, 0 if no)

[51] There are a number of other proposed estimators that also have a smaller bias than the MLE. The Burrows (1987) estimator of $\hat{\hat{\pi}}_B = 1 - [1 - t/(K+c)]^{1/I}$, where $c = (I-1)/(2I)$, is the most often used of these estimators.

- `Contact`: Contact with someone known to have a sexually transmitted disease (1 if yes, 0 if no)
- `Group.numb`: Group number
- `Group.test`: Group test outcome (1 if positive, 0 if negative)
- `Retest`: Individual retest outcome (1 if positive, 0 if negative, NA if no retest performed)

(a) What was the total number of tests used? Compare this value to the number of tests needed if each individual were tested separately.

(b) Fit a binary regression model for the probability of being positive for chlamydia using `Age`, `Symptoms`, `New.partner`, `Multiple.Partners`, and `Contact` as explanatory variables without interactions or transformations. Perform this estimation two separate ways: (1) with the group tests alone and (2) with both the group tests and retests. Use a sensitivity of 0.947 and a specificity of 0.989 for these tests.

(c) Compare the regression parameter estimates along with the corresponding standard errors and Wald tests for the two models estimated in (b).

(d) Examine the model with the group tests and retests more closely as described below.

   i. Interpret the effect of each explanatory variable on the odds of being positive.

   ii. Determine the combinations of explanatory variable values that would result in having the smallest and the largest probability of being positive.

(e) Suppose the NA values for `Retest` are actually 0's. Fit the same model as given in (b) but using `Retest` as a response variable that came about through individual testing. Compare the resulting model and inferences to those given in (b) and (d). This estimation requires the use of methods from Section 6.1.2.

13. Tebbs and Bilder (2004) investigate the true confidence levels for the Wald and Wilson intervals for $\tilde{\pi}$ when there is no testing error. The purpose of this problem is to reproduce some of their results.

(a) Below is code that shows the true confidence level of the Wilson interval is 0.9626 when $\alpha = 0.05$, $K = 40$, $I = 10$, and $\tilde{\pi} = 0.02$. Modify this code to show that the true confidence level for the Wald interval is 0.9415 with these same values of $\alpha, K, I,$ and $\tilde{\pi}$.

```
library(package = binGroup2)

K <- 40  # Number of groups
sum.z <- 0:K  # Potential number of positive groups
I <- 10  # Group size
pi.tilde <- 0.02  # Probability positive
alpha <- 0.05

# New function that can be used with apply()
new.propCI <- function(y, n, s, conf.level, alternative,
   method) {
  propCI(x = y, m = I, n = n, ci.method = method, conf.level
     = 1 - alpha)$conf.int
}
```

```
# Calculate intervals for each sum.z
wilson.int <- apply(X = matrix(data = sum.z, nrow =
    length(sum.z), ncol = 1), MARGIN = 1, FUN = new.propCI,
    n = K, s = I, conf.level = 1 - alpha, method = "score")

pmf <- dbinom(x = sum.z, size = K, prob = 1 - (1 -
    pi.tilde)^I)
save.wilson <- pi.tilde > wilson.int[1,] & pi.tilde <
    wilson.int[2,]
sum(save.wilson*pmf)  # True confidence level
```

(b) Figure 1 in the paper contains a plot with the true confidence levels of both the Wald and Wilson intervals where $\alpha = 0.05$, $n = 40$, $I = 10$, and $\tilde{\pi}$ within the range of 0.001 to 0.1 by 0.0005. Construct the same plot for this exercise. Comment on which interval performs better.

(c) Exercise 17 of Chapter 1 discusses the expected length of a confidence interval when estimating the probability of success for individual testing data. Provide an expression for the expected length when group testing is implemented. Why is the expected length an important consideration?

(d) Construct a plot that examines the expected length of the Wilson and Wald interval for the same settings as given in part (b). Because the Wald interval cannot be calculated when all groups are positive, you may assume that the expected length of the interval to be 1 in this situation. After taking into account the true confidence levels and the expected lengths, which interval is better to use in practice? Explain.

Splines and additive models

1. A cubic spline is a piecewise polynomial as given in Equation 6.34. To show this relationship, consider the truncated power version of a cubic spline given by $f(x) = \beta_0 + \beta_1 x + \beta_2 x^2 + \beta_3 x^3 + \beta_4 (x - k_1)^3_+$ for one knot k_1 and the piecewise polynomial in Equation 6.34.

   (a) For $x > k_1$, express the truncated power spline so that terms are grouped by $x^0, \ldots, x^3$. Thus, multiply the expression out and group like terms by powers of x.

   (b) Equate the coefficients on the x-terms with those of the corresponding polynomial in Equation 6.34 for all values of x.

2. Equation 6.37 shows how to find the log of the odds ratio for x using the link function of a logistic regression model that includes a spline. Using a similar process, show how the log-mean ratio ($\log(\mu_a/\mu_b)$ for two values of x, a and b) can be formed using the link function for a Poisson regression model that involves a spline for x. Show how this also leads to an expression for the percentage change in the mean (PC).

3. Examine a different number of knots with B-splines for the placekicking data example (Placekick.csv). Does the number of knots matter? How do your results compare to the same type of analysis performed with natural splines?

4. The example on alcohol consumption from Section 4.2.2 examined the effect of negative/positive event experiences on the number of alcoholic drinks that an individual

consumes. The purpose of this exercise is to continue this examination by using splines with the negative events variable.

(a) Estimate Poisson regression models using B- and natural splines for negative events. Use knots equal to the 0.25, 0.50, and 0.75 quantiles by specifying `df = 6` within `bs()` and `df = 4` within `ns()`. Construct a scatter plot of the data with the models included. Add first- and third-order models to the plot as well. Are there differences among the models?

(b) Use LRTs to judge the importance of including negative events in the models with the B- and natural splines.

(c) Are other knot choices better for the B- and natural splines? Use AIC and BIC to help make this decision.

(d) Fit the Poisson regression model using a smoothing spline for negative events. Construct a scatter plot of the data with the model included. Add the previous spline models in (a) along with the first- and third-order models to the plot. Are there differences among the models?

(e) Compare the effective degrees of freedom from the smoothing spline to the degrees of freedom used in (a).

(f) Compute the percentage change in the mean with a one-unit increase in negative events. Use the natural spline model from (a) and the smoothing spline model from (d) for these calculations. Examine the percentage change in the mean at a few different values of the negative events. Interpret the corresponding estimates and confidence intervals. Compare your results to those found for the first-order model in Section 4.2.2.

5. Continuing Exercise 4, a Poisson regression model with negative events, positive events, and their interaction was examined in Section 4.2.2 as well. The purpose here is to introduce how this type of model can be made more flexible by using smoothing splines.

(a) Splines for one variable can be generalized to model interactions between variables through tensor product smoothers. These smoothers use basis functions similar to Equation 6.36 but the terms in the sum now involve products of basis terms from each variable. The function fit is

$$f(x_1, x_2) = \sum_j \sum_k \beta_{jk} h_j(x_1) g_k(x_2),$$

where $h_j(x_1)$ is a basis function for explanatory variable x_1, $g_k(x_2)$ is a basis function for explanatory variable x_2, β_{jk} is the regression parameter for the cross product between these basis functions, and the sum is carried out over all combinations of the two sets of basis functions. The `ti()` function of the `mgcv` package provides a convenient way to specify tensor product smoothers within the `gam()` function. Fit a Poisson regression model with these smoothing splines using `numall ~ ti(negevent, bs = "cr") + ti(posevent, bs = "cr") + ti(negevent, posevent, bs = "cr")` for the `formula` argument in `gam()`. Specifying the model in this manner provides a main effects and interaction interpretation for the model. Interpret the Wald-like statistics that provide hypothesis tests to judge the importance of the smoothers.

(b) Construct contour and three-dimensional surface plots for the estimated mean number of drinks using the model from part (a). The code to construct these types of plots in Section 4.2.2 can be adapted here with minor modifications. Use these plots to interpret the model and to compare it to the model estimated in Section 4.2.2.

(c) Calculate the percentage change in the mean number of alcoholic drinks for a one-unit increase in negative events. Use the first, second, and third quartiles for positive events with the model from part (a). Compare the resulting values to those obtained in Section 4.2.2.

(d) Discuss the advantages and disadvantages for using the tensor product smoother approach versus using the approach from Section 4.2.2.

6. The wheat kernels data example from Section 3.3 examined the relationship between a kernel's physical properties and its overall condition type. The purpose here is to continue this examination using multinomial regression models when kernel density is the only explanatory variable and this variable is represented by a spline.

   (a) Fit a multinomial regression model using a natural spline for the density. This can be done using the `multinom()` function with `ns()` in the `formula` argument. Use knots equal to the 0.25, 0.50, and 0.75 quantiles by specifying `df = 4` within `ns()`. We recommend increasing the maximum number of iterations to achieve convergence of the regression parameter estimates. Construct a plot of the model and include a first-order model on the plot as well. Are there differences between the models?

   (b) Use AIC and BIC to compare the natural spline and first-order models. Which model is better according to these measures?

   (c) Repeat (a) using a smoothing spline rather than a natural spline. While the `gam()` function of `mgcv` estimates these models, its syntax is different than what has been presented previously. Below is our code.

```
wheat$type1 <- ifelse(wheat$type == "Healthy", yes = 0, no =
    ifelse(wheat$type == "Scab", yes = 1, no = 2))
mod.fit.nom.ps <- gam(formula = list(type1 ~ s(density, bs =
    "cr"), ~ s(density, bs = "cr")), family = multinom(K =
    2), data = wheat)
```

   The response variable for `gam()` needs to have consecutive integer values starting with 0, so this is the reason for using `ifelse()`. The `family` argument specifies a multinomial response. The `K = 2` designation in `multinom()` represents three levels (0, 1, and 2) for the response variable. The `formula` argument states the model in a `list` format to allow for different smoothing splines with the two log-odds.

   (d) For the very large density values, both spline models in this exercise estimate the probability a kernel is healthy or scab quite differently than for the first-order model. Suggest a reason by examining Figure 3.2. Discuss whether this reason alone justifies using spline models rather than the first-order model.

7. Repeat Exercise 6 for a proportional odds regression model with appropriately ordered response categories. The model with the natural spline can be estimated using `polr()` with `ns()`. Because the `mgcv` package cannot use smoothing splines with these models,

use the `vgam()` function of the `VGAM` package instead. The function has a similar syntax as `vglm()` but now with this package's `sm.ps()` function to denote the smoothing spline.

```
vgam(formula = type.order ~ sm.ps(x = density), data = wheat,
   family = cumulative(parallel = TRUE))
```

Note that `type.order` is the ordered version of `type` as given in Section 3.4.

8. Exercise 5 of Chapter 2 examined the field goal kicking data from Berry and Wood (2004). The authors use linear splines for the distance in a logistic regression model to estimated the probability of success. The purpose here is to use cubic splines to investigate these data further.

   (a) Fit a logistic regression model using a natural spline for the distance. Use knots equal to 20 to 60 by five-yard increments as the authors did in their analysis.

      i. Construct a plot of the model and include a first-order model on the plot as well. Are there differences between the models?

      ii. Comment on the wiggliness of the spline model.

      iii. Are there other knot choices that may result in a better fitting model?

   (b) Reconstruct the plot in (a) without the first-order model. Add 95% confidence interval bands for the probability of success to the plot. If you have access to Berry and Wood (2004), compare your plot to Figure 1 of the paper. At least small differences should be expected because of their use of linear splines and a Bayesian modeling approach.

   (c) Fit a logistic regression model using a smoothing spline for the distance.

      i. Construct a plot of the model and include a first-order model on the plot as well.

      ii. Compare the model to the first-order and the original natural spline model from (a).

   (d) Use the smoothing and original natural spline models to interpret distance with odds ratios and 10-yard decreases in the variable. Compare to an odds ratio obtained from a first-order model.

Appendix A

An introduction to R

R is a statistical software package that shares many similarities with the statistical programming language named S. A preliminary version of S was created by Bell Labs in the 1970s that was meant to be a programming language like C but for statistics. John Chambers was one of the primary developers of the language, and he later won the Association for Computing Machinery Award for it in 1999. A nice video interview with John Chambers about the early days of S is available at `https://www.youtube.com/watch?v=jk9S3RTAl38`. In the 1980s, Bell Labs sold an exclusive license to a company that would market S with additional add-ons. The resulting software became known as S-Plus and was sold to users. This software was popular among many individuals in academics and had some niche followings outside of academics.

Ross Ihaka and Robert Gentleman, University of Auckland faculty, started to develop a new *free* software package named R that was "not unlike S" in the early 1990s. The name "R" was chosen because both of them have this as their first letter of their first name, and because it was close to "S" in the English alphabet. Ihaka and Gentleman (1996) formally introduced the world to R, and version 1.0.0 of R was released on February 29, 2000. R quickly became the standard statistical analysis software for those in academics. Its popularity continues to grow in industry, especially as new graduates enter the workplace. This popularity has led many commercial statistical software packages, including SAS and SPSS, to incorporate ways to run R directly from their own software. Discussions about R have even made it into mainstream media outlets, including the *New York Times* article "Data Analysts Captivated by R's Power" by Vance (2009), which is often thought of as a landmark for R's acceptance.

The purpose of this appendix is to help readers begin using R. While we start with its native graphical user interface, we will provide details on how to use R within integrated development environments that allow users to write programs within editors, view output, and perform other tasks that make using R much easier. Finally, we give a basic regression analysis example using R. All R code used in this appendix is within the AppendixInitialExamples.R and GPA.R programs that are available on the textbook's website. For a more comprehensive introduction to R, we refer readers to our "Computing Tools for Statisticians" course website at `www.chrisbilder.com/stat850`. This website provides course notes, teaching videos, and R code to reproduce all examples discussed in that course.

A.1 Basics

R is a free, open-source software package that is available for download through the Comprehensive R Archive Network (CRAN). Below is the step-by-step process to find the executable file to download:

DOI: 10.1201/9781003093091-A

1. Go to the R website at `http://www.r-project.org`. Select the CRAN link on the left side.

2. CRAN is mirrored at a large number of locations around the world. Choose the location closest to you.

3. R is available for Linux, Mac, and Windows operating systems. Choose your desired operating system's link. We chose the Windows operating system, and we will use an R version for Windows throughout the book.

4. Select the "base" link for the base distribution of R to download.

5. Select the "Download R-*.*.* for Windows" link to download the executable file.

After downloading, run the executable file to install R. The default installation settings are satisfactory for most R users.

To begin R, select the icon on the desktop or from the Windows Start or Apps menu. When R is first opened, the graphical user interface (GUI) looks similar to Figure A.1. The cursor is positioned in the R Console window at the ">" command prompt waiting for a command. The simplest commands use the R Console as a calculator.

```
> 2+2
[1] 4
> (2-3)/6
[1] -0.1666667
> 2^2
[1] 4
> pnorm(1.96)
[1] 0.9750021
> sin(pi/2)
[1] 1
> log(1)
[1] 0
```

For each example, the <Enter> key is pressed after the line of code is typed. Each numerical result is preceded by a `[1]` indicating this is the first value in a *vector* result; we will discuss vectors and how multiple values can be produced later.

Results from calculations can be stored in an *object*. Objects are just listings of information, possibly of different types. Simple objects may contain only one number. Complicated objects may contain data, model-fitting information, parameter estimates, residuals, and other relevant output. Below is an example creating an object called `save` to store the results from $2 + 2$.

```
> save <- 2+2
> save
[1] 4
```

The symbol combination `<-` (less than and a dash) makes the storage assignment; notice that it looks like an arrow pointing the computation results to the object name. This symbol combination is read as "gets." Thus, "save gets the results from $2 + 2$." The equal sign can be used to make the storage assignment too, although `<-` is more prevalent among R

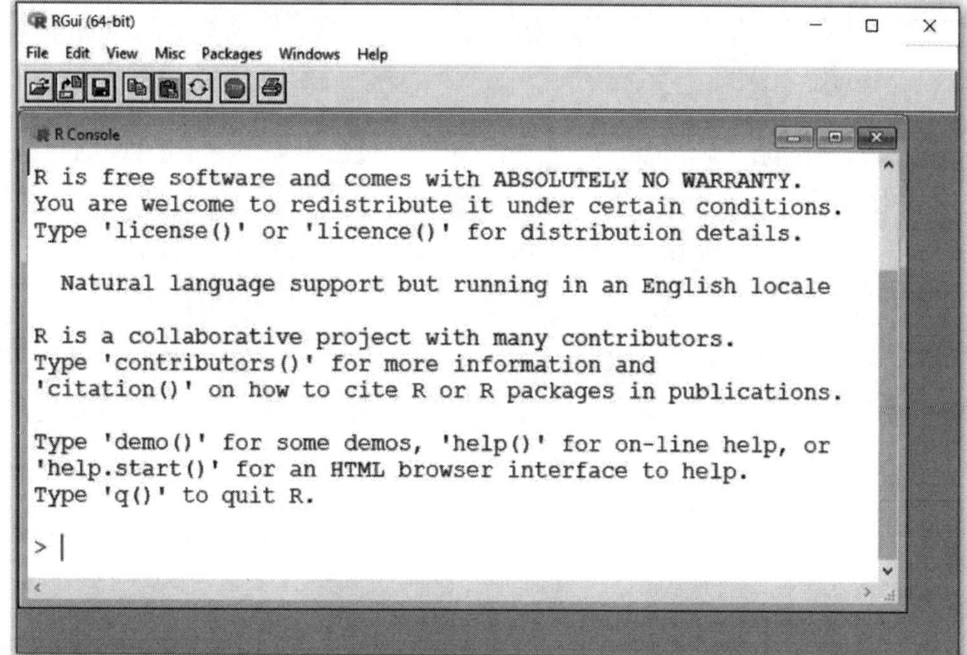

Figure A.1: The R GUI.

users. An object name can be any word or combination of letters, numbers, periods, and underscores, provided a letter is first. Periods and underscores are often used to separate multiple descriptive words. With respect to letters, R is case sensitive. Thus, `save` differs from `Save`.

To see a list of objects created in R, either of the following will work.

```
> ls()
[1] "save"
> objects()
[1] "save"
```

To delete an object, use `rm()` and insert the object name inside the parentheses. All objects are removed automatically from R's *workspace* when R is closed (select File > Exit from the menu bar) unless the user chooses to save them upon on exit. For most applications in this book, saving these objects for use in a future R session will not be necessary.[1]

[1]When exiting R, a window will appear that asks the user if they would like to Save workplace image? If Yes is selected, a .RData file is created that contains all objects from the current R session. These objects can be loaded back into a new R session by selecting File > Load Workspace. A situation when it is useful to retain these objects is when a set of R code takes a long time to execute. Saving the workspace allows one to use the results in a new R session without running the same code again.

A.2 Functions

R performs its calculations using *functions*. For example, the `pnorm()` command used earlier is a function that calculates the probability that a standard normal random variable is less than the specified value within the parentheses. Writing your own function is simple. Suppose you want to write a function to calculate the standard deviation. While the `sd()` function already exists for this purpose, our new function will provide a basic illustration. Below is our new function `sd2()` along with a set of data in a new object named `x`.

```
> x <- c(1,2,3,4,5)
> x
[1] 1 2 3 4 5
> sd2 <- function(numbers) {sqrt(var(numbers))}
> sd2(x)
[1] 1.581139
```

In this example, five observations are saved or "concatenated" into `x` using the `c()` function, which combines the numbers into a vector object type. This vector has five elements inside of it with its first (`[1]`) value equal to 1. A function called `sd2()` is then written to calculate the square root of the variance by taking advantage of the functions `sqrt()` and `var()`. The function requires a single input, or *argument*, which represents the information that will need to be given to the function when it is run. This argument can be any name, subject to the naming rules for objects described earlier, and we call it `numbers` here. The function's body is included within braces, { }, and contains the manipulations of the argument that will lead to the desired result. The `sd2()` function is stored as an object in R's workspace. This function can be run at any time during the R session (i.e., until R is exited). The last line of code runs the function on our object `x`, which is passed into the function as the value for `numbers`.

When a function has multiple lines of code in its body, the last line corresponds to the returned value. For example, the code below shows a new `sd2()` function with two lines of code in its body, but only the last line is saved into an object.

```
> sd2 <- function(numbers) {
+   cat("Print the data \n", numbers, "\n")
+   sqrt(var(numbers))
}
> save <- sd2(x)
Print the data
1 2 3 4 5
> save
[1] 1.581139
```

Included within the function is a call to the `cat()` function. This is used to print text and objects from within inside a function (it works outside a function too). Inside `cat()`, the `\n` is a special escape character that moves printed text to the next line. The + signs in the above code were not actually typed. Rather, at the end of each line in the function body, the <Enter> key is pressed. When R detects that a complete expression has not been included, the cursor moves to the next line and shows the + sign to prompt the user for more code.

A.3 Help

Help available within R is accessed by selecting Help from the R menu bar. From the resulting drop-down menu, there are a number of help options, including answers to frequently asked questions, user manuals, and search tools. We are going to focus on the HTML help here. By selecting HTML help, we obtain a list for many of the same help resources, but now within a web page. In particular, the "Packages" link under "Reference" leads to a list of *packages*, which are groupings of R functions and data. For example, the `base` package contains a number of R's basic functions, including `log()` and `ls()` used earlier. Selecting the "base" link displays functions within the `base` package. Other packages include `stats` (contains functions used for the most common statistical calculations) and `graphics` (contains functions for plotting).

For the remainder of this section, we examine the help available for the `pnorm()` function, which was used earlier to find the probability that a standard normal random variable is less than 1.96. Beginning from the web page that lists the available packages, select the `stats` package and then select the "pnorm" link. Under the "Usage" heading in the resulting help web page for the function, the full function syntax is given as shown below.

```
pnorm(q, mean = 0, sd = 1, lower.tail = TRUE, log.p = FALSE)
```

The function has arguments `q`, `mean`, `sd`, `lower.tail`, and `log.p` that either need to be specified or will take on the stated default values given after the "=" symbol. The help web page further explains that `q` is a quantile, which we set at 1.96 earlier. Values for arguments can appear in any order in the function call as long as their names are specified, or they can be listed in order without names. Using the optional forms of syntax for `pnorm()`, we produce the same probability as before with the following code.

```
> pnorm(1.96)
[1] 0.9750021
> pnorm(q = 1.96)
[1] 0.9750021
> pnorm(1.96, 0, 1)
[1] 0.9750021
> pnorm(mean = 0, sd = 1, q = 1.96)
[1] 0.9750021
```

The first and third lines of code provide examples where the argument names are excluded, but the numerical values are in the same order as in the function syntax. The second and fourth lines of code include the argument names. We strongly recommend including argument names in functions (except for the most basic functions) to avoid possible mistakes and to make code more readable.

The help web page contains other information in addition to syntax for `pnorm()`. For example, it gives usage examples and syntax information for closely related R functions. Finally, the same help for `pnorm()` can be opened by typing `help(topic = pnorm)` and <Enter> at the R Console prompt, or by other methods that depend on the integrated development environment being used (see Appendix A.6).

A.4 Using functions on vectors

Many R functions repeat a calculation when given a vector input. For example, the code below produces two results for each function call.

```
> pnorm(q = c(-1.96,1.96))
[1] 0.02499790 0.97500210
> qt(p = c(0.025, 0.975), df = 9)
[1] -2.262157 2.262157
```

The `pnorm()` function gives the probability of being to the left of -1.96 and 1.96 for a standard normal random variable, and the `qt()` function gives the 0.025 and 0.975 quantiles from a t distribution with degrees of freedom specified by the `df` argument value.

For a more complex example, suppose you would like to find the 95% confidence interval for a population mean using the usual formula based on the t distribution. Below is a set of code demonstrating how this can be done with vectors and a sample contained within an object named `x`.

```
> x <- c(3.68, -3.63, 0.80, 3.03, -9.86, -8.66, -2.38, 8.94, 0.52,
   1.25)
> x
[1] 3.68 -3.63 0.80 3.03 -9.86 -8.66 -2.38 8.94 0.52 1.25
> var.xbar <- var(x)/length(x)
> var.xbar
[1] 3.246608
> mean(x) + qt(p = c(0.025, 0.975), df = length(x)-1) *
   sqrt(var.xbar)
[1] -4.707033  3.445033

> t.test(x = x, mu = 2, conf.level = 0.95)

        One Sample t-test

data:  x
t = -1.4602, df = 9, p-value = 0.1782
alternative hypothesis: true mean is not equal to 2
95 percent confidence interval:
 -4.707033  3.445033
sample estimates:
 mean of x
 -0.631
```

In the code, the `mean()` function calculates the sample mean, the `var()` function calculates the sample variance, the `length()` finds the length of a vector (sample size), and the `sqrt()` function calculates the square root of the sample size. Because the `qt()` function produces a vector of length 2, it may not be immediately clear how R multiplies this vector and `sqrt(var.xbar)`. When R performs arithmetic operations on both a single number and a vector, it performs the operation separately on the single number with each element of the vector. Thus, R multiplies both -2.262157 and 2.262157 by $\sqrt{3.246608}$ resulting in a new vector of length 2. When the mean is eventually added to the vector, both the lower and upper confidence interval limits result. While the ability to work with vectors in this manner can simplify code, we recommend caution to make sure the calculations are done correctly.

The last function call to `t.test()` provides a simpler way to calculate the confidence interval. The results from a hypothesis test (population mean $= 2$ or $\neq 2$) are also given by the function.

A.5 External packages

Outside of the default installation for R, there are many packages, most of which can be downloaded from CRAN. These packages have been created by R core group members,[2] researchers, and everyday users to extend the capabilities of R. This is one of the primary reasons why R has become widespread—anyone can write their own functions to perform tasks and distribute these functions through a package. Next, we go through a specific installation example involving the `binom` package, which is used in Section 1.1.2.

While in the R Console, select Package > Install package(s) from the R menu bar. A number of CRAN mirror locations from around the world then appear in a new window. After a location is selected, all available packages are listed. We scroll down to select the `binom` package and then select OK. The package is downloaded from the CRAN mirror and installed. An alternative installation method is to run the `install.packages()` function from the command prompt. For the `binom` package, we would run `install.packages(pkgs = "binom")`. With either installation method, other packages may be installed automatically if this chosen package depends on them.

To use the installed package, type `library(package = binom)` and <Enter> in the R Console. This needs to be done only once when R is open. If you exit R and reopen it, the package does not need to be installed again, but `library(package = binom)` does need to be executed.

A.6 Programming environments

The R Console automatically executes each line as it is entered. It is often more convenient to type multiple lines of code all at once and then execute them together. To do this, a *program* (or *script*) needs to be typed into a text editor. Notepad or other basic editors can suffice, but there are several built specifically for use with R that have very useful features. Many of these editors are included within integrated development environments (IDEs) to facilitate not just writing a program but also to see its output, to list objects created, to enable dynamic document creation, and to even create packages.

[2]These individuals can modify the R source code; see `http://www.r-project.org/contributors.html`.

A.6.1 R editor

A very limited program editor is available within R. The editor is accessed through File > New script. Code can be typed into the R Editor window and saved as a program using File > Save, while the cursor is within the editor. Note that the file extension used with R programs is .R, and these files contain plain text. A whole program is run by selecting Edit > Run all. Alternatively, a mouse highlighted set of code is run by selecting Edit > Run line or selection.

A.6.2 RStudio

RStudio Desktop (`https://posit.co/products/open-source/rstudio`; more commonly referred to simply as "RStudio") is an IDE produced by Posit (`https://posit.co`). It is by far the most widely used IDE for R. Its market penetration is so dominant that new R users often think of R and RStudio as the same. Instead, Posit is a for-profit company that first released RStudio in 2011. The software is available for free under a public license, but a paid version is available for commercial use. RStudio runs on all major operating systems.

RStudio integrates a program editor, an R Console window, a graphics window, an R-help window, and many other items within one overall window environment. To create a new program, select File > New File > R Script, or open an existing program by selecting File > Open File. Selecting the corresponding icons on the toolbar work for this purpose as well. Note that program files cannot be dragged and dropped into the program editor to open. Figure A.2 shows a screen capture of RStudio where AppendixInitialExamples.R is shown in the program editor.[3] To run a segment of code, highlight it and then select the Run icon in the program editor window (upper right) or just press <Ctrl-Enter>. The code will be automatically transferred to the Console window and executed. Any plots that are created will be sent by default to the Plots tab in the lower-right window. Help can be accessed through the Help or Packages tabs in the lower-right window as well.

One of the most significant benefits to using RStudio rather than R's editor is syntax highlighting. This means that code is colorized according to its purpose. For example, comments are green (these begin with a hashtag "#" symbol). This syntax highlighting can make reading and editing R code much easier. Multiple programs can be open at once in different tabs of the program editor window. Also, the editor can suggest function or package names from any loaded package if <Tab> is pressed at the end of a text string. For example, typing "pn" and pressing <Tab> yields a pop-up window suggesting functions `pnbinom()`, `png()`, and `pnorm()`. Pressing <Tab> where an argument could be given within a function (e.g., after a function name and its opening parenthesis or after a comma following another argument) gives a list of possible arguments for that function.

Below are additional important comments and tips for using RStudio:

- All window sizes can be adjusted by grabbing the borders between them with a mouse and sliding these borders left and right or up and down.
- The program editor can be launched in a completely separate window on the computer by selecting the Show in new window icon in the upper-left portion of the program editor window. This is useful when working in a multiple-monitor environment—one

[3]The program editor window of RStudio will only be present if there is a program open. This can be confusing to new RStudio users who open RStudio for the first time without directly including a program. Simply creating a new program or opening a program will open the program editor window.

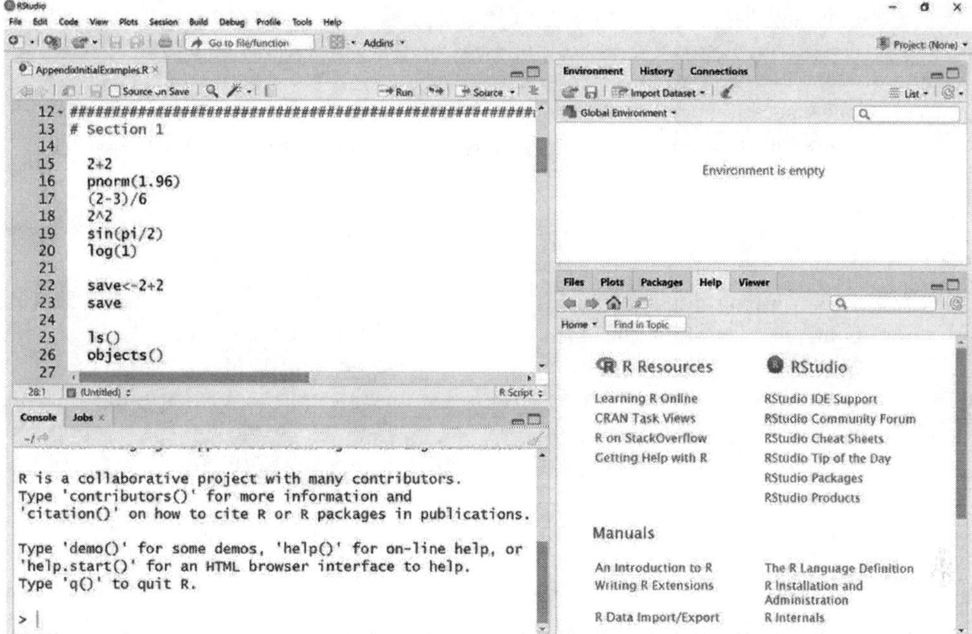

Figure A.2: Screen capture of RStudio.

monitor can be devoted to the program editor while another monitor can be devoted to the console.

- Panes are groups of windows, like for the program editor, within RStudio. Their locations can be interchanged by selecting Tools > Global Options > Pane Layout. For example, some users will have the program editor and the console in the upper-left and right panes, respectively.

- Font sizes for most panes can be changed by selecting Tools > Global Options > Appearance. For example, it may be useful to increase the font size during a presentation (e.g., teaching in a classroom when a laptop screen is mirrored upon a projection screen). However, using larger fonts reduce the number of characters that can be seen within the windows. Viewing the program editor outside of the main RStudio GUI can be quite helpful in this situation—maximize the program editor while discussing code and maximize the console window when discussing output from the code.

- The style of syntax highlighting can be changed by selecting Tools > Global Options > Appearance. There are a number of themes available for syntax highlighting. While there are no built-in ways to make specific changes to a particular theme, users can create their own theme outside of RStudio and load it into the software.[4]

- To keep programs, data, objects, and other items that are part of a central focus together, RStudio provides *projects* as an organizational tool. Projects are created by selecting File > New Project. This will create a .Rproj file in a selected folder that contains information regarding your current settings.

[4]Information is available at `https://support.rstudio.com/hc/en-us/articles/115011846747-Using-RStudio-Themes`.

The positive impact that Posit has had on R is immense. Along with creating an IDE that brought R to the masses, the company has led in the development of dynamic document creation (R Markdown), web-based applications (Shiny), and made it much easier to create packages. Still, there is concern among the R community that the company has too large of an impact on the R open-source project and its influence now and into the future may not be for the best (Matloff, 2023).

A.6.3 Tinn-R

Before RStudio was created, there were a number of program editors available and none had a dominant position among users. One of these editors that continues to be developed is Tinn-R. This project has gone beyond being a program editor and can be considered a full IDE. While Tinn-R does not provide the same level of options as RStudio, this can be beneficial because most R users do not need all of these options. Tinn-R is open source and available for free from `https://tinn-r.org/en`. This software is limited to Windows operating systems.

Figure A.3 shows a screen capture of Tinn-R with the AppendixInitialExamples.R program. Again, syntax highlighting is automatically incorporated to help distinguish the purpose of the code. To run code from the editor, R's GUI needs to be open. This can be opened by selecting the R control: gui/PuTTY (start/close) icon from the R toolbar (see #1 in Figure A.3). Tinn-R subsequently opens R in its *SDI* (single-document interface), which is a little different from R's *MDI* (multiple-document interface) that we saw in Figure A.1. The difference between the two interfaces is simply that MDI uses the R GUI to contain all windows that R opens and SDI does not. Once R is open in its SDI, program code in Tinn-R can be transferred to R by selecting specific icons on Tinn-R's R toolbar. For example, a highlighted portion of code can be transferred to and then run in R by selecting the R send: selection (echo = TRUE) icon. Note that the transfer of code from Tinn-R to R does not work in the MDI.

Below are some additional important comments and tips for using Tinn-R:

- Upon Tinn-R's first use with R's SDI, the `TinnRcom` package is automatically installed within R to allow for the communication between R and Tinn-R. This package is subsequently always loaded for later uses.

- When R code is sent from Tinn-R to R, the default behavior is for Tinn-R to return as the window of focus (i.e., the window location of the cursor) after R completes running the code. If Tinn-R and R are sharing the same location on a monitor, this prevents the user from immediately seeing the results in R because of it being hidden behind the Tinn-R window. To change this behavior, select Options > Application > R > Rgui/PuTTY and uncheck the Return to Tinn-R box. Alternatively, select the Options: return focus after send/control Rgui icon on the Misc toolbar.

- By default, the line containing the cursor is highlighted in yellow. To turn this option off, select Options > Highlighters (settings) and uncheck the Active line (choice) box.

- Long lines of code are wrapped to a new line by default. This behavior can be changed by selecting Options > Application > Editor > Display and then selecting the No radio button for Line wrapping.

- Syntax highlighting can be maintained with code that is copied and pasted into a word processing program. After highlighting the desired code to copy, select Edit > Copy formatted (to export) > RTF. The subsequently pasted code will retain its color.

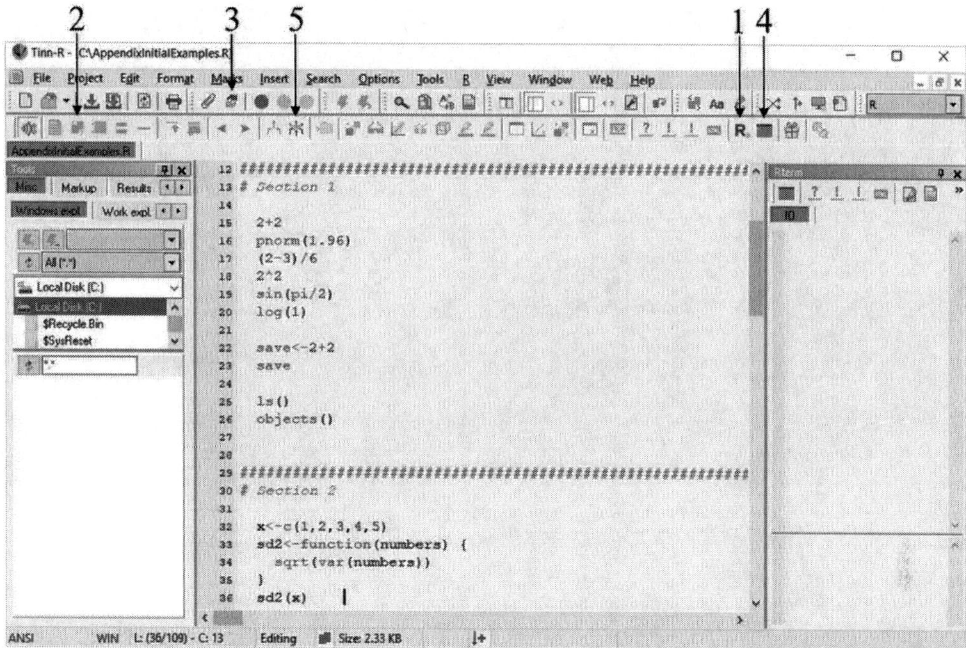

Figure A.3: Screen capture of Tinn-R. Toolbar icons referenced in the text include: (1) R control: gui/PuTTY (start/close), (2) R send: selection (echo = TRUE), (3) Options: return focus after send/control Rgui, (4) R control: term (start/close), and (5) R Send: Knit (file).

- The style of syntax highlighting can be changed by selecting Options > Highlighters (settings). We have found it to be especially helpful to use background colors with particular code types, like comments, to make code quite distinct.
- Tinn-R can run R within its interface by using a link to a terminal version of R rather than R's GUI. To direct code to the Rterm window (located on the right side of Figure A.3), select the R control: term (start/close) icon on the R toolbar. One benefit from using R in this manner is that the syntax highlighting from the program editor is maintained in the R terminal window.

When using Tinn-R and R's GUI, we have found it to be more efficient to view each window side-by-side in a wide-screen monitor or in a multiple monitor environment. This allows for both to be viewable at the same time. Code and output can be side-by-side in large windows without needing to switch back-and-forth between overlaying windows.

A.6.4 Other ways to use R

Using R programs with IDEs is the predominant way that most individuals use R. RStudio and Tinn-R are the main IDEs that we and our students use to write/run R programs. However, many IDEs that are meant for programming in a wide variety of languages (e.g., Visual Studio, Eclipse, and Atom) have plug-ins that make them compatible with writing and running R programs.[5]

[5] A Stack Exchange discussion of IDEs for R is available at `https://datascience.stackexchange.com/questions/5345/ide-alternatives-for-r-programming-rstudio-intellij-idea-eclipse-visual-stud`.

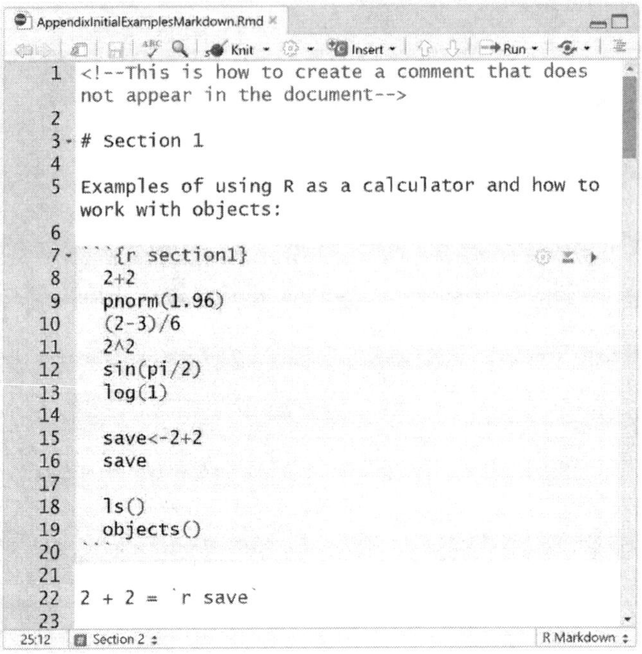

AppendixInitialExamplesMarkdown.Rmd

```
<!--This is how to create a comment that does
not appear in the document-->

# Section 1

Examples of using R as a calculator and how to
work with objects:

```{r section1}
  2+2
  pnorm(1.96)
  (2-3)/6
  2^2
  sin(pi/2)
  log(1)

  save<-2+2
  save

  ls()
  objects()
```


2 + 2 = `r save`

```

25:12 Section 2 R Markdown

**Figure A.4**: Screen capture of using R Markdown in RStudio.

All of these systems merely provide a convenient environment in which to develop code and view output. We next discuss two additional ways to use R beyond running R programs. The first way allows users to embed the program within a report structure. The second way allows users to interact with R in an environment similar to an R console that allows them to edit and rerun past code in place.

### R Markdown

Some R users combine their code, output, and data analysis discussion within one document via the R Markdown language. This language combines the use of the R and the Markdown languages to create a document, like an HTML, PDF,[6] or Word document, when the program is run. The process is known as dynamic document creation. A main advantage to this approach is it allows for the reproducibility of work much more easily and accurately without needing to copy/paste code and output into a document containing the data analysis discussion. Both RStudio and Tinn-R support the use of R Markdown. All R Markdown programs are simply plain text files that use a .Rmd extension.

To create an R Markdown program using RStudio, select File > New File > R Markdown. A window will appear that allows one to specify the title and author of the document along with the type of document to create when the program is run. After selecting OK in this window, RStudio will create a template program that can be edited. To create a similar template program in Tinn-R, select File > Template > R Markdown. Figure A.4 shows the beginning of our own program based on the examples earlier in Appendix A. Our corresponding file is AppendixInitialExamplesMarkdown.Rmd.

[6]PDF files require an installation of LaTeX, which is a typesetting system to produce documents. Distribution systems, such as MikTeX (`https://miktex.org`) and Tex Live (`https://tug.org/texlive`), provide the necessary software for free.

Regular text in the document is included using the Markdown language. For example, a section title for the document is created by using a "#" symbol. Notice that this is different from an R comment that used the same symbol. Regular text within a section is typed as it would be in a word processor. Text attributes, like bolding, are created using code (for bolding, use two asterisks around the corresponding text). R code to be executed is included within *chunks*. These chunks begin with three backquote symbols that are followed by the syntax: `{r <chunk name>, <chunk options separated by commas>}`, where the appropriate name/option is substituted for "< >". The chunk ends with three backquotes. Previews of chunk output in RStudio are available by selecting the right arrow box within a chunk. R code may also be included within the Markdown code by enclosing it within single backquotes. Additional examples of R Markdown code—including equations and plots—are available in our R Markdown file.

When R Markdown code is run, the output from the two languages is combined into one document. This is performed by the `knitr` package, whose name originates from the concept of "knitting together" output from separate languages. Within RStudio, the program is run by selecting the drop-down arrow next to the Knit icon in the program editor and then choosing the appropriate document type. HTML documents are sent to the Viewer window within RStudio. These documents can be exported to a web browser by selecting the Show in new window icon within this window. To create an HTML document from Tinn-R, select the R Send: Knit (file) icon (see #5 in Figure A.3) on the R toolbar. The document is sent to a web browser, and the HTML code is opened into a new tab of Tinn-R. More information on how to use R Markdown is available on RStudio's website at `https://rmarkdown.rstudio.com`.

A more flexible alternative to R Markdown is to use chunks of R code in LaTeX or LyX.[7] When code from these documents are run in combination with `knitr` and R, the R code/output is knit together with the output created by LaTeX/LyX, usually in a PDF form. We include the LyX file, AppendixInitialExamplesLyX.lyx, with our supplementary materials that duplicates the examples given earlier in Appendix A. Additional introductory and more advanced examples of this work flow are provided on our "Computing Tools for Statisticians" course website mentioned at the beginning of Appendix A. This includes the first portion of a short course that we have given on this book.

### Notebooks

Most statistical software packages keep the code and output in separate windows, although the code may be echoed into the output window when it is executed. Another way to combine code and output is within a single-window *notebook* programming environment. This environment is similar to the programming editors described previously in this appendix, except that as soon as any portion of the code is executed, the output for that code is displayed in the same window immediately below it and remains there even when other code portions are run later. Thus, it is like using the R Console window directly, but one can easily edit and rerun previous parts of code by simply going back to where it is located.

RStudio provides access to a notebook environment with its R Markdown Notebook. A template file is created by selecting File > New File > R Notebook. Saved notebooks are simply plain text R Markdown files with small formatting differences. These differences allow the output to be previewed without running the entire code. This output needs to be manually created within the program editor by selecting the right arrow box within a

[7]LaTeX uses a markup-style code that needs to be run to produce documents, like a PDF file. LyX (`www.lyx.org`; available for free) provides an interface similar to Microsoft Word for using LaTeX. This allows individuals to largely avoid coding. Our book is written in LyX!

chunk, like what was described in our earlier R Markdown discussion. The Preview icon in the program editor sends an HTML version of the document to the Viewer window. Whenever the notebook is saved, the Viewer window updates its contents.

Jupyter Notebook provides an alternative to RStudio. This is part of Project Jupyter (`https://jupyter.org`) that provides free, open-source software for individuals to write/run computer programs. The name "Jupyter" is derived from the three languages originally used with it—Julia, Python, and R. While Jupyter Notebook can be downloaded and installed by users on their own computer, in our experience most people will use it through a web-based interface from a central webserver at a university or a company that makes access available. There are number of companies that provide this type of service, including CoCalc at `cocalc.com`. The company provides free, convenient access with an option to upgrade to a paid subscription that provides better access and a larger amount of file storage.

To demonstrate how to use Jupyter Notebook, we logged into a computing cluster at one of our universities via a web browser. We opened Jupyter with R (often referred to as an "R kernel") to begin using the program. New R code is included in a *cell* via the + icon on the main toolbar. The cell's code is run by selecting the right arrow next to the cell. Figure A.5 provides a screen capture showing our code from examples earlier in Appendix A. Cells can be converted to a Markdown format by selecting Markdown from the drop-down menu on the toolbar (shown as "Code" in Figure A.5). To run all code within the notebook at once, select Cell > Run All. Notebook files are saved with a .ipynb extension.[8] Our file for the notebook is AppendixInitialExamplesJupyter.ipynb.

While it is helpful to see the output right after the code, this can also present problems. In particular, code from earlier portions of the program can be changed and rerun, but later portions of the output that depend on these results will not change without rerunning those corresponding cells. Also, depending on how one writes code within a cell, it can be difficult to match the output with the code, as illustrated by the Section 1 example of Figure A.5. Using this programming environment requires careful planning and organization of code.

---

## A.7 Regression example

This section provides a simple example showing how to use R in a linear regression setting. The regression aspect should not be difficult for readers with the needed background for this book. The R aspect will be new to many readers.

### A.7.1 Background

Universities often want to predict how well new freshmen will perform given their background. Suppose a university selects a simple random sample of 20 current students at the end of their first semester. The college grade point average (GPA) and the high school GPA are recorded for each student. A hypothetical data set for this scenario is given in the data file GPA.csv that is available on the book's website. We recommend viewing the structure of the file in a simple text editor, such as Notepad for Windows, to see its format. The file's first line contains the variable names, and subsequent lines contain the observations for the 20 students. For example, the first observation is for a student with a high school GPA

[8] Jupyter was originally part of iPython and retains a file extension from it.

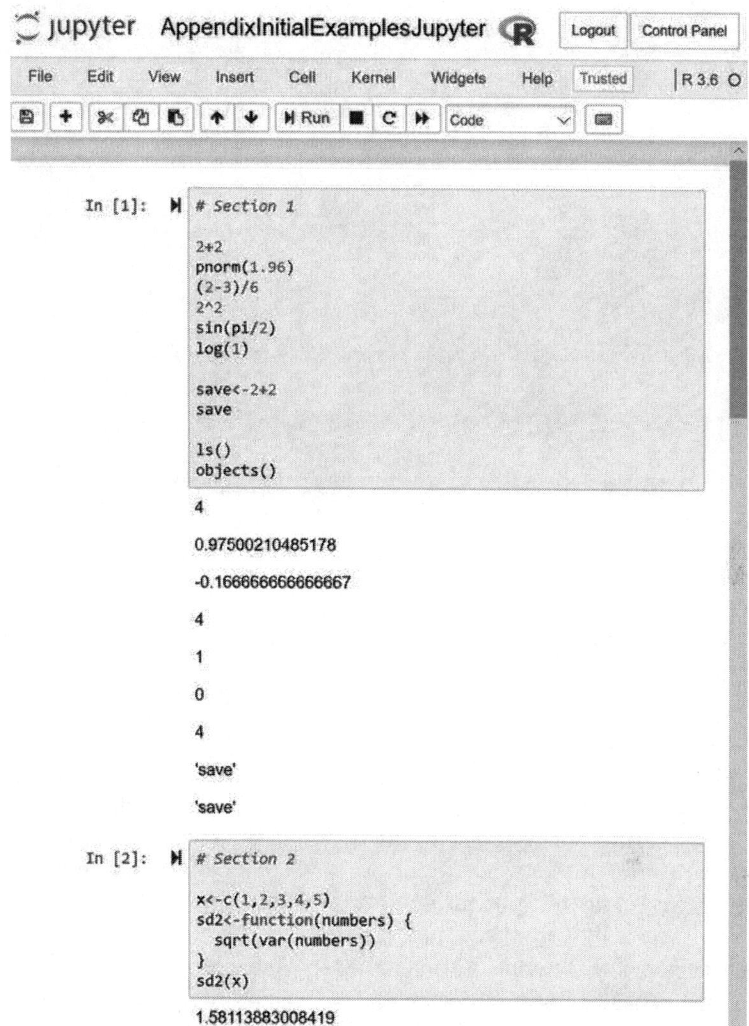


**Figure A.5**: Screen capture of Jupyter Notebook.

of 3.04 and a college GPA of 3.10. Variable names and data values for an observation are separated by commas. Thus, the file format is called "comma-delimited." This file format is a standard way to represent data in an organized manner that can be used by multiple software packages. In particular, Excel can open these files directly, replacing commas with cell borders.

Using the data in GPA.csv, we want to fit a simple linear regression model with college GPA as the response variable and high school GPA as the explanatory variable. The GPA.R program contains all of the R code used for our analysis. Below is the first part of the code after running it in R.

```
> # Read in the data
> gpa <- read.csv(file = "C:\\data\\GPA.csv")
>
> # Print the data
```

```
> gpa
   HSGPA CollegeGPA
1   3.04       3.10
2   2.35       2.30
3   2.70       3.00
4   2.55       2.45
5   2.83       2.50
6   4.32       3.70
7   3.39       3.40
8   2.32       2.60
9   2.69       2.80
10  2.83       3.60
11  2.39       2.00
12  3.65       2.90
13  2.85       3.30
14  3.83       3.20
15  2.22       2.80
16  1.98       2.40
17  2.88       2.60
18  4.00       3.80
19  2.28       2.20
20  2.88       2.60
```

Lines that begin with a # symbol denote comments that are not executed in R. We use comments like these to inform other users or to make notes for ourselves. The first executable line involves the `read.csv()` function that reads (imports) the data file into R stored at the location given as the `file` argument value. This location needs to be updated to the corresponding location on the user's computer. Note the use of double backslashes between folder names. Single forward slashes would work as well. After reading in the data, they are stored as a *data frame* and put into an object named gpa. The contents of this object are displayed by typing its name and then <Enter> at the command prompt.

Data files may come in different formats than we have for GPA.csv. A space-delimited format uses spaces, rather than commas, to separate columns of data. For example, our file GPA.txt contains the same data as in GPA.csv but with this alternative delimiting format. The file can be read into R using the `read.table()` function.

```
> gpa6 <- read.table(file = "C:\\data\\GPA.txt", header = TRUE, sep
   = "")
```

The `header = TRUE` argument indicates that the first line in the file contains the variable names. The `sep = ""` argument specifies that some type of "white space," such as a space, is used to separate each variable name and each data value.

Another commonly used data format is an Excel file. The *R Data Import/Export* manual (select Help > Manuals (in PDF) within R) provides options for how to read in Excel files; however, the manual says "The first piece of advice is to avoid doing so if possible!" This is because of the different Excel file formats (.xls or .xlsx) and 32-bit vs. 64-bit driver issues, all of which can interfere with the import process. The recommended alternative is to save an Excel file in a comma- or space-delimited format instead and use the previously described methods to read in the data. For those readers still wanting to read-in an Excel file into R, we provide examples in our program using the `xlsx` and `readxl` packages.

A.7.2 Data summary

We calculate simple summary statistics for the gpa object.

```
> summary(object = gpa)
     HSGPA          CollegeGPA
Min.   :1.980   Min.   :2.000
1st Qu.:2.380   1st Qu.:2.487
Median :2.830   Median :2.800
Mean   :2.899   Mean   :2.862
3rd Qu.:3.127   3rd Qu.:3.225
Max.   :4.320   Max.   :3.800
```

The `summary()` function summarizes the data for all variables in the object. To create a scatter plot of the data, we use the `plot()` function from the `graphics` package.

```
> plot(x = gpa$HSGPA, y = gpa$CollegeGPA)
```

The `graphics` package, along with a few others, is always available to users so that `library(package = graphics)` does not need to be run prior in the R session.[9] The `plot()` function creates plots in an R Graphics window, like the one shown in Figure A.6. In RStudio, the plot will appear in the plot tab of a pane instead. The `x` and `y` arguments of `plot()` define the data that represent the x and y axes, respectively. To specify a variable within a data frame, the `$` is used to separate the object name from the variable name. For example, `gpa$HSGPA` refers to the `HSGPA` variable in the `gpa` data frame, as shown below.[10]

```
> gpa$HSGPA
 [1] 3.04 2.35 2.70 2.55 2.83 4.32 3.39 2.32 2.69 2.83 2.39
[12] 3.65 2.85 3.83 2.22 1.98 2.88 4.00 2.28 2.88
```

There are other ways to access parts of a data frame. R uses a matrix representation for such objects, with variables in columns and observations in rows. Positions within the matrix are stored by `[row,column]`. For example, `gpa[,1]` produces the values for all rows in the first column of `gpa`, giving the same result as `gpa$HSGPA`. Other examples include: (1) `gpa[1,]`, which gives all variable values for the first observation, (2) `gpa[1,1]`, which gives the value in row 1 and column 1, and (3) `gpa[1:10,1]`, which gives the first 10 observations of variable 1 (`1:10` means 1 to 10 by 1). In addition to numerical values for the columns, variable names can be used too. For example, `gpa[,"HSGPA"]` gives the observations for the first variable.

Optional arguments for `plot()` can be used to improve the scatter plot. Figure A.7 displays a plot using the following code.

---

[9]Use `search()` to see an ordered list of packages that are available within an R session. The ordering is important because there can be times when multiple packages have functions by the same name. When this occurs, R uses the function from the package that appears earlier in this ordering.

[10]If a program editor like those discussed in Appendix A.6 is used, one can simply highlight `gpa$HSGPA` in the `plot()` function call and run the code to produce the same output. This type of highlighting and running can be useful to new users of R, especially for diagnosing errors in complicated function calls.

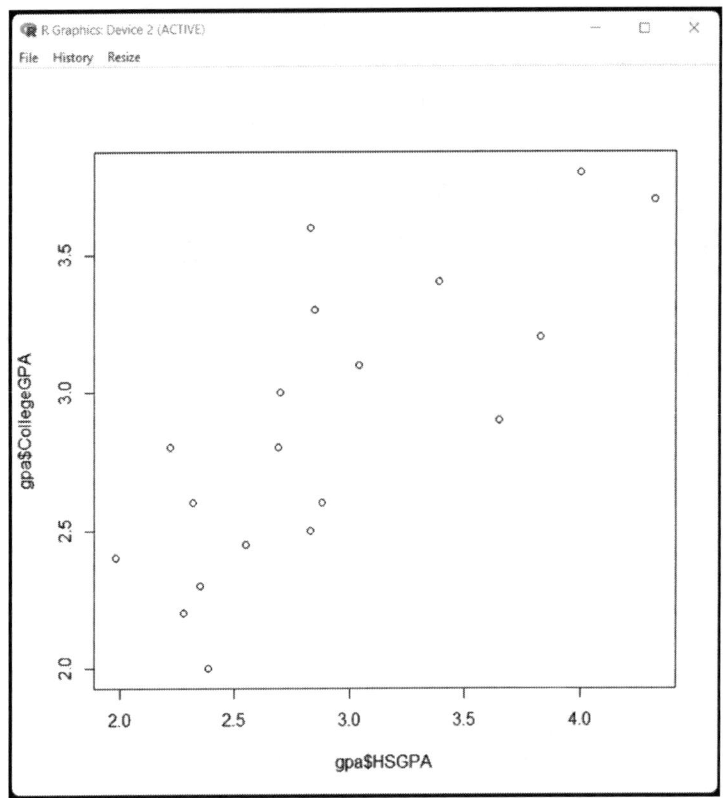

Figure A.6: R Graphics window.

```
> plot(x = gpa$HSGPA, y = gpa$CollegeGPA, xlab = "HS GPA", ylab =
   "College GPA", main = "College GPA vs. HS GPA", xlim = c(0,4.5),
   ylim = c(0,4.0), col = "red", pch = 1, cex = 1.0, panel.first =
   grid(col = "gray", lty = "dotted"))
```

The new arguments are

- `xlab` and `ylab`: The x- and y-axis labels
- `main`: The plot title
- `xlim` and `ylim`: The x- and y-axis limits, where the lower and upper values are combined into a vector
- `col`: The color of the plotting points; colors can be specified by their names (run `colors()` to see all of them or view them at `https://github.com/EarlGlynn/colorchart/wiki/Color-Chart-in-R`) or numbers corresponding to the ordering in the vector produced by `palette()`[11]

[11]Most values represented in the vector are given in the hexadecimal format standard used outside of R. For example, the 2nd color is DF536B, which represents a version of red. Using `col = 2` in `plot()` leads to this red being used by the function.

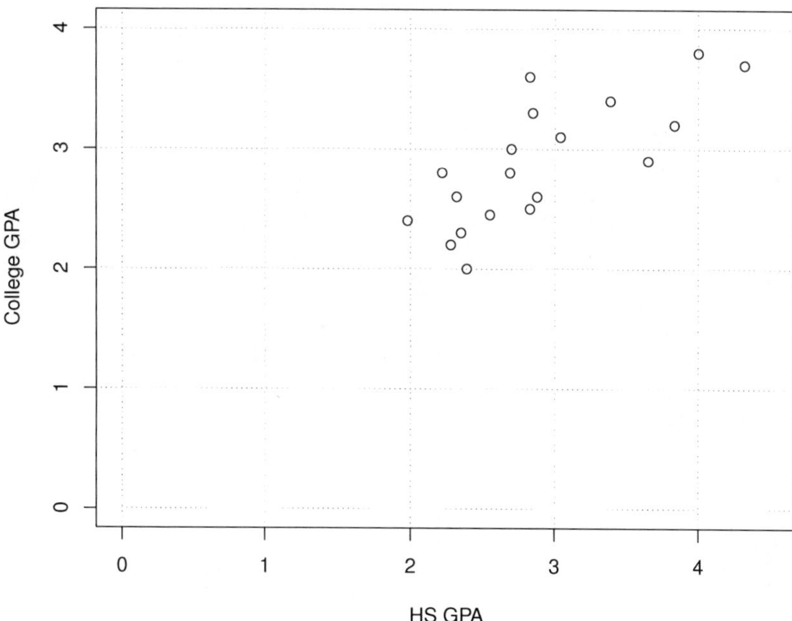

Figure A.7: Scatter plot using optional arguments.

- `pch`: The plotting character, where `pch = 1` denotes an open circle (default); a list of characters can be viewed by running the `pchShow()` function available in the help for `points()`
- `cex`: The plotting character size relative to the default, where 1.0 is the default
- `panel.first`: The plotting of additional items, where gray grid lines with a dotted line type are designated here; the `lty` argument specifies the line type with 1 = "solid", 2 = "dashed", 3 = "dotted", 4 = "dotdash", 5 = "longdash", and 6 = "twodash" (either the number or character string can be used)

The `par()` function, which allows for further control of the graphics parameters, gives additional information on these arguments in its own help.

A popular alternative to the `graphics` package is the `ggplot2` package. Generally, functions from `ggplot2` produce decent plots without much code. However, to obtain a plot of the same quality as displayed in Figure A.7, more code is frequently needed than what would be used with the `graphics` package. We provide an example of using `ggplot2` in our program.

Plots can be copied into word processing software or exported to a file. When the R Graphics window is the current window, select File > Copy to the clipboard > as a Metafile within R (for RStudio, use Export > Copy to Clipboard in the plot window, and then select Metafile and Copy Plot). The plot can be then pasted into a word-processing document. Alternatively, select File > Save as within the R Graphics window to save the plot in a specific format (for RStudio, use Export > Save as Image > Save within the plot window).

A.7.3 Regression modeling

The lm() function

To better understand the relationship between college and high school GPA, we fit a simple linear regression model using the `lm()` function.

```
> mod.fit <- lm(formula = CollegeGPA ~ HSGPA, data = gpa)
> mod.fit
Call: lm(formula = CollegeGPA ~ HSGPA, data = gpa)
Coefficients:
(Intercept)  HSGPA
     1.0869 0.6125
```

The results from `lm()` are saved into an object we have named `mod.fit` to mean "model fit" (a different object name could have been used). The `formula` argument of `lm()` specifies the model in the form `response variable ~ explanatory variable`. If there is more than one explanatory variable, a plus symbol is used to separate each explanatory variable. For example, a second explanatory variable z is included using `formula = CollegeGPA ~ HSGPA + z`. The estimated regression model is $\hat{Y} = 1.0869 + 0.6125x$ where $\hat{Y}$ is the estimated College GPA and x is the high school GPA.

The `mod.fit` object contains much more information than shown so far. Its contents are listed using the `names()` function.

```
> names(mod.fit)
[1] "coefficients"  "residuals"    "effects"      "rank"
[5] "fitted.values" "assign"       "qr"           "df.residual"
[9] "xlevels"       "call"         "terms"        "model"
> mod.fit$coefficients
(Intercept)       HSGPA
1.0868795    0.6124941
> round(mod.fit$residuals[1:5], digits = 2)
   1     2     3     4     5
0.15 -0.23  0.26 -0.20 -0.32
```

The `mod.fit` object is called a *list*. A list has a number of components combined into one object. These components are accessed using the `$` syntax that we saw earlier with the `gpa` data frame. In fact, a data frame is a special type of list where each component is a vector of the same length. Within the `mod.fit` list, the components include `mod.fit$coefficients`, which contains the estimated regression parameters, and `mod.fit$residuals`, which contains the residuals from the model fit. The `round()` function rounds the residuals to two decimal places at the end of the code.

It is often helpful to view the data, the residuals, and the predicted values together in one data set. In the code below, the `data.frame()` function creates a new data frame called `save.fit` containing these items.

```
> save.fit <- data.frame(gpa, C.GPA.hat =
   round(mod.fit$fitted.values, digits = 2), residuals =
   round(mod.fit$residuals, digits = 2))
> head(save.fit)
  HSGPA  CollegeGPA C.GPA.hat residuals
```

```
1    3.04        3.10       2.95      0.15
2    2.35        2.30       2.53     -0.23
3    2.70        3.00       2.74      0.26
4    2.55        2.45       2.65     -0.20
5    2.83        2.50       2.82     -0.32
6    4.32        3.70       3.73     -0.03
```

Inside the data frame, two new variable names (`C.GPA.hat` and `residuals`) are created. Also, we round their numerical values to two decimal places. The `head()` function limits printing to the first six rows of data.

To summarize the information within `mod.fit`, the `summary()` function is used.

```
> summary(object = mod.fit)

Call:
lm(formula = CollegeGPA ~ HSGPA, data = gpa)

Residuals:
     Min       1Q   Median       3Q      Max
-0.55074 -0.25086  0.01633  0.24242  0.77976

Coefficients:
            Estimate Std. Error t value Pr(>|t|)
(Intercept)   1.0869     0.3666   2.965 0.008299 **
HSGPA         0.6125     0.1237   4.953 0.000103 ***
---
Signif. codes: 0 '***' 0.001 '**' 0.01 '*' 0.05 '.' 0.1 ' ' 1

Residual standard error: 0.3437 on 18 degrees of freedom
Multiple R-squared: 0.5768,     Adjusted R-squared: 0.5533
F-statistic: 24.54 on 1 and 18 DF,  p-value: 0.0001027
```

In the `Pr(>|t|)` column of the output, we see that the hypothesis test for a linear relationship between high school and college GPA results in a very small p-value of 0.0001.

Object-oriented language

We have shown that information (i.e., model specifications, estimates, test results) created by a function is stored within an object. Different collections of information are created by functions depending on the types of calculations that are performed. To distinguish objects that contain different collections of information, R assigns each object an attribute called a *class*. We can use the `class()` function to determine an object's class.

```
> class(mod.fit)
[1] "lm"
> class(gpa)
[1] "data.frame"
```

Thus, `mod.fit` has a class of `lm` and `gpa` has a class of `data.frame` (classes do not need to be function names).

Functions are typically designed to operate on only one or very few classes of objects. However, some functions, like summary(), are *generic*, in the sense that different versions of them have been constructed to work with different classes of objects. When a generic function is run with an object as an argument, R first checks the object's class type and then looks to find a *method* function with the name format <generic function>.<class name>. For example, when summary(mod.fit) was run earlier, the function summary.lm() is actually used to summarize the fit from a regression model. Also, when summary(gpa) was run earlier, the function found summary.data.frame(). These two functions produce different output, tailored to suit the type of object on which they are run. If there was no specific method function for an object class, summary.default() would be run. For example, coefficients() is a generic function that gives the same results as mod.fit$coefficients. There is no coefficients.lm() function so the coefficients.default() function is invoked.

The purpose of generic functions is to use a familiar language set with any object. For example, we frequently want to summarize data or a model fit (summary()), plot data (plot()), and find predictions (predict()), so it is convenient to use the same language set no matter the application. This is why R is referred to as an object-oriented language. The object class type determines the function action. Understanding generic functions may be one of the most difficult topics for new R users. The most important point that readers need to know now is where to find help for these functions. For example, if you want help on the results from summary(mod.fit), examine the help for summary.lm() rather than the help for summary() itself.

To see a list of all method functions associated with a class, use methods(class = <class name>). For our regression example, the method functions associated with the lm class are given below.

```
> methods(class = "lm")
 [1] add1          alias          anova           case.names
 [5] coerce        confint        cooks.distance  deviance
 [9] dfbeta        dfbetas        drop1           dummy.coef
 13] effects       extractAIC     family          formula
[17] hatvalues     influence      initialize      kappa
[21] labels        logLik         model.frame     model.matrix
[25] nobs          plot           predict         print
[29] proj          qr             residuals       rstandard
[33] rstudent      show           simulate        slotsFromS3
[37] summary       variable.names vcov
see '?methods' for accessing help and source code
```

To see a list of all method functions for a generic function, use methods(generic.function = <generic function name>). Below are the method functions associated with summary().

```
> methods(generic.function = "summary")
 [1] summary.aov                  summary.aovlist*
 [3] summary.aspell*              summary.check_packages_in_dir*
 [5] summary.connection           summary.data.frame

<OUTPUT EDITED>
```

```
[31] summary.table                        summary.tukeysmooth*
[33] summary.warnings
see '?methods' for accessing help and source code
```

One advantage of using RStudio is that you can type the first few characters of any function name in its help search box to show a list of all functions with these characters, including different method functions.

Graphics

Now that we have the estimated model, we can plot it on a scatter plot. In our demonstration, we make two plots side-by-side in one graphics window. First, we use the `dev.new()` function[12] to open a new graphics window and then use the `par()` function to create a partition of it.

```
> dev.new(width = 8, height = 6, pointsize = 10)
> par(mfrow = c(1,2))
```

The resulting plot window is a 8" × 6" window with a default font size of 10. While opening a new graphics window is not needed (a window will automatically open whenever a plot is constructed in R), we decided to open a new one here to show how to control the window and font size. The `mfrow` argument for `par()` stands for "make frame by row" which results in splitting the plot window into one row and two columns.

Next, we use a similar call to the `plot()` function as earlier, but now add the `abline()` function to plot a line $y = a + bx$ that is solid blue with a thickness twice the default (`lwd = 2`).

```
> plot(x = gpa$HSGPA, y = gpa$CollegeGPA, xlab = "HS GPA", ylab =
    "College GPA", main = "College GPA vs. HS GPA", xlim = c(0,4.5),
    ylim = c(0,4.5), panel.first = grid(col = "gray", lty = "dotted"))
> abline(a = mod.fit$coefficients[1], b = mod.fit$coefficients[2],
    lty = "solid", col = "blue", lwd = 2)
```

As shown earlier, `mod.fit$coefficients` is a vector of length 2. To obtain the estimated y-intercept and slope for the model, we use `mod.fit$coefficients[1]` and `mod.fit$coefficients[2]`, respectively. The resulting plot is given on the left side of Figure A.8.

An alternative to `abline()` is the `curve()` function. This function plots mathematical expressions that vary over one variable, and it allows one to specify the range over which to draw the mathematical function. The right-side plot in Figure A.8 shows the results from rerunning the previous `plot()` function call and following it with a new `curve()` function call as shown below.

[12]There are a number of other functions available to open new graphics windows in R, but they are dependent on the operating system. These functions include `windows()` and `win.graph()` for Windows operating systems and `quartz()` for Mac operating systems. We have found `dev.new()` to be the most general across operating systems despite its Linux-based name. There is one small oddity with using `dev.new()` in RStudio: the function will not open a graphics window on its first run. A second run of the function will open a graphics window outside of the RStudio GUI.

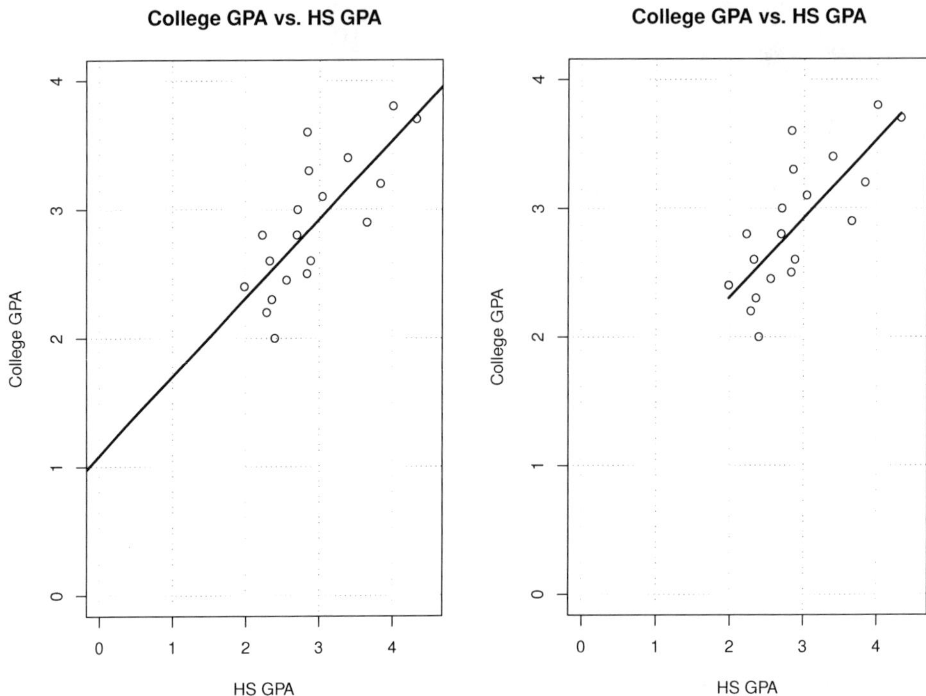

Figure A.8: Scatter plots with estimated model drawn using the `abline()` function (left) and `curve()` function (right).

```
> curve(expr = mod.fit$coefficients[1] + mod.fit$coefficients[2]*x,
   xlim = c(min(gpa$HSGPA), max(gpa$HSGPA)), col= "blue", add = TRUE)
```

The `expr` argument specifies the formula for the curve to be plotted, which in our case is the regression model. The letter `x` is required (instead of `HSGPA` or any other explanatory variable name) for the variable plotted on the x-axis. The `xlim` argument specifies the evaluation limits for the mathematical function, where `min()` and `max()` find the minimum and maximum high school GPAs, respectively. The `add = TRUE` argument value instructs R to add the line to the current plot.

Writing functions

In a simple linear regression analysis, estimation is often followed by plotting. We can partially automate this process by constructing a new R function that does both the estimation and plotting. The code is shown in Figure A.9 from a screen capture of Tinn-R. The function begins by specifying the explanatory variable (`x`), response variable (`y`), and data set (`data`) as its arguments. In the function body, we use almost the same code as earlier; the main difference is the use of `x`, `y`, and `data` rather than `HSGPA`, `CollegeGPA`, and `gpa`, respectively. The last line of code in the function body gives `mod.fit` as the returned value from the function. Below are the results from running the code (excluding the plot).

```
my.reg.func <- function(x, y, data) {

  # Fit simple linear regression model and save results in mod.fit
  mod.fit <- lm(formula = y ~ x, data = data)

  # Open a new graphics window - do not need to
  dev.new(width = 6, height = 6, pointsize = 10)

  # Same scatter plot as before
  plot(x = x, y = y, xlab = "x", ylab = "y", main = "y vs. x",
    panel.first = grid(col = "gray", lty = "dotted"))

  # Include regression model
  curve(expr = mod.fit$coefficients[1] + mod.fit$coefficients[2]*x,
    xlim = c(min(x), max(x)), col = "blue", add = TRUE, lwd = 2)

  # This is the object returned
  mod.fit
}
```

Figure A.9: The `my.reg.func()` in Tinn-R.

```
> save.it <- my.reg.func(x = gpa$HSGPA, y = gpa$CollegeGPA, data =
    gpa)
> names(save.it)
[1] "coefficients"  "residuals"     "effects"       "rank"
[5] "fitted.values" "assign"        "qr"            "df.residual"
[9] "xlevels"       "call"          "terms"         "model"
> summary(save.it)
Call: lm(formula = CollegeGPA ~ HSGPA, data = gpa)

<OUTPUT EDITED>

F-statistic: 24.54 on 1 and 18 DF,  p-value: 0.0001027
```

The function returns the same list as before, and the numerical summary is identical to the previous output. This example shows that it is easy to extend R's capabilities to suit your own needs.

A.7.4 Additional items

There are a number of other plot enhancements that can be used to create better and more descriptive plots. In particular, mathematical expressions can be included in labels and titles on plots using the `expression()` function. For example, including `main = expression(hat(Y) = hat(beta[0]) + hat(beta[1])*x)` in `plot()` prints $\hat{Y} = \hat{\beta}_0 + \hat{\beta}_1 x$ as the plot title. Running `demo(plotmath)` at the R Console prompt provides syntax examples. Another plot enhancement allows one to have finer control of axis tick marks by using the `axis()` function. For example, the code below changes the x-axis tick marks from the default to produce a modified version of the scatter plots given in Figure A.8.

```
> plot(x = gpa$HSGPA, y = gpa$CollegeGPA, xaxt = "n", xlim = c(0,
   4.5), ylim = c(0,4.5))
> # Major tick marks
> axis(side = 1, at = seq(from = 0, to = 4.5, by = 0.5))
> # Minor tick marks
> axis(side = 1, at = seq(from = 0, to = 4.5, by = 0.1), tck = 0.01,
   labels = FALSE)
```

The `xaxt = "n"` argument value in `plot()` prevents the default x-axis tick marks from being plotted. The `axis()` function is used twice to create a specific sequence of numbers (`seq()`) for the major and minor tick marks on `side = 1` (bottom x-axis). The `tck` argument sets a specific tick mark length. More details on each of these arguments are available in the R help.

One of the reasons R has become popular is it facilitates the sharing of code in an open-source software environment. For example, newly published statistical methodology frequently has corresponding R code available for immediate implementation. This code may be downloadable from an author's website or incorporated into a package. Furthermore, all R code is available for viewing. For example, one can type a function name, such as `lm`, at an R Console prompt to see its code. This does not work, however, for functions that are not meant to be run directly by the user. Instead, their code may still be seen by using `getAnywhere(<function name>)` or in the package source code at CRAN.

The open-source environment has also led to the development of a large community for R users to go to when in need of help. A simple web search frequently leads to question/answer websites, such as Stack Overflow, that provide the needed help. In particular for Stack Overflow, a web page is available at `http://stackoverflow.com/tags/r/info` that lists questions/answers tagged with R. There are also a large number of blogs devoted to the use of R. The R-bloggers website at `http://www.r-bloggers.com` serves as an aggregator for many of these blogs. Finally, there are active listservs (`http://www.r-project.org/mail.html`) where even members of the R core group at times answer submitted questions. Before posting a question to the listserv though, search its archives because the same question may have been asked before.

Some new R users may be used to a completely point-and-click process when it comes to using a statistical software package. For those individuals, the `Rcmdr` (R Commander) package may serve as a gentler way to begin using R, because this package provides a GUI for data management and for statistical functions. Once installed, type `library(package = Rcmdr)` and <Enter> at the R Console prompt to open R Commander. In its first use, additional packages are installed automatically to enable all drop-down menu options. This can take a significant amount of time. Once the R Commander window is open, a data set already in the R workspace can be selected through the <No active dataset> icon. Depending on the data type, the drop-down menus allow for editing the data, creating data summaries, fitting models, and plotting. For example, suppose the `gpa` data frame is the active data set within R Commander. We select Statistics > Fit Models > Linear Regression from the main menu bar to open a window that allows us to choose `CollegeGPA` as the response variable and `HSGPA` as the explanatory variable. We also type `mod.fit` in the Enter name for model: box as the object name for saving the results. Figure A.10 shows what happens after selecting OK. The code and output in the script and output windows, respectively, both appear and they are the same as we used earlier in this section. For other applications, R Commander may produce different code from what we might enter. This is because the R Commander package has a number of its own functions built into it to

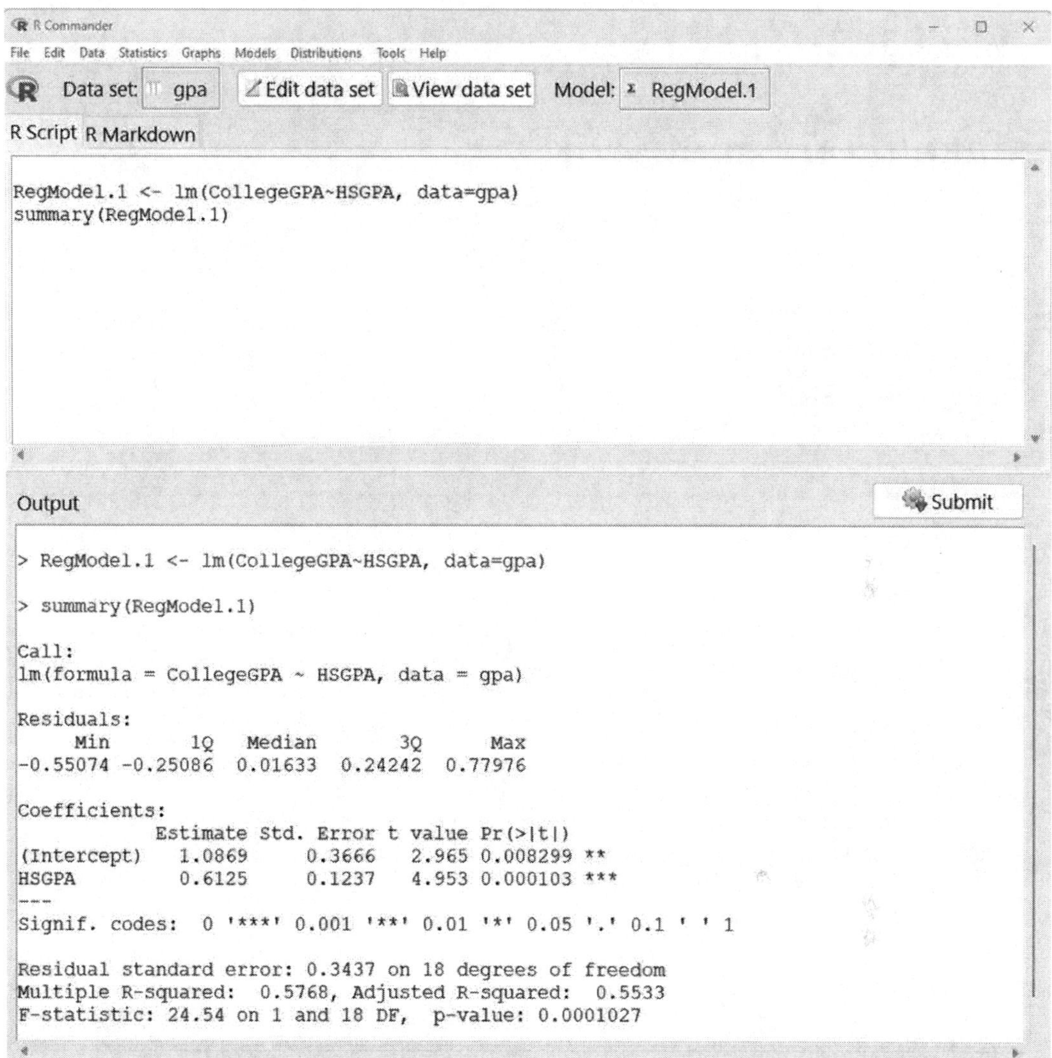

Figure A.10: R Commander window.

facilitate the computational or graphical processes. However, we believe that this package still serves as a useful tool to help new users learn to write R code.

Appendix B

Likelihood methods

B.1 Introduction

Likelihood methods are commonly used in statistical analyses. In particular, they provide the basis for most of the categorical data analysis techniques that are covered in this book. The main body of this book is written assuming that readers are already familiar with likelihood-based methods, at least enough to be able to apply them with some confidence. However, we recognize that readers may come from a broad range of disciplines and may not have had any previous exposure to likelihood methods. Therefore, we offer a brief primer on this important class of procedures. We suggest that readers who have no past exposure to likelihood methods at least look over this appendix before reading the book's main body. We cross-reference this appendix with the sections where likelihood methods are used, so that readers and instructors can also refer to this material as needed.

The scope of this appendix is deliberately limited. An understanding of the theory and asymptotics of likelihood methods is not at all required for this book, although a heuristic appreciation of them is helpful. Accordingly, the descriptions in this appendix are light in mathematics. For a more complete treatment of likelihood methods, please see texts such as Casella and Berger (2002) and Severini (2000).

B.1.1 Model and parameters

The goal of a statistical analysis is to learn something about a population (what we might call the "truth"). To achieve this goal, data are gathered, but data contain variability (or "noise") that prevents us from seeing the truth clearly. To extract the truth from data, it helps if we start with some idea of what the truth might look like and if we know something about the origin of the noise. For example, rather than just looking at a plot of a response variable against an explanatory variable, it may help to speculate that the relationship should be a straight line and that the deviations around the straight line should be independent and approximately normally distributed. A *statistical model* is an assumed structure for the truth and the noise (i.e., it is an educated guess). The features of the model are combined into a probability distribution, such as a normal, Bernoulli, or Poisson, that is intended to serve as a useful approximation to reality.

Models generally contain *parameters*. Model parameters relate to the structure of the truth and/or noise and hold the place of unknown or flexible features of the model. They represent population quantities, some of which may be of direct interest to the researcher—such as the mean of a normal distribution, or the probability of success in a binomial distribution—but not all (we often do not care to investigate the intercept parameter in many linear regression problems). The goal of a statistical analysis is to learn about the model parameters or about some functions of the model parameters (like predictions in a regression, which are a function of the slope and intercept parameters). Functions of model parameters are

DOI: 10.1201/9781003093091-B

also parameters, since they are also unknown population quantities, so our discussion will not distinguish between parameters in the model and other parameters.

B.1.2 The role of maximum likelihood estimation

A statistical model serves to relate the data to the parameters. We still need to find values for the parameters and use them to learn about the population. The first step—finding values for the parameters—is called *estimation.* The second step—using them to draw conclusions—is called *inference.*

There are many ways to estimate parameters from a statistical model, but the one that has been adopted almost universally because of its quality and flexibility is maximum likelihood (ML) estimation. This is because the procedure is adaptable to nearly any statistical model, leading to an automatic process for estimation. It also has associated with it a variety of tools that can be used for inference. The procedure and its tools have strengths and weaknesses. We discuss these as they arise in different settings throughout the text.

B.2 Likelihood

B.2.1 Definition

Statistical models are generally described in terms of a *probability mass function* or a *probability density function.* A probability mass function (PMF) for a discrete random variable provides the probabilities for each possible outcome. For a continuous random variable, the corresponding probability density function (PDF) is a little more complicated to interpret because it assumes that measurements are made to an infinite number of decimal places. Loosely speaking, a PDF describes the relative chances of observing values from different areas of the stated probability distribution. The familiar "normal curve" is an example of a PDF. In both the discrete and continuous cases, higher values of the PMF or PDF are produced by ranges of values that are "more likely" to occur. The *joint* PMF or PDF for a sample of observations can be interpreted as how likely we are to observe the entire sample, given the distribution and its parameters.

In practice, we do not know the values of the parameters, but we do know the data. We therefore cannot know the joint mass or density for our sample exactly. We *can* calculate this quantity if we assume certain values for the parameters. If we change the values of the parameters, we get a different value for the PDF/PMF of the sample, because the data are more or less likely to occur under different parameters. For example, it is very unlikely that we would observe 5 successes in 10 Bernoulli trials when the true probability of success is 0.01. This outcome would be somewhat more likely to happen if the probability of success is 0.30, and even more likely if it is 0.50.

This is the nature of the *likelihood function*: we consider the PMF or PDF in reverse. We observe how the function changes for different values of parameters, while holding the data fixed. We can then make use of this to judge which values of the parameters lead to greater relative chances for the sample to occur. Formally, if we define the joint PMF or PDF of a sample to be $f(\mathbf{y}|\boldsymbol{\theta})$, where $\mathbf{y} = (y_1, \ldots, y_n)$ represents a vector[1] containing the

[1] A vector can be thought of as a simple way to express a group of values using one symbol.

n sampled values and $\boldsymbol{\theta} = (\theta_1, \ldots, \theta_p)$ represents a vector for p different parameters[2], then the likelihood function is

$$L(\boldsymbol{\theta}|\mathbf{y}) = f(\mathbf{y}|\boldsymbol{\theta}).$$

Larger values of the likelihood (shortened name for "likelihood function") correspond to values of the parameters that are relatively better supported by the data.

A likelihood is not a probability, because the only random part, $\mathbf{y}$, is considered fixed in its construction. In particular, the likelihood is not expected to add to 1 across all values of $\boldsymbol{\theta}$. The actual numerical values of a likelihood are unimportant. The *relative* sizes of the likelihoods for different parameter values are all that matters.

When the observations are drawn independently, the likelihood function is simply the product of the PDFs or PMFs evaluated for each observation,

$$L(\boldsymbol{\theta}|\mathbf{y}) = \prod_{i=1}^{n} f(y_i|\boldsymbol{\theta}),$$

where we use the $\prod$ symbol to denote multiplying indexed terms together. Thus, likelihoods are very easy to construct for simple random samples, which is the setting for most problems in this book. Also, notice that the value of the likelihood depends on the sample, so that likelihoods—and any features calculated from them—are statistics. This means that such features are random and have probability distributions.

B.2.2 Examples

Below are some simple examples of likelihoods that are commonly used in categorical data analysis.

Example: Bernoulli

Suppose that the random variable Y takes on only two possible values. The Bernoulli PMF for Y is $f(y|\pi) = \pi^y(1-\pi)^{(1-y)}$ with probability-of-success parameter π ($0 < \pi < 1$) and $y = 1$ or 0 denoting a "success" or a "failure," respectively. Let $y_1, \ldots, y_n$ be observations of independent Bernoulli random variables from this PMF. The likelihood for the parameter π is

$$L(\pi|\mathbf{y}) = \prod_{i=1}^{n} \pi^{y_i}(1-\pi)^{(1-y_i)} = \pi^w(1-\pi)^{n-w}, \tag{B.1}$$

where $\mathbf{y} = (y_1, \ldots, y_n)$ and $w = \sum_{i=1}^{n} y_i$. This likelihood is used in Section 1.1.1.

Example: Binomial

An alternative form of the Bernoulli case occurs when the total number of successes, w, is observed instead of the individual trial results. In this case, the joint PMF of $y_1, \ldots, y_n$ cannot be found because we do not know which y_i should be 1 and which should be 0. However, we know that there are $\binom{n}{w} = n!/\left[w!(n-w)!\right]$ ways for the w successes to be observed among the n observations, so the PMF of w given π is

$$f(w|\pi) = \frac{n!}{w!(n-w)!}\pi^w(1-\pi)^{n-w}.$$

[2]The actual symbols used for the parameters in a given problem vary depending on the context of the problem; see examples later.

If only one set of n trials is run, so that only one total number of successes w is observed, then $L(\pi|w) = f(w|\pi)$. Notice that this is very similar to the Bernoulli likelihood.

Example: Poisson

The Poisson PMF for a random variable Y is $f(y|\mu) = e^{-\mu}\mu^y/y!$ with parameter $\mu > 0$ and y taking on integer values 0, 1, 2, ..., such as for a count of something. Let $y_1, \ldots, y_n$ be observations of independent Poisson random variables. The likelihood for the parameter μ is

$$L(\mu|\mathbf{y}) = \prod_{i=1}^{n} \frac{e^{-\mu}\mu^{y_i}}{y_i!}, \tag{B.2}$$

where $\mathbf{y} = (y_1, \ldots, y_n)$. This likelihood is used in Section 4.1.2.

Example: Multinomial

Consider a random variable Y with responses consisting of one of c categories, labeled 1, 2, ..., c, with respective category probabilities $\pi_1, \pi_2, \ldots, \pi_c$ ($\sum_{k=1}^{c} \pi_k = 1$). Let $y_1, \ldots, y_n$ be observations of Y measured on independent trials of this type; that is, the possible values for each y_i are the categories $1, 2, \ldots, c$. Likelihoods can be constructed as for the Bernoulli and binomial cases above, depending on whether individual trial results $y_1, \ldots, y_n$ are observed or just the summary counts for each category, $w_1, \ldots, w_c$. The multinomial distribution is based on the summary counts, for which the PMF is

$$f(w_1, \ldots, w_c|\pi_1, \ldots, \pi_c) = \frac{n!}{w_1!w_2!\ldots w_c!}\pi_1^{w_1}\pi_2^{w_2}\ldots\pi_c^{w_c}. \tag{B.3}$$

If only one set of n trials is run, so that only one set of category counts $w_1, \ldots, w_c$ is observed, then we have $L(\pi_1, \ldots, \pi_c|w_1, \ldots, w_c) = f(w_1, \ldots, w_c|\pi_1, \ldots, \pi_c)$. This likelihood is used in Section 3.1.

B.3 Maximum likelihood estimates

We define the maximum likelihood estimate (MLE), $\hat{\boldsymbol{\theta}}$, of a parameter as the value of the parameter at which the likelihood function from the given sample is maximized: $L(\hat{\boldsymbol{\theta}}|\mathbf{y}) \geq L(\tilde{\boldsymbol{\theta}}|\mathbf{y})$ for any possible value $\tilde{\boldsymbol{\theta}}$ of the parameter. The next example illustrates how to find this MLE through simple evaluation of a likelihood function.

Example: MLE for a sample of Bernoulli random variables (LikelihoodFunction.R)

Suppose $w = 4$ successes are observed out of $n = 10$ trials. Given this information, we want to determine the most plausible value for π. The likelihood function is

Table B.1: $L(\pi|\mathbf{y})$ evaluated at different values of π.

π	$L(\pi\|\mathbf{y})$
0.20	0.000419
0.30	0.000953
0.35	0.001132
0.39	0.001192
0.40	0.001194
0.41	0.001192
0.50	0.000977

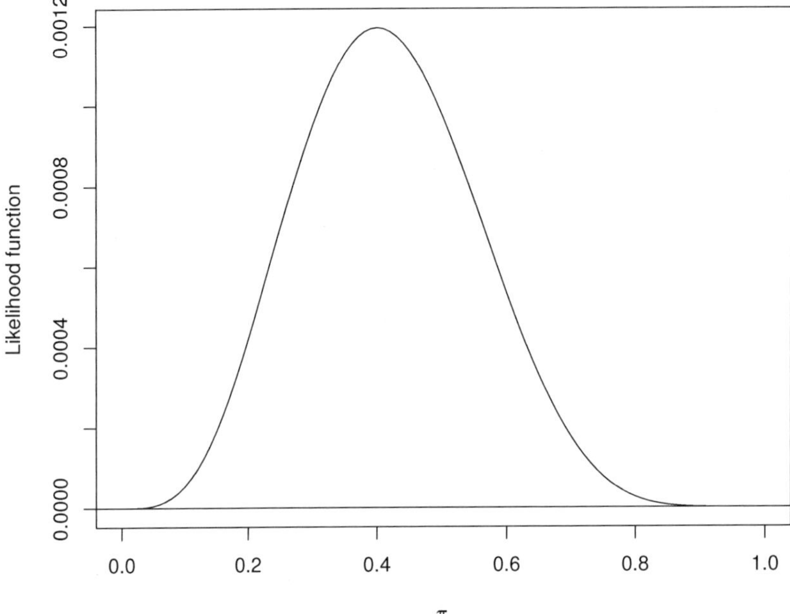

Figure B.1: Bernoulli likelihood function evaluated at $w = 4$ and $n = 10$.

$L(\pi|\mathbf{y}) = \pi^4(1-\pi)^6$. Table B.1 shows the likelihood function evaluated at a few different values of π, and Figure B.1 plots the function (R code for the table and plot is included in our program for this example). We can see that the likelihood function reaches its maximum value when $\pi = 0.4$. Therefore, the most plausible value of π, given the observed data, is 0.4, which makes sense because it is the observed proportion of successes. Formally, we say 0.4 is the MLE of π, and we denote it as $\hat{\pi} = 0.4$.

For various mathematical reasons, it turns out to be easier to work with the natural log of the likelihood, $\log[L(\boldsymbol{\theta}|\mathbf{y})]$, when we attempt to find MLEs. This causes no problem because the log transformation does not change the ordering of likelihood values across different values of $\boldsymbol{\theta}$, so the value that maximizes $\log[L(\boldsymbol{\theta}|\mathbf{y})]$ also maximizes the likelihood.

B.3.1 Mathematical maximization of the log-likelihood function

For simple models with a single parameter, finding the value of $\boldsymbol{\theta}$ that maximizes the log-likelihood function is easily done using calculus. The standard technique is to differentiate the log-likelihood function with respect to the parameter, set the result equal to 0, and solve for the parameter. This process is demonstrated on some of the examples from earlier.

Example: Bernoulli

From Equation B.1, we obtain

$$\log[L(\pi|\mathbf{y})] = w\log\pi + (n-w)\log(1-\pi).$$

Differentiating, we find

$$\frac{d}{d\pi}\log[L(\pi|\mathbf{y})] = \frac{w}{\pi} - \frac{n-w}{1-\pi}.$$

Setting this equal to 0 and solving for π leads to $\hat{\pi} = w/n = \sum_{i=1}^{n} y_i/n$. Thus, the MLE for π is the sample proportion of successes, which was shown in the previous example for $w = 4$ and $n = 10$.

Example: Poisson

From Equation B.2, we obtain

$$\log[L(\mu|\mathbf{y})] = -n\mu + \sum_{i=1}^{n} y_i \log\mu - \sum_{i=1}^{n}\log(y_i!).$$

Differentiating, we find

$$\frac{d}{d\mu}\log[L(\mu|\mathbf{y})] = -n + \frac{1}{\mu}\sum_{i=1}^{n} y_i.$$

Setting this equal to 0 and solving for μ leads to $\hat{\mu} = \sum_{i=1}^{n} y_i/n$. Thus, the MLE for μ is the sample mean.

Models with several parameters

Many models rely on more than one parameter. The multinomial model in Equation B.3 is an example, because there is a different probability parameter for each category. Other models use regression functions to describe the relationship between population means or probabilities and one or more explanatory variables. In these models, the maximization is carried out exactly as described above, with a separate derivative taken with respect to each parameter. Setting each derivative equal to zero results in a system of equations that needs to be solved simultaneously to find the MLE. Theoretically, this presents no difficulty. Practically, the process can be challenging to carry out manually, and the equations may not have closed-form solutions. For example, see the logistic regression likelihood from Section 2.2.

B.3.2 Computational maximization of the log-likelihood function

When the system of equations has no closed-form solution, it is solved using an educated version of trial and error. The general idea is to start with some initial guess at the parameter value, calculate the log likelihood for that value, and then iteratively find parameter values with larger and larger log-likelihoods until no further improvement can be achieved. Improving the log-likelihood can be done, for example, by calculating the slope of the log-likelihood at the current guess and then moving the next guess some distance in the direction leading to larger log-likelihood values. Alternatively, one can work with the first derivative of the log-likelihood function and seek values of the parameters that cause it to be zero. There are a variety of fast, reliable computational algorithms for carrying out these procedures. One of the most widely implemented is the Newton-Raphson algorithm.

Example: MLE for a sample of Bernoulli random variables (NewtonRaphson.R)

We demonstrate the Newton-Raphson algorithm in a simple setting to find the MLE of π for the previous Bernoulli example where $\sum_{i=1}^{n} y_i = w = 4$ in $n = 10$ trials. We found earlier that $\hat{\pi} = w/n = 0.4$.

In this context, the algorithm uses the following equation to obtain an estimate $\pi^{(i+1)}$ using a previous estimate $\pi^{(i)}$:

$$\begin{aligned}\pi^{(i+1)} &= \pi^{(i)} - \frac{\left.\dfrac{d}{d\pi}\log[L(\pi|\mathbf{y})]\right|_{\pi=\pi^{(i)}}}{\left.\dfrac{d^2}{d\pi^2}\log[L(\pi|\mathbf{y})]\right|_{\pi=\pi^{(i)}}} \\ &= \pi^{(i)} - \frac{\dfrac{w}{\pi^{(i)}} - \dfrac{n-w}{1-\pi^{(i)}}}{-\dfrac{w}{\left(\pi^{(i)}\right)^2} - \dfrac{n-w}{\left(1-\pi^{(i)}\right)^2}} \qquad \text{(B.4)}\end{aligned}$$

Equation B.4 comes from a first-order Taylor series expansion about $\pi^{(i)}$.[3] To begin using the algorithm, we choose a starting value $\pi^{(0)}$ that we think may be close to $\hat{\pi}$ and substitute this into Equation B.4 for $\pi^{(i)}$ to obtain $\pi^{(1)}$. If $\pi^{(1)}$ is "close enough" to $\pi^{(0)}$, we stop and use $\pi^{(1)}$ as $\hat{\pi}$; otherwise, we substitute $\pi^{(1)}$ into Equation B.4 for $\pi^{(i)}$ to find $\pi^{(2)}$. This process continues until $\left|\pi^{(i+1)} - \pi^{(i)}\right| < \varepsilon$ for some small number $\epsilon > 0$, and we say that *convergence* has been reached at iteration $i + 1$.

Using the observed data, suppose we guess $\pi^{(0)} = 0.3$, and we feel that $\varepsilon = 0.0001$ represents "close enough." Table B.2 shows the iteration history, where convergence is achieved after five iterations. The R code that produced this table is available in the program for this example. We have included in this program the code to create plots illustrating the Newton-Raphson algorithm.

Generally, readers of this book will not need to implement a Newton-Raphson procedure like this directly. Instead, we will use R functions that take care of these details. Also, note that there are other algorithms besides Newton-Raphson that are used to find maximum likelihood estimates. Many of these are implemented in R's `optim()` function.

[3]Suppose we would like to approximate a function $f(x)$ at a point x_0. The first-order Taylor series expansion approximates $f(x)$ with $f(x_0) + (x - x_0)f'(x_0)$, where $f'(\cdot)$ is the first derivative of $f(\cdot)$ with respect to x.

Table B.2: Iterations for the Newton-Raphson algorithm.

Iteration	$\pi^{(i)}$
1	0.3000000
2	0.3840000
3	0.3997528
4	0.3999999
5	0.4000000

B.3.3 Large-sample properties of the MLE

To use an MLE in confidence intervals and tests, we need to know its probability distribution (the probability distribution of a statistic is also known as its "sampling distribution"). It can be shown that *all* of the MLEs we will use share certain properties relating to their sampling distributions that make them very appealing bases for inference procedures. These properties generally hold for large samples; in other words, these properties generally hold *asymptotically*, which means, as the sample size grows toward ∞. Below is a list of the properties:

1. MLEs ARE ASYMPTOTICALLY NORMALLY DISTRIBUTED. This result is analogous to the central limit theorem for sample means. The fact that normality holds asymptotically means that in any given sample, the normal distribution is typically an *approximation* to the correct sampling distribution of $\hat{\boldsymbol{\theta}}$, and the approximation is better with larger sample sizes.

2. MLEs ARE CONSISTENT. This means essentially that if you sample the whole population (or sample infinitely), the MLE will be exactly the same as the population parameter. In particular, any bias in the estimate (the difference between the expected value of the MLE and the true value of the parameter) vanishes as the sample size grows, and the variance shrinks to 0.

3. MLEs ARE ASYMPTOTICALLY EFFICIENT. This means that as the sample size grows toward ∞, they achieve the smallest variance possible for estimates of their type (e.g., among all asymptotically normal estimates). An important implication of this result is that confidence intervals based on MLEs have the potential to be shorter and tests more powerful than those based on other forms of estimates.

These properties are not guaranteed to hold in samples that are not "large." The normal distribution approximation is generally very good in "large" samples, but may be very poor in "small" samples. Other estimates may have smaller variance than the MLE in finite samples. Unfortunately, there is no uniform way to define "large" or "small." However, it is often not too difficult to simulate data from the chosen model and check whether the MLE has a distribution that appears roughly normal and has an acceptably small bias.

B.3.4 Variance of the MLE

The variance of any MLE is related to the curvature of the log-likelihood function in the neighborhood of the MLE. If the log-likelihood is very flat near the maximum, then there is much uncertainty in the data regarding the location of the parameter (many different values of $\boldsymbol{\theta}$ lead to similarly large likelihoods). Conversely, if the log-likelihood has a sharp peak, then the data show little doubt about the region in which the parameter must lie.

Because the curvature of a function is determined by its second derivative, the asymptotic variance of the MLE is based upon the second derivative of the log-likelihood function.

When there is only one parameter θ ($\boldsymbol{\theta}$ has only one element), this variance is

$$Var(\hat{\theta}) = -E\left(\frac{\partial^2}{\partial\theta^2}\log[L(\theta|\mathbf{Y})]\right)^{-1},$$

where the expected value is taken with respect to the distribution of Y. The same approach is used to find the variance of an MLE when there is more than one parameter. This requires computing second derivatives with respect to a vector,

$$\frac{\partial^2}{\partial\boldsymbol{\theta}^2}\log[L(\boldsymbol{\theta}|\mathbf{y})],$$

which is referred to as the *Hessian* matrix.[4] From this matrix, the asymptotic variance is

$$Var(\hat{\boldsymbol{\theta}}) = -E\left(\frac{\partial^2}{\partial\boldsymbol{\theta}^2}\log[L(\boldsymbol{\theta}|\mathbf{Y})]\right)^{-1}.$$

This expression involves finding the expected value for each element of the Hessian matrix and then inverting the resulting matrix.[5]

The estimated asymptotic variance (we will refer to this more simply as the "estimated variance") results from replacing $\boldsymbol{\theta}$ with its corresponding estimate $\hat{\boldsymbol{\theta}}$,

$$\widehat{Var}(\hat{\boldsymbol{\theta}}) = -E\left(\frac{\partial^2}{\partial\boldsymbol{\theta}^2}\log[L(\boldsymbol{\theta}|\mathbf{Y})]\right)^{-1}\Bigg|_{\boldsymbol{\theta}=\hat{\boldsymbol{\theta}}}. \tag{B.5}$$

The estimated variance can be approximated instead by

$$\widehat{Var}(\hat{\boldsymbol{\theta}}) \approx -\left(\frac{\partial^2}{\partial\boldsymbol{\theta}^2}\log[L(\boldsymbol{\theta}|\mathbf{y})]\right)^{-1}\Bigg|_{\boldsymbol{\theta}=\hat{\boldsymbol{\theta}}}, \tag{B.6}$$

which is often easier to compute. Equation B.6 is *asymptotically equivalent* to Equation B.5, meaning that these estimated variances will essentially be the same in very large samples. Other asymptotically equivalent variants of these formulas are sometimes used as well.

When $p > 1$, the form of the variance is a $p \times p$ matrix,

$$\widehat{Var}(\hat{\boldsymbol{\theta}}) = \begin{bmatrix} \widehat{Var}(\hat{\theta}_1) & \widehat{Cov}(\hat{\theta}_1,\hat{\theta}_2) & \cdots & \widehat{Cov}(\hat{\theta}_1,\hat{\theta}_p) \\ \widehat{Cov}(\hat{\theta}_1,\hat{\theta}_2) & \widehat{Var}(\hat{\theta}_2) & \cdots & \widehat{Cov}(\hat{\theta}_2,\hat{\theta}_p) \\ \vdots & \vdots & \ddots & \vdots \\ \widehat{Cov}(\hat{\theta}_1,\hat{\theta}_p) & \widehat{Cov}(\hat{\theta}_2,\hat{\theta}_p) & \cdots & \widehat{Var}(\hat{\theta}_p) \end{bmatrix}.$$

This is known as an estimated *variance-covariance matrix* (sometimes shortened to *covariance matrix* or *variance matrix*). The diagonal elements of the matrix (where the row and column numbers are the same) are the estimated variances of the MLEs. The estimated standard deviation of a statistic (i.e., the standard error) is computed by finding the

[4]The second derivative of a function $f(\boldsymbol{\theta})$ with respect to a vector of length p produces a $p \times p$ matrix. The jth diagonal element is $\frac{\partial^2}{\partial\theta_j^2}f(\boldsymbol{\theta})$, while the off-diagonal element in position (i, j) is $\frac{\partial^2}{\partial\theta_i\partial\theta_j}f(\boldsymbol{\theta})$, for $i = 1, \ldots, p$; $j = 1, \ldots, p$; $i \neq j$.

[5]The inverse of a matrix is a generalization of the inverse of a single number. For example, the inverse of a number like 3 is $3^{-1} = 1/3$ because $3 \times 3^{-1} = 1$. With respect to a matrix $\mathbf{A}$ with p rows and p columns, the inverse of $\mathbf{A}$ is denoted by $\mathbf{A}^{-1}$ and results in $\mathbf{A}\mathbf{A}^{-1} = \mathbf{I}$, where $\mathbf{I}$ is a matrix with 1's as diagonal elements and 0's as off-diagonal elements. Matrix algebra books, like Hartman (2011) and Fieller (2016), provide more details about the inverse of a matrix.

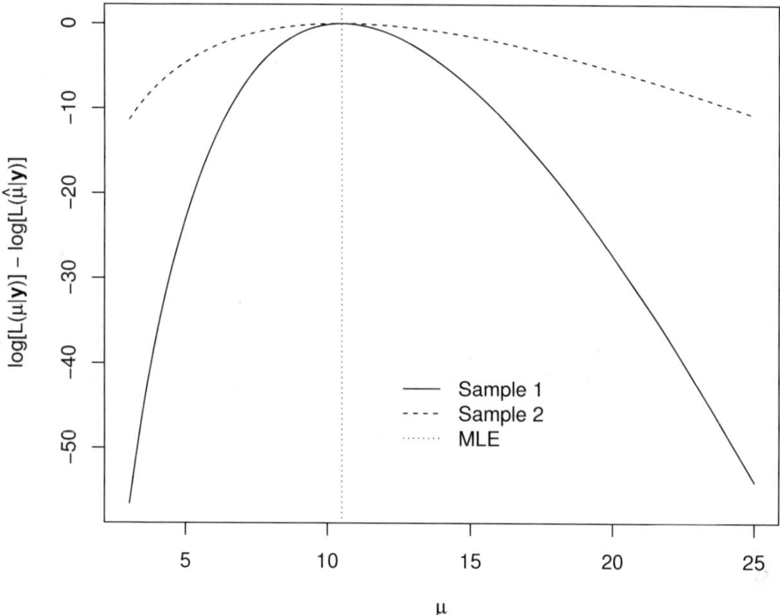

Figure B.2: Poisson log-likelihoods for samples of size $n = 10$ (sample 1) and $n = 2$ (sample 2) with common MLE $\hat{\mu} = 10.5$. Note that the two curves have been shifted vertically so that the log-likelihoods are both 0 at the MLE.

square root of its corresponding diagonal term. For example, the standard error of $\hat{\theta}_1$ is $\widehat{Var}(\hat{\theta}_1)^{1/2}$. The off-diagonal elements of the matrix are the covariances between pairs of MLEs. These estimated covariances measure the dependence between the MLEs, and they can be useful for finding the variances of functions of MLEs (see Appendix B.4). Because $\widehat{Cov}(\hat{\theta}_i, \hat{\theta}_j) = \widehat{Cov}(\hat{\theta}_j, \hat{\theta}_i)$, the matrix is symmetric, so that the element in row i, column j equals the element in row j, column i for any $i \neq j$.

Example: Poisson (PoissonLikelihood.R)

The purpose of this example is to find the variance for $\hat{\mu}$ in the Poisson example. Figure B.2 shows the curvature of the log-likelihood function for two different samples, where sample 1 has $\mathbf{y} = (3, 5, 6, 6, 7, 10, 13, 15, 18, 22)$ and sample 2 has $\mathbf{y} = (9, 12)$. For both samples, $\hat{\mu} = 10.5$. We can see that sample 2's log-likelihood function is relatively flat because of its small sample size, while sample 1's log-likelihood function has much more curvature because of its larger sample size. Because the variance for $\hat{\mu}$ is based on the inverse of this curvature, we would expect the variance for sample 1 to be much less than the variance for sample 2.

Formally, we calculate the variance as

$$\widehat{Var}(\hat{\mu}) = -\left(\frac{\partial^2}{\partial \mu^2}\log[L(\mu|\mathbf{y})]\right)^{-1}\Bigg|_{\mu=\hat{\mu}} = \frac{\hat{\mu}^2}{\sum_{i=1}^n y_i} = \frac{\hat{\mu}}{n}.$$

For sample 1, $\widehat{Var}(\hat{\mu}) = 10.5/10 = 1.05$, while for sample 2, $\widehat{Var}(\hat{\mu}) = 10.5/2 = 5.25$. As expected, the variance for $\hat{\mu}$ is larger for sample 2 than for sample 1. The R code for these calculations is included in the R program for this example.

B.4 Functions of parameters

B.4.1 Invariance property of MLEs

Often, our interest is not limited to the model parameters that are directly estimated using the procedures in Appendix B.3.1 or B.3.2. Instead, we may want to estimate parameters that are *functions* of model parameters. As examples, (1) the odds ratio (e.g., Sections 1.2.5, 2.2.3, 3.3.1, and 4.2.3) can be written as a function of probabilities from two binomials, or from multinomial probabilities, or as a function of Poisson means; and (2) predicted values in any regression model can be written as functions of regression coefficients. The *invariance property of MLEs* states that if $\hat{\boldsymbol{\theta}}$ is the MLE for $\boldsymbol{\theta}$ and $g(\boldsymbol{\theta})$ is any function of $\boldsymbol{\theta}$ then $g(\hat{\boldsymbol{\theta}})$ is the MLE for $g(\boldsymbol{\theta})$. In words, the MLE of any function of parameters is just the same function of the parameters' MLE. The important implication of this result is that the large-sample properties listed in Appendix B.3.3 hold for functions of MLEs, and all of the inference procedures in Appendix B.5 can be applied to these functions.[6]

B.4.2 Delta method for variances of functions

The delta method is a very general and useful procedure to estimate the variance for a function of random variables. We present it here to estimate the variance for a function of an MLE, which is not simply the same function of the MLE's variance.

Suppose that $\boldsymbol{\theta}$ consists of p parameters, $\theta_1, \theta_2, \ldots, \theta_p$ and define

$$g_1'(\hat{\boldsymbol{\theta}}) = \frac{\partial}{\partial \theta_1} g(\boldsymbol{\theta})\bigg|_{\boldsymbol{\theta}=\hat{\boldsymbol{\theta}}},\ g_2'(\hat{\boldsymbol{\theta}}) = \frac{\partial}{\partial \theta_2} g(\boldsymbol{\theta})\bigg|_{\boldsymbol{\theta}=\hat{\boldsymbol{\theta}}}, \ldots,\ g_p'(\hat{\boldsymbol{\theta}}) = \frac{\partial}{\partial \theta_p} g(\boldsymbol{\theta})\bigg|_{\boldsymbol{\theta}=\hat{\boldsymbol{\theta}}}.$$

Then, from a Taylor series approximation to $g(\boldsymbol{\theta})$ centered at $\hat{\boldsymbol{\theta}}$, one can show that

$$\widehat{Var}(g(\hat{\boldsymbol{\theta}})) \approx \sum_{i=1}^{p} [g_i'(\hat{\boldsymbol{\theta}})]^2 \widehat{Var}(\hat{\theta}_i) + 2\sum\sum_{i>j} [g_i'(\hat{\boldsymbol{\theta}})][g_j'(\hat{\boldsymbol{\theta}})]\widehat{Cov}(\hat{\theta}_i, \hat{\theta}_j), \quad \text{(B.7)}$$

where $\widehat{Var}(\hat{\theta}_i)$, $i = 1, \ldots, p$ are the estimated variances of the MLEs and $\widehat{Cov}(\hat{\theta}_i, \hat{\theta}_j)$, $i, j = 1, \ldots, p$, $i \neq j$ are the estimated covariances described in Appendix B.3.4.

Example: Variance of an odds ratio

Suppose there are two independent binomial variables: W_1 with n_1 trials, probability of success π_1, and w_1 observed successes; and W_2 with n_2 trials, probability of success π_2, and w_2 observed successes. As described in Section 1.2.5, inference on the odds ratio $OR = [\pi_1/(1-\pi_1)]/[\pi_2/(1-\pi_2)]$ is done based on the sampling distribution of $\log(\widehat{OR})$. The variance of $\log(\widehat{OR})$ is found using the delta method as follows.

First, note that we have $p = 2$ parameters here, so that $\boldsymbol{\theta} = (\pi_1, \pi_2)$, and $g(\boldsymbol{\theta}) = \log([\pi_1/(1-\pi_1)]/[\pi_2/(1-\pi_2)])$. Then

$$g_1'(\hat{\boldsymbol{\theta}}) = \frac{\partial}{\partial \pi_1} \log([\pi_1/(1-\pi_1)]/[\pi_2/(1-\pi_2)])\bigg|_{(\pi_1,\pi_2)=(\hat{\pi}_1,\hat{\pi}_2)} = \frac{1}{\hat{\pi}_1(1-\hat{\pi}_1)},$$

[6]These properties are again approximate in finite samples. The sample sizes required to make them hold satisfactorily for a given function of parameters may be similar to or quite different from those required for the parameters themselves. This can be checked by simulation.

and similarly $g_2'(\hat{\boldsymbol{\theta}}) = 1/[\hat{\pi}_2(1-\hat{\pi}_2)]$, where $\hat{\pi}_i = w_i/n_i$. Also, we have that $\widehat{Var}(\hat{\pi}_i) = \hat{\pi}_i(1-\hat{\pi}_i)/n_i$, and because W_1 and W_2 are independent, $\widehat{Cov}(\hat{\pi}_1, \hat{\pi}_2) = 0$. Thus,

$$\begin{aligned}\widehat{Var}(g(\hat{\boldsymbol{\theta}})) &\approx \textstyle\sum_{i=1}^{p}[g_i'(\hat{\boldsymbol{\theta}})]^2\widehat{Var}(\hat{\theta}_i) + 2\sum\sum_{i>j}[g_i'(\hat{\boldsymbol{\theta}})][g_j'(\hat{\boldsymbol{\theta}})]\widehat{Cov}(\hat{\theta}_i, \hat{\theta}_j) \\ &= \left[\frac{1}{\hat{\pi}_1(1-\hat{\pi}_1)}\right]^2 \frac{\hat{\pi}_1(1-\hat{\pi}_1)}{n_1} + \left[\frac{1}{\hat{\pi}_2(1-\hat{\pi}_2)}\right]^2 \frac{\hat{\pi}_2(1-\hat{\pi}_2)}{n_2} \\ &= \frac{1}{n_1\hat{\pi}_1(1-\hat{\pi}_1)} + \frac{1}{n_2\hat{\pi}_2(1-\hat{\pi}_2)} \\ &= \frac{1}{w_1} + \frac{1}{n_1 - w_1} + \frac{1}{w_2} + \frac{1}{n_2 - w_2}.\end{aligned}$$

This is the formula given in Section 1.2.5.

B.5 Inference with MLEs

Throughout this section, we consider the problem of conducting inference on a single parameter θ, which may be a model parameter or some function of model parameters. Where appropriate, we mention briefly extensions to multiple parameters.

B.5.1 Tests for parameters

Consider the hypotheses

$$\begin{aligned}H_0 &: \theta = \theta_0 \\ H_a &: \theta \neq \theta_0\end{aligned},$$

where θ_0 is some special value of interest. Several different procedures based on likelihood principles can be used to perform this test. All of them are approximate procedures in the sense that they may not achieve the stated type I error rate α exactly. These approximations are very accurate in large samples, but may provide poor results in small samples.

Wald tests

The Wald test (Wald, 1943) for a single parameter is the most familiar likelihood-based test procedure, because it uses the same ideas as the standard normal test that is part of every introductory statistics course. Because the MLE is asymptotically normally distributed with variance estimated as given in Equation B.5 or Equation B.6, we have that $Z_W = (\hat{\theta} - \theta_0)/\sqrt{\widehat{Var}(\hat{\theta})} \dot{\sim} N(0,1)$ for a large sample when the null hypothesis is true.[7] Thus, we reject H_0 if

$$|Z_W| = \frac{|(\hat{\theta} - \theta_0)|}{\sqrt{\widehat{Var}(\hat{\theta})}} > Z_{1-\alpha/2}, \tag{B.8}$$

where the critical value $Z_{1-\alpha/2}$ is a $1-\alpha/2$ quantile from a standard normal distribution. Alternatively, a p-value is calculated as $2P(Z > |Z_W|)$, where Z has a standard normal distribution.[8]

[7]The symbol $\dot{\sim}$ means "approximately distributed as," and N(0,1) represents a standard normal distribution.

[8]Note that the critical value is from the normal distribution and not a t distribution, even though we are estimating the variance in the denominator of Z_W. This is because the t distribution arises specifically when the variance in the denominator of the test statistic is based on a sum-of-squares calculation on data from a normal distribution. The variance in Z_W above is based on Equation B.5 or Equation B.6, which most often is not a sum-of-squares calculation.

A Wald test is usually simple to conduct, but it is not always very effective. In particular, the $Z_{1-\alpha/2}$ critical value is guaranteed to be close to the correct critical value only in very large samples. In small samples, the normal approximation may be poor and $Z_{1-\alpha/2}$ may not be an appropriate approximation to the true critical value of the test. Thus, this test is recommended only when the sample size is large (in the context of the problem) or when no other test is possible.

A version of the Wald test is available to test hypotheses involving more than one parameter. This can be useful in regression problems with a categorical explanatory variable, which is represented in the regression by several indicator variables, each with a separate parameter. A null hypothesis of no association between the explanatory variable and the response implies that these parameters must all be zero together. Let $\boldsymbol{\theta}_0$ be the hypothesized value of a p-dimensional parameter $\boldsymbol{\theta}$. Let $\hat{\boldsymbol{\theta}}$ be the parameter estimate, and $\widehat{Var}(\hat{\boldsymbol{\theta}})$ be its estimated variance. Then $H_0 : \boldsymbol{\theta} = \boldsymbol{\theta}_0$ vs. $H_a : \boldsymbol{\theta} \neq \boldsymbol{\theta}_0$ is tested using the Wald statistic

$$Q = (\hat{\boldsymbol{\theta}} - \boldsymbol{\theta}_0)'[\widehat{Var}(\hat{\boldsymbol{\theta}})]^{-1}(\hat{\boldsymbol{\theta}} - \boldsymbol{\theta}_0),$$

which has an approximate χ^2_p distribution in large samples. The expression for Q is based on matrix calculations, where we use $'$ to represent a transpose (not a derivative). The p degrees of freedom here result from the null hypothesis constraints placed on each value of $\boldsymbol{\theta}$. In general application, the degrees of freedom for any test (not only Wald tests) are found by calculating

$$\text{(Number of parameters under } H_a) - \text{(Number of parameters under } H_0).$$

Likelihood ratio tests

Values of θ that have likelihood functions near the maximum are more plausible guesses for the true value of the parameter than those whose likelihoods are much lower than the maximum. A comparison of the likelihood's value at its peak against the largest value of the likelihood when parameters are constrained to follow the null hypothesis is therefore a measure of evidence against the null. It turns out that the best way to make this comparison is through the likelihood ratio (LR) statistic

$$\Lambda = \frac{L(\theta_0|\mathbf{y})}{L(\hat{\theta}|\mathbf{y})}. \tag{B.9}$$

Note that $0 \leq \Lambda \leq 1$ because $L(\hat{\theta}|\mathbf{y})$ is the maximum value of the likelihood for the given data. The value of Λ is near 1 when $L(\theta_0|\mathbf{y})$ is close to $L(\hat{\theta}|\mathbf{y})$, indicating that θ_0 is plausible given the data. Closeness is judged by the fact that $-2\log(\Lambda) = 2[\log L(\hat{\theta}|\mathbf{y}) - \log L(\theta_0|\mathbf{y})]$ has an asymptotic χ^2_1 distribution when the null hypothesis is true. Thus, the *likelihood ratio test* (LRT) for $H_0 : \theta = \theta_0$ vs. $H_a : \theta \neq \theta_0$ rejects H_0 when $-2\log(\Lambda) > \chi^2_{1,1-\alpha}$, where $\chi^2_{1,1-\alpha}$ is the $1-\alpha$ quantile from a χ^2 distribution with 1 degree of freedom. Because $-2\log(\Lambda)$ is used to perform the test, rather than Λ alone, this transformed statistic is often referred to as the LRT statistic.

Example: LR for Poisson (LRplots.R)

The left plot in Figure B.3 gives a plot of the log-likelihood function for the previous Poisson example with $n = 10$ (sample 1). Evidence against $\mu = 9$ is fairly light, because the log-likelihood at $\mu = 9$ is fairly close to the peak at $\hat{\mu} = 10.5$. However, there is more evidence against $\mu = 5$, whose log-likelihood is much farther from the peak.

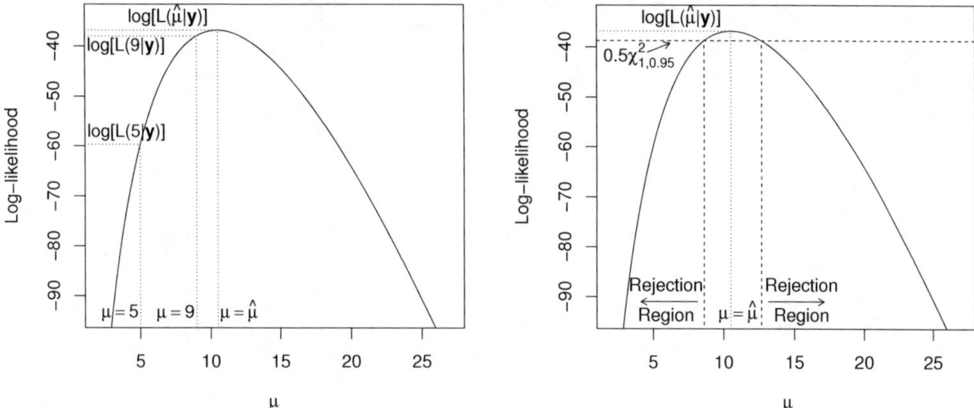

Figure B.3: Left: Poisson log-likelihood for the sample of size 10 with three values of μ indicated. Right: The rejection region for a likelihood ratio test using $\alpha = 0.05$. Any value of μ that lies in the rejection region has its null hypothesis rejected.

An LRT of $\mathrm{H_0}: \mu = 9$ vs. $\mathrm{H_a}: \mu \neq 9$ leads to

$$\begin{aligned} -2\log(\Lambda) &= -2\log\left(\frac{L(\mu = 9|\mathbf{y})}{L(\mu = 10.5|\mathbf{y})}\right) \\ &= -2\log\left(L(\mu = 9|\mathbf{y})\right) + 2\log\left(L(\mu = 10.5|\mathbf{y})\right) \\ &= 2.37. \end{aligned}$$

With $\chi^2_{1,0.95} = 3.84$, we do not reject $H_0: \mu = 9$. In a similar manner for $H_0: \mu = 5$ vs. $H_a: \mu \neq 5$, we calculate $-2\log(\Lambda) = 45.81$, leading to a rejection of the null hypothesis.

The right plot in Figure B.3 demonstrates the results from the two hypothesis tests differently. We represent the rejection region here by finding the set of all possible values of μ such that $-2\log(\Lambda) < \chi^2_{1,1-\alpha}$.[9] The figure again shows that $H_0: \mu = 9$ is not rejected, but $H_0: \mu = 5$ is rejected.

The LRT generalizes to a broad range of problems involving multiple parameters and hypotheses involving intervals or other constraints for the parameters. The general approach is that the numerator of Equation B.9 is replaced by the maximum value of the likelihood function across all parameters that satisfy H_0. Additionally, the denominator is replaced by the maximum value of the likelihood function without constraints. The degrees of freedom for the χ^2 distribution are determined in the same way as described for Wald tests. This is explained further throughout the book where LRTs are used.

Many problems use models that contain extra parameters that are not involved in the null hypothesis (for example, hypotheses about the mean of a normal model generally do not place constraints on the variance). In these cases, the extra parameters not specified in H_0, say ϕ, are set at their MLEs under the conditions on θ from H_0. That is, in the denominator of Equation B.9, the likelihood is maximized with respect to θ and ϕ simultaneously. In the numerator, the likelihood is again maximized for both parameters simultaneously, but subject to the constraint that θ satisfies H_0. Thus, the values for ϕ that yield the largest likelihoods in the numerator and denominator may be different.

[9] Solve for μ in $-2\log\left(L(\mu|\mathbf{y})\right) + 2\log\left(L(\mu = 10.5|\mathbf{y})\right) = \chi^2_{1,0.95}$ in order to find the rejection regions.

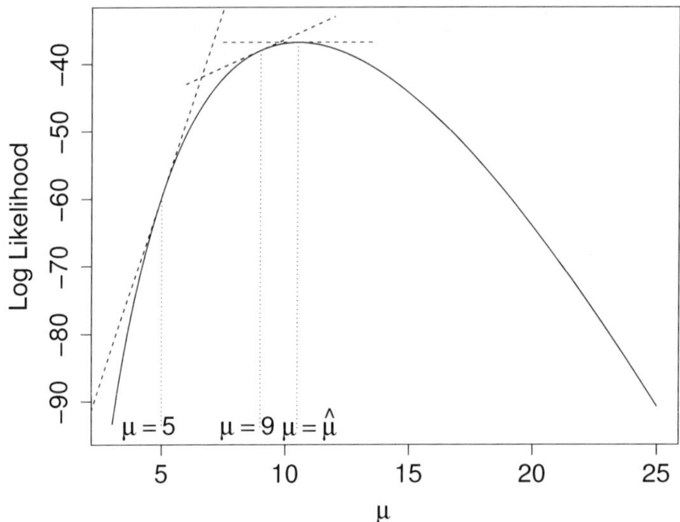

Figure B.4: Poisson log likelihood for artificial sample of size 10 showing the score function (slope) at three values of θ. The score is 0 at the MLE. The magnitude is larger for the θ farther from the MLE ($\theta = 5$) than for the one closer ($\theta = 9$). Code for the plot is in ScorePlot.R.

The LRT is often not simple to do by hand, but there are very good computational techniques that can efficiently maximize likelihoods subject to constraints. This makes LRTs broadly applicable to a wide range of testing problems. Also, the accuracy of the LRT critical value is generally much better for a given sample size than that of the Wald test critical value, so it is generally preferred over Wald tests when both are available.

Score test

An alternative approach to testing hypotheses using likelihoods comes from examining properties of the likelihood function at the null hypothesis. As Figure B.4 shows, the slope of the log-likelihood near the peak should have a smaller magnitude than the slope far from the peak. Therefore, we use this slope as a measure of the evidence in the data against H_0. This slope, also called the *score*, is just the first derivative of the log likelihood, evaluated at θ_0.

In particular, for independent random samples, the score is

$$U_0 = \frac{\partial}{\partial\theta}\log[L(\theta|\mathbf{y})]\bigg|_{\theta=\theta_0}.$$

The central limit theorem ensures that U_0 is asymptotically normally distributed. Under the null hypothesis, the average slope across all possible data sets is zero: $E(U_0) = 0$. It can be shown that the asymptotic variance of the score—measuring how variable the slopes would be at H_0 from likelihood functions calculated based on different data sets—can be estimated analogously to Equation B.5. Specifically,

$$\widehat{Var}(U_0) = -E\left(\frac{\partial^2}{\partial\theta^2}\log[L(\theta|\mathbf{y})]\right)\bigg|_{\theta=\theta_0}.$$

A score test is then carried out much like the Wald test, comparing $Z_S = U_0/\sqrt{\widehat{Var}(U_0)}$ to $Z_{1-\alpha/2}$. Multiparameter extensions are carried out just as in the Wald test.

The score test is also based on large-sample arguments, so that the critical value is approximate for any finite sample. It generally performs better than the Wald test, but not necessarily better than the LRT. Its main advantage is that it uses the likelihood only at the null hypotheses. In some complicated problems, the null hypothesis represents a considerable simplification to the general model—for example, by setting certain parameters to 0—so that calculations are much easier to carry out at θ_0 than anywhere else.

B.5.2 Confidence intervals for parameters

Wald

Like the Wald test, the Wald confidence interval is based on familiar relationships using the normal distribution. We can write $(\hat{\theta} - \theta)/\sqrt{\widehat{Var}(\hat{\theta})} \dot{\sim} N(0,1)$, where $\hat{\theta}$ and $\widehat{Var}(\hat{\theta})$ are the same as defined in Appendix B.3. Then

$$P\left(Z_{\alpha/2} < (\hat{\theta} - \theta)/\sqrt{\widehat{Var}(\hat{\theta})} < Z_{1-\alpha/2}\right) \approx 1 - \alpha,$$

where $Z_{1-\alpha/2}$ is a $1 - \alpha/2$ quantile from a standard normal distribution. After rearranging terms, we obtain

$$P\left(\hat{\theta} - Z_{1-\alpha/2}\sqrt{\widehat{Var}(\hat{\theta})} < \theta < \hat{\theta} - Z_{\alpha/2}\sqrt{\widehat{Var}(\hat{\theta})}\right) \approx 1 - \alpha.$$

Recognizing that $-Z_{\alpha/2} = Z_{1-\alpha/2}$, this leads to the familiar form of a $(1-\alpha)100\%$ confidence interval for θ as

$$\hat{\theta} \pm Z_{1-\alpha/2}\sqrt{\widehat{Var}(\hat{\theta})}. \tag{B.10}$$

An alternative approach to finding a confidence interval is to "invert" a test. That is, we seek the set of values for θ_0 for which $H_0 : \theta = \theta_0$ is *not* rejected. This procedure is easily carried out for the Wald test starting from Equation B.8 and leads to the same interval as Equation B.10.

Likelihood ratio

Likelihood ratio confidence intervals are found by inverting the LRT. A $(1 - \alpha)100\%$ confidence interval for θ is the set of all possible values of θ such that

$$-2[L(\theta|\mathbf{y})/L(\hat{\theta}|\mathbf{y})] \leq \chi^2_{1,1-\alpha}. \tag{B.11}$$

In Figure B.3, this is the interval between the two areas labeled "Rejection Region." As was the case with tests, likelihood ratio confidence intervals tend to be more accurate than Wald in the sense of having a true confidence level closer to the stated $1 - \alpha$ level for a given sample size. However, there is rarely a closed form for the solution, so iterative numerical procedures are needed to locate the endpoints of the interval where equality holds in Equation B.11. These computations can be difficult to carry out in some more complex problems, so not all software packages compute them.

In situations with multiple parameters, such as regression models, we frequently find confidence intervals for individual parameters. For example, suppose that a model contains two parameters, θ_1 and θ_2, and that we want to find a $(1 - \alpha)100\%$ LR confidence interval for θ_1. Because $L(\theta_1, \theta_2|\mathbf{y})$ changes as a function of both parameters, we need to account somehow for θ_2 in the process of computing Equation B.11. One approach is to fix θ_2 at some value, say d, and then find the values of θ_1 that satisfy $-2[L(\theta_1, d|\mathbf{y})/L(\tilde{\theta}_1(d), d|\mathbf{y})] \leq \chi^2_{1,1-\alpha}$, where $\tilde{\theta}_1(d)$ is the MLE of θ_1 when $\theta_2 = d$. However, this may lead to a different confidence interval for each value of d. As an alternative, we can let the value of θ_2 be its MLE for each value of θ_1 that we consider. That is, we fix the denominator of the LR statistic to be the overall maximum, $L(\hat{\theta}_1, \hat{\theta}_2|\mathbf{y})$, and for each different θ_1 we try in the numerator, we set $L(\theta_1, \theta_2|\mathbf{y})$ at the maximum that it achieves across all values θ_2. If we let $\tilde{\theta}_2(c)$ be the MLE of θ_2 when we fix $\theta_1 = c$, then the profile LR confidence interval is the set of values for c that satisfy $-2[L(c, \tilde{\theta}_2(c)|\mathbf{y})/L(\hat{\theta}_1, \hat{\theta}_2|\mathbf{y})] \leq \chi^2_{1,1-\alpha}$.

Score

Score confidence intervals are also found by inverting the score test. This is not as easy to do as it is for the Wald test, however. In the latter case, the standard error in the denominator of the test statistic does not change as one examines different values of θ_0, so that the rearrangement of Equation B.8 is easy. For the score test, the denominator changes with θ_0, and so the rearrangement can result in complicated mathematics unless the form of $\widehat{Var}(U_0)$ is fairly simple. Instead, the interval typically needs to be found using iterative numerical procedures similar to those described in Appendix B.3.2. An example where the mathematics *can* be worked out with relative ease is the Wilson score interval in Section 1.1.2.

B.5.3 Tests for models

Many forms of regression are used in categorical data analysis. Their use in practice often requires comparing several models involving different subsets of explanatory variables or comparing models with and without groups of explanatory variables (e.g., groups of indicator variables representing a categorical explanatory variable). When model parameters are estimated using maximum likelihood estimation, standard model comparison techniques are available based on likelihood ratio tests.

Consider two models: a *full model*, M_1, consisting of a set of p_1 explanatory variables, and the *reduced model*, M_0, containing a proper subset of p_0 explanatory variables. It is important that the reduced model does not contain any variables that are not also in the full model. Comparing M_0 and M_1 is equivalent to a hypothesis test specifying M_0 as the null hypothesis and M_1 as the alternative. Fit both M_0 and M_1 to the data, and let L_{M_1} and L_{M_0} be their respective maximized likelihoods. If we use $\Lambda(M_0, M_1)$ to denote the likelihood ratio L_{M_0}/L_{M_1}, then the LRT for $H_0 : M_0$ vs. $H_a : M_1$ rejects the null hypothesis if

$$-2\log\left(\Lambda(M_0, M_1)\right) > \chi^2_{(p_1-p_0),1-\alpha}. \tag{B.12}$$

A rejection of H_0 means that at least one of the parameters (thus, one of the variables) that make up the difference between M_0 and M_1 is important to include in a model already containing M_0. Failure to reject H_0 suggests that this simpler model may suffice and that the extra variables do not contribute significantly to the explanatory power of the model. Of course, we can never conclude that the null hypothesis is true for any hypothesis test, so it is not possible to say that M_0 is a "significantly better model" than M_1.

Deviance

In linear regression with normally distributed errors and constant variance, the sum of squared errors (SSE) for any model measures in aggregate how close the predictions are to the observed responses. In more general regression models involving different distributions, the common measure of a model's fit is the *deviance*. The deviance is just

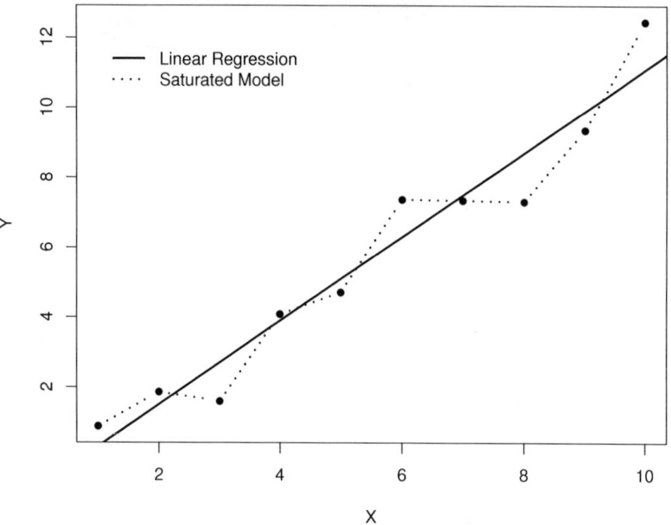

Figure B.5: Comparison of a simple linear regression with the saturated model. Code for the plot is in SaturatedModelExample.R.

$D_M = \Lambda(M, M_{SAT})$ from Equation B.12 where M is the model being fit and M_{SAT} is the *saturated model*, which fits a separate parameter for each observation. The saturated model provides perfect prediction for the observed data, but is often not considered a viable model for predicting new data, because new data is unlikely to have the same random "errors" as the current data (see Figure B.5 for a simple example of a saturated model). The reason to use a saturated model in computing deviances is that any other model is guaranteed to be a proper subset of the saturated model, and hence $\Lambda(M, M_{SAT})$ can always be computed. Thus, the deviance is often part of the default output for many modeling procedures in categorical data analysis.[10] If we have the deviances from the two models M_0 and M_1 that we want to compare as above, we can compute $\Lambda(M_0, M_1) = D_{M_0} - D_{M_1}$ to make this comparison, and carry out the test as in Equation B.12.

[10]For reasons described in Chapter 5, the comparison against $\chi^2_{(p_1-p_0),1-\alpha}$ is not always done.

Bibliography

Abdalhamid, B., Bilder, C., McCutchen, E., Hinrichs, S., Koepsell, S., and Iwen, P. (2020). Assessment of specimen pooling to conserve SARS CoV-2 testing resources. *American Journal of Clinical Pathology*, 153:715–718.

Adams, J. (2004). Ginkgo biloba and acetazolamide for acute mountain sickness: Exclusion of high risk, low status groups perpetuates discrimination and inequalities. *BMJ*, 329:171.

Agresti, A. (2002). *Categorical Data Analysis.* Wiley, second edition.

Agresti, A. (2007). *An Introduction to Categorical Data Analysis.* Wiley, second edition.

Agresti, A. (2013). *Categorical Data Analysis.* Wiley, third edition.

Agresti, A. (2019). *An Introduction to Categorical Data Analysis.* Wiley, third edition.

Agresti, A. and Coull, B. (1998). Approximate is better than "exact" for interval estimation of binomial proportions. *The American Statistician*, 52:119–126.

Agresti, A. and Liu, I. (1999). Modeling a categorical variable allowing arbitrarily many category choices. *Biometrics*, 55:936–943.

Agresti, A. and Min, Y. (2005a). Frequentist performance of Bayesian confidence intervals for comparing proportions in 2×2 contingency tables. *Biometrics*, 61:515–523.

Agresti, A. and Min, Y. (2005b). Simple improved confidence intervals for comparing matched proportions. *Statistics in Medicine*, 24:729–740.

Aguilera-Cogley, V. and Vicent, A. (2020). Sensitivity of *Zasmidium citri-griseum* to fenbuconazole and control of citrus greasy spot in Panama. *Agronomy*, 10:1577.

Akaike, H. (1974). A new look at the statistical model identification. *IEEE Transactions on Automatic Control*, 19:716–723.

American Red Cross (2023). Infectious Disease Testing. `https://www.redcrossblood.org/biomedical-services/blood-diagnostic-testing/blood-testing.html`.

Ando, N., Mizushima, D., Watanabe, K., Takano, M., Shiojiri, D., Uemura, H., Takahiro, A., Yanagawa, Y., Kikuchi, Y., Oka, S., and Gatanaga, H. (2021). Modified self-obtained pooled sampling to screen for *Chlamydia trachomatis* and *Neisseria gonorrhoeae* infections in men who have sex with men. *Sexually Transmitted Infections*, 97:324–328.

Aseffa, A., Ishak, A., Stevens, R., Fergussen, E., Giles, M., Yohannes, G., and Kidan, K. (1998). Prevalence of HIV, syphilis and genital chlamydial infection among women in north-west Ethiopia. *Epidemiology and Infection*, 120:171–177.

Austin, P. and Leckie, G. (2020). Bootstrapped inference for variance parameters, measures of heterogeneity and random effects in multilevel logistic regression models. *Journal of Statistical Computation and Simulation*, 90:3175–3199.

Barsalou, M. (2013). Keep your redshirt on: A Bayesian exploration. *Significance*.

Bates, D. (2010). *lme4: Mixed-Effects Modeling with R*. http://lme4.r-forge.r-project.org/book.

Becker, C., Loughin, T., and Santander, T. (2008). Identification of forest-obligate birds by mist netting and strip counts in Andean Ecuador. *Journal of Field Ornithology*, 79:229–244.

Beller, E. (2009). Bringing intergenerational social mobility research into the twenty-first century: Why mothers matter. *American Sociological Review*, 74:507–528.

Belsley, D., Kuh, E., and Welsch, R. (1980). *Regression Diagnostics: Identifying Influential Data and Sources of Collinearity*. Wiley.

Berry, K. and Mielke, P. (2003). Permutation analysis of data with multiple binary category choices. *Psychological Reports*, 92:91–98.

Berry, S. and Wood, C. (2004). The cold-foot effect. *Chance*, 17:47–51.

Biggerstaff, B. (2008). Confidence intervals for the difference of two proportions estimated from pooled samples. *Journal of Agricultural, Biological and Environmental Statistics*, 13:478–496.

Bilder, C. (2009). Human or Cylon? Group testing on 'Battlestar Galactica'. *Chance*, 22:46–50.

Bilder, C. (2019a). Group testing for estimation. *Wiley StatsRef: Statistics Reference Online*. https://doi.org/10.1002/9781118445112.stat08231.

Bilder, C. (2019b). Group testing for identification. *Wiley StatsRef: Statistics Reference Online*. https://doi.org/10.1002/9781118445112.stat08227.

Bilder, C. (2020). In or out? The new flagstick dilemma for putting in golf. *Chance*, 33:56–61.

Bilder, C., Iwen, P., Abdalhamid, B., Tebbs, J., and McMahan, C. (2020). Tests in short supply? Try group testing. *Significance*, 17:15–16.

Bilder, C. and Loughin, T. (1998). "It's Good!" An analysis of the probability of success for placekicks. *Chance*, 11:20–24.

Bilder, C. and Loughin, T. (2002). Testing for conditional multiple marginal independence. *Biometrics*, 58:200–208.

Bilder, C. and Loughin, T. (2004). Testing for marginal independence between two categorical variables with multiple responses. *Biometrics*, 60:241–248.

Bilder, C. and Loughin, T. (2007). Modeling association between two or more categorical variables that allow for multiple category choices. *Communications in Statistics: Theory and Methods*, 36:433–451.

Bilder, C., Loughin, T., and Nettleton, D. (2000). Multiple marginal independence testing for pick any/c variables. *Communications in Statistics: Simulation and Computation*, 29:1285–1316.

Bilder, C. and Tebbs, J. (2005). Empirical Bayesian estimation of the disease transmission probability in multiple-vector-transfer designs. *Biometrical Journal*, 47:502–516.

Bilder, C. and Tebbs, J. (2009). Bias, efficiency, and agreement for group-testing regression models. *Journal of Statistical Computation and Simulation*, 79:67–80.

Bilder, C. and Tebbs, J. (2012). Pooled-testing procedures for screening high volume clinical specimens in heterogeneous populations. *Statistics in Medicine*, 31:3261–3268.

Bilder, C., Tebbs, J., and Chen, P. (2010). Informative retesting. *Journal of the American Statistical Association*, 105:942–955.

Bilder, C., Tebbs, J., and McMahan, C. (2019). Informative group testing for multiplex assays. *Biometrics*, 75:278–288.

Bilder, C., Tebbs, J., and McMahan, C. (2021a). Discussion on 'Is group testing ready for prime-time in disease identification'. *Statistics in Medicine*, 40:3881–3886.

Bilder, C., Tebbs, J., and McMahan, C. (2021b). Informative array testing with multiplex assays. *Statistics in Medicine*, 40:3021–3034.

Binder, D. (1983). On the variances of asymptotically normal estimators from complex surveys. *International Statistical Review*, 51:279–292.

Binder, D. and Roberts, G. (2003). Statistical inference for survey data analysis. In *ASA Proceedings of the Joint Statistical Meetings*, pages 568–572. American Statistical Association.

Binder, D. and Roberts, G. (2009). Design- and model-based inference for model parameters. In Pfeffermann, D. and Rao, C., editors, *Handbook of Statistics*, volume 29. Elsevier.

Black, M., Bilder, C., and Tebbs, J. (2012). Group testing in heterogeneous populations by using halving algorithms. *Journal of the Royal Statistical Society: Series C (Applied Statistics)*, 61:277–290.

Black, M., Bilder, C., and Tebbs, J. (2015). Optimal retesting configurations for hierarchical group testing. *Journal of the Royal Statistical Society. Series C: Applied Statistics*, 64:693–710.

Blaker, H. (2000). Confidence curves and improved exact confidence intervals for discrete distributions. *The Canadian Journal of Statistics*, 28:783–798.

Blaker, H. (2001). Corrigenda: Confidence curves and improved exact confidence intervals for discrete distributions. *The Canadian Journal of Statistics*, 29:681.

Bolker, B. (2009). Dealing with quasi-models in R. `http://cran.r-project.org/web/packages/bbmle/vignettes/quasi.pdf`.

Bonett, D. and Price, R. (2006). Confidence intervals for a ratio of binomial proportions based on paired data. *Statistics in Medicine*, 25:3039–3047.

Box, G. (1976). Science and statistics. *Journal of the American Statistical Association*, 356:791–799.

Breslow, N. and Lin, X. (1995). Bias correction in generalised linear mixed models with a single component of dispersion. *Biometrika*, 82:81–91.

Bretz, F., Hothorn, T., and Westfall, P. (2011). *Multiple Comparisons Using R*. CRC Press.

Broach, D. and Schroeder, D. (2006). Air traffic control specialist age and en route operational errors. *The International Journal of Aviation Psychology*, 16:363–373.

Brown, L., Cai, T., and DasGupta, A. (2001). Interval estimation for a binomial proportion. *Statistical Science*, 16:101–133.

Brown, P., Stone, J., and Ord-Smith, C. (1983). Toxaemic signs during pregnancy. *Applied Statistics*, 32:69–72.

Brownlee, K. (1955). Statistics of the 1954 polio vaccine trials. *Journal of the American Statistical Association*, 50:1005–1013.

Buonaccorsi, J. (2010). *Measurement Error: Models, Methods, and Applications*. CRC Press.

Burden, R., Faires, J., and Burden, A. (2015). *Numerical Analysis*. Cengage learning.

Burnham, K. and Anderson, D. (2002). *Model Selection and Multimodel Inference: A Practical Information-Theoretic Approach*. Springer, second edition.

Burrows, P. (1987). Improved estimation of pathogen transmission rates by group testing. *Phytopathology*, 77:363–365.

Carlin, B. and Louis, T. (2008). *Bayesian Methods for Data Analysis*. CRC Press, third edition.

Carpenter, B., Gelman, A., Hoffman, M., Lee, D., Goodrich, B., Betancourt, M., Brubaker, M., Guo, J., Li, P., and Riddell, A. (2017). Stan: A probabilistic programming language. *Journal of Statistical Software*, 76:1–32.

Carroll, R., Ruppert, D., Stefanski, L., and Crainiceanu, C. (2006). *Measurement Error in Nonlinear Models: A Modern Perspective*. CRC Press.

Casella, G. and Berger, R. (2002). *Statistical Inference*. Duxbury Press, second edition.

Chankov, E. (2017). Improved implementation of the marginal independence test. `https://github.com/e-chankov/MItest_improvements`.

Cheskin, L., Miday, R., Zorich, N., and Filloon, T. (1998). Gastrointestinal symptoms following consumption of Olestra or regular triglyceride potato chips: a controlled comparison. *Journal of the American Medical Association*, 279:150–152.

Christensen, R., Johnson, W., Branscum, A., and Hanson, T. (2010). *Bayesian Ideas and Data Analysis: An Introduction for Scientists and Statisticians*. CRC Press.

Chung, H., Jian, M., Chang, C., Lin, J., Yeh, K., Chen, C., Yang, Y., Hsieh, S., Chen, E., Yang, M., et al. (2022). Multicenter study evaluating novel multi-specimen pooling assay for the detection of SARS-CoV-2: High sensitivity and high throughput testing. *Journal of Microbiology, Immunology and Infection*, 55:1069–1075.

Clarke, T. (2009). Trial shows first HIV protection from vaccines. *PBS NewsHour*. `https://www.pbs.org/newshour/science/health-july-dec09-hiv_09-24`.

Clogg, C. and Eliason, S. (1987). Some common problems in log-linear analysis. *Sociological Methods & Research*, 16:8–44.

Clopper, C. and Pearson, E. (1934). The use of confidence or fiducial limits illustrated in the case of the binomial. *Biometrika*, 26:404–413.

Cognard, C., Gobin, Y., Pierot, L., Bailly, A., Houdart, E., Casasco, A., Chiras, J., and Merland, J. (1995). Cerebral dural arteriovenous fistulas: Clinical and angiographic correlation with a revised classification of venous drainage. *Radiology*, 194:671–680.

Coombs, C. (1964). *A Theory of Data*. Wiley.

Cowles, M. and Carlin, B. (1996). Markov chain Monte Carlo convergence diagnostics: A comparative review. *Journal of the American Statistical Association*, 91:883–904.

Curtis, C., Laudner, K., McLoda, T., and McCaw, S. (2008). The role of shoe design in ankle sprain rates among collegiate basketball players. *Journal of Athletic Training*, 43:230–233.

Dalal, S., Fowlkes, E., and Hoadley, B. (1989). Risk analysis of the space shuttle: Pre-Challenger prediction of failure. *Journal of the American Statistical Association*, 84:945–957.

Davison, A. and Hinkley, D. (1997). *Bootstrap Methods and their Application*. Cambridge University Press.

Dawson, L. (2004). The Salk polio vaccine clinical trial of 1954: Risks, randomization and public involvement in research. *Clinical Trials*, 1:122–130.

Deb, P. and Trivedi, P. (1997). Demand for medical care by the elderly: A finite mixture approach. *Journal of Applied Econometrics*, 12:313–336.

DeHart, T., Tennen, H., Armeli, S., Todd, M., and Affleck, G. (2008). Drinking to regulate romantic relationship interactions: The moderating role of self-esteem. *Journal of Experimental Social Psychology*, 44:527–538.

Dodd, R. (2019). Case study: West Nile virus. In Shan, H. and Dodd, R., editors, *Blood Safety*. Springer.

Dorfman, R. (1943). The detection of defective members of large populations. *The Annals of Mathematical Statistics*, 14:436–440.

Dres, K., Hepworth, G., and Watson, R. (2015). Exact confidence intervals for proportions estimated by group testing with different group sizes. *Australian & New Zealand Journal of Statistics*, 57:501–516.

Elphick, H. and Elphick, D. (2004). Ginkgo biloba and acetazolamide for acute mountain sickness: Bias in participants may underestimate effectiveness of agents. *BMJ*, 329:172.

Fagerland, M. and Hosmer, D. (2016). Tests for goodness of fit in ordinal logistic regression models. *Journal of Statistical Computation and Simulation*, 86:3398–3418.

Fagerland, M., Hosmer, D., and Bofin, A. (2008). Multinomial goodness-of-fit tests for logistic regression models. *Statistics in Medicine*, 27:4238–4253.

Fang, L. and Loughin, T. (2012). Analyzing binomial data in a split-plot design: Classical approach or modern techniques? *Communications in Statistics: Simulation and Computation*, 42:727–740.

Faraway, J. (2016). Does data splitting improve prediction? *Statistical Computing*, 26:49–60.

Fieller, N. (2016). *Basics of Matrix Algebra for Statistics with R*. CRC Press.

Firth, D. (1993). Bias reduction of maximum likelihood estimates. *Biometrika*, 80:27–38.

Fitzmaurice, G., Davidian, M., Verbeke, G., and Molenberghs, G. (2009). *Longitudinal Data Analysis.* CRC Press.

Fitzpatrick, M., Pandey, A., Wells, C., Sah, P., and Galvani, A. (2021). Buyer beware: inflated claims of sensitivity for rapid COVID-19 tests. *The Lancet*, 397:24–25.

Flaherty, S., Patenaude, G., Close, A., and Lurz, P. (2012). The impact of forest stand structure on red squirrel habitat use. *Forestry: An International Journal of Forest Research*, 85:437–444.

Fox, J. (2008). *Applied Regression Analysis and Generalized Linear Models.* Sage Publications, second edition.

Foxman, B., Marsh, J., Gillespie, B., Rubin, N., Koopman, J., and Spear, S. (1997). Condom use and first-time urinary tract infection. *Epidemiology*, 8:637–641.

Francis, T., Korns, R., Voight, R., Boisen, M., Hemphill, F., Napier, J., and Tolchinsky, E. (1955). An evaluation of the 1954 poliomyelitis vaccine trials. *American Journal of Public Health*, 45:1–63.

Freeman, R., Zinner, S., Müller-Vahl, K., Fase, D., Burd, L., Kano, Y., Rothenberger, A., Roessner, V., Kerbeshian, J., Stern, J., et al. (2009). Coprophenomena in Tourette syndrome. *Developmental Medicine and Child Neurology*, 51:218–227.

Friendly, M. (1992). Graphical methods for categorical data. In *SAS User Group International Conference Proceedings*, volume 17, pages 190–200.

Gange, S. (1995). Generating multivariate categorical variates using the iterative proportional fitting algorithm. *The American Statistician*, 49:134–138.

Gelman, A., Carlin, J., Stern, H., Dunson, D., Vehtari, A., and Rubin, D. (2013). *Bayesian Data Analysis.* CRC Press, third edition.

Gelman, A. and Rubin, D. (1992). Inference from iterative simulation using multiple sequences. *Statistical Science*, 7:457–472.

Gelman, A. and Shirley, K. (2011). Inference from simulations and monitoring convergence. In Brooks, S., Gelman, A., Jones, G., and Meng, X.-L., editors, *Handbook of Markov Chain Monte Carlo.* CRC Press.

Gertsch, J., Basnyat, B., Johnson, E., and Holck, P. (2004a). Ginkgo biloba and acetazolamide for acute mountain sickness: Authors' reply. *BMJ*, 329:172.

Gertsch, J., Basnyat, B., Johnson, E., Onopa, J., and Holck, P. (2004b). Randomised, double blind, placebo controlled comparison of ginkgo biloba and acetazolamide for prevention of acute mountain sickness among Himalayan trekkers: The prevention of high altitude illness trial (PHAIT). *BMJ*, 328:797–801.

Gildow, F., Shah, D., Sackett, W., Butzler, T., Nault, B., and Fleischer, S. (2008). Transmission efficiency of *Cucumber mosaic virus* by aphids associated with virus epidemics in snap bean. *Phytopathology*, 98:1233–1241.

Givens, G. and Hoeting, J. (2013). *Computational Statistics.* Wiley.

Grechanovsky, E. (1987). Stepwise regression procedures: Overview, problems, results, and suggestions. *Annals of the New York Academy of Sciences*, 491:197–232.

Gregory, K., Groth, P., Scharnhorst, A., and Wyatt, S. (2020). Lost or found? Discovering data needed for research. *Harvard Data Science Review*, 2. `https://hdsr.mitpress.mit.edu/pub/gw3r97ht`.

Greven, S. and Kneib, T. (2010). On the behaviour of marginal and conditional AIC in linear mixed models. *Biometrika*, 97:773–789.

Gustafson, C., Rakhmatullaeva, N., Beckford, S., Ammachathram, A., Cristobal, A., and Koehler, K. (2018). Exercise and the timing of snack choice: Healthy snack choice is reduced in the post-exercise state. *Nutrients*, 10:1–11.

Gustafson, P. (2004). *Measurement Error and Misclassification in Statistics and Epidemiology: Impacts and Bayesian Adjustments.* CRC Press.

Halekoh, U., Højsgaard, S., and Yan, J. (2006). The R package `geepack` for generalized estimating equations. *Journal of Statistical Software*, 15:1–11.

Hammer, S., Sobieszczyk, M., Janes, H., Karuna, S., Mulligan, M., Grove, D., Koblin, B., Buchbinder, S., Keefer, M., Tomaras, G., et al. (2013). Efficacy trial of a DNA/rAd5 HIV-1 preventive vaccine. *New England Journal of Medicine*, 369:2083–2092.

Hartman, G. (2011). *Fundamentals of Matrix Algebra.* Open Textbook Library, third edition.

Hastie, T. and Tibshirani, R. (1986). Generalized additive models. *Statistical Science*, 1:297–318.

Hastie, T. and Tibshirani, R. (1990). *Generalized Additive Models.* CRC Press.

Hastie, T., Tibshirani, R., and Friedman, J. (2009). *The Elements of Statistical Learning.* Springer, second edition.

Heath, P., Galiza, E., Baxter, D., Boffito, M., Browne, D., Burns, F., Chadwick, D., Clark, R., Cosgrove, C., Galloway, J., et al. (2021). Safety and efficacy of NVX-CoV2373 Covid-19 vaccine. *New England Journal of Medicine*, 385:1172–1183.

Heeringa, S., West, B., and Berglund, P. (2010). *Applied Survey Data Analysis.* CRC Press.

Heinze, G. (2006). A comparative investigation of methods for logistic regression with separated or nearly separated data. *Statistics in Medicine*, 25:4216–4226.

Heinze, G. and Schemper, M. (2002). A solution to the problem of separation in logistic regression. *Statistics in Medicine*, 21:2409–2419.

Heinze, G., Wallisch, C., and Dunkler, D. (2018). Variable selection–A review and recommendations for the practicing statistician. *Biometrical Journal*, 60:431–449.

Henao-Restrepo, A., Camacho, A., Longini, I., Watson, C., Edmunds, W., Egger, M., Carroll, M., Dean, N., Diatta, I., Doumbia, M., et al. (2017). Efficacy and effectiveness of an rVSV-vectored vaccine in preventing Ebola virus disease: Final results from the Guinea ring vaccination, open-label, cluster-randomised trial (Ebola Ça Suffit!). *The Lancet*, 389:505–518.

Hepworth, G. (2005). Confidence intervals for proportions estimated by group testing with groups of unequal size. *Journal of Agricultural, Biological and Environmental Statistics*, 10:478–497.

Hepworth, G. and Biggerstaff, B. (2017). Bias correction in estimating proportions by pooled testing. *Journal of Agricultural, Biological and Environmental Statistics*, 22:602–614.

Hirji, K., Mehta, C., and Patel, N. (1987). Computing distributions for exact logistic regression. *Journal of the American Statistical Association*, 82:1110–1117.

Hitt, B., Bilder, C., Schaarschmidt, F., Biggerstaff, B., Tebbs, J., and McMahan, C. (2023). `binGroup2`: Identification and estimation using group testing. `https://cran.r-project.org/web/packages/binGroup2/index.html`.

Hitt, B., Bilder, C., Tebbs, J., and McMahan, C. (2019). The objective function controversy for group testing: Much ado about nothing? *Statistics in Medicine*, 38:4912–4923.

Hoeting, J., Madigan, D., Raftery, A., and Volinsky, C. (1999). Bayesian model averaging: A tutorial. *Statistical Science*, 14:382–417.

Hosmer, D. and Hjort, N. (2002). Goodness-of-fit processes for logistic regression: simulation results. *Statistics in Medicine*, 21:2723–2738.

Hosmer, D., Hosmer, T., le Cessie, S., and Lemeshow, S. (1997). A comparison of goodness-of-fit tests for the logistic regression model. *Statistics in Medicine*, 16:965–980.

Hosmer, D. and Lemeshow, S. (1980). Goodness-of-fit tests for the multiple logistic regression model. *Communications in Statistics: Theory and Methods*, 9:1043–1069.

Hosmer, D. and Lemeshow, S. (2000). *Applied Logistic Regression*. Wiley, second edition.

Hou, P., Tebbs, J., Bilder, C., and McMahan, C. (2017). Hierarchical group testing for multiple infections. *Biometrics*, 73:656–665.

Hubert, J. (1992). *Bioassay*. Kendall Hunt Publishing Company, third edition.

Hughes-Oliver, J. (2006). Pooling experiments for blood screening and drug discovery. In Dean, A. and Lewis, S., editors, *Screening: Methods for Experimentation in Industry, Drug Discovery, and Genetics*. Springer.

Ihaka, R. and Gentleman, R. (1996). R: A language for data analysis and graphics. *Journal of Computational and Graphical Statistics*, 3:299–314.

Imori, S., Yanagihara, H., and Wakaki, H. (2014). Simple Formula for Calculating Bias-corrected AIC in Generalized Linear Models. *Scandinavian Journal of Statistics*, 41:535–555.

Imrey, P., Koch, G., Stokes, M., in collaboration with Darroch, J., Freeman, D., and Tolley, H. (1982). Some reflections on the log-linear model and logistic regression. Part II: Data analysis. *International Statistical Review*, 50:35–63.

James, G., Witten, D., Hastie, T., and Tibshirani, R. (2013). *An Introduction to Statistical Learning*. Springer.

Johnson & Johnson (2021). Clinical trial data for the Janssen COVID-19 vaccine. `https://web.archive.org/web/20210308143536/https://www.janssencovid19vaccine.com/hcp/clinical-trial-data.html`.

Joyner, C., McMahan, C., Tebbs, J., and Bilder, C. (2020). From mixed effects modeling to spike and slab variable selection: A Bayesian regression model for group testing data. *Biometrics*, 76:913–923.

Kainkaryam, R. and Woolf, P. (2009). Pooling in high-throughput drug screening. *Current Opinion in Drug Discovery & Development*, 12:339–350.

Kim, H., Hudgens, M., Dreyfuss, J., Westreich, D., and Pilcher, C. (2007). Comparison of group testing algorithms for case identification in the presence of test error. *Biometrics*, 63:1152–1163.

Korall, A., Loughin, T., Feldman, F., Cameron, I., Leung, P., Sims-Gould, J., Godin, J., and Robinovitch, S. (2018). Determinants of commitment to hip protectors among long-term care providers of a regional health authority in British Columbia, Canada: A cross-sectional survey. *International Journal of Nursing Studies*, 82:139–148.

Korn, E. and Graubard, B. (1999). *Analysis of Health Surveys*. Wiley.

Kosmidis, I., Pagui, E., and Sartori, N. (2020). Mean and median bias reduction in generalized linear models. *Statistics and Computing*, 30:43–59.

Kott, P. and Carr, D. (1997). Developing an estimation strategy for a pesticide data program. *Journal of Official Statistics*, 13:367–383.

Koziol, N. and Bilder, C. (2014). `MRCV`: A package for analyzing categorical variables with multiple response options. *R Journal*, 6:144–150.

Küchenhoff, H., Mwalili, S., and Lesaffre, E. (2006). A general method for dealing with misclassification in regression: The misclassification SIMEX. *Biometrics*, 62:85–96.

Kuchibhotla, A., Kolassa, J., and Kuffner, T. (2022). Post-selection inference. *Annual Review of Statistics and Its Application*, 9:505–527.

Kuonen, D. (1999). Saddlepoint approximations for distributions of quadratic forms in normal variables. *Biometrika*, 86:929–935.

Kupper, L. and Haseman, J. (1978). The use of a correlated binomial model for the analysis of certain toxicological experiments. *Biometrics*, 34:69–76.

Kutner, M., Nachtsheim, C., and Neter, J. (2004). *Applied Linear Regression Models*. McGraw-Hill/Irwin, fourth edition.

Lambert, D. (1992). Zero-inflated Poisson regression, with an application to defects in manufacturing. *Technometrics*, 34:1–14.

Landsman, V., Landsman, D., Li, C., and Bang, H. (2019). Overdispersion models for correlated multinomial data: Applications to blinding assessment. *Statistics in Medicine*, 38:4963–4976.

Larntz, K. (1978). Small-sample comparisons of exact levels for chi-squared goodness-of-fit statistics. *Journal of the American Statistical Association*, 73:253–263.

Laurin, E., Thakur, K., Mohr, P., Hick, P., Crane, M., Gardner, I., Moody, N., Colling, A., and Ernst, I. (2019). To pool or not to pool? Guidelines for pooling samples for use in surveillance testing of infectious diseases in aquatic animals. *Journal of Fish Diseases*, 42:1471–1491.

Lederer, W. and Küchenhoff, H. (2006). A short introduction to the SIMEX and MCSIMEX. *R News*, 6:26–31.

Leeb, H. and Poetscher, B. (2005). Model selection and inference: Facts and fiction. *Econometric Theory*, 21:21–59.

Lesaffre, E. and Albert, A. (1989). Multi-group logistic regression diagnostics. *Journal of the Royal Statistical Society, Series C*, 38:425–440.

Lewis, J., Lockary, V., and Kobic, S. (2012). Cost savings and increased efficiency using a stratified specimen pooling strategy for *Chlamydia trachomatis* and *Neisseria gonorrhoeae*. *Sexually Transmitted Diseases*, 39:46–48.

Liang, K. and Zeger, S. (1986). Longitudinal data analysis using generalized linear models. *Biometrika*, 73:13–22.

Lipsitz, S., Fitzmaurice, G., Sinha, D., and Hevelone, N. (2015). Testing for Independence in JxK contingency tables with complex sample survey data. *Biometrics*, 71:832–840.

Littell, R., Milliken, G., Stroup, W., Wolfinger, R., and Schabenberger, O. (2006). *SAS for Mixed Models*. SAS Institute, second edition.

Litvak, E., Tu, X., and Pagano, M. (1994). Screening for the presence of a disease by pooling sera samples. *Journal of the American Statistical Association*, 89:424–434.

Liu, A., Liu, C., Zhang, Z., and Albert, P. (2012). Optimality of group testing in the presence of misclassification. *Biometrika*, 99:245–251.

Liu, P., Shi, Z., Zhang, Y., Xu, Z., Shu, H., and Zhang, X. (1997). A prospective study of a serum-pooling strategy in screening blood donors for antibody to hepatitis C virus. *Transfusion*, 37:732–736.

Lo, C., Bai, Y., Liu, M., and Lynch, J. (2013). Efficient sensor fault detection using combinatorial group testing. In *2013 IEEE International Conference on Distributed Computing in Sensor Systems*, pages 199–206. IEEE.

Lohr, S. (2022). *Sampling: Design and Analysis*. CRC Press, third edition.

Long, J. (1990). The origins of sex differences in science. *Social Forces*, 68:1297–1316.

Loughin, T. (2004). A systematic comparison of methods for combining p-values from independent tests. *Computational Statistics and Data Analysis*, 47:467–485.

Loughin, T. and Bilder, C. (2010). On the use of a log-rate model for survey-weighted categorical data. *Communications in Statistics: Theory and Methods*, 40:2661–2669.

Loughin, T., Roediger, M., Milliken, G., and Schmidt, J. (2007). On the analysis of long-term experiments. *Journal of the Royal Statistical Society, Series A*, 170:29–42.

Loughin, T. and Scherer, P. (1998). Testing for association in contingency tables with multiple column responses. *Biometrics*, 54:630–637.

Louis, T. (1982). Finding the observed information matrix when using the EM algorithm. *Journal of the Royal Statistical Society, Series B*, 44:226–233.

Lui, K. and Lin, C. (2003). A revisit on comparing the asymptotic interval estimators of odds ratio in a single 2x2 table. *Biometrical Journal*, 45:226–237.

Lumley, T. (2011). *Complex Surveys: A Guide to Analysis using R*. Wiley.

Lumley, T. and Scott, A. (2015). AIC and BIC for modeling with complex survey data. *Journal of Survey Statistics and Methodology*, 3:1–18.

Margolin, B., Kaplan, N., and Zeiger, E. (1981). Statistical analysis of the Ames *Salmonella*/microsome test. *Proceedings of the National Academy of Sciences USA*, 78:3779–3783.

Martin, C., Herrman, T., Loughin, T., and Oentong, S. (1998). Micropycnometer measurement of single-kernel density of healthy, sprouted, and scab-damaged wheats. *Cereal Chemistry*, 75:177–180.

Matloff, N. (2023). Teaching R in a kinder, gentler, more effective manner: Teach base-R, not just the tidyverse. https://github.com/matloff/TidyverseSkeptic.

Maugh, T. (2009). Results of AIDS vaccine trial 'weak' in second analysis. *Los Angeles Times.* October 20; https://www.latimes.com/nation/la-sci-aids-vaccine20-2009oct20-story.html.

McCann, M. and Tebbs, J. (2007). Pairwise comparisons for proportions estimated by pooled testing. *Journal of Statistical Planning and Inference*, 137:1278–1290.

McCullagh, P. and Nelder, J. (1989). *Generalized Linear Models.* CRC Press, second edition.

McElreath, R. (2020). *Statistical Rethinking: A Bayesian Course with Examples in R and Stan.* CRC Press.

McLean, R., Sanders, W., and Stroup, W. (1991). A unified approach to mixed linear models. *The American Statistician*, 45:54–64.

McMahan, C., Tebbs, J., and Bilder, C. (2012). Informative Dorfman screening. *Biometrics*, 68:287–296.

McMahan, C., Tebbs, J., Hanson, T., and Bilder, C. (2017). Bayesian regression for group testing data. *Biometrics*, 73:1443–1452.

McNeil, D. (2013). Agency halts trial for AIDS vaccine. *New York Times.* April 26; https://www.nytimes.com/2013/04/26/health/agency-halts-trial-for-aids-vaccine.html.

McNemar, Q. (1947). Note on the sampling error of the difference between correlated proportions or percentages. *Pyschometrika*, 12:153–157.

Mebane, W. and Sekhon, J. (2004). Robust estimation and outlier detection for overdispersed multinomial models of count data. *American Journal of Political Science*, 48:392–411.

Mehta, C. and Patel, N. (1995). Exact logistic regression: Theory and examples. *Statistics in Medicine*, 14:2143–2160.

Mehta, C., Patel, N., and Senchaudhuri, P. (2000). Efficient Monte Carlo methods for conditional logistic regression. *Journal of the American Statistical Association*, 95:99–108.

Meinshausen, N. (2007). Relaxed lasso. *Computational Statistics and Data Analysis*, 52:374–393.

Merck (2021). Merck and Ridgeback's investigational oral antiviral Molnupiravir reduced the risk of hospitalization or death by approximately 50 percent compared to placebo for patients with mild or moderate COVID-19 in positive interim analysis of phase 3 study. https://web.archive.org/web/20211001100425/https://www.merck.com/news/merck-and-ridgebacks-investigational-oral-antiviral-molnupiravir-reduced-the-risk-of-hospitalization-or-death-by-approximately-50-percent-compared-to-placebo-for-patients-with-mild-or-moderat.

Mester, P., Witte, A., Robben, C., Streit, E., Fister, S., Schoder, D., and Rossmanith, P. (2017). Optimization and evaluation of the qPCR-based pooling strategy DEP-pooling in dairy production for the detection of Listeria monocytogenes. *Food Control*, 82:298–304.

Meulders, M. and De Bruecker, P. (2017). `plfm`: Probabilistic latent feature analysis. `https://cran.r-project.org/web/packages/plfm/index.html`.

Michalewicz, Z. (1996). *Genetic Algorithms + Data Structures = Evolution Programs*. Springer, third edition.

Miller, A. (1984). Selection of subsets of regression variables (with discussion). *Journal of the Royal Statistical Society, Series A*, 147:389–425.

Milliken, G. and Johnson, D. (2001). *Analysis of Messy Data, Volume III: Analysis of Covariance*. CRC Press.

Milliken, G. and Johnson, D. (2004). *Analysis of Messy Data Volume I: Designed Experiments*. CRC Press, second edition.

Moderna (2021). Clinical trial data. `https://web.archive.org/web/20210926031407/https://www.modernatx.com/covid19vaccine-eua/providers/clinical-trial-data`.

Molenberghs, G. and Verbeke, G. (2005). *Models for Discrete Longitudinal Data*. Springer.

Montgomery, D., Peck, E., and Vining, G. (2021). *Introduction to Linear Regression Analysis*. Wiley, sixth edition.

Moore, D. and Notz, W. (2009). *Statistics: Concepts and Controversies*. W.H. Freeman & Company, 7th edition.

Mullahy, J. (1986). Specification and testing of some modified count data models. *Journal of Econometrics*, 33:341–365.

Nattino, G., Pennell, M., and Lemeshow, S. (2020). Assessing the goodness of fit of logistic regression models in large samples: A modification of the Hosmer-Lemeshow test. *Biometrics*, 76:549–560.

NIAID (2020). Experimental HIV vaccine regimen ineffective in preventing HIV. `https://www.niaid.nih.gov/news-events/experimental-hiv-vaccine-regimen-ineffective-preventing-hiv`.

Ornaghi, J., Boito, G., Marinelli, A., Beviacqua, J., Giuggia, J., Lenardon, S., and March, G. (1999). Infectivity in natural populations of *Delphacodes kuscheli* vector of Mal de Rio Cuarto virus. *Maydica*, 44:219–223.

Osius, G. and Rojek, D. (1992). Normal goodness-of-fit tests for multinomial models with large degrees of freedom. *Journal of the American Statistical Association*, 87:1145–1152.

Oxford-AstraZeneca (2021). Summary of product characteristics for Vaxzevria (previously COVID-19 Vaccine AstraZeneca). `https://web.archive.org/web/20210828143137/https://www.gov.uk/government/publications/regulatory-approval-of-covid-19-vaccine-astrazeneca`.

Pasteur, R. and David, J. (2017). Evaluation of quarterbacks and kickers. In Albert, J., Glickman, M., Swartz, T., and Koning, R., editors, *Handbook of Statistical Methods and Analyses in Sports*. CRC Press.

Paul, P., Pennell, M., and Lemeshow, S. (2013). Standardizing the power of the Hosmer-Lemeshow goodness of fit test in large data sets. *Statistics in Medicine*, 32:67–80.

Pena, W., De Massaguer, P., Zuniga, A., and Saraiva, S. (2011). Modeling the growth limit of *Alicyclobacillus acidoterrestris* CRA7152 in apple juice: Effect of pH, Brix, temperature and nisin concentration. *Journal of Food Processing and Preservation*, 35:509–517.

Perperoglou, A., Sauerbrei, W., Abrahamowicz, M., and Schmid, M. (2019). A review of spline function procedures in R. *BMC Medical Research Methodology*, 19:1–16.

Plummer, M., Best, N., Cowles, K., and Vines, K. (2006). `CODA`: Convergence diagnosis and output analysis for MCMC. *R News*, 6:7–11.

Polack, F., Thomas, S., Kitchin, N., Absalon, J., Gurtman, A., Lockhart, S., Perez, J., Pérez Marc, G., Moreira, E., Zerbini, C., et al. (2020). Safety and efficacy of the BNT162b2 mRNA COVID-19 vaccine. *New England Journal of Medicine*, 383:2603–2615.

Potter, D. (2005). A permutation test for inference in logistic regression with small- and moderate-sized data sets. *Statistics in Medicine*, 24:693–708.

Pregibon, D. (1981). Logistic regression diagnostics. *The Annals of Statistics*, 9:705–724.

Raftery, A. (1995). Bayesian model selection in social research. *Sociological Methodology*, 25:111–163.

Rao, J. and Scott, A. (1981). The analysis of categorical data from complex sample surveys: Chi-squared tests for goodness of fit and independence in two-way tables. *Journal of the American Statistical Association*, 76:221–230.

Rao, J. and Scott, A. (1984). On chi-squared tests for multiway contingency tables with cell proportions estimated from survey data. *The Annals of Statistics*, 12:46–60.

Rao, J. and Thomas, D. (2003). Analysis of categorical response data from complex surveys: An appraisal and update. In Chambers, R. and Skinner, C., editors, *Analysis of Survey Data*. Wiley.

Raudenbush, S. and Bryk, A. (2002). *Hierarchical Linear Models: Applications and Data Analysis Methods*. Sage Publications, second edition.

Rerks-Ngarm, S., Pitisuttithum, P., Nitayaphan, S., Kaewkungwal, J., Chiu, J., Paris, R., Premsri, N., Namwat, C., de Souza, M., Adams, E., et al. (2009). Vaccination with ALVAC and AIDSVAX to prevent HIV-1 infection in Thailand. *New England Journal of Medicine*, 361:2209–2220.

Richardson, J. (2021). A closer look at the lady tasting tea. *Significance*, 18:34–37.

Richardson, M. and Haller, S. (2002). What is the probability of a kiss? (It's not what you think). *Journal of Statistics Education*, 10.

Richert, B., Tokach, M., Goodband, R., and Nelssen, J. (1995). Assessing producer awareness of the impact of swine production on the environment. *Journal of Extension*, 33.

Riemer, S., Wright, B., and Brown, R. (2011). Food habits of Steller sea lions (*Eumetopias jubatus*) off Oregon and northern California, 1986–2007. *Fishery Bulletin*, 109:369–381.

Robert, C. (2007). *The Bayesian Choice: From Decision-Theoretic Foundations to Computational Implementation.* Springer, second edition.

Robert, C. and Casella, G. (2004). *Monte Carlo Statistical Methods.* Springer.

Robert, C. and Casella, G. (2010). *Introducing Monte Carlo Methods with R.* Springer.

Rogan, W. and Gladen, B. (1978). Estimating the prevalence from the results of a screening test. *American Journal of Epidemiology*, 107:71–76.

Root, R. (1967). The niche exploitation pattern of the blue-gray gnatcatcher. *Ecological Monographs*, 37:317–350.

Rours, G., Verkooyen, R., Willemse, H., van der Zwaan, E., van Belkum, A., de Groot, R., Verbrugh, H., and Ossewaarde, J. (2005). Use of pooled urine samples and automated DNA isolation to achieve improved sensitivity and cost-effectiveness of large-scale testing for *Chlamydia trachomatis* in pregnant women. *Journal of Clinical Microbiology*, 43:4684–4690.

Rue, H., Martino, S., and Chopin, N. (2009). Approximate Bayesian inference for latent Gaussian models by using integrated nested Laplace approximations. *Journal of the Royal Statistical Society, Series B*, 71:319–392.

Rust, K. and Rao, J. (1996). Variance estimation for complex surveys using replication techniques. *Statistical Methods in Medical Research*, 5:293–310.

Saefken, B., Kneib, T., van Waveren, C.-S., and Greven, S. (2014). A unifying approach to the estimation of the conditional Akaike information in generalized linear mixed models. *Electronic Journal of Statistics*, 8:201–225.

Salsburg, D. (2001). *The Lady Tasting Tea: How Statistics Revolutionized Science in the Twentieth Century.* Henry Holt and Company, LLC.

Satterthwaite, F. (1946). An approximate distribution of estimates of variance components. *Biometrics Bulletin*, 2:110–114.

Schonnop, R., Yang, Y., Feldman, F., Robinson, E., Loughin, M., and Robinovitch, S. (2013). Prevalence of and factors associated with head impact during falls in older adults in long-term care. *Canadian Medical Association Journal*, 185:E803–E810.

Schrage, S. (2019). Pre-emptive decision on post-workout snack may fight indulgence. *Nebraska Today.* `https://news.unl.edu/newsrooms/today/article/pre-emptive-decision-on-post-workout-snack-may-fight-indulgence`.

Schwartz, C. and Mare, R. (2005). Trends in educational assortative marriage from 1940 to 2003. *Demography*, 42:621–646.

Schwarz, G. (1978). Estimating the dimension of a model. *The Annals of Statistics*, 6:461–464.

Scott, A. (2007). Rao-Scott corrections and their impact. In *Proceedings of the Section on Survey Research Methods*, pages 3514–3518. American Statistical Association.

Scott, A. and Rao, J. (1981). Chi-squared tests for contingency tables with proportions estimated from survey data. In Krewski, J., Platek, R., and Rao, J., editors, *Current Topics in Survey Sampling.* Academic Press.

Seeber, G. (2005). Poisson regression. In Armitage, P. and Colton, T., editors, *Encyclopedia of Biostatistics*. Wiley.

Severini, T. (2000). *Likelihood Methods in Statistics*. Oxford University Press.

Sherlock, M., Zetola, N., and Klausner, J. (2007). Routine detection of acute HIV infection through RNA pooling: Survey of current practice in the United States. *Sexually Transmitted Diseases*, 34:314–316.

Shtatland, E., Kleinman, K., and Cain, E. (2003). Stepwise methods using SAS Proc Logistic and SAS Enterprise Miner for prediction. In *SAS Users Group International*, volume 28, paper 258. SAS Institute.

Shumway, R. and Stoffer, D. (2017). *Time Series Analysis and Its Applications*. Springer, fourth edition.

Siino, M., Fasola, S., and Muggeo, V. (2018). Inferential tools in penalized logistic regression for small and sparse data: A comparative study. *Statistical Methods in Medical Research*, 27:1365–1375.

Singer, J. and Willett, J. (2003). *Applied Longitudinal Data Analysis: Modeling Change and Event Occurrence*. Oxford University Press.

Skinner, C. (2019). Analysis of categorical data for complex surveys. *International Statistical Review*, 87, S1:S64–S78.

Skinner, C. and Vallet, L. (2010). Fitting log-linear models to contingency tables from surveys with complex sampling designs: An investigation of the Clogg-Eliason approach. *Sociological Methods & Research*, 39:83–108.

Snee, R. (1974). Graphical display of two-way contingency tables. *The American Statistician*, 28:9–12.

Stroup, W. (2013). *Generalized Linear Mixed Models: Modern Concepts, Methods, and Applications*. CRC Press.

Stroup, W., Milliken, G., Claassen, E., and Wolfinger, R. (2018). *SAS for Mixed Models: Introduction and Basic Applications*. SAS Institute Inc.

Stukel, T. (1988). Generalized logistic models. *Journal of the American Statistical Association*, 83:426–431.

Sugiura, N. (1978). Further analysis of the data by Akaike's information criterion and the finite corrections. *Communications in Statistics: Theory and Methods*, 7:13–26.

Surendra, T. (2016). Marginal loglinear models for three multiple-response categorical variables. Master's thesis, Simon Fraser University.

Surjanovic, N., Lockhart, R., and Loughin, T. (2023). A generalized Hosmer-Lemeshow goodness-of-fit test for a family of generalized linear models. *TEST*. `https://doi.org/10.1007/s11749-023-00912-8`.

Surjanovic, N. and Loughin, T. (2023). Improving the Hosmer-Lemeshow goodness-of-fit test in large models with replicated trials. *Journal of Applied Statistics*. `https://doi.org/10.1080/02664763.2023.2272223`.

Swallow, W. (1985). Group testing for estimating infection rates and probabilities of disease transmission. *Phytopathology*, 75:882–889.

Swift, M. (2009). Comparison of confidence intervals for a Poisson mean—further considerations. *Communications in Statistics: Theory and Methods*, 238:748–759.

Tanner, M. (1996). *Tools for Statistical Inference: Methods for the Exploration of Posterior Distributions and Likelihood Functions*. Springer, third edition.

Tauber, M., Tauber, C., and Nechols, J. (1996). Life history of *Galerucella nymphaeae* and implications of reproductive diapause for rearing univoltine chrysomelids. *Physiological Entomology*, 21:317–324.

Tebbs, J. and Bilder, C. (2004). Confidence interval procedures for the probability of disease transmission in multiple-vector-transfer designs. *Journal of Agricultural, Biological and Environmental Statistics*, 9:75–90.

Tebbs, J., McMahan, C., and Bilder, C. (2013). Two-stage hierarchical group testing for multiple infections with application to the Infertility Prevention Project. *Biometrics*, 69:1064–1073.

Thomas, D. and Decady, Y. (2004). Testing for association using multiple response survey data: Approximate procedures based on the Rao-Scott approach. *International Journal of Testing*, 4:43–59.

Thomas, D. and Rao, J. (1987). Small-sample comparisons of level and power for simple goodness-of-fit statistics under cluster sampling. *Journal of the American Statistical Association*, 82:630–636.

Thomas, D., Singh, A., and Roberts, G. (1996). Tests of independence on two-way tables under cluster sampling: An evaluation. *International Statistical Review*, 64:295–311.

Thomas, K. (2020). New Pfizer results: Coronavirus vaccine is safe and 95% effective. *The New York Times*. November 18; `https://www.nytimes.com/2020/11/18/health/pfizer-covid-vaccine.html`.

Thomas, S. and Tu, W. (2021). Learning Hamiltonian Monte Carlo in R. *The American Statistician*, 75:403–413.

Thompson, S. (2002). *Sampling*. Wiley.

Thorburn, D., Dundas, D., McCruden, E., Cameron, S., Goldberg, D., Symington, I., Kirk, A., and Mills, P. (2001). A study of hepatitis C prevalence in healthcare workers in the west of Scotland. *Gut*, 48:116–120.

Tibshirani, R. (1996). Regression shrinkage and selection via the LASSO. *Journal of the Royal Statistical Society, Series B*, 58:267–288.

Turner, D., Ralphs, M., and Evans, J. (1992). Logistic analysis for monitoring and assessing herbicide efficacy. *Weed Technology*, 6:424–430.

Uldbjerg, C., Miller, J., Burgner, D., Pedersen, L., and Bech, B. (2021). Antibiotic exposure during pregnancy and childhood asthma: A national birth cohort study investigating timing of exposure and mode of delivery. *Archives of Disease in Childhood*, 106:888–894.

Vance, A. (2009). Data analysts captivated by R's power. *The New York Times*. January 6; `http://www.nytimes.com/2009/01/07/technology/business-computing/07program.html`.

Vansteelandt, S., Goetghebeur, E., and Verstraeten, T. (2000). Regression models for disease prevalence with diagnostic tests on pools of serum samples. *Biometrics*, 56:1126–1133.

Ver Hoef, J. and Boveng, P. (2007). Quasi-Poisson vs. negative binomial regression: How should we model overdispersed count data? *Ecology*, 11:2766–2772.

Vermunt, J. and Magidson, J. (2007). Latent class analysis with sampling weights: A maximum-likelihood approach. *Sociological Methods & Research*, 36:87–111.

Verstraeten, T., Farah, B., Duchateau, L., and Matu, R. (1998). Pooling sera to reduce the cost of HIV surveillance: A feasibility study in a rural Kenyan district. *Tropical Medicine and International Health*, 3:747–750.

Wald, A. (1943). Tests of statistical hypotheses concerning several parameters when the number of observations is large. *Transactions of the American Mathematical Society*, 54:426–482.

Warasi, M., McMahan, C., Tebbs, J., and Bilder, C. (2017). Group testing regression models with dilution submodels. *Statistics in Medicine*, 36:4860–4872.

Wardrop, R. (1995). Simpson's paradox and the hot hand in basketball. *The American Statistician*, 49:24–28.

Wedderburn, R. (1974). Quasi-likelihood, generalized linear models, and the Gauss-Newton method. *Biometrika*, 61:439–447.

Weisberg, S. (2013). *Applied Linear Regression.* Wiley, fourth edition.

Westfall, P. and Young, S. (1993). *Resampling-Based Multiple Testing: Examples and Methods for P-Value Adjustment.* Wiley.

White, D., Kramer, L., Backenson, P., Lukacik, G., Johnson, G., Oliver, J., Howard, J., Means, R., Eidson, M., Gotham, I., et al. (2001). Mosquito surveillance and polymerase chain reaction detection of West Nile virus, New York State. *Emerging Infectious Diseases*, 7:643.

Wilkins, T., Malcolm, J., Raina, D., and Schade, R. (2010). Hepatitis C: Diagnosis and treatment. *American Family Physician*, 81:1351–1357.

Williams, D. (1975). The analysis of binary responses from toxicological experiments involving reproduction and teratogenicity. *Biometrics*, 31:949–952.

Wilson, E. (1927). Probable inference, the law of succession, and statistical inference. *Journal of the American Statistical Association*, 22:209–212.

Wood, S. (2013). On p-values for smooth components of an extended generalized additive model. *Biometrika*, 100:221–228.

Wood, S. (2017). *Generalized Additive Models: An Introduction with R.* CRC Press.

Wright, B. (2010). Use of chi-square tests to analyze scat-derived diet composition data. *Marine Mammal Science*, 26:395–401.

Xie, M. (2001). Regression analysis of group testing samples. *Statistics in Medicine*, 20:1957–1969.

Yee, T. (2015). *Vector Generalized Linear and Additive Models: With an Implementation in R.* Springer.

Yee, T. (2022). On the Hauck–Donner effect in Wald tests: Detection, tipping points, and parameter space characterization. *Journal of the American Statistical Association*, 117:1763–1774.

Yi, G. (2017). *Statistical Analysis with Measurement Error or Misclassification: Strategy, Method and Application*. Springer.

Yuan, M. and Lin, Y. (2006). Model selection and estimation in regression with grouped variables. *Journal of the Royal Statistical Society, Series B*, 68:49–67.

Zamar, D., McNeney, B., and Graham, J. (2007). `elrm`: Software implementing exact-like inference for logistic regression models. *Journal of Statistical Software*, 21:1–18.

Zeger, S. and Liang, K. (1986). Longitudinal data analysis for discrete and continuous outcomes. *Biometrics*, 42:121–130.

Zeileis, A., Kleiber, C., and Jackman, S. (2008). Regression models for count data in R. *Journal of Statistical Software*, 27:1–25.

Zhang, B., Bilder, C., and Tebbs, J. (2013). Group testing regression model estimation when case identification is a goal. *Biometrical Journal*, 55:173–189.

Zhou, X. and Qin, G. (2005). A new confidence interval for the difference between two binomial proportions of paired data. *Journal of Statistical Planning and Inference*, 128:527–542.

Zou, H. (2006). The adaptive LASSO and its oracle properties. *Journal of the American Statistical Association*, 101:1418–1429.

Index

Printed in the United States
by Baker & Taylor Publisher Services